A TEXTBOOK OF ENGINEERING MATHEMATICS

A TEXTBOOK OF

ENGINEERING MATHEMATICS

For

B.E. Ist Year (*1st Semester*)
M.D.U., G.J.U., K.U.K., and DCRUST, Haryana

(Strictly According to Latest Syllabus)

By

N.P. Bali

Formerly, Principal
S.B. College, Gurgaon
Haryana

LAXMI PUBLICATIONS (P) LTD

(An ISO 9001:2008 Certified Company)

BENGALURU • CHENNAI • COCHIN • GUWAHATI • HYDERABAD
JALANDHAR • KOLKATA • LUCKNOW • MUMBAI • RANCHI • NEW DELHI
INDIA • USA • GHANA • KENYA

A TEXTBOOK OF ENGINEERING MATHEMATICS

Printed and bound in India
Typeset at Goswami Associates, Delhi
First Edition: 2002; Second Edition : 2004; Third Edition: 2005; Fourth Edition: 2007; Fifth Edition : 2008; Sixth Edition: 2010; Seventh Edition: 2011; Eight Edition: 2012; Ninth Edition: 2013; Tenth Edition : 2014; Eleventh Edition: 2015
ISBN 978-81-318-0808-5

Branches

	Branch	Phone	
✆	Bengaluru	080-26 75 69 30	
✆	Chennai	044-24 34 47 26,	24 35 95 07
✆	Cochin	0484-237 70 04,	405 13 03
✆	Guwahati	0361-254 36 69,	251 38 81
✆	Hyderabad	040-27 55 53 83,	27 55 53 93
✆	Jalandhar	0181-222 12 72	
✆	Kolkata	033-22 27 43 84	
✆	Lucknow	0522-220 99 16	
✆	Mumbai	022-24 91 54 15,	24 92 78 69
✆	Ranchi	0651-220 44 64	

PUBLISHED IN INDIA BY

Laxmi Publications (P) Ltd.
(An ISO 9001:2008 Certified Company)
113, GOLDEN HOUSE, DARYAGANJ,
NEW DELHI - 110002, INDIA
Telephone : 91-11-4353 2500, 4353 2501
Fax : 91-11-2325 2572, 4353 2528
www.laxmipublications.com info@laxmipublications.com

C—R/015/08
Printed at: Sanjay Printer, Delhi

Contents

---

*Chapter 1 is not included in the syllabus of KU, Kurukshetra.

**Chapters 2, 3 and 4 are not included in the syllabus of KU, Kurukshetra.

***Chapter 13 is not included in the syllabus of MDU, Rohtak.

Preface to the Eleventh Edition

With the grace of the Almighty and the overwhelming response given by the teachers and the students, the book is gaining increasing popularity. I express my sincerest thanks and sense of gratitude for their appreciation of the book.

The new edition has been revised with lot of care, dedication and patience. Many new types of questions have been added, in examples as well as in exercises, to enhance the utility of the book. The book presents the subject matter in a very systematic, simple and lucid style, so that students can solve the questions on their own. To make the students familiar with the university pattern, all the questions set in university papers (2008 onwards) have been included.

The most outstanding and distinguishing feature of the book is the large number of typical solved examples followed by well graded problems.

I thank my student Mr. Sandeep Yadav for checking all answers and pointing out some printing errors.

All suggestions for further improvement of the book will be highly appreciated and gratefully acknowledged.

—Author

Preface to the First Edition

This book is a part of the original book (*Vol. I and Vol. II with 28 chapters and covering the syllabi of engineering courses of all semesters of all the Indian Universities*) running its seventh edition and very well received by the teachers and students of all Indian Universities. The rapid sale of the seventh edition bears testimony to the overwhelming response. I thank them all for the appreciation.

The present form of the book contains only thirteen chapters and covers the entire syllabus for the students of *first semester* of MDU, GJU, and KUK, Haryana.

There is no dearth of books on Engineering Mathematics but the students find it difficult to solve most of the problems in the exercise in the absence of an adequate number of solved examples. An outstanding and distinguishing feature of the book is the large number of typical solved examples followed by well-graded problems. Many examples and problems have been selected from recent paper (2000 onwards) of various engineering examinations.

I have endeavoured to present the fundamental concepts in a comprehensive and lucid manner. I am indebted to all authors, Indian and Foreign, whose works I have frequently consulted.

All efforts have been made to keep the book free from errors. Answers to all exercises have been re-checked. All suggestions for improvement will be highly appreciated and gratefully acknowledged.

—Author

Syllabus

(MDU, Rohtak)
MATHEMATICS–I

		Class-work	: 50 Marks
		Exam.	: 100 Marks
Total	150 Marks	Duration of Exam.	: 3 Hours

L T P
4 1 0

Section-A

Infinite Series

Convergence and divergence, comparison tests, D'Alembert's ratio test, integral test, Raabe's test, logarithmic and Cauchy root tests, Gauss's Test, alternating series, absolute and conditional convergence.

Section-B

Matrices and Its Applications

Rank of a matrix, elementary transformations, elementary matrices, inverse using elementary transformations, normal form of a matrix, linear dependence and independence of vectors, consistency of linear system of equations, linear and orthogonal transformations, eigen values and eigen vectors, properties of eigen values, Cayley-Hamilton theorem and its applications, diagonalization of matrices, similar matrices, quadratic forms.

Section-C

Differential Calculus

Successive differentiation, Leibnitz theorem and applications, Taylor's and Maclaurin's series, Curvature, Asymptotes, Curve tracing. Functions of two or more variables, limit and continuity, partial derivatives, total differential and differentiability, derivatives of composite and implicit functions, Jacobians, higher order partial derivatives, homogeneous functions, Euler's theorem and applications. Taylor's series for functions of two variables (without proof), Maxima-Minima of function of two variables, Lagrange's method of undetermined multipliers, differentiation under integral sign (Leibnitz rule).

Section-D

Integral Calculus

Beta and gamma functions and relationship between them. Applications of single integration to find volume of solids and surface area of solids of revolution. Double integral, change of order of integration, double integral in polar coordinates, applications of double integral to find area enclosed by plane curves, triple integral, change of variables, volume of solids, Dirichlet's integral.

Note. Examiner will set 9 questions in total, with two questions from each section and one question covering all sections which will be Q.1. This Q.1 is compulsory and of short answer type. Each question carries equal mark (20 marks). Students have to attempt 5 questions in total.

Syllabus

(KU, Kurukshetra)
MATHEMATICS–I

Theory	: 100 Marks
Sessional	: 50 Marks
Total	: 150 Marks
Duration of Exam.	: 3 Hours

Unit–I

Applications of Differentiation

Taylor's and Maclaurin's series, Expansion by use of known series, Expansion by forming a differential equation, Asymptotes. Curvature, Radius of Curvature for Cartesian, parametric and Polar Curves, Centre of curvature and Chord of curvature, Tracing of Cartesian and Polar curves (standard curves).

Unit–II

Partial Differentiation and Its Applications

Functions of two or more variables, Partial derivatives, Total differential and differentiability, Derivatives of composite and implicit functions, Change of variables.

Homogeneous functions, Euler's theorem, Jacobian, Taylor's and Maclaurin's series for functions of two variables (without proof), Errors and approximations, Maxima-Minima of functions of two variables, Lagrange's method of undetermined multipliers, Differentiation under the integral sign.

Unit–III

Multiple Integrals and Their Applications

Double integral, change of order of integration, Double integral in polar coordinates, Applications of double integral to find area enclosed by plane curves and volume of solids of revolution.

Triple integration, volume of solids, change of variables, Beta and Gamma functions and relationship between them.

Unit–IV

Vector Calculus

Differentiation of vectors, Scalar and Vector point function, Gradient of a Scalar field and directional derivative, divergence and curl of a vector field and their physical interpretations. Del applied twice to point functions. Del applied to product of point functions.

Integration of vectors, Line integral, surface integral, volume integral, Green's, Stoke's and Gauss divergence theorems (without proof), and their simple applications.

Note. The Examiners will set eight questions, taking two from each unit. The students are required to attempt five questions in all selecting at least one from each unit. All questions will carry equal marks.

Syllabus

(DCRUST, Murthal)
MATHEMATICS–I
B.Tech. Semester–I (Common for all Branches)

L	T	P	Credits		
3	2	–	5	Class Work	: 50 Marks
				Examination	: 100 Marks
				Total	: 150 Marks
				Duration of Exam.	: 3 Hours

Part–A

Infinite series: Convergence and divergence, Comparison, D' Alembert's ratio, Integral, Rabbes, Logrithmic and Cauchy root tests, Alternating series, Absolute and conditional convergence.

Applications of Differentiation: Taylor's and Maclaurin's series, Asymptotes, Curvature.

Partial Differentiation and Its Applications: Function of two or more variables; partial derivatives, Total differential and differentiability, Derivatives of composite and implicit functions, Jacobians, Higher order partial derivatives.

Homogeneous function, Euler's theorem, Taylor's series for functions of two variables (without proof), maxima-minima of function of two variables, Lagrange's method of undetermined multipliers, Differentiation under integral sign.

Part–B

Applications of Single and Multiple Integration: Applications of single integration to find volume of solids and surface area of solids of revolution. Double integral, change of order of integration, Double integral in polar coordinates, Applications of double integral to find area enclosed by plane curves and volume of solids of revolution.

Triple integral, volume of solids, change of variables, Beta and gamma functions and relationship between them.

Vector Calculus: Differentiation of vectors, scalar and vector point function Gradient of a scalar field and directional derivative, divergence and curl of a vector field and their physical interpretations.

Integration of vectors, line integral, surface integral, volume integral, Green, Stoke's and Gauss theorems (without proof) and their simple applications.

Basic Concepts

1. Greek Letters Used

α	alpha	ϕ	phi	κ	kappa	τ	tau
β	beta	ψ	psi	μ	mu	χ	chi
γ	gamma	ξ	xi	ν	nu	ω	omega
δ	delta	η	eta	π	pi	Δ	cap. delta
ε	epsilon	ζ	zeta	ρ	rho		
i	iota	λ	lambda	σ	sigma		
θ	theta	Γ	cap. gamma	Σ	cap. sigma		

2. Some Notations

$\in$	belongs to	$\Leftrightarrow$	implies and implied by
$\notin$	does not belong to	iff	if and only if
$\Rightarrow$	implies	$\cup$	union
		$\cap$	intersection

3. Useful Data

$\sqrt{2} = 1.4142$	$\frac{1}{\pi} = 0.3183$	1 rad. = 57° 17′ 45″	$\log_{10} e = 0.4343$
$\sqrt{3} = 1.732$	$e = 2.7183$	1° = 0.0174 rad.	$\log_e 2 = 0.6931$
$\pi = 3.1416$	$\frac{1}{e} = 0.3679$	$\log_e 10 = 2.3026$	$\log_e 3 = 1.0986$

4. Quadratic Equation

Roots of quadratic equation $ax^2 + bx + c = 0, \quad a \neq 0$

are $\frac{-b \pm \sqrt{D}}{2a}$, where $D = b^2 - 4ac$ is called discriminant

Sum of roots $= -\frac{b}{a}$, Product of roots $= \frac{c}{a}$

If $D > 0$, roots are real and distinct.

If $D = 0$, roots are equal.

If $D < 0$, roots are imaginary.

5. Progressions

(*i*) For the A.P. (Arithmetic Progression) $a, a + d, a + 2d, \cdots$

$$T_n = a + (n-1)d, \quad S_n = \frac{n}{2}\,[2a + (n-1)d\,] = \frac{n}{2}\,(a + l).$$

(*ii*) For the G.P. (Geometric Progression) $a, ar, ar^2, \cdots$

$$T_n = ar^{n-1}, \quad S_n = \begin{cases} \dfrac{a(1-r^n)}{1-r}, & \text{when } r \neq 1 \\ na, & \text{when } r = 1 \end{cases}$$

$$S_{\infty} = \frac{a}{1-r} \text{ provided } |r| < 1 \quad i.e., \quad -1 < r < 1$$

(*iii*) A sequence is said to be in H.P. (Harmonic Progression) if the reciprocals of its terms are in A.P.

For the H.P. $\frac{1}{a}, \frac{1}{a+d}, \frac{1}{a+2d}, \ldots, T_n = \frac{1}{a+(n-1)d}$

(*iv*) For two numbers a and b,

$$\text{A.M.} = \frac{a+b}{2}, \quad \text{G.M.} = \sqrt{ab}, \quad \text{H.M.} = \frac{2ab}{a+b}$$

(*v*) For natural numbers 1, 2, 3, ..., n

$$\Sigma n = \frac{n(n+1)}{2}, \ \Sigma n^2 = \frac{n(n+1)(2n+1)}{6}, \ \Sigma n^3 = \left[\frac{n(n+1)}{2}\right]^2$$

6. Permutations and Combinations

$$^nP_r = \frac{n!}{(n-r)!}, \quad ^nC_r = \frac{^nP_r}{r!} = \frac{n!}{r!(n-r)!}$$

$$^nP_n = n!, \ ^nC_r = {^nC_{n-r}}, \quad ^nC_n = {^nC_0} = 1$$

7. Binomial Theorem

(*i*) When n is a positive integer

$$(a+b)^n = {^nC_0}\, a^n + {^nC_1}\, a^{n-1} b + {^nC_2}\, a^{n-2} b^2 + \cdots + {^nC_r}\, a^{n-r} b^r + \cdots + {^nC_n}\, b^n$$

$$(1+x)^n = 1 + {^nC_1}\, x + {^nC_2}\, x^2 + \cdots + {^nC_n}\, x^n$$

(*ii*) When n is a negative integer or a fraction

$$(1+x)^n = 1 + nx + \frac{n(n-1)}{2!} x^2 + \frac{n(n-1)(n-2)}{3!} x^3 + \cdots \infty$$

provided $|x| < 1$.

8. Logarithms

(*i*) Natural logarithm of a positive real number x is denoted by $\log_e x$ or simply $\log x$ or $\ln x$. It is the inverse of e^x.

Common logarithm of a positive real number x is denoted by $\log_{10} x$.

Relation: (*i*) $\log_{10} x = 0.4343 \log_e x$

(*ii*) $\log_a 1 = 0, \qquad \log_a a = 1, \qquad \log_a 0 = -\infty \ (a > 1)$

(*iii*) $\log(mn) = \log m + \log n, \quad \log\left(\frac{m}{n}\right) = \log m - \log n$

$\log(m^n) = n \log m, \quad \log_n m \times \log_m n = 1$

9. Matrices and Determinants

(*i*) Two matrices $A = [a_{ij}]$ and $B = [b_{ij}]$ are equal if they have same order and corresponding elements are equal. *i.e.*, $a_{ij} = b_{ij}$ for all i and j.

(*ii*) If $A = [a_{ij}]$ and $B = [b_{ij}]$ are two matrices of the **same order**, then $A + B$ is defined and $A + B = [a_{ij} + b_{ij}]$, *i.e.*, add corresponding elements.

(*iii*) If $A = [a_{ij}]$ is a matrix and k is a scalar, then kA is another matrix obtained by multiplying each element of A by the scalar k. Thus, $kA = [ka_{ij}]$.

(*iv*) The product AB of two matrices A and B is defined if the number of columns in A is equal to the number of rows in B. If A is an $m \times n$ matrix and B is an $n \times p$ matrix then AB is a matrix of order $m \times p$. The $(i, j)^{th}$ element of AB is obtained by multiplying the corresponding elements of ith row of A and j^{th} column of B and adding all these products.

Matrix multiplication is not commutative, *i.e.*, $AB \neq BA$ in general.

If the product of two matrices is a zero matrix, it is not necessary that one of the matrices is a zero matrix.

$$(AB)C = A(BC), \quad A(B + C) = AB + AC$$

whenever both sides of equality are defined.

(*v*) For every square matrix A, there exists an identity matrix I of same order such that $AI = IA = A$.

(*vi*) The matrix obtained by interchanging the rows and columns of a matrix A is called the transpose of A. Transpose of A is denoted by A' or A^T.

If $A = [a_{ij}]_{m \times n}$, then $A' = [a_{ji}]_{n \times m}$

$$(A')' = A, \qquad (kA)' = kA', \quad (A + B)' = A' + B',$$

$$(AB)' = B'A'.$$

(*vii*) A square matrix A is called symmetric if $A' = A$ and skew symmetric if $A' = -A$. All the diagonal elements of a skew symmetric matrix are zero.

(*viii*) A square matrix A is said to be invertible if there exists a square matrix B of same order as A, such that $AB = BA = I$. The matrix B is called the inverse of A and it is denoted by A^{-1}.

Thus, $$AA^{-1} = A^{-1}A = I$$

Also, $$(A^{-1})^{-1} = A, \qquad (AB)^{-1} = B^{-1}A^{-1}.$$

(*ix*) A determinant is a function which associates each square matrix with a unique number (real or complex). The determinant of a square matrix A is denoted by $|A|$ or det. A or Δ.

(*x*) $$\begin{vmatrix} a & b \\ c & d \end{vmatrix} = ad - cb.$$

$$\begin{vmatrix} a_1 & a_2 & a_3 \\ b_1 & b_2 & b_3 \\ c_1 & c_2 & c_3 \end{vmatrix} = a_1\begin{vmatrix} b_2 & b_3 \\ c_2 & c_3 \end{vmatrix} - a_2\begin{vmatrix} b_1 & b_3 \\ c_1 & c_3 \end{vmatrix} + a_3\begin{vmatrix} b_1 & b_2 \\ c_1 & c_2 \end{vmatrix}$$

$$= a_1(b_2c_3 - b_3c_2) - a_2(b_1c_3 - b_3\,c_1) + a_3\,(b_1c_2 - b_2c_1)$$

$$\begin{vmatrix} a_1 & a_2 & a_3 \\ 0 & b_2 & b_3 \\ 0 & 0 & c_3 \end{vmatrix} = \begin{vmatrix} a_1 & 0 & 0 \\ b_1 & b_2 & 0 \\ c_1 & c_2 & c_3 \end{vmatrix} = a_1b_2c_3 = \text{product of diagonal elements.}$$

(*xi*) If A is a square matrix of order n, then $|kA| = k^n |A|$.

(*xii*) The value of a determinant remains unchanged if its rows and columns are interchanged, *i.e.*, $|A'| = |A|$.

(*xiii*) If any two-rows (or columns) of a determinant are interchanged, then sign of determinant changes.

(xiv) If any two rows (or columns) of a determinant are identical or proportional, then value of determinant is zero.

(xv) If each element of a row (or column) of a determinant is multiplied by a constant k, then its value gets multiplied by k. By this property, we can take out any common factor from any one row (or column) of a given determinant.

(xvi) If each element of a row (or column) of a determinant is the sum of m terms, then the determinant can be expressed as the sum of m determinants.

(xvii) If, to each element of a row (or column) of a determinant, be added equi-multiples of the corresponding elements of some other row (or column), then value of determinant remains the same.

(xviii) If $A = \begin{bmatrix} a_{11} & a_{12} & a_{13} \\ a_{21} & a_{22} & a_{23} \\ a_{31} & a_{32} & a_{33} \end{bmatrix}$ then adj. $A = \begin{bmatrix} A_{11} & A_{12} & A_{13} \\ A_{21} & A_{22} & A_{23} \\ A_{31} & A_{32} & A_{33} \end{bmatrix}'$, where A_{ij} is co-factor of a_{ij}.

(xix) For a square matrix of order n,

$$A\ (\text{adj. } A) = (\text{adj. } A) = |\ A\ |\ I$$

where I is the identity matrix of order n.

(xx) If A is a non-singular matrix of order n, then $|\ \text{adj. } A\ | = |\ A\ |^{n-1}$

(xxi) A square matrix A is invertible if and only if A is a non-singular matrix.

$$A^{-1} = \frac{1}{|A|} \text{ adj. } A$$

10. Trigonometry

(i)

$x°$	0°	30°	45°	60°	90°
$\sin x$	0	$\frac{1}{2}$	$\frac{1}{\sqrt{2}}$	$\frac{\sqrt{3}}{2}$	1
$\cos x$	1	$\frac{\sqrt{3}}{2}$	$\frac{1}{\sqrt{2}}$	$\frac{1}{2}$	0
$\tan x$	0	$\frac{1}{\sqrt{3}}$	1	$\sqrt{3}$	∞

(ii) The figure shows a unit circle with centre at origin. If $\angle XOP = x$, then $P = (\cos x, \sin x)$.

$\therefore$

$$\cos 0 = 1, \qquad \sin 0 = 0$$

$$\cos \frac{\pi}{2} = 0, \qquad \sin \frac{\pi}{2} = 1$$

$$\cos \pi = -1, \qquad \sin \pi = 0$$

$$\cos \frac{3\pi}{2} = 0, \qquad \sin \frac{3\pi}{2} = -1$$

$$\cos 2\pi = 1, \qquad \sin 2\pi = 0$$

(iii) Any t-ratio of $(n.\ 90° \pm x) = \pm$ same t-ratio of x, when n is even $= \pm$ co-ratio of x, when n is odd.

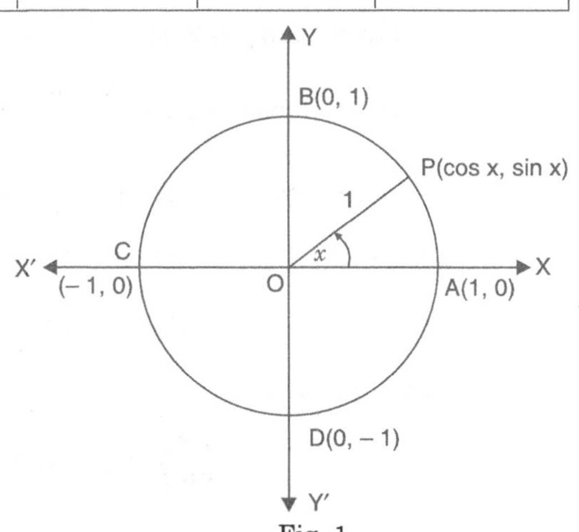

Fig. 1

where co-ratio of x is obtained by dropping co if present and adding co if absent. Thus, sin $\rightleftharpoons$ cos, tan $\rightleftharpoons$ cot, sec $\rightleftharpoons$ cosec.

The sign $\pm$ or $-$ is decided from the quadrant in which $n.\ 90° \pm x$ lies.

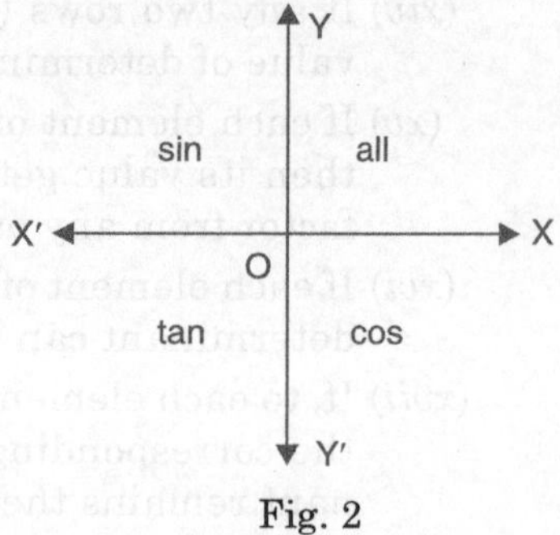

Fig. 2

(iv) Signs of t-ratios in different quadrants (Fig. 2)

(v) $\cos^2 x + \sin^2 x = 1$, $\sec^2 x - \tan^2 x = 1$, $\text{cosec}^2\, x - \cot^2 x = 1$

(vi) $$\sin (x \pm y) = \sin x \cos y \pm \cos x \sin y$$

$$\cos (x \pm y) = \cos x \cos y \mp \sin x \sin y$$

$$\tan (x \pm y) = \frac{\tan x \pm \tan y}{1 \mp \tan x \tan y}$$

(vii) $$\sin 2x = 2 \sin x \cos x = \frac{2 \tan x}{1 + \tan^2 x}$$

(viii) $$\cos 2x = \cos^2 x - \sin^2 x = 2 \cos^2 x - 1 = 1 - 2 \sin^2 x = \frac{1 - \tan^2 x}{1 + \tan^2 x}$$

(ix) $$\sin 3x = 3 \sin x - 4 \sin^3 x,\ \cos 3x = 4 \cos^3 x - 3 \cos x$$

$$\tan 3x = \frac{3 \tan x - \tan^3 x}{1 - 3 \tan^2 x}$$

(x) $$\sin x \cos y = \frac{1}{2} [\sin (x + y) + \sin (x - y)]$$

$$\cos x \sin y = \frac{1}{2} [\sin (x + y) - \sin (x - y)]$$

$$\cos x \cos y = \frac{1}{2} [\cos (x + y) + \cos (x - y)]$$

$$\sin x \sin y = \frac{1}{2} [\cos (x - y) - \cos (x + y)]$$

(xi) $$\sin x + \sin y = 2 \sin \frac{x + y}{2} \cos \frac{x - y}{2}$$

$$\sin x - \sin y = 2 \cos \frac{x + y}{2} \sin \frac{x - y}{2}$$

$$\cos x + \cos y = 2 \cos \frac{x + y}{2} \cos \frac{x - y}{2}$$

$$\cos x - \cos y = -2 \sin \frac{x + y}{2} \sin \frac{x - y}{2} = 2 \sin \frac{x + y}{2} \sin \frac{y - x}{2}$$

(xii) $a \sin x + b \cos x = r \sin (x + \theta)$, $a \cos x + b \sin x = r \cos (x - \theta)$, where $a = r \cos \theta$, $b = r \sin \theta$

so that $r = \sqrt{a^2 + b^2}$, $\theta = \tan^{-1} \left(\frac{b}{a}\right)$.

(xiii) In any ΔABC:

$$\frac{a}{\sin A} = \frac{b}{\sin B} = \frac{c}{\sin C} \qquad \text{(sine formula)}$$

$$\cos A = \frac{b^2 + c^2 - a^2}{2bc} \quad \text{(cosine formula)}$$

$$a = b \cos C + c \cos B \quad \text{(projection formula)}$$

Area of $\Delta\, ABC = \frac{1}{2}\, bc \sin A = \sqrt{s(s-a)(s-b)(s-c)}$, where $s = \frac{1}{2}(a + b + c)$

$\frac{a}{\sin A} = \frac{b}{\sin B} = \frac{c}{\sin C} = 2R$, where R is the circum-radius of Δ ABC.

$$R = \frac{abc}{4\Delta}$$

$r = \frac{\Delta}{s}$, where r is the radius of inscribed circle of Δ ABC.

11. Demoivre's Theorem: $(\cos\theta + i \sin\theta)^n = \cos n\theta + i \sin n\theta$

Euler's Theorem: $\cos\theta + i\sin\theta = e^{i\theta}$.

12. Hyperbolic Functions

$$\sinh x = \frac{e^x - e^{-x}}{2}, \ \cosh x = \frac{e^x + e^{-x}}{2}, \ \tanh x = \frac{e^x - e^{-x}}{e^x + e^{-x}}$$

$$\text{cosech}\, x = \frac{1}{\sinh x}, \ \text{sech}\, x = \frac{1}{\cosh x}, \ \coth x = \frac{1}{\tanh x}$$

$$\tanh x = \frac{\sinh x}{\cosh x}, \ \coth x = \frac{\cosh x}{\sinh x}, \ \cosh^2 x - \sinh^2 x = 1$$

$$\sin ix = i \sinh x, \ \cos ix = \cosh x, \ \tan ix = i \tanh x$$

$$\sinh^{-1} x = \log\left(x + \sqrt{x^2 + 1}\right), \ \cosh^{-1} x = \log\left(x + \sqrt{x^2 - 1}\right),$$

$$\tanh^{-1} x = \frac{1}{2} \log \frac{1+x}{1-x}.$$

13. Series

$$e^x = 1 + \frac{x}{1!} + \frac{x^2}{2!} + \frac{x^3}{3!} + \cdots \infty$$

$$\sin x = x - \frac{x^3}{3!} + \frac{x^5}{5!} - \cdots \infty, \qquad \sinh x = x + \frac{x^3}{3!} + \frac{x^5}{5!} + \cdots \infty$$

$$\cos x = 1 - \frac{x^2}{2!} + \frac{x^4}{4!} - \cdots \infty, \qquad \cosh x = 1 + \frac{x^2}{2!} + \frac{x^4}{4!} + \cdots \infty$$

$$\log(1 + x) = x - \frac{x^2}{2} + \frac{x^3}{3} - \cdots \infty, \quad \log(1 - x) = -\left(x + \frac{x^2}{2} + \frac{x^3}{3} + \cdots \infty\right)$$

$$\tan^{-1} x = x - \frac{x^3}{3} + \frac{x^5}{5} - \cdots \infty \quad \text{(Gregory series)}$$

$$\tanh^{-1} x = x + \frac{x^3}{3} + \frac{x^5}{5} + \cdots \infty$$

14. Calculus

(a) Standard Limits

(i) $\lim\limits_{x \to a} \dfrac{x^n - a^n}{x - a} = na^{n-1}, n \in Q$

(ii) $\lim\limits_{x \to 0} \dfrac{\sin x}{x} = 1$

(iii) $\lim\limits_{x \to 0} \dfrac{\sin^{-1} x}{x} = 1$

(iv) $\lim\limits_{x \to 0} \dfrac{\tan x}{x} = 1$

(v) $\lim\limits_{x \to 0} \dfrac{\tan^{-1} x}{x} = 1$

(vi) $\lim\limits_{x \to 0} (1+x)^{\frac{1}{x}} = e$

(vii) $\lim\limits_{x \to 0} \dfrac{a^x - 1}{x} = \log_e a$

(viii) $\lim\limits_{x \to \infty} x^{1/x} = 1$

(ix) If $f(a) = g(a) = 0$, then

$\lim\limits_{x \to a} \dfrac{f(x)}{g(x)} = \lim\limits_{x \to a} \dfrac{f'(x)}{g'(x)}$ (L' Hospital's Rule)

(Differentiate the numerator and denominator separately)

(b) Differentiation

(i) $\dfrac{d}{dx}(uv) = u\dfrac{dv}{dx} + v\dfrac{du}{dx}$ (Product Rule)

(ii) $\dfrac{d}{dx}\left(\dfrac{u}{v}\right) = \dfrac{v\dfrac{du}{dx} - u\dfrac{dv}{dx}}{v^2}$ (Quotient Rule)

(iii) $\dfrac{dy}{dx} = \dfrac{dy}{du} \times \dfrac{du}{dx}$ (Chain Rule)

(iv) If $x = f(t)$, $y = g(t)$, then $\dfrac{dy}{dx} = \dfrac{\dfrac{dy}{dt}}{\dfrac{dx}{dt}}$

(v) $\dfrac{d}{dx}(c) = 0$

(vi) $\dfrac{d}{dx}(x^n) = nx^{n-1}$

(vii) $\dfrac{d}{dx}(e^x) = e^x$

(viii) $\dfrac{d}{dx}(a^x) = a^x \log_e a$

(ix) $\dfrac{d}{dx}(\log_e x) = \dfrac{1}{x}$

(x) $\dfrac{d}{dx}(\log_a x) = \dfrac{1}{x \log a}$

(xi) $\dfrac{d}{dx}(\sin x) = \cos x$

(xii) $\dfrac{d}{dx}(\cos x) = -\sin x$

(xiii) $\dfrac{d}{dx}(\tan x) = \sec^2 x$

(xiv) $\dfrac{d}{dx}(\cot x) = -\text{cosec}^2 x$

(xv) $\dfrac{d}{dx}(\sec x) = \sec x \tan x$

(xvi) $\dfrac{d}{dx}(\text{cosec } x) = -\text{cosec } x \cot x$

(xvii) $\dfrac{d}{dx}(\sin^{-1} x) = \dfrac{1}{\sqrt{1-x^2}}$

(xviii) $\dfrac{d}{dx}(\cos^{-1} x) = \dfrac{-1}{\sqrt{1-x^2}}$

(xix) $\frac{d}{dx}(\tan^{-1} x) = \frac{1}{1+x^2}$

(xx) $\frac{d}{dx}(\cot^{-1} x) = \frac{-1}{1+x^2}$

(xxi) $\frac{d}{dx}(\sec^{-1} x) = \frac{1}{x\sqrt{x^2-1}}$

(xxii) $\frac{d}{dx}(\text{cosec}^{-1} x) = \frac{-1}{x\sqrt{x^2-1}}$

(xxiii) $\frac{d}{dx}(\sinh x) = \cosh x$

(xxiv) $\frac{d}{dx}(\cosh x) = \sinh x.$

(c) **Integration**

(i) $\int x^n\, dx = \frac{x^{n+1}}{n+1} \quad (n \neq -1)$

(ii) $\int \frac{1}{x}\, dx = \log_e x$

(iii) $\int e^x\, dx = e^x$

(iv) $\int a^x\, dx = \frac{a^x}{\log_e a}$

(v) $\int \sin x\, dx = -\cos x$

(vi) $\int \cos x\, dx = \sin x$

(vii) $\int \tan x\, dx = -\log \cos x$

(viii) $\int \cot x\, dx = \log \sin x$

(ix) $\int \sec x \tan x\, dx = \sec x$

(x) $\int \text{cosec}\, x \cot x\, dx = -\text{cosec}\, x$

(xi) $\int \sec^2 x\, dx = \tan x$

(xii) $\int \text{cosec}^2 x\, dx = -\cot x$

(xiii) $\int \sec x\, dx = \log(\sec x + \tan x)$

(xiv) $\int \text{cosec}\, x\, dx = \log(\text{cosec}\, x - \cot x) = \log \tan \frac{x}{2}$

(xv) $\int \frac{dx}{a^2+x^2} = \frac{1}{a}\tan^{-1}\frac{x}{a}$

(xvi) $\int \frac{dx}{\sqrt{a^2-x^2}} = \sin^{-1}\frac{x}{a}$

(xvii) $\int \frac{dx}{a^2-x^2} = \frac{1}{2a}\log\frac{a+x}{a-x}$

(xviii) $\int \frac{dx}{\sqrt{a^2+x^2}} = \sinh^{-1}\frac{x}{a}$

(xix) $\int \frac{dx}{x^2-a^2} = \frac{1}{2a}\log\frac{x-a}{x+a}$

(xx) $\int \frac{dx}{\sqrt{x^2-a^2}} = \cosh^{-1}\frac{x}{a}$

(xxi) $\int \sqrt{a^2-x^2}\, dx = \frac{x\sqrt{a^2-x^2}}{2} + \frac{a^2}{2}\sin^{-1}\frac{x}{a}$

(xxii) $\int \sqrt{a^2+x^2}\, dx = \frac{x\sqrt{a^2+x^2}}{2} + \frac{a^2}{2}\sinh^{-1}\frac{x}{a}$

(xxiii) $\int \sqrt{x^2-a^2}\, dx = \frac{x\sqrt{x^2-a^2}}{2} - \frac{a^2}{2}\cosh^{-1}\frac{x}{a}$

(xxiv) $\int e^{ax} \sin bx\, dx = \dfrac{e^{ax}}{a^2+b^2}(a \sin bx - b \cos bx) = \dfrac{e^{ax}}{\sqrt{a^2+b^2}} \sin\left(bx - \tan^{-1}\dfrac{b}{a}\right)$

(xxv) $\int e^{ax} \cos bx\, dx = \dfrac{e^{ax}}{a^2+b^2}(a \cos bx + b \sin bx) = \dfrac{e^{ax}}{\sqrt{a^2+b^2}} \cos\left(bx - \tan^{-1}\dfrac{b}{a}\right)$

(xxvi) $\int \sinh x\, dx = \cosh x$

(xxvii) $\int \cosh x\, dx = \sinh x$

(xxviii) $\int \tanh x\, dx = \log \cosh x$

(xxix) $\int \coth x\, dx = \log \sinh x$

(xxx) $\int \text{sech}^2\, x\, dx = \tanh x$

(xxxi) $\int \text{cosech}^2 x\, dx = -\coth x$

(xxxii) $\int_a^b f(x)\, dx = -\int_b^a f(x)\, dx$

(xxxiii) $\int_a^b f(x)\, dx = \int_a^c f(x)\, dx + \int_c^b f(x)\, dx, \quad a < c < b$

(xxxiv) $\int_0^a f(x)\, dx = \int_0^a f(a-x)\, dx$

(xxxv) $\int_a^b f(x)\, dx = \int_a^b f(a+b-x)\, dx$

(xxxvi) $\int_{-a}^a f(x)\, dx = \begin{cases} 0, & \text{if } f \text{ is an odd function} \\ 2\int_0^a f(x)\, dx, & \text{if } f \text{ is an even function} \end{cases}$

(xxxvii) $\int_0^{2a} f(x)\, dx = \begin{cases} 0, & \text{if } f(2a-x) = -f(x) \\ 2\int_0^a f(x)\, dx, & \text{if } f(2a-x) = f(x) \end{cases}$

(xxxviii) $\int_0^{\pi/2} \sin^n x\, dx = \int_0^{\pi/2} \cos^n x\, dx \quad (n > 1)$

$$= \frac{(n-1)(n-3)(n-5)\ldots}{n(n-2)(n-4)\ldots} \times \left(\frac{\pi}{2}, \text{only if } n \text{ is even}\right)$$

(xxxix) $\int_0^{\pi/2} \sin^m x \cos^n x\, dx$

$$= \frac{(m-1)(m-3)\ldots \times (n-1)(n-3)\ldots}{(m+n)(m+n-2)(m+n-4)\ldots} \times \left(\frac{\pi}{2}, \text{only if both } m \text{ and } n \text{ are even}\right).$$

Abbreviations

M.D.U.	Mahrishi Dayanand University, Rohtak
M.T.U.	Mahamaya Technical University, Noida
K.U.K.	Kurukshetra University, Kurukshetra
P.T.U.	Punjab Technical University, Jalandhar
V.T.U.	Vivekanand Technical University
J.N.T.U.	Jawaharlal Nehru Technical University
G.B.T.U.	Gautam Budh Technical University

SECTION A
Infinite Series

1*

Infinite Series

1.1. SEQUENCE

A sequence is a function whose domain is the set N of all natural numbers whereas the range may be any set S. In other words, a sequence in a set S is a rule which assigns to each natural number a unique element of S.

1.2. REAL SEQUENCE

A real sequence is a function whose domain is the set N of all natural numbers and range a subset of the set R of real numbers.

Symbolically $f : \mathrm{N} \to \mathrm{R}$ (or $x : \mathrm{N} \to \mathrm{R}$ or $a : \mathrm{N} \to \mathrm{R}$)

is a real sequence.

Note. If $x : \mathrm{N} \to \mathrm{R}$ be a sequence, the image of $n \in \mathrm{N}$ instead of denoting it by $x(n)$, we shall generally denote it by x_n. Thus x_1, x_2, x_3 etc. are the real numbers associated to 1, 2, 3, etc. by this mapping. Also, the sequence $x : \mathrm{N} \to \mathrm{R}$ is denoted by $\{x_n\}$ or $< x_n >$.

$x_1, x_2, \ldots$ are called the first, second ... terms of the sequence. The mth and nth terms x_m and x_n for $m \neq n$ are treated as distinct event if $x_m = x_n$ *i.e.*, the terms occurring at different positions are treated as distinct terms even if they have the same value.

1.3. RANGE OF A SEQUENCE

The set of all **distinct** terms of a sequence is called its range.

Note. In a sequence $\{x_n\}$, since $n \in \mathrm{N}$ and N is an infinite set, **the number of terms of a sequence is always infinite.** The range of a sequence may be a finite set. *e.g.* if $x_n = (-1)^n$ then $\{x_n\} = \{-1, 1, -1, 1, \ldots\}$

The range of sequence $\{x_n\} = \{-1, 1\}$ which is a finite set.

1.4. CONSTANT SEQUENCE

A sequence $\{x_n\}$ defined by $x_n = c$, where c is a fixed real number, $\forall\ n \in \mathrm{N}$ is called a constant sequence.

e.g. $\{x_n\} = \{c, c, c, \ldots\}$ is a constant sequence with range $= \{c\}$.

1.5. BOUNDED AND UNBOUNDED SEQUENCES

Bounded above sequence. A sequence $\{a_n\}$ is said to be bounded above if there exists a real number K such that $a_n \leq \mathrm{K} \quad \forall\ n \in \mathrm{N}$.

*This chapter is not included in the syllabus of KU, Kurukshetra.

Bounded below sequence. A sequence $\{a_n\}$ is said to be bounded below if there exists a real number k such that $a_n \geq k \quad \forall \ n \in \text{N}$.

Bounded sequence. A sequence $\{a_n\}$ is said to be bounded when it is bounded both above and below.

$\Rightarrow$ A sequence $\{a_n\}$ is bounded if there exist two real numbers k and K ($k \leq \text{K}$) such that

$$k \leq a_n \leq \text{K} \quad \forall \ n \in \text{N}.$$

Choosing M = maximum $\{|\ k\ |, |\ \text{K}\ |\}$, we can also define a sequence $\{a_n\}$ to be bounded if

$$|\ a_n\ | \leq \text{M} \quad \forall \quad n \in \text{N}.$$

Unbounded sequence. If there exists no real number M such that $|\ a_n\ | \leq \text{M} \ \forall \ n \in \text{N}$, then the sequence $\{a_n\}$ is said to be unbounded.

Examples (1). The sequence $\{a_n\}$ defined by $a_n = \dfrac{1}{n}$.

Here $$\{a_n\} = \left\{\frac{1}{1}, \frac{1}{2}, \frac{1}{3}, \ldots\right\}$$

$\because$ $$0 < a_n \leq 1 \quad \forall \quad n \in \text{N}$$

$\therefore$ $\{a_n\}$ is bounded.

(2) The sequence $\{a_n\}$ defined by $a_n = 2^{n-1}$

Here $$\{a_n\} = \{1, 2, 2^2, 2^3, \cdots\}.$$

Although $a_n \geq 1, \quad \forall \quad n \in \text{N}$, there exists no real number K such that $a_n \leq \text{K}$.

$\therefore$ The sequence is unbounded above.

1.6. CONVERGENT, DIVERGENT, OSCILLATORY SEQUENCES

Convergent sequence. A sequence $\{a_n\}$ is said to be convergent if $\lim\limits_{n \to \infty} a_n$ is finite.

For example, consider the sequence $\dfrac{1}{2}, \dfrac{1}{2^2}, \dfrac{1}{2^3}, \ldots, \dfrac{1}{2^n} \ \ldots$

Here $a_n = \dfrac{1}{2^n}$, $\lim\limits_{n \to \infty} a_n = \lim\limits_{n \to \infty} \dfrac{1}{2^n} = 0$, which is finite.

$\Rightarrow$ the sequence $\{a_n\}$ is convergent.

Divergent sequence. A sequence $\{a_n\}$ is said to be divergent if $\lim\limits_{n \to \infty} a_n$ is not finite,

i.e., if $$\lim_{n \to \infty} a_n = +\infty \text{ or } -\infty.$$

For example

(*i*) Consider the sequence $\{n^2\}$

Here $a_n = n^2$, $\lim\limits_{n \to \infty} a_n = +\infty \quad \Rightarrow \quad$ The sequence $\{n^2\}$ is divergent.

(*ii*) Consider the sequence $\{-2^n\}$.

Here $$a_n = -2^n, \lim_{n \to \infty} a_n = \lim_{n \to \infty} (-2^n) = -\infty$$

$\Rightarrow$ the sequence $\{-2^n\}$ is divergent.

Oscillatory sequence. If a sequence $\{a_n\}$ neither converges to a finite number nor diverges to $+\infty$ or $-\infty$, it is called an oscillatory sequence. Oscillatory sequences are of two types:

(*i*) A bounded sequence which does not converge is said to **oscillate finitely.**

For example, consider the sequence $\{(-1)^n\}$.

Here $$a_n = (-1)^n$$

It is a bounded sequence. $\lim_{n\to\infty} a_{2n} = \lim_{n\to\infty} (-1)^{2n} = 1$

$$\lim_{n\to\infty} a_{2n+1} = \lim_{n\to\infty} (-1)^{2n+1} = -1.$$

Thus $\lim_{n\to\infty} a_n$ does not exist $\Rightarrow$ the sequence does not converge.

Hence this sequence oscillates finitely.

(*ii*) An unbounded sequence which does not diverge is said to **oscillate infinitely.**

For example, consider the sequence $\{(-1)^n\, n\}$.

Here $$a_n = (-1)^n\, n.$$

It is an unbounded sequence.

$$\lim_{n\to\infty} a_{2n} = \lim_{n\to\infty} (-1)^{2n} \,.\, 2n = \lim_{n\to\infty} 2n = +\infty$$

$$\lim_{n\to\infty} a_{2n+1} = \lim_{n\to\infty} (-1)^{2n+1}\,(2n+1) = \lim_{n\to\infty} -(2n+1) = -\infty.$$

Thus the sequence does not diverge.

Hence this sequence oscillates infinitely.

Note. When we say $\lim_{n\to\infty} a_n = l$, it means $\lim_{n\to\infty} a_{2n} = \lim_{n\to\infty} a_{2n+1} = l$

Similarly $\lim_{n\to\infty} a_n = +\infty$ means $\lim_{n\to\infty} a_{2n} = \lim_{n\to\infty} a_{2n+1} = +\infty.$

1.7. MONOTONIC SEQUENCES

(*i*) A sequence $\{a_n\}$ is said to be **monotonically increasing** if $a_{n+1} \geq a_n \;\; \forall \;\; n \in \mathrm{N}$.

i.e., if $$a_1 \leq a_2 \leq a_3 \leq \ldots \leq a_n \leq a_{n+1} \leq \ldots$$

(*ii*) A sequence $\{a_n\}$ is said to be **monotonically decreasing** if $a_{n+1} \leq a_n \;\; \forall \;\; n \in \mathrm{N}$.

i.e., if $$a_1 \geq a_2 \geq a_3 \geq \ldots \geq a_n \geq a_{n+1} \geq \ldots$$

(*iii*) A sequence $\{a_n\}$ is said to be **monotonic** if it is either monotonically increasing or monotonically decreasing.

(*iv*) A sequence $\{a_n\}$ is said to be **strictly monotonically increasing** if

$$a_{n+1} > a_n \;\; \forall \;\; n \in \mathrm{N}.$$

(*v*) A sequence $\{a_n\}$ is said to be **strictly monotonically decreasing** if

$$a_{n+1} < a_n \;\; \forall \;\; n \in \mathrm{N}.$$

(*vi*) A sequence $\{a_n\}$ is said to be **strictly monotonic** if it is either strictly monotonically increasing or strictly monotonically decreasing.

1.8. LIMIT OF A SEQUENCE

A sequence $\{a_n\}$ is said to approach the limit l (say) when $n \to \infty$, if for each $\varepsilon > 0$, there exists a +ve integer m (depending upon ε) such that $|\, a_n - l \,| < \varepsilon \;\; \forall \;\; n \geq m$.

In symbols, we write $\lim_{n\to\infty} a_n = l$.

Note. $|\, a_n - l \,| < \varepsilon \; \forall \; n \geq m \;\; \Rightarrow \;\; l - \varepsilon < a_n < l + \varepsilon \;$ for $\; n = m, m+1, m+2, \cdots$

1.9. EVERY CONVERGENT SEQUENCE IS BOUNDED

Let the sequence $\{a_n\}$ be convergent. Let it tend to the limit l.

Then given $\varepsilon > 0$, there exists a +ve integer m, such that

$$| a_n - l | < \varepsilon \quad \forall \quad n \geq m$$

$$\Rightarrow \qquad l - \varepsilon < a_n < l + \varepsilon \quad \forall \quad n \geq m.$$

Let k and K be the least and the greatest of $a_1, a_2, a_3, \ldots, a_{m-1}, l - \varepsilon, l + \varepsilon$

Then $\qquad k \leq a_n \leq K \quad \forall \quad n \in N,$

$\Rightarrow$ the sequence $\{a_n\}$ is bounded.

The converse is not always true *i.e.*, a sequence may be bounded, yet it may not be convergent. *e.g.*, Consider $a_n = (-1)^n$, then the sequence $\{a_n\}$ is bounded but not convergent since it does not have a unique limit.

1.10. CONVERGENCE OF MONOTONIC SEQUENCES

Theorem I. *The necessary and sufficient condition for the convergence of a monotonic sequence is that it is bounded.*

A monotonic increasing sequence which is bounded above converges.

A monotonic decreasing sequence which is bounded below converges.

Theorem II. *If a monotonic increasing sequence is not bounded above, it diverges to* $+\infty$.

Theorem III. *If a monotonic decreasing sequence is not bounded below, it diverges to* $-\infty$.

Theorem IV. *If* $\{a_n\}$ *and* $\{b_n\}$ *are two convergent sequences, then sequence* $\{a_n + b_n\}$ *is also convergent.*

Or

If $\quad \lim a_n = A \quad$ *and* $\quad \lim b_n = B,\quad$ *then* $\quad \lim (a_n + b_n) = A + B.$

Theorem V. *If* $\{a_n\}$ *and* $\{b_n\}$ *are two convergent sequences such that* $\lim a_n = A$ *and* $\lim b_n = B$, *then*

(*i*) *sequence* $\{a_n b_n\}$ *is also convergent and converges to AB.*

(*ii*) *sequence* $\left\{\dfrac{a_n}{b_n}\right\}$ *is also convergent and converges to* $\dfrac{A}{B}$, $(B \neq 0)$.

Theorem VI. *The sequence* $\{| a_n |\}$ *converges to zero if and only if the sequence* $\{a_n\}$ *converges to zero.*

Theorem VII. *If a sequence* $\{a_n\}$ *converges to a and* $a_n \geq 0 \quad \forall\ n$, *then* $a \geq 0$.

Theorem VIII. *If* $a_n \to a$, $b_n \to b$ *and* $a_n \leq b_n \quad \forall\ n$, *then* $a \leq b$.

Theorem IX. *If* $a_n \to l$, $b_n \to l$, *and* $a_n \leq c_n \leq b_n$, $\forall\ n$, *then* $c_n \to l$. (*Squeeze Principle*)

ILLUSTRATIVE EXAMPLES

Example 1. *Give an example of a monotonic increasing sequence which is (i) convergent, (ii) divergent.*

Sol. (*i*) Consider the sequence $\frac{1}{2}, \frac{2}{3}, \frac{3}{4}, \ldots, \frac{n}{n+1}, \ldots$

Since $\frac{1}{2} < \frac{2}{3} < \frac{3}{4} < \ldots$ the sequence is monotonic increasing.

$$a_n = \frac{n}{n+1}, \quad \lim_{n\to\infty} a_n = \lim_{n\to\infty} \frac{n}{n+1} = \lim_{n\to\infty} \frac{1}{1+\frac{1}{n}} = 1$$

which is finite. $\therefore$ The sequence is convergent.

(*ii*) Consider the sequence 1, 2, 3, ···, n, ···

Since $1 < 2 < 3 < \cdots$, the sequence is monotonic increasing,

$$a_n = n, \quad \lim_{n\to\infty} a_n = \lim_{n\to\infty} n = \infty$$

$\therefore$ The sequence diverges to $+\infty$.

Example 2. *Give an example of a monotonic decreasing sequence which is (i) convergent, (ii) divergent.*

Sol. (*i*) Consider the sequence $1, \frac{1}{2}, \frac{1}{3}, \ldots, \frac{1}{n}, \ldots$

Since $1 > \frac{1}{2} > \frac{1}{3} > \ldots$, the sequence is monotonic decreasing.

$$a_n = \frac{1}{n}, \quad \lim_{n\to\infty} a_n = \lim_{n\to\infty} \frac{1}{n} = 0.$$

$\therefore$ The sequence converges to 0.

(*ii*) Consider the sequence $-1, -2, -3, \cdots, -n, \cdots$

Since $-1 > -2 > -3 > \ldots$, the sequence is monotonic decreasing.

$$a_n = -n, \quad \lim_{n\to\infty} a_n = \lim_{n\to\infty} (-n) = -\infty$$

$\therefore$ The sequence diverges to $-\infty$.

Example 3. *Discuss the convergence of the sequence* $\{a_n\}$ *where*

(*i*) $a_n = \frac{n+1}{n}$ (*ii*) $a_n = \frac{n}{n^2+1}$ (*iii*) $a_n = 1 + \frac{1}{3} + \frac{1}{3^2} + \cdots + \frac{1}{3^n}$.

Sol. (*i*) Here $a_n = \frac{n+1}{n} = 1 + \frac{1}{n}$

$$\therefore \quad a_{n+1} - a_n = \left(1 + \frac{1}{n+1}\right) - \left(1 + \frac{1}{n}\right) = \frac{-1}{n(n+1)} < 0 \quad \forall\, n$$

$\Rightarrow$ $a_{n+1} < a_n \ \forall\, n$

$\Rightarrow$ $\{a_n\}$ is a decreasing sequence.

Also $a_n = \frac{n+1}{n} = 1 + \frac{1}{n} > 1 \quad \forall\, n$

$\Rightarrow$ $\{a_n\}$ is bounded below by 1,

$\because$ $\{a_n\}$ is decreasing and bounded below, it is convergent.

$$\lim_{n\to\infty} a_n = \lim_{n\to\infty}\left(1+\frac{1}{n}\right) = 1.$$

$\therefore$ The sequence $\{a_n\}$ converges to 1.

(*ii*) Here $\quad a_n = \dfrac{n}{n^2+1}$

$$\therefore \quad a_{n+1} - a_n = \frac{n+1}{(n+1)^2+1} - \frac{n}{n^2+1} = \frac{(n+1)(n^2+1) - n(n^2+2n+2)}{(n^2+2n+2)(n^2+1)}$$

$$= \frac{-n^2-n+1}{(n^2+2n+2)(n^2+1)} < 0 \quad \forall \; n \;\Rightarrow\; a_{n+1} < a_n \;\; \forall\, n$$

$\Rightarrow$ $\{a_n\}$ is a decreasing sequence.

Also $\quad a_n = \dfrac{n}{n^2+1} > 0 \;\; \forall \; n \;\Rightarrow\;$ $\{a_n\}$ is bounded below by 0.

$\because$ $\{a_n\}$ is decreasing and bounded below, it is convergent.

$$\lim_{n\to\infty} a_n = \lim_{n\to\infty} \frac{n}{n^2+1} = \lim_{n\to\infty} \frac{\frac{1}{n}}{1+\frac{1}{n^2}} = 0.$$

$\therefore$ The sequence $\{a_n\}$ converges to 0.

(*iii*) Here $a_n = 1 + \dfrac{1}{3} + \dfrac{1}{3^2} + \cdots + \dfrac{1}{3^n}$

= sum of $(n + 1)$ terms of a G.P. whose first term is 1 and common ratio is $\frac{1}{3}$

$$= \frac{1\left(1-\frac{1}{3^{n+1}}\right)}{1-\frac{1}{3}} \qquad \left|\; S_n = \frac{a(1-r^n)}{1-r}\right.$$

$$= \frac{3}{2}\left(1-\frac{1}{3^{n+1}}\right)$$

Now $\quad a_{n+1} = 1 + \dfrac{1}{3} + \dfrac{1}{3^2} + \cdots + \dfrac{1}{3^n} + \dfrac{1}{3^{n+1}}$

$$\therefore \quad a_{n+1} - a_n = \frac{1}{3^{n+1}} > 0 \;\; \forall\, n \;\Rightarrow\; a_{n+1} > a_n \;\; \forall\, n$$

$\Rightarrow$ $\{a_n\}$ is an increasing sequence.

Also $\quad a_n = \dfrac{3}{2}\left(1-\dfrac{1}{3^{n+1}}\right) < \dfrac{3}{2} \;\; \forall\, n \;\Rightarrow\;$ $\{a_n\}$ is bounded above by $\dfrac{3}{2}$.

$\because$ $\{a_n\}$ is increasing and bounded above, it is convergent.

$$\lim_{n\to\infty} a_n = \lim_{n\to\infty} \frac{3}{2}\left(1-\frac{1}{3^{n+1}}\right) = \frac{3}{2}(1-0) = \frac{3}{2}.$$

$\therefore$ The sequence $\{a_n\}$ converges to $\dfrac{3}{2}$.

1.11. INFINITE SERIES

If $\{u_n\}$ is a sequence of real numbers, then the expression $u_1 + u_2 + u_3 + \cdots + u_n \cdots$ [*i.e.*, the sum of the terms of the sequence, which are infinite in number] is called an infinite series.

The infinite series $u_1 + u_2 + \cdots + u_n + \cdots$ is denoted by $\sum_{n=1}^{\infty} u_n$ or more briefly, by Σu_n.

1.12. SERIES OF POSITIVE TERMS

If all the terms of the series $\Sigma u_n = u_1 + u_2 + \cdots + u_n + \cdots$ are positive *i.e.*, if $u_n > 0, \forall\, n$, then the series Σu_n is called a series of positive terms.

1.13. ALTERNATING SERIES

A series in which the terms are alternately positive and negative is called an alternating series. Thus, the series $\Sigma\, (-1)^{n-1}\, u_n = u_1 - u_2 + u_3 - u_4 + \cdots + (-1)^{n-1}\, u_n + \cdots$ where $u_n > 0\ \forall\, n$ is an alternating series.

1.14. PARTIAL SUMS

If $\Sigma u_n = u_1 + u_2 + u_3 + \cdots + u_n + \cdots$ is an infinite series where the terms may be +ve or –ve, then

$S_n = u_1 + u_2 + \cdots + u_n$ is called the nth partial sum of Σu_n. Thus, the nth partial sum of an infinite series is the sum of its first n terms.

$S_1, S_2, S_3, \ldots$ are the first, second, third, ... partial sums of the series.

Since $n \in N$, $\{S_n\}$ is a sequence called the **sequence of partial sums** of the infinite series Σu_n.

$\therefore$ To every infinite series Σu_n, there corresponds a sequence $\{S_n\}$ of its partial sums.

1.15. BEHAVIOUR OF AN INFINITE SERIES (Convergence, Divergence and Oscillation)

An infinite series Σu_n converges, diverges or oscillates (finitely or infinitely) according as the sequence $\{S_n\}$ of its partial sums converges, diverges or oscillates (finitely or infinitely).

(*i*) The series Σu_n converges (or is said to be convergent) if the sequence $\{S_n\}$ of its partial sums converges.

Thus, Σu_n is convergent if $\lim\limits_{n \to \infty} S_n$ = a finite quantity.

(*ii*) The series Σu_n diverges (or is said to be divergent) if the sequence $\{S_n\}$ of its partial sums diverges.

Thus, Σu_n is divergent if $\lim\limits_{n \to \infty} S_n = +\infty$ or $-\infty$

(*iii*) The series Σu_n oscillates finitely if the sequence $\{S_n\}$ of its partial sums oscillates finitely.

Thus, Σu_n oscillates finitely if $\{S_n\}$ is bounded and neither converges nor diverges.

(*iv*) The series Σu_n oscillates infinitely if the sequence $\{S_n\}$ of its partial sums oscillates infinitely.

Thus, Σu_n oscillates infinitely if $\{S_n\}$ is unbounded and neither converges nor diverges.

ILLUSTRATIVE EXAMPLES

Example 1. *Discuss the convergence or otherwise of the series*

$$\frac{1}{1\cdot 2}+\frac{1}{2\cdot 3}+\frac{1}{3\cdot 4}+\ldots+\frac{1}{n(n+1)}+\ldots\infty.$$

Sol. Here $\quad u_n=\dfrac{1}{n(n+1)}=\dfrac{(n+1)-n}{n(n+1)}=\dfrac{1}{n}-\dfrac{1}{n+1}$

Putting $n = 1, 2, 3, \cdots, n$, we get

$$u_1=\frac{1}{1}-\frac{1}{2}$$

$$u_2=\frac{1}{2}-\frac{1}{3}$$

$$u_3=\frac{1}{3}-\frac{1}{4}$$

....

....

$$u_n=\frac{1}{n}-\frac{1}{n+1}$$

Adding, $\quad S_n=1-\dfrac{1}{n+1}$

$$\lim_{n\to\infty} S_n=1-0=1$$

$\Rightarrow$ $\{S_n\}$ converges to 1 $\Rightarrow$ Σu_n converges to 1.

Note. For another method, see Comparison Test.

Example 2. *Show that the series* $\displaystyle\sum_{n=1}^{\infty}\frac{1}{n(n+2)}$ *converges to* $\dfrac{3}{4}$.

Sol. Let $\quad u_n=\dfrac{1}{n(n+2)}=\dfrac{1}{2}\cdot\dfrac{(n+2)-n}{n(n+2)}=\dfrac{1}{2}\left(\dfrac{1}{n}-\dfrac{1}{n+2}\right)$

$$=\frac{1}{2}\left[\left(\frac{1}{n}-\frac{1}{n+1}\right)+\left(\frac{1}{n+1}-\frac{1}{n+2}\right)\right]$$

Putting $n = 1, 2, 3, \ldots, n$, we get

$$u_1=\frac{1}{2}\left[\left(1-\frac{1}{2}\right)+\left(\frac{1}{2}-\frac{1}{3}\right)\right]$$

$$u_2=\frac{1}{2}\left[\left(\frac{1}{2}-\frac{1}{3}\right)+\left(\frac{1}{3}-\frac{1}{4}\right)\right]$$

$$u_3=\frac{1}{2}\left[\left(\frac{1}{3}-\frac{1}{4}\right)+\left(\frac{1}{4}-\frac{1}{5}\right)\right]$$

....

....

$$u_n=\frac{1}{2}\left[\left(\frac{1}{n}-\frac{1}{n+1}\right)+\left(\frac{1}{n+1}-\frac{1}{n+2}\right)\right]$$

Adding, $\quad S_n=\dfrac{1}{2}\left[\left(1-\dfrac{1}{n+1}\right)+\left(\dfrac{1}{2}-\dfrac{1}{n+2}\right)\right]$

$$\lim_{n\to\infty} S_n = \frac{1}{2}\left(1-0+\frac{1}{2}-0\right) = \frac{3}{4}, \text{ a finite quantity.}$$

$\Rightarrow$ the sequence $< S_n >$ converges to $\frac{3}{4}$.

$\Rightarrow$ the infinite series $\sum_{n=1}^{\infty} u_n$ converges to $\frac{3}{4}$.

[**Note.** For another method, see Comparison Test.]

Example 3. *Show that the series* $\sum_{n=1}^{\infty}\left(\frac{3}{4}\right)^{n-1}$ *converges to 4.*

Sol. Let $$u_n = \left(\frac{3}{4}\right)^{n-1}$$

then
$$S_n = u_1 + u_2 + u_3 + \cdots + u_n$$
$$= 1 + \frac{3}{4} + \left(\frac{3}{4}\right)^2 + \cdots + \left(\frac{3}{4}\right)^{n-1}$$
$$= \frac{1 \times \left[1-\left(\frac{3}{4}\right)^n\right]}{1-\frac{3}{4}} \qquad \left[\because \text{ In a G.P. } S_n = \frac{a(1-r^n)}{1-r}\right]$$
$$= 4\left[1-\left(\frac{3}{4}\right)^n\right]$$
$$\lim_{n\to\infty} S_n = 4\,[1-0] \qquad \left[\because \text{ if } |x|<1, \text{ then } x^n \to 0 \text{ as } n\to\infty\right]$$
$$= 4, \text{ a finite quantity.}$$

$\Rightarrow$ the sequence $< S_n >$ converges to 4.

$\Rightarrow$ the infinite series $\sum_{n=1}^{\infty} u_n$ converges to 4.

Example 4. *Show that the series* $1^2 + 2^2 + 3^2 + \cdots + n^2 + \cdots$ *diverges to* $+\infty$.

Sol. Here $$S_n = 1^2 + 2^2 + 3^2 + \cdots + n^2 = \frac{n(n+1)(2n+1)}{6}$$
$$\lim_{n\to\infty} S_n = +\infty$$

$\Rightarrow$ $< S_n >$ diverges to $+\infty$.

$\Rightarrow$ the given series diverges to $+\infty$.

Example 5. *Show that the series* $-1-2-3-\cdots-n-\cdots$ *diverges to* $-\infty$.

Sol. Here $$S_n = -1-2-3-\cdots-n$$
$$= -(1+2+3+\cdots+n) = -\frac{n(n+1)}{2}$$
$$\lim_{n\to\infty} S_n = -\infty \quad \Rightarrow \quad < S_n > \text{ diverges to } -\infty$$

$\Rightarrow$ the given series diverges to $-\infty$.

Example 6. *Test the convergence or otherwise of* $\sum_{n=1}^{\infty} (-1)^{n-1}$.

Sol. Here $S_n = 1 - 1 + 1 - 1 + 1 - 1 + \cdots$ to n terms

$= 1$ or 0 according as n is odd or even.

The subsequence $< S_{2n-1} >$ converges to 1, while the subsequence $< S_{2n} >$ converges to 0.

$\Rightarrow$ $< S_n >$ is not convergent.

Since $< S_n >$ is bounded.

$\therefore$ $< S_n >$ oscillates finitely. $\Rightarrow$ $\sum_{n=1}^{\infty} (-1)^{n-1}$ oscillates finitely.

Example 7. *Test the convergence of the series*

$$5 - 4 - 1 + 5 - 4 - 1 + 5 - 4 - 1 + \cdots \text{ to } \infty.$$

Sol. Here $S_n = 5 - 4 - 1 + 5 - 4 - 1 + 5 - 4 - 1 + \cdots$ to n terms $= 0, 5$ or 1 according as the number of terms is $3m$, $3m + 1$ or $3m + 2$ respectively.

Clearly S_n does not tend to a unique limit. Since $< S_n >$ is bounded, it oscillates finitely.

$\Rightarrow$ The given series oscillates finitely.

Example 8. *Show that the series* $\sum_{n=1}^{\infty} n(-1)^n$ *oscillates infinitely.*

Sol. Here $S_n = -1 + 2 - 3 + 4 - 5 + 6 + \cdots$ to n terms

$$= \begin{cases} -\left(\dfrac{n+1}{2}\right), & \text{if } n \text{ is odd} \\ \dfrac{n}{2}, & \text{if } n \text{ is even} \end{cases}$$

The subsequence $< S_{2n-1} >$ diverges to $-\infty$, while the subsequence $< S_{2n} >$ diverges to $+\infty$.

$\therefore$ $< S_n >$ oscillates infinitely.

$\Rightarrow$ $\sum_{n=1}^{\infty} n(-1)^n$ oscillates infinitely.

EXERCISE 1.1

Test the nature of the following series:

1. $1 + 2 + 3 + 4 + \cdots \infty$

2. $1 + 3 + 5 + 7 + \cdots \infty$

3. $1 + \dfrac{1}{2} + \dfrac{1}{4} + \dfrac{1}{8} + \cdots \infty$

4. $1 - \dfrac{1}{5} + \dfrac{1}{5^2} - \dfrac{1}{5^3} + \cdots \infty$

5. $1^2 + 3^2 + 5^2 + 7^2 + \cdots \infty$

6. $7 - 4 - 3 + 7 - 4 - 3 + 7 - 4 - 3 + \cdots \infty$

7. $\dfrac{1}{1.3} + \dfrac{1}{3.5} + \dfrac{1}{5.7} + \cdots \infty$

8. $-1 - 8 - 27 - 64 - \cdots \infty$.

Answers

1. Divergent **2.** Divergent **3.** Convergent **4.** Convergent

5. Divergent **6.** Oscillatory **7.** Convergent **8.** Divergent.

1.16. THE GEOMETRIC SERIES $1 + x + x^2 + x^3 + \cdots \infty$ (*M.D.U. Dec. 2010*)

(*i*) converges if $-1 < x < 1$, *i.e.*, $|x| < 1$ (*ii*) diverges if $x \geq 1$

(*iii*) oscillates finitely if $x = -1$ (*iv*) oscillates infinitely if $x < -1$.

Proof. (*i*) When $|x| < 1$

Since $|x| < 1, x^n \to 0$ as $n \to \infty$

$$S_n = 1 + x + x^2 + \cdots \text{ to } n \text{ terms} = \frac{1(1-x^n)}{1-x} = \frac{1}{1-x} - \frac{x^n}{1-x}$$

$$\lim_{n\to\infty} S_n = \frac{1}{1-x} \Rightarrow \text{ the sequence } \{S_n\} \text{ is convergent}$$

$\Rightarrow$ the given series is convergent.

(*ii*) When $x \geq 1$

Sub-case I. When $x = 1$

$$S_n = 1 + 1 + 1 + \cdots \text{ to } n \text{ terms} = n$$

$$\lim_{n\to\infty} S_n = \infty \Rightarrow \text{ the sequence } \{S_n\} \text{ diverges to } \infty.$$

$\Rightarrow$ the given series diverges to ∞.

Sub-case II. When $x > 1$, $x^n \to \infty$ as $n \to \infty$

$$S_n = 1 + x + x^2 + \cdots \text{ to } n \text{ terms} = \frac{1(x^n - 1)}{x-1}$$

$$\lim_{n\to\infty} S_n = \infty \Rightarrow \text{ the sequence } \{S_n\} \text{ diverges to } \infty$$

$\Rightarrow$ the given series diverges to ∞.

(*iii*) When $x = -1$

$$S_n = 1 - 1 + 1 - 1 + \cdots \text{ to } n \text{ terms}$$

$$= 1 \text{ or } 0 \text{ according as } n \text{ is odd or even.}$$

$$\Rightarrow \lim_{n\to\infty} S_n = 1 \text{ or } 0 \Rightarrow \text{ the sequence } \{S_n\} \text{ oscillates finitely.}$$

$\Rightarrow$ the given series oscillates finitely.

(*iv*) When $x < -1$

$x < -1 \Rightarrow -x > 1$

Let $r = -x$, then $r > 1$

$\therefore$ $r^n \to \infty$ as $n \to \infty$

$$S_n = 1 + x + x^2 + x^3 + \cdots \text{ to } n \text{ terms} = \frac{1-x^n}{1-x} = \frac{1-(-r)^n}{1+r} \qquad [\because x = -r]$$

$$= \frac{1-r^n}{1+r} \text{ or } \frac{1+r^n}{1+r} \text{ according as } n \text{ is even or odd}$$

$$\lim_{n\to\infty} S_n = \frac{1-\infty}{1+r} \text{ or } \frac{1+\infty}{1+r} = -\infty \text{ or } +\infty$$

$\Rightarrow$ the sequence $\{S_n\}$ oscillates infinitely

$\Rightarrow$ the given series oscillates infinitely.

Examples. 1. Consider $1-\frac{3}{4}+\frac{9}{16}-\frac{27}{64}+\cdots$

It is a geometric series with common ratio $r=-\frac{3}{4}$.

Since $|r|=\left|-\frac{3}{4}\right|=\frac{3}{4}<1$, the given series is convergent.

2. Consider $2+3+\frac{9}{2}+\frac{27}{4}+\cdots$

It is a geometric series with $r=\frac{3}{2}>1$

$\Rightarrow$ the given series is divergent.

3. Consider $\frac{3}{4}+\frac{3}{4}+\frac{3}{4}+\frac{3}{4}+\cdots$

It is a geometric series with $r=1$

$\Rightarrow$ the given series is divergent.

4. Consider $\frac{3}{4}-\frac{3}{4}+\frac{3}{4}-\frac{3}{4}+\cdots$

It is a geometric series with $r=-1$

$\Rightarrow$ the given series oscillates finitely.

5. Consider $1-\frac{5}{2}+\frac{25}{4}-\frac{125}{8}+\cdots$

It is a geometric series with $r=-\frac{5}{2}<-1$

$\Rightarrow$ the given series oscillates infinitely.

1.17. NECESSARY CONDITION FOR CONVERGENCE *(M.D.U. Dec. 2009)*

If a series Σu_n is convergent, then $\lim_{n\to\infty} u_n = 0$.

Proof. Let S_n denote the nth partial sum of the series Σu_n.

Then Σu_n is convergent $\Rightarrow$ $\{S_n\}$ is convergent

$\Rightarrow$ $\lim_{n\to\infty} S_n$ is finite and unique $= s$ (say). $\Rightarrow$ $\lim_{n\to\infty} S_{n-1} = s$

Now $S_n - S_{n-1} = u_n$

$\therefore$ $\lim_{n\to\infty} u_n = \lim_{n\to\infty} (S_n - S_{n-1}) = \lim_{n\to\infty} S_n - \lim_{n\to\infty} S_{n-1} = s - s = 0.$

Hence Σu_n is convergent $\Rightarrow$ $\lim_{n\to\infty} u_n = 0$.

The converse of the above theorem is not always true, *i.e.*, the nth term may tend to zero as $n \to \infty$ even if the series is not convergent.

For example, consider the series $1+\frac{1}{\sqrt{2}}+\frac{1}{\sqrt{3}}+\cdots+\frac{1}{\sqrt{n}}+\cdots$

Here $$S_n = 1+\frac{1}{\sqrt{2}}+\frac{1}{\sqrt{3}}+\cdots+\frac{1}{\sqrt{n}}$$

$$> \frac{1}{\sqrt{n}} + \frac{1}{\sqrt{n}} + \frac{1}{\sqrt{n}} + \ldots + \frac{1}{\sqrt{n}} \qquad \left(\because\ m < n \Rightarrow \frac{1}{\sqrt{m}} > \frac{1}{\sqrt{n}}\right)$$

$$= \frac{n}{\sqrt{n}} = \sqrt{n}$$

$$\therefore \qquad \lim_{n\to\infty} S_n = \lim_{n\to\infty} \sqrt{n} \to \infty$$

$\Rightarrow$ the series is divergent, though $\lim_{n\to\infty} u_n = \lim_{n\to\infty} \frac{1}{\sqrt{n}} = 0$

Thus $\lim_{n\to\infty} u_n = 0$ is a necessary condition but not a sufficient condition for convergence of Σu_n.

Note 1. Σu_n is convergent $\Rightarrow \lim_{n\to\infty} u_n = 0$

Note 2. $\lim_{n\to\infty} u_n = 0 \Rightarrow \Sigma u_n$ may or may not be convergent

Note 3. $\lim_{n\to\infty} u_n \neq 0 \Rightarrow \Sigma u_n$ is not convergent.

1.18. A POSITIVE TERM SERIES EITHER CONVERGES OR DIVERGES TO +∞

Proof. Let Σu_n be a positive term series and S_n be its nth partial sum.

Then $\qquad S_{n+1} = u_1 + u_2 + \cdots + u_n + u_{n+1} = S_n + u_{n+1}$

$\Rightarrow \qquad S_{n+1} - S_n = u_{n+1} > 0 \quad \forall\ n \qquad [\because\ u_n > 0\ \forall\ n]$

$\Rightarrow \qquad S_{n+1} > S_n \quad \forall\ n$

$\Rightarrow$ $\{S_n\}$ is a monotonic increasing sequence.

Two cases arise. The sequence $\{S_n\}$ may be bounded or unbounded above.

Case I. *When $\{S_n\}$ is bounded above.*

Since $\{S_n\}$ is monotonic increasing and bounded above, it is convergent $\Rightarrow \Sigma u_n$ is convergent.

Case II. *When $\{S_n\}$ is not bounded above.*

Since $\{S_n\}$ is monotonic increasing and not bounded above, it diverges to $+\infty \Rightarrow \Sigma u_n$ diverges to $+\infty$.

Hence a positive term series either converges or diverges to +∞.

Cor. If $\quad u_n > 0 \ \forall\ n$ and $\lim_{n\to\infty} u_n \neq 0$, then the series Σu_n diverges to $+\infty$.

Proof. $\quad u_n > 0 \ \forall\ n \Rightarrow \Sigma u_n$ is a series of +ve terms.

$\Rightarrow \Sigma u_n$ either converges or diverges to $+\infty$

Since $\qquad \lim_{n\to\infty} u_n \neq 0$ (given)

$\therefore\ \Sigma u_n$ does not converge.

Hence Σu_n must diverge to $+\infty$.

1.19. (a) THE NECESSARY AND SUFFICIENT CONDITION FOR THE CONVERGENCE OF A POSITIVE TERM SERIES Σu_n IS THAT THE SEQUENCE $\{S_n\}$ OF ITS PARTIAL SUMS IS BOUNDED ABOVE

Proof. (*i*) Suppose the sequence $\{S_n\}$ is bounded above. Since the series Σu_n is of positive terms, the sequence $\{S_n\}$ is monotonically increasing. Since every monotonically increasing sequence which is bounded above, converges, therefore $\{S_n\}$ and hence Σu_n converges.

(*ii*) Conversely, suppose Σu_n converges. Then the sequence $\{S_n\}$ of its partial sums also converges. Since every convergent sequence is bounded, $\{S_n\}$ is bounded. In particular, $\{S_n\}$ is bounded above.

1.19. (b) CAUCHY'S GENERAL PRINCIPLE OF CONVERGENCE OF SERIES

The necessary and sufficient condition for the infinite series $\sum_{n=1}^{\infty} u_n$ to converge is that given $\varepsilon > 0$, however small, there exists a positive integer m such that

$$| u_{m+1} + u_{m+2} + \cdots + u_n | < \varepsilon \quad \forall \quad n > m.$$

Example. *Prove with the help of Cauchy's general principle of convergence that the series*

$$\sum_{n=1}^{\infty} \frac{1}{n} = 1 + \frac{1}{2} + \frac{1}{3} + \cdots + \frac{1}{n} + \cdots \textit{ does not converge.}$$

Sol. If possible, suppose the given series is convergent.

Take $\varepsilon = \dfrac{1}{2}$

By Cauchy's general principle of convergence, there exists a positive integer m such that

$$\left| \frac{1}{m+1} + \frac{1}{m+2} + \cdots + \frac{1}{n} \right| < \frac{1}{2} \quad \forall \quad n > m$$

or
$$\frac{1}{m+1} + \frac{1}{m+2} + \cdots + \frac{1}{n} < \frac{1}{2} \quad \forall \quad n > m \qquad \ldots(1)$$

By taking $n = 2m$, we observe that $\dfrac{1}{m+1} + \dfrac{1}{m+2} + \cdots + \dfrac{1}{n}$

$$= \frac{1}{m+1} + \frac{1}{m+2} + \cdots + \frac{1}{2m} > \frac{1}{2m} + \frac{1}{2m} + \cdots + \frac{1}{2m} = \frac{m}{2m} = \frac{1}{2}$$

i.e.,
$$\frac{1}{m+1} + \frac{1}{m+2} + \cdots + \frac{1}{n} > \frac{1}{2}$$

where $n = 2m > m$. This contradicts (1).

$\Rightarrow$ our supposition is wrong.

$\Rightarrow$ the given series does not converge.

1.20. IF m IS A GIVEN POSITIVE INTEGER, THEN THE TWO SERIES $u_1 + u_2 + \cdots + u_{m+1} + u_{m+2} + \cdots$ AND $u_{m+1} + u_{m+2} + \cdots$ CONVERGE OR DIVERGE TOGETHER

Proof. Let S_n and s_n denote the nth partial sums of the two series.

Then $\qquad S_n = u_1 + u_2 + \cdots + u_n$

$$s_n = u_{m+1} + u_{m+2} + \cdots + u_{m+n}$$
$$= (u_1 + u_2 + \cdots + u_{m+n}) - (u_1 + u_2 + \cdots + u_m)$$
$$= S_{m+n} - S_m \Rightarrow s_n = S_{m+n} - S_m \qquad ...(1)$$

S_m being the sum of a finite number of terms of Σu_n is a fixed finite quantity.

(i) If $S_{m+n} \to$ a finite limit as $n \to \infty$, then from (1), so does s_n.

(ii) If $S_{m+n} \to +\infty$ as $n \to \infty$, so does s_n.

(iii) If $S_{m+n} \to -\infty$ as $n \to \infty$, so does s_n.

(iv) If S_{m+n} does not tend to any limit (finite or infinite), so does s_n.

$\Rightarrow$ The sequences $\{S_n\}$ and $\{s_n\}$ converge or diverge together.

$\Rightarrow$ The two given series converge or diverge together. Hence the result.

Note. The above theorem shows that the convergence, divergence or oscillation of a series is not affected by addition or omission of a finite number of its terms.

1.21. IF Σu_n AND Σv_n CONVERGE TO u AND v RESPECTIVELY, THEN $\Sigma(u_n + v_n)$ CONVERGES TO (u + v)

Proof. Let $U_n = u_1 + u_2 + \cdots + u_n$

$V_n = v_1 + v_2 + \cdots + v_n$

and $S_n = (u_1 + v_1) + (u_2 + v_2) + \cdots + (u_n + v_n)$

Then $S_n = (u_1 + u_2 + \cdots + u_n) + (v_1 + v_2 + \cdots + v_n) = U_n + V_n$.

Since Σu_n converges to u, $\lim\limits_{n\to\infty} U_n = u$

Σv_n converges to v, $\lim\limits_{n\to\infty} V_n = v$

$$\therefore \quad \lim_{n\to\infty} S_n = \lim_{n\to\infty} (U_n + V_n) = \lim_{n\to\infty} U_n + \lim_{n\to\infty} V_n = u + v.$$

$\Rightarrow$ $\Sigma(u_n + v_n)$ converges to $(u + v)$.

1.22. COMPARISON TESTS

Test I. *If Σu_n and Σv_n are series of positive terms and Σv_n is convergent and there is a positive constant k such that $u_n \le kv_n$, $\forall\, n > m$, then Σu_n is also convergent.*

Proof. Let $U_n = u_1 + u_2 + \cdots + u_n$ and $V_n = v_1 + v_2 + \cdots + v_n$

Now $u_n \le kv_n \ \forall\, n > m$

$\Rightarrow$ $u_{m+1} \le kv_{m+1}$

$u_{m+2} \le kv_{m+2}$

....

Adding $u_{m+1} + u_{m+2} + \ldots + u_n \le k(v_{m+1} + v_{m+2} + \cdots + v_n)$

$\Rightarrow$ $U_n - U_m \le k(V_n - V_m) \quad \forall \ n > m$

$\Rightarrow$ $U_n \le kV_n + (U_m - kV_m) \quad \forall \ n > m$

$\Rightarrow$ $U_n \le kV_n + k_0 \quad \forall \ n > m \qquad ...(1)$

where $k_0 = U_m - kV_m$ is a fixed number. Since Σv_n is convergent, the sequence $\{V_n\}$ is convergent and hence bounded above.

$\therefore$ From (1), the sequence $\{U_n\}$ is bounded above.

$\because$ Σu_n is a series of +ve terms, $\{U_n\}$ is monotonic increasing.

$\because$ $\{U_n\}$ is monotonic increasing sequence and is bounded above.

$\therefore$ It is convergent.

$\Rightarrow$ Σu_n is convergent.

Test II. *If Σu_n and Σv_n are two series of positive terms and Σv_n is divergent and there is a positive constant k such that $u_n > kv_n$, $\forall\ n > m$, then Σu_n is also divergent.*

Proof. Let $\quad U_n = u_1 + u_2 + \cdots + u_n$

and $\quad V_n = v_1 + v_2 + \cdots + v_n$

Now $\quad u_n > kv_n \quad \forall \quad n > m$

$$\Rightarrow \quad u_{m+1} > kv_{m+1}$$
$$u_{m+2} > kv_{m+2}$$
$$\cdots\cdots\cdots\cdots$$
$$\cdots\cdots\cdots\cdots$$
$$u_n > kv_n$$

Adding $u_{m+1} + u_{m+2} + \cdots + u_n > k\,(v_{m+1} + v_{m+2} + \cdots + v_n)$

$$\Rightarrow \quad U_n - U_m > k\,(V_n - V_m) \quad \forall \quad n > m$$
$$\Rightarrow \quad U_n > kV_n + (U_m - kV_m) \quad \forall \quad n > m$$
$$\Rightarrow \quad U_n > kV_n + k_0 \quad \forall \quad n > m \qquad \ldots(1)$$

where $k_0 = U_m - kV_m$ is a fixed number.

Since Σv_n is divergent, the sequence $\{V_n\}$ is divergent.

$\Rightarrow$ for each positive real number k_1, however large, there exists a +ve integer m' such that

$$V_n > k_1 \quad \forall \quad n > m'$$

Let $\quad m^* = \max.\ \{m, m'\}$, then $V_n > k_1 \quad \forall \quad n > m^*$

From (1), $\quad U_n > kk_1 + k_0 = K \quad \forall \quad n > m^*$

$\Rightarrow$ $\{U_n\}$ is divergent

$\Rightarrow$ Σu_n is divergent.

Test III. *If Σu_n and Σv_n are two positive term series and there exist two positive constants H and K (independent of n) and a positive integer m such that $H < \dfrac{u_n}{v_n} < K\ \forall\ n > m$, then the two series Σu_n and Σv_n converge or diverge together.*

Proof. Since Σv_n is a series of + ve terms, $v_n > 0$, $\ \forall\ n$

$$\therefore \quad H < \frac{u_n}{v_n} < K \quad \forall\ n > m$$
$$\Rightarrow \quad Hv_n < u_n < Kv_n \quad \forall\ n > m \qquad \ldots(1)$$

Case I. *When Σv_n is convergent*

From (1), $\quad u_n < Kv_n \quad \forall\ n > m$ and Σv_n is convergent

$\Rightarrow$ Σu_n is convergent. [See Test I]

Case II. *When Σv_n is divergent*

From (1), $\quad u_n > Hv_n \quad \forall\ n > m$ and Σv_n is divergent.

$\Rightarrow$ Σu_n is divergent. [See Test II]

Case III. *When Σu_n is convergent*

From (1), $\quad Hv_n < u_n \quad \forall\ n > m$

$$\Rightarrow \quad v_n < \frac{1}{H}\,u_n \quad \forall\ n > m \qquad (\because \quad H > 0)$$

Since Σu_n is convergent $\ \therefore\ $ Σv_n is convergent. [See Test I]

Case IV. *When Σu_n is divergent*

From (1), $\quad Kv_n > u_n \quad \forall\ n > m$

$\Rightarrow \qquad v_n > \frac{1}{K} u_n. \qquad \forall\, n > m \qquad (\because \ K > 0)$

Since Σu_n is divergent $\therefore$ Σv_n is divergent. [See Test II]

Particular Case of Test III (When m = 0)

If Σu_n and Σv_n are two positive term series and there exist two positive constants H and K (independent of n) such that $H < \frac{u_n}{v_n} < K \ \forall\, n$,

then the two series Σu_n and Σv_n converge or diverge together.

Test IV. *Let Σu_n and Σv_n be two positive term series.*

(i) If $\lim\limits_{n\to\infty} \frac{u_n}{v_n} = l$ (finite and non-zero), then Σu_n and Σv_n both converge or diverge together.

(ii) If $\lim\limits_{n\to\infty} \frac{u_n}{v_n} = 0$ and Σv_n converges, then Σu_n also converges.

(iii) If $\lim\limits_{n\to\infty} \frac{u_n}{v_n} = \infty$ and Σv_n diverges, then Σu_n also diverges.

Proof. (*i*) Since $u_n > 0,\ v_n > 0 \ \therefore\ \frac{u_n}{v_n} > 0$

$$\therefore \qquad \lim_{n\to\infty} \frac{u_n}{v_n} \ge 0$$

$$\text{But} \qquad \lim_{n\to\infty} \frac{u_n}{v_n} = l \ne 0 \ \Rightarrow \ l > 0$$

$$\text{Now} \qquad \lim_{n\to\infty} \frac{u_n}{v_n} = l$$

$\Rightarrow$ given $\varepsilon > 0$, there exists a +ve integer m such that $\left| \frac{u_n}{v_n} - l \right| < \varepsilon \ \ \forall \ \ n > m$

$$\Rightarrow \qquad l - \varepsilon < \frac{u_n}{v_n} < l + \varepsilon \qquad \forall\, n > m$$

$$\Rightarrow \qquad (l - \varepsilon)\, v_n < u_n < (l + \varepsilon)\, v_n \qquad \forall\, n > m \qquad (\because \ v_n > 0)$$

Choose $\varepsilon > 0$ such that $l - \varepsilon > 0$.

Let $l - \varepsilon = H,\ l + \varepsilon = K$, where H, K are > 0

$$\therefore \qquad Hv_n < u_n < Kv_n \qquad \forall\, n > m \qquad \text{...(1)}$$

Case I. *When Σu_n is convergent*

From (1), $\qquad Hv_n < u_n \qquad \forall\, n > m$

$$\Rightarrow \qquad v_n < \frac{1}{H} u_n \qquad \forall\, n > m \qquad (\because \ H > 0)$$

Since Σu_n is convergent, Σv_n is also convergent.

Case II. *When Σu_n is divergent*

From (1), $\qquad Kv_n > u_n \qquad \forall\, n > m$

$$\Rightarrow \qquad v_n > \frac{1}{K} u_n \qquad \forall\, n > m \qquad (\because \ K > 0)$$

Since Σu_n is divergent, Σv_n is also divergent.

Case III. *When Σv_n is convergent*

From (1), $u_n < Kv_n \quad \forall\ n > m$

Since Σv_n is convergent, Σu_n is also convergent.

Case IV. *When Σv_n is divergent*

From (1), $u_n > Hv_n \quad \forall\ n > m$

Since Σv_n is divergent, Σu_n is also divergent.

Hence Σu_n and Σv_n converge or diverge together.

(*ii*) Here $\lim\limits_{n\to\infty} \dfrac{u_n}{v_n} = 0$

$\therefore$ Given $\varepsilon > 0$, there exists a +ve integer m such that $\left|\dfrac{u_n}{v_n} - 0\right| < \varepsilon \quad \forall\ n > m$

$$\Rightarrow \quad -\varepsilon < \frac{u_n}{v_n} < \varepsilon \quad \forall\ n > m$$

$$\Rightarrow \quad u_n < \varepsilon\, v_n \quad \forall\ n > m \qquad (\because\ v_n > 0)$$

Since Σv_n is convergent, Σu_n is also convergent.

(*iii*) Here $\lim\limits_{n\to\infty} \dfrac{u_n}{v_n} = \infty$

$\therefore$ Given M > 0, however large, there exists a +ve integer m such that $\dfrac{u_n}{v_n} > M \ \forall\ n > m$

$$\Rightarrow \quad u_n > Mv_n \quad \forall\ n > m$$

Since Σv_n is divergent, Σu_n is also divergent.

Test V. *Let Σu_n and Σv_n be two positive term series.*

(*i*) *If* $\dfrac{u_n}{u_{n+1}} > \dfrac{v_n}{v_{n+1}}$ $\forall\ n > m$ *and Σv_n is convergent, then Σu_n is also convergent.*

(*ii*) *If* $\dfrac{u_n}{u_{n+1}} < \dfrac{v_n}{v_{n+1}}$ $\forall\ n > m$ *and Σv_n is divergent, then Σu_n is also divergent.*

Proof. (*i*) $\dfrac{u_n}{u_{n+1}} > \dfrac{v_n}{v_{n+1}} \quad \forall\ n > m$

$$\Rightarrow \quad \frac{u_{m+1}}{u_{m+2}} > \frac{v_{m+1}}{v_{m+2}}$$

$$\frac{u_{m+2}}{u_{m+3}} > \frac{v_{m+2}}{v_{m+3}}$$

$$\frac{u_{m+3}}{u_{m+4}} > \frac{v_{m+3}}{v_{m+4}}$$

....

....

$$\frac{u_{n-1}}{u_n} > \frac{v_{n-1}}{v_n}$$

Multiplying the corresponding sides of the above inequalities, we have

$$\frac{u_{m+1}}{u_n} > \frac{v_{m+1}}{v_n} \quad \forall\ n > m$$

$$\Rightarrow \qquad u_n < \left(\frac{u_{m+1}}{v_{m+1}}\right) v_n \qquad \forall\, n > m$$

$$\Rightarrow \qquad u_n < kv_n \qquad \forall\, n > m,$$

where $k = \dfrac{u_{m+1}}{v_{m+1}}$ is a fixed +ve quantity.

Since Σv_n is convergent, so is Σu_n.

(*ii*) Using $\qquad \dfrac{u_n}{u_{n+1}} < \dfrac{v_n}{v_{n+1}} \qquad \forall\, n > m$

and proceeding as in part (*i*), we have $\dfrac{u_{m+1}}{u_n} < \dfrac{v_{m+1}}{v_n} \qquad \forall\, n > m$

$$\Rightarrow \qquad u_n > \left(\frac{u_{m+1}}{v_{m+1}}\right) v_n \qquad \forall\, n > m$$

$$\Rightarrow \qquad u_n > kv_n \qquad \forall\, n > m,$$

where $k = \dfrac{u_{m+1}}{v_{m+1}}$ is a fixed +ve quantity.

Since Σv_n is divergent, so is Σu_n.

1.23. AN IMPORTANT TEST FOR COMPARISON $\sum \frac{1}{n^p}$ [HYPER HARMONIC SERIES OR P-SERIES]

Statement: The series $\sum \frac{1}{n^p} = \frac{1}{1^p} + \frac{1}{2^p} + \frac{1}{3^p} + \dots + \frac{1}{n^p} + \dots \infty$ converges if p > 1 and diverges if p ≤ 1. (M.D.U. Dec. 2013)

Proof. Case I. *When p > 1*

$$\frac{1}{1^p} = 1$$

$$\frac{1}{2^p} + \frac{1}{3^p} < \frac{1}{2^p} + \frac{1}{2^p} = \frac{2}{2^p} = \frac{1}{2^{p-1}} \qquad \left[\because\ \frac{1}{3^p} < \frac{1}{2^p}\right]$$

$$\frac{1}{4^p} + \frac{1}{5^p} + \frac{1}{6^p} + \frac{1}{7^p} < \frac{1}{4^p} + \frac{1}{4^p} + \frac{1}{4^p} + \frac{1}{4^p}$$

$$= \frac{4}{4^p} = \frac{1}{4^{p-1}} = \frac{1}{(2^{p-1})^2} \qquad \left[\because\ \frac{1}{5^p} < \frac{1}{4^p}, \frac{1}{6^p} < \frac{1}{4^p} \text{ etc.}\right]$$

Similarly, the sum of next eight terms

$$= \frac{1}{8^p} + \frac{1}{9^p} + \dots + \frac{1}{15^p} < \frac{1}{8^p} + \frac{1}{8^p} + \dots + \frac{1}{8^p} = \frac{8}{8^p}$$

$$= \frac{1}{8^{p-1}} = \frac{1}{(2^{p-1})^3} \text{ and so on.}$$

Now $\sum \dfrac{1}{n^p} = \dfrac{1}{1^p} + \dfrac{1}{2^p} + \dfrac{1}{3^p} + \dots + \dfrac{1}{n^p} + \dots$

$$= \frac{1}{1^p} + \left(\frac{1}{2^p} + \frac{1}{3^p}\right) + \left(\frac{1}{4^p} + \frac{1}{5^p} + \frac{1}{6^p} + \frac{1}{7^p}\right) + \left(\frac{1}{8^p} + \frac{1}{9^p} + \dots + \frac{1}{15^p}\right) + \dots \qquad \dots(1)$$

$\therefore$ Each term of (1) after the first is less than the corresponding term in

$$1+\frac{1}{2^{p-1}}+\frac{1}{(2^{p-1})^2}+\frac{1}{(2^{p-1})^3} \quad ...(2)$$

But (2) is a G.P. whose common ratio $= \dfrac{1}{2^{p-1}} < 1$ $\quad (\because \; p > 1)$

$\therefore$ (2) is convergent $\Rightarrow$ (1) is convergent.

Hence the given series is convergent.

Case II. *When p = 1*

$$\sum \frac{1}{n^p} = \sum \frac{1}{n} = 1+\frac{1}{2}+\frac{1}{3}+\frac{1}{4}+\cdots$$

$$1+\frac{1}{2} = 1+\frac{1}{2}$$

$$\frac{1}{3}+\frac{1}{4} > \frac{1}{4}+\frac{1}{4} = \frac{2}{4} = \frac{1}{2}$$

$$\frac{1}{5}+\frac{1}{6}+\frac{1}{7}+\frac{1}{8} > \frac{1}{8}+\frac{1}{8}+\frac{1}{8}+\frac{1}{8} = \frac{4}{8} = \frac{1}{2} \quad \text{and so on.}$$

Now $\quad \sum \dfrac{1}{n} = 1+\dfrac{1}{2}+\dfrac{1}{3}+\dfrac{1}{4}+\cdots+\dfrac{1}{n}+\cdots$

$$= 1+\frac{1}{2}+\left(\frac{1}{3}+\frac{1}{4}\right)+\left(\frac{1}{5}+\frac{1}{6}+\frac{1}{7}+\frac{1}{8}\right)+\cdots \quad ...(1)$$

Each term of (1) after the second is greater than the corresponding term in

$$1+\frac{1}{2}+\frac{1}{2}+\frac{1}{2}+\cdots \quad ...(2)$$

But after the second term (2) is a G.P. whose common ratio = 1.

$\therefore$ (2) is divergent. $\Rightarrow$ (1) is divergent.

Hence the given series is divergent.

Case III. *When p < 1*

$$p < 1 \;\Rightarrow\; n^p < n \;\Rightarrow\; \frac{1}{n^p} > \frac{1}{n} \quad \forall \; n$$

But the series $\sum \dfrac{1}{n}$ is divergent (Case II).

Hence $\sum \dfrac{1}{n^p}$ is also divergent.

Note. The above test can also be proved by using Cauchy's Integral test.

ILLUSTRATIVE EXAMPLES

Example 1. *Examine the convergence of the series:*

(*i*) $\dfrac{3}{5}+\dfrac{4}{5^2}+\dfrac{3}{5^3}+\dfrac{4}{5^4}+\cdots\infty$ $\qquad$ (*ii*) $1+\dfrac{1}{4^{2/3}}+\dfrac{1}{9^{2/3}}+\dfrac{1}{16^{2/3}}+\cdots$

Sol. (*i*) $\dfrac{3}{5}+\dfrac{4}{5^2}+\dfrac{3}{5^3}+\dfrac{4}{5^4}+\cdots\infty$

$$= \left(\frac{3}{5}+\frac{3}{5^3}+\cdots\infty\right)+\left(\frac{4}{5^2}+\frac{4}{5^4}+\cdots\infty\right) = \Sigma u_n + \Sigma v_n \text{ (say)}$$

Now Σu_n is a G.P. with common ratio $= \frac{1}{5^2}$ which is numerically less than 1.

$\therefore$ Σu_n is convergent.

Σv_n is also a G.P. with common ratio $= \frac{1}{5^2}$ which is numerically less than 1.

$\therefore$ Σv_n is convergent.

$\therefore$ The given series *viz.* $\Sigma(u_n + v_n)$ is also convergent.

(ii) $1 + \frac{1}{4^{2/3}} + \frac{1}{9^{2/3}} + \frac{1}{16^{2/3}} + \cdots \infty = 1 + \frac{1}{(2^2)^{2/3}} + \frac{1}{(3^2)^{2/3}} + \frac{1}{(4^2)^{2/3}} + \cdots \infty$

$$= \frac{1}{1^{4/3}} + \frac{1}{2^{4/3}} + \frac{1}{3^{4/3}} + \frac{1}{4^{4/3}} + \cdots \infty = \sum \frac{1}{n^{4/3}} = \sum \frac{1}{n^p} \text{ with } p = \frac{4}{3} > 1$$

$\therefore$ By p-series test, the given series is convergent.

Example 2. *Test the convergence of the series:* $\frac{1}{1\cdot2\cdot3} + \frac{3}{2\cdot3\cdot4} + \frac{5}{3\cdot4\cdot5} + \cdots$

Sol. Here $u_n = \frac{T_n \text{ of } 1, 3, 5, \ldots}{n(n+1)(n+2)} = \frac{2n-1}{n(n+1)(n+2)} = \frac{n\left(2 - \frac{1}{n}\right)}{n^3\left(1+\frac{1}{n}\right)\left(1+\frac{2}{n}\right)}$

$$= \frac{2 - \frac{1}{n}}{n^2\left(1+\frac{1}{n}\right)\left(1+\frac{2}{n}\right)}$$

Let us compare Σu_n with Σv_n, where $v_n = \frac{1}{n^2}$

$$\frac{u_n}{v_n} = \frac{2 - \frac{1}{n}}{\left(1+\frac{1}{n}\right)\left(1+\frac{2}{n}\right)}$$

$$\lim_{n\to\infty} \frac{u_n}{v_n} = \lim_{n\to\infty} \frac{2 - \frac{1}{n}}{\left(1+\frac{1}{n}\right)\left(1+\frac{2}{n}\right)} = \frac{2}{(1)(1)} = 2 \text{ which is finite and non-zero.}$$

$\therefore$ Σu_n and Σv_n converge or diverge together.

Since $\Sigma v_n = \Sigma \frac{1}{n^2}$ is of the form $\Sigma \frac{1}{n^p}$ with $p = 2 > 1$.

$\therefore$ Σv_n is convergent $\Rightarrow$ Σu_n is convergent.

Example 3. *Test the convergence of the following series:*

(i) $\frac{1}{\sqrt{1}+\sqrt{2}} + \frac{1}{\sqrt{2}+\sqrt{3}} + \frac{1}{\sqrt{3}+\sqrt{4}} + \cdots$ *(ii)* $\sqrt{\frac{1}{4}} + \sqrt{\frac{2}{6}} + \sqrt{\frac{3}{8}} + \cdots + \sqrt{\frac{n}{2(n+1)}} + \cdots$

Sol. (*i*) Here $$u_n = \frac{1}{\sqrt{n}+\sqrt{n+1}} = \frac{1}{\sqrt{n}\left[1+\sqrt{1+\frac{1}{n}}\right]}$$

Let us compare Σu_n with Σv_n, where $v_n = \frac{1}{\sqrt{n}}$

$$\frac{u_n}{v_n} = \frac{1}{1+\sqrt{1+\frac{1}{n}}}$$

$$\lim_{n\to\infty}\frac{u_n}{v_n} = \lim_{n\to\infty}\frac{1}{1+\sqrt{1+\frac{1}{n}}} = \frac{1}{1+1} = \frac{1}{2}$$ which is finite and non-zero.

$\therefore$ Σu_n and Σv_n converge or diverge together.

Since $\Sigma v_n = \Sigma \frac{1}{n^{1/2}}$ is of the form $\Sigma \frac{1}{n^p}$ with $p = \frac{1}{2} < 1$

$\therefore$ Σv_n is divergent $\Rightarrow$ Σu_n is divergent.

(*ii*) Here $$u_n = \sqrt{\frac{n}{2(n+1)}} = \sqrt{\frac{1}{2\left(1+\frac{1}{n}\right)}}$$

$$\lim_{n\to\infty} u_n = \lim_{n\to\infty}\frac{1}{\sqrt{2\left(1+\frac{1}{n}\right)}} = \frac{1}{\sqrt{2}} \neq 0$$

$\Rightarrow$ Σu_n does not converge.

Since the given series is a series of +ve terms, it either converges or diverges. Since it does not converge, it must diverge.

Hence the given series is divergent.

Example 4. *Test the convergence of the series:* $1 + \frac{1}{2^2} + \frac{2^2}{3^3} + \frac{3^3}{4^4} + \frac{4^4}{5^5} + \cdots$

(M.D.U. Dec. 2008)

Sol. *Leaving aside the first term* ($\because$ *addition or deletion of a finite number of terms does not alter the nature of the series*), we have $u_n = \frac{n^n}{(n+1)^{n+1}} = \frac{n^n}{n^{n+1}\left(1+\frac{1}{n}\right)^{n+1}} = \frac{1}{n\left(1+\frac{1}{n}\right)^{n+1}}$

Take $$v_n = \frac{1}{n}.$$

$$\lim_{n\to\infty}\frac{u_n}{v_n} = \lim_{n\to\infty}\frac{1}{\left(1+\frac{1}{n}\right)^{n+1}} = \lim_{n\to\infty}\frac{1}{\left(1+\frac{1}{n}\right)^{n}} \times \frac{1}{\left(1+\frac{1}{n}\right)}$$

$$= \frac{1}{e}\cdot\frac{1}{1} \qquad \left[\because \lim_{n\to\infty}\left(1+\frac{1}{n}\right)^n = e\right]$$

$$= \frac{1}{e} \text{ which is finite and non-zero.}$$

$\therefore$ Σu_n and Σv_n converge or diverge together.

Since $\Sigma v_n = \Sigma \frac{1}{n}$ is of the form $\Sigma \frac{1}{n^p}$ with $p = 1$

$\therefore$ Σv_n is divergent. $\Rightarrow$ Σu_n is divergent.

Example 5. *Examine the convergence of the series:* $\frac{\sqrt{2}-1}{3^3-1}+\frac{\sqrt{3}-1}{4^3-1}+\frac{\sqrt{4}-1}{5^3-1}+\cdots$

(M.D.U. Dec. 2013)

Sol. Here
$$u_n = \frac{\sqrt{n+1}-1}{(n+2)^3-1} = \frac{\sqrt{n}\left(\sqrt{1+\frac{1}{n}}-\frac{1}{\sqrt{n}}\right)}{n^3\left[\left(1+\frac{2}{n}\right)^3-\frac{1}{n^3}\right]} = \frac{\sqrt{1+\frac{1}{n}}-\frac{1}{\sqrt{n}}}{n^{5/2}\left[\left(1+\frac{2}{n}\right)^3-\frac{1}{n^3}\right]}$$

Take
$$v_n = \frac{1}{n^{5/2}}$$

$$\lim_{n\to\infty}\frac{u_n}{v_n} = \lim_{n\to\infty}\frac{\sqrt{1+\frac{1}{n}}-\frac{1}{\sqrt{n}}}{\left(1+\frac{2}{n}\right)^3-\frac{1}{n^3}} = \frac{\sqrt{1+0}-0}{(1-0)^3-0} = 1 \quad \text{which is finite and non-zero.}$$

$\therefore$ Σu_n and Σv_n converge or diverge together.

Since $\Sigma v_n = \Sigma \frac{1}{n^{5/2}}$ is of the form $\Sigma \frac{1}{n^p}$ with $p = \frac{5}{2} > 1$.

Σv_n is convergent. $\Rightarrow$ Σu_n is convergent.

Example 6. *Examine the convergence of the series:*

(*i*) $\sum \frac{\sqrt{n+1}-\sqrt{n}}{n^p}$ (*ii*) $\sum\left(\sqrt[3]{n^3+1}-n\right)$.

Sol. (*i*) Here
$$u_n = \frac{\sqrt{n+1}-\sqrt{n}}{n^p} = \frac{\sqrt{n+1}-\sqrt{n}}{n^p}\times\frac{\sqrt{n+1}+\sqrt{n}}{\sqrt{n+1}+\sqrt{n}} \quad \text{(Rationalising)}$$

$$= \frac{(n+1)-n}{n^p\cdot\sqrt{n}\left(\sqrt{1+\frac{1}{n}}+1\right)} = \frac{1}{n^{p+1/2}\left(\sqrt{1+\frac{1}{n}}+1\right)}$$

Take
$$v_n = \frac{1}{n^{p+1/2}}$$

$$\lim_{n\to\infty}\frac{u_n}{v_n} = \lim_{n\to\infty}\frac{1}{\sqrt{1+\frac{1}{n}}+1} = \frac{1}{1+1} = \frac{1}{2} \quad \text{which is finite and non-zero.}$$

$\therefore$ Σu_n and Σv_n converge or diverge together.

Since $\sum v_n = \sum \frac{1}{n^{p+1/2}}$ is convergent if $p+\frac{1}{2} > 1$ and divergent if $p+\frac{1}{2} \le 1$.

i.e., convergent if $p > \frac{1}{2}$ and divergent if $p \le \frac{1}{2}$.

$\therefore$ Σu_n is convergent if $p > \frac{1}{2}$ and divergent if $p \le \frac{1}{2}$.

(*ii*) Here
$$u_n = (n^3 + 1)^{1/3} - n = \left[n^3\left(1+\frac{1}{n^3}\right)\right]^{1/3} - n$$
$$= n\left(1+\frac{1}{n^3}\right)^{1/3} - n = n\left[\left(1+\frac{1}{n^3}\right)^{1/3} - 1\right]$$
$$= n\left[1+\frac{1}{3}\cdot\frac{1}{n^3}+\frac{\frac{1}{3}(\frac{1}{3}-1)}{2!}\cdot\left(\frac{1}{n^3}\right)^2 + \dots - 1\right]$$
$$= \frac{n}{n^3}\left[\frac{1}{3}-\frac{1}{9}\cdot\frac{1}{n^3}\cdots\right] = \frac{1}{n^2}\left[\frac{1}{3}-\frac{1}{9}\cdot\frac{1}{n^3}\cdots\right]$$

Take
$$v_n = \frac{1}{n^2}.$$
$$\lim_{n\to\infty}\frac{u_n}{v_n} = \lim_{n\to\infty}\left(\frac{1}{3}-\frac{1}{9}\cdot\frac{1}{n^3}\cdots\right) = \frac{1}{3}$$

which is finite and non-zero.

$\therefore$ Σu_n and Σv_n converge or diverge together.

Since $\sum v_n = \sum \frac{1}{n^2}$ is of the form $\sum \frac{1}{n^p}$ with $p = 2 > 1$

$\therefore$ Σv_n is convergent $\Rightarrow$ Σu_n is convergent.

Note. Rationalisation is effective only when square roots are involved whereas this method of Binomial Expansion is general.

Example 7. *Test the convergence of the series* $\sum_{n=1}^{\infty} \frac{1}{x^n + x^{-n}}$, $x > 0$.

Sol. Here
$$u_n = \frac{1}{x^n + x^{-n}} = \frac{x^n}{x^{2n}+1}$$

Case I. *When* $0 < x < 1$, *take* $v_n = x^n$

$$\therefore \quad \lim_{n\to\infty}\frac{u_n}{v_n} = \lim_{n\to\infty}\frac{1}{x^{2n}+1} = \frac{1}{0+1} = 1 \qquad [\because \; x^{2n} \to 0 \text{ as } n \to \infty]$$

which is non-zero, finite.

$\therefore$ Σu_n and Σv_n converge or diverge together.

But $\Sigma v_n = \Sigma x^n = x + x^2 + x^3 + \cdots$ is an infinite geometric series with common ratio $x < 1$.

$\therefore$ Σv_n is convergent. $\Rightarrow$ Σu_n is convergent.

Case II. *When* $x > 1$ *so that* $0 < \frac{1}{x} < 1$

$$u_n = \frac{x^n}{x^{2n}+1} = \frac{\frac{1}{x^n}}{1+\frac{1}{x^{2n}}}$$

Take
$$v_n = \frac{1}{x^n}$$

$\therefore \qquad \lim_{n\to\infty} \frac{u_n}{v_n} = \lim_{n\to\infty} \frac{1}{1+\frac{1}{x^{2n}}} = \frac{1}{1+0} = 1 \qquad \left[\because \frac{1}{x^{2n}} \to 0 \text{ as } n \to \infty\right]$

which is non-zero, finite.

$\therefore$ Σu_n and Σv_n converge or diverge together.

But $\Sigma v_n = \Sigma \frac{1}{x^n}$ is an infinite geometric series with common ratio $\frac{1}{x} < 1$.

$\therefore$ Σv_n is convergent $\Rightarrow$ Σu_n is convergent.

Case III. *When $x = 1$, $u_n = \frac{1}{2}$*

$$\lim_{n\to\infty} u_n = \frac{1}{2} \neq 0. \text{ Therefore, } \Sigma u_n \text{ is divergent.}$$

Hence Σu_n converges for $x < 1$ and $x > 1$ but diverges for $x = 1$.

Example 8. *Show that $\sum_{n=1}^{\infty} \frac{1}{n} \sin \frac{1}{n}$ converges.* (M.D.U. Dec. 2012)

Sol. Here $\qquad u_n = \frac{1}{n} \sin \frac{1}{n} = \frac{1}{n}\left[\frac{1}{n} - \frac{1}{3!\,n^3} + \frac{1}{5!\,n^5} \cdots\right]$

Take $v_n = \frac{1}{n^2}$ $\qquad \left(\because \sin x = x - \frac{x^3}{3!} + \frac{x^5}{5!} \cdots\right)$

Then $\qquad \lim_{n\to\infty} \frac{u_n}{v_n} = \lim_{n\to\infty} \left[1 - \frac{1}{3!\,n^2} + \frac{1}{5!\,n^4} - \cdots\right] = 1$

which is finite and non-zero.

$\therefore$ Σu_n and Σv_n converge or diverge together.

Since $\Sigma v_n = \sum \frac{1}{n^2}$ is of the form $\sum \frac{1}{n^p}$ with $p = 2 > 1$

$\therefore$ Σv_n is convergent $\Rightarrow$ Σu_n is convergent.

EXERCISE 1.2

Test the convergence or divergence of the following series:

1. $1 - \frac{1}{3} + \frac{1}{2} - \frac{1}{3^3} + \frac{1}{2^2} - \frac{1}{3^5} + \cdots$ (*M.D.U. Dec. 2011*)
2. $\frac{1}{1.2} + \frac{1}{2.3} + \frac{1}{3.4} + \cdots \infty$
3. $\frac{1}{1.4} + \frac{1}{2.5} + \frac{1}{3.6} + \cdots \infty$
4. $\frac{1}{1.2} + \frac{2}{3.4} + \frac{3}{5.6} + \cdots \infty$
5. $\frac{1}{1.3} + \frac{2}{3.5} + \frac{3}{5.7} + \cdots \infty$
6. $\frac{1}{\sqrt{1.2}} + \frac{1}{\sqrt{2.3}} + \frac{1}{\sqrt{3.4}} + \cdots \infty$

7. $\frac{1}{1^p}+\frac{1}{3^p}+\frac{1}{5^p}+\cdots\infty$

8. $\frac{2}{1^p}+\frac{3}{2^p}+\frac{4}{3^p}+\cdots\infty$ *(M.D.U. May 2008)*

9. $\sum_{n=1}^{\infty}\frac{n+1}{n(2n-1)}$

10. $\sum_{n=1}^{\infty}\frac{1}{n^p(n+1)^p}$

11. $\frac{2^p}{1^q}+\frac{3^p}{2^q}+\frac{4^p}{3^q}+\cdots\infty$ (p and q are positive numbers)

12. $\sum_{n=1}^{\infty}\frac{\sqrt{n}}{n^2+1}$

13. $\sum\frac{2n^3+5}{4n^5+1}$

14. $\sum\frac{\sqrt{n^2-1}}{n^3+1}$

15. (*i*) $\frac{1}{5}+\frac{\sqrt{2}}{7}+\frac{\sqrt{3}}{9}+\frac{\sqrt{4}}{11}+\cdots\infty$ (*ii*) $\frac{1}{1\cdot2\cdot3}+\frac{1}{2\cdot3\cdot4}+\frac{1}{3\cdot4\cdot5}+\cdots\infty$

16. $\sum\left(\sqrt{n^2+1}-n\right)$

17. $\sum\left(\sqrt{n^3+1}-\sqrt{n^3}\right)$

18. $\sum\left(\sqrt{n^4+1}-\sqrt{n^4-1}\right)$ *(M.D.U. May 2009, Dec. 2012; Kerala 2010)*

19. $\frac{\sqrt{2}-\sqrt{1}}{1}+\frac{\sqrt{3}-\sqrt{2}}{2}+\frac{\sqrt{4}-\sqrt{3}}{3}+\cdots$

20. $\sum\left\{(\sqrt[3]{n+1}-\sqrt[3]{n})\right\}$

21. $\sum\frac{1}{\sqrt{n}}\sin\frac{1}{n}$

22. $\sum\cot^{-1}n^2$ *(M.D.U. Dec. 2009)*

23. $\sum\left(\frac{1}{n}-\log\frac{n+1}{n}\right)$

24. $\sum\frac{\sqrt[3]{3n^2+1}}{\sqrt[4]{4n^3+2n+7}}$

25. $\sum\frac{x^{n-1}}{1+x^n},\ x>0$

26. $\frac{1}{1^2}+\frac{1+2}{1^2+2^2}+\frac{1+2+3}{1^2+2^2+3^2}+\cdots$

27. $\frac{1}{1+\sqrt{2}}+\frac{2}{1+2\sqrt{3}}+\frac{3}{1+3\sqrt{4}}+\cdots$

28. $\sum_{n=1}^{\infty}\left(1-\cos\frac{\pi}{n}\right)$

29. $\frac{1}{a(a+b)}+\frac{1}{(a+2b)(a+3b)}+\frac{1}{(a+4b)(a+5b)}+\cdots\infty,\ a>0,\ b>0.$

Answers

1. Convergent 2. Convergent 3. Convergent 4. Divergent
5. Divergent 6. Divergent 7. Convergent for $p > 1$, divergent for $p \le 1$
8. Convergent for $p > 2$, divergent for $p \le 2$ 9. Divergent
10. Convergent for $p > \frac{1}{2}$, divergent for $p \le \frac{1}{2}$ 11. Convergent for $q > p + 1$, divergent for $q \le p + 1$
12. Convergent 13. Convergent 14. Convergent
15. (*i*) Divergent (*ii*) Convergent 16. Divergent 17. Convergent
18. Convergent 19. Convergent 20. Divergent 21. Convergent
22. Convergent 23. Convergent 24. Divergent
25. Convergent for $x < 1$, divergent for $x \ge 1$ 26. Divergent 27. Divergent
28. Convergent 29. Convergent.

1.24. D'ALEMBERT'S RATIO TEST

Statement. *If* Σu_n *is a positive term series, and* $\lim_{n \to \infty} \frac{u_{n+1}}{u_n} = l$, *then*

(*i*) Σu_n *is convergent if* $l < 1$. (*ii*) Σu_n *is divergent if* $l > 1$. (M.D.U. Dec. 2012)

Note. If $l = 1$, the test fails, *i.e.*, no conclusion can be drawn about the convergence or divergence of the series. The series may converge, it may diverge.

Proof. Since $\lim_{n \to \infty} \frac{u_{n+1}}{u_n} = l$,

$\therefore$ Given $\varepsilon > 0$, however small, there exists a positive integer m such that

$$\left| \frac{u_{n+1}}{u_n} - l \right| < \varepsilon \quad \forall \quad n \geq m$$

$$\Rightarrow \qquad l - \varepsilon < \frac{u_{n+1}}{u_n} < 1 + \varepsilon \quad \forall \quad n \geq m \qquad ...(i)$$

(*i*) **When $l < 1$**

Choose $\varepsilon > 0$ such that $l < l + \varepsilon < 1$

Put $l + \varepsilon = r$, then $0 < r < 1$.

From (*i*), $\frac{u_{n+1}}{u_n} < r \ \forall \ n \geq m \Rightarrow u_{n+1} < ru_n \ \forall \ n \geq m$ $\qquad (\because \ u_n > 0)$

Putting $n = m, m + 1, m + 2, \cdots$, we get

$$u_{m+1} < ru_m$$
$$u_{m+2} < ru_{m+1} < r^2 u_m$$
$$u_{m+3} < ru_{m+2} < r^3 u_m \quad \text{and so on.}$$

Adding $u_{m+1} + u_{m+2} + u_{m+3} + \cdots < u_m (r + r^2 + r^3 + \cdots)$

$\Rightarrow$ each term of the given series Σu_n after leaving the first m terms (*i.e.* a finite number of terms) is less than the corresponding term of a geometric series which is convergent ($\because$ its common ratio $r < 1$). Hence the given series is also convergent.

(*ii*) **When $l > 1$**

Choose $\varepsilon > 0$ such that $l - \varepsilon > 1$

Put $l - \varepsilon = \text{R}$, then $\text{R} > 1$.

From (*i*), $\frac{u_{n+1}}{u_n} > \text{R} \ \forall \ n \geq m \Rightarrow u_{n+1} > \text{R}u_n \ \forall \ n \geq m$ $\qquad (\because \ u_n > 0)$

Putting $n = m, m + 1, m + 2, \cdots$, we get

$$u_{m+1} > \text{R}u_m$$
$$u_{m+2} > \text{R}u_{m+1} > \text{R}^2 u_m$$
$$u_{m+3} > \text{R}u_{m+2} > \text{R}^3 u_m \quad \text{and so on.}$$

Adding $u_{m+1} + u_{m+2} + u_{m+3} + \cdots > u_m (\text{R} + \text{R}^2 + \text{R}^3 + \cdots)$

$\Rightarrow$ each term of the given series Σu_n after leaving the first m terms (*i.e.* a finite number of terms) is greater than the corresponding term of a geometric series which is divergent ($\because$ its common ratio $R > 1$). Hence the given series is also divergent.

Practical Form of D'Alembert's Ratio Test

In practice, Ratio Test is used in the following form:

If Σu_n is a positive term series, and $\lim_{n\to\infty} \frac{u_n}{u_{n+1}} = l$, then

(*i*) Σu_n is convergent if $l > 1$ (*ii*) Σu_n is divergent if $l < 1$.

ILLUSTRATIVE EXAMPLES

Example 1. *Discuss the convergence of the following series:*

(*i*) $1 + \frac{2^p}{2!} + \frac{3^p}{3!} + \frac{4^p}{4!} + \cdots, (p > 0)$ (*ii*) $\frac{1}{2} + \frac{1}{3} + \frac{1}{5} + \cdots + \frac{1}{2^{n-1}+1} + \cdots$

(*iii*) $\frac{1^2 \cdot 2^2}{1!} + \frac{2^2 \cdot 3^2}{2!} + \frac{3^2 \cdot 4^2}{3!} + \frac{4^2 \cdot 5^2}{4!} + \cdots$

Sol. (*i*) Here $u_n = \frac{n^p}{n!}$ $\left[\because \ 1 = \frac{1^p}{1!}\right]$

$\therefore$
$$u_{n+1} = \frac{(n+1)^p}{(n+1)!}$$

$$\frac{u_n}{u_{n+1}} = \frac{n^p}{n!} \frac{(n+1)!}{(n+1)^p} = \frac{n^p \cdot (n+1)\, n!}{n!\,(n+1)^p} = \frac{n^p}{(n+1)^{p-1}}$$

$$= \frac{n^p}{n^{p-1}\left(1+\frac{1}{n}\right)^{p-1}} = \frac{n}{\left(1+\frac{1}{n}\right)^{p-1}}$$

$$\lim_{n\to\infty} \frac{u_n}{u_{n+1}} = \lim_{n\to\infty} \frac{n}{\left(1+\frac{1}{n}\right)^{p-1}} = \infty > 1$$

$\therefore$ By D'Alembert's Ratio Test, Σu_n is convergent.

(*ii*) Here $u_n = \frac{1}{2^{n-1}+1}$ $\therefore$ $u_{n+1} = \frac{1}{2^n+1}$

$$\frac{u_n}{u_{n+1}} = \frac{2^n+1}{2^{n-1}+1} = \frac{2^n\left(1+\frac{1}{2^n}\right)}{2^{n-1}\left(1+\frac{1}{2^{n-1}}\right)} = 2 \cdot \frac{1+\frac{1}{2^n}}{1+\frac{1}{2^{n-1}}}$$

$$\lim_{n\to\infty} \frac{u_n}{u_{n+1}} = \lim_{n\to\infty} 2 \cdot \frac{1+\frac{1}{2^n}}{1+\frac{1}{2^{n-1}}} = 2 > 1$$

$\therefore$ By D'Alembert's Ratio Test, Σu_n is convergent.

(*iii*) Here $u_n = \dfrac{n^2 (n+1)^2}{n!}$ $\therefore$ $u_{n+1} = \dfrac{(n+1)^2 (n+2)^2}{(n+1)!}$

$$\frac{u_n}{u_{n+1}} = \frac{n^2 (n+1)^2}{n!} \cdot \frac{(n+1)!}{(n+1)^2 (n+2)^2} = \frac{n^2 \cdot (n+1)\, n!}{n!\,(n+2)^2}$$

$$= \frac{n^3\left(1+\frac{1}{n}\right)}{n^2\left(1+\frac{2}{n}\right)^2} = n \cdot \frac{1+\frac{1}{n}}{\left(1+\frac{2}{n}\right)^2}$$

$$\lim_{n\to\infty} \frac{u_n}{u_{n+1}} = \lim_{n\to\infty} n \cdot \frac{1+\frac{1}{n}}{\left(1+\frac{2}{n}\right)^2} = \infty > 1$$

$\therefore$ By D'Alembert's Ratio Test, Σu_n is convergent.

Example 2. *Test the convergence of the following series:*

(*i*) $\sum \dfrac{n^3+a}{2^n+a}$ (*ii*) $\sum \dfrac{n!\,3^n}{n^n}$.

Sol. (*i*) Here $u_n = \dfrac{n^3+a}{2^n+a}$ $\therefore$ $u_{n+1} = \dfrac{(n+1)^3+a}{2^{n+1}+a}$

$$\frac{u_n}{u_{n+1}} = \frac{n^3+a}{(n+1)^3+a} \cdot \frac{2^{n+1}+a}{2^n+a}$$

$$= \frac{n^3\left(1+\frac{a}{n^3}\right)}{n^3\left[\left(1+\frac{1}{n}\right)^3+\frac{a}{n^3}\right]} \cdot \frac{2^{n+1}\left(1+\frac{a}{2^{n+1}}\right)}{2^n\left(1+\frac{a}{2^n}\right)} = \frac{1+\frac{a}{n^3}}{\left(1+\frac{1}{n}\right)^3+\frac{a}{n^3}} \cdot \frac{2\left(1+\frac{a}{2^{n+1}}\right)}{1+\frac{a}{2^n}}$$

$$\lim_{n\to\infty} \frac{u_n}{u_{n+1}} = \frac{1+0}{1+0} \cdot 2 \cdot \frac{1+0}{1+0} = 2 > 1$$

$\therefore$ By D'Alembert's Ratio Test, Σu_n is convergent.

(*ii*) Here $u_n = \dfrac{n!\,3^n}{n^n}$ $\therefore$ $u_{n+1} = \dfrac{(n+1)!\,3^{n+1}}{(n+1)^{n+1}}$

$$\frac{u_n}{u_{n+1}} = \frac{n!\,3^n}{(n+1)!\,3^{n+1}} \cdot \frac{(n+1)^{n+1}}{n^n} = \frac{1}{3(n+1)} \cdot \frac{(n+1)^{n+1}}{n^n}$$

$$= \frac{1}{3} \cdot \frac{(n+1)^n}{n^n} = \frac{1}{3} \cdot \left(\frac{n+1}{n}\right)^n = \frac{1}{3}\left(1+\frac{1}{n}\right)^n$$

$$\lim_{n\to\infty} \frac{u_n}{u_{n+1}} = \lim_{n\to\infty} \frac{1}{3}\left(1+\frac{1}{n}\right)^n = \frac{e}{3} < 1 \qquad (\because \; e = 2.7)$$

$\therefore$ Σu_n is divergent.

Example 3. *Discuss the convergence of the series:* $\sum \frac{n!}{n^n}$.

Sol. Here $$u_n = \frac{n!}{n^n}$$

$$u_{n+1} = \frac{(n+1)!}{(n+1)^{n+1}} = \frac{(n+1)\,n!}{(n+1)^{n+1}} = \frac{n!}{(n+1)^n}$$

$$\frac{u_n}{u_{n+1}} = \frac{(n+1)^n}{n^n} = \left(\frac{n+1}{n}\right)^n = \left(1+\frac{1}{n}\right)^n$$

$$\lim_{n\to\infty} \frac{u_n}{u_{n+1}} = \lim_{n\to\infty}\left(1+\frac{1}{n}\right)^n = e > 1 \qquad (\because \quad e = 2.7)$$

$\therefore$ By D'Alembert's Ratio Test, Σu_n is convergent.

Example 4. *Test the convergence of the following series:*

(*i*) $1 + \frac{1^2 \cdot 2^2}{1 \cdot 3 \cdot 5} + \frac{1^2 \cdot 2^2 \cdot 3^2}{1 \cdot 3 \cdot 5 \cdot 7 \cdot 9} + \cdots \infty$ (*ii*) $\sum_{n=1}^{\infty} \frac{3 \cdot 6 \cdot 9 \cdots 3n}{4 \cdot 7 \cdot 10 \cdots (3n+1)} \cdot \frac{5^n}{3n+2}$

Sol. (*i*) Here $$u_n = \frac{1^2 \cdot 2^2 \cdot 3^2 \ldots n^2}{1 \cdot 3 \cdot 5 \ldots (4n-5)(4n-3)}$$

[Note that two new factors are added with every term in the denominator and u_n has $(2n-1)$ factors in the denominator.]

$$u_{n+1} = \frac{1^2 \cdot 2^2 \cdot 3^2 \ldots n^2 (n+1)^2}{1 \cdot 3 \cdot 5 \ldots (4n-5)(4n-3)(4n-1)(4n+1)}$$

$$\therefore \quad \frac{u_n}{u_{n+1}} = \frac{(4n-1)(4n+1)}{(n+1)^2} = \frac{16n^2-1}{(n+1)^2} = \frac{16 - \frac{1}{n^2}}{\left(1+\frac{1}{n}\right)^2}$$

$$\lim_{n\to\infty} \frac{u_n}{u_{n+1}} = \lim_{n\to\infty} \frac{16 - \frac{1}{n^2}}{\left(1+\frac{1}{n}\right)^2} = \frac{16-0}{(1+0)^2} = 16 > 1$$

$\therefore$ By D'Alembert's Ratio Test, Σu_n is convergent.

(*ii*) Here $$u_n = \frac{3 \cdot 6 \cdot 9 \ldots 3n}{4 \cdot 7 \cdot 10 \ldots (3n+1)} \cdot \frac{5^n}{3n+2}$$

$$u_{n+1} = \frac{3 \cdot 6 \cdot 9 \ldots (3n)(3n+3)}{4 \cdot 7 \cdot 10 \ldots (3n+1)(3n+4)} \cdot \frac{5^{n+1}}{3n+5}$$

$$\therefore \quad \frac{u_n}{u_{n+1}} = \frac{3n+4}{3n+3} \cdot \frac{3n+5}{3n+2} \cdot \frac{1}{5} = \frac{\left(1+\frac{4}{3n}\right)\left(1+\frac{5}{3n}\right)}{\left(1+\frac{1}{n}\right)\left(1+\frac{2}{3n}\right)} \cdot \frac{1}{5}$$

$$\lim_{n\to\infty} \frac{u_n}{u_{n+1}} = \lim_{n\to\infty} \frac{\left(1+\frac{4}{3n}\right)\left(1+\frac{5}{3n}\right)}{\left(1+\frac{1}{n}\right)\left(1+\frac{2}{3n}\right)} \cdot \frac{1}{5} = \frac{1\times 1}{1 \times 1} \cdot \frac{1}{5} = \frac{1}{5} < 1$$

$\therefore$ By D'Alembert's Ratio Test, Σu_n is divergent.

Example 5. *Discuss the convergence of the series:* $\sum \frac{\sqrt{n}}{\sqrt{n^2+1}} x^n \cdot (x > 0)$.

Sol. Here
$$u_n = \sqrt{\frac{n}{n^2+1}}\, x^n$$

$\therefore$
$$u_{n+1} = \sqrt{\frac{n+1}{(n+1)^2+1}} \cdot x^{n+1}$$

$$\frac{u_n}{u_{n+1}} = \sqrt{\frac{n}{n+1} \cdot \frac{n^2+2n+2}{n^2+1}} \cdot \frac{1}{x} = \sqrt{\frac{1}{1+\frac{1}{n}} \cdot \frac{1+\frac{2}{n}+\frac{2}{n^2}}{1+\frac{1}{n^2}}} \cdot \frac{1}{x}$$

$$\lim_{n\to\infty} \frac{u_n}{u_{n+1}} = \lim_{n\to\infty} \sqrt{\frac{1}{1+\frac{1}{n}} \cdot \frac{1+\frac{2}{n}+\frac{2}{n^2}}{1+\frac{1}{n^2}}} \cdot \frac{1}{x} = \frac{1}{x}$$

$\therefore$ By D'Alembert's Ratio Test, $\sum u_n$ converges if $\frac{1}{x} > 1$ *i.e.*, $x < 1$ and diverges if $\frac{1}{x} < 1$ *i.e.*, $x > 1$

When $x = 1$, the Ratio Test fails.

When $x = 1$, $u_n = \sqrt{\frac{n}{n^2+1}} = \frac{n}{\sqrt{n^2\left(1+\frac{1}{n^2}\right)}} = \frac{1}{\sqrt{n}} \cdot \frac{1}{\sqrt{1+\frac{1}{n^2}}}$

Take $v_n = \frac{1}{\sqrt{n}}$, $\lim_{n\to\infty} \frac{u_n}{v_n} = \lim_{n\to\infty} \frac{1}{\sqrt{1+\frac{1}{n^2}}} = 1$

which is finite and non-zero.

$\therefore$ By Comparison Test, Σu_n and Σv_n converge or diverge together.

Since $\sum v_n = \sum \frac{1}{\sqrt{n}}$ is of the form $\sum \frac{1}{n^p}$ with $p = \frac{1}{2} < 1$

Σv_n diverges $\Rightarrow$ Σu_n diverges.

Hence the given series Σu_n converges if $x < 1$ and diverges if $x \geq 1$.

Example 6. *Examine the convergence or divergence of the following series for $x > 0$:*

$$\frac{1}{2\sqrt{1}} + \frac{x^2}{3\sqrt{2}} + \frac{x^4}{4\sqrt{3}} + \frac{x^6}{5\sqrt{4}} + \cdots$$ (M.D.U. Dec. 2008)

Sol. Here
$$u_n = \frac{x^{2n-2}}{(n+1)\sqrt{n}} \quad \therefore \quad u_{n+1} = \frac{x^{2n}}{(n+2)\sqrt{n+1}}$$

$$\frac{u_n}{u_{n+1}} = \frac{(n+2)\sqrt{n+1}}{(n+1)\sqrt{n}} \cdot \frac{1}{x^2} = \frac{1+\frac{2}{n}}{1+\frac{1}{n}} \cdot \sqrt{1+\frac{1}{n}} \cdot \frac{1}{x^2}$$

$$\lim_{n\to\infty} \frac{u_n}{u_{n+1}} = \frac{1}{x^2}$$

By D'Alembert's Ratio Test, $\sum u_n$ converges if $\frac{1}{x^2} > 1$, *i.e.*, $x^2 < 1$

and diverges if $\frac{1}{x^2} < 1$ *i.e.* $x^2 > 1$.

When $x^2 = 1$, $u_n = \frac{1}{(n+1)\sqrt{n}} = \frac{1}{n^{3/2}\left(1+\frac{1}{n}\right)}$

Take $v_n = \frac{1}{n^{3/2}}$, $\lim_{n\to\infty} \frac{u_n}{v_n} = \lim_{n\to\infty} \frac{1}{1+\frac{1}{n}} = 1$ which is finite and non-zero. By comparison test Σu_n is convergent. $(\because \; p > 1)$

Hence Σu_n is convergent if $x^2 \le 1$ and divergent if $x^2 > 1$.

Example 7. *Examine the convergence or divergence of the following series:*

$$1 + \frac{2}{5}x + \frac{6}{9}x^2 + \frac{14}{17}x^3 + \dots + \frac{2^{n+1}-2}{2^{n+1}+1}x^n + \cdots \; (x > 0).$$

Sol. Here, leaving the first term, $u_n = \frac{2^{n+1}-2}{2^{n+1}+1}x^n$

$$\therefore \qquad u_{n+1} = \frac{2^{n+2}-2}{2^{n+2}+1}x^{n+1}$$

$$\frac{u_n}{u_{n+1}} = \frac{2^{n+1}-2}{2^{n+1}+1} \cdot \frac{2^{n+2}+1}{2^{n+2}-2} \cdot \frac{1}{x} = \frac{2^{n+1}\left(1-\frac{2}{2^{n+1}}\right)}{2^{n+1}\left(1+\frac{1}{2^{n+1}}\right)} \cdot \frac{2^{n+2}\left(1+\frac{1}{2^{n+2}}\right)}{2^{n+2}\left(1-\frac{2}{2^{n+2}}\right)} \cdot \frac{1}{x}$$

$$\lim_{n\to\infty} \frac{u_n}{u_{n+1}} = \lim_{n\to\infty} \frac{1-\frac{1}{2^n}}{1+\frac{1}{2^{n+1}}} \cdot \frac{1+\frac{1}{2^{n+2}}}{1-\frac{1}{2^{n+1}}} \cdot \frac{1}{x} = \frac{1}{x}$$

$\therefore$ By D'Alembert's Ratio Test, $\sum u_n$ converges if $\frac{1}{x} > 1$, *i.e.*, $x < 1$

and diverges if $\frac{1}{x} < 1$, *i.e.*, $x > 1$.

When $x = 1$, $u_n = \frac{2^{n+1}-2}{2^{n+1}+1} = \frac{2^{n+1}\left(1-\frac{2}{2^{n+1}}\right)}{2^{n+1}\left(1+\frac{1}{2^{n+2}}\right)} = \frac{1-\frac{1}{2^n}}{1+\frac{1}{2^{n+1}}}$

$\lim_{n\to\infty} u_n = 1 \neq 0 \;\Rightarrow\; \Sigma u_n$ does not converge. Being a series of + ve terms, it must diverge.

Hence Σu_n is convergent if $x < 1$ and divergent if $x \ge 1$.

Example 8. *Test for convergence the positive term series:*

$$1+\frac{\alpha+1}{\beta+1}+\frac{(\alpha+1)(2\alpha+1)}{(\beta+1)(2\beta+1)}+\frac{(\alpha+1)(2\alpha+1)(3\alpha+1)}{(\beta+1)(2\beta+1)(3\beta+1)}+\cdots$$ (M.D.U. Dec. 2013)

Sol. Leaving the first term $u_n=\dfrac{(\alpha+1)(2\alpha+1)\dots(n\alpha+1)}{(\beta+1)(2\beta+1)\dots(n\beta+1)}$

$$\therefore \qquad u_{n+1}=\frac{(\alpha+1)(2\alpha+1)\dots(n\alpha+1)[(n+1)\alpha+1]}{(\beta+1)(2\beta+1)\dots(n\beta+1)[(n+1)\beta+1]}$$

$$\frac{u_n}{u_{n+1}}=\frac{(n+1)\beta+1}{(n+1)\alpha+1}=\frac{\left(1+\frac{1}{n}\right)\beta+\frac{1}{n}}{\left(1+\frac{1}{n}\right)\alpha+\frac{1}{n}}$$

$$\lim_{n\to\infty}\frac{u_n}{u_{n+1}}=\lim_{n\to\infty}\frac{\left(1+\frac{1}{n}\right)\beta+\frac{1}{n}}{\left(1+\frac{1}{n}\right)\alpha+\frac{1}{n}}=\frac{\beta}{\alpha}$$

$\therefore$ By D'Alembert's Ratio Test, Σu_n converges if $\dfrac{\beta}{\alpha}>1$ *i.e.,* $\beta>\alpha>0$

and diverges if $\dfrac{\beta}{\alpha}<1$ *i.e.,* $\beta<\alpha$ or $\alpha>\beta>0$

When $\alpha=\beta$, the Ratio Test fails.

When $\alpha=\beta$, $u_n=1$ $\lim\limits_{n\to\infty}u_n=1\neq 0$

$\Rightarrow$ Σu_n does not converge. Being a series of +ve terms, it must diverge.

Hence the given series is convergent if $\beta>\alpha>0$ and divergent if $\alpha\geq\beta>0$.

EXERCISE 1.3

Discuss the convergence of the following series:

1. $1+\dfrac{2^2}{2!}+\dfrac{3^2}{3!}+\dfrac{4^2}{4!}+\cdots\infty$

2. $1+\dfrac{2!}{2^2}+\dfrac{3!}{3^3}+\dfrac{4!}{4^4}+\cdots\infty$

3. $\dfrac{1}{1+2}+\dfrac{2}{1+2^2}+\dfrac{3}{1+2^3}+\cdots\infty$

4. $\dfrac{2}{1}+\dfrac{2\cdot5\cdot8}{1\cdot5\cdot9}+\dfrac{2\cdot5\cdot8\cdot11}{1\cdot5\cdot9\cdot13}+\cdots\infty$

5. $\dfrac{2!}{3}+\dfrac{3!}{3^2}+\dfrac{4!}{3^3}+\cdots\infty$

6. $\left(\dfrac{1}{3}\right)^2+\left(\dfrac{1\cdot2}{3\cdot5}\right)^2+\left(\dfrac{1\cdot2\cdot3}{3\cdot5\cdot7}\right)^2+\left(\dfrac{1\cdot2\cdot3\cdot4}{3\cdot5\cdot7\cdot9}\right)^2+\cdots\infty$

7. $\sum\dfrac{2^{n-1}}{3^n+1}$

8. $\sum\dfrac{1}{n!}$

9. $\dfrac{1}{1\cdot2\cdot3}+\dfrac{x}{4\cdot5\cdot6}+\dfrac{x^2}{7\cdot8\cdot9}+\cdots\infty\ (x>0)$

10. $\sum\dfrac{x^n}{3^n\cdot n^2}$, $x>0$

11. $\sum \frac{x^n}{n}, x > 0$

12. $\sum \frac{n}{n^2+1} x^n, x > 0$

13. $\sum \left(\frac{n^2}{2^n} + \frac{1}{n^2}\right)$

14. $\sum \frac{n^3 - n + 1}{n!}$

15. $\sum \sqrt{\frac{n+1}{n^3+1}} \cdot x^n, x > 0$

16. $x + 2x^2 + 3x^3 + 4x^4 + \cdots \infty$ *(M.D.U. Dec. 2009)*

17. $1 + \frac{x}{2} + \frac{x^2}{5} + \frac{x^3}{10} + \cdots + \frac{x^n}{n^2+1} + \cdots \infty$

18. $\frac{x}{1.3} + \frac{x^2}{3.5} + \frac{x^3}{5.7} + \cdots \infty$

19. $x + \frac{3}{5}x^2 + \frac{8}{10}x^3 + \frac{15}{17}x^4 + \cdots + \frac{n^2-1}{n^2+1}x^n + \cdots \infty$

20. $\frac{x}{2\sqrt{3}} + \frac{x^2}{3\sqrt{4}} + \frac{x^3}{4\sqrt{5}} + \cdots \infty$

21. $\sum \frac{3^n - 2}{3^n + 1} \cdot x^{n-1}, x > 0$

22. $\sum_{n=1}^{\infty} \frac{n^2}{2^n}$ *(M.D.U. Dec. 2011)*

23. $\sum_{n=1}^{\infty} \frac{x^n}{(2n)!}$

24. $\frac{x}{1+\sqrt{1}} + \frac{x^2}{2+\sqrt{2}} + \frac{x^3}{3+\sqrt{3}} + \cdots \infty$

25. $\sum_{n=1}^{\infty} \frac{n\,x^n}{(n+1)(n+2)}$

26. $\sum_{n=1}^{\infty} \frac{x^n}{(2n-1)^2 \cdot 2^n}$

27. $1 + \frac{3}{2!} + \frac{5}{3!} + \frac{7}{4!} + \cdots \infty$ *(M.D.U. Dec. 2011)*

28. $1 + \frac{x^2}{2} + \frac{x^4}{4} + \frac{x^6}{6} + \cdots \infty$

29. $\frac{4}{18} + \frac{4 \cdot 12}{18 \cdot 27} + \frac{4 \cdot 12 \cdot 20}{18 \cdot 27 \cdot 36} + \cdots \infty$

30. $\sum_{0}^{\infty} \frac{(3x+5)^n}{(n+1)!}$

31. $\sum_{n=1}^{\infty} \frac{x^{n+1}}{(n+1)\sqrt{n}}$ *(M.D.U. Dec. 2012)*

Answers

1. Convergent **2.** Convergent **3.** Convergent **4.** Convergent

5. Divergent **6.** Convergent **7.** Convergent **8.** Convergent

9. Convergent if $x \le 1$, Divergent if $x > 1$ **10.** Convergent for $x \le 3$, divergent for $x > 3$

11. Convergent for $x < 1$, divergent for $x \ge 1$ **12.** Convergent for $x < 1$, divergent for $x \ge 1$

13. Convergent **14.** Convergent **15.** Convergent for $x < 1$, divergent for $x \ge 1$

16. Convergent for $x < 1$, divergent for $x \ge 1$ **17.** Convergent for $x \le 1$, divergent for $x > 1$

18. Convergent for $x \le 1$, divergent for $x > 1$ **19.** Convergent for $x < 1$, divergent for $x \ge 1$

20. Convergent for $x \le 1$, divergent for $x > 1$ **21.** Convergent for $x < 1$, divergent for $x \ge 1$

22. Convergent **23.** Convergent **24.** Convergent for $x < 1$, divergent for $x \ge 1$

25. Convergent for $x < 1$, divergent for $x \ge 1$ **26.** Convergent for $x \le 2$, divergent for $x > 2$

27. Convergent **28.** Convergent for $x < 1$, divergent for $x \ge 1$

29. Convergent **30.** Convergent for all x **31.** Convergent for $x \le 1$, divergent for $x > 1$.

1.25. CAUCHY'S ROOT TEST

Statement. *If* Σu_n *is a positive term series and* $\lim_{n \to \infty} (u_n)^{1/n} = l$, *then*

(*i*) Σu_n *is convergent if* $l < 1$ (*ii*) Σu_n *is divergent if* $l > 1$.

Note. If $l = 1$, the test fails *i.e.* no conclusion can be drawn about the convergence or divergence of the series. The series may converge, it may diverge.

Proof. Since $\lim_{n \to \infty} (u_n)^{1/n} = l$,

$\therefore$ Given $\varepsilon > 0$, however small, there exists a +ve integer m such that

$$| (u_n)^{1/n} - l | < \varepsilon \qquad \forall \ n \geq m$$

$$\Rightarrow \qquad l - \varepsilon < (u_n)^{1/n} < l + \varepsilon \qquad \forall \ n \geq m$$

$$\Rightarrow \qquad (l - \varepsilon)^n < u_n < (l + \varepsilon)^n \qquad \forall \ n \geq m \qquad ...(1)$$

(*i*) **When $l < 1$**

Choose $\varepsilon > 0$ such that $l < l + \varepsilon < 1$

Put $l + \varepsilon = r$, then $0 < r < 1$

From (1), $u_n < r^n \quad \forall \quad n \geq m$

Putting $n = m, m + 1, m + 2, \ldots$, we get $u_m < r^m$, $u_{m+1} < r^{m+1}$, $u_{m+2} < r^{m+2}$, ... and so on.

Adding $u_m + u_{m+1} + u_{m+2} + \cdots + < r^m + r^{m+1} + r^{m+2} + \cdots$

$\Rightarrow$ each term of the given series Σu_n after leaving the first $(m - 1)$ terms, (*i.e.* a finite number of terms) is less than the corresponding term of a geometric series which is convergent ($\because$ its common ratio $r < 1$). Hence the given series is also convergent.

(*ii*) **When $l > 1$**

Choose $\varepsilon > 0$ such that $l - \varepsilon > 1$

Put $l - \varepsilon = \text{R}$, then $\text{R} > 1$

From (1), $u_n > \text{R}^n \quad \forall \quad n \geq m$

Putting $n = m, m + 1, m + 2, \ldots$, we get $u_m > \text{R}^m$, $u_{m+1} > \text{R}^{m+1}$, $u_{m+2} > \text{R}^{m+2}$, ... and so on.

Adding $u_m + u_{m+1} + u_{m+2} + \cdots > \text{R}^m + \text{R}^{m+1} + \text{R}^{m+2} + \cdots$

$\Rightarrow$ each term of the series Σu_n after leaving the first $(m - 1)$ terms, (*i.e.* a finite number of terms) is greater than the corresponding term of a geometric series which is divergent. ($\because$ its common ratio $\text{R} > 1$). Hence the given series is also divergent.

ILLUSTRATIVE EXAMPLES

Example 1. *Test the convergence of the following series:*

$$\sum \left(\frac{n}{n+1}\right)^{n^2} \quad or \quad \sum \left(1 + \frac{1}{n}\right)^{-n^2}. \qquad \text{(M.D.U. Dec. 2010)}$$

Sol. Here $u_n = \left(\dfrac{n}{n+1}\right)^{n^2}$

$$\therefore \qquad (u_n)^{1/n} = \left[\left(\frac{n}{n+1}\right)^{n^2}\right]^{1/n} = \left(\frac{n}{n+1}\right)^n = \left(\frac{n+1}{n}\right)^{-n} = \left[\left(1 + \frac{1}{n}\right)^n\right]^{-1}$$

$$\lim_{n\to\infty} (u_n)^{1/n} = \lim_{n\to\infty} \left[\left(1+\frac{1}{n}\right)^n\right]^{-1} = e^{-1} = \frac{1}{e} < 1 \qquad (\because \ e = 2.7)$$

$\therefore$ By Cauchy's Root Test, the given series Σu_n is convergent.

Example 2. *Examine the convergence of the series:*

$$\left(\frac{2^2}{1^2}-\frac{2}{1}\right)^{-1} + \left(\frac{3^3}{2^3}-\frac{3}{2}\right)^{-2} + \left(\frac{4^4}{3^4}-\frac{4}{3}\right)^{-3} + \cdots \qquad \text{(Kerala 2010)}$$

Sol. Here
$$u_n = \left[\left(\frac{n+1}{n}\right)^{n+1} - \frac{n+1}{n}\right]^{-n}$$

$$(u_n)^{1/n} = \left[\left(\frac{n+1}{n}\right)^{n+1} - \frac{n+1}{n}\right]^{-1} = \left[\left(1+\frac{1}{n}\right)^{n+1} - \left(1+\frac{1}{n}\right)\right]^{-1}$$

$$= \left[\left(1+\frac{1}{n}\right)^{n} \cdot \left(1+\frac{1}{n}\right) - \left(1+\frac{1}{n}\right)\right]^{-1}$$

$$\lim_{n\to\infty} (u_n)^{1/n} = \lim_{n\to\infty} \left[\left(1+\frac{1}{n}\right)^{n} \left(1+\frac{1}{n}\right) - \left(1+\frac{1}{n}\right)\right]^{-1}$$

$$= (e.1 - 1)^{-1} = \frac{1}{e-1} < 1 \qquad (\because \ e = 2.7)$$

$\therefore$ By Cauchy's Root Test, Σu_n is convergent.

Example 3. *Test the following series for convergence:*

(*i*) $\frac{1}{2} + \frac{2}{3}x + \left(\frac{3}{4}\right)^2 x^2 + \left(\frac{4}{5}\right)^3 x^3 + \cdots \ (x > 0)$ (M.D.U. Dec. 2013) (*ii*) $\sum_{n=1}^{\infty} \frac{(n+1)^n \, x^n}{n^{n+1}}$. (M.D.U. Dec. 2012)

Sol. (*i*) Neglecting the first term, we have

$$u_n = \left(\frac{n+1}{n+2}\right)^n x^n$$

$$\Rightarrow \qquad (u_n)^{1/n} = \left(\frac{n+1}{n+2}\right) x = \left(\frac{1+\frac{1}{n}}{1+\frac{2}{n}}\right) x$$

$$\Rightarrow \qquad \lim_{n\to\infty} (u_n)^{1/n} = \lim_{n\to\infty} \left(\frac{1+\frac{1}{n}}{1+\frac{2}{n}}\right) x = x$$

$\therefore$ By Cauchy's root test, the series is convergent if $x < 1$ and divergent if $x > 1$. The test fails when $x = 1$.

When $x = 1$, $$u_n = \left(\frac{n+1}{n+2}\right)^n = \left(\frac{1+\frac{1}{n}}{1+\frac{2}{n}}\right)^n = \frac{\left(1+\frac{1}{n}\right)^n}{\left(1+\frac{2}{n}\right)^n}$$

$$\Rightarrow \quad \lim_{n\to\infty} u_n = \lim_{n\to\infty} \frac{\left(1+\frac{1}{n}\right)^n}{\left[\left(1+\frac{2}{n}\right)^{n/2}\right]^2} = \frac{e}{e^2} = \frac{1}{e} \neq 0$$

Since $\lim\limits_{n\to\infty} u_n \neq 0$, Σu_n cannot converge. Being a series of positive terms, it must diverge.

Hence the given series is convergent if $x < 1$ and divergent if $x \geq 1$.

(*ii*) Here $$u_n = \frac{(n+1)^n \cdot x^n}{n^{n+1}} = \left[\frac{(n+1)\,x}{n}\right]^n \cdot \frac{1}{n}$$

$$\Rightarrow \quad (u_n)^{1/n} = \frac{(n+1)\,x}{n} \cdot \frac{1}{n^{1/n}} = \left(1+\frac{1}{n}\right) x \cdot \frac{1}{n^{1/n}}$$

$$\Rightarrow \quad \lim_{n\to\infty} (u_n)^{1/n} = \left[\lim_{n\to\infty}\left(1+\frac{1}{n}\right) x\right]\left[\lim_{n\to\infty} \frac{1}{n^{1/n}}\right]$$

$$= (1+0)\,x \cdot \frac{1}{1} = x \qquad \left[\because \lim_{n\to\infty} n^{1/n} = 1\right]$$

$\therefore$ By Cauchy's root test, Σu_n is convergent if $x < 1$ and divergent if $x > 1$. The test fails when $x = 1$.

When $x = 1$, $$u_n = \frac{(n+1)^n}{n^{n+1}} = \frac{1}{n} \cdot \frac{(n+1)^n}{n^n} = \frac{1}{n}\left(1+\frac{1}{n}\right)^n$$

Taking $$v_n = \frac{1}{n}, \quad \frac{u_n}{v_n} = \left(1+\frac{1}{n}\right)^n$$

$$\lim_{n\to\infty} \frac{u_n}{v_n} = \lim_{n\to\infty}\left(1+\frac{1}{n}\right)^n = e \quad \text{which is finite and non-zero.}$$

$\therefore$ By comparison test, Σu_n and Σv_n converge or diverge together.

Since $\Sigma v_n = \sum \frac{1}{n}$ is of the form $\sum \frac{1}{n^p}$ with $p = 1$, Σv_n is divergent.

$\Rightarrow$ Σu_n is divergent.

Hence Σu_n is convergent if $x < 1$ and divergent if $x \geq 1$.

EXERCISE 1.4

Discuss the convergence of the following series:

1. $\sum\left(1+\frac{1}{n}\right)^{n^2}$ (*M.D.U. Dec. 2011*) 2. $\sum_{n=2}^{\infty}\frac{1}{(\log n)^n}$ 3. $\sum\left(\frac{n+1}{3n}\right)^n$

4. $\sum\frac{(n-\log n)^n}{2^n \cdot n^n}$ 5. $\sum\left(1+\frac{1}{\sqrt{n}}\right)^{-n^{3/2}}$ 6. $\sum\left(\frac{nx}{n+1}\right)^n$

7. $\sum 5^{-n-(-1)^n}$ 8. $\sum\frac{(1+nx)^n}{n^n}$ 9. $\sum\frac{(x+2)^n}{3^n \cdot n}$

10. $1+\frac{x}{2}+\frac{x^2}{3^2}+\frac{x^3}{4^3}+\cdots\ (x>0)$ 11. $\frac{2}{1^2}x+\frac{3^2}{2^3}x^2+\frac{4^3}{3^4}x^3+\dots+\frac{(n+1)^n}{n^{n+1}}x^n+\cdots$

Answers

1. Divergent 2. Convergent 3. Convergent 4. Convergent
5. Convergent 6. Convergent for $x<1$, divergent for $x\ge 1$ 7. Convergent
8. Convergent for $x<1$, divergent for $x\ge 1$ 9. Convergent for $x<1$, divergent for $x\ge 1$
10. Convergent 11. Convergent for $x<1$, divergent for $x\ge 1$.

1.26. RAABE'S TEST

Statement. *If Σu_n is a series of positive terms and* $\lim_{n\to\infty} n\left(\frac{u_n}{u_{n+1}}-1\right)=l$, *then the series is convergent if $l>1$ and divergent if $l<1$.*

Proof. Let us compare the given series Σu_n with an auxiliary series $\Sigma v_n = \Sigma\frac{1}{n^p}$ which we know converges if $p>1$ and diverges if $p\le 1$.

Now
$$\frac{v_n}{v_{n+1}}=\frac{\frac{1}{n^p}}{\frac{1}{(n+1)^p}}=\left(\frac{n+1}{n}\right)^p=\left(1+\frac{1}{n}\right)^p=1+\frac{p}{n}+\frac{p(p-1)}{2!}\cdot\frac{1}{n^2}+\cdots$$

Case I. *Let* $\Sigma v_n=\Sigma\frac{1}{n^p}$ *be convergent, so that $p>1$.*

Then Σu_n will also converge if $\frac{u_n}{u_{n+1}}>\frac{v_n}{v_{n+1}}$

or if
$$\frac{u_n}{u_{n+1}}>1+\frac{p}{n}+\frac{p(p-1)}{2!}\cdot\frac{1}{n^2}+\cdots$$

or if $$n\left(\frac{u_n}{u_{n+1}}-1\right)>p+\frac{p(p-1)}{2!}\cdot\frac{1}{n}+\cdots$$

or if $$\lim_{n\to\infty} n\left(\frac{u_n}{u_{n+1}}-1\right)>p \quad \text{or} \quad \text{if } l>p$$

But p is itself greater than 1, $\therefore$ Σu_n is convergent if $l > 1$.

Case II. *Let Σv_n be divergent, so that $p \le 1$*

Then Σu_n will also diverge if $\dfrac{u_n}{u_{n+1}} < \dfrac{v_n}{v_{n+1}}$

or if $$\frac{u_n}{u_{n+1}}<1+\frac{p}{n}+\frac{p(p-1)}{2!}\cdot\frac{1}{n^2}+\cdots$$

or if $$n\left(\frac{u_n}{u_{n+1}}-1\right)<p+\frac{p(p-1)}{2!}\cdot\frac{1}{n}+\cdots$$

or if $$\lim_{n\to\infty} n\left(\frac{u_n}{u_{n+1}}-1\right)<p \quad \text{or} \quad \text{if } l<p.$$

But p itself ≤ 1. Thus the given series Σu_n diverges if $l < 1$. This proves the result.

Note 1. If $\lim\limits_{n\to\infty} n\left(\dfrac{u_n}{u_{n+1}}-1\right)=1$, then Raabe's test fails.

Note 2. Raabe's test is used when D'Alembert's Ratio test fails and when in the ratio test, $\dfrac{u_n}{u_{n+1}}$ does not involve the number e. When $\dfrac{u_n}{u_{n+1}}$ involves e, we apply logarithmic test after the ratio test and not Raabe's test.

1.27. LOGARITHMIC TEST

Statement. *A positive term series Σu_n converges or diverges according as*

$$\lim_{n\to\infty} n \log \frac{u_n}{u_{n+1}} > 1 \text{ or } < 1.$$

Proof. Let us compare the given series Σu_n with an auxiliary series $\Sigma v_n = \Sigma \dfrac{1}{n^p}$ which we know converges if $p > 1$ and diverges if $p \le 1$.

Now $$\frac{v_n}{v_{n+1}}=\frac{(n+1)^p}{n^p}=\left(1+\frac{1}{n}\right)^p$$

Case I. *Let Σv_n be convergent, so that $p > 1$*

Then Σu_n will also be convergent if $\dfrac{u_n}{u_{n+1}} > \dfrac{v_n}{v_{n+1}}$

or if $$\frac{u_n}{u_{n+1}}>\left(1+\frac{1}{n}\right)^p$$

or if $$\log\frac{u_n}{u_{n+1}}>\log\left(1+\frac{1}{n}\right)^p=p\log\left(1+\frac{1}{n}\right)$$

or if $\log \dfrac{u_n}{u_{n+1}} > p\left[\dfrac{1}{n} - \dfrac{1}{2n^2} + \dfrac{1}{3n^3} - \dots\right]$ $\left[\because \log(1+x) = x - \dfrac{x^2}{2} + \dfrac{x^3}{3} - \dots\right]$

or if $n \log \dfrac{u_n}{u_{n+1}} > p\left[1 - \dfrac{1}{2n} + \dfrac{1}{3n^2} - \dots\right]$ or if $\lim\limits_{n\to\infty} n \log \dfrac{u_n}{u_{n+1}} > p$

or if $\lim\limits_{n\to\infty} n \log \dfrac{u_n}{u_{n+1}} > 1$ | $\because$ $p > 1$ (given)

Case II. *Let Σv_n be divergent, so that $p \le 1$*

Then Σu_n also diverges if $\dfrac{u_n}{u_{n+1}} < \dfrac{v_n}{v_{n+1}}$ or if $\log \dfrac{u_n}{u_{n+1}} < \log \dfrac{v_n}{v_{n+1}}$

or if $\log \dfrac{u_n}{u_{n+1}} < \log\left(1 + \dfrac{1}{n}\right)^p = p \log\left(1 + \dfrac{1}{n}\right)$

or if $\log \dfrac{u_n}{u_{n+1}} < p\left(\dfrac{1}{n} - \dfrac{1}{2n^2} + \dfrac{1}{3n^3} - \dots\right)$

or if $n \log \dfrac{u_n}{u_{n+1}} < p\left[1 - \dfrac{1}{2n} + \dfrac{1}{3n^2} - \dots\right]$ or if $\lim\limits_{n\to\infty} n \log \dfrac{u_n}{u_{n+1}} < p$

or if $\lim\limits_{n\to\infty} n \log \dfrac{u_n}{u_{n+1}} < 1$ | $\because$ $p \le 1$

Thus the series Σu_n converges or diverges according as $\lim\limits_{n\to\infty} n \log \dfrac{u_n}{u_{n+1}} > 1$ or < 1.

Note 1. The test fails if $\lim\limits_{n\to\infty} n \log \dfrac{u_n}{u_{n+1}} = 1$.

Note 2. The test is applied after the failure of Ratio test and generally when in Ratio test, $\dfrac{u_n}{u_{n+1}}$ involves '*e*'.

1.28. GAUSS TEST

Statement. *If for the series Σu_n of positive terms, $\dfrac{u_n}{u_{n+1}}$ can be expanded in the form*

$$\frac{u_n}{u_{n+1}} = 1 + \frac{\lambda}{n} + O\left(\frac{1}{n^2}\right)$$

then Σu_n converges if $\lambda > 1$ and diverges if $\lambda \le 1$. (M.D.U. Dec. 2012)

Note. The test never fails as we know that the series diverges for $\lambda = 1$. Moreover the test is applied after the failure of Ratio test and when it is possible to expand $\dfrac{u_n}{u_{n+1}}$ in powers of $\dfrac{1}{n}$ by Binomial Theorem or by any other method.

ILLUSTRATIVE EXAMPLES

Example 1. *Discuss the convergence of the series:* $\frac{1}{2}+\frac{1\cdot 3}{2\cdot 4}+\frac{1\cdot 3\cdot 5}{2\cdot 4\cdot 6}+\cdots$

Sol. Here $$u_n=\frac{1\cdot 3\cdot 5\cdots(2n-1)}{2\cdot 4\cdot 6\cdots 2n}.$$

$$\therefore \quad u_{n+1}=\frac{1\cdot 3\cdot 5\cdots(2n-1)(2n+1)}{2\cdot 4\cdot 6\cdots(2n)(2n+2)}$$

$$\therefore \quad \frac{u_n}{u_{n+1}}=\frac{1\cdot 3\cdot 5\cdots(2n-1)}{2\cdot 4\cdot 6\cdots 2n}\times\frac{2\cdot 4\cdot 6\cdots 2n(2n+2)}{1\cdot 3\cdot 5\cdots(2n-1)(2n+1)}$$

$$=\frac{2n+2}{2n+1}=\frac{1+\frac{1}{n}}{1+\frac{1}{2n}}\to 1 \text{ as } n\to\infty.$$

$\therefore$ D'Alembert's Ratio test fails.

$$n\left[\frac{u_n}{u_{n+1}}-1\right]=n\left[\frac{2n+2}{2n+1}-1\right]=\frac{n}{2n+1}=\frac{1}{2+\frac{1}{n}}$$

$$\therefore \quad \lim_{n\to\infty} n\left[\frac{u_n}{u_{n+1}}-1\right]=\frac{1}{2}<1.$$

$\therefore$ By Raabe's test, Σu_n diverges.

Example 2. *Discuss the convergence of the series:* $\frac{1^2}{2^2}+\frac{1^2\cdot 3^2}{2^2\cdot 4^2}+\frac{1^2\cdot 3^2\cdot 5^2}{2^2\cdot 4^2\cdot 6^2}+\ldots\ldots$

(M.D.U. Dec. 2009, Dec. 2010)

Sol. Here $$u_n=\frac{1^2\cdot 3^2\cdot 5^2\cdots(2n-1)^2}{2^2\cdot 4^2\cdot 6^2\cdots(2n)^2}$$

and $$u_{n+1}=\frac{1^2\cdot 3^2\cdot 5^2\cdots(2n-1)^2\,(2n+1)^2}{2^2\cdot 4^2\cdot 6^2\cdots(2n)^2\,(2n+2)^2}$$

$$\therefore \quad \frac{u_n}{u_{n+1}}=\frac{(2n+2)^2}{(2n+1)^2}=\frac{4n^2\left(1+\frac{1}{n}\right)^2}{4n^2\left(1+\frac{1}{2n}\right)^2}=\frac{\left(1+\frac{1}{n}\right)^2}{\left(1+\frac{1}{2n}\right)^2}$$

$$\therefore \quad \lim_{n\to\infty}\frac{u_n}{u_{n+1}}=1. \quad \text{Hence the ratio test fails}$$

$$n\left(\frac{u_n}{u_{n+1}}-1\right)=n\left[\frac{(2n+2)^2}{(2n+1)^2}-1\right]$$

$$=n\left[\frac{4n^2+8n+4-(4n^2+4n+1)}{(2n+1)^2}\right]=n\,\frac{(4n+3)}{(2n+1)^2}=\frac{4n^2+3n}{(2n+1)^2}$$

$$=\frac{1+\frac{3}{4n}}{\left(1+\frac{1}{2n}\right)^2}\to 1 \text{ as } n\to\infty$$

$\therefore$ Raabe's test also fails.

When D'Alembert ratio test fails, we can directly apply Gauss test.

Now, $$\frac{u_n}{u_{n+1}} = \frac{(2n+2)^2}{(2n+1)^2} = \frac{\left(1+\frac{1}{n}\right)^2}{\left(1+\frac{1}{2n}\right)^2} = \left(1+\frac{1}{n}\right)^2\left(1+\frac{1}{2n}\right)^{-2}$$

$$= \left(1+\frac{2}{n}+\frac{1}{n^2}\right)\left(1-\frac{2}{2n}+\frac{3}{4n^2}-\cdots\right) = 1+\frac{1}{n}+\frac{1}{n^2}\left(1-2+\frac{3}{4}\right)+\cdots$$

$$= 1+\frac{1}{n}-\frac{1}{4n^2}+\cdots = 1+\frac{1}{n}+O\left(\frac{1}{n^2}\right)$$

Comparing it with $\frac{u_n}{u_{n+1}} = 1+\frac{\lambda}{n}+O\left(\frac{1}{n^2}\right)$

we have $\lambda = 1$. Thus by Gauss test, the series Σu_n diverges.

Example 3. *Discuss the convergence of the series:* $\frac{x}{1}+\frac{1}{2}\cdot\frac{x^3}{3}+\frac{1\cdot 3}{2\cdot 4}\cdot\frac{x^5}{5}+\frac{1\cdot 3\cdot 5}{2\cdot 4\cdot 6}\cdot\frac{x^7}{7}+$ $\cdots$ *($x > 0$).* (M.D.U. May 2008, Dec. 2012)

Sol. Neglecting the first term, we have $u_n = \frac{1\cdot 3\cdot 5\ldots(2n-1)}{2\cdot 4\cdot 6\ldots(2n)}\cdot\frac{x^{2n+1}}{2n+1}$

and $$u_{n+1} = \frac{1\cdot 3\cdot 5\cdots(2n-1)}{2\cdot 4\cdot 6\cdots(2n)}\cdot\frac{x^{2n+1}}{2n+1}$$

$\therefore$ $$\frac{u_n}{u_{n+1}} = \frac{2n+2}{2n+1}\cdot\frac{2n+3}{2n+1}\cdot\frac{1}{x^2} = \frac{2n\left(1+\frac{1}{n}\right)\cdot 2n\left(1+\frac{3}{2n}\right)}{2n\left(1+\frac{1}{2n}\right)\cdot 2n\left(1+\frac{1}{2n}\right)}\cdot\frac{1}{x^2}$$

$$= \frac{\left(1+\frac{1}{n}\right)\left(1+\frac{3}{2n}\right)}{\left(1+\frac{1}{2n}\right)^2}\cdot\frac{1}{x^2}$$

$$\lim_{n\to\infty}\frac{u_n}{u_{n+1}} = \lim_{n\to\infty}\frac{\left(1+\frac{1}{n}\right)\left(1+\frac{3}{2n}\right)}{\left(1+\frac{1}{2n}\right)^2}\cdot\frac{1}{x^2} = \frac{1}{x^2}$$

$\therefore$ By Ratio Test, Σu_n is convergent if $\frac{1}{x^2} > 1$ *i.e.*, $x^2 < 1$ and divergent if $\frac{1}{x^2} < 1$ *i.e.*, $x^2 > 1$.

If $x^2 = 1$, then Ratio Test fails.

When $x^2 = 1$, we have $\frac{u_n}{u_{n+1}} = \frac{(2n+2)(2n+3)}{(2n+1)^2} = \frac{4n^2+10n+6}{4n^2+4n+1}$

$$\lim_{n\to\infty} n\left(\frac{u_n}{u_{n+1}}-1\right) = \lim_{n\to\infty} n\left(\frac{4n^2+10n+6}{4n^2+4n+1}-1\right)$$

$$= \lim_{n\to\infty} \frac{6n^2+5n}{4n^2+4n+1} = \lim_{n\to\infty} \frac{6+\frac{5}{n}}{4+\frac{4}{n}+\frac{1}{n^2}} = \frac{6}{4} = \frac{3}{2} > 1.$$

$\therefore$ By Raabe's Test, the series converges.

Hence Σu_n is convergent if $x^2 \le 1$ *i.e.*, $0 < x \le 1$ and divergent if $x^2 > 1$ *i.e.*, $x > 1$.

Example 4. *Discuss the convergence of the series:* $1 + \frac{x}{2} + \frac{2!}{3^2}x^2 + \frac{3!}{4^3}x^3 + \frac{4!}{5^4}x^4 + \cdots$

Sol. Neglecting the first term, we have

$$u_n = \frac{n!}{(n+1)^n}x^n \quad \text{and} \quad u_{n+1} = \frac{(n+1)!}{(n+2)^{n+1}}\cdot x^{n+1}.$$

$$\therefore \qquad \lim_{n\to\infty}\frac{u_n}{u_{n+1}} = \lim_{n\to\infty}\frac{n!}{(n+1)^n}\cdot\frac{(n+2)^{n+1}}{(n+1)!}\cdot\frac{1}{x}$$

$$= \lim_{n\to\infty}\frac{1}{n^n\left(1+\frac{1}{n}\right)^n}\cdot\frac{n^{n+1}\left(1+\frac{2}{n}\right)^{n+1}}{(n+1)}\cdot\frac{1}{x}$$

$$= \lim_{n\to\infty}\frac{\left(1+\frac{2}{n}\right)^n\left(1+\frac{2}{n}\right)}{\left(1+\frac{1}{n}\right)^n\left(1+\frac{1}{n}\right)}\cdot\frac{1}{x} = \frac{e^2}{e}\cdot\frac{1}{x} = \frac{e}{x}$$

$$\left[\because \lim_{n\to\infty}\left(1+\frac{a}{n}\right)^n = \lim_{n\to\infty}\left[\left(1+\frac{a}{n}\right)^{n/a}\right]^a = e^a\right]$$

$\therefore$ By D'Alembert's ratio test, the series converges if $\frac{e}{x} > 1$ or if $x < e$ and diverges if $\frac{e}{x} < 1$ or if $x > e$.

If $x = e$, the ratio test fails, $\because \lim_{n\to\infty}\frac{u_n}{u_{n+1}} = 1$.

Now when $x = e$

$$\frac{u_n}{u_{n+1}} = \frac{\left(1+\frac{2}{n}\right)^{n+1}}{\left(1+\frac{1}{n}\right)^{n+1}}\cdot\frac{1}{e}.$$

Since the expression $\frac{u_n}{u_{n+1}}$ involves the number e, so we do not apply Raabe's test but apply logarithmic test.

$$\therefore \qquad \log\frac{u_n}{u_{n+1}} = (n+1)\log\left(1+\frac{2}{n}\right) - (n+1)\log\left(1+\frac{1}{n}\right) - \log e$$

$$= (n+1)\left[\log\left(1+\frac{2}{n}\right) - \log\left(1+\frac{1}{n}\right)\right] - 1$$

$$= (n+1)\left[\left(\frac{2}{n} - \frac{1}{2}\cdot\frac{4}{n^2} + \frac{1}{3}\cdot\frac{8}{n^3} - \cdots\right) - \left(\frac{1}{n} - \frac{1}{2n^2} + \frac{1}{3n^3} + \cdots\right)\right] - 1$$

$$= (n+1)\left[\frac{1}{n} - \frac{3}{2n^2} + \cdots\right] - 1$$

$$= 1 - \frac{3}{2n} + \frac{1}{n} - \frac{3}{2n^2} + \cdots - 1 = -\frac{1}{2n} - \frac{3}{2n^2} + \cdots$$

$$\therefore \quad \lim_{n\to\infty} n \log\frac{u_n}{u_{n+1}} = \lim_{n\to\infty} n\left[-\frac{1}{2n} - \frac{3}{2n^2} + \cdots\right] = \lim_{n\to\infty}\left(-\frac{1}{2} - \frac{3}{2n} + \cdots\right) = -\frac{1}{2} < 1$$

$\therefore$ By log test, the series *diverges*.

Hence the given series Σu_n converges if $x < e$ and diverges if $x \geq e$.

Example 5. *Test the following series for convergence:*

$$\frac{a+x}{1!} + \frac{(a+2x)^2}{2!} + \frac{(a+3x)^3}{3!} + \cdots \infty.$$

Sol. Here $\quad u_n = \dfrac{(a+nx)^n}{n!} \quad \therefore \quad u_{n+1} = \dfrac{[a+(n+1)x]^{n+1}}{(n+1)!}$

$$\frac{u_n}{u_{n+1}} = \frac{(a+nx)^n}{[a+(n+1)x]^{n+1}}\cdot\frac{(n+1)!}{n!} = \frac{n^n x^n\left(1+\dfrac{a}{nx}\right)^n}{(n+1)^{n+1}x^{n+1}\left[1+\dfrac{a}{(n+1)x}\right]^{n+1}}\cdot(n+1)$$

$$= \frac{n^n\cdot\left(1+\dfrac{a}{nx}\right)^n}{(n+1)^n\left[1+\dfrac{a}{(n+1)x}\right]^{n+1}}\cdot\frac{1}{x} = \frac{\left(1+\dfrac{a}{nx}\right)^n}{\left(1+\dfrac{1}{n}\right)^n\left[1+\dfrac{a}{(n+1)x}\right]^{n+1}}\cdot\frac{1}{x} \qquad \ldots(1)$$

$$\therefore \quad \lim_{n\to\infty}\frac{u_n}{u_{n+1}} = \lim_{n\to\infty}\frac{\left[\left(1+\dfrac{a}{nx}\right)^{\frac{nx}{a}}\right]^{a/x}}{\left(1+\dfrac{1}{n}\right)^n\left[\left\{1+\dfrac{a}{(n+1)x}\right\}^{\frac{(n+1)x}{a}}\right]^{a/x}}\cdot\frac{1}{x} = \frac{e^{a/x}}{e\,.\,e^{a/x}}\cdot\frac{1}{x} = \frac{1}{ex}$$

$\therefore$ By D'Alembert's ratio test, the series converges if $\dfrac{1}{ex} > 1$, *i.e.*, if $x < \dfrac{1}{e}$ and diverges if $\dfrac{1}{ex} < 1$, *i.e.*, if $x > \dfrac{1}{e}$.

The test fails when $\quad x = \dfrac{1}{e}$.

When $\quad x = \dfrac{1}{e}$, we have from (1)

$$\frac{u_n}{u_{n+1}} = \frac{e\left(1+\dfrac{ae}{n}\right)^n}{\left(1+\dfrac{1}{n}\right)^n\left(1+\dfrac{ae}{n+1}\right)^{n+1}}$$

Since $\frac{u_n}{u_{n+1}}$ involves e, we use logarithmic test after the failure of ratio test.

$$\log \frac{u_n}{u_{n+1}} = \log \left[\frac{e\left(1+\frac{ae}{n}\right)^n}{\left(1+\frac{1}{n}\right)^n \left(1+\frac{ae}{n+1}\right)^{n+1}} \right]$$

$$= \log e + n \log \left(1+\frac{ae}{n}\right) - n \log \left(1+\frac{1}{n}\right) - (n+1) \log \left(1+\frac{ae}{n+1}\right)$$

$$= 1 + n \left(\frac{ae}{n} - \frac{a^2e^2}{2n^2} + \frac{a^3e^3}{3n^3} - \ldots\right)$$

$$- n\left(\frac{1}{n} - \frac{1}{2n^2} + \frac{1}{3n^3} - \ldots\right) - (n+1)\left[\frac{ae}{n+1} - \frac{a^2e^2}{2(n+1)^2} + \frac{a^3e^3}{3(n+1)^3} - \ldots\right]$$

$$= 1 + \left(ae - \frac{a^2e^2}{2n} + \frac{a^3e^3}{3n^2} - \ldots\right) - \left(1 - \frac{1}{2n} + \frac{1}{3n^2} - \ldots\right)$$

$$- \left[ae - \frac{a^2e^2}{2(n+1)} + \frac{a^3e^3}{3(n+1)^2} - \ldots\right]$$

$$= \left(-\frac{a^2e^2}{2n} + \frac{a^3e^3}{3n^2} - \ldots\right) + \left(\frac{1}{2n} - \frac{1}{3n^2} + \ldots\right)$$

$$+ \left[\frac{a^2e^2}{2(n+1)} - \frac{a^3e^3}{3(n+1)^2} + \ldots\right]$$

$$\Rightarrow n \log \frac{u_n}{u_{n+1}} = \left(-\frac{a^2e^2}{2} + \frac{a^3e^3}{3n} - \ldots\right) + \left(\frac{1}{2} - \frac{1}{3n} + \ldots\right)$$

$$+ \left[\frac{a^2e^2}{2} \cdot \frac{n}{n+1} - \frac{a^3e^3}{3(n+1)} \cdot \frac{n}{n+1} + \ldots\right]$$

$$= \left(-\frac{a^2e^2}{2} + \frac{a^3e^3}{3n} - \ldots\right) + \left(\frac{1}{2} - \frac{1}{3n} + \ldots\right)$$

$$+ \left[\frac{a^2e^2}{2} \cdot \frac{1}{1+\frac{1}{n}} - \frac{a^3e^3}{3(n+1)} \cdot \frac{1}{1+\frac{1}{n}} + \ldots\right]$$

$$\Rightarrow \qquad \lim_{n\to\infty} n \log \frac{u_n}{u_{n+1}} = \left(-\frac{a^2e^2}{2}\right) + \left(\frac{1}{2}\right) + \left(\frac{a^2e^2}{2}\right) = \frac{1}{2} < 1$$

$\therefore$ By Logarithmic Test, the series is divergent.

Hence the given series converges for $x < \frac{1}{e}$ and diverges for $x \geq \frac{1}{e}$.

Example 6. *Test the following series for convergence:*

$$x^2 (\log 2)^q + x^3 (\log 3)^q + x^4 (\log 4)^q + \cdots \infty.$$

Sol. Here $u_n = x^{n+1} [\log (n+1)]^q \quad \therefore \quad u_{n+1} = x^{n+2} [\log (n+2)]^q$

$$\frac{u_n}{u_{n+1}} = \frac{1}{x} \cdot \left[\frac{\log (n+1)}{\log (n+2)}\right]^q = \frac{1}{x}\left[\frac{\log n\left(1+\frac{1}{n}\right)}{\log n\left(1+\frac{2}{n}\right)}\right]^q$$

$$= \frac{1}{x}\left[\frac{\log n + \log\left(1+\frac{1}{n}\right)}{\log n + \log\left(1+\frac{2}{n}\right)}\right]^q = \frac{1}{x}\left[\frac{\log n + \frac{1}{n} - \frac{1}{2n^2} + \cdots}{\log n + \frac{2}{n} - \frac{1}{2}\left(\frac{2}{n}\right)^2 + \cdots}\right]^q$$

$$= \frac{1}{x}\left[\frac{1+\frac{1}{n\log n} - \frac{1}{2n^2 \log n} + \cdots}{1+\frac{2}{n\log n} - \frac{2}{n^2\log n} + \cdots}\right] \qquad \ldots(1)$$

$$\Rightarrow \qquad \lim_{n\to\infty} \frac{u_n}{u_{n+1}} = \lim_{n\to\infty} \frac{1}{x}\left[\frac{1+\frac{1}{n\log n} - \frac{1}{2n^2 \log n} + \cdots}{1+\frac{2}{n\log n} - \frac{2}{n^2\log n} + \cdots}\right]^q = \frac{1}{x}$$

$\therefore$ By Ratio Test, the series Σu_n converges if $\frac{1}{x} > 1$, *i.e.*, if $x < 1$ and diverges if $\frac{1}{x} < 1$, *i.e.*, if $x > 1$.

Ratio test fails when $x = 1$.

When $x = 1$, from (1) we have

$$\frac{u_n}{u_{n+1}} = \left[\frac{1+\frac{1}{n\log n} - \frac{1}{2n^2 \log n} + \cdots}{1+\frac{2}{n\log n} - \frac{2}{n^2\log n} + \cdots}\right]^q$$

$$= \left[1+\left(\frac{1}{n\log n} - \frac{1}{2n^2\log n} + \cdots\right)\right]^q \left[1+\left(\frac{2}{n\log n} - \frac{2}{n^2\log n} + \cdots\right)\right]^{-q}$$

Expanding by Binomial Theorem

$$= \left[1+q\left(\frac{1}{n\log n} - \frac{1}{2n^2\log n} + \cdots\right) + \cdots\right]\left[1-q\left(\frac{2}{n\log n} - \frac{2}{n^2\log n} + \cdots\right) + \cdots\right]$$

$$= 1 + q\left(\frac{1}{n\log n} - \frac{2}{n\log n}\right) + \cdots$$

$\Rightarrow \quad \frac{u_n}{u_{n+1}} - 1 = -\frac{q}{n\log n}$ + (terms containing squares and higher power of n and $\log n$ in the denominator)

$$\Rightarrow \quad n\left(\frac{u_n}{u_{n+1}}-1\right)=-\frac{q}{\log n}$$ + (terms containing n, n^2, ..., $(\log n)^2$, $(\log n)^3$ ... in the denominator)

$$\Rightarrow \qquad \lim_{n\to\infty} n\left(\frac{u_n}{u_{n+1}}-1\right)=0<1$$

$\therefore$ By Raabe's test, the series Σu_n is divergent.

Hence the given series Σu_n is convergent for $x < 1$ and divergent for $x \geq 1$.

Example 7. *Discuss the convergence of the hyper-geometric series:*

$$1+\frac{\alpha\beta}{1\cdot\gamma}x+\frac{\alpha(\alpha+1)\,\beta\,(\beta+1)}{1\cdot 2\cdot\gamma(\gamma+1)}x^2+\frac{\alpha(\alpha+1)(\alpha+2)\,\beta(\beta+1)(\beta+2)}{1\cdot 2\cdot 3\cdot\gamma(\gamma+1)(\gamma+2)}x^3+\cdots$$

Sol. Neglecting the first term,

$$u_n=\frac{\alpha(\alpha+1)\cdots(\alpha+n-1)\,\beta(\beta+1)\cdots(\beta+n-1)}{1\cdot 2\cdot 3\cdots n\cdot\gamma(\gamma+1)\cdots(\gamma+n-1)}\,.\,x^n$$

$$u_{n+1}=\frac{\alpha(\alpha+1)\cdots(\alpha+n-1)\,(\alpha+n)\,\beta(\beta+1)\cdots(\beta+n-1)(\beta+n)}{1\cdot 2\cdot 3\cdots n(n+1)\,\gamma(\gamma+1)(\gamma+2)\cdots(\gamma+n-1)(\gamma+n)}\cdot x^{n+1}$$

$$\therefore \qquad \frac{u_n}{u_{n+1}}=\frac{(n+1)(\gamma+n)}{(\alpha+n)(\beta+n)}\cdot\frac{1}{x}=\frac{\left(1+\frac{1}{n}\right)\left(1+\frac{\gamma}{n}\right)}{\left(1+\frac{\alpha}{n}\right)\left(1+\frac{\beta}{n}\right)}\cdot\frac{1}{x}$$

$$\Rightarrow \qquad \lim_{n\to\infty}\frac{u_n}{u_{n+1}}=\frac{1}{x}\,.$$

$\therefore$ By D'Alembert's Ratio test the series Σu_n converges if $\frac{1}{x} > 1$ *i.e.*, if $x < 1$ and diverges if $\frac{1}{x} < 1$ or if $x > 1$.

If $\qquad x = 1,\ \lim_{n\to\infty}\frac{u_n}{u_{n+1}}=1 \quad \therefore$ Ratio test fails.

Putting $x = 1$ in $\frac{u_n}{u_{n+1}}$, we have $\frac{u_n}{u_{n+1}}=\frac{\left(1+\frac{1}{n}\right)\left(1+\frac{\gamma}{n}\right)}{\left(1+\frac{\alpha}{n}\right)\left(1+\frac{\beta}{n}\right)}$

$$=\left(1+\frac{1}{n}\right)\left(1+\frac{\gamma}{n}\right)\left(1+\frac{\alpha}{n}\right)^{-1}\left(1+\frac{\beta}{n}\right)^{-1}$$ [Expand by Binomial Theorem]

$$=\left(1+\frac{1}{n}\right)\left(1+\frac{\gamma}{n}\right)\left(1-\frac{\alpha}{n}+\frac{\alpha^2}{n^2}+\cdots\right)\left(1-\frac{\beta}{n}+\frac{\beta^2}{n^2}\cdots\right)$$

$$=\left(1+\frac{1}{n}+\frac{\gamma}{n}+\frac{\gamma}{n^2}\right)\left(1-\frac{\alpha}{n}-\frac{\beta}{n}+\frac{\alpha\beta}{n^2}+\frac{\alpha^2}{n^2}+\frac{\beta^2}{n^2}\cdots\right)$$

$$= 1 + \frac{1}{n}(1 + \gamma - \alpha - \beta) + O\left(\frac{1}{n^2}\right).$$

$\therefore$ By Gauss test, the series Σu_n converges if $1 + \gamma - \alpha - \beta > 1$ *i.e.,* if $\gamma > \alpha + \beta$ and diverges if $1 + \gamma - \alpha - \beta \le 1$ *i.e.,* if $\gamma \le \alpha + \beta$.

Thus the given series *converges if $x < 1$ and diverges if $x > 1$. If $x = 1$, then the series converges if $\gamma > \alpha + \beta$ and diverges if $\gamma \le \alpha + \beta$.*

EXERCISE 1.5

Discuss the convergence of the following series:

1. $1 + \frac{2^2}{3^2} + \frac{2^2 \cdot 4^2}{3^2 \cdot 5^2} + \frac{2^2 \cdot 4^2 \cdot 6^2}{3^2 \cdot 5^2 \cdot 7^2} + \cdots \infty$

2. $\frac{1^2}{4^2} + \frac{1^2 \cdot 5^2}{4^2 \cdot 8^2} + \frac{1^2 \cdot 5^2 \cdot 9^2}{4^2 \cdot 8^2 \cdot 12^2} + \cdots \infty$

3. $1 + \frac{3}{7}x + \frac{3 \cdot 6}{7 \cdot 10}x^2 + \frac{3 \cdot 6 \cdot 9}{7 \cdot 10 \cdot 13}x^3 + \cdots \infty$

4. $1 + \frac{2}{1} \cdot \frac{1}{2} + \frac{2 \cdot 4}{1 \cdot 3} \cdot \frac{1}{3} + \frac{2 \cdot 4 \cdot 6}{1 \cdot 3 \cdot 5} \cdot \frac{1}{4} + \cdots \infty$

5. $\frac{1^2}{2^2} + \frac{1^2 \cdot 3^2}{2^2 \cdot 4^2}x + \frac{1^2 \cdot 3^2 \cdot 5^2}{2^2 \cdot 4^2 \cdot 6^2}x^2 + \ldots \infty$

6. $x + \frac{2^2 x^2}{2!} + \frac{3^3 x^3}{3!} + \frac{4^4 x^4}{4!} + \frac{5^5 x^5}{5!} + \cdots \infty$ *(M.D.U. May 2009, Dec. 2010, Dec. 2013)*

7. $1 + \frac{2x}{2!} + \frac{3^2 x^2}{3!} + \frac{4^3 x^3}{4!} + \cdots \infty$

8. $1 + \frac{1}{2} \cdot \frac{x^2}{4} + \frac{1 \cdot 3 \cdot 5}{2 \cdot 4 \cdot 6} \cdot \frac{x^4}{8} + \frac{1 \cdot 3 \cdot 5 \cdot 7 \cdot 9}{2 \cdot 4 \cdot 6 \cdot 8 \cdot 10} \cdot \frac{x^6}{12} + \cdots \infty$

9. $1^p + \left(\frac{1}{2}\right)^p + \left(\frac{1 \cdot 3}{2 \cdot 4}\right)^p + \left(\frac{1.3.5}{2.4.6}\right)^p + \cdots \infty$

10. $\frac{a}{b} + \frac{a(a+1)}{b(b+1)} + \frac{a(a+1)(a+2)}{b(b+1)(b+2)} + \cdots \infty$

11. $\frac{a}{b} + \frac{a(a+d)}{b(b+d)}x + \frac{a(a+d)(a+2d)}{b(b+d)(b+2d)}x^2 + \cdots \infty$ $(a > 0,\ b > 0,\ x > 0)$

12. $\sum \frac{n!}{x(x+1)(x+2)\ldots(x+n-1)}$

13. $1 + \frac{1}{2}x + \frac{1 \cdot 3}{2 \cdot 4}x^2 + \frac{1 \cdot 3 \cdot 5}{2 \cdot 4 \cdot 6}x^3 + \cdots \infty$ $(x > 0)$

14. $\frac{x}{1.2} + \frac{x^2}{3.4} + \frac{x^3}{5.6} + \frac{x^4}{7.8} + \cdots \infty$ $(x > 0)$

15. $\frac{x}{1 \cdot 2} + \frac{x^2}{2 \cdot 3} + \frac{x^3}{3 \cdot 4} + \frac{x^4}{4 \cdot 5} + \cdots \infty$

16. $1 + \frac{a}{1!} + \frac{a(a+1)}{2!} + \frac{a(a+1)(a+2)}{3!} + \cdots \infty$ $(a > 0)$

17. $\sum \frac{1 \cdot 3 \cdot 5 \cdots (2n-1)}{2 \cdot 4 \cdot 6 \cdots (2n)} \cdot \frac{x^{2n}}{2n}$, $x > 0$

18. $\sum \frac{(n!)^2}{(2n)!} x^{2n}$, $x > 0$. *(M.D.U. Dec. 2010)*

19. $\sum \frac{4 \cdot 7 \cdot 10 \cdots (3n+1)}{1 \cdot 2 \cdot 3 \cdots n} x^n$

20. $1 + \frac{(1!)^2}{2!}x^2 + \frac{(2!)^2}{4!}x^4 + \frac{(3!)^2}{6!}x^6 + \cdots \infty$ $(x > 0)$

21. $\frac{1}{2}x + x^2 + \frac{9}{8}x^3 + x^4 + \frac{25}{32}x^5 + \cdots \infty$

22. $\frac{x^2}{2\log 2} + \frac{x^3}{3\log 3} + \frac{x^4}{4\log 4} + \cdots \infty$ *(M.D.U. Dec. 2011)*

23. $1+\frac{2}{3}x+\frac{2\cdot 3}{3\cdot 5}x^2+\frac{2\cdot 3\cdot 4}{3\cdot 5\cdot 7}x^3+\cdots\infty$

24. $1+\frac{2^2}{3^2}x+\frac{2^2\cdot 4^2}{3^2\cdot 5^2}x^2+\frac{2^2\cdot 4^2\cdot 6^2}{3^2\cdot 5^2\cdot 7^2}x^3+\cdots\infty$

25. $\sum_{n=1}^{\infty}\frac{(2n)!}{(n!)^2}x^n$

26. $1+\frac{a}{b}x+\frac{a(a+1)}{b(b+1)}x^2+\frac{a(a+1)(a+2)}{b(b+1)(b+2)}x^3+\cdots\infty.$

27. $\sum\frac{n^2(n+1)^2}{n!}$

28. $x^2+\frac{2^2}{3\cdot 4}x^4+\frac{2^2\cdot 4^2}{3\cdot 4\cdot 5\cdot 6}x^6+\frac{2^2\cdot 4^2\cdot 6^2}{3\cdot 4\cdot 5\cdot 6\cdot 7\cdot 8}x^8+\cdots\infty\ (x>0)$

29. $\frac{1}{(\log 2)^k}+\frac{1}{(\log 3)^k}+\frac{1}{(\log 4)^k}+\cdots\infty.$

Answers

1. Divergent
2. Convergent
3. Convergent for $x \le 1$, divergent for $x > 1$
4. Divergent
5. Convergent for $x < 1$, divergent for $x \ge 1$
6. Convergent for $x < \frac{1}{e}$, divergent for $x \ge \frac{1}{e}$
7. Convergent for $x \le \frac{1}{e}$, divergent for $x > \frac{1}{e}$
8. Convergent for $x^2 \le 1$, divergent for $x^2 > 1$

$\left[\textbf{Hint.}\text{ Here } u_n=\frac{1\cdot 3\cdot 5\cdots(4n-3)}{2\cdot 4\cdot 6\cdots(4n-2)}\cdot\frac{x^{2n}}{4n}\ \text{ and }\ u_{n+1}=\frac{1\cdot 3\cdot 5\cdots(4n-3)(4n-1)(4n+1)}{2\cdot 4\cdot 6\cdots(4n-2)(4n)(4n+2)}\cdot\frac{x^{2n+2}}{4n+4}\right]$

9. Convergent for $p > 2$, divergent for $p \le 2$
10. Convergent for $b > a + 1$, divergent for $b \le a + 1$
11. Convergent for $x < 1$ or $x = 1$ and $b > a + d$, divergent for $x > 1$ or $x = 1$ and $b \le a + d$
12. Convergent for $x > 2$, divergent for $x \le 2$
13. Convergent for $x < 1$, divergent for $x \ge 1$
14. Convergent for $x \le 1$, divergent for $x > 1$
15. Convergent for $x \le 1$, divergent for $x > 1$
16. Divergent
17. Convergent for $x^2 \le 1$, divergent for $x^2 > 1$
18. Convergent for $x^2 < 4$, divergent for $x^2 \ge 4$
19. Convergent for $x < \frac{1}{3}$, divergent for $x \ge \frac{1}{3}$
20. Convergent for $x^2 < 4$, divergent for $x^2 \ge 4$
21. Convergent for $x < 2$, divergent for $x \ge 2$
22. Convergent for $x < 1$, divergent for $x \ge 1$
23. Convergent for $x < 2$, divergent for $x \ge 2$
24. Convergent for $x < 1$, divergent for $x \ge 1$
25. Convergent for $x < \frac{1}{4}$, divergent for $x \ge \frac{1}{4}$
26. Convergent for $x < 1$, divergent for $x > 1$, when $x = 1$, convergent for $b - a > 1$, divergent for $b - a \le 1$
27. Convergent
28. Convergent if $x^2 \le 1$, divergent if $x^2 > 1$
29. Divergent.

1.29. CAUCHY'S INTEGRAL TEST

Statement. *If for $x \ge 1$, $f(x)$ is a non-negative, monotonic decreasing function of x such that $f(n) = u_n$ for all positive integral values of n, then the series Σu_n and the integral $\int_1^{\infty} f(x)\,dx$ converge or diverge together.*

Proof. Let r be a +ve integer. Choose x such that $r + 1 \ge x \ge r \ge 1$

Since $f(x)$ is a monotonic decreasing function of x.

$\therefore \qquad f(r+1) \le f(x) \le f(r) \ \Rightarrow\ u_{r+1} \le f(x) \le u_r \qquad [\because\ f(n) = u_n,\ n \in N]$

$$\Rightarrow \quad \int_r^{r+1} u_{r+1}\, dx \le \int_r^{r+1} f(x)\, dx \le \int_r^{r+1} u_r\, dx$$

$$\Rightarrow \quad u_{r+1} \int_r^{r+1} dx \le \int_r^{r+1} f(x)\, dx \le u_r \int_r^{r+1} dx$$

$$\Rightarrow \quad u_{r+1} \Big[x \Big]_r^{r+1} \le \int_r^{r+1} f(x)\, dx \le u_r \Big[x \Big]_r^{r+1}$$

$$\Rightarrow \quad u_{r+1} \le \int_r^{r+1} f(x)\, dx \le u_r \qquad ...(1)$$

Putting $r = 1, 2, 3, ..., n$ in succession in (1), we have

$$u_2 \le \int_1^2 f(x)\, dx \le u_1$$

$$u_3 \le \int_2^3 f(x)\, dx \le u_2$$

....

....

$$u_{n+1} \le \int_n^{n+1} f(x)\, dx \le u_n$$

Adding the above inequalities, we have

$$u_2 + u_3 + \cdots + u_{n+1} \le \int_1^2 f(x)\, dx + \int_2^3 f(x)\, dx + \cdots + \int_n^{n+1} f(x)\, dx \le u_1 + u_2 + \cdots + u_n$$

$$\Rightarrow \quad S_{n+1} - u_1 \le \int_1^{n+1} f(x)\, dx \le S_n, \text{ where } S_n = \sum_1^n u_n = u_1 + u_2 + \cdots + u_n.$$

Proceeding to the limit as $n \to \infty$

$$\lim_{n\to\infty} S_{n+1} - u_1 \le \lim_{n\to\infty} \int_1^{n+1} f(x)\, dx \le \lim_{n\to\infty} S_n$$

$$\Rightarrow \quad \lim_{n\to\infty} S_{n+1} - u_1 \le \int_1^{\infty} f(x)\, dx \le \lim_{n\to\infty} S_n \qquad ...(2)$$

(*i*) **If $\int_1^\infty$ f(x) dx converges,** then $\int_1^\infty f(x)\, dx$ = a fixed finite number = I (say).

Then from (2), we have $\lim_{n\to\infty} S_{n+1} - u_1 \le I$

$$\Rightarrow \quad \lim_{n\to\infty} S_{n+1} \le I + u_1 = \text{a fixed finite number}$$

$\Rightarrow$ $\{S_n\}$ is a convergent sequence

$\Rightarrow$ **the series Σu_n is convergent.**

(*ii*) If $\int_1^\infty f(x)\, dx$ diverges, then $\int_1^\infty f(x)\, dx = +\infty$

From (2), $\lim_{n\to\infty} S_n \ge \int_1^\infty f(x)\, dx = +\infty$

$\Rightarrow$ $\{S_n\}$ is a divergent sequence

$\Rightarrow$ the series Σu_n is divergent.

Hence Σu_n and $\int_1^\infty f(x)\, dx$ converge or diverge together.

Note. If $x \ge k$, then Σu_n and $\int_k^\infty f(x)\, dx$ converge or diverge together.

ILLUSTRATIVE EXAMPLES

Example 1. *Test for convergence the series:* $\sum \frac{1}{n^2+1}$.

Sol. Here $$u_n = \frac{1}{n^2+1} = f(n)$$

$\therefore$ $$f(x) = \frac{1}{x^2+1}$$

For $x \geq 1$, $f(x)$ is +ve and monotonic decreasing.

$\therefore$ Cauchy's Integral Test is applicable.

Now $$\int_1^\infty f(x)\,dx = \int_1^\infty \frac{dx}{x^2+1} = \left[\tan^{-1} x\right]_1^\infty = \frac{\pi}{2} - \frac{\pi}{4} = \frac{\pi}{4} = \text{finite}$$

$\Rightarrow$ $\int_1^\infty f(x)\,dx$ converges and hence by Integral Test, Σu_n also converges.

Example 2. *Show that the series* $\sum_1^\infty \frac{1}{n^p}$ *converges if* $p > 1$ *and diverges if* $0 < p \leq 1$.

(M.D.U. Dec. 2010)

Sol. Here $$u_n = \frac{1}{n^p} = f(n)$$

$\therefore$ $$f(x) = \frac{1}{x^p}$$

For $x \geq 1$, $f(x)$ is + ve and monotonic decreasing.

$\therefore$ Cauchy's Integral Test is applicable.

Case I. *When* $p \neq 1$

$$\int_1^\infty f(x)\,dx = \int_1^\infty \frac{1}{x^p}\,dx = \int_1^\infty x^{-p}\,dx = \left[\frac{x^{-p+1}}{-p+1}\right]_1^\infty$$

Sub-Case 1. When $p > 1$, $p - 1$ is +ve, so that $\int_1^\infty f(x)\,dx = -\frac{1}{p-1}\left[\frac{1}{x^{p-1}}\right]_1^\infty$

$$= -\frac{1}{p-1}[0-1] = \frac{1}{p-1} = \text{finite}$$

$\Rightarrow$ $\int_1^\infty f(x)\,dx$ converges $\Rightarrow$ Σu_n is convergent.

Sub-Case 2. When $0 < p < 1$, $1 - p$ is +ve, so that

$$\int_1^\infty f(x)\,dx = \frac{1}{1-p}\left[x^{1-p}\right]_1^\infty = \frac{1}{1-p}(\infty - 1) = \infty$$

$\Rightarrow$ $\int_1^\infty f(x)\,dx$ diverges $\Rightarrow$ Σu_n is divergent.

Case II. *When* $p = 1$, $f(x) = \frac{1}{x}$

$$\int_1^\infty f(x)dx = \int_1^\infty \frac{1}{x}\,dx = \left[\log x\right]_1^\infty = \infty - \log 1 = \infty - 0 = \infty$$

$\Rightarrow \quad \int_1^\infty f(x)\, dx$ diverges $\quad \Rightarrow \quad \Sigma u_n$ is divergent.

Hence Σu_n converges if $p > 1$ and diverges if $p \le 1$.

Example 3. *Discuss the convergence of the series:* $\sum_{n=2}^{\infty} \frac{1}{n(\log n)^p}$, $(p > 0)$.

(M.D.U. Dec. 2009)

Sol. Here $\quad u_n = \frac{1}{n(\log n)^p} = f(n) \quad \therefore \quad f(x) = \frac{1}{x(\log x)^p}$

For $x \ge 2$, $p > 0$, $f(x)$ is +ve and monotonic decreasing.

$\therefore$ By Cauchy's Integral Test, $\sum_{n=2}^{\infty} u_n$ and $\int_2^\infty f(x)\, dx$ converge or diverge together.

Case I. *When* $p \ne 1$

$$\int_2^\infty f(x)\, dx = \int_2^\infty (\log x)^{-p} \cdot \frac{1}{x}\, dx = \left[\frac{(\log x)^{-p+1}}{-p+1}\right]_2^\infty$$

Sub-Case 1. When $p > 1$, $p - 1$ is +ve, so that $\int_2^\infty f(x)\, dx = -\frac{1}{p-1}\left[\frac{1}{(\log x)^{p-1}}\right]_2^\infty$

$$= -\frac{1}{p-1}\left[0 - \frac{1}{(\log 2)^{p-1}}\right] = \frac{1}{(p-1)(\log 2)^{p-1}} = \text{finite}$$

$\Rightarrow \quad \int_2^\infty f(x)\, dx$ converges $\quad \Rightarrow \quad \sum_{n=2}^{\infty} u_n$ converges.

Sub-Case 2. When $p < 1$, $1 - p$ is +ve, so that

$$\int_2^\infty f(x)\, dx = \frac{1}{1-p}\left[(\log x)^{1-p}\right]_2^\infty = \frac{1}{1-p}\,[\infty - (\log 2)^{1-p}] = \infty$$

$\Rightarrow \quad \int_2^\infty f(x)\, dx$ diverges $\quad \Rightarrow \quad \sum_{n=2}^{\infty} u_n$ diverges.

Case II. *When* $\quad p = 1,\ f(x) = \frac{1}{x \log x}$

$$\int_2^\infty f(x)\, dx = \int_2^\infty \frac{dx}{x \log x} = \int_2^\infty \frac{\frac{1}{x}}{\log x}\, dx = \Big[\log \log x\Big]_2^\infty = \infty - \log \log 2 = \infty$$

$\Rightarrow \quad \int_2^\infty f(x)\, dx$ diverges $\quad \Rightarrow \quad \sum_{n=2}^{\infty} u_n$ diverges.

Hence Σu_n converges if $p > 1$ and diverges if $0 < p \le 1$.

EXERCISE 1.6

Using the integral test, discuss the convergence of the following series:

1. $\sum \frac{1}{2n+3}$ 2. $\sum \frac{1}{n(n+1)}$ 3. $\sum \frac{1}{\sqrt{n}}$

4. $\sum \frac{1}{(n+1)^2}$ (*M.D.U. May 2008*) 5. $\sum \frac{2n^3}{n^4+3}$ 6. $\sum \frac{n}{(n^2+1)^2}$

7. $\sum_{n=2}^{\infty} \frac{1}{n\sqrt{n^2-1}}$ 8. $\sum n e^{-n^2}$ 9. $\sum_{n=3}^{\infty} \frac{1}{n \log n (\log \log n)^p}$, $p > 0$

10. $\sum_{n=2}^{\infty} \frac{1}{n(\log n)^2}$ 11. $\sum_{n=1}^{\infty} \frac{1}{(n+1)\log(n+1)}$ 12. $\sum_{n=1}^{\infty} \frac{8\tan^{-1} n}{1+n^2}$.

Answers

1. Divergent 2. Convergent 3. Divergent 4. Convergent
5. Divergent 6. Convergent 7. Convergent 8. Convergent
9. Convergent for $p > 1$, divergent for $p \le 1$ 10. Convergent 11. Divergent
12. Convergent.

1.30. LEIBNITZ' TEST ON ALTERNATING SERIES

Statement. *The alternating series* $\Sigma(-1)^{n-1} u_n = u_1 - u_2 + u_3 - u_4 + \cdots$ $(u_n > 0\ \forall\ n)$ *converges if*

(i) $u_n > u_{n+1}$ $\forall$ *n and* *(ii)* $\lim_{n\to\infty} u_n = 0$.

Proof. Let S_n denote the *n*th partial sum of the series $\Sigma (-1)^{n-1} . u_n$.

$$S_{2n} = u_1 - u_2 + u_3 - u_4 + u_5 - \cdots - u_{2n-2} + u_{2n-1} - u_{2n}$$
$$= u_1 - (u_2 - u_3) - (u_4 - u_5) \cdots - (u_{2n-2} - u_{2n-1}) - u_{2n}$$
$$= u_1 - [(u_2 - u_3) + (u_4 - u_5) + \cdots + (u_{2n-2} - u_{2n-1}) + u_{2n}]$$
$$< u_1 \qquad [\because\ u_n > u_{n+1} \text{ and } u_n > 0 \text{ for all } n]$$

$\Rightarrow$ The sequence $\{S_{2n}\}$ is bounded above.

Also $S_{2n+2} = S_{2n} + u_{2n+1} - u_{2n+2}$

$\Rightarrow$ $S_{2n+2} - S_{2n} = u_{2n+1} - u_{2n+2} > 0$ for all n $\Rightarrow$ $S_{2n+2} > S_{2n}$

$\Rightarrow$ The sequence $\{S_{2n}\}$ is monotonically increasing.

Since every monotonically increasing sequence which is bounded above converges, therefore, the sequence $\{S_{2n}\}$ converges. Let it converge to S, then $\lim_{n\to\infty} S_{2n} = S$

Now $\lim_{n\to\infty} S_{2n+1} = \lim_{n\to\infty} (S_{2n} + u_{2n+1}) = \lim_{n\to\infty} S_{2n} + \lim_{n\to\infty} u_{2n+1}$

$$= S + 0 \qquad [\because\ \lim_{n\to\infty} u_n = 0]$$
$$= S$$

$\Rightarrow$ The sequences $\{S_{2n}\}$ and $\{S_{2n+1}\}$ converge to the same real number S.

$\therefore$ Given $\varepsilon > 0$, there exist positive integers m_1 and m_2 such that

$$| S_{2n} - S | < \varepsilon \ \ \forall\ \ 2n > m_1 \ \text{ and } \ | S_{2n+1} - S | < \varepsilon \ \ \forall\ \ 2n + 1 > m_2$$

Let $m = \max. \{m_1, m_2\}$, then

$$| S_{2n} - S | < \varepsilon \ \ \forall\ \ n > m \ \text{ and } \ | S_{2n+1} - S | < \varepsilon \ \ \forall\ \ n > m$$

$\Rightarrow$ $| S_n - S | < \varepsilon \ \ \forall\ \ n > m$

$\therefore$ The sequence $\{S_n\}$ converges to S.

Hence the given series is convergent.

Note 1. The alternating series will not be convergent if any one of the two conditions is not satisfied.

Note 2. When $\lim_{n\to\infty} u_n \neq 0$, the alternating series is oscillatory.

ILLUSTRATIVE EXAMPLES

Example 1. *Examine the convergence of the series:* $1 - \frac{1}{2} + \frac{1}{3} - \frac{1}{4} + \cdots$

Sol. It is an alternating series $\Sigma(-1)^{n-1} u_n$.

(*i*) $u_n = \frac{1}{n}, \quad u_{n+1} = \frac{1}{n+1}$

$$\because \quad \frac{1}{n} > \frac{1}{n+1} \ \forall\, n \quad \therefore \quad u_n > u_{n+1} \ \forall\, n$$

(*ii*) $\lim_{n\to\infty} u_n = \lim_{n\to\infty} \frac{1}{n} = 0$

$\therefore$ Both the conditions of Leibnitz's Test are satisfied.

Hence the given series is convergent.

Example 2. *Examine the convergence of the series:*

(*a*) $2 - \frac{3}{2} + \frac{4}{3} - \frac{5}{4} + \cdots$

(*b*) $\frac{1}{2^3} - \frac{1}{3^3}(1+2) + \frac{1}{4^3}(1+2+3) - \frac{1}{5^3}(1+2+3+4) + \cdots$

Sol. (*a*) It is an alternating series $\Sigma(-1)^{n-1} u_n$.

(*i*) $u_n = \frac{n+1}{n}, \quad u_{n+1} = \frac{n+2}{n+1}$

$$u_n - u_{n+1} = \frac{n+1}{n} - \frac{n+2}{n+1} = \frac{(n+1)^2 - n(n+2)}{n(n+1)} = \frac{1}{n(n+1)} > 0 \quad \forall\, n$$

$$\Rightarrow \quad u_n > u_{n+1} \quad \forall \quad n$$

(*ii*) $\lim_{n\to\infty} u_n = \lim_{n\to\infty} \frac{n+1}{n} = \lim_{n\to\infty}\left(1 + \frac{1}{n}\right) = 1 \neq 0.$

Since the second condition of Leibnitz's Test is not satisfied, the series is not convergent.

(*b*) It is an alternating series $\Sigma(-1)^{n-1} u_n$.

(*i*) $u_n = \frac{1}{(n+1)^3}[1+2+3+\cdots+n] = \frac{1}{(n+1)^3}\cdot\frac{n(n+1)}{2} = \frac{1}{2}\cdot\frac{n}{(n+1)^2}.$

$u_{n+1} = \frac{1}{2}\cdot\frac{n+1}{(n+2)^2}$

$$u_n - u_{n+1} = \frac{1}{2}\left[\frac{n}{(n+1)^2} - \frac{n+1}{(n+2)^2}\right] = \frac{1}{2}\,\frac{n(n+2)^2 - (n+1)^3}{(n+1)^2(n+2)^2}$$

$$= \frac{1}{2}\cdot\frac{n^2+n-1}{(n+1)^2(n+2)^2} > 0 \quad \forall \quad n$$

$$\Rightarrow \quad u_n > u_{n+1} \quad \forall \quad n$$

(ii) $\lim\limits_{n\to\infty} u_n = \lim\limits_{n\to\infty} \dfrac{n}{2(n+1)^2} = \lim\limits_{n\to\infty} \dfrac{\frac{1}{n}}{2\left(1+\frac{1}{n}\right)^2} = 0.$

Since both the conditions of Leibnitz's Test are satisfied, the given series is convergent.

Example 3. *Test the convergence of the series:* $\sum\limits_{n=1}^{\infty} \dfrac{(-1)^{n-1} . n}{2n-1}$.

Sol. The given series is $\sum\limits_{n=1}^{\infty} \dfrac{(-1)^{n-1} \cdot n}{2n-1} = \dfrac{1}{1} - \dfrac{2}{3} + \dfrac{3}{5} - \dfrac{4}{7} + \ldots$

It is an alternating series $\Sigma(-1)^{n-1} u_n$.

(i) $u_n = \dfrac{n}{2n-1}, \quad u_{n+1} = \dfrac{n+1}{2n+1}$

$$u_n - u_{n+1} = \frac{1}{4n^2-1} > 0 \quad \forall \; n \;\Rightarrow\; u_n > u_{n+1} \quad \forall \; n$$

(ii) $\lim\limits_{n\to\infty} u_n = \lim\limits_{n\to\infty} \dfrac{n}{2n-1} = \lim\limits_{n\to\infty} \dfrac{1}{2-\frac{1}{n}} = \dfrac{1}{2} \neq 0.$

Here the second condition of Leibnitz's Test is not satisfied. Hence the given series is not convergent.

Example 4. *Show that the series* $\dfrac{\log 2}{2^2} - \dfrac{\log 3}{3^2} + \dfrac{\log 4}{4^2} - \ldots \infty$ *converges.*

Sol. The given series is an alternating series $\Sigma(-1)^{n-1} u_n$ with

$$u_n = \frac{\log(n+1)}{(n+1)^2} \quad \forall \; n \geq 1.$$

Now consider the function $f(x) = \dfrac{\log x}{x^2}$, $x > 0$

$$f'(x) = \frac{x^2 \cdot \frac{1}{x} - 2x \log x}{x^4} = \frac{1 - 2\log x}{x^3} < 0 \text{ if } 1 - 2\log x < 0$$

i.e., if $\log x > \frac{1}{2}$ or if $x > e^{1/2}$

$\Rightarrow$ $f(x)$ is monotonically decreasing for $x > \sqrt{e} = 1.65$

$\Rightarrow$ $f(2) > f(3) > f(4) > \ldots$

In particular $f(n+1) > f(n+2) \quad \forall \; n \geq 1$ *i.e.,* $u_n > u_{n+1} \quad \forall \; n.$

Also $\lim\limits_{n\to\infty} u_n = \lim\limits_{n\to\infty} \dfrac{\log(n+1)}{(n+1)^2}$ $\left[\text{Form } \dfrac{\infty}{\infty}\right]$

$$= \lim_{n\to\infty} \frac{\frac{1}{n+1}}{2(n+1)} = \lim_{n\to\infty} \frac{1}{2(n+1)^2} = 0$$

Thus both the conditions of Leibnitz's Test are satisfied. Hence the given series is convergent.

Example 5. *Discuss the convergence of the series*

$$\sum_{n=2}^{\infty} \frac{(-1)^{n-1} x^n}{n(n-1)}, \; 0 < x < 1.$$

Sol. The given series is an alternating series $\sum_{n=2}^{\infty} (-1)^{n-1} u_n$ with

$$u_n = \frac{x^n}{n(n-1)},\ 0 < x < 1,\ n \geq 2$$

$$\Rightarrow \qquad u_{n+1} = \frac{x^{n+1}}{(n+1)n}$$

$$\therefore \qquad u_n - u_{n+1} = \frac{x^n}{n}\left(\frac{1}{n-1} - \frac{x}{n+1}\right) = \frac{x^n}{n}\left[\frac{(n+1)-(n-1)x}{(n-1)(n+1)}\right]$$

$$= \frac{x^n}{n}\left[\frac{n(1-x)+(1+x)}{n^2-1}\right]$$

Since $0 < x < 1$ $\therefore$ x^n, $(1 - x)$ and $(1 + x)$ are all positive.

Since $n \geq 2$ $\therefore$ $n^2 - 1 > 0$

Thus $u_n - u_{n+1} > 0 \Rightarrow u_n > u_{n+1}\ \forall\ n \geq 2$

Also $\lim_{n\to\infty} u_n = \lim_{n\to\infty} \frac{1}{n(n-1)} \cdot x^n = 0 \times 0 = 0$ $\qquad [\because\ 0 < x < 1]$

Both the conditions of Leibnitz's Test are satisfied. Hence the given series is convergent.

EXERCISE 1.7

Discuss the convergence of the following series:

1. $1 - \frac{1}{\sqrt{2}} + \frac{1}{\sqrt{3}} - \frac{1}{\sqrt{4}} + \cdots \infty$

2. $1 - \frac{1}{2^2} + \frac{1}{3^2} - \frac{1}{4^2} + \cdots \infty$

3. $\frac{1}{\log 2} - \frac{1}{\log 3} + \frac{1}{\log 4} - \frac{1}{\log 5} + \cdots \infty$

4. $1 - \frac{1}{2} + \frac{1}{4} - \frac{1}{8} + \cdots \infty$

5. $1 - \frac{1}{2!} + \frac{1}{4!} - \frac{1}{6!} + \cdots \infty$

6. $\frac{1}{6} - \frac{2}{11} + \frac{3}{16} - \frac{4}{21} + \cdots \infty$

7. $\frac{5}{2} - \frac{7}{4} + \frac{9}{6} - \frac{11}{8} + \cdots \infty$

8. $\frac{1}{1.2} - \frac{1}{3.4} + \frac{1}{5.6} - \frac{1}{7.8} + \cdots \infty$

9. $\frac{1}{\sqrt{2}+1} - \frac{1}{\sqrt{3}+1} + \frac{1}{\sqrt{4}+1} - \frac{1}{\sqrt{5}+1} + \cdots \infty$

10. $\frac{1}{a} - \frac{1}{a+b} + \frac{1}{a+2b} - \frac{1}{a+3b} + \cdots \infty \quad (a > 0,\ b > 0)$

11. $\frac{x}{1+x} - \frac{x^2}{1+x^2} + \frac{x^3}{1+x^3} - \frac{x^4}{1+x^4} + \cdots \infty \quad (0 < x < 1)$

12. $\sum_{n=1}^{\infty} \frac{(-1)^{n-1}}{3n-2}$

13. $\sum_{n=1}^{\infty} \frac{n(-1)^{n-1}}{5^n}$

14. $\sum_{n=1}^{\infty} (-1)^{n-1} \frac{n}{n^2+1}$

15. $\sum_{n=1}^{\infty} \frac{\cos n\pi}{n^2+1}$ [**Hint.** $\cos n\pi = (-1)^n$]

16. $\log\left(\frac{1}{2}\right) - \log\left(\frac{2}{3}\right) + \log\left(\frac{3}{4}\right) - \log\left(\frac{4}{5}\right) + \cdots \infty$

[The given series $\sum_{n=1}^{\infty} (-1)^{n-1} \log\left(\frac{n}{n+1}\right)$ is not an alternating series because $0 < \frac{n}{n+1} < 1$

and $u_n = \log\left(\frac{n}{n+1}\right) < 0$.

Re-write the given series as $-\log\left(\frac{2}{1}\right) + \log\left(\frac{3}{2}\right) - \log\left(\frac{4}{3}\right) + \log\left(\frac{5}{4}\right) - \ldots = \sum_{n=1}^{\infty} (-1)^n \log\left(\frac{n+1}{n}\right)$

Since $\quad \frac{n+1}{n} = 1 + \frac{1}{n} > 1,\ u_n = \log\left(\frac{n+1}{n}\right) > 0$]

Answers

1. Convergent 2. Convergent 3. Convergent
4. $\sum_{n=1}^{\infty} \frac{(-1)^{n-1}}{2^{n-1}}$, Convergent 5. Convergent 6. $\sum_{n=1}^{\infty} \frac{n(-1)^{n-1}}{5n+1}$, not convergent
7. Not convergent 8. Convergent 9. Convergent 10. Convergent
11. Convergent 12. Convergent 13. Convergent 14. Convergent
15. Convergent 16. Convergent.

1.31. ABSOLUTE CONVERGENCE OF A SERIES *(M.D.U. Dec. 2012)*

Def. *If a convergent series whose terms are not all positive, remains convergent when all its terms are made positive, then it is called an* **absolutely convergent series**, *i.e.,*

The series Σu_n is said to be absolutely convergent if $\Sigma \mid u_n \mid$ is a convergent series.

A series is said to be conditionally convergent if it is convergent but does not converge absolutely.

Example. *Test whether the following series are absolutely convergent or conditionally convergent ?*

(a) $1 - \frac{1}{2^2} + \frac{1}{3^2} - \frac{1}{4^2} + \ldots$ (b) $\sum_{n=1}^{\infty} \frac{(-1)^{n-1}}{2n-1}$.

Sol. (a) The series is $1 - \frac{1}{2^2} + \frac{1}{3^2} - \frac{1}{4^2} + \ldots$

(i) In this alternating series, each term is less than the preceding term numerically.

(ii) Moreover $u_n = \frac{1}{n^2}$ which $\to 0$ as $n \to \infty$.

Hence the series satisfies both the conditions of the test on alternating series and so the given series *converges*.

Again when all the term of the series are made positive, the series becomes

$$\sum \mid u_n \mid = 1 + \frac{1}{2^2} + \frac{1}{3^2} + \frac{1}{4^2} + \cdots = \Sigma \frac{1}{n^2}$$

which we know is a convergent series. ($\because$ here $p = 2 > 1$)

Thus the given series converges absolutely.

(b) $\sum_{n=1}^{\infty} \frac{(-1)^{n-1}}{2n-1} = \Sigma (-1)^{n-1} u_n$ (say)

Putting $n = 1, 2, 3, \cdots$, the series becomes $1 - \frac{1}{3} + \frac{1}{5} + \frac{1}{7} + \cdots$

The series is clearly an alternating series.

(i) The terms go on decreasing numerically and

(ii) $\lim_{n \to \infty} u_n = \lim_{n \to \infty} \frac{1}{2n-1} = 0$

∴ By Leibnitz's Test, the series *converges.*

But when all terms are made positive, the series becomes,

$$\sum | u_n | = 1 + \frac{1}{3} + \frac{1}{5} + \frac{1}{7} + \cdots$$

Here $| u_n | = \frac{1}{2n-1}$. Take $v_n = \frac{1}{n}$

∴ $$\lim_{n\to\infty} \frac{|u_n|}{v_n} = \lim_{n\to\infty} \frac{n}{2n-1} = \lim_{n\to\infty} \left[\frac{1}{2 - \frac{1}{n}} \right] = \frac{1}{2} = \text{finite} \neq 0$$

Hence by comparison test series $\Sigma | u_n |$ and Σv_n behave alike.

But $\Sigma v_n = \Sigma \frac{1}{n}$ is a divergent series (∵ here $p = 1$), ∴ $\Sigma | u_n |$ also diverges.

Since the given series converges, and the series of absolute terms diverges, therefore the given series converges conditionally.

1.32. EVERY ABSOLUTELY CONVERGENT SERIES IS CONVERGENT

Proof. Let $\sum_{n=1}^{\infty} u_n$ be an absolutely convergent series.

Then $\sum_{n=1}^{\infty} | u_n |$ is convergent.

By Cauchy's general principle of convergence, given $\varepsilon > 0$, there exists a positive integer m such that

$$\| u_{m+1} | + | u_{m+2} | + \cdots + | u_n \| < \varepsilon \quad \forall \quad n > m$$

or $$| u_{m+1} | + | u_{m+1} | + \cdots + | u_n | < \varepsilon \quad \forall \quad n > m \qquad ...(1)$$

Now, by triangle inequality, we have

$$| u_{m+1} + u_{m+2} + \cdots + u_n | \leq | u_{m+1} | + | u_{m+2} | + \cdots + | u_n |$$
$$< \varepsilon \quad \forall \quad n > m \qquad \text{[Using (1)]}$$

∴ By Cauchy's general principle of convergence, the series $\sum_{n=1}^{\infty} u_n$ is convergent.

Hence $\sum | u_n |$ is convergent ⇒ Σu_n is convergent.

Note 1. Absolute convergence ⇒ Convergence, but convergence need not imply absolute convergence *i.e.,* the converse of above theorem need not be true.

For example, consider the series $\sum \frac{(-1)^{n-1}}{n} = 1 - \frac{1}{2} + \frac{1}{3} - \frac{1}{4} + \cdots$

It is convergent. [See Example 1 with Leibnitz's Test]

But the series $\sum \left| \frac{(-1)^{n-1}}{n} \right| = \sum \frac{1}{n}$ is divergent.

Note 2. The divergence of $\Sigma \mid u_n \mid$ does not imply the divergence of Σu_n.

For example, $\sum \left| \frac{(-1)^{n-1}}{n} \right| = \sum \frac{1}{n}$ is divergent whereas $\sum \frac{(-1)^{n-1}}{n}$ is convergent.

ILLUSTRATIVE EXAMPLES

Example 1. *Test the following series for convergence and absolute convergence:*

(*i*) $1 - \frac{1}{2\sqrt{2}} + \frac{1}{3\sqrt{3}} - \frac{1}{4\sqrt{4}} + \ldots$

(*ii*) $\frac{1}{1.3} - \frac{1}{2.4} + \frac{1}{3.5} - \frac{1}{4.6} + \ldots$ (*iii*) $1 - \frac{1}{2^3}(1+2) + \frac{1}{3^3}(1+2+3) - \frac{1}{4^3}(1+2+3+4) + \ldots$

Sol. (*i*) The given series is

$$\sum_{n=1}^{\infty} u_n = \sum_{n=1}^{\infty} (-1)^{n-1} \cdot \frac{1}{n\sqrt{n}} = \sum_{n=1}^{\infty} (-1)^{n-1} \cdot \frac{1}{n^{3/2}} = \sum_{n=1}^{\infty} (-1)^{n-1} a_n$$

It is an alternating series.

Here $\quad a_n = \frac{1}{n^{3/2}} > 0 \quad \forall\, n$

$$a_{n+1} = \frac{1}{(n+1)^{3/2}}$$

Since $\quad \frac{1}{n^{3/2}} > \frac{1}{(n+1)^{3/2}} \quad \forall\, n \quad \therefore \quad a_n > a_{n+1} \quad \forall\, n$

Also $\quad \lim_{n\to\infty} a_n = \lim_{n\to\infty} \frac{1}{n^{3/2}} = 0$

$\therefore$ By Leibnitz's test, the series Σu_n is convergent.

Now $\quad | u_n | = \frac{1}{n^{3/2}}$

$\therefore \quad \sum | u_n | = \sum \frac{1}{n^{3/2}}$ is convergent $\quad (\because \; p = 3/2 > 1)$

Hence the series Σu_n is absolutely convergent.

(*ii*) The given series is

$$\sum_{n=1}^{\infty} u_n = \sum_{n=1}^{\infty} (-1)^{n-1} \cdot \frac{1}{n(n+2)} = \sum_{n=1}^{\infty} (-1)^{n-1} \cdot a_n$$

It is an alternating series

Here $\quad a_n = \frac{1}{n(n+2)} > 0 \quad \forall\, n$

$$a_{n+1} = \frac{1}{(n+1)(n+3)}$$

$$\therefore \quad a_n - a_{n+1} = \frac{1}{n(n+2)} - \frac{1}{(n+1)(n+3)} = \frac{2n+3}{n(n+1)(n+2)(n+3)} > 0 \quad \forall\, n$$

$\Rightarrow \quad a_n > a_{n+1} \quad \forall\, n$

Also $\quad \lim_{n\to\infty} a_n = \lim_{n\to\infty} \frac{1}{n(n+2)} = 0$

$\therefore$ By Leibnitz's test, the series Σu_n is convergent.

Now $\quad | u_n | = \frac{1}{n(n+2)} = \frac{1}{n^2\left(1+\frac{2}{n}\right)}$

Take $\quad v_n = \frac{1}{n^2}$

$$\lim_{n\to\infty} \frac{|u_n|}{v_n} = \lim_{n\to\infty} \frac{1}{1+\frac{2}{n}} = 1, \text{ which is non-zero and finite.}$$

$\therefore$ By comparison test, $\Sigma | u_n |$ and Σv_n converge or diverge together. Since $\Sigma v_n = \sum \frac{1}{n^2}$ converges $(\because p = 2 > 1)$, $\Sigma | u_n |$ also converges.

Hence the given series Σu_n is absolutely convergent.

(*iii*) The given series is

$$\sum_{n=1}^{\infty} u_n = \sum_{n=1}^{\infty} (-1)^{n-1} . \frac{1+2+3+\dots+n}{n^3}$$

$$= \sum_{n=1}^{\infty} (-1)^{n-1} . \frac{\frac{n(n+1)}{2}}{n^3} = \sum_{n=1}^{\infty} (-1)^{n-1} . \frac{n+1}{2n^2} = \sum_{n=1}^{\infty} (-1)^{n-1} a_n$$

It is an alternating series.

Here $\quad a_n = \frac{n+1}{2n^2}, \quad a_{n+1} = \frac{n+2}{2(n+1)^2}$

$$a_n - a_{n+1} = \frac{n+1}{2n^2} - \frac{n+2}{2(n+1)^2} = \frac{(n+1)^3 - n^2(n+2)}{2n^2(n+1)^2} = \frac{n^2+3n+1}{2n^2(n+1)^2} > 0 \quad \forall\, n$$

$\Rightarrow \quad a_n > a_{n+1} \quad \forall\, n$

Also $\quad \lim_{n\to\infty} a_n = \lim_{n\to\infty} \frac{n+1}{2n^2} = \lim_{n\to\infty} \frac{1}{2}\left(\frac{1}{n} + \frac{1}{n^2}\right) = 0$

$\therefore$ By Leibnitz's test, the series Σu_n is convergent.

Now $\quad | u_n | = \frac{n+1}{2n^2} = \frac{n\left(1+\frac{1}{n}\right)}{2n^2} = \frac{1+\frac{1}{n}}{2n}$

Take $\quad v_n = \frac{1}{n}$

$$\lim_{n\to\infty} \frac{|u_n|}{v_n} = \lim_{n\to\infty} \frac{1}{2}\left(1+\frac{1}{n}\right) = \frac{1}{2}, \text{ which is non-zero and finite.}$$

$\therefore$ By comparison test, $\Sigma | u_n |$ and Σv_n converge or diverge together.

Since $\sum v_n = \sum \frac{1}{n}$ diverges ($\because$ $p = 1$), $\Sigma |u_n|$ also diverges.

Hence the given series Σu_n is conditionally convergent.

Example 2. *Test* $\sum_{n=2}^{\infty} \frac{(-1)^n}{n(\log n)^2}$ *for convergence and absolute convergence.*

Sol. The given series is $\sum_{n=2}^{\infty} u_n = \sum_{n=2}^{\infty} \frac{(-1)^n}{n(\log n)^2} = \sum_{n=2}^{\infty} (-1)^n a_n$

It is an alternating series.

Here $$a_n = \frac{1}{n(\log n)^2}, \quad a_{n+1} = \frac{1}{(n+1)[\log(n+1)]^2}$$

Clearly, $a_n > a_{n+1} \quad \forall\, n \geq 2$

Also $$\lim_{n\to\infty} a_n = \lim_{n\to\infty} \frac{1}{n(\log n)^2} = 0$$

$\therefore$ By Leibnitz's test, the series $\sum_{n=2}^{\infty} u_n$ is convergent.

Now $$|u_n| = \frac{1}{n(\log n)^2} = f(n) \quad \therefore \quad f(x) = \frac{1}{x(\log x)^2}$$

For $x \geq 2$, $f(x)$ is positive and monotonic decreasing.

$\therefore$ By Cauchy's Integral Test, $\sum_{n=2}^{\infty} |u_n|$ and $\int_2^{\infty} f(x)\,dx$ converge or diverge together.

Now $$\int_2^{\infty} f(x)\,dx = \int_2^{\infty} \frac{dx}{x(\log x)^2} = \int_2^{\infty} (\log x)^{-2} \cdot \frac{1}{x}\,dx$$

$$= \left[\frac{(\log x)^{-1}}{-1}\right]_2^{\infty} = \left[-\frac{1}{\log x}\right]_2^{\infty} = \frac{1}{\log 2}$$

which is a finite quantity.

$\Rightarrow$ The integral $\int_2^{\infty} f(x)\,dx$ is convergent.

$\Rightarrow$ The series $\sum_{n=2}^{\infty} |u_n|$ is convergent.

Hence the given series $\sum_{n=2}^{\infty} u_n$ is absolutely convergent.

Example 3. *Test the convergence and absolute convergence of the series:*

(*i*) $\sum_{n=1}^{\infty} \frac{(-1)^{n-1} \cdot n}{5^n}$ (*ii*) $\sum_{n=1}^{\infty} (-1)^{n-1} \left[\frac{1}{n^2} + \frac{1}{(n+1)^2}\right]$

(*iii*) $\frac{1}{2(\log 2)^p} - \frac{1}{3(\log 3)^p} + \frac{1}{4(\log 4)^p} - \dots \infty \ (p > 0)$. (M.D.U. Dec. 2012)

Sol. (*i*) The given series is $\sum_{n=1}^{\infty} u_n = \sum_{n=1}^{\infty} (-1)^{n-1} \cdot \frac{n}{5^n} = \sum_{n=1}^{\infty} (-1)^{n-1} a_n$

It is an alternating series.

Here $a_n = \frac{n}{5^n}, \quad a_{n+1} = \frac{n+1}{5^{n+1}}$

$$a_n - a_{n+1} = \frac{n}{5^n} - \frac{n+1}{5^{n+1}} = \frac{5n-(n+1)}{5^{n+1}} = \frac{4n-1}{5^{n+1}} > 0 \quad \forall\, n$$

$\Rightarrow$ $a_n > a_{n+1} \ \forall\, n$

Also $\lim_{n\to\infty} a_n = \lim_{n\to\infty} \frac{n}{5^n}$ $\left[\text{Form } \frac{\infty}{\infty}\right]$

$= \lim_{n\to\infty} \frac{1}{5^n \log 5}$ | Hospital's Rule

$= 0$

$\therefore$ By Leibnitz's test, the series $\sum_{n=1}^{\infty} u_n$ is convergent.

Now $|u_n| = \frac{n}{5^n}, \quad |u_{n+1}| = \frac{n+1}{5^{n+1}}$

$$\lim_{n\to\infty} \frac{|u_n|}{|u_{n+1}|} = \lim_{n\to\infty} \frac{5n}{n+1} = \lim_{n\to\infty} \frac{5}{1+\frac{1}{n}} = 5 > 1$$

$\therefore$ By Ratio test, the series $\Sigma |u_n|$ is convergent.

Hence the given series Σu_n is absolutely convergent.

(*ii*) The given series is $\sum_{n=1}^{\infty} u_n = \sum_{n=1}^{\infty} (-1)^{n-1} \left[\frac{1}{n^2} + \frac{1}{(n+1)^2}\right] = \sum_{n=1}^{\infty} (-1)^{n-1} a_n$

It is an alternating series

Here $a_n = \frac{1}{n^2} + \frac{1}{(n+1)^2} > 0 \quad \forall\, n$

$$a_{n+1} = \frac{1}{(n+1)^2} + \frac{1}{(n+2)^2}$$

$$a_n - a_{n+1} = \frac{1}{n^2} - \frac{1}{(n+2)^2} = \frac{4n+4}{n^2(n+2)^2} > 0 \quad \forall\, n$$

$\Rightarrow$ $a_n > a_{n+1} \quad \forall\, n$

Also $\lim_{n\to\infty} a_n = \lim_{n\to\infty} \left[\frac{1}{n^2} + \frac{1}{(n+1)^2}\right] = 0$

$\therefore$ By Leibnitz's test, the series $\sum_{n=1}^{\infty} u_n$ is convergent.

Now $$|u_n| = \frac{1}{n^2} + \frac{1}{(n+1)^2} < \frac{1}{n^2} + \frac{1}{n^2} = \frac{2}{n^2} \ \forall\ n$$

Since the series $\sum \frac{2}{n^2} = 2\sum \frac{1}{n^2}$ is convergent ($\because p = 2 > 1$), by comparison test, $\Sigma |u_n|$ is also convergent. Hence the given series is absolutely convergent.

(*iii*) The given series is $\sum_{n=2}^{\infty} u_n = \sum_{n=2}^{\infty} (-1)^n \cdot \frac{1}{n(\log n)^p} = \sum_{n=2}^{\infty} (-1)^n a_n$

It is an alternating series.

Here $|u_n| = \frac{1}{n(\log n)^p}$ for all $n \geq 2$ and $\sum_{n=2}^{\infty} |u_n|$ is convergent if $p > 1$ and divergent if $0 < p \leq 1$.

[See Example 3 with Cauchy's Integral Test.]

Hence the given series is absolutely convergent if $p > 1$. Now we discuss the convergence of $\sum_{n=2}^{\infty} u_n = \sum_{n=2}^{\infty} (-1)^n a_n$ for $0 < p \leq 1$.

Here $$a_n = \frac{1}{n(\log n)^p}$$

$$a_{n+1} = \frac{1}{(n+1)[\log(n+1)]^p}$$

Since $n < n + 1$ for all $n \geq 2$

$\therefore$ $$n(\log n)^p < (n+1)[\log(n+1)]^p \quad \text{for all } n \geq 2$$

Taking reciprocals, we get

$$\frac{1}{n(\log n)^p} > \frac{1}{(n+1)[\log(n+1)]^p} \quad \text{for all } n \geq 2$$

$\Rightarrow$ $$a_n > a_{n+1} \quad \text{for all } n \geq 2$$

Also $$\lim_{n\to\infty} a_n = \lim_{n\to\infty} \frac{1}{n(\log n)^p} = 0$$

$\therefore$ By Leibnitz's test, $\sum_{n=2}^{\infty} (-1)^n a_n = \sum_{n=2}^{\infty} u_n$ is convergent.

Hence the given series is absolutely convergent if $p > 1$ and conditionally convergent if $0 < p \leq 1$.

Example 4. *Prove that the series* $\frac{\sin x}{1^3} - \frac{\sin 2x}{2^3} + \frac{\sin 3x}{3^3} - \cdots$ *converges absolutely.*

(Kerala 2009)

Sol. The given series is $\sum_{n=1}^{\infty} u_n = \sum_{n=1}^{\infty} \left[\frac{(-1)^{n-1}\sin nx}{n^3}\right]$

Since $\quad | u_n | = \dfrac{|\sin nx|}{n^3} \le \dfrac{1}{n^3} \; \forall \, n$ and $\sum \dfrac{1}{n^3}$ converges.

$\therefore$ By comparison test, the series $\sum | u_n |$ converges.

$\Rightarrow$ the given series converges absolutely.

Example 5. *Discuss the convergence of the series* $1 + \dfrac{x}{1!} + \dfrac{x^2}{2!} + \cdots$ *for all values of* x.

Sol. The given series is $\sum\limits_{n=1}^{\infty} u_n = \sum\limits_{n=1}^{\infty} \dfrac{x^{n-1}}{(n-1)!}$

Here $\quad u_n = \dfrac{x^{n-1}}{(n-1)!}$ and $u_{n+1} = \dfrac{x^n}{n!}$

$$\therefore \quad \frac{|u_n|}{|u_{n+1}|} = \frac{|x|^{n-1}}{(n-1)!} \times \frac{n!}{|x|^n} = \frac{n}{|x|} \text{ for } x \neq 0$$

$$\lim_{n \to \infty} \frac{|u_n|}{|u_{n+1}|} = \lim_{n \to \infty} \frac{n}{|x|} = \infty \text{ for } x \neq 0$$

$\therefore$ By ratio test, the series $\sum\limits_{n=1}^{\infty} |u_n|$ is convergent for $x \neq 0$.

When $x = 0$, the series becomes $1 + 0 + 0 + \cdots$ and is convergent.

Thus $\Sigma |u_n|$ is convergent for all x.

$\Rightarrow$ The given series is absolutely convergent for all x.

Since every absolutely convergent series is convergent, therefore, the given series is convergent for all x.

Note. In the above example, Σu_n is convergent for all x

$$\Rightarrow \quad \lim_{n \to \infty} u_{n+1} = 0 \;\Rightarrow\; \lim_{n \to \infty} \frac{x^n}{n!} = 0 \text{ for all } x.$$

Example 6. *Discuss the convergence and absolute convergence of the series* $x - \dfrac{x^3}{3} + \dfrac{x^5}{5} - \cdots$, *x being real.*

Sol. The given series is $\sum\limits_{n=1}^{\infty} u_n = \sum\limits_{n=1}^{\infty} (-1)^{n-1} \cdot \dfrac{x^{2n-1}}{2n-1} = \sum\limits_{n=1}^{\infty} (-1)^{n-1} v_n$

Here $\quad u_n = (-1)^{n-1} \cdot \dfrac{x^{2n-1}}{2n-1}$ and $u_{n+1} = (-1)^n \cdot \dfrac{x^{2n+1}}{2n+1}$

$$\therefore \quad \frac{|u_n|}{|u_{n+1}|} = \frac{|x|^{2n-1}}{2n-1} \times \frac{2n+1}{|x|^{2n+1}} = \frac{2n+1}{2n-1} \cdot \frac{1}{|x|^2} = \frac{1+\frac{1}{2n}}{1-\frac{1}{2n}} \cdot \frac{1}{x^2}$$

$(\because \; |x|^2 = x^2)$

$$\lim_{n \to \infty} \frac{|u_n|}{|u_{n+1}|} = \lim_{n \to \infty} \frac{1+\frac{1}{2n}}{1-\frac{1}{2n}} \cdot \frac{1}{x^2} = \frac{1}{x^2}$$

$\therefore$ By ratio test, the series $\sum_{n=1}^{\infty} |u_n|$ is convergent if $\frac{1}{x^2} > 1$ *i.e.*, if $x^2 < 1$ *i.e.*, if $-1 < x < 1$,

and divergent if $\frac{1}{x^2} < 1$ *i.e.*, if $x^2 > 1$ *i.e.*, if $x > 1$ or $x < -1$.

Ratio test fail when $x^2 = 1$.

When $x^2 = 1$, we have $\frac{|u_n|}{|u_{n+1}|} = \frac{1+\frac{1}{2n}}{1-\frac{1}{2n}} = \left(1+\frac{1}{2n}\right)\left(1-\frac{1}{2n}\right)^{-1}$

$$= \left(1+\frac{1}{2n}\right)\left[1+\frac{1}{2n}+O\left(\frac{1}{n^2}\right)\right] = 1+\frac{1}{n}+O\left(\frac{1}{n^2}\right)$$

$\therefore$ By Gauss test, the series $\Sigma\,|u_n|$ is divergent.

Hence the given series is absolutely convergent when $-1 < x < 1$.

Since absolute convergence implies convergence, the series $\sum_{n=1}^{\infty} u_n$ is convergent when $-1 < x < 1$.

When x = 1, the series becomes $\Sigma u_n = \Sigma(-1)^{n-1} \cdot \frac{1}{2n-1} = \Sigma(-1)^{n-1} v_n$

It is an alternating series.

Here $\quad v_n = \frac{1}{2n-1}, \quad v_{n+1} = \frac{1}{2n+1}$

Since $\quad 2n-1 < 2n+1 \quad \forall\, n$

$\therefore \quad \frac{1}{2n-1} > \frac{1}{2n+1} \quad \forall\, n$ or $v_n > v_{n+1} \quad \forall\, n$

Also $\quad \lim_{n\to\infty} v_n = \lim_{n\to\infty} \frac{1}{2n-1} = 0$

$\therefore$ By Leibnitz's test, the series is convergent.

When x = – 1, the series becomes

$$\Sigma u_n = \Sigma(-1)^{n-1} \cdot \frac{(-1)^{2n-1}}{2n-1} = \Sigma(-1)^{n-1} \cdot \frac{-1}{2n-1} = -\Sigma(-1)^{n-1} \cdot \frac{1}{2n-1}$$

Since $\Sigma(-1)^{n-1} \cdot \frac{1}{2n-1}$ is convergent, Σu_n is convergent.

When $x^2 > 1$, *i.e.*, $|x| > 1$, $\lim_{n\to\infty} v_n \neq 0$. Therefore $\sum_{n=1}^{\infty} u_n$ is not convergent.

Hence Σu_n is convergent if $-1 \le x \le 1$ and absolutely convergent if $-1 < x < 1$.

Example 7. *Test the convergence of the series* $\sum_{n=1}^{\infty} \frac{(-1)^n (x+1)^n}{2^n \cdot n^2}$. (M.D.U. Dec. 2010)

Sol. Here $u_n = \dfrac{(-1)^n (x+1)^n}{2^n \cdot n^2}$ and $u_{n+1} = \dfrac{(-1)^{n+1} \cdot (x+1)^{n+1}}{2^{n+1} \cdot (n+1)^2}$

$$\frac{|u_n|}{|u_{n+1}|} = \frac{|x+1|^n}{2^n \cdot n^2} \times \frac{2^{n+1} \cdot (n+1)^2}{|x+1|^{n+1}}$$

$$= 2\left(\frac{n+1}{n}\right)^2 \cdot \frac{1}{|x+1|} = 2\left(1+\frac{1}{n}\right)^2 \cdot \frac{1}{|x+1|}$$

$\therefore$ $$\lim_{n\to\infty} \frac{|u_n|}{|u_{n+1}|} = \lim_{n\to\infty} 2\left(1+\frac{1}{n}\right)^2 \cdot \frac{1}{|x+1|} = \frac{2}{|x+1|}$$

$\therefore$ By ratio test, the series $\Sigma\,|u_n|$ is convergent

if $\dfrac{2}{|x+1|} > 1$ *i.e.,* if $|x+1| < 2$

i.e., if $-2 < x+1 < 2$ *i.e.,* if $-3 < x < 1$

Also $\Sigma\,|u_n|$ is divergent if $\dfrac{2}{|x+1|} < 1$ *i.e.,* if $|x+1| > 2$ *i.e.,* if $x+1 > 2$

or $x+1 < -2$ *i.e.,* if $x > 1$ or $x < -3$.

Ratio test fails when $x = 1$ or -3.

When x = 1 $\Sigma u_n = \sum \dfrac{(-1)^n \cdot 2^n}{2^n \cdot n^2} = \sum \dfrac{(-1)^n}{n^2} = \Sigma(-1)^n \cdot v_n$

It is an alternating series.

Here $v_n = \dfrac{1}{n^2}$, $v_{n+1} = \dfrac{1}{(n+1)^2}$

Clearly, $v_n > v_{n+1}\ \forall\ n$

Also $$\lim_{n\to\infty} v_n = \lim_{n\to\infty} \frac{1}{n^2} = 0$$

$\therefore$ By Leibnitz's test, Σu_n is convergent.

When x = – 3 $\Sigma u_n = \sum \dfrac{(-1)^n \cdot (-2)^n}{2^n \cdot n^2} = \sum \dfrac{(-1)^{2n} \cdot 2^n}{2^n \cdot n^2} = \sum \dfrac{1}{n^2}$

which is convergent.

Thus $\Sigma\,|u_n|$ is convergent when $-3 \le x \le 1$. Since absolute convergence implies convergence, $\Sigma\,u_n$ is convergent when $-3 \le x \le 1$.

If $|x+1| > 2$, $\lim_{n\to\infty} v_n \ne 0$ so that $\Sigma\,u_n$ is not convergent.

Hence the given series is convergent if $-3 \le x \le 1$.

EXERCISE 1.8

Test the convergence and absolute convergence of the series (1 – 9):

1. $\dfrac{1}{5} - \dfrac{1}{7} + \dfrac{1}{9} - \dfrac{1}{11} + \cdots$

2. $1 - \dfrac{1}{2^p} + \dfrac{1}{3^p} - \dfrac{1}{4^p} + \cdots$ $(p > 0)$

3. $1 - \dfrac{1}{3!} + \dfrac{1}{5!} - \dfrac{1}{7!} + \cdots$

4. $\displaystyle\sum_{n=1}^{\infty} (-1)^{n-1} \cdot \sin \frac{1}{n}$

5. $\sum_{n=1}^{\infty} \frac{(-1)^{n-1} \cdot n}{5n-7}$

6. $\frac{1}{\sqrt{2}+1} - \frac{1}{\sqrt{3}+1} + \frac{1}{\sqrt{4}+1} - \frac{1}{\sqrt{5}+1} + \dots$

7. $1 - \frac{1}{4 \cdot 3} + \frac{1}{4^2 \cdot 5} - \frac{1}{4^3 \cdot 7} + \dots$

8. $\sum_{n=1}^{\infty} \frac{(-1)^{n-1} \cdot n}{n^2+1}$

9. $\sum_{n=1}^{\infty} (-1)^{n-1} \cdot \frac{n+2}{2^n+5}$

10. Test whether the series $1 + \frac{1}{2^2} - \frac{1}{3^2} - \frac{1}{4^2} + \frac{1}{5^2} + \frac{1}{6^2} - \frac{1}{7^2} - \frac{1}{8^2} + \dots$ is convergent or not? *(M.D.U. Dec. 2011)*

11. Prove that the following series converge absolutely:

(*i*) $\sum_{n=1}^{\infty} \frac{(-1)^{n-1} \sin n\alpha}{n^2}$, α real *(M.D.U. Dec. 2010)*

(*ii*) $\sum_{n=1}^{\infty} \frac{(-1)^{n-1} \cdot \cos^2 n\alpha}{n\sqrt{n}}$, α real

(*iii*) $\sum_{n=1}^{\infty} (-1)^{n-1} \cdot \frac{2^n}{n!}$

(*iv*) $\sum_{n=1}^{\infty} (-1)^{n-1} \cdot \frac{n^3}{(n+1)!}$.

12. For what values of x are the following series convergent:

(*i*) $x - \frac{x^2}{2} + \frac{x^3}{3} - \frac{x^4}{4} + \frac{x^5}{5} - \dots$ *(M.D.U. May 2009)*

(*ii*) $x - \frac{x^2}{\sqrt{2}} + \frac{x^3}{\sqrt{3}} - \frac{x^4}{\sqrt{4}} + \dots$ *(M.D.U. Dec. 2008)*

(*iii*) $x - \frac{x^2}{2^2} + \frac{x^3}{3^2} - \frac{x^4}{4^2} + \dots$

(*iv*) $1 - \frac{x^2}{2} + \frac{x^4}{4} - \frac{x^6}{6} + \dots$

(*v*) $\frac{1}{1-x} + \frac{1}{2(1-x)^2} + \frac{1}{3(1-x)^3} + \dots$ *(M.D.U. Dec. 2009)*

13. For what values of x does the series $\sum_{n=1}^{\infty} (-1)^{n-1}(4x+1)^n$ converge absolutely?

14. Discuss the convergence of the series $\sum_{n=1}^{\infty} (-1)^n \frac{\tan^{-1} n}{1+n^2}$.

15. Test for absolute/conditional convergence of the series

$\frac{1}{6}x^2 - \frac{2}{11}x^3 + \frac{3}{16}x^4 - \frac{4}{21}x^5 + \frac{5}{26}x^6 - \dots\dots$ *(M.D.U. Dec. 2013)*

Answers

1. Conditionally convergent
2. Absolutely convergent if $p > 1$, conditionally convergent if $0 < p \leq 1$
3. Absolutely convergent
4. Conditionally convergent
5. Not convergent
6. Conditionally convergent
7. Absolutely convergent
8. Conditionally convergent
9. Absolutely convergent
10. Absolutely convergent and hence convergent
12. (*i*) $-1 < x \leq 1$ (*ii*) $-1 < x \leq 1$ (*iii*) $-1 \leq x \leq 1$ (*iv*) $-1 \leq x \leq 1$ (*v*) $x < 0$ and $x \geq 2$
13. $-\frac{1}{2} < x < 0$
14. Convergent
15. Absolutely convergent when $-1 < x < 1$.

SECTION B

Matrices and Its Applications

2

Matrices and Its Applications*

2.1. ELEMENTARY TRANSFORMATIONS (OR OPERATIONS)

Any one of the following operations on a matrix is called an elementary transformation (or E-operation).

(i) Interchange of two rows or two columns.

The interchange of ith and j^{th} rows is denoted by R_{ij} or $R_i \leftrightarrow R_j$.

The interchange of ith and j^{th} columns is denoted by C_{ij} or $C_i \leftrightarrow C_j$.

(ii) Multiplication of (each element of) a row or column by a non-zero number k.

The multiplication of ith row by k is denoted by kR_i.

The multiplication of ith column by k is denoted by kC_i.

(iii) Addition of k times the elements of a row (or column) to the corresponding elements of another row (or column), k ≠ 0.

The addition of k times the jth row to the ith row is denoted by $R_i + kR_j$.

The addition of k times the jth column to the ith column is denoted by $C_i + kC_j$.

If a matrix B is obtained from a matrix A by one or more E-operations, then B is said to be equivalent to A.

Two equivalent matrices A and B are written as A ~ B.

2.2. ELEMENTARY MATRICES

The matrix obtained from a unit matrix I by subjecting it to one of the E-operations is called an elementary matrix.

For example, let $I = \begin{bmatrix} 1 & 0 & 0 \\ 0 & 1 & 0 \\ 0 & 0 & 1 \end{bmatrix}$

(i) Operating R_{23} or C_{23} on I, we get the same elementary matrix $\begin{bmatrix} 1 & 0 & 0 \\ 0 & 0 & 1 \\ 0 & 1 & 0 \end{bmatrix}$

It is denoted by E_{23}. [illegible]

(ii) Operating $3R_2$ or $3C_2$ on I, we get the same elementary matrix $\begin{bmatrix} 1 & 0 & 0 \\ 0 & 3 & 0 \\ 0 & 0 & 1 \end{bmatrix}$.

*This chapter is not included in the syllabus of KU, Kurukshetra and DCRUST, Murthal.

2*

Matrices and Its Applications

2.1. ELEMENTARY TRANSFORMATIONS (OR OPERATIONS)

Any one of the following operations on a matrix is called an elementary transformation (or E-operation).

(*i*) **Interchange of two rows or two columns.**

The interchange of ith and j^{th} rows is denoted by R_{ij} or $R_i \leftrightarrow R_j$

The interchange of ith and j^{th} columns is denoted by C_{ij} or $C_i \leftrightarrow C_j$.

(*ii*) **Multiplication of (each element of) a row or column by a non-zero number k.**

The multiplication of ith row by k is denoted by kR_i.

The multiplication of ith column by k is denoted by kC_i.

(*iii*) **Addition of k times the elements of a row (or column) to the corresponding elements of another row (or column), $k \neq 0$.**

The addition of k times the jth row to the ith row is denoted by $R_i + kR_j$.

The addition of k times the jth column to the ith column is denoted by $C_i + kC_j$.

If a matrix B is obtained from a matrix A by one or more E-operations, then B is said to be **equivalent** to A.

Two equivalent matrices A and B are written as A ~ B.

2.2. ELEMENTARY MATRICES *(M.D.U. Dec. 2012)*

The matrix obtained from a unit matrix I by subjecting it to one of the E-operations is called an elementary matrix.

For example, let $I = \begin{bmatrix} 1 & 0 & 0 \\ 0 & 1 & 0 \\ 0 & 0 & 1 \end{bmatrix}$

(*i*) Operating R_{23} or C_{23} on I, we get the same elementary matrix $\begin{bmatrix} 1 & 0 & 0 \\ 0 & 0 & 1 \\ 0 & 1 & 0 \end{bmatrix}$.

It is denoted by E_{23}. Thus, the E-matrix obtained by either of the operations R_{ij} or C_{ij} on I is denoted by E_{ij}.

(*ii*) Operating $5R_2$ or $5C_2$ on I, we get the same elementary matrix $\begin{bmatrix} 1 & 0 & 0 \\ 0 & 5 & 0 \\ 0 & 0 & 1 \end{bmatrix}$.

*This chapter is not included in the syllabus of KU, Kurukshetra and DCRUST, Murthal.

It is denoted by $5E_2$. Thus, the E-matrix obtained by either of the operations kR_i or kC_i is denoted by kE_i.

(*iii*) Operating $R_2 + 4R_3$ on I, we get the elementary matrix $\begin{bmatrix} 1 & 0 & 0 \\ 0 & 1 & 4 \\ 0 & 0 & 1 \end{bmatrix}$.

It is denoted by $E_{23}(4)$. Thus, the E-matrix obtained by the operation $R_i + kR_j$ is denoted by $E_{ij}(k)$.

(*iv*) Operating $C_2 + 4C_3$ on I, we get the elementary matrix $\begin{bmatrix} 1 & 0 & 0 \\ 0 & 1 & 0 \\ 0 & 4 & 1 \end{bmatrix}$, which is the transpose of $\begin{bmatrix} 1 & 0 & 0 \\ 0 & 1 & 4 \\ 0 & 0 & 1 \end{bmatrix} = E_{23}(4)$ and is, therefore, denoted by $E'_{23}(4)$. Thus, the E-matrix obtained by the operation $C_i + kC_j$ is denoted by $E'_{ij}(k)$.

2.3. THE FOLLOWING THEOREMS ON THE EFFECT OF E-OPERATIONS ON MATRICES HOLD GOOD

(*a*) *Any E-row operation on the product of two matrices is equivalent to the same E-row operation on the pre-factor.*

If the E-row operation is denoted by R, then R(AB) = R(A).B.

(*b*) *Any E-column operation on the product of two matrices is equivalent to the same E-column operation on the post-factor.*

If the E-column operation is denoted by C, then C(AB) = A . C(B).

(*c*) *Every E-row operation on a matrix is equivalent to pre-multiplication by the corresponding E-matrix.*

Thus, the effect of E-row operation R_{ij} on $A = E_{ij} . A$

The effect of E-row operation kR_i on $A = kE_i . A$

The effect of E-row operation $R_i + kR_j$ on $A = E_{ij}(k) . A$.

(*d*) *Every E-column operation on a matrix is equivalent to post-multiplication by the corresponding E-matrix.*

Thus, the effect of E-column operation C_{ij} on $A = A . E_{ij}$

The effect of *E*-column operation kC_i on $A = A . (kE_i)$

The effect of *E*-column operation $C_i + kC_j$ on $A = A . E'_{ij}(k)$.

2.4. INVERSE OF A MATRIX BY E-OPERATIONS (GAUSS-JORDAN METHOD)

The elementary row transformations which reduce a square matrix A to the unit matrix, when applied to the unit matrix, give the inverse matrix A^{-1}.

Let A be a non-singular square matrix. Then A = IA

Apply suitable E-row operations to A on the left hand side so that A is reduced to I.

Simultaneously, apply the same E-row operations to the pre-factor I on right hand side. Let I reduce to B, so that

$$I = BA$$

Post-multiplying by A^{-1}, we get

$$IA^{-1} = BAA^{-1} \quad \Rightarrow \quad A^{-1} = B(AA^{-1}) = BI = B$$

$$\therefore \qquad B = A^{-1}.$$

Note. In practice, to find the inverse of A by E-row operations, we write A and I side by side in the form [A : I] and the same E-row operations are performed on both. As soon as A is reduced to I, I will reduce to A^{-1}.

ILLUSTRATIVE EXAMPLES

Example 1. *Find the inverse of* $A = \begin{bmatrix} 0 & 1 & 2 \\ 1 & 2 & 3 \\ 3 & 1 & 1 \end{bmatrix}$ *by elementary row operations.*

Sol. Writing the given matrix side by side with unit matrix I_3, we get

$$[A : I_3] = \left[\begin{array}{ccc:ccc} 0 & 1 & 2 & 1 & 0 & 0 \\ 1 & 2 & 3 & 0 & 1 & 0 \\ 3 & 1 & 1 & 0 & 0 & 1 \end{array}\right] \qquad \text{operating } R_1 \leftrightarrow R_2$$

$$\sim \left[\begin{array}{ccc:ccc} 1 & 2 & 3 & 0 & 1 & 0 \\ 0 & 1 & 2 & 1 & 0 & 0 \\ 3 & 1 & 1 & 0 & 0 & 1 \end{array}\right] \qquad \text{operating } R_3 \to R_3 - 3R_1$$

$$\sim \left[\begin{array}{ccc:ccc} 1 & 2 & 3 & 0 & 1 & 0 \\ 0 & 1 & 2 & 1 & 0 & 0 \\ 0 & -5 & -8 & 0 & -3 & 1 \end{array}\right] \qquad \text{operating } R_1 \to R_1 - 2R_2,\ R_3 \to R_3 + 5R_2$$

$$\sim \left[\begin{array}{ccc:ccc} 1 & 0 & -1 & -2 & 1 & 0 \\ 0 & 1 & 2 & 1 & 0 & 0 \\ 0 & 0 & 2 & 5 & -3 & 1 \end{array}\right] \qquad \text{operating } R_3 \to \tfrac{1}{2}R_3$$

$$\sim \left[\begin{array}{ccc:ccc} 1 & 0 & -1 & -2 & 1 & 0 \\ 0 & 1 & 2 & 1 & 0 & 0 \\ 0 & 0 & 1 & \frac{5}{2} & -\frac{3}{2} & \frac{1}{2} \end{array}\right] \qquad \text{operating } R_1 \to R_1 + R_3,\ R_2 \to R_2 - 2R_3$$

$$\sim \left[\begin{array}{ccc:ccc} 1 & 0 & 0 & \frac{1}{2} & -\frac{1}{2} & \frac{1}{2} \\ 0 & 1 & 0 & -4 & 3 & -1 \\ 0 & 0 & 1 & \frac{5}{2} & -\frac{3}{2} & \frac{1}{2} \end{array}\right] = [I_3 : A^{-1}]$$

$$\therefore \qquad A^{-1} = \begin{bmatrix} \frac{1}{2} & -\frac{1}{2} & \frac{1}{2} \\ -4 & 3 & -1 \\ \frac{5}{2} & -\frac{3}{2} & \frac{1}{2} \end{bmatrix}$$

Note. Check the answer.

$$A^{-1}A = \begin{bmatrix} \frac{1}{2} & -\frac{1}{2} & \frac{1}{2} \\ -4 & 3 & -1 \\ \frac{5}{2} & -\frac{3}{2} & \frac{1}{2} \end{bmatrix} \begin{bmatrix} 0 & 1 & 2 \\ 1 & 2 & 3 \\ 3 & 1 & 1 \end{bmatrix}$$

$$= \begin{bmatrix} 0 - \frac{1}{2} + \frac{3}{2} & \frac{1}{2} - 1 + \frac{1}{2} & 1 - \frac{3}{2} + \frac{1}{2} \\ 0 + 3 - 3 & -4 + 6 - 1 & -8 + 9 - 1 \\ 0 - \frac{3}{2} + \frac{3}{2} & \frac{5}{2} - 3 + \frac{1}{2} & 5 - \frac{9}{2} + \frac{3}{2} \end{bmatrix}$$

$$= \begin{bmatrix} 1 & 0 & 0 \\ 0 & 1 & 0 \\ 0 & 0 & 1 \end{bmatrix} = I_3$$

Example 2. *Using Gauss-Jordan method, find the inverse of the matrix*

$$A = \begin{bmatrix} 2 & 4 & 3 & 2 \\ 3 & 6 & 5 & 2 \\ 2 & 5 & 2 & -3 \\ 4 & 5 & 14 & 14 \end{bmatrix}.$$

Sol. Writing the given matrix A and the unit matrix I_4 side by side, we get

$$[A : I_4] = \left[\begin{array}{cccc:cccc} 2 & 4 & 3 & 2 & 1 & 0 & 0 & 0 \\ 3 & 6 & 5 & 2 & 0 & 1 & 0 & 0 \\ 2 & 5 & 2 & -3 & 0 & 0 & 1 & 0 \\ 4 & 5 & 14 & 14 & 0 & 0 & 0 & 1 \end{array}\right]$$

Operating $R_2 \to R_2 - R_1$, $R_3 \to R_3 - R_1$, $R_4 \to R_4 - 2R_1$

$$\sim \left[\begin{array}{cccc:cccc} 2 & 4 & 3 & 2 & 1 & 0 & 0 & 0 \\ 1 & 2 & 2 & 0 & -1 & 1 & 0 & 0 \\ 0 & 1 & -1 & -5 & -1 & 0 & 1 & 0 \\ 0 & -3 & 8 & 10 & -2 & 0 & 0 & 1 \end{array}\right]$$

Operating $R_1 \longleftrightarrow R_2$

$$\sim \left[\begin{array}{cccc:cccc} 1 & 2 & 2 & 0 & -1 & 1 & 0 & 0 \\ 2 & 4 & 3 & 2 & 1 & 0 & 0 & 0 \\ 0 & 1 & -1 & -5 & -1 & 0 & 1 & 0 \\ 0 & -3 & 8 & 10 & -2 & 0 & 0 & 1 \end{array}\right]$$

Operating $R_2 \to R_2 - 2R_1$

$$\sim \left[\begin{array}{cccc:cccc} 1 & 2 & 2 & 0 & -1 & 1 & 0 & 0 \\ 0 & 0 & -1 & 2 & 3 & -2 & 0 & 0 \\ 0 & 1 & -1 & -5 & -1 & 0 & 1 & 0 \\ 0 & -3 & 8 & 10 & -2 & 0 & 0 & 1 \end{array}\right]$$

Operating $R_1 \to R_1 - 2R_3$, $R_4 \to R_4 + 3R_3$

$$\sim \left[\begin{array}{cccc:cccc} 1 & 0 & 4 & 10 & 1 & 1 & -2 & 0 \\ 0 & 0 & -1 & 2 & 3 & -2 & 0 & 0 \\ 0 & 1 & -1 & -5 & -1 & 0 & 1 & 0 \\ 0 & 0 & 5 & -5 & -5 & 0 & 3 & 1 \end{array}\right]$$

Operating $R_2 \longleftrightarrow R_3$

$$\sim \left[\begin{array}{cccc:cccc} 1 & 0 & 4 & 10 & 1 & 1 & -2 & 0 \\ 0 & 1 & -1 & -5 & -1 & 0 & 1 & 0 \\ 0 & 0 & -1 & 2 & 3 & -2 & 0 & 0 \\ 0 & 0 & 5 & -5 & -5 & 0 & 3 & 1 \end{array}\right]$$

Operating $R_1 \to R_1 + 4R_3$, $R_2 \to R_2 - R_3$, $R_4 \to R_4 + 5R_3$

$$\sim \left[\begin{array}{cccc:cccc} 1 & 0 & 0 & 18 & 13 & -7 & -2 & 0 \\ 0 & 1 & 0 & -7 & -4 & 2 & 1 & 0 \\ 0 & 0 & -1 & 2 & 3 & -2 & 0 & 0 \\ 0 & 0 & 0 & 5 & 10 & -10 & 3 & 1 \end{array}\right]$$

Operating $R_3 \to (-1)\, R_3$, $R_4 \to 1/5 R_4$

$$\sim \left[\begin{array}{cccc:cccc} 1 & 0 & 0 & 18 & 13 & -7 & -2 & 0 \\ 0 & 1 & 0 & -7 & -4 & 2 & 1 & 0 \\ 0 & 0 & 1 & -2 & -3 & 2 & 0 & 0 \\ 0 & 0 & 0 & 1 & 2 & -2 & 3/5 & 1/5 \end{array}\right]$$

Operating $R_1 \to R_1 - 18R_4$, $R_2 \to R_2 + 7R_4$, $R_3 \to R_3 + 2R_4$

$$\sim \left[\begin{array}{cccc:cccc} 1 & 0 & 0 & 0 & -23 & 29 & -64/5 & -18/5 \\ 0 & 1 & 0 & 0 & 10 & -12 & 26/5 & 7/5 \\ 0 & 0 & 1 & 0 & 1 & -2 & 6/5 & 2/5 \\ 0 & 0 & 0 & 1 & 2 & -2 & 3/5 & 1/5 \end{array}\right] = [I_4 \vdots A^{-1}]$$

$$\therefore \qquad A^{-1} = \begin{bmatrix} -23 & 29 & -64/5 & -18/5 \\ 10 & -12 & 26/5 & 7/5 \\ 1 & -2 & 6/5 & 2/5 \\ 2 & -2 & 3/5 & 1/5 \end{bmatrix}$$

Note. Check the answer.

2.5. RANK OF A MATRIX *(M.D.U. Dec. 2010 ; P.T.U. May 2008)*

Let A be any $m \times n$ matrix. It has square sub-matrices of different orders. The determinants of these square sub-matrices are called minors of A. If **all** minors of order $(r + 1)$ are zero but there **is at least one** non-zero minor of order r, then r is called the rank of A. Symbolically, rank of A is written as $\rho(A)$.

From the definition of the rank of a matrix A, it follows that :

(*i*) If A is a null matrix, then $\rho(A) = 0$.

[$\because$ every minor of A has zero value.]

(*ii*) If A is not a null matrix, then $\rho(A) \geq 1$.

(*iii*) If A is a non-singular $n \times n$ matrix, then $\rho(A) = n$

[$\because$ $|A| \neq 0$ is the largest minor of A.]

If I_n is the $n \times n$ unit matrix, then $|I_n| = 1 \neq 0 \Rightarrow \rho(I_n) = n$.

(*iv*) If A is an $m \times n$ matrix, then $\rho(A) \leq$ minimum of m and n.

(*v*) If all minors of order r are equal to zero, then $\rho(A) < r$.

To determine the rank of a matrix A, we adopt the following different methods:

(*i*) Start with the highest order minor (or minors) of A. Let their order be r. **If any one of them** is non-zero, then $\rho(A) = r$.

If **all of them** are zero, start with minors of next lower order $(r - 1)$ and so on till you get a non-zero minor. The order of that minor is the rank of A.

This method **usually** involves a lot of computational work since we have to evaluate several determinants.

(*ii*) Convert the given matrix A into lower triangular matrix by elementary column operations or upper triangular matrix by elementary row operations.

$$\begin{bmatrix} a & 0 & 0 \\ b & c & 0 \\ d & e & f \end{bmatrix}$$

Then $\rho(A)$ = number of non-zero columns in lower triangular matrix or number of non-zero rows in upper triangular matrix.

$$\begin{bmatrix} a & b & c \\ 0 & d & e \\ 0 & 0 & f \end{bmatrix}$$

A non-zero row or column has at least one non-zero element.

(*iii*) If A is an $m \times n$ matrix and by a series of elementary (*row or column or both*) operations, it can be put into one of the following forms (called normal forms) :

$\begin{bmatrix} I_r & \vdots & O \\ \ldots & \ldots & \ldots \\ O & \vdots & O \end{bmatrix}$, $\begin{bmatrix} I_r \\ \\ O \end{bmatrix}$, $[I_r \vdots O]$, $[I_r]$, where I_r is the unit matrix of order r.

Since the rank of a matrix is not changed as a result of elementary transformations, it follows that

$$\rho(A) = r. \qquad [\because \ r\text{th order minor } |I_r| = 1 \neq 0]$$

Note. *For an $m \times n$ matrix A of rank r, to find square matrices P and Q of orders m and n respectively, such that PAQ is in the normal form* $\begin{bmatrix} I_r & O \\ O & O \end{bmatrix}$.

Method. Write A = IAI.

Reduce the matrix on L.H.S. to normal form by affecting elementary row and/or column transformations.

Every elementary row transformation on A must be accompanied by the same transformation on the pre-factor on R.H.S.

Every elementary column transformation on A must be accompanied by the same transformation on the post-factor on R.H.S.

Example 3. *Find the rank of the matrix*

(*i*) $\begin{bmatrix} 1 & 2 & 3 & 4 \\ -2 & 0 & 5 & 7 \end{bmatrix}$ (*ii*) $\begin{bmatrix} 1 & 2 & 3 \\ 1 & 4 & 2 \\ 2 & 6 & 5 \end{bmatrix}$ (M.D.U. Dec. 2010)

(*iii*) $\begin{bmatrix} 2 & 3 & 4 & -1 \\ 5 & 2 & 0 & -1 \\ -4 & 5 & 12 & -1 \end{bmatrix}$ (*iv*) $\begin{bmatrix} 1 & 2 & 3 & 0 \\ 2 & 4 & 3 & 2 \\ 3 & 2 & 1 & 3 \\ 6 & 8 & 7 & 5 \end{bmatrix}$.

Sol. (*i*) Here $A = \begin{bmatrix} 1 & 2 & 3 & 4 \\ -2 & 0 & 5 & 7 \end{bmatrix}$ is a 2×4 matrix.

$\therefore \quad \rho(A) \le 2$, the smaller of 2 and 4.

The second order minor $\begin{vmatrix} 1 & 2 \\ -2 & 0 \end{vmatrix} = 4 \ne 0 \qquad \therefore \quad \rho(A) = 2.$

(*ii*) Here $A = \begin{bmatrix} 1 & 2 & 3 \\ 1 & 4 & 2 \\ 2 & 6 & 5 \end{bmatrix}$ is a 3×3 matrix. $\qquad \therefore \quad \rho(A) \le 3$

Operating $R_2 \to R_2 - R_1$, $R_3 \to R_3 - 2R_1$

$$A \sim \begin{bmatrix} 1 & 2 & 3 \\ 0 & 2 & -1 \\ 0 & 2 & -1 \end{bmatrix} \quad \text{operating } R_3 \to R_3 - R_2$$

$$\sim \begin{bmatrix} 1 & 2 & 3 \\ 0 & 2 & -1 \\ 0 & 0 & 0 \end{bmatrix}$$

The only third order minor is zero, but the second order minor

$$\begin{vmatrix} 1 & 2 \\ 0 & 2 \end{vmatrix} = 2 \ne 0 \quad \therefore \quad \rho(A) = 2.$$

(*iii*) Here $A = \begin{bmatrix} 2 & 3 & 4 & -1 \\ 5 & 2 & 0 & -1 \\ -4 & 5 & 12 & -1 \end{bmatrix}$ is a 3×4 matrix. $\quad \therefore \quad \rho(A) \le 3$

Operating C_{14}

$$A \sim \begin{bmatrix} -1 & 3 & 4 & 2 \\ -1 & 2 & 0 & 5 \\ -1 & 5 & 12 & -4 \end{bmatrix} \quad \text{operating } R_2 \to R_2 - R_1,\ R_3 \to R_3 - R_1$$

$$\sim \begin{bmatrix} -1 & 3 & 4 & 2 \\ 0 & -1 & -4 & 3 \\ 0 & 2 & 8 & -6 \end{bmatrix} \quad \text{operating } R_1 \to R_1 + 3R_2,\ R_3 \to R_3 + 2R_2$$

$$\sim \begin{bmatrix} -1 & 0 & -8 & 11 \\ 0 & -1 & -4 & 3 \\ 0 & 0 & 0 & 0 \end{bmatrix}$$

All the third order minors are zero but the second order minor

$$\begin{vmatrix} -1 & 0 \\ 0 & -1 \end{vmatrix} = 1 \ne 0 \quad \therefore \quad \rho(A) = 2.$$

(*iv*) Here $\qquad A = \begin{bmatrix} 1 & 2 & 3 & 0 \\ 2 & 4 & 3 & 2 \\ 3 & 2 & 1 & 3 \\ 6 & 8 & 7 & 5 \end{bmatrix}$ is a 4×4 matrix $\quad \therefore \quad \rho(A) \le 4$

Operating $R_4 \to R_4 - (R_3 + R_2 + R_1)$

$$A \sim \begin{bmatrix} 1 & 2 & 3 & 0 \\ 2 & 4 & 3 & 2 \\ 3 & 2 & 1 & 3 \\ 0 & 0 & 0 & 0 \end{bmatrix} \quad \text{operating } R_2 \to R_2 - 2R_1,\ R_3 \to R_3 - 3R_1$$

$$\sim \begin{bmatrix} 1 & 2 & 3 & 0 \\ 0 & 0 & -3 & 2 \\ 0 & -4 & -8 & 3 \\ 0 & 0 & 0 & 0 \end{bmatrix} \quad \text{operating } R_2 \leftrightarrow R_3$$

$$\sim \begin{bmatrix} 1 & 2 & 3 & 0 \\ 0 & -4 & -8 & 3 \\ 0 & 0 & -3 & 2 \\ 0 & 0 & 0 & 0 \end{bmatrix}$$

The only fourth order minor is zero. Since the third order minor

$$\begin{vmatrix} 1 & 2 & 3 \\ 0 & -4 & -8 \\ 0 & 0 & -3 \end{vmatrix} = (1)(-4)(-3) = 12 \neq 0 \quad \therefore \quad \rho(A) = 3.$$

Example 4. *Reduce each of the following matrices to normal form and hence, find their ranks :*

(*i*) $\begin{bmatrix} 2 & 3 & -1 & -1 \\ 1 & -1 & -2 & -4 \\ 3 & 1 & 3 & -2 \\ 6 & 3 & 0 & -7 \end{bmatrix}$ (*ii*) $\begin{bmatrix} 8 & 1 & 3 & 6 \\ 0 & 3 & 2 & 2 \\ -8 & -1 & -3 & 4 \end{bmatrix}$.

(M.D.U. Dec. 2009)

Sol. (*i*) Let

$$A = \begin{bmatrix} 2 & 3 & -1 & -1 \\ 1 & -1 & -2 & -4 \\ 3 & 1 & 3 & -2 \\ 6 & 3 & 0 & -7 \end{bmatrix} \quad \text{operating } R_4 \to R_4 - (R_1 + R_2 + R_3)$$

$$\sim \begin{bmatrix} 2 & 3 & -1 & -1 \\ 1 & -1 & -2 & -4 \\ 3 & 1 & 3 & -2 \\ 0 & 0 & 0 & 0 \end{bmatrix} \quad \text{operating } R_1 \leftrightarrow R_2$$

$$\sim \begin{bmatrix} 1 & -1 & -2 & -4 \\ 2 & 3 & -1 & -1 \\ 3 & 1 & 3 & -2 \\ 0 & 0 & 0 & 0 \end{bmatrix} \quad \text{operating } C_2 \to C_2 + C_1,\ C_3 \to C_3 + 2C_1,\ C_4 \to C_4 + 4C_1$$

$$\sim \begin{bmatrix} 1 & 0 & 0 & 0 \\ 2 & 5 & 3 & 7 \\ 3 & 4 & 9 & 10 \\ 0 & 0 & 0 & 0 \end{bmatrix} \quad \text{operating } R_2 \to R_2 - 2R_1,\ R_3 \to R_3 - 3R_1$$

$$\sim \begin{bmatrix} 1 & 0 & 0 & 0 \\ 0 & 5 & 3 & 7 \\ 0 & 4 & 9 & 10 \\ 0 & 0 & 0 & 0 \end{bmatrix} \quad \text{operating } R_2 \to R_2 - R_3$$

$$\sim \begin{bmatrix} 1 & 0 & 0 & 0 \\ 0 & 1 & -6 & -3 \\ 0 & 4 & 9 & 10 \\ 0 & 0 & 0 & 0 \end{bmatrix} \quad \text{operating } C_3 \to C_3 + 6C_2,\ C_4 \to C_4 + 3C_2$$

$$\sim \begin{bmatrix} 1 & 0 & 0 & 0 \\ 0 & 1 & 0 & 0 \\ 0 & 4 & 33 & 22 \\ 0 & 0 & 0 & 0 \end{bmatrix} \quad \text{operating } R_3 \to R_3 - 4R_2$$

$$\sim \begin{bmatrix} 1 & 0 & 0 & 0 \\ 0 & 1 & 0 & 0 \\ 0 & 0 & 33 & 22 \\ 0 & 0 & 0 & 0 \end{bmatrix} \quad \text{operating } C_3 \to \frac{1}{33} C_3,\ C_4 \to \frac{1}{22} C_4$$

$$\sim \begin{bmatrix} 1 & 0 & 0 & 0 \\ 0 & 1 & 0 & 0 \\ 0 & 0 & 1 & 1 \\ 0 & 0 & 0 & 0 \end{bmatrix} \quad \text{operating } C_4 \to C_4 - C_3$$

$$\sim \left[\begin{array}{ccc|c} 1 & 0 & 0 & 0 \\ 0 & 1 & 0 & 0 \\ 0 & 0 & 1 & 0 \\ \hline 0 & 0 & 0 & 0 \end{array}\right] = \left[\begin{array}{c|c} I_3 & 0 \\ \hline 0 & 0 \end{array}\right] \quad \text{which is the required normal form}$$

$\therefore \qquad \rho(A) = 3.$

(*ii*) Let
$$A = \begin{bmatrix} 8 & 1 & 3 & 6 \\ 0 & 3 & 2 & 2 \\ -8 & -1 & -3 & 4 \end{bmatrix} \quad \text{operating } C_1 \to \frac{1}{8} C_1$$

$$\sim \begin{bmatrix} 1 & 1 & 3 & 6 \\ 0 & 3 & 2 & 2 \\ -1 & -1 & -3 & 4 \end{bmatrix} \quad \text{operating } R_3 \to R_3 + R_1$$

$$\sim \begin{bmatrix} 1 & 1 & 3 & 6 \\ 0 & 3 & 2 & 2 \\ 0 & 0 & 0 & 10 \end{bmatrix} \quad \text{operating } C_2 \to C_2 - C_1,\ C_3 \to C_3 - 3C_1,\ C_4 \to C_4 - 6C_1$$

$$\sim \begin{bmatrix} 1 & 0 & 0 & 0 \\ 0 & 3 & 2 & 2 \\ 0 & 0 & 0 & 10 \end{bmatrix} \quad \text{operating } C_2 \to \frac{1}{3} C_2$$

$$\sim \begin{bmatrix} 1 & 0 & 0 & 0 \\ 0 & 1 & 2 & 2 \\ 0 & 0 & 0 & 10 \end{bmatrix} \quad \text{operating } C_3 \to C_3 - 2C_2,\ C_4 \to C_4 - 2C_2$$

$$\sim \begin{bmatrix} 1 & 0 & 0 & 0 \\ 0 & 1 & 0 & 0 \\ 0 & 0 & 0 & 10 \end{bmatrix} \quad \text{operating } C_4 \to \frac{1}{10} C_4$$

$$\sim \begin{bmatrix} 1 & 0 & 0 & 0 \\ 0 & 1 & 0 & 0 \\ 0 & 0 & 0 & 1 \end{bmatrix} \quad \text{operating } C_3 \leftrightarrow C_4$$

$$\sim \begin{bmatrix} 1 & 0 & 0 & \vdots & 0 \\ 0 & 1 & 0 & \vdots & 0 \\ 0 & 0 & 1 & \vdots & 0 \end{bmatrix} = [I_3 : O]$$

which is the required normal form

$$\rho(A) = 3.$$

Example 5. *For the matrix* $A = \begin{bmatrix} 1 & 1 & 2 \\ 1 & 2 & 3 \\ 0 & -1 & -1 \end{bmatrix}$, *find non-singular matrices P and Q such that PAQ is in the normal form. Hence find the rank of A.* (M.D.U., 2010, 2011; K.U.K. 2013)

Sol. We write A = IAI

$$\Rightarrow \quad \begin{bmatrix} 1 & 1 & 2 \\ 1 & 2 & 3 \\ 0 & -1 & -1 \end{bmatrix} = \begin{bmatrix} 1 & 0 & 0 \\ 0 & 1 & 0 \\ 0 & 0 & 1 \end{bmatrix} A \begin{bmatrix} 1 & 0 & 0 \\ 0 & 1 & 0 \\ 0 & 0 & 1 \end{bmatrix}$$

Operating $C_2 \to C_2 - C_1$, $C_3 \to C_3 - 2C_1$ (subjecting the *post-factor* on R.H.S. to same operations)

$$\begin{bmatrix} 1 & 0 & 0 \\ 1 & 1 & 1 \\ 0 & -1 & -1 \end{bmatrix} = \begin{bmatrix} 1 & 0 & 0 \\ 0 & 1 & 0 \\ 0 & 0 & 1 \end{bmatrix} A \begin{bmatrix} 1 & -1 & -2 \\ 0 & 1 & 0 \\ 0 & 0 & 1 \end{bmatrix}$$

Operating $R_2 \to R_2 - R_1$ (subjecting the *pre-factor* on R.H.S. to same operation)

$$\begin{bmatrix} 1 & 0 & 0 \\ 0 & 1 & 1 \\ 0 & -1 & -1 \end{bmatrix} = \begin{bmatrix} 1 & 0 & 0 \\ -1 & 1 & 0 \\ 0 & 0 & 1 \end{bmatrix} A \begin{bmatrix} 1 & -1 & -2 \\ 0 & 1 & 0 \\ 0 & 0 & 1 \end{bmatrix}$$

Operating $C_3 \to C_3 - C_2$ (subjecting the *post-factor* on R.H.S. to same operation)

$$\begin{bmatrix} 1 & 0 & 0 \\ 0 & 1 & 0 \\ 0 & -1 & 0 \end{bmatrix} = \begin{bmatrix} 1 & 0 & 0 \\ -1 & 1 & 0 \\ 0 & 0 & 1 \end{bmatrix} A \begin{bmatrix} 1 & -1 & -1 \\ 0 & 1 & -1 \\ 0 & 0 & 1 \end{bmatrix}$$

Operating $R_3 \to R_3 + R_2$ (subjecting the *pre-factor* on R.H.S. to same operation)

$$\begin{bmatrix} 1 & 0 & 0 \\ 0 & 1 & 0 \\ 0 & 0 & 0 \end{bmatrix} = \begin{bmatrix} 1 & 0 & 0 \\ -1 & 1 & 0 \\ -1 & 1 & 0 \end{bmatrix} A \begin{bmatrix} 1 & -1 & -1 \\ 0 & 1 & -1 \\ 0 & 0 & 1 \end{bmatrix}$$

$$\Rightarrow \quad \begin{bmatrix} I_2 & O \\ O & O \end{bmatrix} = PAQ, \quad \text{where} \quad P = \begin{bmatrix} 1 & 0 & 0 \\ -1 & 1 & 0 \\ -1 & 1 & 1 \end{bmatrix} \text{ and } Q = \begin{bmatrix} 1 & -1 & -1 \\ 0 & 1 & -1 \\ 0 & 0 & 1 \end{bmatrix}.$$

Also, rank of $A = \rho(A) = 2$

Remarks 1. The matrices P and Q are not unique.

2. If the normal form of PAQ involves I_r, then $\rho(A) = r$.

EXERCISE 2.1

1. Reduce to triangular form (*i*) $\begin{bmatrix} 3 & -4 & -5 \\ -9 & 1 & 4 \\ -5 & 3 & 1 \end{bmatrix}$ (*ii*) $\begin{bmatrix} 2 & 3 & -1 \\ 0 & -3 & 2 \\ -5 & 3 & 7 \end{bmatrix}$

2. (*a*) Use Gauss-Jordan method to find the inverse of each of the following matrices:

(*i*) $\begin{bmatrix} 8 & 4 & 3 \\ 2 & 1 & 1 \\ 1 & 2 & 1 \end{bmatrix}$ (*ii*) $\begin{bmatrix} 2 & 0 & -1 \\ 5 & 1 & 0 \\ 0 & 1 & 3 \end{bmatrix}$ (*iii*) $\begin{bmatrix} 2 & 1 & -1 \\ 0 & 2 & 1 \\ 5 & 2 & -3 \end{bmatrix}$

(*K.U.K. 2008, 2011, 2013*)

(*b*) Compute the inverse of the following matrices by using Gauss-Jordan method:

(*i*) $\begin{bmatrix} 1 & 2 & 1 \\ -1 & 1 & 2 \\ 2 & -1 & 1 \end{bmatrix}$ (*ii*) $\begin{bmatrix} 2 & -6 & -2 & -3 \\ 5 & -13 & -4 & -7 \\ -1 & 4 & 1 & 2 \\ 0 & 1 & 0 & 1 \end{bmatrix}$

(*K.U.K. May 2012*)

3. By using elementary transformations, find the inverse of each of the following matrices:

(*i*) $\begin{bmatrix} 1 & 1 & 3 \\ 1 & 3 & -3 \\ -2 & -4 & -4 \end{bmatrix}$ (*ii*) $\begin{bmatrix} 3 & -3 & 4 \\ 2 & -3 & 4 \\ 0 & -1 & 1 \end{bmatrix}$ (*iii*) $\begin{bmatrix} 1 & 2 & 5 \\ 2 & 3 & 1 \\ -1 & 1 & 1 \end{bmatrix}$

(*G.B.T.U. 2010*)

4. Find the ranks of the matrices

(*i*) $\begin{bmatrix} 2 & -1 & 0 & 5 \\ 0 & 3 & 1 & 4 \end{bmatrix}$ (*ii*) $\begin{bmatrix} 1 & 2 & 3 \\ 2 & 4 & 7 \\ 3 & 6 & 10 \end{bmatrix}$ (*iii*) $\begin{bmatrix} 1 & 3 & 4 & 5 \\ 1 & 2 & 6 & 7 \\ 1 & 5 & 0 & 10 \end{bmatrix}$

(*iv*) $\begin{bmatrix} 1 & 3 & 4 & 3 \\ 3 & 9 & 12 & 3 \\ 1 & 3 & 4 & 1 \end{bmatrix}$ (*v*) $\begin{bmatrix} 1 & 4 & 5 \\ 2 & 6 & 8 \\ 3 & 7 & 22 \end{bmatrix}$ (*vi*) $\begin{bmatrix} 2 & -4 & 6 \\ -1 & 2 & -3 \\ 3 & -6 & 9 \end{bmatrix}$.

5. (*i*) Find rank of the matrix $\begin{bmatrix} 3 & -1 & 2 \\ -6 & 2 & 4 \\ -3 & 1 & 2 \end{bmatrix}$ by reducing it in its normal form.

(*ii*) Determine the rank of the matrix $A = \begin{bmatrix} 6 & 1 & 3 & 8 \\ 4 & 2 & 6 & -1 \\ 10 & 3 & 9 & 7 \\ 16 & 4 & 12 & 15 \end{bmatrix}$ by reducing it to the normal form.

6. Find the ranks of the following matrices:

(*i*) $\begin{bmatrix} 3 & -1 & 2 \\ -6 & 2 & 4 \\ -3 & 1 & 2 \end{bmatrix}$ (*ii*) $\begin{bmatrix} 1 & 2 & 3 & 2 \\ 2 & 3 & 5 & 1 \\ 1 & 3 & 4 & 5 \end{bmatrix}$ (*iii*) $\begin{bmatrix} 1 & -1 & 2 & -3 \\ 4 & 1 & 0 & 2 \\ 0 & 3 & 1 & 4 \\ 0 & 1 & 0 & 2 \end{bmatrix}$

(*U.K.T.U. 2012*)

(*iv*) $\begin{bmatrix} 0 & 1 & -3 & -1 \\ 1 & 0 & 1 & 1 \\ 3 & 1 & 0 & 2 \\ 1 & 1 & -2 & 0 \end{bmatrix}$ (*v*) $\begin{bmatrix} 5 & 6 & 7 & 8 \\ 6 & 7 & 8 & 9 \\ 11 & 12 & 13 & 14 \\ 16 & 17 & 18 & 19 \end{bmatrix}$ (*vi*) $\begin{bmatrix} 1^2 & 2^2 & 3^2 & 4^2 \\ 2^2 & 3^2 & 4^2 & 5^2 \\ 3^2 & 4^2 & 5^2 & 6^2 \\ 4^2 & 5^2 & 6^2 & 7^2 \end{bmatrix}$

(*K.U.K. 2008*)

7. Find the ranks of the following matrices:

(i) $\begin{bmatrix} 3 & -4 & -1 & 2 \\ 1 & 7 & 3 & 1 \\ 5 & -2 & 5 & 4 \\ 9 & -3 & 7 & 7 \end{bmatrix}$ (ii) $\begin{bmatrix} -1 & 2 & 3 & -2 \\ 2 & -5 & 1 & 2 \\ 3 & -8 & 5 & 2 \\ 5 & -12 & -1 & 6 \end{bmatrix}$

(iii) $\begin{bmatrix} 2 & 4 & 3 & -2 \\ -3 & -2 & -1 & 4 \\ 6 & -1 & 7 & 2 \end{bmatrix}$ (iv) $\begin{bmatrix} 1 & 2 & 3 & 4 \\ 2 & 3 & 4 & 5 \\ 3 & 4 & 5 & 6 \\ 4 & 5 & 6 & 7 \end{bmatrix}$

8. Reduce each of the following matrices to normal form and hence, find their ranks:

(i) $\begin{bmatrix} 0 & 2 & 3 & 4 \\ 2 & 3 & 5 & 4 \\ 4 & 8 & 13 & 12 \end{bmatrix}$ (ii) $\begin{bmatrix} 1 & 2 & 0 & -1 \\ 3 & 4 & 1 & 2 \\ -2 & 3 & 2 & 5 \end{bmatrix}$

(iii) $\begin{bmatrix} 1 & 2 & 1 & 0 \\ 3 & 2 & 1 & 2 \\ 2 & -1 & 2 & 5 \\ 5 & 6 & 3 & 2 \\ 1 & 3 & -1 & -3 \end{bmatrix}$ (iv) $\begin{bmatrix} 2 & 1 & -3 & -6 \\ 3 & -3 & 1 & 2 \\ 1 & 1 & 1 & 2 \end{bmatrix}$

9. Find non-singular matrices P and Q such that PAQ is in the normal form for the matrices:

(i) $A = \begin{bmatrix} 1 & -1 & -1 \\ 1 & 1 & 1 \\ 3 & 1 & 1 \end{bmatrix}$ (ii) $A = \begin{bmatrix} 1 & 2 & 3 \\ 3 & 2 & 1 \\ 1 & 3 & 2 \\ 2 & 1 & 3 \end{bmatrix}$ (iii) $A = \begin{bmatrix} 1 & 2 & 3 & -2 \\ 2 & -2 & 1 & 3 \\ 3 & 0 & 4 & 1 \end{bmatrix}$

(M.D.U. Dec. 2013)

10. For the matrix

$$A = \begin{bmatrix} 2 & 1 & -3 & 6 \\ 3 & -3 & 1 & 2 \\ 1 & 1 & 1 & 2 \end{bmatrix}$$

find non-singular matrices P and Q such that PAQ is in the normal form. Hence find the rank of A.

(M.D.U. May 2008, Dec. 2012)

11. If $A = \begin{bmatrix} 3 & -3 & 4 \\ 2 & -3 & 4 \\ 0 & -1 & 1 \end{bmatrix}$ find A^{-1}. Also find two non-singular matrices P and Q such that PAQ = I, where I is the unit matrix and verify that $A^{-1} = QP$.

Answers

2. (a) (i) $\begin{bmatrix} \frac{1}{3} & -\frac{2}{3} & -\frac{1}{3} \\ \frac{1}{3} & -\frac{5}{3} & \frac{2}{3} \\ -1 & 4 & 0 \end{bmatrix}$ (ii) $\begin{bmatrix} 3 & -1 & 1 \\ -15 & 6 & -5 \\ 5 & -2 & 2 \end{bmatrix}$ (iii) $\begin{bmatrix} 8 & -1 & -3 \\ -5 & 1 & 2 \\ 10 & -1 & -4 \end{bmatrix}$

(b) (i) $\frac{1}{12}\begin{bmatrix} 3 & -3 & 3 \\ 5 & -1 & -3 \\ -1 & 5 & 3 \end{bmatrix}$ (ii) $\begin{bmatrix} -2 & 1 & 0 & 1 \\ 1 & 0 & 2 & -1 \\ -4 & 1 & -3 & 1 \\ -1 & 0 & -2 & 2 \end{bmatrix}$

3. (i) $\begin{bmatrix} 3 & 1 & \frac{3}{2} \\ -\frac{5}{4} & -\frac{1}{4} & -\frac{3}{4} \\ -\frac{1}{4} & -\frac{1}{4} & -\frac{1}{4} \end{bmatrix}$ (ii) $\begin{bmatrix} 1 & -1 & 0 \\ -2 & 3 & -4 \\ -2 & 3 & -3 \end{bmatrix}$ (iii) $\begin{bmatrix} \frac{2}{21} & \frac{1}{7} & -\frac{13}{21} \\ -\frac{1}{7} & \frac{2}{7} & \frac{3}{7} \\ \frac{5}{21} & -\frac{1}{7} & -\frac{1}{21} \end{bmatrix}$.

4. (i) 2 (ii) 2 (iii) 3 (iv) 2
(v) 3 (vi) 1 5. (i) 2 (ii) 2

6. (i) 2 (ii) 2 (iii) 4 (iv) 2
(v) 2 (vi) 3

7. (i) 3 (ii) 2 (iii) 3 (iv) 2

8. (i) $\begin{bmatrix} I_2 & O \\ O & O \end{bmatrix}$, 2 (ii) $[I_3 \;\; O]$, 3 (iii) $\begin{bmatrix} I_3 & O \\ O & O \end{bmatrix}$, 3 (iv) $[I_3 \;\; O]$, 3

10. $PAQ = [I_3 \;\; O]$, $\rho(A) = 3$ 11. $A^{-1} = \begin{bmatrix} 1 & -1 & 0 \\ -2 & 3 & -4 \\ -2 & 3 & -3 \end{bmatrix}$.

2.6. SOLUTION OF A SYSTEM OF LINEAR EQUATIONS

Consider the system of equations $\left.\begin{aligned} a_1x + b_1y + c_1z &= d_1 \\ a_2x + b_2y + c_2z &= d_2 \\ a_3x + b_3y + c_3z &= d_3 \end{aligned}\right\}$ (3 equations in 3 unknowns)

In matrix notation, these equations can be written as

$$\begin{bmatrix} a_1x + b_1y + c_1z \\ a_2x + b_2y + c_2z \\ a_3x + b_3y + c_3z \end{bmatrix} = \begin{bmatrix} d_1 \\ d_2 \\ d_3 \end{bmatrix} \quad \text{or} \quad \begin{bmatrix} a_1 & b_1 & c_1 \\ a_2 & b_2 & c_2 \\ a_3 & b_3 & c_3 \end{bmatrix} \begin{bmatrix} x \\ y \\ z \end{bmatrix} = \begin{bmatrix} d_1 \\ d_2 \\ d_3 \end{bmatrix}$$

or
$$AX = B$$

where $A = \begin{bmatrix} a_1 & b_1 & c_1 \\ a_2 & b_2 & c_2 \\ a_3 & b_3 & c_3 \end{bmatrix}$ is called the co-efficient matrix,

$X = \begin{bmatrix} x \\ y \\ z \end{bmatrix}$ is the column matrix of unknowns

$B = \begin{bmatrix} d_1 \\ d_2 \\ d_3 \end{bmatrix}$ is the column matrix of constants.

If $d_1 = d_2 = d_3 = 0$, then $B = O$ and the matrix equation $AX = B$ reduces to $AX = O$. Such a system of equations is called a system of **homogeneous linear equations.** If at least one of d_1, d_2, d_3 is non-zero, then $B \neq O$.
Such a system of equations is called a system of **non-homogeneous linear equations.**

Solving the matrix equation $AX = B$ means finding X, *i.e.*, finding a column matrix $\begin{bmatrix} \alpha \\ \beta \\ \gamma \end{bmatrix}$ such that

$$X = \begin{bmatrix} x \\ y \\ z \end{bmatrix} = \begin{bmatrix} \alpha \\ \beta \\ \gamma \end{bmatrix}. \text{ Then } x = \alpha, y = \beta, z = \gamma.$$

The matrix equation AX = B need not always have a solution. It may have no solution or a unique solution or an infinite number of solutions.

A system of equations having **no solution** is called an **inconsistent** system of equations.

A system of equations having **one or more solutions** is called **a consistent** system of equations.

For a system of non-homogeneous linear equations AX = B

(*i*) if $\rho[A : B] \neq \rho(A)$, the system is inconsistent.

(*ii*) if $\rho[A : B] = \rho(A)$ = number of unknowns, the system has a unique solution.

(*iii*) $\rho[A : B] = \rho(A)$ < number of unknowns, the system has an infinite number of solutions.

The matrix [A : B] in which the elements of A and B are written side by side is called the **augmented matrix.**

For a system of homogeneous linear equations AX = O

(*i*) X = O is always a solution. This solution in which each unknown has the value zero is called the **Null Solution** or the **Trivial Solution.** Thus a homogeneous system is always consistent.

A system of homogeneous linear equations has either the trivial solution or an infinite number of solutions.

(*ii*) if $\rho(A)$ = number of unknowns, the system has only the trivial solution.

(*iii*) if $\rho(A)$ < number of unknowns, the system has an infinite number of non-trivial solutions.

2.7. IF A IS A NON-SINGULAR MATRIX, THEN THE MATRIX EQUATION AX = B HAS A UNIQUE SOLUTION

The given equation is $AX = B$ ...(1)

$\because$ A is a non-singular matrix, $\therefore$ A^{-1} exists.

Pre-multiplying both sides of (1) by A^{-1}, we get

$$A^{-1}AX = A^{-1}B \quad \text{or} \quad (A^{-1}A)X = A^{-1}B$$

or $$IX = A^{-1}B \quad \text{or} \quad X = A^{-1}B$$

which is the required unique solution (since A^{-1} is unique).

Another Method

Write the augmented matrix [A : B]. By **E-row operations** on A and B, reduce A to a diagonal matrix, thus getting

$$[A : B] \sim \begin{bmatrix} p_1 & 0 & 0 & \vdots & q_1 \\ 0 & p_2 & 0 & \vdots & q_2 \\ 0 & 0 & p_3 & \vdots & q_3 \end{bmatrix}$$

Then $$p_1x = q_1,\ p_2y = q_2,\ p_3z = q_3.$$

ILLUSTRATIVE EXAMPLES

Example 1. *Solve, with the help of matrices, the simultaneous equations:*

$$x + y + z = 3, \quad x + 2y + 3z = 4, \quad x + 4y + 9z = 6.$$

Sol. [In this question, there is no restriction that the solution must be obtained by finding A^{-1}].

Augmented matrix $[A : B] = \left[\begin{array}{ccc:c} 1 & 1 & 1 & 3 \\ 1 & 2 & 3 & 4 \\ 1 & 4 & 9 & 6 \end{array}\right]$ operating $R_2 \to R_2 - R_1$, $R_3 \to R_3 - R_1$

$$\sim \left[\begin{array}{ccc:c} 1 & 1 & 1 & 3 \\ 0 & 1 & 2 & 1 \\ 0 & 3 & 8 & 3 \end{array}\right] \quad \text{operating } R_3 \to R_3 - 3R_2$$

$$\sim \left[\begin{array}{ccc:c} 1 & 1 & 1 & 3 \\ 0 & 1 & 2 & 1 \\ 0 & 0 & 2 & 0 \end{array}\right] \quad \text{operating } R_3 \to \tfrac{1}{2} R_3$$

$$\sim \left[\begin{array}{ccc:c} 1 & 1 & 1 & 3 \\ 0 & 1 & 2 & 1 \\ 0 & 0 & 1 & 0 \end{array}\right] \quad \text{operating } R_1 \to R_1 - R_3,\ R_2 \to R_2 - 2R_3$$

$$\sim \left[\begin{array}{ccc:c} 1 & 1 & 0 & 3 \\ 0 & 1 & 0 & 1 \\ 0 & 0 & 1 & 0 \end{array}\right] \quad \text{operating } R_1 \to R_1 - R_2$$

$$\sim \left[\begin{array}{ccc:c} 1 & 0 & 0 & 2 \\ 0 & 1 & 0 & 1 \\ 0 & 0 & 1 & 0 \end{array}\right]$$

$\therefore \qquad x = 2,\ y = 1,\ z = 0.$

Example 2. *Solve the system of equations:*

$$2x_1 + x_2 + 2x_3 + x_4 = 6, \qquad 6x_1 - 6x_2 + 6x_3 + 12x_4 = 36$$
$$4x_1 + 3x_2 + 3x_3 - 3x_4 = -1, \qquad 2x_1 + 2x_2 - x_3 + x_4 = 10.$$

Sol. In matrix notation, the given system of equations can be written as AX = B

where $A = \begin{bmatrix} 2 & 1 & 2 & 1 \\ 6 & -6 & 6 & 12 \\ 4 & 3 & 3 & -3 \\ 2 & 2 & -1 & 1 \end{bmatrix}, X = \begin{bmatrix} x_1 \\ x_2 \\ x_3 \\ x_4 \end{bmatrix}, B = \begin{bmatrix} 6 \\ 36 \\ -1 \\ 10 \end{bmatrix}$

Augmented matrix

$$[A : B] = \left[\begin{array}{cccc:c} 2 & 1 & 2 & 1 & 6 \\ 6 & -6 & 6 & 12 & 36 \\ 4 & 3 & 3 & -3 & -1 \\ 2 & 2 & -1 & 1 & 10 \end{array}\right]$$

Operating $R_2 \to R_2 - 3R_1$, $R_3 \to R_3 - 2R_1$, $R_4 \to R_4 - R_1$

$$\sim \left[\begin{array}{cccc:c} 2 & 1 & 2 & 1 & 6 \\ 0 & -9 & 0 & 9 & 18 \\ 0 & 1 & -1 & -5 & -13 \\ 0 & 1 & -3 & 0 & 4 \end{array}\right] \quad \text{operating } R_2 \to -\frac{1}{9} R_2$$

$$\sim \begin{bmatrix} 2 & 1 & 2 & 1 & \vdots & 6 \\ 0 & 1 & 0 & -1 & \vdots & -2 \\ 0 & 1 & -1 & -5 & \vdots & -13 \\ 0 & 1 & -3 & 0 & \vdots & 4 \end{bmatrix}$$

Operating $R_1 \to R_1 - R_2$, $R_3 \to R_3 - R_2$, $R_4 \to R_4 - R_2$

$$\sim \begin{bmatrix} 2 & 0 & 2 & 2 & \vdots & 8 \\ 0 & 1 & 0 & -1 & \vdots & -2 \\ 0 & 0 & -1 & -4 & \vdots & -11 \\ 0 & 0 & -3 & 1 & \vdots & 6 \end{bmatrix} \quad \text{operating } R_4 \to R_4 - 3R_3,\ R_1 \to \tfrac{1}{2}R_1$$

$$\sim \begin{bmatrix} 1 & 0 & 1 & 1 & \vdots & 4 \\ 0 & 1 & 0 & -1 & \vdots & -2 \\ 0 & 0 & -1 & -4 & \vdots & -11 \\ 0 & 0 & 0 & 13 & \vdots & 39 \end{bmatrix} \quad \text{operating } R_1 \to R_1 + R_3,\ R_4 \to \frac{1}{13} R_4$$

$$\sim \begin{bmatrix} 1 & 0 & 0 & -3 & \vdots & 7 \\ 0 & 1 & 0 & -1 & \vdots & -2 \\ 0 & 0 & -1 & -4 & \vdots & -11 \\ 0 & 0 & 0 & 1 & \vdots & 3 \end{bmatrix} \quad \text{operating } R_1 \to R_1 + 3R_4,\ R_2 \to R_2 + R_4,\ R_3 \to R_3 + 4R_4$$

$$\sim \begin{bmatrix} 1 & 0 & 0 & 0 & \vdots & 2 \\ 0 & 1 & 0 & 0 & \vdots & 1 \\ 0 & 0 & -1 & 0 & \vdots & 1 \\ 0 & 0 & 0 & 1 & \vdots & 3 \end{bmatrix} \quad \text{operating } R_3 \to (-1)\, R_3$$

$$\sim \begin{bmatrix} 1 & 0 & 0 & 0 & \vdots & 2 \\ 0 & 1 & 0 & 0 & \vdots & 1 \\ 0 & 0 & 1 & 0 & \vdots & -1 \\ 0 & 0 & 0 & 1 & \vdots & 3 \end{bmatrix}$$

Hence $x_1 = 2,\ x_2 = 1,\ x_3 = -1,\ x_4 = 3.$

Example 3. *Using matrix method, show that the equations $3x + 3y + 2z = 1$, $x + 2y = 4$, $10y + 3z = -2$, $2x - 3y - z = 5$ are consistent and hence obtain the solutions for x, y and z.*

(M.D.U. Dec., 2012; U.K.T.U. 2010; G.B.T.U. 2010)

Sol. In matrix notation, the given system of equations can be written as AX = B

where $A = \begin{bmatrix} 3 & 3 & 2 \\ 1 & 2 & 0 \\ 0 & 10 & 3 \\ 2 & -3 & -1 \end{bmatrix},\ X = \begin{bmatrix} x \\ y \\ z \end{bmatrix},\ B = \begin{bmatrix} 1 \\ 4 \\ -2 \\ 5 \end{bmatrix}$

Augmented matrix

$$[A : B] = \begin{bmatrix} 3 & 3 & 2 & \vdots & 1 \\ 1 & 2 & 0 & \vdots & 4 \\ 0 & 10 & 3 & \vdots & -2 \\ 2 & -3 & -1 & \vdots & 5 \end{bmatrix} \quad \text{operating } R_1 \leftrightarrow R_2$$

$$\sim \begin{bmatrix} 1 & 2 & 0 & \vdots & 4 \\ 3 & 3 & 2 & \vdots & 1 \\ 0 & 10 & 3 & \vdots & -2 \\ 2 & -3 & -1 & \vdots & 5 \end{bmatrix} \quad \text{operating } R_2 \to R_2 - 3R_1,\ R_4 \to R_4 - 2R_1$$

$$\sim \begin{bmatrix} 1 & 2 & 0 & : & 4 \\ 0 & -3 & 2 & : & -11 \\ 0 & 10 & 3 & : & -2 \\ 0 & -7 & -1 & : & -3 \end{bmatrix} \quad \text{operating } R_3 \to R_3 + 3R_2,\ R_4 \to R_4 - 2R_2$$

$$\sim \begin{bmatrix} 1 & 2 & 0 & : & 4 \\ 0 & -3 & 2 & : & -11 \\ 0 & 1 & 9 & : & -35 \\ 0 & -1 & -5 & : & 19 \end{bmatrix} \quad \text{operating } R_1 \to R_1 - 2R_3,\ R_2 \to R_2 + 3R_3,\ R_4 \to R_4 + R_3$$

$$\sim \begin{bmatrix} 1 & 0 & -18 & : & 74 \\ 0 & 0 & 29 & : & -116 \\ 0 & 1 & 9 & : & -35 \\ 0 & 0 & 4 & : & -16 \end{bmatrix} \quad \text{operating } R_2 \leftrightarrow R_3,\ R_4 \to \tfrac{1}{4} R_4$$

$$\sim \begin{bmatrix} 1 & 0 & -18 & : & 74 \\ 0 & 1 & 9 & : & -35 \\ 0 & 0 & 29 & : & -116 \\ 0 & 0 & 1 & : & -4 \end{bmatrix} \quad \text{operating } R_1 \to R_1 + 18R_4,\ R_2 \to R_2 - 9R_4,\ R_3 \to R_3 - 29R_4$$

$$\sim \begin{bmatrix} 1 & 0 & 0 & : & 2 \\ 0 & 1 & 0 & : & 1 \\ 0 & 0 & 0 & : & 0 \\ 0 & 0 & 1 & : & -4 \end{bmatrix} \quad \text{operating } R_3 \leftrightarrow R_4$$

$$\sim \begin{bmatrix} 1 & 0 & 0 & : & 2 \\ 0 & 1 & 0 & : & 1 \\ 0 & 0 & 1 & : & -4 \\ 0 & 0 & 0 & : & 0 \end{bmatrix}$$

$\rho(A) = \rho(A : B) = 3$ = number of unknowns

$\Rightarrow$ The given system of equations is consistent and the unique solution is

$$x = 2,\ y = 1,\ z = -4.$$

Example 4. *For what values of parameters λ and μ do the system of equations*

$$x + y + z = 6,\ x + 2y + 3z = 10,\ x + 2y + \lambda z = \mu$$

have (i) no solution (ii) unique solution (iii) more than one solution ?

(Kerala May 2010; M.T.U. 2013; K.U.K. 2008, 2012)

Sol. In matrix notation, the given system of equations can be written as AX = B

$$\text{where} \quad A = \begin{bmatrix} 1 & 1 & 1 \\ 1 & 2 & 3 \\ 1 & 2 & \lambda \end{bmatrix},\ X = \begin{bmatrix} x \\ y \\ z \end{bmatrix},\ B = \begin{bmatrix} 6 \\ 10 \\ \mu \end{bmatrix}$$

$$\text{Augmented matrix } [A : B] = \begin{bmatrix} 1 & 1 & 1 & : & 6 \\ 1 & 2 & 3 & : & 10 \\ 1 & 2 & \lambda & : & \mu \end{bmatrix} \quad \text{operating } R_2 \to R_2 - R_1,\ R_3 \to R_3 - R_1$$

$$\sim \begin{bmatrix} 1 & 1 & 1 & : & 6 \\ 0 & 1 & 2 & : & 4 \\ 0 & 1 & \lambda - 1 & : & \mu - 6 \end{bmatrix} \quad \text{operating } R_1 \to R_1 - R_2,\ R_3 \to R_3 - R_2$$

$$\sim \begin{bmatrix} 1 & 0 & -1 & \vdots & 2 \\ 0 & 1 & 2 & \vdots & 4 \\ 0 & 0 & \lambda - 3 & \vdots & \mu - 10 \end{bmatrix}$$

Case I. *If* $\lambda = 3$, $\mu \neq 10$

$$\rho(A) = 2, \rho(A : B) = 3$$

$\because$ $\rho(A) \neq \rho(A : B)$

$\therefore$ The system has no solution.

Case II. *If* $\lambda \neq 3$, μ *may have any value*

$$\rho(A) = \rho(A : B) = 3 = \text{number of unknowns}$$

$\therefore$ The system has unique solution.

Case III. *If* $\lambda = 3$, $\mu = 10$

$$\rho(A) = \rho(A : B) = 2 < \text{number of unknowns}$$

$\therefore$ The system has an infinite number of solutions.

Example 5. *Solve the equations :*

$x_1 + 3x_2 + 2x_3 = 0,\ 2x_1 - x_2 + 3x_3 = 0,\ 3x_1 - 5x_2 + 4x_3 = 0,\ x_1 + 17x_2 + 4x_3 = 0.$

Sol. In matrix notation, the given system of equations can be written as AX = O

where $A = \begin{bmatrix} 1 & 3 & 2 \\ 2 & -1 & 3 \\ 3 & -5 & 4 \\ 1 & 17 & 4 \end{bmatrix}, X = \begin{bmatrix} x_1 \\ x_2 \\ x_3 \end{bmatrix}$

Operating $R_2 \to R_2 - 2R_1$, $R_3 \to R_3 - 3R_1$, $R_4 \to R_4 - R_1$

$$A \sim \begin{bmatrix} 1 & 3 & 2 \\ 0 & -7 & -1 \\ 0 & -14 & -2 \\ 0 & 14 & 2 \end{bmatrix} \quad \text{operating } R_3 \to R_3 - 2R_2,\ R_4 \to R_4 + 2R_2$$

$$\sim \begin{bmatrix} 1 & 3 & 2 \\ 0 & -7 & -1 \\ 0 & 0 & 0 \\ 0 & 0 & 0 \end{bmatrix} \quad \text{operating } R_1 \to R_1 + 2R_2$$

$$\sim \begin{bmatrix} 1 & -11 & 0 \\ 0 & -7 & -1 \\ 0 & 0 & 0 \\ 0 & 0 & 0 \end{bmatrix}$$

$\therefore$ $\rho(A) = 2 <$ number of unknowns

$\Rightarrow$ The system has an infinite number of non-trivial solutions given by

$$x_1 - 11x_2 = 0, \qquad -7x_2 - x_3 = 0$$

i.e., $x_1 = 11k, \quad x_2 = k, \quad x_3 = -7k$, where k is arbitrary.

Different values of k give different solutions.

Example 6. *Show that the system of equations:*

$$2x_1 - 2x_2 + x_3 = \lambda x_1$$
$$2x_1 - 3x_2 + 2x_3 = \lambda x_2$$
$$-x_1 + 2x_2 = \lambda x_3$$

can possess a non-trivial solution only if $\lambda = 1$ *or* -3. *Obtain the general solution in each case.*

Sol. The given system of equation is

$$(2-\lambda)x_1 - 2x_2 + x_3 = 0$$
$$2x_1 - (3+\lambda)x_2 + 2x_3 = 0$$
$$-x_1 + 2x_2 - \lambda x_3 = 0$$

In matrix notation, it can be written as AX = O

$$\text{where}\quad A = \begin{bmatrix} 2-\lambda & -2 & 1 \\ 2 & -(3+\lambda) & 2 \\ -1 & 2 & -\lambda \end{bmatrix}, X = \begin{bmatrix} x_1 \\ x_2 \\ x_3 \end{bmatrix}$$

For non-trivial solution, $|A| = 0$

$$\Rightarrow \quad \begin{vmatrix} 2-\lambda & -2 & 1 \\ 2 & -(3+\lambda) & 2 \\ -1 & 2 & -\lambda \end{vmatrix} = 0$$

$$\Rightarrow \quad (2-\lambda)[\lambda(3+\lambda) - 4] + 2(-2\lambda + 2) + [4 - (3+\lambda)] = 0$$

$$\Rightarrow \quad \lambda^3 + \lambda^2 - 5\lambda + 3 = 0$$

$$\Rightarrow \quad (\lambda - 1)^2(\lambda + 3) = 0$$

$$\therefore \quad \lambda = 1 \quad \text{or} \quad -3.$$

When $\lambda = 1$, the equations become

$$x_1 - 2x_2 + x_3 = 0$$
$$2x_1 - 4x_2 + 2x_3 = 0$$
$$-x_1 + 2x_2 - x_3 = 0$$

which are identical.

The given system is equivalent to a single equation

$$x_1 - 2x_2 + x_3 = 0$$

Taking $x_2 = t$, $x_3 = s$, we get $x_1 = 2t - s$

$$\therefore \quad x_1 = 2t - s,\ x_2 = t,\ x_3 = s$$

which give an infinite number of non-trivial solutions, t and s being the parameters.

When $\lambda = -3$, the equations become

$$5x_1 - 2x_2 + x_3 = 0$$
$$2x_1 + 2x_3 = 0$$
$$-x_1 + 2x_2 + 3x_3 = 0$$

Solving the first two, we have

$$\frac{x_1}{-4} = \frac{x_2}{2-10} = \frac{x_3}{4} \quad \text{or} \quad x_1 = \frac{x_2}{2} = \frac{x_3}{-1}$$

$$\therefore \quad x_1 = t,\quad x_2 = 2t,\quad x_3 = -t$$

which give an infinite number of non-trivial solutions, t being the parameter.

EXERCISE 2.2

1. Write the following equations in matrix form AX = B and solve for X:

(i) $4x - 5y + z = 2$, $3x + y - 2z = 9$, $x + 4y + z = 5$

(ii) $x_1 + x_2 + x_3 = 1$, $2x_1 - x_2 + x_3 = 4$, $x_1 - 3x_2 - 2x_3 = 2$

2. Using the loop current method on a circuit, the following equations were obtained:

$$7i_1 - 4i_2 = 12, \ - 4i_1 + 12i_2 - 6i_3 = 0, \ - 6i_2 + 14i_3 = 0.$$

By matrix method, solve for i_1, i_2 and i_3.

3. Solve the following systems of equations by matrix method:

(i) $x + y + z = 8,\quad x - y + 2z = 6,\quad 3x + 5y - 7z = 14$

(ii) $x + y + z = 6,\quad x - y + 2z = 5,\quad 3x + y + z = 8$

(iii) $x + 2y + 3z = 1,\quad 2x + 3y + 2z = 2,\quad 3x + 3y + 4z = 1.$

4. Show that the equations $x + 2y - z = 3$, $3x - y + 2z = 1$, $2x - 2y + 3z = 2$, $x - y + z = - 1$ are consistent and solve them. *(Calicut 2008)*

5. (a) Test for consistency the equations $2x - 3y + 7z = 5$, $3x + y - 3z = 13$, $2x + 19y - 47z = 32$.

(b) Examine if the following equations are consistent and solve them if they are consistent:

$2x + 6y + 11 = 0,\quad 6x + 20y - 6z + 3 = 0,\quad 6y - 18z + 1 = 0$ *(U.K.T.U. 2011)*

(c) Test for consistency the following equations and if consistent, obtain the solution:

$2x + 3y + 4z = 11,\quad x + 5y + 7z = 15,\quad 3x + 11y + 13z = 25$

(d) Test for consistency and solve

$5x + 3y + 7z = 4$, $3x + 26y + 2z = 9$, $7x + 2y + 10z = 5$.

6. (a) For what values of a and b do the equations $x + 2y + 3z = 6$, $x + 3y + 5z = 9$, $2x + 5y + az = b$ have (i) no solution (ii) a unique solution and (iii) more than one solution?

(b) For what values of a and b, the equations : $x + y + 5z = 0$, $x + 2y + 3az = b$ and $x + 3y + az = 1$ have (i) no solution (ii) unique solution and (iii) infinitely many solutions. *(K.U.K. May 2013)*

7. Find the value of k so that the equations $x + ky + 3z = 0$, $4x + 3y + kz = 0$, $2x + y + 2z = 0$ have a non-trivial solution.

8. Show that the equations $3x + 4y + 5z = a$, $4x + 5y + 6z = b$, $5x + 6y + 7z = c$ do not have a solution unless $a + c = 2b$. *(U.P.T.U. 2008; M.T.U. 2009)*

9. Investigate the values of λ and μ so that the equations $2x + 3y + 5z = 9$, $7x + 3y - 2z = 8$, $2x + 3y + \lambda z = \mu$ have (i) no solution, (ii) a unique solution and (iii) an infinite number of solutions. *(M.D.U. Dec. 2011)*

10. Determine the values of λ for which the following system of equations may possess non-trivial solution. $3x_1 + x_2 - \lambda x_3 = 0$, $4x_1 - 2x_2 - 3x_3 = 0$, $2\lambda x_1 + 4x_2 + \lambda x_3 = 0$. For each permissible value of λ, determine the general solution.

11. Investigate for consistency the following equations and if possible, find the solutions:

$4x - 2y + 6z = 8$, $x + y - 3z = - 1$, $15x - 3y + 9z = 21$.

12. For what values of k the equations $x + y + z = 1$, $2x + y + 4z = k$, $4x + y + 10z = k^2$ have a solution and solve them completely in each case. *(K.U.K. 2013; M.D.U. Dec. 2010; Kerala 2013)*

13. Show that if $\lambda \neq - 5$, the system of equations $3x - y + 4z = 3$, $x + 2y - 3z = - 2$, $6x + 5y + \lambda z = - 3$ have a unique solution. If $\lambda = - 5$, show that the equations are consistent. Determine the solutions in each case.

14. Find the values of a and b for which the equations $x + ay + z = 3$, $x + 2y + 2z = b$, $x + 5y + 3z = 9$ are consistent. When will these equations have a unique solution? *(M.D.U. Dec. 2009; K.U.K. 2008)*

15. Find the value of λ for which the system of equations $x + y + 4z = 1$, $x + 2y - 2z = 1$ and $\lambda x + y + z = 1$ will have unique solution.

16. Find the value of k such that the system of equations $4x + 9y + z = 0$, $kx + 3y + kz = 0$ and $x + 4y + 2z = 0$ has non-trivial solution. Hence find the solution of the system of equations.

17. Determine the values of a and b for which the system $\begin{bmatrix} 3 & -2 & 1 \\ 5 & -8 & 9 \\ 2 & 1 & a \end{bmatrix}\begin{bmatrix} x \\ y \\ z \end{bmatrix} = \begin{bmatrix} b \\ 3 \\ -1 \end{bmatrix}$ has

(*i*) a unique solution (*ii*) no solution (*iii*) an infinite number of solutions.

18. Solve the equations

$$x + y + z + w = 0, x + 3y + 2z + w = 0, \quad 2x + z - w = 0.$$

19. Discuss the consistency of the system of equations :

$$2x - 3y + 6z - 5w = 3,\ y - 4z + w = 1,\ 4x - 5y + 8z - 9w = \lambda$$

for various values of λ. If consistent, find the solution. (*M.D.U. May 2008*)

20. Find the values of λ for which the equations

$$(2 - \lambda)x + 2y + 3 = 0, \quad 2x + (4 - \lambda)y + 7 = 0, \quad 2x + 5y + (6 - \lambda) = 0$$

are consistent and find the values of x and y corresponding to each of these values of λ.

21. Find the values of λ for which the equations

$$(\lambda - 1)x + (3\lambda + 1)y + 2\lambda z = 0$$
$$(\lambda - 1)x + (4\lambda - 2)y + (\lambda + 3)z = 0$$
$$2x + (3\lambda + 1)y + 3(\lambda - 1)z = 0$$

have non-trivial solution and find the ratios $x : y : z$, when λ has the smallest of these values. What happens when λ has the greatest of these values?

Answers

1. (*i*) $x = 2, y = 1, z = -1$ (*ii*) $x_1 = 1, x_2 = -1, x_3 = 1$ **2.** $i_1 = \frac{396}{175}, i_2 = \frac{24}{25}, i_3 = \frac{72}{175}$

3. (*i*) $x = 5, y = \frac{5}{3}, z = \frac{4}{3}$ (*ii*) $x = 1, y = 2, z = 3$ (*iii*) $x = -\frac{3}{7}, y = \frac{8}{7}, z = -\frac{2}{7}$

4. $x = -1, y = 4, z = 4$ **5.** (*a*) Inconsistent (*b*) Inconsistent

(*c*) Consistent; $x = 2, y = -3, z = 4$ (*d*) Consistent; $x = \frac{1}{11}(7 - 16k), y = \frac{1}{11}(3 + k), z = k$

6. (*a*) (*i*) $a = 8, b \neq 15$ (*ii*) $a \neq 8$, b may have any value (*iii*) $a = 8, b = 15$

(*b*) (*i*) $a = 1, b \neq \frac{1}{2}$ (*ii*) $a \neq 1$, b may have any value (*iii*) $a = 1, b = \frac{1}{2}$

7. $k = 0, 9/2$ **9.** (*i*) $\lambda = 5, \mu \neq 9$ (*ii*) $\lambda \neq 5$, μ arbitrary (*iii*) $\lambda = 5, \mu = 9$

10. $\lambda = 1, -9$. For $\lambda = 1$, solution is $x = k, y = -k, z = 2k$. For $\lambda = -9$, solution is $x = 3k, y = 9k, z = -2k$

11. Consistent ; $x = 1, y = 3k - 2, z = k$, where k is arbitrary

12. $k = 1, x = -3\lambda, y = 1 + 2\lambda, z = \lambda$; $k = 2, x = 1 - 3\lambda, y = 2\lambda, z = \lambda$, where λ is arbitrary

13. $\lambda \neq -5, x = \frac{4}{7}, y = -\frac{9}{7}, z = 0$; $\lambda = -5, x = \frac{4 - 5k}{7}, y = \frac{13k - 9}{7}, z = k$, where k is arbitrary

14. If $a = -1, b = 6$, equations will be consistent and have infinitely many solutions

If $a = -1, b \neq 6$, equations will be inconsistent

If $a \neq -1$, b is arbitrary, equations will be consistent and have a unique solution

15. $\lambda \neq \frac{7}{10}$ **16.** $k = 1$; $x = 2\lambda, y = -\lambda, z = \lambda$

17. (*i*) $a \neq -3$ (*ii*) $a = -3, b \neq \frac{1}{3}$ (*iii*) $a = -3, b = \frac{1}{3}$

18. $x = k, y = k, z = -2k, w = 0$ where k is arbitrary

19. Inconsistent for $\lambda \neq 7$. When $\lambda = 7$, the number of solutions is infinite $x = 3k_1 + k_2 + 3, y = 4k_1 - k_2 + 1, z = k_1, w = k_2$ where k_1, k_2 are arbitrary

20. $\lambda = -1, 1, 12$; $x = -\frac{1}{11}, y = -\frac{15}{11}$; $x = -5, y = 1$; $x = \frac{1}{2}, y = 1$

21. $\lambda = 0, 3$; $x : y : z = 1 : 1 : 1$; equations become identical.

2.8. VECTORS

Any ordered n-tuple of numbers is called an n-vector. By an ordered n-tuple, we mean a set consisting of n numbers in which the place of each number is fixed. If $x_1, x_2, \ldots\ldots, x_n$ be any n numbers then the ordered n-tuple $X = (x_1, x_2, \ldots, x_n)$ is called an n-vector. Thus the coordinates of a point in space can be represented by a 3-vector (x, y, z). Similarly $(1, 0, 2, -1)$ and $(2, 7, 5, -3)$ are 4-vectors. The n numbers $x_1, x_2, \ldots\ldots, x_n$ are called the components of the n-vector $X = (x_1, x_2, \ldots\ldots, x_n)$. A vector may be written either as a *row vector* or as a *column vector*. If A be a matrix of order $m \times n$, then each row of A will be an n-vector and each column of A will be an m-vector. A vector whose components are all zero is called a zero vector and is denoted by O. Thus $O = (0, 0, 0, \ldots\ldots, 0)$.

Let $X = (x_1, x_2, \ldots\ldots, x_n)$ and $Y = (y_1, y_2, \ldots\ldots, y_n)$ be two vectors.

Then X = Y if and only if their corresponding components are equal.

i.e., if $x_i = y_i$, for $i = 1, 2, \ldots., n$

$$X + Y = (x_1 + y_1, x_2 + y_2, \ldots\ldots, x_n + y_n)$$

If k be a scalar, then $kX = (kx_1, kx_2, \ldots\ldots, kx_n)$.

2.9. LINEAR DEPENDENCE AND LINEAR INDEPENDENCE OF VECTORS

A set of r n-vectors $X_1, X_2, \ldots\ldots, X_r$ is said to be *linearly dependent* if there exist r scalars (numbers) $k_1, k_2, \ldots\ldots, k_r$ not all zero, such that

$$k_1X_1 + k_2X_2 + \ldots. + k_rX_r = O$$

If $k_i \neq 0$, then $k_iX_i = -k_1X_1 - k_2X_2 \ldots - k_rX_r$

$$\Rightarrow \quad X_i = -\frac{k_1}{k_i}X_1 - \frac{k_2}{k_i}X_2 - \ldots\ldots - \frac{k_r}{k_i}X_r$$

or

$$X_i = \lambda_1X_1 + \lambda_2X_2 + \ldots + \lambda_rX_r$$

The vector X_i is called a linear combination of the remaining vectors $X_1, X_2, \ldots, X_r$.

If a set of vectors is linearly dependent, then at least one member of the set can be expressed as a linear combination of the remaining vectors.

A set of r n-vectors $X_1, X_2, \ldots.., X_r$ is said to be *linearly independent* if every relation of the type

$$k_1X_1 + k_2X_2 + \ldots\ldots + k_rX_r = O \text{ implies } k_1 = k_2 = \ldots\ldots = k_r = 0$$

To test the linear dependence of r given vectors, write them as row vectors. Add suitable multiples of one vector to the others so that the resulting $(r - 1)$ vectors have their first component zero. Choose any one of these $(r - 1)$ vectors and add its multiples to the others so that the resulting $(r - 2)$ vectors have their second component zero. In this way continue, reducing the successive components to zero. If the final reduction gives a vector all of whose components are zero, then the original vectors are linearly dependent. However, if the final reduction gives a vector all of whose components are not zero, then the original vectors are linearly independent.

Example. *Show that the vectors $x_1 = (1, 2, 4)$, $x_2 = (2, -1, 3)$, $x_3 = (0, 1, 2)$ and $x_4 = (-3, 7, 2)$ are linearly dependent and find the relation between them.* (U.K.T.U. 2012)

Sol. Adding suitable multiples of x_1 to x_2 and x_4 so that the first component reduces to zero, we have

$$x_2 - 2x_1 = (2, -1, 3) - (2, 4, 8) = (0, -5, -5)$$
$$x_4 + 3x_1 = (-3, 7, 2) + (3, 6, 12) = (0, 13, 14)$$

Also $\quad x_3 = (0, 1, 2).$

Adding suitable multiples of x_3 to the above vectors so that the second component reduces to zero, we have

$$(x_2 - 2x_1) + 5x_3 = (0, -5, -5) + (0, 5, 10) = (0, 0, 5)$$
$$(x_4 + 3x_1) - 13x_3 = (0, 13, 14) - (0, 13, 26) = (0, 0, -12)$$

To reduce the third component to zero, multiplying the above vectors by 12 and 5 respectively and adding, we have

$$12(x_2 - 2x_1 + 5x_3) + 5(x_4 + 3x_1 - 13x_3) = (0, 0, 60) + (0, 0, -60)$$
$$\Rightarrow \quad -9x_1 + 12x_2 - 5x_3 + 5x_4 = (0, 0, 0)$$
$$\Rightarrow \quad 9x_1 - 12x_2 + 5x_3 - 5x_4 = O \quad \text{...(1)}$$

Thus, there exist numbers $k_1 = 9$, $k_2 = -12$, $k_3 = 5$, $k_4 = -5$ which are not all zero such that

$$k_1x_1 + k_2x_2 + k_3x_3 + k_4x_4 = O$$

Hence the vectors x_1, x_2, x_3 and x_4 are linearly dependent. Also, (1) is the relation between them.

Second Method

Consider the relation $\quad k_1x_1 + k_2x_2 + k_3x_3 + k_4x_4 = O$

i.e., $\quad k_1(1, 2, 4) + k_2(2, -1, 3) + k_3(0, 1, 2) + k_4(-3, 7, 2) = O$

or $\quad (k_1, 2k_1, 4k_1) + (2k_2, -k_2, 3k_2) + (0, k_3, 2k_3) + (-3k_4, 7k_4, 2k_4) = O$

or $\quad (k_1 + 2k_2 - 3k_4, 2k_1 - k_2 + k_3 + 7k_4, 4k_1 + 3k_2 + 2k_3 + 2k_4) = (0, 0, 0)$

Equating corresponding components on both sides, we have

$$k_1 + 2k_2 - 3k_4 = 0 \quad \text{...(1)}$$
$$2k_1 - k_2 + k_3 + 7k_4 = 0 \quad \text{...(2)}$$
$$4k_1 + 3k_2 + 2k_3 + 2k_4 = 0 \quad \text{...(3)}$$

We have a homogeneous system of 3 linear equations in 4 unknowns. The coefficient matrix

$$A = \begin{bmatrix} 1 & 2 & 0 & -3 \\ 2 & -1 & 1 & 7 \\ 4 & 3 & 2 & 2 \end{bmatrix} \text{ is } 3 \times 4.$$

$\rho(A) \le \min\{3, 4\} = 3$. Since rank of A is less than the number of unknowns, the homogeneous system has infinitely many non-zero solution. Thus there exist scalars k_1, k_2, k_3, k_4, not all zero, such that

$$k_1x_1 + k_2x_2 + k_3x_3 + k_4x_4 = O \quad \text{...(4)}$$

$\therefore$ The vectors x_1, x_2, x_3, x_4 are linearly dependent.

To find the relation between the given vectors

In matrix notation, the system of equations (1), (2) and (3) can be written as

$$\begin{bmatrix} 1 & 2 & 0 & -3 \\ 2 & -1 & 1 & 7 \\ 4 & 3 & 2 & 2 \end{bmatrix} \begin{bmatrix} k_1 \\ k_2 \\ k_3 \\ k_4 \end{bmatrix} = \begin{bmatrix} 0 \\ 0 \\ 0 \end{bmatrix}$$

Operating $R_2 \to R_2 - 2R_1$ and $R_3 \to R_3 - 4R_1$

$$\sim \begin{bmatrix} 1 & 2 & 0 & -3 \\ 0 & -5 & 1 & 13 \\ 0 & -5 & 2 & 14 \end{bmatrix} \begin{bmatrix} k_1 \\ k_2 \\ k_3 \\ k_4 \end{bmatrix} = \begin{bmatrix} 0 \\ 0 \\ 0 \end{bmatrix}$$

Operating $R_3 \to R_3 - R_2$

$$\sim \begin{bmatrix} 1 & 2 & 0 & -3 \\ 0 & -5 & 1 & 13 \\ 0 & 0 & 1 & 1 \end{bmatrix} \begin{bmatrix} k_1 \\ k_2 \\ k_3 \\ k_4 \end{bmatrix} = \begin{bmatrix} 0 \\ 0 \\ 0 \end{bmatrix}$$

Operating $R_2 \to R_2 - R_3$

$$\sim \begin{bmatrix} 1 & 2 & 0 & -3 \\ 0 & -5 & 0 & 12 \\ 0 & 0 & 1 & 1 \end{bmatrix} \begin{bmatrix} k_1 \\ k_2 \\ k_3 \\ k_4 \end{bmatrix} = \begin{bmatrix} 0 \\ 0 \\ 0 \end{bmatrix}$$

$$\Rightarrow \quad k_1 + 2k_2 - 3k_4 = 0, \quad -5k_2 + 12\, k_4 = 0, \quad k_3 + k_4 = 0$$

$$\Rightarrow \quad k_1 + \frac{24}{5}\, k_4 - 3k_4 = 0, \quad k_2 = \frac{12}{5}\, k_4, \quad k_3 = -\, k_4$$

$$\Rightarrow \quad k_1 = -\frac{9}{5} k_4, \quad k_2 = \frac{12}{5}\, k_4, \quad k_3 = -\, k_4$$

For non-zero solutions, $k_4 \neq 0$

Putting these values in (4), we get

$$-\frac{9}{5}\, k_4\, x_1 + \frac{12}{5}\, k_4\, x_2 - k_4\, x_3 + k_4\, x_4 = O$$

$$\Rightarrow \quad 9x_1 - 12x_2 + 5x_3 - 5x_4 = O$$

Since k_1, k_2, k_3, k_4 are all non-zero, we can express each of the given vectors as a linear combination of the remaining three.

e.g., $$9x_1 = 12x_2 - 5x_3 + 5x_4$$

$$\therefore \quad x_1 = \frac{4}{3} x_2 - \frac{5}{9} x_3 + \frac{5}{9} x_4 .$$

Note 1. The elements in the four columns of A are the components of the four given vectors.

Note 2. Any super set of a linearly dependent set of vectors is linearly dependent.

Thus if X_1, X_2, X_3 are three linearly dependent vectors, then there exist scalars k_1, k_2, k_3 (not all zero) such that

$$k_1 X_1 + k_2\, X_2 + k_3\, X_3 = O$$

$$\Rightarrow \quad k_1 X_1 + k_2\, X_2 + k_3\, X_3 + 0X_4 + 0X_5 + \cdots = O$$

$\Rightarrow$ Vectors X_1, X_2, X_3, X_4, X_5, ... are linearly dependent.

2.10. LINEAR TRANSFORMATIONS

Let a point $P(x, y)$ in a plane transform to the point $P'(x', y')$ under reflection in the co-ordinate axes, or reflection in the line $y = x \tan \theta$ or rotation of OP through an angle θ about the origin or rotation of axes through an angle θ etc.

Then the co-ordinates of P′ can be expressed in terms of those of P by the linear relations of the form

$$\left.\begin{aligned} x' &= a_1x + b_1y \\ y' &= a_2x + b_2y \end{aligned}\right]$$

which in matrix notation is $\begin{bmatrix} x' \\ y' \end{bmatrix} = \begin{bmatrix} a_1 & b_1 \\ a_2 & b_2 \end{bmatrix}\begin{bmatrix} x \\ y \end{bmatrix}$ or $X' = AX$

Such transformations are called linear transformations in two dimensions.

Similarly, relations of the form $\left.\begin{aligned} x' &= a_1x + b_1y + c_1z \\ y' &= a_2x + b_2y + c_2z \\ z' &= a_3x + b_3y + c_3z \end{aligned}\right]$

which in matrix notation is $\begin{bmatrix} x' \\ y' \\ z' \end{bmatrix} = \begin{bmatrix} a_1 & b_1 & c_1 \\ a_2 & b_2 & c_2 \\ a_3 & b_3 & c_3 \end{bmatrix}\begin{bmatrix} x \\ y \\ z \end{bmatrix}$ or $X' = AX$ gives a linear transformation $(x, y, z) \longrightarrow (x', y', z')$ in three dimensions.

In general, the relation $Y = AX$, where $Y = \begin{bmatrix} y_1 \\ y_2 \\ \vdots \\ y_n \end{bmatrix}$, $A = \begin{bmatrix} a_{11} & a_{12} & \cdots & a_{1n} \\ a_{21} & a_{22} & \cdots & a_{2n} \\ \dots & \dots & \dots & \dots \\ a_{n1} & a_{n2} & \cdots & a_{nn} \end{bmatrix}$, $X = \begin{bmatrix} x_1 \\ x_2 \\ \vdots \\ x_n \end{bmatrix}$

defines a linear transformation which carries any vector X into another vector Y over the matrix A which is called the linear operator of the transformation.

This transformation is called linear because $Y_1 = AX_1$ and $Y_2 = AX_2$ implies $aY_1 + bY_2 = A(aX_1 + bX_2)$ for all values of a and b.

Thus, if $X = \begin{bmatrix} 2 \\ -3 \end{bmatrix}$ and $A = \begin{bmatrix} 1 & -1 \\ 2 & 3 \end{bmatrix}$, then $Y = \begin{bmatrix} y_1 \\ y_2 \end{bmatrix} = \begin{bmatrix} 1 & -1 \\ 2 & 3 \end{bmatrix}\begin{bmatrix} 2 \\ -3 \end{bmatrix} = \begin{bmatrix} 5 \\ -5 \end{bmatrix}$

so that $(2, -3) \to (5, -5)$ under the transformation defined by A.

If the transformation matrix A is non-singular, *i.e.*, if $|A| \neq 0$, then the linear transformation is called *non-singular* or *regular*.

If the transformation matrix A is singular, *i.e.*, if $|A| = 0$, then the linear transformation is also called *singular*.

For a non-singular transformation $Y = AX$, since A is non-singular, A^{-1} exists and we can write the inverse transformation, which carries the vector Y back into the vector X, as $X = A^{-1}Y$.

Note. If a transformation from $(x_1, x_2, \ldots\ldots, x_n)$ to $(y_1, y_2, \ldots\ldots, y_n)$ is given by $Y = AX$ and another transformation from $(y_1, y_2, \ldots\ldots, y_n)$ to $(z_1, z_2, \ldots\ldots, z_n)$ is given by $Z = BY$, then the transformation from $(x_1, x_2, \ldots\ldots, x_n)$ to $(z_1, z_2, \ldots\ldots, z_n)$ is given by $Z = BY = B(AX) = (BA)X$.

2.11. ORTHOGONAL TRANSFORMATION

The linear transformation $Y = AX$, where

$$Y = \begin{bmatrix} y_1 \\ y_2 \\ \vdots \\ y_n \end{bmatrix}, A = \begin{bmatrix} a_{11} & a_{12} & \cdots & a_{1n} \\ a_{21} & a_{22} & \cdots & a_{2n} \\ \dots & \dots & \dots & \dots \\ a_{n1} & a_{n2} & \cdots & a_{nn} \end{bmatrix}, X = \begin{bmatrix} x_1 \\ x_2 \\ \vdots \\ x_n \end{bmatrix}$$

is said to be *orthogonal* if it transforms $y_1^2 + y_2^2 + \cdots\cdots + y_n^2$ into $x_1^2 + x_2^2 + \cdots\cdots + x_n^2$.

The matrix A of this transformation is called an *orthogonal matrix.*

Now $$X'X = [x_1\ x_2\ \ldots\ldots\ x_n]\begin{bmatrix} x_1 \\ x_2 \\ \vdots \\ x_n \end{bmatrix} = x_1^2 + x_2^2 + \ldots\ldots + x_n^2$$

and similarly $$Y'Y = y_1^2 + y_2^2 + \ldots\ldots + y_n^2$$

$\therefore$ If Y = AX is an orthogonal transformation, then

$$X'X = x_1^2 + x_2^2 + \ldots\ldots + x_n^2 = y_1^2 + y_2^2 + \ldots\ldots + y_n^2$$
$$= Y'Y = (AX)'(AX) = (X'A')(AX) \qquad [\because\ (AB)' = B'A']$$
$$= X'(A'A)X$$

which holds only when $A'A = I$ or when $A'A = A^{-1}A$ $\qquad [\because\ A^{-1}A = I]$

or when $A' = A^{-1}$.

Hence *a square matrix A is said to be orthogonal if AA′ = A′A = I*

Also, *for an orthogonal matrix A, $A' = A^{-1}$.*

ILLUSTRATIVE EXAMPLES

Example 1. *Show that the transformation $y_1 = 2x_1 + x_2 + x_3$, $y_2 = x_1 + x_2 + 2x_3$, $y_3 = x_1 - 2x_3$ is regular. Write down the inverse transformation.*

Sol. In matrix notation, the given transformation is Y = AX,

where $$Y = \begin{bmatrix} y_1 \\ y_2 \\ y_3 \end{bmatrix},\ A = \begin{bmatrix} 2 & 1 & 1 \\ 1 & 1 & 2 \\ 1 & 0 & -2 \end{bmatrix},\ X = \begin{bmatrix} x_1 \\ x_2 \\ x_3 \end{bmatrix}$$

Since $$|A| = \begin{vmatrix} 2 & 1 & 1 \\ 1 & 1 & 2 \\ 1 & 0 & -2 \end{vmatrix} = -1 \neq 0,$$

the matrix A is non-singular and hence, the given transformation is non-singular or regular.

As usual, $$A^{-1} = \begin{bmatrix} 2 & -2 & -1 \\ -4 & 5 & 3 \\ 1 & -1 & -1 \end{bmatrix}$$

The inverse transformation is given by $X = A^{-1}Y$

$$\Rightarrow \qquad \begin{bmatrix} x_1 \\ x_2 \\ x_3 \end{bmatrix} = \begin{bmatrix} 2 & -2 & -1 \\ -4 & 5 & 3 \\ 1 & -1 & -1 \end{bmatrix}\begin{bmatrix} y_1 \\ y_2 \\ y_3 \end{bmatrix}$$

$$\Rightarrow \qquad x_1 = 2y_1 - 2y_2 - y_3,\ x_2 = -4y_1 + 5y_2 + 3y_3,\ x_3 = y_1 - y_2 - y_3.$$

Example 2. *Prove that the matrix* $\dfrac{1}{3}\begin{bmatrix} 1 & 2 & 2 \\ 2 & 1 & -2 \\ 2 & -2 & 1 \end{bmatrix}$ *is orthogonal.*

Sol. Denoting the given matrix by A, we have

$$A' = \frac{1}{3}\begin{bmatrix} 1 & 2 & 2 \\ 2 & 1 & -2 \\ 2 & -2 & 1 \end{bmatrix}$$

Now $$AA' = \frac{1}{3}\begin{bmatrix} 1 & 2 & 2 \\ 2 & 1 & -2 \\ 2 & -2 & 1 \end{bmatrix} \times \frac{1}{3}\begin{bmatrix} 1 & 2 & 2 \\ 2 & 1 & -2 \\ 2 & -2 & 1 \end{bmatrix}$$

$$= \frac{1}{9}\begin{bmatrix} 9 & 0 & 0 \\ 0 & 9 & 0 \\ 0 & 0 & 9 \end{bmatrix} = \begin{bmatrix} 1 & 0 & 0 \\ 0 & 1 & 0 \\ 0 & 0 & 1 \end{bmatrix} = I$$

Since AA′ = I, A is an orthogonal matrix.

Example 3. *If A is an orthogonal matrix, prove that* $|A| = \pm 1$.

Sol. A is an orthogonal matrix

$\Rightarrow \quad AA' = I \quad \Rightarrow \quad |AA'| = |I|$

$\Rightarrow \quad |A||A'| = 1 \quad \Rightarrow \quad |A||A| = 1 \qquad [\because \ |A'| = |A|]$

$\Rightarrow \quad |A|^2 = 1 \quad \therefore \quad |A| = \pm 1.$

Example 4. *Prove that the inverse of an orthogonal matrix is orthogonal and its transpose is also orthogonal.*

Sol. (*i*) A is orthogonal $\Rightarrow AA' = I$

$\Rightarrow \quad (AA')^{-1} = I^{-1} \quad \Rightarrow \quad (A')^{-1} A^{-1} = I \qquad [\because \ (AB)^{-1} = B^{-1}A^{-1}]$

$\Rightarrow \quad (A^{-1})' A^{-1} = I \qquad [\because \ (A')^{-1} = (A^{-1})']$

$\Rightarrow$ Product of A^{-1} and its transpose $(A^{-1})'$ is equal to I

$\Rightarrow$ A^{-1} is also orthogonal.

(*ii*) A is orthogonal $\Rightarrow AA' = I \Rightarrow (AA')' = I' \Rightarrow (A')' A' = I$

$\Rightarrow$ Product of A′ and its transpose (A′)′ is equal to I.

$\Rightarrow$ A′ is also orthogonal.

Example 5. *If* $A = \frac{1}{3}\begin{bmatrix} 1 & 2 & a \\ 2 & 1 & b \\ 2 & -2 & c \end{bmatrix}$ *is orthogonal, find a, b, c and* A^{-1}.

Sol. Given $A = \frac{1}{3}\begin{bmatrix} 1 & 2 & a \\ 2 & 1 & b \\ 2 & -2 & c \end{bmatrix} \Rightarrow A' = \frac{1}{3}\begin{bmatrix} 1 & 2 & 2 \\ 2 & 1 & -2 \\ a & b & c \end{bmatrix}$

Since A is orthogonal, AA′ = I

$$\Rightarrow \quad \frac{1}{3}\begin{bmatrix} 1 & 2 & a \\ 2 & 1 & b \\ 2 & -2 & c \end{bmatrix} \times \frac{1}{3}\begin{bmatrix} 1 & 2 & 2 \\ 2 & 1 & -2 \\ a & b & c \end{bmatrix} = \begin{bmatrix} 1 & 0 & 0 \\ 0 & 1 & 0 \\ 0 & 0 & 1 \end{bmatrix}$$

$$\Rightarrow \quad \begin{bmatrix} 5 + a^2 & 4 + ab & -2 + ac \\ 4 + ab & 5 + b^2 & 2 + bc \\ -2 + ac & 2 + bc & 8 + c^2 \end{bmatrix} = \begin{bmatrix} 9 & 0 & 0 \\ 0 & 9 & 0 \\ 0 & 0 & 9 \end{bmatrix}$$

Equating corresponding elements

$5 + a^2 = 9,\quad 4 + ab = 0,\quad -2 + ac = 0,$

$5 + b^2 = 9,\quad 2 + bc = 0,\quad 8 + c^2 = 9$

$\Rightarrow \quad a^2 = 4 \quad ...(1) \quad b^2 = 4 \quad ...(2) \quad c^2 = 1 \quad ...(3)$

$ab = -4 \quad ...(4) \quad ac = 2 \quad ...(5) \quad bc = -2 \quad ...(6)$

From (1), $a = 2$ or -2

When $a = 2$, from (4) $b = -2$, from (5) $c = 1$

These values satisfy all other equations.

$$\therefore \quad A = \frac{1}{3}\begin{bmatrix} 1 & 2 & 2 \\ 2 & 1 & -2 \\ 2 & -2 & 1 \end{bmatrix} \text{ and } A^{-1} = A' = \frac{1}{3}\begin{bmatrix} 1 & 2 & 2 \\ 2 & 1 & -2 \\ 2 & -2 & 1 \end{bmatrix}$$

When $a = -2$, from (4) $b = 2$, from (5) $c = -1$

These values satisfy all other equations.

$$\therefore \quad A = \frac{1}{3}\begin{bmatrix} 1 & 2 & -2 \\ 2 & 1 & 2 \\ 2 & -2 & -1 \end{bmatrix} \text{ and } A^{-1} = A' = \frac{1}{3}\begin{bmatrix} 1 & 2 & 2 \\ 2 & 1 & -2 \\ -2 & 2 & -1 \end{bmatrix}.$$

EXERCISE 2.3

1. (*a*) Are the following vectors linearly dependent? If so, find a relation between them.

(*i*) $x_1 = (1, 2, 1), x_2 = (2, 1, 4), x_3 = (4, 5, 6)$

(*ii*) $x_1 = (2, -1, 4), x_2 = (0, 1, 2), x_3 = (6, -1, 16)$

(*iii*) $x_1 = (3, 2, 7), x_2 = (2, 4, 1), x_3 = (1, -2, 6)$

(*iv*) $x_1 = (2, -1, 3, 2), x_2 = (1, 3, 4, 2), x_3 = (3, -5, 2, 2)$

(*v*) $x_1 = (2, 3, 1, -1), x_2 = (2, 3, 1, -2), x_3 = (4, 6, 2, 1)$

(*vi*) $x_1 = (1, 1, 1, 3), x_2 = (1, 2, 3, 4), x_3 = (2, 3, 4, 9)$

(*vii*) $X_1 = (1, 1, -1, 1), X_2 = (1, -1, 2, -1), X_3 = (3, 1, 0, 1)$.

(*b*) Verify that the following set of vectors in R^3 is linearly dependent : (1, 0, 1), (1, 1, 1), (1, 1, 2) and (1, 2, 1).

Also find the number of linearly independent vectors. (*M.D.U. May 2008*)

(*c*) Show that the row vectors of the matrix $\begin{bmatrix} 1 & 2 & -2 \\ -1 & 3 & 0 \\ 0 & -2 & 1 \end{bmatrix}$ are linearly independent.

(*G.B.T.U. 2010*)

(*d*) Are the following vectors linearly dependent? If so, find the relation between them.

$x_1 = (1, 2, 1), x_2 = (2, 1, 4), x_3 = (4, 5, 6), x_4 = (1, 8, -3)$ (*M.D.U. Dec. 2013*)

[**Hint.** Apply the result of Note 2 with 2.9 to question 1(*a*) (*i*)]

2. Show that the transformation $y_1 = x_1 - x_2 + x_3$, $y_2 = 3x_1 - x_2 + 2x_3$, $y_3 = 2x_1 - 2x_2 + 3x_3$ is non-singular. Find the inverse transformation.

3. Represent each of the transformations $x_1 = 3y_1 + 2y_2$, $x_2 = -y_1 + 4y_2$, $y_1 = z_1 + 2z_2$ and $y_2 = 3z_1$ by the use of matrices and find the composite transformation which expresses x_1, x_2 in terms of z_1, z_2.

4. A transformation from the variables x_1, x_2, x_3 to y_1, y_2, y_3 is given by Y = AX, and another transformation from y_1, y_2, y_3 to z_1, z_2, z_3 is given by Z = BY, where $A = \begin{bmatrix} 2 & 1 & 0 \\ 0 & 1 & -2 \\ -1 & 2 & 1 \end{bmatrix}, B = \begin{bmatrix} 1 & 1 & 1 \\ 1 & 2 & 3 \\ 1 & 3 & 5 \end{bmatrix}$.

Obtain the transformation from x_1, x_2, x_3 to z_1, z_2, z_3.

5. Which of the following matrices are orthogonal ?

(*i*) $\frac{1}{9}\begin{bmatrix} -8 & 4 & 1 \\ 1 & 4 & -8 \\ 4 & 7 & 4 \end{bmatrix}$ (*ii*) $\begin{bmatrix} 2 & -3 & 1 \\ 4 & 3 & 1 \\ -3 & 1 & 9 \end{bmatrix}$.

6. (*a*) Prove that the matrix $\begin{bmatrix} -\frac{2}{3} & \frac{1}{3} & \frac{2}{3} \\ \frac{2}{3} & \frac{2}{3} & \frac{1}{3} \\ \frac{1}{3} & -\frac{2}{3} & \frac{2}{3} \end{bmatrix}$ is orthogonal.

(*b*) Show that the transformation

$$y_1 = \frac{1}{3}x_1 + \frac{2}{3}x_2 + \frac{2}{3}x_3$$

$$y_2 = \frac{2}{3}x_1 + \frac{1}{3}x_2 - \frac{2}{3}x_3$$

$$y_3 = \frac{2}{3}x_1 - \frac{2}{3}x_2 + \frac{1}{3}x_3$$

is orthogonal.

7. (*i*) Verify that the matrix $\begin{bmatrix} \cos\theta & 0 & \sin\theta \\ 0 & 1 & 0 \\ -\sin\theta & 0 & \cos\theta \end{bmatrix}$ is orthogonal

(*ii*) Prove that $\begin{bmatrix} l & m & n & 0 \\ 0 & 0 & 0 & -1 \\ n & l & -m & 0 \\ -m & n & -l & 0 \end{bmatrix}$ is orthogonal when $l = \frac{2}{7}, m = \frac{3}{7}, n = \frac{6}{7}$.

(*iii*) Show that $A = \begin{bmatrix} \cos\phi & 0 & \sin\phi \\ \sin\theta\sin\phi & \cos\theta & -\sin\theta\cos\phi \\ -\cos\theta\sin\phi & \sin\theta & \cos\theta\cos\phi \end{bmatrix}$ is an orthogonal matrix.

8. If A and B are orthogonal matrices, prove that AB is also orthogonal.

9. Find the values of *a*, *b*, *c* if $A = \begin{bmatrix} 0 & 2b & c \\ a & b & -c \\ a & -b & c \end{bmatrix}$ is orthogonal.

Answers

1. (*a*) (*i*) Yes ; $x_3 = 2x_1 + x_2$ (*ii*) Yes ; $x_3 = 3x_1 + 2x_2$

(*iii*) Yes ; $x_1 = x_2 + x_3$ (*iv*) Yes ; $2x_1 - x_2 - x_3 = O$ (*v*) Yes ; $5x_1 - 3x_2 - x_3 = O$ (*vi*) No

(*vii*) Yes ; $2X_1 + X_2 - X_3 = O$ (*b*) 3 (*d*) Yes; $2x_1 + x_2 - x_3 + 0x_4 = O$

2. $x_1 = \frac{1}{2}(y_1 + y_2 - y_3), x_2 = \frac{1}{2}(-5y_1 + y_2 + y_3), x_3 = -2y_1 + y_3$

3. $\begin{bmatrix} x_1 \\ x_2 \end{bmatrix} = \begin{bmatrix} 9 & 6 \\ 11 & -2 \end{bmatrix} \begin{bmatrix} z_1 \\ z_2 \end{bmatrix}$

4. Z = (BA)X, where $BA = \begin{bmatrix} 1 & 4 & -1 \\ -1 & 9 & -1 \\ -3 & 14 & -1 \end{bmatrix}$

5. (*i*) Orthogonal (*ii*) Not orthogonal

9. $a = \pm \frac{1}{\sqrt{2}}, b = \pm \frac{1}{\sqrt{6}}, c = \pm \frac{1}{\sqrt{3}}$.

2.12. CHARACTERISTIC EQUATION

If A is a square matrix of order n, we can form the matrix $A - \lambda I$, where λ is a scalar and I is the unit matrix of order n. The determinant of this matrix equated to zero, *i.e.*,

$$|A - \lambda I| = \begin{vmatrix} a_{11} - \lambda & a_{12} & \cdots a_{1n} \\ a_{21} & a_{22} - \lambda & \cdots a_{2n} \\ \cdots & \cdots & \cdots \\ a_{n1} & a_{n2} & \cdots a_{nn} - \lambda \end{vmatrix} = 0$$ is called the *characteristic equation* of A.

On expanding the determinant, the characteristic equation can be written as a polynomial equation of degree n in λ of the form $(-1)^n \lambda^n + k_1\lambda^{n-1} + k_2\lambda^{n-2} + \ldots\ldots + k_n = 0$.

The roots of this equation are called the *characteristic roots* or *latent roots* or *eigen values* of A. The set of all the characteristic roots of A is called *spectrum* of A.

2.13. EIGEN VECTORS

Consider the linear transformation Y = AX ...(1)

which transforms the column vector X into the column vector Y. In practice, we are often required to find those vectors X which transform into scalar multiples of themselves.

Let X be such a vector which transforms into λX(λ being a non-zero scalar) by the transformation (1).

Then $Y = \lambda X$ ...(2)

From (1) and (2), $AX = \lambda X \Rightarrow AX - \lambda IX = O \Rightarrow (A - \lambda I)X = O$ ...(3)

This matrix equation gives n homogeneous linear equations

$$\left.\begin{array}{llll} (a_{11} - \lambda)x_1 & + a_{12}x_2 & \cdots\cdots + & \cdots\cdots + a_{1n} \; x_n = 0 \\ a_{21}x_1 & + (a_{22} - \lambda)x_2 & + & \cdots\cdots + a_{2n} \; x_n = 0 \\ \cdots\cdots & \cdots\cdots & \cdots\cdots & \cdots\cdots \\ a_{n1}x_1 & + a_{n2}x_2 + & \cdots\cdots + & (a_{nn} - \lambda) \; x_n = 0 \end{array}\right] \quad \ldots(4)$$

These equations will have a non-trivial solution only if the co-efficient matrix $A - \lambda I$ is singular

i.e., if $|A - \lambda I| = 0$ ...(5)

This is the characteristic equation of the matrix A and has n roots which are the eigen values of A. Corresponding to each root of (5), the homogeneous system (3) has a non-zero solution

$$X = \begin{bmatrix} x_1 \\ x_2 \\ \vdots \\ x_n \end{bmatrix}$$ which is called an *eigen vector* or *latent vector*.

Note. If X is a solution of (3), then so is kX, where k is an arbitrary constant. Thus, the eigen vector corresponding to an eigen value is not unique.

ILLUSTRATIVE EXAMPLES

Example 1. *Find the eigen values and eigen vectors of the matrix* $A = \begin{bmatrix} 1 & -2 \\ -5 & 4 \end{bmatrix}$.

Sol. The characteristic equation of the given matrix is

$$| A - \lambda I | = 0 \quad \text{or} \quad \begin{vmatrix} 1-\lambda & -2 \\ -5 & 4-\lambda \end{vmatrix} = 0$$

$$\text{or} \quad (1 - \lambda)(4 - \lambda) - 10 = 0 \quad \text{or} \quad \lambda^2 - 5\lambda - 6 = 0$$

$$\text{or} \quad (\lambda - 6)(\lambda + 1) = 0 \quad \therefore \quad \lambda = 6, -1.$$

Thus, the eigen values of A are 6, – 1.

Corresponding to $\lambda = 6$, the eigen vectors are given by $(A - 6I)X = O$

$$\text{or} \quad \begin{bmatrix} 1-6 & -2 \\ -5 & 4-6 \end{bmatrix}\begin{bmatrix} x_1 \\ x_2 \end{bmatrix} = O \quad \text{or} \quad \begin{bmatrix} -5 & -2 \\ -5 & -2 \end{bmatrix}\begin{bmatrix} x_1 \\ x_2 \end{bmatrix} = O$$

Operating $R_2 \to R_2 - R_1 \sim \begin{bmatrix} -5 & -2 \\ 0 & 0 \end{bmatrix}\begin{bmatrix} x_1 \\ x_2 \end{bmatrix} = \begin{bmatrix} 0 \\ 0 \end{bmatrix}$

We get only one independent equation $-5x_1 - 2x_2 = 0$, since rank of co-efficient matrix is 1.

$\therefore \quad \dfrac{x_1}{2} = \dfrac{x_2}{-5}$ giving the eigen vector (2, – 5).

Corresponding to $\lambda = -1$, the eigen vectors are given by $\begin{bmatrix} 2 & -2 \\ -5 & 5 \end{bmatrix}\begin{bmatrix} x_1 \\ x_2 \end{bmatrix} = O$

Operating $R_1 \to \dfrac{1}{2} R_1$ and $R_2 \to \dfrac{1}{5} R_2$

$$\sim \begin{bmatrix} 1 & -1 \\ -1 & 1 \end{bmatrix}\begin{bmatrix} x_1 \\ x_2 \end{bmatrix} = O$$

Operating $R_2 \to R_2 + R_1$

$$\sim \begin{bmatrix} 1 & -1 \\ 0 & 0 \end{bmatrix}\begin{bmatrix} x_1 \\ x_2 \end{bmatrix} = \begin{bmatrix} 0 \\ 0 \end{bmatrix}.$$

We get only one independent equation $2x_1 - 2x_2 = 0$, since rank of co-efficient matrix is 1.

$\therefore \quad x_1 = x_2$ giving the eigen vector (1, 1).

Example 2. *Find the eigen values and eigen vectors of the matrix* $A = \begin{bmatrix} -2 & 2 & -3 \\ 2 & 1 & -6 \\ -1 & -2 & 0 \end{bmatrix}$.

(K.U.K. 2013; M.D.U. 2012; U.K.T.U. 2011)

Sol. The characteristic equation of the given matrix is $|A - \lambda I| = 0$

or $$\begin{vmatrix} -2-\lambda & 2 & -3 \\ 2 & 1-\lambda & -6 \\ -1 & -2 & -\lambda \end{vmatrix} = 0$$

or $$(-2-\lambda)[-\lambda(1-\lambda) - 12] - 2[-2\lambda - 6] - 3[-4 + 1(1-\lambda)] = 0$$

or $$\lambda^3 + \lambda^2 - 21\lambda - 45 = 0$$

By trial, $\lambda = -3$ satisfies it.

$\therefore \quad (\lambda + 3)(\lambda^2 - 2\lambda - 15) = 0 \Rightarrow (\lambda + 3)(\lambda + 3)(\lambda - 5) = 0 \Rightarrow \lambda = -3, -3, 5$

Thus, the eigen values of A are $-3, -3, 5$.

Corresponding to $\lambda = -3$, the eigen vectors are given by

$$(A + 3I)\,X = O \quad \text{or} \quad \begin{bmatrix} 1 & 2 & -3 \\ 2 & 4 & -6 \\ -1 & -2 & 3 \end{bmatrix}\begin{bmatrix} x_1 \\ x_2 \\ x_3 \end{bmatrix} = O$$

Operating $R_2 \to R_2 - 2R_1$ and $R_3 \to R_3 + R_1$

$$\sim \begin{bmatrix} 1 & 2 & -3 \\ 0 & 0 & 0 \\ 0 & 0 & 0 \end{bmatrix}\begin{bmatrix} x_1 \\ x_2 \\ x_3 \end{bmatrix} = \begin{bmatrix} 0 \\ 0 \\ 0 \end{bmatrix}$$

We get only one independent equation $x_1 + 2x_2 - 3x_3 = 0$, since rank of co-efficient matrix is 1.

Choosing $x_2 = 0$, we have $x_1 - 3x_3 = 0$

$\therefore \quad \dfrac{x_1}{3} = \dfrac{x_2}{0} = \dfrac{x_3}{1}$ giving the eigen vector (3, 0, 1)

Choosing $x_3 = 0$, we have $x_1 + 2x_2 = 0$

$\therefore \quad \dfrac{x_1}{2} = \dfrac{x_2}{-1} = \dfrac{x_3}{0}$ giving the eigen vector (2, – 1, 0)

Any other eigen vector corresponding to $\lambda = -3$ will be a linear combination of these two.

Corresponding to $\lambda = 5$, the eigen vectors are given by $\begin{bmatrix} -7 & 2 & -3 \\ 2 & -4 & -6 \\ -1 & -2 & -5 \end{bmatrix}\begin{bmatrix} x_1 \\ x_2 \\ x_3 \end{bmatrix} = O$

Operating $R_1 \to R_1 - 7R_3$ and $R_2 \to R_2 + 2R_3$

$$\sim \begin{bmatrix} 0 & 16 & 32 \\ 0 & -8 & -16 \\ -1 & -2 & -5 \end{bmatrix}\begin{bmatrix} x_1 \\ x_2 \\ x_3 \end{bmatrix} = O$$

Operating $R_1 \to \dfrac{1}{16} R_1$ and $R_2 \to -\dfrac{1}{8} R_2$

$$\sim \begin{bmatrix} 0 & 1 & 2 \\ 0 & 1 & 2 \\ -1 & -2 & -5 \end{bmatrix}\begin{bmatrix} x_1 \\ x_2 \\ x_3 \end{bmatrix} = O$$

Operating $R_1 \to R_1 - R_2$, $R_3 \to R_3 + 2R_2$

$$\sim \begin{bmatrix} 0 & 0 & 0 \\ 0 & 1 & 2 \\ -1 & 0 & -1 \end{bmatrix} \begin{bmatrix} x_1 \\ x_2 \\ x_3 \end{bmatrix} = O$$

Operating $R_1 \leftrightarrow R_3$

$$\sim \begin{bmatrix} -1 & 0 & -1 \\ 0 & 1 & 2 \\ 0 & 0 & 0 \end{bmatrix} \begin{bmatrix} x_1 \\ x_2 \\ x_3 \end{bmatrix} = \begin{bmatrix} 0 \\ 0 \\ 0 \end{bmatrix}$$

We get only two independent equations $-x_1 - x_3 = 0$ and $x_2 + 2x_3 = 0$, since rank of co-efficient matrix is 2.

$\Rightarrow \qquad x_1 = -x_3$ and $x_2 = -2x_3$.

$\Rightarrow \qquad \dfrac{x_1}{1} = \dfrac{x_2}{2} = \dfrac{x_3}{-1}$ giving the eigen vector (1, 2, – 1).

2.14. PROPERTIES OF EIGEN VALUES

(a) The eigen values of a square matrix A and its transpose A′ are the same.

Let
$$A = \begin{bmatrix} a_{11} & a_{12} & \cdots\cdots & a_{1n} \\ a_{21} & a_{22} & \cdots\cdots & a_{2n} \\ \cdots & \cdots & \cdots & \cdots \\ a_{n1} & a_{n2} & \cdots\cdots & a_{nn} \end{bmatrix}$$

then
$$A' = \begin{bmatrix} a_{11} & a_{21} & \cdots\cdots & a_{n1} \\ a_{12} & a_{22} & \cdots\cdots & a_{n2} \\ \cdots & \cdots & \cdots & \cdots \\ a_{1n} & a_{2n} & \cdots\cdots & a_{nn} \end{bmatrix}$$

The characteristic equation of A is given by

$$|A - \lambda I| = \begin{vmatrix} a_{11}-\lambda & a_{12} & \cdots\cdots & a_{1n} \\ a_{21} & a_{22}-\lambda & \cdots\cdots & a_{2n} \\ \cdots & \cdots & \cdots & \cdots \\ a_{n1} & a_{n2} & \cdots\cdots & a_{nn}-\lambda \end{vmatrix} = 0 \qquad \text{...(1)}$$

Also, the characteristic equation of A′ is given by

$$|A' - \lambda I| = \begin{vmatrix} a_{11}-\lambda & a_{21} & \cdots\cdots & a_{n1} \\ a_{12} & a_{22}-\lambda & \cdots\cdots & a_{n2} \\ \cdots & \cdots & \cdots & \cdots \\ a_{1n} & a_{2n} & \cdots\cdots & a_{nn}-\lambda \end{vmatrix} = 0 \qquad \text{...(2)}$$

Since the value of a determinant remains unchanged when its rows and columns are interchanged, therefore, the characteristic equations (1) and (2) are same. Hence the eigen values of A and A′ are the same.

(b) The sum of the eigen values of a matrix is the sum of the elements on the principal diagonal. (M.D.U. May 2008, Dec. 2011)

We prove the property for a matrix of order 3. However, the result is true for a matrix of any order.

Let $$A = \begin{bmatrix} a_{11} & a_{12} & a_{13} \\ a_{21} & a_{22} & a_{23} \\ a_{31} & a_{32} & a_{33} \end{bmatrix}$$

so that $$|A - \lambda I| = \begin{vmatrix} a_{11}-\lambda & a_{12} & a_{13} \\ a_{21} & a_{22}-\lambda & a_{23} \\ a_{31} & a_{32} & a_{33}-\lambda \end{vmatrix}$$

$$= -\lambda^3 + \lambda^2(a_{11} + a_{22} + a_{33}) - \ldots\ldots \qquad \ldots(1)$$

If $\lambda_1, \lambda_2, \lambda_3$ be the eigen values of A, then

$$|A - \lambda I| = (-1)^3 (\lambda - \lambda_1)(\lambda - \lambda_2)(\lambda - \lambda_3)$$

$$= -\lambda^3 + \lambda^2 (\lambda_1 + \lambda_2 + \lambda_3) - \ldots\ldots \qquad \ldots(2)$$

From (1) and (2), we have

$$-\lambda^3 + \lambda^2(a_{11} + a_{22} + a_{33}) \ldots\ldots = -\lambda^3 + \lambda^2(\lambda_1 + \lambda_2 + \lambda_3) \ldots\ldots$$

Comparing co-efficients of λ^2, we get

$$\lambda_1 + \lambda_2 + \lambda_3 = a_{11} + a_{22} + a_{33}$$

= The sum of the elements on the principal diagonal.

(c) *The product of the eigen values of a matrix A is equal to* $|A|$. (M.D.U. Dec. 2011)

If $\lambda_1, \lambda_2, \ldots\ldots, \lambda_n$ be the eigen values of a square matrix A of order n, then

$$|A - \lambda I| = (-1)^n (\lambda - \lambda_1)(\lambda - \lambda_2) \ldots\ldots (\lambda - \lambda_n)$$

Putting $\lambda = 0$, we get

$$|A| = (-1)^n \cdot (-1)^n \lambda_1\lambda_2 \ldots\ldots \lambda_n$$

$$\Rightarrow \qquad \lambda_1\lambda_2 \ldots\ldots \lambda_n = |A| \qquad [\because \ (-1)^{2n} = 1]$$

(d) *If* λ *is an eigen value of a non-singular matrix A, then* $\dfrac{1}{\lambda}$ *is an eigen value of* A^{-1}.

λ is an eigen value of A (Kerala 2009)

$\Rightarrow$ There exists a non-zero vector X such that $AX = \lambda X$

Pre-multiplying both sides by A^{-1}, we get

$$A^{-1}(AX) = A^{-1}(\lambda X)$$

$$\Rightarrow \qquad (A^{-1}A)X = \lambda(A^{-1}X) \qquad \Rightarrow \quad X = \lambda(A^{-1}X)$$

$$\Rightarrow \qquad \frac{1}{\lambda} X = A^{-1}X \ \Rightarrow \ A^{-1}X = \frac{1}{\lambda} X$$

$\Rightarrow$ $\dfrac{1}{\lambda}$ is an eigen value of A^{-1}.

(e) *If* λ *is an eigen value of an orthogonal matrix, then* $\dfrac{1}{\lambda}$ *is also its eigen value.*

(Kerala 2010)

Let λ be an eigen value of an orthogonal matrix A, then $\frac{1}{\lambda}$ is an eigen value of A^{-1}.

[Part (*d*) above]

But $$A^{-1} = A' \quad (\because \text{ A is an orthogonal matrix})$$

$\therefore$ $\frac{1}{\lambda}$ is an eigen value of A′.

But the matrices A and A′ have the same eigen values. [Part (*a*) above]

$\therefore$ $\frac{1}{\lambda}$ is also an eigen value of A.

(*f*) *If $\lambda_1, \lambda_2, \ldots\ldots, \lambda_n$ are the eigen values of a matrix A, then A^m has the eigen values $\lambda_1^m, \lambda_2^m, \ldots\ldots, \lambda_n^m$ (m being a positive integer).* (Kerala 2012)

Let λ be an eigen value of a matrix A, then there exists a non-zero vector X such that

$$AX = \lambda X \quad \ldots(1)$$

Now, $$A(AX) = A(\lambda X) \Rightarrow A^2X = \lambda(AX)$$

$$\Rightarrow \quad A^2 X = \lambda(\lambda X) \quad \text{[by (1)]}$$

$$\Rightarrow \quad A^2X = \lambda^2X$$

Similarly, $A^3X = \lambda^3X$. In general $A^mX = \lambda^mX$

$\Rightarrow$ λ^m is an eigen value of A^m.

Since λ is an eigen value of A $\Rightarrow$ λ^m is an eigen value of A^m.

$\therefore$ If $\lambda_1, \lambda_2, \ldots\ldots, \lambda_n$ are the eigen values of A, then $\lambda_1^m, \lambda_2^m, \ldots\ldots, \lambda_n^m$ *are the eigen values of* A^m.

(*g*) *The eigen values of an idempotent matrix are either zero or unity.*

Let A be an idempotent matrix, then $A^2 = A$.

Let λ be an eigen value of A, then there exists a non-zero vector X such that

$$AX = \lambda X \quad \ldots(1)$$

$$\therefore \quad A(AX) = A(\lambda X) \quad \text{or} \quad A^2X = \lambda(AX)$$

or $$AX = \lambda(\lambda X) \quad [\because \; A^2 = A \text{ and } AX = \lambda X]$$

or $$AX = \lambda^2 X \quad \ldots(2)$$

From (1) and (2), we have $\lambda^2X = \lambda X$ or $(\lambda^2 - \lambda)X = O$

or $$\lambda^2 - \lambda = 0 \quad (\because \; X \neq O)$$

or $$\lambda(\lambda - 1) = 0 \quad \therefore \; \lambda = 0 \text{ or } 1.$$

Example 3. *If λ be an eigen value of a non-singular matrix A, show that $\frac{|A|}{\lambda}$ is an eigen value of the matrix adj. A.*

Sol. λ is an eigen value of A.

$\Rightarrow$ There exists a non-zero matrix X such that $AX = \lambda X$

$$\Rightarrow \quad (\text{adj A})(AX) = (\text{adj A})(\lambda X)$$

$\Rightarrow$ $((\text{adj } A)A)\, X = \lambda\, (\text{adj } A)\, X$

$\Rightarrow$ $|A|\, IX = \lambda\, (\text{adj } A)\, X$ $[\because\ (\text{adj } A)\, A = |A|\, I]$

$\Rightarrow$ $|A|\, X = \lambda\, (\text{adj } A)\, X \Rightarrow \frac{|A|}{\lambda} X = (\text{adj } A)\, X$

$\Rightarrow$ $(\text{adj } A)\, X = \frac{|A|}{\lambda} X \Rightarrow \frac{|A|}{\lambda}$ is an eigen value of adj A.

2.15. CAYLEY HAMILTON THEOREM

Every square matrix satisfies its characteristic equation.

i.e., if the characteristic equation of the nth order square matrix A is

$$|A - \lambda I| = (-1)^n \lambda^n + k_1 \lambda^{n-1} + k_2 \lambda^{n-2} + \ldots.. + k_n = 0 \qquad \ldots(1)$$

then $$(-1)^n A^n + k_1 A^{n-1} + k_2 A^{n-2} + \ldots\ldots + k_n I = O$$

Let $$P = \text{adj } (A - \lambda I)$$

Since the elements of $A - \lambda I$ are at most of first degree in λ, the elements of $P = \text{adj } (A - \lambda I)$ are polynomials in λ of degree $(n - 1)$ or less. We can, therefore, split up P into a number of matrices each containing the same power of λ and write

$$P = P_1 \lambda^{n-1} + P_2 \lambda^{n-2} + P_{n-1} \lambda + P_n \qquad \ldots(2)$$

Also, we know that if M is a square matrix, then $M\, (\text{adj } M) = |M| \times I$

$\therefore$ $$(A - \lambda I)P = |A - \lambda I| \times I$$

By (1) and (2), we have

$$(A - \lambda I)(P_1 \lambda^{n-1} + P_2 \lambda^{n-2} + \ldots.. + P_{n-1} \lambda + P_n) = [(-1)^n \lambda^n + k_1 \lambda^{n-1} + k_2 \lambda^{n-2} + \ldots.. + k_{n-1} \lambda + k_n] I$$

Equating co-efficients of like powers of λ on both sides, we have

$$-P_1 = (-1)^n I \qquad [\because\ IP_1 = P_1]$$

$$AP_1 - P_2 = k_1 I$$

$$AP_2 - P_3 = k_2 I$$

....

$$AP_{n-1} - P_n = k_{n-1} I$$

$$AP_n = k_n I$$

Pre-multiplying these equations by $A^n, A^{n-1}, A^{n-2}, \ldots\ldots, A, I$ respectively and adding, we get

$$O = (-1)^n A^n + k_1 A^{n-1} + k_2 A^{n-2} + \ldots.. + k_{n-1} A + k_n I$$

[terms on the L.H.S. cancel in pairs]

or $$(-1)^n A^n + k_1 A^{n-1} + k_2 A^{n-2} + \ldots.. + k_{n-1} A + k_n I = O \qquad \ldots(3)$$

which proves the theorem.

Note 1. Multiplying (3) by A^{-1}, we have $(-1)^n A^{n-1} + k_1 A^{n-2} + \ldots\ldots + k_{n-1} I + k_n A^{-1} = O$

$\Rightarrow$ $$A^{-1} = -\frac{1}{k_n} [(-1)^n A^{n-1} + k_1 A^{n-2} + \ldots\ldots + k_{n-1} I]$$

Thus Cayley-Hamilton theorem gives another method for computing the inverse of a matrix. Since this method expresses the inverse of a matrix of order n in terms of $(n-1)$ powers of A, it is most suitable for computing inverses of large matrices.

Note 2. If m be a positive integer such that $m > n$, then multiplying (3) by A^{m-n}, we get

$$(-1)^n A^m + k_1 A^{m-1} + k_2 A^{m-2} + \ldots\ldots + k_{n-1} A^{m-n+1} + k_n A^{m-n} = O$$

showing that any positive integral power A^m $(m > n)$ of A is linearly expressible in terms of those of lower degree.

Example 4. *Verify Cayley-Hamilton theorem for the matrix* $A = \begin{bmatrix} 2 & -1 & 1 \\ -1 & 2 & -1 \\ 1 & -1 & 2 \end{bmatrix}$.

Hence compute A^{-1}. (K.U.K. 2011; M.D.U. Dec. 2012; G.B.T.U. 2012; U.K.T.U. 2011)

Sol. The characteristic equation of A is

$$|A - \lambda I| = 0 \quad i.e., \quad \begin{vmatrix} 2-\lambda & -1 & 1 \\ -1 & 2-\lambda & 1 \\ 1 & -1 & 2-\lambda \end{vmatrix} = 0$$

or $$\lambda^3 - 6\lambda^2 + 9\lambda - 4 = 0 \quad \text{(on simplification)}$$

To verify Cayley-Hamilton theorem, we have to show that

$$A^3 - 6A^2 + 9A - 4I = O \qquad \ldots(1)$$

Now $$A^2 = \begin{bmatrix} 2 & -1 & 1 \\ -1 & 2 & -1 \\ 1 & -1 & 2 \end{bmatrix} \times \begin{bmatrix} 2 & -1 & 1 \\ -1 & 2 & -1 \\ 1 & -1 & 2 \end{bmatrix} = \begin{bmatrix} 6 & -5 & 5 \\ -5 & 6 & -5 \\ 5 & -5 & 6 \end{bmatrix}$$

$$A^3 = A^2 \times A = \begin{bmatrix} 6 & -5 & 5 \\ -5 & 6 & -5 \\ 5 & -5 & 6 \end{bmatrix} \begin{bmatrix} 2 & -1 & 1 \\ -1 & 2 & -1 \\ 1 & -1 & 2 \end{bmatrix} = \begin{bmatrix} 22 & -21 & 21 \\ -21 & 22 & -21 \\ 21 & -21 & 22 \end{bmatrix}$$

$$\therefore \quad A^3 - 6A^2 + 9A - 4I = \begin{bmatrix} 22 & -21 & 21 \\ -21 & 22 & -21 \\ 21 & -21 & 22 \end{bmatrix} - 6 \begin{bmatrix} 6 & -5 & 5 \\ -5 & 6 & -5 \\ 5 & -5 & 6 \end{bmatrix}$$

$$+ 9 \begin{bmatrix} 2 & -1 & 1 \\ -1 & 2 & -1 \\ 1 & -1 & 2 \end{bmatrix} - 4 \begin{bmatrix} 1 & 0 & 0 \\ 0 & 1 & 0 \\ 0 & 0 & 1 \end{bmatrix}$$

$$= \begin{bmatrix} 0 & 0 & 0 \\ 0 & 0 & 0 \\ 0 & 0 & 0 \end{bmatrix} = O$$

This verifies Cayley-Hamilton theorem.

Now, multiplying both sides of (1) by A^{-1}, we have

$$A^2 - 6A + 9I - 4A^{-1} = O \quad \Rightarrow \quad 4A^{-1} = A^2 - 6A + 9I$$

$$\Rightarrow \qquad 4A^{-1} = \begin{bmatrix} 6 & -5 & 5 \\ -5 & 6 & -5 \\ 5 & -5 & 6 \end{bmatrix} - 6\begin{bmatrix} 2 & -1 & 1 \\ -1 & 2 & -1 \\ 1 & -1 & 2 \end{bmatrix} + 9\begin{bmatrix} 1 & 0 & 0 \\ 0 & 1 & 0 \\ 0 & 0 & 1 \end{bmatrix} = \begin{bmatrix} 3 & 1 & -1 \\ 1 & 3 & 1 \\ -1 & 1 & 3 \end{bmatrix}$$

$$\therefore \qquad A^{-1} = \frac{1}{4}\begin{bmatrix} 3 & 1 & -1 \\ 1 & 3 & 1 \\ -1 & 1 & 3 \end{bmatrix}.$$

Example 5. *Using Cayley-Hamilton theorem, find A^8 if $A = \begin{bmatrix} 1 & 2 \\ 2 & -1 \end{bmatrix}$*

Sol. The characteristic equation of A is

$$|A - \lambda I| = 0 \quad i.e., \quad \begin{vmatrix} 1-\lambda & 2 \\ 2 & -1-\lambda \end{vmatrix} = 0$$

or $\qquad -(1 - \lambda^2) - 4 = 0 \quad$ or $\quad \lambda^2 = 5 \qquad$ ...(1)

By Cayley-Hamilton theorem. A satisfies its characteristic equation (1)

$$\therefore \qquad A^2 = 5I$$

$$\Rightarrow \qquad (A^2)^4 = (5I)^4 \quad \Rightarrow \quad A^8 = 625\ I \qquad (\because\ I^4 = I)$$

Example 6. *Find the characteristic equation of the matrix $A = \begin{bmatrix} 2 & 1 & 1 \\ 0 & 1 & 0 \\ 1 & 1 & 2 \end{bmatrix}$ and, hence, find the matrix represented by $A^8 - 5A^7 + 7A^6 - 3A^5 + A^4 - 5A^3 + 8A^2 - 2A + I$.*

(M.T.U. 2013; U.K.T.U. 2010)

Sol. The characteristic equation of A is

$$|A - \lambda I| = \begin{vmatrix} 2-\lambda & 1 & 1 \\ 0 & 1-\lambda & 0 \\ 1 & 1 & 2-\lambda \end{vmatrix} = 0 \quad \text{or} \quad \lambda^3 - 5\lambda^2 + 7\lambda - 3 = 0$$

By Cayley-Hamilton theorem, $A^3 - 5A^2 + 7A - 3I = O$ ...(1)

Now $A^8 - 5A^7 + 7A^6 - 3A^5 + A^4 - 5A^3 + 8A^2 - 2A + I$

$$= A^5(A^3 - 5A^2 + 7A - 3I) + A(A^3 - 5A^2 + 7A - 3I) + (A^2 + A + I)$$

$$= A^2 + A + I \qquad \text{[Using (1)]}$$

$$= \begin{bmatrix} 2 & 1 & 1 \\ 0 & 1 & 0 \\ 1 & 1 & 2 \end{bmatrix}\begin{bmatrix} 2 & 1 & 1 \\ 0 & 1 & 0 \\ 1 & 1 & 2 \end{bmatrix} + \begin{bmatrix} 2 & 1 & 1 \\ 0 & 1 & 0 \\ 1 & 1 & 2 \end{bmatrix} + \begin{bmatrix} 1 & 0 & 0 \\ 0 & 1 & 0 \\ 0 & 0 & 1 \end{bmatrix}$$

$$= \begin{bmatrix} 5 & 4 & 4 \\ 0 & 1 & 0 \\ 4 & 4 & 5 \end{bmatrix} + \begin{bmatrix} 2 & 1 & 1 \\ 0 & 1 & 0 \\ 1 & 1 & 2 \end{bmatrix} + \begin{bmatrix} 1 & 0 & 0 \\ 0 & 1 & 0 \\ 0 & 0 & 1 \end{bmatrix} = \begin{bmatrix} 8 & 5 & 5 \\ 0 & 3 & 0 \\ 5 & 5 & 8 \end{bmatrix}.$$

EXERCISE 2.4

1. Find the sum and product of the eigen values of

(i) $\begin{bmatrix} 2 & 1 & 1 \\ 0 & 1 & 0 \\ 1 & 1 & 2 \end{bmatrix}$ (ii) $\begin{bmatrix} -2 & 2 & -3 \\ 2 & 1 & -6 \\ -1 & -2 & 0 \end{bmatrix}$

(iii) $\begin{bmatrix} 6 & -2 & 2 \\ -2 & 3 & -1 \\ 2 & -1 & 3 \end{bmatrix}$ (iv) $\begin{bmatrix} 1 & 2 & 3 & 4 \\ 2 & 1 & 5 & 6 \\ 7 & 4 & 3 & 2 \\ 4 & 3 & 0 & 5 \end{bmatrix}$. (*M.D.U. May 2008*)

2. (*a*) Find the product of the eigen values of $\begin{bmatrix} 7 & 2 & 2 \\ -6 & -1 & 2 \\ 6 & 2 & -1 \end{bmatrix}$.

(*b*) Find the sum of eigen values of the inverse of the matrix

$$A = \begin{bmatrix} 3 & 0 & 0 \\ 8 & 4 & 0 \\ 6 & 2 & 5 \end{bmatrix}$$ (*Calicut 2008*)

[**Hint.** Eigen values of A are 3, 4, 5 ⇒ Eigen values of A^{-1} are 3^{-1}, 4^{-1}, 5^{-1}]

(*c*) Find the eigen values of $2A^2$, if $A = \begin{bmatrix} 4 & 1 \\ 3 & 2 \end{bmatrix}$. (*Calicut 2010*)

[**Hint.** Eigen values of A are 1, 5 ⇒ Eigen values of A^2 are 1^2, 5^2
⇒ Eigen values of $2A^2$ are 2×1^2, 2×5^2]

3. Find the eigen values and the corresponding eigen vectors of the following matrices:

(i) $\begin{bmatrix} 5 & 4 \\ 1 & 2 \end{bmatrix}$ (ii) $\begin{bmatrix} 1 & 2 \\ 2 & 4 \end{bmatrix}$ (iii) $\begin{bmatrix} 1 & 4 \\ 3 & 2 \end{bmatrix}$.

4. Find the eigen values and the corresponding eigen vectors of the following matrices:

(i) $\begin{bmatrix} 8 & -6 & 2 \\ -6 & 7 & -4 \\ 2 & -4 & 3 \end{bmatrix}$ (*M.D.U. Dec. 2013; G.B.T.U. 2011*)

(ii) $\begin{bmatrix} 6 & -2 & 2 \\ -2 & 3 & -1 \\ 2 & -1 & 3 \end{bmatrix}$ (*M.D.U. Dec. 2010 ; K.U.K. 2008, 2011*)

(iii) $\begin{bmatrix} 2 & 0 & 1 \\ 0 & 2 & 0 \\ 1 & 0 & 2 \end{bmatrix}$ (iv) $\begin{bmatrix} 2 & 1 & 1 \\ 1 & 2 & 1 \\ 0 & 0 & 1 \end{bmatrix}$ (v) $\begin{bmatrix} 3 & 1 & 4 \\ 0 & 2 & 0 \\ 0 & 0 & 5 \end{bmatrix}$

(vi) $\begin{bmatrix} 1 & 1 & 3 \\ 1 & 5 & 1 \\ 3 & 1 & 1 \end{bmatrix}$ (vii) $\begin{bmatrix} -9 & 2 & 6 \\ 5 & 0 & -3 \\ -16 & 4 & 11 \end{bmatrix}$ (viii) $\begin{bmatrix} 1 & 0 & -2 \\ 0 & 0 & 0 \\ -2 & 0 & 4 \end{bmatrix}$

(ix) $\begin{bmatrix} 3 & -1 & 1 \\ -1 & 5 & -1 \\ 1 & -1 & 3 \end{bmatrix}$ (x) $\begin{bmatrix} 1 & 0 & -1 \\ 1 & 2 & 1 \\ 2 & 2 & 3 \end{bmatrix}$ (xi) $\begin{bmatrix} -1 & 2 & -2 \\ 1 & 2 & 1 \\ -1 & -1 & 0 \end{bmatrix}$.

5. Show that the eigen values of a triangular matrix A are equal to the elements of the principal diagonal of A.

6. Show that 0 is a characteristic root of a matrix if and only if the matrix is singular.

7. Show that if λ is a characteristic root of a matrix A, then $\lambda + k$ is a characteristic root of the matrix $A + kI$.

8. If $\lambda_1, \lambda_2, \ldots\ldots, \lambda_n$ are the eigen values of a matrix A, then A^2 has the eigen values $\lambda_1^2, \lambda_2^2, \ldots\ldots, \lambda_n^2$.

9. Prove that the matrices A and $P^{-1}AP$ have the same eigen values, P being an invertible matrix of same order as A. *(M.D.U. 2011)*

10. (*a*) Find the characteristic equation of the matrix $A = \begin{bmatrix} 1 & 3 & 7 \\ 4 & 2 & 3 \\ 1 & 2 & 1 \end{bmatrix}$ *(M.D.U. Dec. 2013)*

Show that the equation is satisfied by A and hence obtain the inverse of the given matrix.

(*b*) Find the characteristic equation of the matrix $A = \begin{bmatrix} 1 & 2 & 3 \\ 2 & 1 & 4 \\ 3 & 1 & 1 \end{bmatrix}$

Show that the equation is satisfied by A. Also find A^{-1}.

11. Verify Cayley-Hamilton theorem for the matrix A and find A^{-1} when A is

(*i*) $\begin{bmatrix} 1 & 0 & 2 \\ 0 & 2 & 1 \\ 2 & 0 & 3 \end{bmatrix}$ *(K.U.K. May 2012)* (*ii*) $\begin{bmatrix} 2 & 6 & 1 \\ 0 & 1 & -6 \\ 3 & 4 & -2 \end{bmatrix}$ (*iii*) $\begin{bmatrix} 7 & 2 & -2 \\ -6 & -1 & 2 \\ 6 & 2 & -1 \end{bmatrix}$. *(M.D.U. 2009; U.K.T.U. 2012)*

12. Using Cayley-Hamilton theorem, find the inverse of

(*i*) $\begin{bmatrix} 5 & 3 \\ 3 & 2 \end{bmatrix}$ *(M.D.U. Dec. 2010)* (*ii*) $\begin{bmatrix} 7 & -1 & 3 \\ 6 & 1 & 4 \\ 2 & 4 & 8 \end{bmatrix}$ (*iii*) $\begin{bmatrix} 1 & 0 & 3 \\ 2 & 1 & -1 \\ 1 & -1 & 1 \end{bmatrix}$.

13. Using Cayley-Hamilton theorem, find A^4 if $A = \begin{bmatrix} 2 & 1 \\ 5 & -2 \end{bmatrix}$. *(M.D.U. Dec. 2013)*

14. Find the characteristic equation of the matrix $A = \begin{bmatrix} 1 & 4 \\ 2 & 3 \end{bmatrix}$ and use it to find the matrix represented by

$$A^5 + 5A^4 - 6A^3 + 2A^2 - 4A + 7I.$$ *(M.D.U. Dec. 2011)*

Also express $A^5 - 4A^4 - 7A^3 + 11A^2 - A - 10I$ as a linear polynomial in A.

15. Show that the matrix $A = \begin{bmatrix} 0 & r & -q \\ -r & 0 & p \\ q & -p & 0 \end{bmatrix}$ satisfies Cayley-Hamilton theorem.

16. If $A = \begin{bmatrix} 1 & 2 \\ -1 & 3 \end{bmatrix}$, express $A^6 - 4A^5 + 8A^4 - 12A^3 + 14A^2$ as a linear polynomial in A.

Answers

1. (*i*) 5, 3 (*ii*) – 1, 45 (*iii*) 12, 32 (*iv*) 10, 262

2. (*a*) – 21 (*b*) 47/60 (*c*) 2, 50

3. (*i*) 1, 6 ; (1, – 1), (4, 1) (*ii*) 5, 0 ; (1, 2), (2, – 1) (*iii*) 5, – 2 ; (1, 1), (4, – 3)

4. (*i*) 0, 3, 15 ; (1, 2, 2), (2, 1, – 2), (2, – 2, 1) (*ii*) 2, 2, 8 ; (1, 0, – 2), (1, 2, 0), (2, – 1, 1)
(*iii*) 1, 2, 3 ; (1, 0, – 1), (0, 1, 0), (1, 0, 1) (*iv*) 1, 1, 3 ; (1, – 1, 0), (1, 0, – 1), (1, 1, 0)
(*v*) 2, 3, 5 ; (1, – 1, 0), (1, 0, 0), (2, 0, 1) (*vi*) – 2, 3, 6 ; (– 1, 0, 1), (1, – 1, 1), (1, 2, 1)
(*vii*) 1, – 1, 2 ; (1, – 1, 2), (2, – 1, 3), (2, – 1, 4) (*viii*) 0, 0, 5 ; (0, 1, 0), (2, 0, 1), (1, 0, – 2)
(*ix*) 2, 3, 6 ; (– 1, 0, 1), (1, 1, 1), (1, – 2, 1) (*x*) 1, 2, 3 ; (1, – 1, 0), (2, – 1, – 2), (1, – 1, – 2)
(*xi*) 1, $-\sqrt{5}$, $\sqrt{5}$; (– 1, 0, 1), ($1 + \sqrt{5}$, – 1, 1), ($1 - \sqrt{5}$, – 1, 1).

9. **[Hint.** $|P^{-1}AP - \lambda I| = |P^{-1}AP - \lambda P^{-1}P| = |P^{-1}(A - \lambda I)P|$
$= |P^{-1}||A - \lambda I||P| = |A - \lambda I||P^{-1}P| = |A - \lambda I|$ **]**

10. (a) $\lambda^3 - 4\lambda^2 - 20\lambda - 35 = 0$, $\frac{1}{35}\begin{bmatrix} -4 & 11 & -5 \\ -1 & -6 & 25 \\ 6 & 1 & -10 \end{bmatrix}$ (b) $\lambda^3 - 3\lambda^2 - 14\lambda - 14 = 0$, $\frac{1}{14}\begin{bmatrix} -3 & 1 & 5 \\ 10 & -8 & 2 \\ -1 & 5 & -3 \end{bmatrix}$

11. (i) $\begin{bmatrix} -3 & 0 & 2 \\ -1 & \frac{1}{2} & \frac{1}{2} \\ 2 & 0 & -1 \end{bmatrix}$ (ii) $\frac{1}{67}\begin{bmatrix} -22 & -16 & 37 \\ 18 & 7 & -12 \\ 3 & -10 & -2 \end{bmatrix}$ (iii) $\frac{1}{3}\begin{bmatrix} -3 & -2 & 2 \\ 6 & 5 & -2 \\ -6 & -2 & 5 \end{bmatrix}$

12. (i) $\begin{bmatrix} 2 & -3 \\ -3 & 5 \end{bmatrix}$ (ii) $\frac{1}{50}\begin{bmatrix} -8 & 20 & -7 \\ 40 & 50 & -10 \\ 22 & -30 & 13 \end{bmatrix}$ (iii) $\frac{1}{9}\begin{bmatrix} 0 & 3 & 3 \\ 3 & 2 & -7 \\ 3 & -1 & -1 \end{bmatrix}$

13. 81I **14.** $\begin{bmatrix} 1861 & 3676 \\ 1838 & 3699 \end{bmatrix}$; A + 5I **16.** – 4A + 5I.

2.16. REDUCTION OF A MATRIX TO DIAGONAL FORM

If a square matrix A of order n has n linearly independent eigen vectors, then a matrix B can be found such that $B^{-1}AB$ is a diagonal matrix.

[We prove the result for a square matrix of order 3. The proof can be easily extended to matrices of higher order.]

Let A be a square matrix of order 3. Let $\lambda_1, \lambda_2, \lambda_3$ be its eigen values and

$X_1 = \begin{bmatrix} x_1 \\ y_1 \\ z_1 \end{bmatrix}, X_2 = \begin{bmatrix} x_2 \\ y_2 \\ z_2 \end{bmatrix}, X_3 = \begin{bmatrix} x_3 \\ y_3 \\ z_3 \end{bmatrix}$ be the corresponding eigen vectors.

Since eigen vectors are non-trivial solutions of the matrix equation $AX = \lambda X$, we have

$$AX_1 = \lambda_1 X_1, AX_2 = \lambda_2 X_2, AX_3 = \lambda_3 X_3$$

Let
$$B = \begin{bmatrix} x_1 & x_2 & x_3 \\ y_1 & y_2 & y_3 \\ z_1 & z_2 & z_3 \end{bmatrix} = [X_1 X_2 X_3]$$

Consider $AB = A[X_1, X_2, X_3] = [AX_1, AX_2, AX_3] = [\lambda_1 X_1, \lambda_2 X_2, \lambda_3 X_3]$

$$= \begin{bmatrix} \lambda_1 x_1 & \lambda_2 x_2 & \lambda_3 x_3 \\ \lambda_1 y_1 & \lambda_2 y_2 & \lambda_3 y_3 \\ \lambda_1 z_1 & \lambda_2 z_2 & \lambda_3 z_3 \end{bmatrix} = \begin{bmatrix} x_1 & x_2 & x_3 \\ y_1 & y_2 & y_3 \\ z_1 & z_2 & z_3 \end{bmatrix}\begin{bmatrix} \lambda_1 & 0 & 0 \\ 0 & \lambda_2 & 0 \\ 0 & 0 & \lambda_3 \end{bmatrix}$$

$= BD$, where D is the diagonal matrix $\begin{bmatrix} \lambda_1 & 0 & 0 \\ 0 & \lambda_2 & 0 \\ 0 & 0 & \lambda_3 \end{bmatrix}$

$\Rightarrow$ $AB = BD \Rightarrow B^{-1}AB = B^{-1}BD = ID = D$ which proves the theorem.

Note. The matrix B which diagonalises A is called the **modal matrix** of A and is obtained by grouping the eigen vectors of A into a square matrix. The diagonal matrix D is called the **spectral matrix** of A and has the eigen values of A as its diagonal elements.

2.17. CALCULATION OF POWERS OF A MATRIX

Given a square matrix A, it is quite tedious to find A^n (n being a positive integer) when n is large. On the other hand, it is quite easy to obtain any positive integral power of a diagonal matrix D, since

$$D = \begin{bmatrix} a & 0 \\ 0 & b \end{bmatrix} \quad \Rightarrow \quad D^n = \begin{bmatrix} a^n & 0 \\ 0 & b^n \end{bmatrix}$$

Therefore, diagonalisation of a matrix is useful for finding positive integral powers of a square matrix.

Let A be a square matrix. Then we know that a non-singular matrix B can be found such that

$$B^{-1}AB = D \qquad \text{...(1)}$$

where D is a diagonal matrix.

From (1), $D^2 = (B^{-1}AB)(B^{-1}AB)$

$= B^{-1}A^2B$ $\qquad [\because BB^{-1} = I]$

Similarly, $D^3 = B^{-1}A^3B$

In general, $D^n = B^{-1}A^nB$ $\qquad$ ...(2)

To find A^n from (2), pre-multiply (2) by B and post-multiply by B^{-1}. Then

$$BD^nB^{-1} = BB^{-1}A^nBB^{-1}$$
$$= A^n$$

Hence $A^n = BD^nB^{-1}$.

ILLUSTRATIVE EXAMPLES

Example 1. *Show the the matrix* $A = \begin{bmatrix} 3 & 1 & -1 \\ -2 & 1 & 2 \\ 0 & 1 & 2 \end{bmatrix}$ *is diagonalizable. Hence find P such that* $P^{-1}AP$ *is a diagonal matrix.* (M.T.U. 2012; P.T.U. May 2008)

Sol. The characteristic equation of A is

$$|A - \lambda I| = \begin{vmatrix} 3-\lambda & 1 & -1 \\ -2 & 1-\lambda & 2 \\ 0 & 1 & 2-\lambda \end{vmatrix} = 0$$

or $\lambda^3 - 6\lambda^2 + 11\lambda - 6 = 0$ $\qquad$ (On simplification)

Solving it, $\lambda = 1, 2, 3$

Since the matrix A has three distinct eigen values, it has three linearly independent eigen vectors and hence A is diagonalizable.

When $\lambda = 1$, the corresponding eigen vector is given by

$$(A - I)X = O \quad \text{or} \quad \begin{bmatrix} 2 & 1 & -1 \\ -2 & 0 & 2 \\ 0 & 1 & 1 \end{bmatrix}\begin{bmatrix} x_1 \\ x_2 \\ x_3 \end{bmatrix} = \begin{bmatrix} 0 \\ 0 \\ 0 \end{bmatrix}$$

Operating $R_1 \to R_1 + R_2$

$$\sim \begin{bmatrix} 0 & 1 & 1 \\ -2 & 0 & 2 \\ 0 & 1 & 1 \end{bmatrix} \begin{bmatrix} x_1 \\ x_2 \\ x_3 \end{bmatrix} = \begin{bmatrix} 0 \\ 0 \\ 0 \end{bmatrix}$$

Operating $R_2 \to \frac{1}{2} R_2$ and $R_3 \to R_3 - R_1$

$$\sim \begin{bmatrix} 0 & 1 & 1 \\ -1 & 0 & 1 \\ 0 & 0 & 0 \end{bmatrix} \begin{bmatrix} x_1 \\ x_2 \\ x_3 \end{bmatrix} = \begin{bmatrix} 0 \\ 0 \\ 0 \end{bmatrix}$$

We get only two independent equations $x_2 + x_3 = 0$ and $- x_1 + x_3 = 0$, since rank of co-efficient matrix is 2.

$\Rightarrow \qquad x_2 = -x_3$ and $x_1 = x_3$

$\Rightarrow \qquad \frac{x_1}{1} = \frac{x_2}{-1} = \frac{x_3}{1}$ giving the eigen vector (1, –1, 1).

When $\lambda = 2$, the corresponding eigen vector is given by

$$(A - 2I)\, X = O \quad \text{or} \quad \begin{bmatrix} 1 & 1 & -1 \\ -2 & -1 & 2 \\ 0 & 1 & 0 \end{bmatrix} \begin{bmatrix} x_1 \\ x_2 \\ x_3 \end{bmatrix} = \begin{bmatrix} 0 \\ 0 \\ 0 \end{bmatrix}$$

Operating $R_1 \to R_1 - R_3$ and $R_2 \to R_2 + R_3$

$$\sim \begin{bmatrix} 1 & 0 & -1 \\ -2 & 0 & 2 \\ 0 & 1 & 0 \end{bmatrix} \begin{bmatrix} x_1 \\ x_2 \\ x_3 \end{bmatrix} = \begin{bmatrix} 0 \\ 0 \\ 0 \end{bmatrix}$$

Operating $R_2 \to R_2 + 2R_1$

$$\sim \begin{bmatrix} 1 & 0 & -1 \\ 0 & 0 & 0 \\ 0 & 1 & 0 \end{bmatrix} \begin{bmatrix} x_1 \\ x_2 \\ x_3 \end{bmatrix} = \begin{bmatrix} 0 \\ 0 \\ 0 \end{bmatrix}$$

Operating $R_2 \leftrightarrow R_3$

$$\sim \begin{bmatrix} 1 & 0 & -1 \\ 0 & 1 & 0 \\ 0 & 0 & 0 \end{bmatrix} \begin{bmatrix} x_1 \\ x_2 \\ x_3 \end{bmatrix} = \begin{bmatrix} 0 \\ 0 \\ 0 \end{bmatrix}$$

We get two independent equations $x_1 - x_3 = 0$ and $x_2 = 0$, since rank of co-efficient matrix is 2.

$\Rightarrow \quad x_1 = x_3$ and $x_2 = 0$ giving the eigen vector (1, 0, 1)

When $\lambda = 3$, the corresponding eigen vector is given by

$$(A - 3I)\, X = O \quad \text{or} \quad \begin{bmatrix} 0 & 1 & -1 \\ -2 & -2 & 2 \\ 0 & 1 & -1 \end{bmatrix} \begin{bmatrix} x_1 \\ x_2 \\ x_3 \end{bmatrix} = \begin{bmatrix} 0 \\ 0 \\ 0 \end{bmatrix}$$

Operating $R_2 \to R_2 + 2R_1$ and $R_3 \to R_3 - R_1$

$$\sim \begin{bmatrix} 0 & 1 & -1 \\ -2 & 0 & 0 \\ 0 & 0 & 0 \end{bmatrix} \begin{bmatrix} x_1 \\ x_2 \\ x_3 \end{bmatrix} = \begin{bmatrix} 0 \\ 0 \\ 0 \end{bmatrix}$$

We get two independent equations $x_2 - x_3 = 0$ and $-2x_1 = 0$, since rank of co-efficient matrix is 2.

$\Rightarrow$ $x_1 = 0$ and $x_2 = x_3$ giving the eigen vector (0, 1, 1)

$\therefore$ The modal matrix $P = [X_1 \quad X_2 \quad X_3] = \begin{bmatrix} 1 & 1 & 0 \\ -1 & 0 & 1 \\ 1 & 1 & 1 \end{bmatrix}$

and $$P^{-1} AP = D = \begin{bmatrix} \lambda_1 & 0 & 0 \\ 0 & \lambda_2 & 0 \\ 0 & 0 & \lambda_3 \end{bmatrix} \begin{bmatrix} 1 & 0 & 0 \\ 0 & 2 & 0 \\ 0 & 0 & 3 \end{bmatrix}$$

is the required diagonal matrix.

Example 2. *Diagonalise the matrix* $A = \begin{bmatrix} -1 & 2 & -2 \\ 1 & 2 & 1 \\ -1 & -1 & 0 \end{bmatrix}$ *and obtain the modal matrix.*

(M.D.U. Dec. 2009; Kerala 2011; U.K.T.U. 2011)

Sol. The characteristic equation of A is

$$|A - \lambda I| = \begin{vmatrix} -1-\lambda & 2 & -2 \\ 1 & 2-\lambda & 1 \\ -1 & -1 & -\lambda \end{vmatrix} = 0 \text{ or } \lambda^3 - \lambda^2 - 5\lambda + 5 = 0 \quad \text{(on simplification)}$$

Solving it, we have $\lambda = 1, \pm\sqrt{5}$.

When $\lambda = 1$, the corresponding eigen vector is given by

$$-2x_1 + 2y_1 - 2z_1 = 0$$
$$x_1 + y_1 + z_1 = 0$$
$$x_1 - y_1 + z_1 = 0$$

Solving the last two $\frac{x_1}{2} = \frac{y_1}{0} = \frac{z_1}{-2}$ giving the eigen vector (1, 0, – 1).

When $\lambda = \sqrt{5}$, the corresponding eigen vector is given by

$$(-1-\sqrt{5})x_1 + 2y_1 - 2z_1 = 0$$
$$x_1 + (2-\sqrt{5})y_1 + z_1 = 0$$
$$-x_1 - y_1 - \sqrt{5}z_1 = 0$$

Solving the last two $\frac{x_1}{6-2\sqrt{5}} = \frac{y_1}{-1+\sqrt{5}} = \frac{z_1}{1-\sqrt{5}}$

or $\frac{x_1}{(\sqrt{5}-1)^2} = \frac{y_1}{\sqrt{5}-1} = \frac{z_1}{1-\sqrt{5}}$ or $\frac{x_1}{\sqrt{5}-1} = \frac{y_1}{1} = \frac{z_1}{-1}$ giving the eigen vector $(\sqrt{5} - 1, 1, -1)$

Similarly, the eigen vector corresponding to $\lambda = -\sqrt{5}$ is $(\sqrt{5} + 1, -1, 1)$.

$\therefore$ The modal matrix $B = \begin{bmatrix} 1 & \sqrt{5}-1 & -\sqrt{5}+1 \\ 0 & 1 & 1 \\ -1 & -1 & 1 \end{bmatrix}$

and $$B^{-1}AB = D = \begin{bmatrix} \lambda_1 & 0 & 0 \\ 0 & \lambda_2 & 0 \\ 0 & 0 & \lambda_3 \end{bmatrix} = \begin{bmatrix} 1 & 0 & 0 \\ 0 & \sqrt{5} & 0 \\ 0 & 0 & -\sqrt{5} \end{bmatrix}$$

is the required diagonal matrix.

Example 3. *Diagonalise the matrix* $A = \begin{bmatrix} 1 & 6 & 1 \\ 1 & 2 & 0 \\ 0 & 0 & 3 \end{bmatrix}$ *and hence find* A^4.

(M.D.U. Dec. 2012)

Sol. The characteristic equation of A is

$$|A - \lambda I| = \begin{vmatrix} 1-\lambda & 6 & 1 \\ 1 & 2-\lambda & 0 \\ 0 & 0 & 3-\lambda \end{vmatrix} = 0$$

or $\lambda^3 - 6\lambda^2 + 5\lambda + 12 = 0$ (On simplification)

or $(\lambda + 1)(\lambda - 3)(\lambda - 4) = 0$

$\therefore$ $\lambda = -1, 3, 4$

When $\lambda = -1$, the corresponding eigen vector is given by

$$2x_1 + 6y_1 + z_1 = 0$$
$$x_1 + 3y_1 = 0$$
$$4z_1 = 0$$

Solving the last two, $\frac{x_1}{3} = \frac{y_1}{-1}$, $z_1 = 0$ giving the eigen vector (3, – 1, 0).

When $\lambda = 3$, the corresponding eigen vector is given by

$$-2x_1 + 6y_1 + z_1 = 0$$
$$x_1 - y_1 = 0$$

Solving $\frac{x_1}{1} = \frac{y_1}{1} = \frac{z_1}{-4}$ giving the eigen vector (1, 1, – 4).

When $\lambda = 4$, the corresponding eigen vector is given by

$$-3x_1 + 6y_1 + z_1 = 0$$
$$x_1 - 2y_1 = 0$$
$$z_1 = 0$$

Solving the last two, $\frac{x_1}{2} = \frac{y_1}{1}$, $z_1 = 0$ giving the eigen vector (2, 1, 0).

$\therefore$ The modal matrix $B = \begin{bmatrix} 3 & 1 & 2 \\ -1 & 1 & 1 \\ 0 & -4 & 0 \end{bmatrix}$

and $$B^{-1}AB = D = \begin{bmatrix} \lambda_1 & 0 & 0 \\ 0 & \lambda_2 & 0 \\ 0 & 0 & \lambda_3 \end{bmatrix} = \begin{bmatrix} -1 & 0 & 0 \\ 0 & 3 & 0 \\ 0 & 0 & 4 \end{bmatrix}$$

is the required diagonal matrix.

To find B^{-1}

$$|B| = \begin{vmatrix} 3 & 1 & 2 \\ -1 & 1 & 1 \\ 0 & -4 & 0 \end{vmatrix} = \begin{vmatrix} a_1 & b_1 & c_1 \\ a_2 & b_2 & c_2 \\ a_3 & b_3 & c_3 \end{vmatrix} \quad \text{(say)}$$

$$A_1 = 4,\ B_1 = 0,\ C_1 = 4,\ A_2 = -8,\ B_2 = 0,\ C_2 = 12,\ A_3 = -1,\ B_3 = -5,\ C_3 = 4$$

$$\therefore \quad |B| = a_1A_1 + b_1B_1 + c_1C_1 = (3)(4) + (1)(0) + 2(4) = 20$$

$$\text{adj. } B = \begin{bmatrix} A_1 & A_2 & A_3 \\ B_1 & B_2 & B_3 \\ C_1 & C_2 & C_3 \end{bmatrix} = \begin{bmatrix} 4 & -8 & -1 \\ 0 & 0 & -5 \\ 4 & 12 & 4 \end{bmatrix}$$

$$B^{-1} = \frac{1}{|B|} \text{adj } B = \frac{1}{20}\begin{bmatrix} 4 & -8 & -1 \\ 0 & 0 & -5 \\ 4 & 12 & 4 \end{bmatrix}$$

Now $$D = \begin{bmatrix} -1 & 0 & 0 \\ 0 & 3 & 0 \\ 0 & 0 & 4 \end{bmatrix} \Rightarrow D^4 = \begin{bmatrix} (-1)^4 & 0 & 0 \\ 0 & 3^4 & 0 \\ 0 & 0 & 4^4 \end{bmatrix} = \begin{bmatrix} 1 & 0 & 0 \\ 0 & 81 & 0 \\ 0 & 0 & 256 \end{bmatrix}$$

$$\therefore \quad A^4 = BD^4B^{-1} = \frac{1}{20}\begin{bmatrix} 3 & 1 & 2 \\ -1 & 1 & 1 \\ 0 & -4 & 0 \end{bmatrix}\begin{bmatrix} 1 & 0 & 0 \\ 0 & 81 & 0 \\ 0 & 0 & 256 \end{bmatrix}\begin{bmatrix} 4 & -8 & -1 \\ 0 & 0 & -5 \\ 4 & 12 & 4 \end{bmatrix}$$

$$= \frac{1}{20}\begin{bmatrix} 3 & 1 & 2 \\ -1 & 1 & 1 \\ 0 & -4 & 0 \end{bmatrix}\begin{bmatrix} 4 & -8 & -1 \\ 0 & 0 & -405 \\ 1024 & 3072 & 1024 \end{bmatrix}$$

$$= \frac{1}{20}\begin{bmatrix} 2060 & 6120 & 1640 \\ 1020 & 3080 & 620 \\ 0 & 0 & 1620 \end{bmatrix} = \begin{bmatrix} 103 & 306 & 82 \\ 51 & 154 & 31 \\ 0 & 0 & 81 \end{bmatrix}$$

EXERCISE 2.5

1. Reduce the matrix $A = \begin{bmatrix} -19 & 7 \\ -42 & 16 \end{bmatrix}$ to the diagonal form.

2. By diagonalising the matrix $A = \begin{bmatrix} -1 & 3 \\ -2 & 4 \end{bmatrix}$, find A^4.

3. Diagonalise each of the following matrices :

(*i*) $\begin{bmatrix} -1 & 1 & 2 \\ 0 & -2 & 1 \\ 0 & 0 & -3 \end{bmatrix}$ (*ii*) $\begin{bmatrix} 8 & -6 & 2 \\ -6 & 7 & -4 \\ 2 & -4 & 3 \end{bmatrix}$ (*iii*) $\begin{bmatrix} 1 & 0 & -1 \\ 1 & 2 & 1 \\ 2 & 2 & 3 \end{bmatrix}$ (*Kerala 2010*)

4. Diagonalise $A = \begin{bmatrix} 1 & -1 & 2 \\ 0 & 2 & -1 \\ 0 & 0 & 3 \end{bmatrix}$ and hence find A^3.

5. Diagonalise the matrix $A = \begin{bmatrix} 3 & -1 & 1 \\ -1 & 5 & -1 \\ 1 & -1 & 3 \end{bmatrix}$ and hence find A^4. (*M.T.U. 2013*)

6. Diagonalise $A = \begin{bmatrix} 1 & 1 & 1 \\ 0 & 2 & 1 \\ -4 & 4 & 3 \end{bmatrix}$ and hence find A^8.

7. Find a matrix P which transforms the matrix $A = \begin{bmatrix} 1 & 1 & 3 \\ 1 & 5 & 1 \\ 3 & 1 & 1 \end{bmatrix}$ to its diagonal form. Hence calculate A^3.

8. Show that the linear transformation $T = \begin{bmatrix} \cos\theta & \sin\theta \\ -\sin\theta & \cos\theta \end{bmatrix}$, where $\theta = \frac{1}{2}\tan^{-1}\left(\frac{2h}{b-a}\right)$ changes the matrix $A = \begin{bmatrix} a & h \\ h & b \end{bmatrix}$ to diagonal form $D = T^{-1}AT$.

Answers

1. $\begin{bmatrix} -5 & 0 \\ 0 & 2 \end{bmatrix}$ 2. $\begin{bmatrix} -29 & 45 \\ -30 & 46 \end{bmatrix}$.

3. (*i*) $\begin{bmatrix} -1 & 0 & 0 \\ 0 & -2 & 0 \\ 0 & 0 & -3 \end{bmatrix}$ (*ii*) $\begin{bmatrix} 0 & 0 & 0 \\ 0 & 3 & 0 \\ 0 & 0 & 15 \end{bmatrix}$ (*iii*) $\begin{bmatrix} 1 & 0 & 0 \\ 0 & 2 & 0 \\ 0 & 0 & 3 \end{bmatrix}$

4. $\begin{bmatrix} 1 & -7 & 32 \\ 0 & 8 & -19 \\ 0 & 0 & 27 \end{bmatrix}$ 5. $\begin{bmatrix} 2 & 0 & 0 \\ 0 & 3 & 0 \\ 0 & 0 & 6 \end{bmatrix}, \begin{bmatrix} 251 & -405 & 235 \\ -405 & 891 & -405 \\ 235 & -405 & 251 \end{bmatrix}$

6. $\begin{bmatrix} -12099 & 12355 & 6305 \\ -12100 & 12356 & 6305 \\ -13120 & 13120 & 6561 \end{bmatrix}$ 7. $P = \begin{bmatrix} -1 & 1 & 1 \\ 0 & -1 & 2 \\ 1 & 1 & 1 \end{bmatrix}, A^3 = \begin{bmatrix} 49 & 63 & 41 \\ 63 & 153 & 63 \\ 41 & 63 & 49 \end{bmatrix}$.

2.18. SIMILAR MATRICES (*M.D.U. Dec. 2010*)

Let A and B be square matrices of the same order. The matrix A is said to be similar to the matrix B if there exists an invertible square matrix P such that $A = P^{-1}BP$.

The matrix P is called the **similarity matrix.**

Note. The matrices on the left and right of B are inverse of each other.

2.19. THEOREMS ON SIMILAR MATRICES

Theorem 1. *A square matrix A is similar to a square matrix B if and only if B is similar to A.*

Proof. A is similar to B if and only if there exists a non-singular matrix P such that

$$A = P^{-1}BP \quad \text{or} \quad PA = BP$$

Post-multiplying both sides by P^{-1}, we get

$$PAP^{-1} = B \quad \text{or} \quad B = PAP^{-1}$$

showing B is similar to A.

Therefore, A is similar to B if and only if B is similar to A.

Theorem 2. *If A is similar to B and B is similar to C, then A is similar to C.* (Transitivity)

Proof. A is similar to B $\Rightarrow$ There exists an invertible matrix P such that $A = P^{-1}BP$ ...(1)

B is similar to C $\Rightarrow$ There exists an invertible matrix Q such that $B = Q^{-1}CQ$ ...(2)

Substituting for B from (2) in (1), we get

$$A = P^{-1}Q^{-1}CQP = (QP)^{-1}\,C\,(QP) = R^{-1}CR \text{ where } R = QP$$

$\Rightarrow$ A is similar to C.

Theorem 3. *Similar matrices have the same characteristic equation* (*and hence the same eigen values*).

Proof. Let A and B be two similar matrices. Then there exists a non-singular matrix P such that

$$B = P^{-1}AP$$

Now,

$$\begin{aligned} |B - \lambda I| &= |P^{-1}AP - \lambda I| \\ &= |P^{-1}AP - \lambda P^{-1}P| && (\because\ P^{-1}P = I) \\ &= |P^{-1}(A - \lambda I)P| = |P^{-1}|\ |A - \lambda I|\ |P| \\ &= |A - \lambda I|\ |P^{-1}|\ |P| \\ &= |A - \lambda I|\ |P^{-1}P| = |A - \lambda I|\ |I| \\ &= |A - \lambda I| && (\because\ |I| = 1) \end{aligned}$$

$\therefore$ Matrices A and B have the same characteristic polynomial and hence the same characteristic equation.

Example. *Show that the matrices* $A = \begin{bmatrix} 5 & 5 \\ -2 & 0 \end{bmatrix}$ *and* $B = \begin{bmatrix} 1 & 2 \\ -3 & 4 \end{bmatrix}$ *are similar.*

Sol. A and B are similar if there exists a non-singular matrix P such that

$$A = P^{-1}BP \quad \text{or} \quad PA = BP$$

Let $P = \begin{bmatrix} a & b \\ c & d \end{bmatrix}$. We shall find a, b, c, d such that PA = BP and then check whether P is non-singular.

Now, $PA = BP$

$$\Rightarrow \quad \begin{bmatrix} a & b \\ c & d \end{bmatrix}\begin{bmatrix} 5 & 5 \\ -2 & 0 \end{bmatrix} = \begin{bmatrix} 1 & 2 \\ -3 & 4 \end{bmatrix}\begin{bmatrix} a & b \\ c & d \end{bmatrix}$$

$$\Rightarrow \quad \begin{bmatrix} 5a - 2b & 5a \\ 5c - 2d & 5c \end{bmatrix} = \begin{bmatrix} a + 2c & b + 2d \\ -3a + 4c & -3b + 4d \end{bmatrix}$$

Equating corresponding elements, we have

$$5a - 2b = a + 2c \quad \text{or} \quad 4a - 2b - 2c = 0$$

or

$$2a - b - c = 0 \qquad ...(1)$$

$$5a = b + 2d \quad \text{or} \quad 5a - b - 2d = 0 \qquad ...(2)$$

$$5c - 2d = -3a + 4c \quad \text{or} \quad 3a + c - 2d = 0 \qquad ...(3)$$

$$5c = -3b + 4d \quad \text{or} \quad 3b + 5c - 4d = 0 \qquad ...(4)$$

The co-efficient matrix $\begin{bmatrix} 2 & -1 & -1 & 0 \\ 5 & -1 & 0 & -2 \\ 3 & 0 & 1 & -2 \\ 0 & 3 & 5 & -4 \end{bmatrix}$ of this system of homogeneous equations has rank 2. A solution to this system of equations is $a = b = c = 1, d = 2$

$\therefore \quad P = \begin{bmatrix} 1 & 1 \\ 1 & 2 \end{bmatrix}$ which is non-singular. Hence the matrices A and B are similar.

EXERCISE 2.6

1. Examine whether A is similar to B, where

$$A = \begin{bmatrix} 1 & 0 \\ 0 & 1 \end{bmatrix} \text{ and } B = \begin{bmatrix} 1 & 1 \\ 0 & 1 \end{bmatrix}$$ **(Ans.** No.)

2. Show that the matrices

$$A = \begin{bmatrix} 2 & 1 & -1 \\ 1 & 3 & 2 \\ -1 & 2 & 2 \end{bmatrix} \text{ and } B = \begin{bmatrix} -13 & -47 & -58 \\ 4 & 14 & 15 \\ 1 & 4 & 6 \end{bmatrix} \text{ are similar if}$$

$$P = \begin{bmatrix} 1 & 3 & 3 \\ 1 & 4 & 3 \\ 1 & 3 & 4 \end{bmatrix} \text{ and } P^{-1} = \begin{bmatrix} 7 & -3 & -3 \\ -1 & 1 & 0 \\ -1 & 0 & 1 \end{bmatrix}$$

[Hint. $P^{-1}AP = B$, $PBP^{-1} = A$]

3. Show that 'similarity of matrices' is an equivalence relation.
4. Let A be similar to B. Then show that

   (*i*) A^{-1} is similar to B^{-1} (*ii*) $|A| = |B|$

2.20. QUADRATIC FORMS

Definition. A homogeneous polynomial of second degree in any number of variables is called a quadratic form. For example,

(*i*) $ax^2 + 2hxy + by^2$ (*ii*) $ax^2 + by^2 + cz^2 + 2hxy + 2gyz + 2fzx$ and

(*iii*) $ax^2 + by^2 + cz^2 + dw^2 + 2hxy + 2gyz + 2fzx + 2lxw + 2myw + 2nzw$

are quadratic forms in two, three and four variables.

In *n*-variables $x_1, x_2, \ldots\ldots, x_n$, the general quadratic form is $\sum_{j=1}^{n} \sum_{i=1}^{n} b_{ij} x_i x_j$

In the expansion, the co-efficient of $x_i x_j = (b_{ij} + b_{ji})$.

Suppose $2a_{ij} = b_{ij} + b_{ij}$ where $a_{ij} = a_{ji}$ and $a_{ii} = b_{ii}$

$$\therefore \quad \sum_{j=1}^{n} \sum_{i=1}^{n} b_{ij} x_i x_j = \sum_{j=1}^{n} \sum_{i=1}^{n} a_{ij} x_i x_j \text{ where } a_{ij} = \frac{1}{2}(b_{ij} + b_{ji}).$$

Hence every quadratic form can be written as $\sum_{j=1}^{n}\sum_{i=1}^{n} a_{ij}x_i x_j = X'AX$, so that the matrix A is always symmetric, where $A = [a_{ij}]$ and $X' = [x_1, x_2, ..., x_n]$.

Rule to write the matrix A of quadratic form

$$A = \begin{bmatrix} \text{co-eff. of } x_1^2 & \frac{1}{2}\text{co-eff. of } x_1x_2 & \frac{1}{2}\text{co-eff. of } x_1x_3 & \dots \\ \frac{1}{2}\text{co-eff. of } x_2x_1 & \text{co-eff. of } x_2^2 & \frac{1}{2}\text{co-eff. of } x_2x_3 & \dots \\ \frac{1}{2}\text{co-eff. of } x_3x_1 & \frac{1}{2}\text{co-eff. of } x_3x_2 & \text{co-eff. of } x_3^2 & \dots \\ \dots & \dots & \dots & \dots \end{bmatrix}$$

Now writing the above said examples of quadratic forms in matrix form, we get

(*i*) $ax^2 + 2hxy + by^2 = [x\ y]\begin{bmatrix} a & h \\ h & b \end{bmatrix}\begin{bmatrix} x \\ y \end{bmatrix}$

(*ii*) $ax^2 + by^2 + cz^2 + 2hxy + 2gyz + 2fzx = [x\ y\ z]\begin{bmatrix} a & h & f \\ h & b & g \\ f & g & c \end{bmatrix}\begin{bmatrix} x \\ y \\ z \end{bmatrix}$

and (*iii*) $ax^2 + by^2 + cz^2 + dw^2 + 2hxy + 2gyz + 2fzx + 2lxw + 2myw + 2nzw$

$$= [x\ y\ z\ w]\begin{bmatrix} a & h & f & l \\ h & b & g & m \\ f & g & c & n \\ l & m & n & d \end{bmatrix}\begin{bmatrix} x \\ y \\ z \\ w \end{bmatrix}.$$

2.21. LINEAR TRANSFORMATION OF A QUADRATIC FORM

Let X′AX be a quadratic form in n-variables and let X = PY ...(1)

where P is a non-singular matrix, be the non-singular transformation.

From (1), X′ = (PY)′ = Y′P′ and hence

X′AX = Y′P′APY = Y′(P′AP)Y = Y′BY ...(2)

where B = P′AP. Therefore Y′BY is also a quadratic form in n-variables. Hence it is a linear transformation of the quadratic form X′AX under the linear transformation X = PY and B = P′AP.

2.22. CANONICAL FORM

If a real quadratic form be expressed as a sum or difference of the squares of new variables by means of any real non-singular linear transformation, then the latter quadratic expression is called a *canonical form* of the given quadratic form.

i.e., if the quadratic form $X'AX = \sum_{j=1}^{n}\sum_{i=1}^{n} a_{ij}x_i x_j$ can be reduced to the quadratic form

$$Y'BY = \sum_{i=1}^{n} \lambda_i y_i^2 \text{ by a non-singular linear transformation}$$

X = PY then Y′BY is called the canonical form of the given one.

$$\therefore \quad \text{If } B = P'AP = \text{diag}(\lambda_1, \lambda_2, \ldots\ldots, \lambda_n) \text{ then } X'AX = Y'BY = \sum_{j=1}^{n} \lambda_i y_i^2 .$$

Note 1. Here some of λ_i (eigen values) may be positive or negative or zero.

2. A quadratic form is said to be real if the elements of the symmetric matrix are real.

3. If $\rho(A) = r$, then the quadratic form X′AX will contain only r terms.

2.23. INDEX AND SIGNATURE OF THE QUADRATIC FORM

The number p of positive terms in the canonical form is called the *index* of the quadratic form.

(The number of positive terms) – (The number of negative terms) *i.e.*, $p - (r - p) = 2p - r$ is called *signature* of the quadratic form, where $\rho(A) = r$.

2.24. DEFINITE, SEMI-DEFINITE AND INDEFINITE REAL QUADRATIC FORMS

Let X′AX be a real quadratic form in n-variables $x_1, x_2, \ldots\ldots, x_n$ with rank r and index p. Then we say that the quadratic form is

(*i*) *positive definite* if $r = n, p = r$

(*ii*) *negative definite* if $r = n, p = 0$

(*iii*) *positive semi-definite* if $r < n, p = r$

and (*iv*) *negative semi-definite* if $r < n, p = 0$

If the canonical form has both positive and negative terms, the quadratic form is said to be *indefinite*.

Note. If X′AX is positive definite then $|A| > 0$.

2.25. LAW OF INERTIA OF QUADRATIC FORM

The index of a real quadratic form is invariant under real non-singular transformations.

ILLUSTRATIVE EXAMPLES

Example 1. *Reduce* $3x^2 + 3z^2 + 4xy + 8xz + 8yz$ *into canonical form.*

Sol. The given quadratic form can be written as X′AX where $X' = [x, y, z]$ and the symmetric matrix

$$A = \begin{bmatrix} 3 & 2 & 4 \\ 2 & 0 & 4 \\ 4 & 4 & 3 \end{bmatrix}.$$

Let us reduce A into diagonal matrix. We know that $A = I_3 A I_3$

i.e.,
$$\begin{bmatrix} 3 & 2 & 4 \\ 2 & 0 & 4 \\ 4 & 4 & 3 \end{bmatrix} = \begin{bmatrix} 1 & 0 & 0 \\ 0 & 1 & 0 \\ 0 & 0 & 1 \end{bmatrix} A \begin{bmatrix} 1 & 0 & 0 \\ 0 & 1 & 0 \\ 0 & 0 & 1 \end{bmatrix}$$

Operating $R_2 \to R_2 - \frac{2}{3} R_1$, $R_3 \to R_3 - \frac{4}{3} R_1$ (for A on L.H.S. and pre-factor on R.H.S.), we get

$$\begin{bmatrix} 3 & 2 & 4 \\ 0 & -\frac{4}{3} & \frac{4}{3} \\ 0 & \frac{4}{3} & -\frac{7}{3} \end{bmatrix} = \begin{bmatrix} 1 & 0 & 0 \\ -\frac{2}{3} & 1 & 0 \\ -\frac{4}{3} & 0 & 1 \end{bmatrix} A \begin{bmatrix} 1 & 0 & 0 \\ 0 & 1 & 0 \\ 0 & 0 & 1 \end{bmatrix}$$

Operating $C_2 \to C_2 - \frac{2}{3} C_1$, $C_3 \to C_3 - \frac{4}{3} C_1$ (for A on L.H.S. and post-factor on R.H.S.), we get

$$\begin{bmatrix} 3 & 0 & 0 \\ 0 & -\frac{4}{3} & \frac{4}{3} \\ 0 & \frac{4}{3} & -\frac{7}{3} \end{bmatrix} = \begin{bmatrix} 1 & 0 & 0 \\ -\frac{2}{3} & 1 & 0 \\ -\frac{4}{3} & 0 & 1 \end{bmatrix} A \begin{bmatrix} 1 & -\frac{2}{3} & -\frac{4}{3} \\ 0 & 1 & 0 \\ 0 & 0 & 1 \end{bmatrix}$$

Operating $R_3 \to R_3 + R_2$, we get

$$\begin{bmatrix} 3 & 0 & 0 \\ 0 & -\frac{4}{3} & \frac{4}{3} \\ 0 & 0 & -1 \end{bmatrix} = \begin{bmatrix} 1 & 0 & 0 \\ -\frac{2}{3} & 1 & 0 \\ -2 & 1 & 1 \end{bmatrix} A \begin{bmatrix} 1 & -\frac{2}{3} & -\frac{4}{3} \\ 0 & 1 & 0 \\ 0 & 0 & 1 \end{bmatrix}$$

Operating $C_3 \to C_3 + C_2$, we get

$$\begin{bmatrix} 3 & 0 & 0 \\ 0 & -\frac{4}{3} & 0 \\ 0 & 0 & -1 \end{bmatrix} = \begin{bmatrix} 1 & 0 & 0 \\ -\frac{2}{3} & 1 & 0 \\ -2 & 1 & 1 \end{bmatrix} A \begin{bmatrix} 1 & -\frac{2}{3} & -2 \\ 0 & 1 & 1 \\ 0 & 0 & 1 \end{bmatrix}$$

or
$$\text{Diag}\left(3, -\frac{4}{3}, -1\right) = P'AP$$

$\therefore$ The canonical form of the given quadratic form is

$$Y'(P'AP)Y = [y_1, y_2, y_3] \begin{bmatrix} 3 & 0 & 0 \\ 0 & -\frac{4}{3} & 0 \\ 0 & 0 & -1 \end{bmatrix} \begin{bmatrix} y_1 \\ y_2 \\ y_3 \end{bmatrix} = 3y_1^2 - \frac{4}{3} y_2^2 - y_3^2.$$

Here $\rho(A) = 3$, index = 1, signature = 1 – 2 = – 1.

Note. In this problem the non-singular transformation which reduces the given quadratic form into the canonical form is X = PY *i.e.*, $\begin{bmatrix} x \\ y \\ z \end{bmatrix} = \begin{bmatrix} 1 & -\frac{2}{3} & -2 \\ 0 & 1 & 1 \\ 0 & 0 & 1 \end{bmatrix} \begin{bmatrix} y_1 \\ y_2 \\ y_3 \end{bmatrix}$. *i.e.*, $x = y_1 - \frac{2}{3} y_2 - 2y_3,\ y = y_2 + y_3,\ z = y_3.$

Note. The above example can also be questioned as 'Diagonalise' the quadratic form $3x^2 + 3z^2 + 4xy + 8xz + 8yz$ by linear transformation and write the linear transformation.

Or

Reduce the quadratic form $3x^2 + 3z^2 + 4xy + 8xz + 8yz$ into the sum of squares.

Example 2. *Reduce the quadratic form* $x^2 - 4y^2 + 6z^2 + 2xy - 4xz + 2w^2 - 6zw$ *into sum of squares.*

Sol. The matrix form of the given quadratic form is X′AX, where X′ = [$x\ y\ z\ w$]

and $$A = \begin{bmatrix} 1 & 1 & -2 & 0 \\ 1 & -4 & 0 & 0 \\ -2 & 0 & 6 & -3 \\ 0 & 0 & -3 & 2 \end{bmatrix}$$

Let us reduce A to the diagonal matrix. We know that $A = I_4 A I_4$.

$$\begin{bmatrix} 1 & 1 & -2 & 0 \\ 1 & -4 & 0 & 0 \\ -2 & 0 & 6 & -3 \\ 0 & 0 & -3 & 2 \end{bmatrix} = \begin{bmatrix} 1 & 0 & 0 & 0 \\ 0 & 1 & 0 & 0 \\ 0 & 0 & 1 & 0 \\ 0 & 0 & 0 & 1 \end{bmatrix} A \begin{bmatrix} 1 & 0 & 0 & 0 \\ 0 & 1 & 0 & 0 \\ 0 & 0 & 1 & 0 \\ 0 & 0 & 0 & 1 \end{bmatrix}$$

Operating $R_2 \to R_2 - R_1$, $R_3 \to R_3 + 2R_1$, we get

$$\begin{bmatrix} 1 & 1 & -2 & 0 \\ 0 & -5 & 2 & 0 \\ 0 & 2 & 2 & -3 \\ 0 & 0 & -3 & 2 \end{bmatrix} = \begin{bmatrix} 1 & 0 & 0 & 0 \\ -1 & 1 & 0 & 0 \\ 2 & 0 & 1 & 0 \\ 0 & 0 & 0 & 1 \end{bmatrix} A \begin{bmatrix} 1 & 0 & 0 & 0 \\ 0 & 1 & 0 & 0 \\ 0 & 0 & 1 & 0 \\ 0 & 0 & 0 & 1 \end{bmatrix}$$

Operating $C_2 \to C_2 - C_1$, $C_3 \to C_3 + 2C_1$, we get

$$\begin{bmatrix} 1 & 0 & 0 & 0 \\ 0 & -5 & 2 & 0 \\ 0 & 2 & 2 & -3 \\ 0 & 0 & -3 & 2 \end{bmatrix} = \begin{bmatrix} 1 & 0 & 0 & 0 \\ -1 & 1 & 0 & 0 \\ 2 & 0 & 1 & 0 \\ 0 & 0 & 0 & 1 \end{bmatrix} A \begin{bmatrix} 1 & -1 & 2 & 0 \\ 0 & 1 & 0 & 0 \\ 0 & 0 & 1 & 0 \\ 0 & 0 & 0 & 1 \end{bmatrix}$$

Operating $R_3 \to R_3 + (2/5)R_2$, we get

$$\begin{bmatrix} 1 & 0 & 0 & 0 \\ 0 & -5 & 2 & 0 \\ 0 & 0 & \frac{14}{5} & -3 \\ 0 & 0 & -3 & 2 \end{bmatrix} = \begin{bmatrix} 1 & 0 & 0 & 0 \\ -1 & 1 & 0 & 0 \\ \frac{8}{5} & \frac{2}{5} & 1 & 0 \\ 0 & 0 & 0 & 1 \end{bmatrix} A \begin{bmatrix} 1 & -1 & 2 & 0 \\ 0 & 1 & 0 & 0 \\ 0 & 0 & 1 & 0 \\ 0 & 0 & 0 & 1 \end{bmatrix}$$

Operating $C_3 \to C_3 + (2/5)\, C_2$, we get

$$\begin{bmatrix} 1 & 0 & 0 & 0 \\ 0 & -5 & 0 & 0 \\ 0 & 0 & \frac{14}{5} & -3 \\ 0 & 0 & -3 & 2 \end{bmatrix} = \begin{bmatrix} 1 & 0 & 0 & 0 \\ -1 & 1 & 0 & 0 \\ \frac{8}{5} & \frac{2}{5} & 1 & 0 \\ 0 & 0 & 0 & 1 \end{bmatrix} A \begin{bmatrix} 1 & -1 & \frac{8}{5} & 0 \\ 0 & 1 & \frac{2}{5} & 0 \\ 0 & 0 & 1 & 0 \\ 0 & 0 & 0 & 1 \end{bmatrix}$$

Operating $R_4 \to R_4 + \frac{15}{14} R_3$, we get

$$\begin{bmatrix} 1 & 0 & 0 & 0 \\ 0 & -5 & 0 & 0 \\ 0 & 0 & \frac{14}{5} & -3 \\ 0 & 0 & 0 & -\frac{17}{14} \end{bmatrix} = \begin{bmatrix} 1 & 0 & 0 & 0 \\ -1 & 1 & 0 & 0 \\ \frac{8}{5} & \frac{2}{5} & 1 & 0 \\ \frac{12}{7} & \frac{3}{7} & \frac{15}{14} & 1 \end{bmatrix} A \begin{bmatrix} 1 & -1 & \frac{8}{5} & 0 \\ 0 & 1 & \frac{2}{5} & 0 \\ 0 & 0 & 1 & 0 \\ 0 & 0 & 0 & 1 \end{bmatrix}$$

Operating $C_4 \to C_4 + \frac{15}{14} C_3$, we get

$$\begin{bmatrix} 1 & 0 & 0 & 0 \\ 0 & -5 & 0 & 0 \\ 0 & 0 & \frac{14}{5} & 0 \\ 0 & 0 & 0 & -\frac{17}{14} \end{bmatrix} = \begin{bmatrix} 1 & 0 & 0 & 0 \\ -1 & 1 & 0 & 0 \\ \frac{8}{5} & \frac{2}{5} & 1 & 0 \\ \frac{12}{7} & \frac{3}{7} & \frac{15}{14} & 1 \end{bmatrix} A \begin{bmatrix} 1 & -1 & \frac{8}{5} & \frac{12}{7} \\ 0 & 1 & \frac{2}{5} & \frac{3}{7} \\ 0 & 0 & 1 & \frac{15}{14} \\ 0 & 0 & 0 & 1 \end{bmatrix}$$

i.e., $$\text{diag}\left(1, -5, \frac{14}{5}, -\frac{17}{14}\right) = P'AP$$

$\therefore$ the canonical form of the given quadratic form is

$$Y'(P'AP)Y = Y' \,\text{diag}\left(1, -5, \frac{14}{5}, -\frac{17}{14}\right) Y$$

$$= [y_1 y_2 y_3 y_4] \begin{bmatrix} 1 & 0 & 0 & 0 \\ 0 & -5 & 0 & 0 \\ 0 & 0 & \frac{14}{5} & 0 \\ 0 & 0 & 0 & -\frac{17}{14} \end{bmatrix} \begin{bmatrix} y_1 \\ y_2 \\ y_3 \\ y_4 \end{bmatrix}$$

$$= y_1^2 - 5y_2^2 + \frac{14}{5} y_3^2 - \frac{17}{14} y_4^2,$$ which is the sum of the squares.

Note. Here $\rho(A) = 4$, index $= 2$, signature $= 2 - 2 = 0$.

2.26. REDUCTION TO CANONICAL FORM BY ORTHOGONAL TRANSFORMATION

Let X′AX be a given quadratic form. The modal matrix B of A is that matrix whose columns are characteristic vectors of A. If B represents the orthogonal matrix of A (the normalised modal matrix of A whose column vectors are pairwise orthogonal) then X = BY will reduce X′AX to Y′DY where D = diag $(\lambda_1, \lambda_2, \ldots\ldots, \lambda_n)$ and $\lambda_1, \lambda_2, \ldots\ldots, \lambda_n$ are characteristic roots of A.

Note. This method works successfully if the characteristic vectors of A are linearly independent which are pairwise orthogonal.

Example 3. *Reduce $8x^2 + 7y^2 + 3z^2 - 12xy + 4xz - 8yz$ into canonical form by orthogonal transformation.* (Calicut 2008)

Sol. The matrix of the quadratic form is

$$A = \begin{bmatrix} 8 & -6 & 2 \\ -6 & 7 & -4 \\ 2 & -4 & 3 \end{bmatrix}$$

The characteristic roots of A are given by $| A - \lambda I | = 0$

i.e., $$\begin{vmatrix} 8-\lambda & -6 & 2 \\ -6 & 7-\lambda & -4 \\ 2 & -4 & 3-\lambda \end{vmatrix} = 0 \quad \text{or} \quad \lambda(\lambda - 3)(\lambda - 15) = 0$$

$\therefore$ $$\lambda = 0, 3, 15$$

Characteristic vector for $\lambda = 0$ is given by $[A - (0)I]\, X = O$

i.e., $$8x_1 - 6x_2 + 2x_3 = 0$$
$$-6x_1 + 7x_2 - 4x_3 = 0$$
$$2x_1 - 4x_2 + 3x_3 = 0$$

Solving first two, we get $\dfrac{x_1}{1} = \dfrac{x_2}{2} = \dfrac{x_3}{2}$ giving the eigen vector $X_1 = k_1(1, 2, 2)'$

When $\lambda = 3$, the corresponding characteristic vector is given by $[A - 3I]X = O$

i.e., $$5x_1 - 6x_2 + 2x_3 = 0$$
$$-6x_1 + 4x_2 - 4x_3 = 0$$
$$2x_1 - 4x_2 = 0$$

Solving any two equations, we get $X_2 = k_2(2, 1, -2)'$.

Similarly characteristic vector corresponding to

$$\lambda = 15 \text{ is } X_3 = k_3(2, -2, 1)'.$$

Now X_1, X_2, X_3 are pairwise orthogonal

i.e., $$X_1 \cdot X_2 = X_2 \cdot X_3 = X_3 \cdot X_1 = 0.$$

$\therefore$ The normalised modal matrix is

$$B = \left[\frac{X_1}{\|X_1\|}, \frac{X_2}{\|X_2\|}, \frac{X_3}{\|X_3\|}\right] = \begin{bmatrix} \frac{1}{3} & \frac{2}{3} & \frac{2}{3} \\ \frac{2}{3} & \frac{1}{3} & -\frac{2}{3} \\ \frac{2}{3} & \frac{-2}{3} & \frac{1}{3} \end{bmatrix}$$

Now B is orthogonal matrix and $| B | = 1$

i.e., $$B^{-1} = B^T \text{ and } B^{-1}AB = D = \text{diag}(0, 3, 15)$$

i.e.,
$$\begin{bmatrix} \frac{1}{3} & \frac{2}{3} & \frac{2}{3} \\ \frac{2}{3} & \frac{1}{3} & -\frac{2}{3} \\ \frac{2}{3} & -\frac{2}{3} & \frac{1}{3} \end{bmatrix} A \begin{bmatrix} \frac{1}{3} & \frac{2}{3} & \frac{2}{3} \\ \frac{2}{3} & \frac{1}{3} & \frac{-2}{3} \\ \frac{2}{3} & -\frac{2}{3} & \frac{1}{3} \end{bmatrix} = \begin{bmatrix} 0 & 0 & 0 \\ 0 & 3 & 0 \\ 0 & 0 & 15 \end{bmatrix}$$

$$X'AX = Y'(B^{-1}AB)\,Y = Y'DY$$

$$= [y_1\; y_2\; y_3] \begin{bmatrix} 0 & 0 & 0 \\ 0 & 3 & 0 \\ 0 & 0 & 15 \end{bmatrix} \begin{bmatrix} y_1 \\ y_2 \\ y_3 \end{bmatrix} = 0y_1^2 + 3\,y_2^2 + 15y_3^2$$

which is the required canonical form.

Note. Here the orthogonal transformation is X = BY, rank of the quadratic form = 2; index = 2, signature = 2. It is positive semi-definite.

Example 4. *Reduce* $6x_1^2 + 3x_2^2 + 3x_3^2 - 4x_1x_2 - 2x_2x_3 + 4x_3x_1$ *into canonical form.*

Sol. The matrix of the quadratic form is $A = \begin{bmatrix} 6 & -2 & 2 \\ -2 & 3 & -1 \\ 2 & -1 & 3 \end{bmatrix}$

The characteristic roots are given by $|A - \lambda I| = 0$

i.e.,
$$\begin{vmatrix} 6-\lambda & -2 & 2 \\ -2 & 3-\lambda & -1 \\ 2 & -1 & 3-\lambda \end{vmatrix} = 0$$

or
$$\lambda^3 - 12\lambda^2 + 36\lambda - 32 = 0$$

which on solving gives λ = 8, 2, 2.

The vector corresponding to λ = 8 is given by [A – 8I]X = O

i.e.,
$$\begin{bmatrix} -2 & -2 & 2 \\ -2 & -5 & -1 \\ 2 & -1 & -5 \end{bmatrix} \begin{bmatrix} x_1 \\ x_2 \\ x_3 \end{bmatrix} = \begin{bmatrix} 0 \\ 0 \\ 0 \end{bmatrix}$$

or
$$-2x_1 - 2x_2 + 2x_3 = 0$$
$$-2x_1 - 5x_2 - x_3 = 0$$
$$2x_1 - x_2 - 5x_3 = 0$$

Solving any two of the equations, we get the vector as $k_1[2, -1, 1]'$.

The characteristic vector for λ = 2 is given by [A – 2I]X = O which reduces to single equation

$$2x_1 - x_2 + x_3 = 0.$$

Putting $x_2 = 0$, we get $\frac{x_2}{1} = \frac{x_3}{1}$ or the vector is $k_2\,[0, 1, 1]'$. Again by putting $x_2 = 0$, we get $\frac{x_1}{1} = \frac{x_3}{-2}$ or the vector is $k_3\,[1, 0 - 2]'$.

Now $X_1 = k_1\,[2, -1, 1]'$; $X_2 = k_2\,[0, 1, 1]'$ and $X_3 = k_3\,[1, 0, -2]'$.

Here X_1, X_2, X_3 are not pairwise orthogonal

$\because$ $\quad X_1 . X_2 = 0$; $X_2 . X_3 \neq 0$ and $X_3 . X_1 = 0$

To get X_3 orthogonal to X_2 assume a vector $[u, v, w]'$ orthogonal to X_2 also satisfying

$2x_1 - x_2 + x_3 = 0$ *i.e.,* $2u - v + w = 0$ and $0.u + 1.v + 1.w = 0$

Solving $[u, v, w]' = [1, 1, -1]' = X_3$ so that $X_1 . X_2 = X_2 . X_3 = X_3 . X_1 = 0$

$\therefore$ The normalised modal matrix is

$$B = \begin{bmatrix} \frac{2}{\sqrt{6}} & 0 & \frac{1}{\sqrt{3}} \\ -\frac{1}{\sqrt{6}} & \frac{1}{\sqrt{2}} & \frac{1}{\sqrt{3}} \\ \frac{1}{\sqrt{6}} & \frac{1}{\sqrt{2}} & -\frac{1}{\sqrt{3}} \end{bmatrix}$$

Now B is orthogonal matrix and $|B| = -1$

i.e., $$B' = B^{-1} \text{ and } B^{-1}AB = D \text{ where } D = \begin{bmatrix} 8 & 0 & 0 \\ 0 & 2 & 0 \\ 0 & 0 & 2 \end{bmatrix}$$

$$\therefore \quad X'AX = Y'(B^{-1}AB)Y = Y'DY$$

$$= [y_1\ y_2\ y_3] \begin{bmatrix} 8 & 0 & 0 \\ 0 & 2 & 0 \\ 0 & 0 & 2 \end{bmatrix} \begin{bmatrix} y_1 \\ y_2 \\ y_3 \end{bmatrix} = 8y_1^2 + 2y_2^2 + 2y_3^2$$

which is the required canonical form.

Note. Here rank of the quadratic form is 3, index = 3, signature = 3. It is positive definite.

EXERCISE 2.7

1. Write down the matrices of the following quadratic forms

   (*i*) $2x^2 + 3y^2 + 6xy$ (*ii*) $2x^2 + 5y^2 - 6z^2 - 2xy - yz + 8zx$

   (*iii*) $x_1^2 + 2x_2^2 + 3x_3^2 + 4x_4^2 + 2x_1x_2 + 4x_1x_3 - 6x_1x_4 - 4x_2x_3 - 8x_2x_4 + 12x_3x_4$.

2. Write down the quadratic forms corresponding to the following matrices:

   (*i*) $\begin{bmatrix} 2 & 4 & 5 \\ 4 & 3 & 1 \\ 5 & 1 & 1 \end{bmatrix}$ (*M.D.U. Dec. 2009*) (*ii*) $\begin{bmatrix} 1 & 1 & -2 & 0 \\ 1 & -4 & 0 & 0 \\ -2 & 0 & 6 & -3 \\ 0 & 0 & -3 & 2 \end{bmatrix}$

3. Reduce the following quadratic forms to canonical forms or to sum of squares by linear transformation. Write also the rank, index and signature.

   (*i*) $2x^2 + 2y^2 + 3z^2 + 2xy - 4yz - 4xz$ (*M.D.U. Dec. 2010*) (*ii*) $12x_1^2 + 4x_2^2 + 5x_3^2 - 4x_2x_3 + 6x_1x_3 - 6x_1x_2$

   (*iii*) $2x^2 + 9y^2 + 6z^2 + 8xy + 8yz + 6zx$ (*iv*) $x^2 + 4y^2 + z^2 + 4xy + 6yz + 2zx$.

4. Reduce the following quadratic forms to canonical forms or to sum of squares by orthogonal transformation. Write also rank, index, signature.

   (*i*) $3x^2 + 5y^2 + 3z^2 - 2xy - 2yz + 2zx$ (*K.U.K. 2008*) (*ii*) $2x_1^2 + 2x_2^2 + 2x_3^2 - 2x_1x_2 + 2x_1x_3 - 2x_2x_3$

   (*iii*) $3x^2 - 2y^2 - z^2 - 4xy + 8xz + 12yz$ (*iv*) $x^2 + 3y^2 + 3z^2 - 2yz$. (*Calicut 2009*)

Answers

1. (i) $\begin{bmatrix} 2 & 3 \\ 3 & 3 \end{bmatrix}$ (ii) $\begin{bmatrix} 2 & -1 & 4 \\ -1 & 5 & -\frac{1}{2} \\ 4 & -\frac{1}{2} & -6 \end{bmatrix}$ (iii) $\begin{bmatrix} 1 & 1 & 2 & -3 \\ 1 & 2 & -2 & -4 \\ 2 & -2 & 3 & 6 \\ -3 & -4 & 6 & 4 \end{bmatrix}$

2. (i) $2x^2 + 3y^2 + z^2 + 8xy + 2yz + 10xz$ (ii) $x_1^2 - 4x_2^2 + 6x_3^2 + 2x_4^2 + 2x_1x_2 - 4x_1x_3 - 6x_3x_4$

3. (i) $2y_1^2 + \frac{3}{2}y_2^2 + \frac{1}{3}y_3^2$; Rank = 3, index = 3, Sig. = 3

   (ii) $12y_1^2 + \frac{13}{4}y_2^2 + \frac{49}{13}y_3^2$; Rank = 3, index = 3, Sig. = 3

   (iii) $2y_1^2 + y_2^2 - \frac{5}{2}y_3^2$; Rank = 3, index = 2, Sig. = 1

   (iv) $y_1^2 + 2y_2^2 - \frac{1}{2}y_3^2$; Rank = 3, index = 2, Sig. = 1.

4. (i) $2y_1^2 + 3y_2^2 + 6y_3^2$; Rank = 3 , index = 3, Sig. = 3

   (ii) $4y_1^2 + y_2^2 + y_3^2$; Rank = 3, index = 3, Sig. = 3

   (iii) $3y_1^2 + 6y_2^2 - 9y_3^2$; Rank = 3, index = 2, Sig. = 1

   (iv) $y_1^2 + 2y_2^2 + 4y_3^2$; Rank = 3, index = 3, Sig. = 3.

2.27. COMPLEX MATRICES

If *all* the elements of a matrix are real numbers, then it is called *a real matrix* or a matrix over R. On the other hand, if *at least one* element of a matrix is a complex number $a + ib$, where a, b are real and $i = \sqrt{-1}$, then the matrix is called a *complex matrix.*

The matrix obtained by replacing the elements of a complex matrix A by the corresponding conjugate complex numbers is called the *conjugate of the matrix* A and is denoted by $\overline{A}$.

Thus, if $A = \begin{bmatrix} 2+3i & -7i \\ 5 & 1-i \end{bmatrix}$, then $\overline{A} = \begin{bmatrix} 2-3i & 7i \\ 5 & 1+i \end{bmatrix}$

It is easy to see that the *conjugate of the transpose of* A *i.e.,* $(\overline{A'})$ and the *transposed conjugate of* A *i.e.,* $(\overline{A})'$ are equal. Each of them is denoted by A^*.

Thus, $(\overline{A'}) = (\overline{A})' = A^*$.

A square matrix A is said to be *Hermitian* if $A^* = A$.

A square matrix A is said to be *skew-Hermitian* if $A^* = -A$.

In a Hermitian matrix, the diagonal elements are all real, while every other element is the conjugate complex of the element in the transposed position. For example,

$$A = \begin{bmatrix} 5 & 2+i & -3i \\ 2-i & -3 & 1-i \\ 3i & 1+i & 0 \end{bmatrix} \text{ is a Hermitian matrix.}$$

In a skew-Hermitian matrix, the diagonal elements are zero or purely imaginary numbers of the form $i\beta$, where β is real. Every other element is the negative of the cojugate complex of the element in the transposed position.

For example, $$B = \begin{bmatrix} 3i & 1+i & 7 \\ -1+i & 0 & -2-i \\ -7 & 2-i & -i \end{bmatrix} \text{ is a skew-Hermitian matrix.}$$

2.28. UNITARY MATRIX

A square matrix A is said to be *unitary* if A*A = I = AA*. The determinant of a unitary matrix is of unit modulus. For a matrix to be unitary, it must be non-singular.

Note. The following results hold:

(*i*) $\overline{(\overline{A})} = A$ (*ii*) $\overline{A + B} = \overline{A} + \overline{B}$ (*iii*) $\overline{\lambda A} = \overline{\lambda}\,\overline{A}$

(*iv*) $\overline{AB} = \overline{A}\,\overline{B}$ (*v*) $(A^*)^* = A$ (*vi*) $(A + B)^* = A^* + B^*$

(*vii*) $(\lambda A)^* = \overline{\lambda}A^*$ (*viii*) $(AB)^* = B^*A^*$.

ILLUSTRATIVE EXAMPLES

Example 1. *If* $A = \begin{bmatrix} 2+i & 3 & -1+3i \\ -5 & i & 4-2i \end{bmatrix}$, *verify that A*A is a Hermitian matrix.*

Sol.
$$A' = \begin{bmatrix} 2+i & -5 \\ 3 & i \\ -1+3i & 4-2i \end{bmatrix}$$

$$A^* = \overline{(A')} = \begin{bmatrix} 2-i & -5 \\ 3 & -i \\ -1-3i & 4+2i \end{bmatrix}$$

$\therefore$
$$A^*A = \begin{bmatrix} 2-i & -5 \\ 3 & -i \\ -1-3i & 4+2i \end{bmatrix} \begin{bmatrix} 2+i & 3 & -1+3i \\ -5 & i & 4-2i \end{bmatrix}$$

$$= \begin{bmatrix} 30 & 6-8i & -19-17i \\ 6+8i & 10 & -5+5i \\ -19+17i & -5+5i & 30 \end{bmatrix} = B \text{ (say)}$$

Now
$$B' = \begin{bmatrix} 30 & 6+8i & -19-17i \\ 6-8i & 10 & -5-5i \\ -19+17i & -5+5i & 30 \end{bmatrix}$$

$$B^* = \overline{(B')} = \begin{bmatrix} 30 & 6-8i & -19+17i \\ 6+8i & 10 & -5+5i \\ -19-17i & -5-5i & 30 \end{bmatrix} = B$$

$\Rightarrow$ B = A*A is a Hermitian matrix.

Example 2. *If A and B are Hermitian, show that AB – BA is skew-Hermitian.*

Sol. A and B are Hermitian $\Rightarrow$ A* = A and B* = B

Now (AB – BA)* = (AB)* – (BA)*

= B*A* – A*B* = BA – AB = –(AB – BA)

$\Rightarrow$ AB – BA is skew-Hermitian.

Example 3. *If A is a skew-Hermitian matrix, then show that iA is Hermitian.*

Sol. A is a skew-Hermitian matrix $\Rightarrow$ $A^* = -A$

Now $(iA)^* = \bar{i}\, A^* = (-i)(-A) = iA$

$\Rightarrow$ iA is a Hermitian matrix.

Example 4. *Prove that a unitary matrix over R is an orthogonal matrix.*

Sol. Let A be a unitary matrix over R, then $\bar{A} = A$

$$\therefore \quad A^* = (\bar{A})' = A' \quad \ldots(1)$$

Since A is unitary, $AA^* = I = A^*A$

$$\Rightarrow \quad AA' = I = A'A \quad \text{[Using (1)]}$$

$\Rightarrow$ A is orthogonal.

Example 5. *Prove that the transpose of a unitary matrix is unitary.*

Sol. Let A be a unitary matrix, then

$$AA^* = I = A^*A$$

$$\Rightarrow \quad (AA^*)' = I' = (A^*A)'$$

$$\Rightarrow \quad (A^*)'A' = I = A'(A^*)' \quad [\because \ (AB)' = B'A']$$

$$\Rightarrow \quad ((\overline{A'}))'\ A' = I = A'((\overline{A'}))'$$

$$\Rightarrow \quad (A')^*\ A' = I = A'(A')^* \quad [\because \ (\bar{P})' = P^*]$$

$\Rightarrow$ A' is unitary.

Example 6. *Prove that the conjugate of a unitary matrix is unitary.*

Sol. Let A be a unitary matrix, then

$$AA^* = I = A^*A$$

$$\Rightarrow \quad \overline{AA^*} = \bar{I} = \overline{A^*A}$$

$$\Rightarrow \quad \bar{A}(\overline{A^*}) = I = (\overline{A^*})\bar{A} \quad [\because \ \overline{AB} = \bar{A}\,\bar{B}]$$

$$\Rightarrow \quad \bar{A}(\overline{(\bar{A})'}) = I = (\overline{(\bar{A})'})\bar{A}$$

$$\Rightarrow \quad \bar{A}(\bar{A})^* = I = (\bar{A})^*\bar{A} \quad [\because \ (\bar{P}') = P^*]$$

$\Rightarrow$ $\bar{A}$ is unitary.

Example 7. *Prove that the product of two unitary matrices is a unitary matrix.*

(K.U.K. 2013)

Sol. Let A and B be two unitary matrices, then

$$AA^* = I = A^*A \quad \text{and} \quad BB^* = I = B^*B \quad \ldots(1)$$

Now
$$\begin{aligned}(AB)(AB)^* &= (AB)(B^*A^*)\\ &= A(BB^*)A^*\\ &= AIA^* \quad \text{[Using (1)]}\\ &= (AI)A^*\\ &= AA^*\\ &= I \quad \text{[Using (1)]}\end{aligned}$$

Similarly $(AB)^*(AB) = I$

Since $(AB)(AB)^* = I = (AB)^*(AB)$

The matrix AB is unitary.

Example 8. *Show that every square matrix is expressible as the sum of a Hermitian matrix and a skew-Hermitian matrix.*

Sol. Let A be any square matrix.

Since $(A + A^*)^* = A^* + (A^*)^* = A^* + A = A + A^*$

and $(A - A^*)^* = A^* - (A^*)^* = A^* - A = -(A - A^*)$

$\therefore$ A + A* is Hermitian and A – A* is skew-Hermitian.

Now $$A = \frac{1}{2}(A + A^*) + \frac{1}{2}(A - A^*) = P + Q \text{ (say)}$$

where P is Hermitian and Q is skew-Hermitian. Thus, every square matrix can be expressed as the sum of a Hermitian matrix and a skew-Hermitian matrix.

Example 9. *Show that every square matrix A can be expressed as P + iQ, where P and Q are Hermitian matrices.*

Sol. Let A be any square matrix.

Consider $$A = \frac{1}{2}(A + A^*) + i \cdot \frac{1}{2i}(A - A^*) = P + iQ$$

where $$P = \frac{1}{2}(A + A^*) \quad \text{and} \quad Q = \frac{1}{2i}(A - A^*)$$

Now $$P^* = \left[\frac{1}{2}(A + A^*)\right]^* = \frac{1}{2}(A + A^*)^*$$

$$= \frac{1}{2}(A^* + (A^*)^*) = \frac{1}{2}(A^* + A) = \frac{1}{2}(A + A^*) = P$$

and $$Q^* = \left[\frac{1}{2i}(A - A^*)\right]^* = \overline{\left(\frac{1}{2i}\right)}(A - A^*)^*$$

$$= -\frac{1}{2i}(A^* - (A^*)^*) = -\frac{1}{2i}(A^* - A) = \frac{1}{2i}(A - A^*) = Q$$

Thus P and Q both are Hermitian matrices.

Example 10. *Show that every Hermitian matrix A can be expressed as P + iQ where P is real and symmetric while Q is real and skew symmetric.*

Sol. Let A be a Hermitian matrix so that A* = A

$\Rightarrow \quad (\bar{A})' = A \quad \Rightarrow \quad \bar{A} = A'$

Consider $$A = \frac{1}{2}(A + \bar{A}) + i \cdot \frac{1}{2i}(A - \bar{A}) = P + iQ$$

where $$P = \frac{1}{2}(A + \bar{A}) \quad \text{and} \quad Q = \frac{1}{2i}(A - \bar{A})$$

If $a_{ij} = x + iy$ then $\bar{a}_{ij} = x - iy$ so that $a_{ij} + \bar{a}_{ij} = 2x$

and $a_{ij} - \bar{a}_{ij} = 2iy$

Thus P and Q are real matrices.

Now $$P' = \left[\frac{1}{2}(A + \bar{A})\right]' = \frac{1}{2}(A + \bar{A})' = \frac{1}{2}(A' + (\bar{A})')$$

$$= \frac{1}{2}(A' + A^*) = \frac{1}{2}(\bar{A} + A) = \frac{1}{2}(A + \bar{A}) = P$$

$\Rightarrow$ P is symmetric.

Also
$$Q' = \left[\frac{1}{2i}(A - \bar{A})\right]' = \frac{1}{2i}(A - \bar{A})' = \frac{1}{2i}(A' - (\bar{A})')$$

$$= \frac{1}{2i}(A' - A^*) = \frac{1}{2i}(\bar{A} - A) = -\frac{1}{2i}(A - \bar{A}) = -Q$$

$\Rightarrow$ Q is skew symmetric.

Example 11. *If* $N = \begin{bmatrix} 0 & 1+2i \\ -1+2i & 0 \end{bmatrix}$*, obtain the matrix* $(I - N)(I + N)^{-1}$ *and show that it is unitary.* (M.T.U. 2013; G.B.T.U. 2011)

Sol.
$$I - N = \begin{bmatrix} 1 & 0 \\ 0 & 1 \end{bmatrix} - \begin{bmatrix} 0 & 1+2i \\ -1+2i & 0 \end{bmatrix} = \begin{bmatrix} 1 & -1-2i \\ 1-2i & 1 \end{bmatrix}$$

$$I + N = \begin{bmatrix} 1 & 0 \\ 0 & 1 \end{bmatrix} + \begin{bmatrix} 0 & 1+2i \\ -1+2i & 0 \end{bmatrix} = \begin{bmatrix} 1 & 1+2i \\ -1+2i & 1 \end{bmatrix}$$

$$|I + N| = \begin{vmatrix} 1 & 1+2i \\ -1+2i & 1 \end{vmatrix} = 1 - (4i^2 - 1) = 6$$

$$\text{adj}(I + N) = \begin{bmatrix} 1 & -1-2i \\ 1-2i & 1 \end{bmatrix}$$

$$(I + N)^{-1} = \frac{1}{|I + N|}\text{adj}(I + N) = \frac{1}{6}\begin{bmatrix} 1 & -1-2i \\ 1-2i & 1 \end{bmatrix}$$

$$\therefore \quad (I - N)(I + N)^{-1} = \begin{bmatrix} 1 & -1-2i \\ 1-2i & 1 \end{bmatrix} \frac{1}{6}\begin{bmatrix} 1 & -1-2i \\ 1-2i & 1 \end{bmatrix}$$

$$= \frac{1}{6}\begin{bmatrix} -4 & -2-4i \\ 2-4i & -4 \end{bmatrix} = A \text{ (say)}$$

$$A' = \frac{1}{6}\begin{bmatrix} -4 & 2-4i \\ -2-4i & -4 \end{bmatrix}$$

$$\overline{(A')} = A^* = \frac{1}{6}\begin{bmatrix} -4 & 2+4i \\ -2+4i & -4 \end{bmatrix}$$

$$A^*A = \frac{1}{6}\begin{bmatrix} -4 & 2+4i \\ -2+4i & -4 \end{bmatrix} \frac{1}{6}\begin{bmatrix} -4 & -2-4i \\ 2-4i & -4 \end{bmatrix}$$

$$= \frac{1}{36}\begin{bmatrix} 36 & 0 \\ 0 & 36 \end{bmatrix} = \begin{bmatrix} 1 & 0 \\ 0 & 1 \end{bmatrix} = I$$

$\Rightarrow$ $A = (I - N)(I + N)^{-1}$ is unitary.

EXERCISE 2.8

1. Show that the matrix $A = \begin{bmatrix} 2 & 3-4i \\ 3+4i & 2 \end{bmatrix}$ is Hermitian and iA is skew-Hermitian. *(U.K.T.U. 2012)*

2. If $H = \begin{bmatrix} 3 & 5+2i & -3 \\ 5-2i & 7 & 4i \\ -3 & -4i & 5 \end{bmatrix}$, prove that H is Hermitian.

   Also verify that iH is skew-Hermitian.

3. If $A = \begin{bmatrix} -1 & 2+i & 5-3i \\ 2-i & 7 & 5i \\ 5+3i & -5i & 2 \end{bmatrix}$, show that A is a Hermitian matrix and iA is a skew-Hermitian matrix.

4. If A is any square matrix, prove that $A + A^*$, AA^*, A^*A are all Hermitian and $A - A^*$ is skew-Hermitian.

5. If A, B are Hermitian or skew-Hermitian, then so is $A + B$.

6. Show that the matrix $B^* AB$ is Hermitian or skew-Hermitian according as A is Hermitian or skew-Hermitian.

7. Show that the matrix $\frac{1}{\sqrt{3}}\begin{bmatrix} 1 & 1+i \\ 1-i & -1 \end{bmatrix}$ is unitary. *(G.B.T.U. 2013)*

8. Prove that $\frac{1}{2}\begin{bmatrix} 1+i & -1+i \\ 1+i & 1-i \end{bmatrix}$ is a unitary matrix and find A^{-1}. *(K.U.K., May 2011, May 2012)*

9. If A is a Hermitian matrix, then show that iA is a skew-Hermitian matrix. *(M.T.U. 2013)*

10. Show that the matrix $\begin{bmatrix} \alpha+i\gamma & -\beta+i\delta \\ \beta+i\delta & \alpha-i\gamma \end{bmatrix}$ is unitary if $\alpha^2 + \beta^2 + \gamma^2 + \delta^2 = 1$. *(G.B.T.U. 2011)*

11. Show that $A = \begin{bmatrix} i & 0 & 0 \\ 0 & 0 & i \\ 0 & i & 0 \end{bmatrix}$ is skew-Hermitian as well as unitary. *(M.T.U. 2012)*

12. Express the matrix $A = \begin{bmatrix} i & 2-3i & 4+5i \\ 6+i & 0 & 4-5i \\ -i & 2-i & 2+i \end{bmatrix}$ as the sum of a Hermitian and a skew-Hermitian matrix. *(G.B.T.U. 2010)*

13. Express the Hermitian matrix $A = \begin{bmatrix} 1 & -i & 1+i \\ i & 0 & 2-3i \\ 1-i & 2+3i & 2 \end{bmatrix}$ as $P + iQ$ where P is a real symmetric and Q is a real skew symmetric matrix. *(G.B.T.U. 2012)*

14. Prove that the conjugate transpose of a unitary matrix is unitary.

15. Prove that the inverse of a unitary matrix is unitary.

16. Prove that the determinant of a unitary matrix has absolute value 1.

SECTION C

Differential Calculus

3*

Successive Differentiation

3.1. INTRODUCTION

The process of differentiating a function again and again is called **Successive Differentiation.**

Notation. If y be a function of x, then its successive derivatives are denoted by

$$\frac{dy}{dx}, \frac{d^2y}{dx^2}, \frac{d^3y}{dx^3}, \ldots, \frac{d^ny}{dx^n}$$

or $\quad y_1, y_2, y_3, \ldots, y_n$

or $\quad y', y'', y''', \ldots, y^{(n)}$

ILLUSTRATIVE EXAMPLES

Example 1. *(a) If* $y = \log(\cos x)$*, find* y_3

(b) If $y = \dfrac{x}{\sqrt{1-x^2}}$*, find* $\dfrac{d^3y}{dx^3}$

(c) If $y = e^{-x} \cos x$*, show that* $y_4 + 4y = 0$ (M.T.U. 2011)

(d) If $y = \cot x$*, find* y_4.

Sol. (*a*) We have $\quad y = \log(\cos x)$

$$y_1 = \frac{1}{\cos x} \cdot \frac{d}{dx}(\cos x) = \frac{-\sin x}{\cos x} = -\tan x$$

$$y_2 = -\sec^2 x$$

$$y_3 = -2 \sec x \cdot \frac{d}{dx}(\sec x) = -2 \sec x \sec x \tan x$$

$$= -2 \sec^2 x \tan x.$$

(*b*) We have $\quad y = \dfrac{x}{\sqrt{1-x^2}}$ $\quad$ | Form $y = \dfrac{u}{v}$

$$\frac{dy}{dx} = \frac{\sqrt{1-x^2} \cdot 1 - x \cdot \dfrac{d}{dx}(\sqrt{1-x^2})}{1-x^2} = \frac{\sqrt{1-x^2} - x \dfrac{1}{2} \cdot (1-x^2)^{-1/2}(-2x)}{1-x^2}$$

*This chapter is not included in the syllabus of KU, Kurukshetra and DCRUST, Murthal.

$$= \frac{\sqrt{1-x^2} + \dfrac{x^2}{\sqrt{1-x^2}}}{1-x^2} = \frac{1-x^2+x^2}{(1-x^2)^{3/2}} = \frac{1}{(1-x^2)^{3/2}} = (1-x^2)^{-3/2}$$

$$\frac{d^2y}{dx^2} = -\frac{3}{2}(1-x^2)^{-5/2}(-2x) = 3x(1-x^2)^{-5/2}$$

$$\frac{d^3y}{dx^3} = 3x\left(-\frac{5}{2}\right)(1-x^2)^{-7/2}(-2x) + 3(1-x^2)^{-5/2}$$

$$= 15x^2(1-x^2)^{-7/2} + 3(1-x^2)^{-5/2}$$

$$= \frac{15x^2}{(1-x^2)^{7/2}} + \frac{3}{(1-x^3)^{5/2}} = \frac{15x^2+3(1-x^2)}{(1-x^2)^{7/2}} = \frac{12x^2+3}{(1-x^2)^{7/2}} = \frac{3(4x^2+1)}{(1-x^2)^{7/2}}.$$

(*c*) We have $y = e^{-x}\cos x$ ...(1)

$y_1 = -e^{-x}\cos x + e^{-x}(-\sin x)$

$= -y - e^{-x}\sin x$ ...(2) [Using (1)]

$y_2 = -y_1 - (-e^{-x}\sin x + e^{-x}\cos x)$

$= -y_1 + e^{-x}\sin x - e^{-x}\cos x$

$= -y_1 + (-y - y_1) - y$ [Using (1) and (2)]

$= -2y - 2y_1$ ...(3)

$y_3 = -2y_1 - 2y_2$

$= -2y_1 - 2(-2y - 2y_1)$ [Using (3)]

$= 4y + 2y_1$

$y_4 = 4y_1 + 2y_2$

$= 4y_1 + 2(-2y - 2y_1)$ [Using (3)]

$= -4y$

$\therefore$ $y_4 + 4y = 0.$

(*d*) $y = \cot x$

$y_1 = -\text{cosec}^2 x = -(1+\cot^2 x) = -1 - \cot^2 x$

$y_2 = 0 - 2\cot x(-\text{cosec}^2 x)$

$= 2\cot x\,\text{cosec}^2 x = 2\cot x(1+\cot^2 x) = 2\cot x + 2\cot^3 x$

$y_3 = 2(-\text{cosec}^2 x) + 2.3\cot^2 x(-\text{cosec}^2 x)$

$= -2\,\text{cosec}^2 x - 6\cot^2 x\,\text{cosec}^2 x$

$= -2(1+\cot^2 x) - 6\cot^2 x(1+\cot^2 x)$

$= -2 - 2\cot^2 x - 6\cot^2 x - 6\cot^4 x$

$= -2 - 8\cot^2 x - 6\cot^4 x$

$y_4 = 0 - 8.2\cot x.(-\text{cosec}^2 x) - 6.4\cot^3 x.(-\text{cosec}^2 x)$

$= 16\cot x\,\text{cosec}^2 x + 24\cot^3 x\,\text{cosec}^2 x$

$= 16\cot x(1+\cot^2 x) + 24\cot^3 x(1+\cot^2 x)$

$= 16\cot x + 40\cot^3 x + 24\cot^5 x.$

Example 2. *If* $y = e^{ax} \sin bx$, *prove that* $\frac{d^2y}{dx^2} - 2a \cdot \frac{dy}{dx} + (a^2 + b^2)y = 0$

Sol.
$$y = e^{ax} \sin bx \qquad ...(i)$$

$$\frac{dy}{dx} = e^{ax} \cdot b \cos bx + ae^{ax} \sin bx$$

$$= b \cdot e^{ax} \cos bx + ay \qquad [\because \text{ of } (i)] \quad ...(ii)$$

$$\frac{d^2y}{dx^2} = b[e^{ax} \cdot (-b \sin bx) + ae^{ax} \cdot \cos bx] + a \cdot \frac{dy}{dx}$$

$$= -b^2 \boxed{e^{ax} \sin bx} + a \boxed{be^{ax} \cdot \cos bx} + a \frac{dy}{dx}$$

$$= -b^2 \cdot y + a\left(\frac{dy}{dx} - ay\right) + a \frac{dy}{dx} \qquad [\because \text{ of } (i) \text{ and } (ii)]$$

$$= -b^2y + a \cdot \frac{dy}{dx} - a^2y + a \cdot \frac{dy}{dx} = 2a \frac{dy}{dx} - (a^2 + b^2)\, y$$

$$\therefore \quad \frac{d^2y}{dx^2} - 2a \frac{dy}{dx} + (a^2 + b^2)\, y = 0.$$

Example 3. *(a) If* $y = \frac{ax+b}{cx+d}$, *prove that* $2y_1y_3 = 3y_2^2$.

(b) If $\frac{x^2}{a^2} + \frac{y^2}{b^2} = 1$, *show that* $\frac{d^2y}{dx^2} = -\frac{b^4}{a^2y^3}$

Sol. *(a)*
$$y = \frac{ax+b}{cx+d} \quad \Rightarrow \quad (cx + d)y = ax + b$$

Differentiating w.r.t. x

$$(cx + d)y_1 + cy = a \qquad \therefore \quad (cx + d)y_1 = a - cy$$

Differentiating again w.r.t. x.

$$(cx + d)y_2 + cy_1 = -cy_1 \qquad \therefore \quad (cx + d)y_2 = -2cy_1 \qquad ...(i)$$

Differentiating again w.r.t. x

$$(cx + d)y_3 + cy_2 = -2cy_2 \qquad \therefore \quad (cx + d)y_3 = -3cy_2 \qquad ...(ii)$$

Dividing *(i)* by *(ii)*,

$$\frac{y_2}{y_3} = \frac{2y_1}{3y_2} \qquad \therefore \quad 2y_1y_3 = 3y_2^2.$$

(b)
$$\frac{x^2}{a^2} + \frac{y^2}{b^2} = 1$$

Differentiating w.r.t. x

$$\frac{2x}{a^2} + \frac{2y}{b^2} \cdot \frac{dy}{dx} = 0 \qquad \therefore \quad \frac{dy}{dx} = -\frac{b^2x}{a^2y} \qquad ...(i)$$

Differentiating again w.r.t. x

$$\frac{d^2y}{dx^2} = -\frac{b^2}{a^2}\cdot\frac{d}{dx}\left(\frac{x}{y}\right) = -\frac{b^2}{a^2}\cdot\frac{y\cdot 1 - x\cdot\frac{dy}{dx}}{y^2}$$

$$= -\frac{b^2}{a^2}\cdot\frac{y - x\left(-\frac{b^2x}{a^2y}\right)}{y^2} \qquad [\because \text{ of } (i)]$$

$$= -\frac{b^2}{a^2}\cdot\frac{a^2y^2 + b^2x^2}{a^2y^3} = -\frac{b^2}{a^2}\cdot\frac{a^2b^2}{a^2y^3}$$

$$\Big|\ \because \frac{x^2}{a^2} + \frac{y^2}{b^2} = 1 \Rightarrow a^2y^2 + b^2x^2 = a^2b^2$$

$$\therefore \qquad \frac{d^2y}{dx^2} = -\frac{b^4}{a^2y^3}.$$

Example 4. (*a*) *If* $x = a(\theta + \sin\theta)$, $y = a(1 + \cos\theta)$, *find the value of* $\frac{d^2y}{dx^2}$ *at* $\theta = \frac{\pi}{2}$.

(*b*) *If* $x = \sin t$, $y = \sin pt$, *prove that* $(1 - x^2)\frac{d^2y}{dx^2} - x\frac{dy}{dx} + p^2y = 0$

(*c*) *If* $x = a(\cos\theta + \theta\sin\theta)$, $y = a(\sin\theta - \theta\cos\theta)$, *prove that* $a\theta\frac{d^2y}{dx^2} = \sec^3\theta$.

Sol. (*a*) $\quad x = a(\theta + \sin\theta);\ y = a(1 + \cos\theta)$

$$\frac{dx}{d\theta} = a(1 + \cos\theta);\ \frac{dy}{d\theta} = -a\sin\theta$$

$$\frac{dy}{dx} = \frac{dy}{d\theta}\Big/\frac{dx}{d\theta} = \frac{-a\sin\theta}{a(1+\cos\theta)} = \frac{-2\sin\frac{\theta}{2}\cos\frac{\theta}{2}}{2\cos^2\frac{\theta}{2}} = -\tan\frac{\theta}{2}$$

Differentiating again w.r.t. x,

$$\frac{d^2y}{dx^2} = -\frac{1}{2}\sec^2\frac{\theta}{2}\cdot\frac{d\theta}{dx} = -\frac{1}{2}\sec^2\frac{\theta}{2}\cdot\frac{1}{a(1+\cos\theta)}$$

$$= -\frac{1}{2a}\sec^2\frac{\theta}{2}\cdot\frac{1}{2\cos^2\frac{\theta}{2}} = -\frac{1}{4a}\sec^4\frac{\theta}{2}$$

$$\left[\frac{d^2y}{dx^2}\right]_{\theta=\frac{\pi}{2}} = -\frac{1}{4a}\sec^4\frac{\pi}{4} = -\frac{1}{4a}(\sqrt{2})^4 = -\frac{1}{a}.$$

(*b*) $\quad x = \sin t;\ y = \sin pt \qquad ...(1)$

$$\frac{dx}{dt} = \cos t;\ \frac{dy}{dt} = p\cos pt$$

$$\therefore \qquad \frac{dy}{dx} = \frac{dy}{dt} \Big/ \frac{dx}{dt} = \frac{p \cos pt}{\cos t} \qquad \text{...(2)}$$

Differentiating both sides w.r.t. x,

$$\frac{d^2y}{dx^2} = \frac{\cos t\,(-p^2 \sin pt) - p \cos pt\,(-\sin t)}{\cos^2 t} \cdot \frac{dt}{dx}$$

$$= \frac{-p^2 \cos t \sin pt + p \cos pt \sin t}{1 - \sin^2 t} \cdot \frac{1}{\cos t}$$

or $\qquad (1 - \sin^2 t) \dfrac{d^2y}{dx^2} = -p^2 \sin pt + \dfrac{p \cos pt}{\cos t} \sin t$

or $\qquad (1 - x^2) \dfrac{d^2y}{dx^2} = -p^2 y + \dfrac{dy}{dx} \cdot x \qquad$ [Using (1) and (2)]

$$\therefore \quad (1 - x^2) \frac{d^2y}{dx^2} - x \frac{dy}{dx} + p^2 y = 0.$$

(c) $\qquad x = a(\cos \theta + \theta \sin \theta),\ y = a\,(\sin \theta - \theta \cos \theta)$

$$\frac{dx}{d\theta} = a[-\sin \theta + \theta \cdot \cos \theta + \sin \theta] = a\theta \cos \theta$$

$$\frac{dy}{d\theta} = a\,[\,\cos \theta - (-\theta \sin \theta + \cos \theta)] = a\theta \sin \theta$$

$$\frac{dy}{dx} = \frac{dy}{d\theta} \Big/ \frac{dx}{d\theta} = \tan \theta$$

Differentiating again w.r.t. x,

$$\frac{d^2y}{dx^2} = \sec^2 \theta \frac{d\theta}{dx} = \sec^2 \theta \cdot \frac{1}{a\theta \cos \theta} \qquad \therefore \quad a\theta \cdot \frac{d^2y}{dx^2} = \sec^3 \theta.$$

Example 5. *If $x = a \sin 2\theta\,(1 + \cos 2\theta)$, $y = a \cos 2\theta\,(1 - \cos 2\theta)$, show that*

$$\frac{\left[1 + \left(\dfrac{dy}{dx}\right)^2\right]^{3/2}}{\dfrac{d^2y}{dx^2}} = 4a \cos 3\theta.$$

Sol. Here $\qquad x = a \sin 2\theta\,(1 + \cos 2\theta)$

$$\Rightarrow \qquad \frac{dx}{d\theta} = a\,[2 \cos 2\theta\,(1 + \cos 2\theta) + \sin 2\theta\,(-\,2 \sin 2\theta)]$$

$$= 2a\,(\cos 2\theta + \cos^2 2\theta - \sin^2 2\theta)$$

$$= 2a\,(\cos 2\theta + \cos 4\theta) = 2a\,.\,2 \cos 3\theta \cos \theta$$

$$= 4a \cos \theta \cos 3\theta$$

Also $\qquad y = a \cos 2\theta\,(1 - \cos 2\theta)$

$$\Rightarrow \qquad \frac{dy}{d\theta} = a\,[-\,2 \sin 2\theta\,(1 - \cos 2\theta) + \cos 2\theta\,(2 \sin 2\theta)]$$

$$= 2a\,[\,-\sin 2\theta + 2 \sin 2\theta \cos 2\theta]$$

$$= 2a\,(\sin 4\theta - \sin 2\theta) = 2a \cdot 2 \cos 3\theta \sin \theta$$

$$= 4a \sin \theta \cos 3\theta$$

$\therefore \qquad \dfrac{dy}{dx} = \dfrac{dy}{d\theta} \div \dfrac{dx}{d\theta} = \dfrac{4a \sin\theta \cos 3\theta}{4a \cos\theta \cos 3\theta} = \tan\theta$

$$\frac{d^2y}{dx^2} = \frac{d}{dx}\left(\frac{dy}{dx}\right) = \frac{d}{dx}(\tan\theta) = \sec^2\theta \times \frac{d\theta}{dx}$$

$$= \frac{\sec^2\theta}{4a\cos\theta\cos 3\theta} = \frac{\sec^3\theta}{4a\cos 3\theta}$$

Now $\qquad \left[1+\left(\dfrac{dy}{dx}\right)^2\right]^{3/2} = (1+\tan^2\theta)^{3/2} = (\sec^2\theta)^{3/2} = \sec^3\theta$

$\therefore \qquad \dfrac{\left[1+\left(\dfrac{dy}{dx}\right)^2\right]^{3/2}}{\dfrac{d^2y}{dx^2}} = \sec^3\theta \times \dfrac{4a\cos 3\theta}{\sec^3\theta} = 4a\cos 3\theta.$

Example 6. *(a) If $p^2 = a^2 \cos^2 \theta + b^2 \sin^2 \theta$, prove that* $p + \dfrac{d^2p}{d\theta^2} = \dfrac{a^2b^2}{p^3}$.

(b) If $y^3 - 3ax^2 + x^3 = 0$, prove that $\dfrac{d^2y}{dx^2} + \dfrac{2a^2x^2}{y^5} = 0.$

(c) If $x^3 + y^3 - 3\,axy = 0$, show that $\dfrac{d^2y}{dx^2} = -\dfrac{2a^3xy}{(y^2-ax)^3}$.

Sol. (*a*) $\qquad p^2 = a^2\cos^2\theta + b^2\sin^2\theta$

$$= a^2\left(\frac{1+\cos 2\theta}{2}\right) + b^2\left(\frac{1-\cos 2\theta}{2}\right)$$

or $\qquad 2p^2 = a^2 + b^2 + (a^2 - b^2)\cos 2\theta$

or $\qquad 2p^2 - (a^2+b^2) = (a^2-b^2)\cos 2\theta \qquad$ ...(*i*)

Differentiating both sides w.r.t. θ, we get

$$4p \cdot \frac{dp}{d\theta} = (a^2-b^2)(-2\sin 2\theta) \text{ or } -2p\frac{dp}{d\theta} = (a^2-b^2)\sin 2\theta \qquad \text{...}(ii)$$

Squaring (*i*) and (*ii*) and adding,

$$4p^4 + (a^2+b^2)^2 - 4p^2(a^2+b^2) + 4p^2 \cdot \left(\frac{dp}{d\theta}\right)^2 = (a^2-b^2)^2(\cos^2 2\theta + \sin^2 2\theta)$$

or $\qquad 4p^4 - 4p^2(a^2+b^2) + 4p^2 \cdot \left(\dfrac{dp}{d\theta}\right)^2 + (a^2+b^2)^2 = (a^2-b^2)^2$

or $\qquad 4p^4 - 4p^2(a^2+b^2) + 4p^2 \cdot \left(\dfrac{dp}{d\theta}\right)^2 + (a^2+b^2)^2 - (a^2-b^2)^2 = 0$

or $\qquad 4p^4 - 4p^2(a^2+b^2) + 4p^2\left(\dfrac{dp}{d\theta}\right)^2 + 4a^2b^2 = 0$

or $$p^4 - p^2(a^2 + b^2) + p^2\left(\frac{dp}{d\theta}\right)^2 + a^2b^2 = 0$$

or $$p^2 - (a^2 + b^2) + \left(\frac{dp}{d\theta}\right)^2 + \frac{a^2b^2}{p^2} = 0$$

Differentiating again w.r.t. θ,

$$2p \cdot \frac{dp}{d\theta} - 0 + 2 \cdot \frac{dp}{d\theta} \cdot \frac{d^2p}{d\theta^2} - \frac{2a^2b^2}{p^3} \cdot \frac{dp}{d\theta} = 0$$

Dividing by $2\,\frac{dp}{d\theta}$,

$$p + \frac{d^2p}{d\theta^2} - \frac{a^2b^2}{p^3} = 0 \quad \text{or} \quad p + \frac{d^2p}{d\theta^2} = \frac{a^2b^2}{p^3}.$$

(*b*) $y^3 - 3ax^2 + x^3 = 0$ ...(*i*)

Differentiating w.r.t. *x*,

$$3y^2 \cdot \frac{dy}{dx} - 6ax + 3x^2 = 0$$

or $$y^2 \cdot \frac{dy}{dx} = 2ax - x^2 \quad \text{or} \quad \frac{dy}{dx} = \frac{2ax - x^2}{y^2}$$ ...(*ii*)

Differentiating again w.r.t. *x*,

$$\frac{d^2y}{dx^2} = \frac{y^2(2a - 2x) - (2ax - x^2) \cdot 2y\,\frac{dy}{dx}}{y^4}$$

$$= \frac{y^2(2a - 2x) - 2(2ax - x^2)\,y \cdot \frac{2ax - x^2}{y^2}}{y^4}$$ [∵ of (*ii*)]

$$= \frac{y^2(2a - 2x) - 2(2ax - x^2) \cdot \frac{2ax - x^2}{y}}{y^4}$$

$$= \frac{y^3(2a - 2x) - 2(2ax - x^2)^2}{y^5}$$

$$= \frac{(3ax^2 - x^3)(2a - 2x) - 2(2ax - x^2)^2}{y^5}$$ [∵ of (*i*)]

$$= \frac{6a^2x^2 - 6ax^3 - 2ax^3 + 2x^4 - 8a^2x^2 - 2x^4 + 8ax^3}{y^5} = \frac{-2a^2x^2}{y^5}$$

$$\therefore \quad \frac{d^2y}{dx^2} + \frac{2a^2x^2}{y^5} = 0.$$

(*c*) $x^3 + y^3 - 3axy = 0$ ...(*i*)

Differentiating w.r.t. *x*,

$$3x^2 + 3y^2 \cdot \frac{dy}{dx} - 3a\left(x \cdot \frac{dy}{dx} + 1 \cdot y\right) = 0$$

or $$x^2 + y^2 \cdot \frac{dy}{dx} - ax \cdot \frac{dy}{dx} - ay = 0$$

or $$(y^2 - ax) \cdot \frac{dy}{dx} = ay - x^2 \quad \text{or} \quad \frac{dy}{dx} = \frac{ay - x^2}{y^2 - ax} \qquad ...(ii)$$

Differentiating again w.r.t. x,

$$\frac{d^2y}{dx^2} = \frac{(y^2 - ax)\left[a \cdot \frac{dy}{dx} - 2x\right] - (ay - x^2)\left[2y \cdot \frac{dy}{dx} - a\right]}{(y^2 - ax)^2}$$

$$= \frac{(y^2 - ax)\left[a \cdot \frac{ay - x^2}{y^2 - ax} - 2x\right] - (ay - x^2)\left[2y \cdot \frac{ay - x^2}{y^2 - ax} - a\right]}{(y^2 - ax)^2} \qquad [\because \text{ of } (ii)]$$

$$= \frac{(y^2 - ax)(a^2y - ax^2 - 2xy^2 + 2ax^2) - (ay - x^2)(2ay^2 - 2x^2y - ay^2 + a^2x)}{(y^2 - ax)^3}$$

$$= \frac{(y^2 - ax)(a^2y - 2xy^2 + ax^2) - (ay - x^2)(ay^2 - 2x^2y + a^2x)}{(y^2 - ax)^3}$$

$$= \frac{a^2y^3 - 2xy^4 + ax^2y^2 - a^3xy + 2ax^2y^2 - a^2x^3 - a^2y^3 + 2ax^2y^2 - a^3xy + ax^2y^2 - 2x^4y + a^2x^3}{(y^2 - ax)^3}$$

$$= \frac{-2xy^4 + 6ax^2y^2 - 2a^3xy - 2x^4y}{(y^2 - ax)^3} = \frac{-2xy(y^3 - 3axy + x^3) - 2a^3xy}{(y^2 - ax)^3}$$

$$= \frac{-2xy(0) - 2a^3xy}{(y^2 - ax)^3} \qquad [\because \text{ of } (i)]$$

$$= -\frac{2a^3xy}{(y^2 - ax)^3}.$$

Example 7. *If* $y = x \log\left(\frac{x}{a + bx}\right)$, *prove that* $x^3 \frac{d^2y}{dx^2} = \left(x\frac{dy}{dx} - y\right)^2$.

Sol. $$y = x \log\left(\frac{x}{a + bx}\right)$$

$\Rightarrow$ $$\frac{y}{x} = \log x - \log(a + bx)$$

Differentiating w.r.t. x

$$\frac{x\frac{dy}{dx} - y \cdot 1}{x^2} = \frac{1}{x} - \frac{b}{a + bx}$$

$\Rightarrow$ $$x\frac{dy}{dx} - y = x^2\left[\frac{a + bx - bx}{x(a + bx)}\right]$$

$\Rightarrow$ $$x\frac{dy}{dx} - y = \frac{ax}{a + bx} \qquad ...(1)$$

Differentiating again w.r.t. x

$$\left(x\frac{d^2y}{dx^2}+\frac{dy}{dx}\right)-\frac{dy}{dx}=\frac{(a+bx)\cdot a-ax\cdot b}{(a+bx)^2}$$

$$\Rightarrow \qquad x\frac{d^2y}{dx^2}=\frac{a^2}{(a+bx)^2}$$

Multiplying by x^2, we have

$$x^3\frac{d^2y}{dx^2}=\left(\frac{ax}{a+bx}\right)^2$$

$$\Rightarrow \qquad x^3\frac{d^2y}{dx^2}=\left(x\frac{dy}{dx}-y\right)^2 \qquad \text{| Using (1)}$$

Another Form of above example

If $(a + bx)\, e^{y/x} = x$, then show that $x^3\dfrac{d^2y}{dx^2}=\left(x\dfrac{dy}{dx}-y\right)^2$

[**Hint.** $\quad (a + bx)\, e^{y/x} = x$

$$\Rightarrow \qquad e^{y/x}=\frac{x}{a+bx}$$

$$\Rightarrow \qquad \log e^{y/x}=\log\left(\frac{x}{a+bx}\right)$$

$$\Rightarrow \qquad \frac{y}{x}=\log\left(\frac{x}{a+bx}\right)$$

$$\Rightarrow \qquad y=x\log\left(\frac{x}{a+bx}\right)\Bigg].$$

EXERCISE 3.1

1. If $y = x^3 \log x$, prove that $y_4=\dfrac{6}{x}$.
2. Find the fourth derivative of $\tan x$ at $x=\dfrac{\pi}{4}$.
3. If $y=\dfrac{\log x}{x}$, prove that $\dfrac{d^2y}{dx^2}=\dfrac{2\log x-3}{x^3}$.
4. If $y = \text{A}\sin mx + \text{B}\cos mx$, show that $\dfrac{d^2y}{dx^2}+m^2y=0$.
5. If $y = ae^{mx} + be^{-mx}$, show that $y_2 = m^2y$.
6. If $y = \text{A}\,e^{mx} + \text{B}\,e^{nx}$, prove that $\dfrac{d^2y}{dx^2}-(m+n)\dfrac{dy}{dx}+mny=0$.
7. If $x = 2\cos t - \cos 2t$, $y = 2\sin t - \sin 2t$, find the value of $\dfrac{d^2y}{dx^2}$, when $t=\dfrac{\pi}{2}$.

8. If $y = (\sin^{-1} x)^2$, prove that $(1 - x^2) \dfrac{d^2y}{dx^2} - x \dfrac{dy}{dx} = 2$.

9. If $y = e^{m \sin^{-1}x}$, prove that $(1 - x^2)y_2 - xy_1 = m^2y$.

10. If $\sqrt{x} + \sqrt{y} = \sqrt{a}$, find the value of $\dfrac{d^2y}{dx^2}$ for $x = a$.

11. If $ax^2 + 2hxy + by^2 = 1$, show that $\dfrac{d^2y}{dx^2} = \dfrac{h^2 - ab}{(hx + by)^3}$.

12. If $y = ae^x + be^{2x} + ce^{3x}$, prove that $y_3 - 6y_2 + 11y_1 - 6y = 0$.

Answers

2. 80 **7.** $-3/2$ **10.** $\dfrac{1}{2a}$.

3.2. STANDARD RESULTS

1. To find the nth differential co-efficient of x^m.

Let $y = x^m$

then
$$y_1 = mx^{m-1}$$
$$y_2 = m(m-1)\, x^{m-2}$$
$$y_3 = m(m-1)\,(m-2)x^{m-3}$$
....

....

$$\therefore \quad y_n = [m(m-1)\,(m-2) \ldots \text{upto } n \text{ factors}] \cdot x^{m-n}$$
$$= m(m-1)\,(m-2) \ldots (m-n+1)x^{m-n}, \text{ where } n < m.$$
$$= \frac{m!}{m-n!}\, x^{m-n} \text{ if } m \in \text{N},\ m > n.$$

Cor. If m be a positive integer and $n = m$,

then
$$y_m = m(m-1)\,(m-2) \ldots (m-m+1)x^{m-m}$$
$$= m(m-1)\,(m-2) \ldots 1$$
$$= m\,!$$

As $y_m = m\,!$ is constant

$\therefore$ y_{m+1} which is the derivative of y_m will be zero and so on.

2. To find the nth differential co-efficient of $(ax + b)^m$.

Let $y = (ax + b)^m$

then
$$y_1 = m(ax+b)^{m-1} \cdot a$$
$$y_2 = m(m-1)(ax+b)^{m-2} \cdot a^2$$
$$y_3 = m(m-1)(m-2)\,(ax+b)^{m-3} \cdot a^3$$
....

....

$$y_n = [m(m-1)(m-2) \ldots \text{upto } n \text{ factors}] \times (ax+b)^{m-n} \cdot a^n$$
$$= m(m-1)(m-2) \ldots (m-n+1).\ (ax+b)^{m-n} \cdot a^n.$$

Cor. If m be a positive integer and $n = m$, then

$$y_m = m(m-1)(m-2)\ldots(m-m+1).(ax+b)^{m-m}\cdot a^m$$
$$= m(m-1)(m-2)\ldots 1\cdot a^m$$
$$= m!.a^m.$$

3. To find the nth differential co-efficient of

$$\frac{\mathbf{1}}{\mathbf{ax+b}},\quad \left[\mathbf{x} \neq -\frac{\mathbf{b}}{\mathbf{a}}\right]$$

Let $\qquad y = \dfrac{1}{ax+b} = (ax+b)^{-1}$

then $\qquad y_1 = -1\cdot(ax+b)^{-2}.a$

$$y_2 = (-1)^2\cdot 1\cdot 2\cdot(ax+b)^{-3}\cdot a^2 = (-1)^2\cdot 2!\cdot(ax+b)^{-3}\cdot a^2$$
$$y_3 = (-1)^3\cdot 3\cdot 2!\cdot(ax+b)^{-4}\cdot a^3 = (-1)^3\cdot 3!\cdot(ax+b)^{-4}\cdot a^3$$

....

....

$$y_n = (-1)^n\cdot n!\cdot(ax+b)^{-(n+1)}\cdot a^n$$
$$y_n = \frac{\mathbf{(-1)^n\cdot n!\cdot a^n}}{\mathbf{(ax+b)^{n+1}}}.$$

ILLUSTRATIVE EXAMPLES

Example 1. *Find the nth derivative of* $\dfrac{1}{1-5x+6x^2}$. (M.D.U. Dec. 2009)

Sol. Let $\qquad y = \dfrac{1}{1-5x+6x^2} = \dfrac{1}{6x^2-5x+1}$

$$= \frac{1}{(3x-1)(2x-1)} = \frac{1}{(3x-1)\left(2\cdot\frac{1}{3}-1\right)} + \frac{1}{\left(3\cdot\frac{1}{2}-1\right)(2x-1)}$$

$$= \frac{1}{(3x-1)\left(-\frac{1}{3}\right)} + \frac{1}{\frac{1}{2}(2x-1)} = \frac{-3}{3x-1} + \frac{2}{2x-1}$$

$\therefore \qquad y_n = -3\cdot\dfrac{(-1)^n\cdot n!\,3^n}{(3x-1)^{n+1}} + 2\cdot\dfrac{(-1)^n\cdot n!\cdot 2^n}{(2x-1)^{n+1}}$

$$= (-1)^n\,n!\left[\frac{-3^{n+1}}{(3x-1)^{n+1}} + \frac{2^{n+1}}{(2x-1)^{n+1}}\right]$$

Note 1. Compare $\dfrac{1}{3x-1}$ with $\dfrac{1}{ax+b}$ [Standard form] $a = 3$

$\therefore \qquad y_n = \dfrac{(-1)^n\cdot n!\cdot 3^n}{(3x-1)^{n+1}}.$

Caution. Do not forget to write 3^n, *i.e.,* (co-eff. of $x)^n$ in the numerator.

Note 2. *To find the partial fractions of a rational fraction when the denominator consists of non-repeated linear factors.*

Method. [Rule of *Everywhere but not there.*] Equate each linear in the denominator to zero separately and find the value of x. Put this value of x everywhere in the given fraction **except in the factor from which the value has been obtained.**

For example, to resolve $\dfrac{1}{(3x-1)(2x-1)}$ into partial fractions

equating $3x - 1$ to zero, $x = \dfrac{1}{3}$, equating $2x - 1$ to zero, $x = \dfrac{1}{2}$

$$\therefore \qquad \frac{1}{(3x-1)(2x-1)} = \frac{1}{(3x-1)\left(2\cdot\frac{1}{3}-1\right)} + \frac{1}{\left(3\cdot\frac{1}{2}-1\right)(2x-1)} \text{ etc.}$$

Example 2. *Find the nth derivative of*

(a) $\dfrac{x^2+4x+1}{x^3+2x^2-x-2}$ *(b)* $\dfrac{x^4}{(x-1)(x-2)}$

Sol. (*a*) Let $y = \dfrac{x^2+4x+1}{x^3+2x^2-x-2} = \dfrac{x^2+4x+1}{x^2(x+2)-(x+2)}$

$$= \frac{x^2+4x+1}{(x+2)(x^2-1)} = \frac{x^2+4x+1}{(x+2)(x+1)(x-1)}$$

$$= \frac{(-2)^2+4(-2)+1}{(x+2)(-2+1)(-2-1)} + \frac{(-1)^2+4(-1)+1}{(-1+2)(x+1)(-1-1)} + \frac{1+4+1}{(1+2)(1+1)(x-1)}$$

$$= \frac{-3}{3(x+2)} + \frac{-2}{-2(x+1)} + \frac{6}{6(x-1)}$$

$$= -\frac{1}{x+2} + \frac{1}{x+1} + \frac{1}{x-1}$$

$$y_n = -\frac{(-1)^n\cdot n!}{(x+2)^{n+1}} + \frac{(-1)^n\cdot n!}{(x+1)^{n+1}} + \frac{(-1)^n\cdot n!}{(x-1)^{n+1}}.$$

(*b*) Let $y = \dfrac{x^4}{(x-1)(x-2)} = x^2 + 3x + 7 + \dfrac{15x-14}{(x-1)(x-2)}$

[Dividing num. by denom.]

$$= x^2 + 3x + 7 + \frac{15(1)-14}{(x-1)(1-2)} + \frac{15(2)-14}{(2-1)(x-2)}$$

$$= x^2 + 3x + 7 - \frac{1}{x-1} + \frac{16}{x-2}$$

$$y_n = -\frac{(-1)^n\cdot n!}{(x-1)^{n+1}} + 16\cdot\frac{(-1)^n\cdot n!}{(x-2)^{n+1}}$$

$$= (-1)^n\, n!\left[\frac{16}{(x-2)^{n+1}} - \frac{1}{(x-1)^{n+1}}\right].$$

Example 3. *Prove that the value of the nth differential coefficient of* $\dfrac{x^3}{x^2-1}$ *for* $x = 0$ *is zero, if n is even; and* $-\,n\,!$ *if n is odd and greater than 1.*

Sol. Let $y = \dfrac{x^3}{x^2-1} = x + \dfrac{x}{x^2-1} = x + \dfrac{1}{2}\left(\dfrac{1}{x+1} + \dfrac{1}{x-1}\right)$ | Partial fractions

If $n > 1$, the second and higher derivatives of x are 0.

$$\therefore \quad y_n = 0 + \frac{1}{2}\left[\frac{(-1)^n \cdot n!}{(x+1)^{n+1}} + \frac{(-1)^n \cdot n!}{(x-1)^{n+1}}\right]$$

$$= \frac{1}{2} \cdot (-1)^n \cdot n! \left[\frac{1}{(x+1)^{n+1}} + \frac{1}{(x-1)^{n+1}}\right] \quad \ldots(i)$$

Case I. *When n is even, n + 1 is odd so that* $(-1)^n = 1, (-1)^{n+1} = -1$

Putting $x = 0$ in (i)

$$y_n(0) = \frac{1}{2} \cdot (-1)^n \cdot n! \left[\frac{1}{(1)^{n+1}} + \frac{1}{(-1)^{n+1}}\right]$$

$$= \frac{1}{2} \cdot (1) \cdot n! \left[\frac{1}{1} + \frac{1}{-1}\right] = \frac{n!}{2}[1-1] = 0.$$

Case II. *When n is odd, n + 1 is even so that* $(-1)^n = -1, (-1)^{n+1} = 1$

Putting $x = 0$ in (i),

$$y_n(0) = \frac{1}{2} \cdot (-1)^n \cdot n! \left[\frac{1}{1^{n+1}} + \frac{1}{(-1)^{n+1}}\right]$$

$$= \frac{1}{2} \cdot (-1)\, n! \left[\frac{1}{1} + \frac{1}{1}\right] = \frac{-n!}{2}(2) = -n!.$$

Example 4. *If* $y\sqrt{1+x} + x\sqrt{1+y} = 0$, *prove that*

$$\frac{d^n y}{dx^n} = \frac{(-1)^n\, n!}{(1+x)^{n+1}}.$$

Sol. $\quad y\sqrt{1+x} + x\sqrt{1+y} = 0$

$\Rightarrow \quad y\sqrt{1+x} = -x\sqrt{1+y}$

Squaring $\quad y^2(1+x) = x^2(1+y)$

or $\quad (y^2 - x^2) + xy(y - x) = 0$

or $\quad (y - x)(y + x + xy) = 0$

or $\quad y + x + xy = 0 \qquad [\because \; y \neq x]$

or $\quad y(1 + x) = -x$

or $\quad y = -\dfrac{x}{1+x} = -1 + \dfrac{1}{1+x}$

$$\therefore \quad y_n = \frac{(-1)^n \cdot n!}{(1+x)^{n+1}}.$$

Example 5. *Find the nth derivative of* $\dfrac{1}{(x^2-a^2)(x^2-b^2)}$.

Sol. Let $y = \dfrac{1}{(x^2-a^2)(x^2-b^2)} = \dfrac{1}{(x^2-a^2)(a^2-b^2)} + \dfrac{1}{(b^2-a^2)(x^2-b^2)}$

$$= \frac{1}{a^2-b^2}\left[\frac{1}{x^2-a^2} - \frac{1}{x^2-b^2}\right]$$

$$= \frac{1}{a^2-b^2}\left[\frac{1}{(x+a)(x-a)} - \frac{1}{(x+b)(x-b)}\right]$$

$$= \frac{1}{a^2-b^2}\left[\left\{\frac{1}{(x+a)(-a-a)} + \frac{1}{(a+a)(x-a)}\right\} - \left\{\frac{1}{(x+b)(-b-b)} + \frac{1}{(b+b)(x-b)}\right\}\right]$$

$$= \frac{1}{a^2-b^2}\left[\frac{1}{2a}\left(\frac{1}{x-a} - \frac{1}{x+a}\right) - \frac{1}{2b}\left(\frac{1}{x-b} - \frac{1}{x+b}\right)\right]$$

$$\therefore\ y_n = \frac{1}{a^2-b^2}\left[\frac{1}{2a}\left\{\frac{(-1)^n\, n!}{(x-a)^{n+1}} - \frac{(-1)^n\, n!}{(x+a)^{n+1}}\right\} - \frac{1}{2b}\left\{\frac{(-1)^n\, n!}{(x-b)^{n+1}} - \frac{(-1)^n\, n!}{(x+b)^{n+1}}\right\}\right]$$

$$= \frac{(-1)^n\, n!}{a^2-b^2}\left[\frac{1}{2a}\left\{\frac{1}{(x-a)^{n+1}} - \frac{1}{(x+a)^{n+1}}\right\} - \frac{1}{2b}\left\{\frac{1}{(x-b)^{n+1}} - \frac{1}{(x+b)^{n+1}}\right\}\right].$$

3.3. STANDARD RESULTS (Continued)

4. To find the nth differential co-efficient of log (ax + b).

Let $y = \log (ax + b)$

$$y_1 = \frac{1}{ax+b} \cdot a = a\,(ax+b)^{-1}$$

$$y_2 = (-1)a^2\,(ax+b)^{-2}$$

$$y_3 = (-1)(-2)a^3\,(ax+b)^{-3} = (-1)^2\, 2!\, a^3(ax+b)^{-3}$$

$$y_4 = (-1)^3\, 3 \cdot 2! \cdot a^4(ax+b)^{-4} = (-1)^3 \cdot 3!\, a^4\,(ax+b)^{-4}$$

....

....

$$y_n = (-1)^{n-1} \cdot (n-1)!\, a^n\,(ax+b)^{-n}$$

Hence
$$\mathbf{y_n = \frac{(-1)^{n-1}\,(n-1)! \cdot a^n}{(ax+b)^n}}.$$

Example. *Find the nth differential co-efficient of* $\log \sqrt{\frac{2x+1}{x-2}}$.

Sol. Let
$$y = \log \sqrt{\frac{2x+1}{x-2}} = \frac{1}{2}\log \frac{2x+1}{x-2}$$

$$= \tfrac{1}{2}\,[\log (2x+1) - \log (x-2)]$$

$$\therefore\quad y_n = \frac{1}{2}\left[\frac{(-1)^{n-1} \cdot (n-1)! \cdot 2^n}{(2x+1)^n} - \frac{(-1)^{n-1}\,(n-1)!}{(x-2)^n}\right]$$

$$= \frac{1}{2} \cdot (-1)^{n-1} \cdot (n-1)!\left[\frac{2^n}{(2x+1)^n} - \frac{1}{(x-2)^n}\right].$$

3.4. STANDARD RESULTS (Continued)

5. To find the nth differential co-efficient of a^{mx}.

Let $y = a^{mx}$

$y_1 = ma^{mx} \cdot (\log a)$

$y_2 = m^2 \cdot a^{mx} \cdot (\log a)^2$

$y_3 = m^3 \cdot a^{mx} \cdot (\log a)^3$

....

....

$y_n = m^n \cdot a^{mx} \cdot (\log a)^n$

Cor. If $y = a^x$, $m = 1$ $\therefore$ $y_n = a^x (\log a)^n$.

6. To find the nth differential co-efficient of e^{mx}.

Let $y = e^{mx}$

$y_1 = me^{mx}$

$y_2 = m^2 \cdot e^{mx}$

$y_3 = m^3 \cdot e^{mx}$

....

....

$y_n = m^n \cdot e^{mx}$.

Cor. If $y = e^x$, $m = 1$ $\therefore$ $y_n = e^x$.

7. To find the nth differential co-efficient of sin (ax + b).

Let $y = \sin (ax + b)$

$y_1 = a \cos (ax + b)$

$= a \sin \left(ax + b + \frac{\pi}{2}\right)$ $\quad | \because \sin (\pi/2 + \theta) = \cos \theta$ $\quad$ **| Note this step**

$y_2 = a^2 \cos \left(ax + b + \frac{\pi}{2}\right) = a^2 \sin \left(ax + b + \frac{\pi}{2} + \frac{\pi}{2}\right)$

$= a^2 \sin \left(ax + b + 2 \cdot \frac{\pi}{2}\right)$

$y_3 = a^3 \cos \left(ax + b + 2 \cdot \frac{\pi}{2}\right) = a^3 \sin \left(ax + b + 2 \cdot \frac{\pi}{2} + \frac{\pi}{2}\right)$

$= a^3 \sin \left(ax + b + 3 \cdot \frac{\pi}{2}\right)$

....

....

$$\mathbf{y_n = a^n \sin \left(ax + b + n \cdot \frac{\pi}{2}\right)}$$

8. Similarly, the nth differential co-efficient of

$$\mathbf{\cos (ax + b) = a^n \cos \left(ax + b + n\frac{\pi}{2}\right)}.$$

Example 1. *Find the nth differential co-efficient of*

(a) $\sin^3 x$ *(b)* $\sin^4 x$.

Sol. (*a*) Let $y = \sin^3 x$

Now $\sin 3x = 3 \sin x - 4 \sin^3 x$

$\Rightarrow \quad 4 \sin^3 x = 3 \sin x - \sin 3x$

$$\Rightarrow \quad \sin^3 x = \frac{1}{4}(3 \sin x - \sin 3x)$$

$$\Rightarrow \quad y = \frac{3}{4} \sin x - \frac{1}{4} \sin 3x$$

$$\therefore \quad y_n = \frac{3}{4} \sin\left(x + n \cdot \frac{\pi}{2}\right) - \frac{1}{4} \cdot 3^n \sin\left(3x + n \cdot \frac{\pi}{2}\right)$$

(*b*) Let $y = \sin^4 x = (\sin^2 x)^2$

$$= \left[\frac{1 - \cos 2x}{2}\right]^2 = \frac{1}{4}[1 - 2\cos 2x + \cos^2 2x]$$

$$= \frac{1}{4}\left[1 - 2\cos 2x + \frac{1 + \cos 4x}{2}\right]$$

$$= \frac{1}{8}[2 - 4\cos 2x + 1 + \cos 4x] = \frac{3}{8} - \frac{1}{2}\cos 2x + \frac{1}{8}\cos 4x$$

$$\therefore \quad y_n = 0 - \frac{1}{2} \cdot 2^n \cos\left[2x + n \cdot \frac{\pi}{2}\right] + \frac{1}{8} \cdot 4^n \cdot \cos\left[4x + n \cdot \frac{\pi}{2}\right]$$

$$= -2^{n-1} \cos\left[2x + n \cdot \frac{\pi}{2}\right] + \frac{1}{2^3} \cdot 2^{2n} \cos\left[4x + n \cdot \frac{\pi}{2}\right]$$

$$= -2^{n-1} \cos\left[2x + n \cdot \frac{\pi}{2}\right] + 2^{2n-3} \cos\left[4x + n \cdot \frac{\pi}{2}\right].$$

Example 2. *Find the nth differential co-efficient of*

(a) $\cos 2x \sin 3x$ *(b)* $\sin^3 x \cos^2 x$

(c) $\cos x \cos 2x \cos 3x$

Sol. (*a*) Let $y = \cos 2x \sin 3x$

$$= \frac{1}{2}(2 \sin 3x \cos 2x)$$

| $2 \sin A \cos B = \sin (A + B) + \sin (A - B)$

$$= \frac{1}{2}[\sin 5x + \sin x]$$

$$\therefore \quad y_n = \frac{1}{2}\left[5^n \sin\left(5x + \frac{n\pi}{2}\right) + \sin\left(x + \frac{n\pi}{2}\right)\right]$$

(*b*) Let $y = \sin^3 x \cos^2 x$

$$= \frac{1}{4} \cdot (4 \sin^2 x \cos^2 x) \sin x = \frac{1}{4} \sin^2 2x \sin x$$

$$= \frac{1}{4} \cdot \frac{1 - \cos 4x}{2} \cdot \sin x = \frac{1}{8}(\sin x - \cos 4x \sin x)$$

$$= \frac{1}{16}(2 \sin x - 2 \cos 4x \sin x)$$

$$= \frac{1}{16}[2 \sin x - (\sin 5x - \sin 3x)]$$

$$= \frac{1}{16}(2 \sin x + \sin 3x - \sin 5x)$$

$\therefore \quad y_n = \frac{1}{16}\left[2\sin\left(x+n\cdot\frac{\pi}{2}\right)+3^n\sin\left(3x+n\cdot\frac{\pi}{2}\right)-5^n\sin\left(5x+n\cdot\frac{\pi}{2}\right)\right].$

(c) Let $\quad y = \cos x \cos 2x \cos 3x$

$= \frac{1}{2}(2\cos 3x\cos x)\cdot\cos 2x = \frac{1}{2}(\cos 4x+\cos 2x)\cos 2x$

$= \frac{1}{4}[2\cos 4x\cos 2x + 2\cos^2 2x] = \frac{1}{4}[\cos 6x+\cos 2x + 1 + \cos 4x]$

$= \frac{1}{4}[1+\cos 2x+\cos 4x+\cos 6x]$

$\therefore \quad y_n = \frac{1}{4}\left[2^n\cos\left(2x+n\cdot\frac{\pi}{2}\right)+4^n\cdot\cos\left(4x+n\cdot\frac{\pi}{2}\right)+6^n\cos\left(6x+n\cdot\frac{\pi}{2}\right)\right].$

Example 3. *If $y = \sin mx + \cos mx$, show that* (M.T.U. 2012)

$y_r = m^r\,[1+(-1)^r\sin 2mx]^{1/2}$.

Sol. $\quad y = \sin mx + \cos mx$

$\therefore \quad y_r = m^r\sin\left(mx+r\cdot\frac{\pi}{2}\right)+m^r\cos\left(mx+r\cdot\frac{\pi}{2}\right)$

$= m^r\left[\sin\left(mx+r\cdot\frac{\pi}{2}\right)+\cos\left(mx+r\cdot\frac{\pi}{2}\right)\right]$

$= m^r\left[\left\{\sin\left(mx+\frac{r\pi}{2}\right)+\cos\left(mx+\frac{r\pi}{2}\right)\right\}^2\right]^{1/2}$

$= m^r\left[\sin^2\left(mx+\frac{r\pi}{2}\right)+\cos^2\left(mx+\frac{r\pi}{2}\right)+2\sin\left(mx+\frac{r\pi}{2}\right)\cos\left(mx+\frac{r\pi}{2}\right)\right]^{1/2}$

$= m^r\left[1+\sin 2\left(mx+\frac{r\pi}{2}\right)\right]^{1/2}$ $\quad \left|\ \because \sin^2\theta+\cos^2\theta = 1 \text{ and } 2\sin\theta\cos\theta = \sin 2\theta\right.$

$= m^r\,[1+\sin(2mx+r\pi)]^{1/2}$

$= m^r\,[1+(-1)^r\sin 2mx]^{1/2}$ $\quad [\because \sin(\theta+n\pi) = (-1)^n\sin\theta]$

3.5. STANDARD RESULTS (Continued)

9. To find the nth differential co-efficient of $e^{ax}\sin(bx+c)$.

Let $\quad y = e^{ax}\sin(bx+c)$ $\quad$ | Form $y = uv$

$y_1 = e^{ax}\cdot\cos(bx+c)\cdot b + \sin(bx+c)\cdot e^{ax}\cdot a$

$= e^{ax}\,[a\sin(bx+c)+b\cos(bx+c)]$

Put $a = r\cos\theta,\ b = r\sin\theta$, so that $a^2+b^2 = r^2(\cos^2\theta+\sin^2\theta) = r^2$

$\therefore \quad r = \sqrt{a^2+b^2}$

and $\quad \tan\theta = \frac{b}{a} \quad \therefore \quad \theta = \tan^{-1}\frac{b}{a}$

$\therefore \quad y_1 = e^{ax}\,[r\cos\theta\sin(bx+c)+r\sin\theta\cos(bx+c)]$

$= r\cdot e^{ax}\,[\sin(bx+c)\cos\theta+\cos(bx+c)\sin\theta]$

$= r\cdot e^{ax}\sin(bx+c+\theta).$

Thus, y_1 has been obtained from y by multiplying it by r and increasing the angle by θ. Applying the same rule for the differentiation of y_1, we have

$$y_2 = r^2 \cdot e^{ax} \sin (bx + c + 2\theta).$$

Similarly, $$y_3 = r^3 \cdot e^{ax} \sin (bx + c + 3\theta)$$

....

....

$$y_n = r^n e^{ax} \sin (bx + c + n\theta)$$

Hence $$y_n = (a^2 + b^2)^{n/2} \cdot e^{ax} \cdot \sin \left[bx + c + n \tan^{-1} \frac{b}{a} \right]$$

10. Similarly, the nth differential co-efficient of $\mathbf{e^{ax} \cos (bx + c) = (a^2 + b^2)^{n/2} \cdot e^{ax}}$ $\mathbf{\cos \left(bx + c + n \tan^{-1} \frac{b}{a} \right)}$.

Example. *Find the nth differential co-efficient of*

(a) $e^{3x} \sin^2 (2x)$ *(b)* $e^x \cos^3 x$

(c) $e^x \cos x \cos 2x$ *(d)* $e^{ax} \sin^2 x \sin 2x$

(e) $e^{x \cos \alpha} \cdot \cos (x \sin \alpha)$

Sol. (*a*) Let $$y = e^{3x} \sin^2 (2x) = e^{3x} \cdot \frac{1 - \cos 4x}{2} = \frac{1}{2} [e^{3x} - e^{3x} \cos 4x]$$

$$\therefore \quad y_n = \frac{1}{2} \left[3^n e^{3x} - (3^2 + 4^2)^{n/2} e^{3x} \cos \left(4x + n \tan^{-1} \frac{4}{3} \right) \right]$$

$$= \frac{1}{2} e^{3x} \left[3^n - 5^n \cos \left(4x + n \tan^{-1} \frac{4}{3} \right) \right]$$

(*b*) Let $$y = e^x \cos^3 x$$

Now $$\cos 3x = 4 \cos^3 x - 3 \cos x \quad \therefore \quad 4 \cos^3 x = 3 \cos x + \cos 3x$$

$$\cos^3 x = \frac{1}{4} (3 \cos x + \cos 3x)$$

$$\therefore \quad y = e^x \cdot \frac{1}{4} (3 \cos x + \cos 3x) = \frac{3}{4} \cdot e^x \cos x + \frac{1}{4} e^x \cos 3x$$

$$\therefore \quad y_n = \frac{3}{4} \cdot (1^2 + 1^2)^{n/2} \cdot e^x \cdot \cos \left(x + n \tan^{-1} \frac{1}{1} \right)$$

$$+ \frac{1}{4} (1^2 + 3^2)^{n/2} \cdot e^x \cdot \cos \left(3x + n \tan^{-1} \frac{3}{1} \right)$$

$$= \frac{1}{4} e^x \left[3.2^{n/2} \cdot \cos \left(x + n \cdot \frac{\pi}{4} \right) + 10^{n/2} \cdot \cos (3x + n \tan^{-1} 3) \right]$$

(*c*) Let $$y = e^x \cos x \cos 2x = \tfrac{1}{2} e^x (2 \cos 2x \cos x)$$

$$= \tfrac{1}{2} e^x (\cos 3x + \cos x) = \tfrac{1}{2} (e^x \cos x + e^x \cos 3x)$$

$$y_n = \tfrac{1}{2} [(1^2 + 1^2)^{n/2} \cdot e^x \cos (x + n \tan^{-1} \tfrac{1}{1})] + (1^2 + 3^2)^{n/2} \cdot e^x \cos (3x + n \tan^{-1} \tfrac{3}{1})]$$

$$= \tfrac{1}{2} \cdot e^x [2^{n/2} \cdot \cos (x + n \cdot \pi/4) + 10^{n/2} \cdot \cos (3x + n \tan^{-1} 3)].$$

(d) Let $y = e^{ax} \sin^2 x \sin 2x = e^{ax} \cdot \dfrac{1 - \cos 2x}{2} \cdot \sin 2x$

$$= \tfrac{1}{2} [e^{ax} \sin 2x - e^{ax} \cdot \cos 2x \sin 2x]$$

$$= \tfrac{1}{2} \cdot e^{ax} \sin 2x - \tfrac{1}{4} \cdot e^{ax} \sin 4x$$

$$y_n = \frac{1}{2} \cdot (a^2 + 2^2)^{n/2} \cdot e^{ax} \sin\left(2x + n \tan^{-1}\frac{2}{a}\right) - \frac{1}{4} (a^2 + 4^2)^{n/2} \cdot e^{ax} \sin\left(4x + n \tan^{-1}\frac{4}{a}\right)$$

$$= \frac{1}{2} (a^2 + 4)^{n/2} \cdot e^{ax} \sin\left(2x + n \tan^{-1}\frac{2}{a}\right) - \frac{1}{4} (a^2 + 16)^{n/2} \cdot e^{ax} \sin\left(4x + n \tan^{-1}\frac{4}{a}\right).$$

(e) Let $y = e^{x \cos \alpha} \cdot \cos(x \sin \alpha) = e^{(\cos \alpha) x} \cdot \cos[(\sin \alpha) \cdot x]$

Comparing with $e^{ax} \cos bx$, here $a = \cos \alpha$, $b = \sin \alpha$

$$\therefore \quad y_n = (\cos^2 \alpha + \sin^2 \alpha)^{n/2} \cdot e^{(\cos \alpha) x} \times \cos\left[(\sin \alpha) \cdot x + n \tan^{-1}\left(\frac{\sin \alpha}{\cos \alpha}\right)\right]$$

$$= e^{x \cos \alpha} \cdot \cos[x \sin \alpha + n \tan^{-1}(\tan \alpha)]$$

$$= e^{x \cos \alpha} \cdot \cos[x \sin \alpha + n\alpha].$$

EXERCISE 3.2

1. Find the nth derivative of

(i) $(a^2 - x^2)^{-1}$ (ii) $\dfrac{1}{(x+2)(x-1)}$

(iii) $\dfrac{3x+7}{x+2}$ (iv) $\dfrac{2x-1}{(x-2)(x+1)}$

(v) $\dfrac{x^2}{(x+2)(2x+3)}$ (U.K.T.U. 2011) (vi) $\dfrac{x+1}{6x^2 - 7x - 3}$

(vii) $\dfrac{x}{1 + 3x + 2x^2}$ (viii) $\dfrac{ax+b}{cx+d}$ (M.T.U. 2013)

2. Find the nth derivative of

(i) $\log(ax + x^2)$ (ii) $\log(x^2 - a^2)$

(iii) $\cos^2 x$ (iv) $\cos^3 x$

(v) $\cos^4 x$ (vi) $\sin x \sin 3x$

(vii) $\sin x \sin 2x \sin 3x$ (viii) $\sin^2 x \cos^3 x$

(ix) $\dfrac{1}{x^2 + 5x + 6} + \sin x \sin 2x$

3. Find the nth differential co-efficient of

(i) $e^x \sin x \cos x$ (ii) $e^{ax} \cos^2 x \sin x$.

Answers

1. (i) $\dfrac{n!}{2a}\left[\dfrac{(-1)^n}{(a+x)^{n+1}} + \dfrac{1}{(a-x)^{n+1}}\right]$ (ii) $\dfrac{(-1)^n n!}{3}\left[\dfrac{1}{(x-1)^{n+1}} - \dfrac{1}{(x+2)^{n+1}}\right]$

(iii) $\dfrac{(-1)^n n!}{(x+2)^{n+1}}$ (iv) $(-1)^n n!\left[\dfrac{1}{(x-2)^{n+1}} + \dfrac{1}{(x+1)^{n+1}}\right]$

(v) $(-1)^n \cdot n!\left[\dfrac{9.2^{n-1}}{(2x+3)^{n+1}}-\dfrac{4}{(x+2)^{n+1}}\right]$ (vi) $\dfrac{(-1)^n\, n!}{11}\left[\dfrac{5\cdot 2^n}{(2x-3)^{n+1}}-\dfrac{2\cdot 3^n}{(3x+1)^{n+1}}\right]$

(vii) $(-1)^n \cdot n!\left[\dfrac{1}{(x+1)^{n+1}}-\dfrac{2^n}{(2x+1)^{n+1}}\right]$ (viii) $\dfrac{(-1)^n\, n!(bc-ad)\,c^{n-1}}{(cx+d)^{n+1}}$

2. (i) $(-1)^{n-1}\cdot(n-1)!\left[\dfrac{1}{x^n}+\dfrac{1}{(x+a)^n}\right]$ (ii) $(-1)^{n-1}\cdot(n-1)!\left[\dfrac{1}{(x+a)^n}+\dfrac{1}{(x-a)^n}\right]$

(iii) $2^{n-1}\cos\left(2x+n\cdot\dfrac{\pi}{2}\right)$ (iv) $\dfrac{1}{4}\cdot 3^n\cos\left(3x+\dfrac{n\pi}{2}\right)+\dfrac{3}{4}\cos\left(x+\dfrac{n\pi}{2}\right)$

(v) $2^{n-1}\cos\left(2x+\dfrac{n\pi}{2}\right)+2^{2n-3}\cos\left(4x+\dfrac{n\pi}{2}\right)$

(vi) $\dfrac{1}{2}\left[2^n\cos\left(2x+\dfrac{n\pi}{2}\right)-4^n\cos\left(4x+\dfrac{n\pi}{2}\right)\right]$

(vii) $\dfrac{1}{4}\left[2^n\sin\left(2x+\dfrac{n\pi}{2}\right)+4^n\sin\left(4x+\dfrac{n\pi}{2}\right)-6^n\sin\left(6x+\dfrac{n\pi}{2}\right)\right]$

(viii) $\dfrac{1}{16}\left[2\cos\left(x+\dfrac{n\pi}{2}\right)-3^n\cos\left(3x+\dfrac{n\pi}{2}\right)-5^n\cos\left(5x+\dfrac{n\pi}{2}\right)\right]$

(ix) $(-1)^n\cdot n!\left[\dfrac{1}{(x+2)^{n+1}}-\dfrac{1}{(x+3)^{n+1}}\right]+\dfrac{1}{2}\left[\cos\left(x+\dfrac{n\pi}{2}\right)-3^n\cos\left(3x+\dfrac{n\pi}{2}\right)\right]$

3. (i) $\dfrac{1}{2}\cdot 5^{n/2}\cdot e^x\sin(2x+n\tan^{-1}2)$

(ii) $\dfrac{1}{4}\cdot e^{ax}\left[(a^2+9)^{n/2}\sin\left(3x+n\tan^{-1}\dfrac{3}{a}\right)+(a^2+1)^{n/2}\sin\left(x+n\tan^{-1}\dfrac{1}{a}\right)\right].$

4*

Leibnitz's Theorem and Applications

4.1. INTRODUCTION

Leibnitz's Theorem helps us to find the nth derivative of the product of two functions.

Statement. *If $y = uv$, where u and v are functions of x, having derivatives of nth order, then*

$$y_n = {}^nC_0 u_n v + {}^nC_1 u_{n-1} v_1 + {}^nC_2 u_{n-2} v_2 + \cdots + {}^nC_r u_{n-r} v_r + \cdots + {}^nC_n u v_n$$

where suffixes of u and v denote the number of times they are differentiated.

Note. If x^m, where m is a +ve integer, is one of the factors, taking $v = x^m$ simplifies the process of writing the nth derivative.

ILLUSTRATIVE EXAMPLES

Example 1. *Find the nth derivative of*

(a) $x^3 e^{ax}$ *(b)* $x^2 \sin x$ (U.P.T.U. 2009)

(c) $x^2 e^{3x} \cos 4x$.

Sol. (*a*) Let

$u = e^{ax}$	$v = x^3$
$u_n = a^n e^{ax}$	$v_1 = 3x^2$
$u_{n-1} = a^{n-1} e^{ax}$	$v_2 = 6x$
$u_{n-2} = a^{n-2} e^{ax}$	$v_3 = 6$
$u_{n-3} = a^{n-3} e^{ax}$	$v_4 = 0$

Now by Leibnitz's Theorem, we have

$$\frac{d^n}{dx^n}(uv) = {}^nC_0 u_n v + {}^nC_1 u_{n-1} v_1 + {}^nC_2 u_{n-2} v_2 + \cdots + {}^nC_n u v_n.$$

$$\therefore \quad \frac{d^n}{dx^n}(e^{ax} \cdot x^3) = {}^nC_0 a^n e^{ax} x^3 + {}^nC_1 a^{n-1} e^{ax} \cdot 3x^2 + {}^nC_2 a^{n-2} e^{ax} \cdot 6x + {}^nC_3 \cdot a^{n-3} e^{ax} \cdot 6$$

$$= e^{ax} \cdot a^{n-3}\left[a^3x^3 + 3na^2x^2 + 6\frac{n(n-1)}{2!} \cdot ax + 6 \cdot \frac{n(n-1)(n-2)}{3!}\right]$$

$$= e^{ax} \cdot a^{n-3}\,[a^3x^3 + 3na^2x^2 + 3n(n-1) \cdot ax + n(n-1)(n-2)].$$

(*b*) Let

$u = \sin x$	$v = x^2$
$u_n = \sin\left(x + \frac{n\pi}{2}\right)$	$v_1 = 2x$
$u_{n-1} = \sin\left[x + (n-1)\frac{\pi}{2}\right]$	$v_2 = 2$

*This chapter is not included in the syllabus of KU, Kurukshetra and DCRUST, Murthal.

$$u_{n-2} = \sin\left[x + (n-2)\frac{\pi}{2}\right] \quad \Bigg| \quad v_3 = 0$$

∴ By Leibnitz's Theorem,

$$\frac{d^n}{dx^n}(\sin x \cdot x^2) = {}^nC_0 \sin\left(x + \frac{n\pi}{2}\right)x^2$$
$$+ {}^nC_1 \cdot \sin\left(x + (n-1)\frac{\pi}{2}\right) \cdot 2x + {}^nC_2 \sin\left[x + (n-2)\frac{\pi}{2}\right] 2$$
$$= x^2 \sin\left(x + \frac{n\pi}{2}\right) + 2nx \sin\left(x + (n-1)\frac{\pi}{2}\right)$$
$$+ 2 \cdot \frac{n(n-1)}{2 \cdot 1} \sin\left[x + (n-2)\frac{\pi}{2}\right]$$
$$= x^2 \sin\left(x + \frac{n\pi}{2}\right) + 2nx \sin\left(x + (n-1)\frac{\pi}{2}\right)$$
$$+ n(n-1) \sin\left(x + (n-2)\frac{\pi}{2}\right).$$

(*c*) Let

$$\begin{aligned} u &= e^{3x} \cos 4x & v &= x^2 \\ u_n &= r^n e^{3x} \cos(4x + n\theta) & v_1 &= 2x \\ u_{n-1} &= r^{n-1} e^{3x} \cos[4x + (n-1)\theta] & v_2 &= 2 \\ u_{n-2} &= r^{n-2} e^{3x} \cos[4x + (n-2)\theta] & v_3 &= 0 \end{aligned}$$

where $r = \sqrt{3^2 + 4^2} = 5,\ \theta = \tan^{-1}\frac{4}{3}$

∴ By Leibnitz's Theorem,

$$\frac{d^n}{dx^n}(e^{3x} \cos 4x \cdot x^2) = \frac{d^n}{dx^n}(uv) = {}^nC_0 u_n v + {}^nC_1 u_{n-1} v_1 + {}^nC_2 u_{n-2} v_2$$
$$= r^n e^{3x} \cos(4x + n\theta) \cdot x^2 + nr^{n-1} e^{3x} \cos[4x + (n-1)\theta] \cdot 2x$$
$$+ \frac{n(n-1)}{2!} \cdot r^{n-2} e^{3x} \cos[4x + (n-2)\theta] \cdot 2$$
$$= r^{n-2} e^{3x} [r^2 x^2 \cos(4x + n\theta) + 2nr\, x \cos(4x + (n-1)\theta)$$
$$+ n(n-1) \cos(4x + (n-2)\theta]$$
$$= 5^{n-2} \cdot e^{3x} \left[25\, x^2 \cos\left(4x + n\tan^{-1}\frac{4}{3}\right) + 10\, nx \cos\left(4x + (n-1)\tan^{-1}\frac{4}{3}\right)\right.$$
$$\left. + n(n-1) \cos\left(4x + (n-2)\tan^{-1}\frac{4}{3}\right)\right].$$

Example 2. *If $y = x^2 \sin x$, prove that*

$$\frac{d^n y}{dx^n} = (x^2 - n^2 + n) \sin\left(x + \frac{n\pi}{2}\right) - 2nx \cos\left(x + \frac{n\pi}{2}\right).$$

Sol. Proceeding as in Example 1 (*b*),

$$\frac{d^n y}{dx^n} = x^2 \sin\left(x + \frac{n\pi}{2}\right) + 2nx \sin\left(x + (n-1)\frac{\pi}{2}\right) + n(n-1) \sin\left(x + (n-2)\frac{\pi}{2}\right)$$
$$= x^2 \sin\left(x + \frac{n\pi}{2}\right) + 2nx \sin\left[\left(x + \frac{n\pi}{2}\right) - \frac{\pi}{2}\right] + n(n-1) \sin\left[\left(x + \frac{n\pi}{2}\right) - \pi\right]$$

$$= x^2 \sin\left(x+\frac{n\pi}{2}\right) - 2nx \sin\left[\frac{\pi}{2}-\left(x+\frac{n\pi}{2}\right)\right] - n(n-1)\sin\left[\pi-\left(x+\frac{n\pi}{2}\right)\right]$$

$$= x^2 \sin\left(x+\frac{\pi n}{2}\right) - 2nx \cos\left(x+\frac{n\pi}{2}\right) - n(n-1)\sin\left(x+\frac{n\pi}{2}\right)$$

$$= (x^2 - n^2 + n)\sin\left(x+\frac{n\pi}{2}\right) - 2nx\cos\left(x+\frac{n\pi}{2}\right).$$

Example 3. *Find the nth derivative of*

(a) $x^2 \log x$ *(b)* $e^x \log x$

Sol. (*a*) Let

$u = \log x$	$v = x^2$
$u_n = \dfrac{(-1)^{n-1}(n-1)!}{x^n}$	$v_1 = 2x$
$u_{n-1} = \dfrac{(-1)^{n-2}(n-2)!}{x^{n-1}}$	$v_2 = 2$
$u_{n-2} = \dfrac{(-1)^{n-3}(n-3)!}{x^{n-2}}$	$v_3 = 0$

$\therefore$ By Leibnitz's Theorem,

$$\frac{d^n}{dx^n}(\log x \cdot x^2) = {}^nC_0 \cdot \frac{(-1)^{n-1}(n-1)!}{x^n}\, x^2 + {}^nC_1 \cdot \frac{(-1)^{n-2}(n-2)!}{x^{n-1}} \cdot 2x + {}^nC_2 \cdot \frac{(-1)^{n-3}(n-3)!}{x^{n-2}} \cdot 2$$

$$= \frac{(-1)^{n-1}(n-1)!}{x^{n-2}} + 2n\,\frac{(-1)^{n-2}(n-2)!}{x^{n-2}} + n(n-1)\cdot\frac{(-1)^{n-3}(n-3)!}{x^{n-2}}$$

$$= \frac{(-1)^{n-1}\,n!}{nx^{n-2}} + 2\cdot\frac{(-1)^{n-2}\,n!}{(n-1)x^{n-2}} + \frac{(-1)^{n-3}\,n!}{(n-2)x^{n-2}}$$

$$= \frac{(-1)^{n-1}\,n!}{x^{n-2}}\left[\frac{1}{n} - \frac{2}{n-1} + \frac{1}{n-2}\right].$$

(*b*) Let

$u = e^x$	$v = \log x$
$u_n = e^x$	$v_1 = \dfrac{1}{x}$
$u_{n-1} = e^x$	$v_2 = -\dfrac{1}{x^2}$
$u_{n-2} = e^x$	$v_3 = \dfrac{2}{x^3}$
....	....
$u_2 = u_1 = e^x$	$v_n = \dfrac{(-1)^{n-1}(n-1)!}{x^n}$

$\therefore$ By Leibnitz's Theorem,

$$\frac{d^n}{dx^n}(e^x \log x) = {}^nC_0\, e^x \cdot \log x + {}^nC_1\, e^x \cdot \left(\frac{1}{x}\right) + {}^nC_2 e^x \cdot \left(-\frac{1}{x^2}\right) + \dots + {}^nC_n \cdot e^x \frac{(-1)^{n-1}(n-1)!}{x^n}$$

$$= e^x\left[\log x + \frac{n}{x} - \frac{n(n-1)}{2}\cdot\frac{1}{x^2} + \dots + \frac{(-1)^{n-1}(n-1)!}{x^n}\right].$$

Example 4. *If* $y = x^2e^x$, *show that* $\dfrac{d^n y}{dx^n} = \dfrac{1}{2}\, n(n-1)\dfrac{d^2y}{dx^2} - n(n-2)\dfrac{dy}{dx} + \dfrac{1}{2}(n-1)$ $(n-2)y$.

Sol. $y = x^2e^x$

Let

$$\begin{array}{rl|l} & u = e^x & v = x^2 \\ & u_n = e^x & v_1 = 2x \\ & u_{n-1} = e^x & v_2 = 2 \\ & u_{n-2} = e^x & v_3 = 0 \end{array}$$

$\therefore$ By Leibnitz's Theorem,

$$\frac{d^n y}{dx^n} = e^x \cdot x^2 + n\cdot e^x \cdot 2x + \frac{n(n-1)}{2!}\cdot e^x \cdot 2 = e^x \cdot [x^2 + 2nx + n(n-1)]$$

Putting $n = 1, 2$

$$\frac{dy}{dx} = e^x\,[x^2 + 2x]\ ;\quad \frac{d^2y}{dx^2} = e^x\,[x^2 + 4x + 2]$$

$$\therefore\quad \text{R.H.S.} = \frac{1}{2}\,n(n-1)\frac{d^2y}{dx^2} - n(n-2)\frac{dy}{dx} + \frac{1}{2}(n-1)(n-2)\,y$$

$$= \frac{1}{2}\,n(n-1)\,e^x\,(x^2+4x+2) - n(n-2)\cdot e^x\,(x^2+2x) + \frac{1}{2}(n-1)(n-2)\,x^2e^x$$

$$= \frac{1}{2}\,e^x\,[n(n-1)(x^2+4x+2) - 2n(n-2)(x^2+2x) + (n-1)(n-2)\,x^2]$$

$$= \frac{1}{2}\,e^x\,[x^2(n^2 - n - 2n^2 + 4n + n^2 - 3n + 2) + x\,(4n^2 - 4n - 4n^2 + 8n) + 2n(n-1)]$$

$$= \frac{1}{2}e^x\,[2x^2 + 4nx + 2n(n-1)]$$

$$= e^x\,[x^2 + 2nx + n(n-1)] = \frac{d^n y}{dx^n} = \text{L.H.S.}$$

Example 5. *If* $y = x^n \log x$, *prove that* $y_{n+1} = \dfrac{n!}{x}$.

Sol. $y = x^n \log x$ ...(*i*)

$\therefore$ $y_1 = x^n \cdot \dfrac{1}{x} + nx^{n-1}\cdot \log x$

or $xy_1 = x^n + nx^n \log x$

or $xy_1 = x^n + ny$ [$\because$ of (*i*)]

Differentiating n times by Leibnitz's Theorem,

$$y_{n+1}\,x + n\cdot y_n \cdot 1 = n\,! + ny_n$$

$\therefore$ $xy_{n+1} = n\,!$ or $y_{n+1} = \dfrac{n!}{x}$.

Note. To differentiate $(1 - x^2)\,y_2$, n times by Leibnitz's Theorem.

Let

$u = y_2$	$v = 1 - x^2$
$u_n = y_{n+2}$	$v_1 = -2x$
$u_{n-1} = y_{n+1}$	$v_2 = -2$
$u_{n-2} = y_n$	$v_3 = 0$

$$\therefore \quad \frac{d^n}{dx^n}[(1-x^2)\,y_2] = y_{n+2}\,(1-x^2) + n\,y_{n+1}\,(-2x) + \frac{n(n-1)}{2!}\,y_n\,(-2)$$
$$= (1-x^2)\,y_{n+2} - 2nxy_{n+1} - n(n-1)\,y_n.$$

Example 6. *Differentiate n times the equation* $(1-x^2)\,y_2 - xy_1 + a^2y = 0$

Sol. $(1-x^2)\,y_2 - xy_1 + a^2y = 0$

Differentiating every term n times by Leibnitz's Theorem

$$\left[y_{n+2}\,(1-x^2) + ny_{n+1}\,(-2x) + \frac{n(n-1)}{2!}\cdot y_n\,(-2)\right]$$
$$-[y_{n+1}\cdot x + ny_n\cdot 1] + a^2\,y_n = 0$$

or $$(1-x^2)\,y_{n+2} + (-2nx - x)\,y_{n+1} + (-n^2 + n - n + a^2)\,y_n = 0$$

or $$(1-x^2)\,y_{n+2} - (2n+1)\,xy_{n+1} - (n^2 - a^2)\,y_n = 0.$$

Example 7. *If* $y = a\cos(\log x) + b\sin(\log x)$, *show that* $x^2y_{n+2} + (2n+1)\,xy_{n+1} + (n^2+1)\,y_n = 0$.

Sol. $$y = a\cos(\log x) + b\sin(\log x) \qquad \ldots(i)$$

$\Rightarrow$ $$y_1 = -a\sin(\log x)\cdot\frac{1}{x} + b\cos(\log x)\cdot\frac{1}{x}$$

or $$xy_1 = -a\sin(\log x) + b\cos(\log x)$$

Differentiating again,

$$xy_2 + y_1\cdot 1 = -a\cos(\log x)\cdot\frac{1}{x} - b\sin(\log x)\cdot\frac{1}{x}$$

or $$x^2y_2 + xy_1 = -[a\cos(\log x) + b\sin(\log x)]$$

or $$x^2y_2 + xy_1 = -y \qquad | \because \text{ of } (i)$$

or $$x^2y_2 + xy_1 + y = 0.$$

Differentiating every term n times by Leibnitz's Theorem,

$$\left[y_{n+2}\,x^2 + ny_{n+1}\,2x + \frac{n(n-1)}{2!}\,y_n\,.\,2\right] + [y_{n+1}\cdot x + ny_n\cdot 1] + y_n = 0$$

or $$x^2y_{n+2} + (2nx + x)\,y_{n+1} + (n^2 - n + n + 1)\,y_n = 0$$

or $$x^2y_{n+2} + (2n+1)\,xy_{n+1} + (n^2+1)\,y_n = 0.$$

Example 8. *If* $y = (\sin^{-1}x)^2$, *prove that* $(1-x^2)\,y_{n+2} - (2n+1)\,xy_{n+1} - n^2y_n = 0.$

(U.P.T.U. 2009)

Sol. $$y = (\sin^{-1}x)^2 \qquad \ldots(i)$$

$\Rightarrow$ $$y_1 = 2\sin^{-1}x\cdot\frac{1}{\sqrt{1-x^2}}$$

Squaring and cross-multiplying,

$$(1-x^2)\,y_1^2 = 4(\sin^{-1}x)^2 = 4y \qquad [\because \text{ of } (i)]$$

Differentiating again,

$$(1-x^2)\,.\,2y_1y_2 + y_1^2\,(-2x) = 4y_1$$

Dividing both sides by $2y_1$, $(1 - x^2)\, y_2 - xy_1 = 2$

Differentiating every term n times by Leibnitz's Theorem,

$$\left[y_{n+2}\,(1-x^2) + ny_{n+1}\,(-2x) + \frac{n(n-1)}{2!} \cdot y_n(-2) \right] - (y_{n+1} \cdot x + ny_n \cdot 1) = 0$$

or $$(1 - x^2)\, y_{n+2} - (2n + 1)\, xy_{n+1} + (-\, n^2 + n - n)\, y_n = 0$$

or $$(1 - x^2)\, y_{n+2} - (2n + 1)\, xy_{n+1} - n^2 y_n = 0.$$

Example 9. *If* $y = e^{a \sin^{-1} x}$, *show that* $(1 - x^2)\, y_{n+2} - (2n + 1)\, xy_{n+1} - (n^2 + a^2)\, y_n = 0.$

(M.D.U. Dec. 2010; Kerala 2010)

Sol. $$y = e^{a \sin^{-1} x} \qquad ...(i)$$

$\Rightarrow$ $$y_1 = e^{a \sin^{-1} x} \cdot \frac{a}{\sqrt{1-x^2}} = \frac{ay}{\sqrt{1-x^2}} \qquad [\because \text{ of } (i)]$$

Squaring and cross-multiplying, $(1 - x^2)\, y_1^2 = a^2 y^2$

Differentiating again, $(1 - x^2) 2y_1 y_2 + y_1^2\, (-2x) = a^2 \cdot 2yy_1$

Dividing both sides by $2y_1$, $(1 - x^2) y_2 - xy_1 = a^2 y$

Differentiating every term n times by Leibnitz's Theorem,

$$\left[y_{n+2}\,(1-x^2) + ny_{n+1}\,(-2x) + \frac{n(n-1)}{2!}\, y_n(-2) \right] - [y_{n+1} \cdot x + ny_n \cdot 1] = a^2 y_n$$

or $$(1 - x^2) y_{n+2} - (2n + 1) xy_{n+1} + (-\, n^2 + n - n - a^2)\, y_n = 0$$

or $$(1 - x^2)\, y_{n+2} - (2n + 1)\, xy_{n+1} - (n^2 + a^2) y_n = 0.$$

Example 10. *If* $y = \sin\,(m \sin^{-1} x)$, *prove that* $(1 - x^2)\, y_{n+2} = (2n + 1)\, xy_{n+1} + (n^2 - m^2)\, y_n$.

Also show that $\underset{x \to 0}{Lt} \dfrac{y_{n+2}}{y_n} = n^2 - m^2$.

Sol. $$y = \sin\,(m \sin^{-1} x) \qquad ...(i)$$

$\Rightarrow$ $$y_1 = \cos\,(m \sin^{-1} x) \frac{m}{\sqrt{1-x^2}}$$

Squaring and cross-multiplying,

$$(1 - x^2)\, y_1^2 = m^2 \cos^2 (m \sin^{-1} x) = m^2\, [1 - \sin^2 (m \sin^{-1} x)]$$

$\therefore$ $$(1 - x^2)\, y_1^2 = m^2\, (1 - y^2) \qquad [\because \text{ of } (i)]$$

Differentiating again, $(1 - x^2)\, 2y_1 y_2 + y_1^2\, (-2x) = m^2\, (-\, 2yy_1)$

Dividing both sides by $2y_1$, $(1 - x^2)\, y_2 - xy_1 + m^2 y = 0$

Differentiating every term n times by Leibnitz's Theorem,

$$\left[y_{n+2}\,(1-x^2) + ny_{n+1}(-2x) + \frac{n(n-1)}{2!}\, y_n(-2) \right] - [y_{n+1}\, x + ny_n \cdot 1] + m^2 y_n = 0$$

or $$(1 - x^2)\, y_{n+2} - (2n + 1)\, xy_{n+1} + (-\, n^2 + n - n + m^2)\, y_n = 0$$

or $$(1 - x^2)\, y_{n+2} - (2n + 1)\, xy_{n+1} - (n^2 - m^2)\, y_n = 0$$

or $$(1 - x^2)\, y_{n+2} = (2n + 1)\, xy_{n+1} + (n^2 - m^2)\, y_n.$$

Second Part. Proceeding to the limit as $x \to 0$, we have

$$\underset{x \to 0}{\text{Lt}} (1 - x^2)\, y_{n+2} = \underset{x \to 0}{\text{Lt}} (n^2 - m^2)\, y_n$$

or $$\underset{x \to 0}{\text{Lt}}\, y_{n+2} = (n^2 - m^2) \underset{x \to 0}{\text{Lt}}\, y_n$$

or $$\underset{x \to 0}{\text{Lt}} \frac{y_{n+2}}{y_n} = n^2 - m^2.$$

Example 11. *If* $y = (x^2 - 1)^n$, *prove that* $(x^2 - 1)\, y_{n+2} + 2xy_{n+1} - n(n + 1)y_n = 0.$

Hence if $P_n = \frac{d^n}{dx^n} [(x^2 - 1)^n]$, *show that* $\frac{d}{dx}\left\{(1 - x^2)\frac{dP_n}{dx}\right\} + n(n + 1)P_n = 0.$

(U.K.T.U. 2010)

Sol. $$y = (x^2 - 1)^n \qquad \text{...}(i)$$

$$\Rightarrow \quad y_1 = n(x^2 - 1)^{n-1} \cdot 2x = \frac{2nx\,(x^2 - 1)^n}{x^2 - 1} = \frac{2nxy}{x^2 - 1} \qquad [\because \text{ of } (i)]$$

$$\Rightarrow \quad (x^2 - 1)\, y_1 = 2nxy$$

Differentiating again,

$$(x^2 - 1)y_2 + y_1(2x) = 2n(xy_1 + y \cdot 1)$$

i.e., $$(x^2 - 1)\, y_2 - 2(n - 1)\, xy_1 - 2ny = 0$$

Differentiating every term n times by Leibnitz's Theorem,

$$\left[y_{n+2}\,(x^2 - 1) + n \cdot y_{n+1} \cdot 2x + \frac{n(n-1)}{2!} \cdot y_n \cdot 2\right]$$

$$- 2(n - 1)\,[y_{n+1} \cdot x + ny_n \cdot 1] - 2ny_n = 0$$

$$(x^2 - 1)\, y_{n+2} + (2nx - 2nx + 2x)\, y_{n+1} + (n^2 - n - 2n^2 + 2n - 2n)\, y_n = 0$$

or $$(x^2 - 1)\, y_{n+2} + 2xy_{n+1} - n(n + 1)\, y_n = 0.$$

Second Part. $$\mathrm{P}_n = \frac{d^n}{dx^n} [(x^2 - 1)^n] = \frac{d^n}{dx^n} (y) = y_n$$

$$\Rightarrow \quad \frac{dP_n}{dx} = \frac{d}{dx}(y_n) = y_{n+1}$$

$$\therefore \quad \frac{d}{dx}\left\{(1 - x^2)\frac{d\mathrm{P}_n}{dx}\right\} = \frac{d}{dx} \{(1 - x^2)\, y_{n+1}\} = (1 - x^2)\, y_{n+2} - 2xy_{n+1}$$

$$= -[(x^2 - 1)\, y_{n+2} + 2xy_{n+1}]$$

$$= -n(n + 1)y_n \qquad \text{(from first part)}$$

$$= -n(n + 1)\, \mathrm{P}_n \qquad [\because \; y_n = \mathrm{P}_n]$$

Hence $$\frac{d}{dx}\left\{(1 - x^2)\frac{d\mathrm{P}_n}{dx}\right\} + n(n + 1)\, \mathrm{P}_n = 0.$$

Example 12. *If $y^{1/m} + y^{-1/m} = 2x$, prove that*

$$(x^2 - 1)y_{n+2} + (2n + 1)xy_{n+1} + (n^2 - m^2)y_n = 0.$$

(U.P.T.U. 2008; U.K.T.U. 2011)

Sol. $y^{1/m} + y^{-1/m} = 2x$

Put $y^{1/m} = z$, then we have

$$z + \frac{1}{z} = 2x \quad \text{or} \quad z^2 - 2xz + 1 = 0$$

$$\therefore \quad z = \frac{2x \pm \sqrt{4x^2 - 4}}{2} = x \pm \sqrt{x^2 - 1}$$

i.e., $$y^{1/m} = x \pm \sqrt{x^2 - 1} \quad \text{or} \quad y = (x \pm \sqrt{(x^2 - 1)})^m$$

If $y = (x + \sqrt{(x^2 - 1)})^m$, then

$$y_1 = m(x + \sqrt{x^2 - 1})^{m-1}\left(1 + \frac{1}{2}\frac{1}{\sqrt{x^2 - 1}} \cdot 2x\right)$$

$$= m(x + \sqrt{x^2 - 1})^{m-1}\left(\frac{x + \sqrt{x^2 - 1}}{\sqrt{x^2 - 1}}\right)$$

$$= \frac{m(x + \sqrt{x^2 - 1})^m}{\sqrt{x^2 - 1}} = \frac{my}{\sqrt{x^2 - 1}}$$

If $y = (x - \sqrt{x^2 - 1})^m$, then

$$y_1 = m(x - \sqrt{x^2 - 1})^{m-1}\left(1 - \frac{1}{2} \cdot \frac{1}{\sqrt{x^2 - 1}} \cdot 2x\right)$$

$$= m(x - \sqrt{x^2 - 1})^{m-1}\left(\frac{\sqrt{x^2 - 1} - x}{\sqrt{x^2 - 1}}\right)$$

$$= \frac{-m(x - \sqrt{x^2 - 1})^{m-1}[x - \sqrt{x^2 - 1}]}{\sqrt{x^2 - 1}}$$

$$= \frac{-m(x - \sqrt{x^2 - 1})^m}{\sqrt{x^2 - 1}} = -\frac{my}{\sqrt{x^2 - 1}}$$

Squaring, we have, in either case

$$y_1^2 = \frac{m^2 y^2}{x^2 - 1} \quad \text{or} \quad (x^2 - 1)\, y_1^2 = m^2 y^2$$

Differentiating again, $(x^2 - 1) \cdot 2y_1 y_2 + y_1^2(2x) = m^2 \cdot 2y \cdot y_1$

Dividing both sides by $2y_1$, $(x^2 - 1)y_2 + xy_1 - m^2 y = 0$

Differentiating every term n times by Leibnitz's Theorem,

$$\left[y_{n+2}(x^2 - 1) + n \cdot y_{n+1} \cdot 2x + \frac{n(n-1)}{2!} \cdot y_n \cdot 2\right] + [y_{n+1} \cdot x + ny_n \cdot 1] - m^2 y_n = 0$$

or $$(x^2 - 1)y_{n+2} + (2n + 1)xy_{n+1} + (n^2 - n + n - m^2)y_n = 0$$

or $$(x^2 - 1)y_{n+2} + (2n + 1)xy_{n+1} + (n^2 - m^2)y_n = 0.$$

Example 13. *If* $\cos^{-1}\left(\frac{y}{b}\right) = \log\left(\frac{x}{n}\right)^n$*, prove that* $x^2y_{n+2} + (2n+1)\,xy_{n+1} + 2n^2y_n = 0$.

Sol. $$\cos^{-1}\left(\frac{y}{b}\right) = \log\left(\frac{x}{n}\right)^n = n\log\frac{x}{n}$$

$$\Rightarrow \quad \frac{y}{b} = \cos\left(n\log\frac{x}{n}\right)$$

$$\Rightarrow \quad y = b\cos\left(n\log\frac{x}{n}\right) \qquad \text{...}(i)$$

$$\therefore \quad y_1 = -\,b\sin\left(n\log\frac{x}{n}\right)\cdot n\cdot\frac{1}{x/n}\cdot\frac{1}{n} = -\frac{bn}{x}\sin\left(n\log\frac{x}{n}\right)$$

or $$xy_1 = -\,bn\sin\left(n\log\frac{x}{n}\right)$$

Differentiating again,

$$xy_2 + y_1\cdot 1 = -bn\cos\left(n\log\frac{x}{n}\right)n\cdot\frac{1}{x/n}\cdot\frac{1}{n}$$

or $$xy_2 + y_1 = -\,bn\cdot\frac{n}{x}\cos\left(n\log\frac{x}{n}\right)$$

or $$x^2y_2 + xy_1 = -\,n^2y \qquad |\because \text{ of } (i)$$

Differentiating every term n times by Leibnitz's Theorem,

$$\left[y_{n+2}\cdot x^2 + ny_{n+1}\cdot 2x + \frac{n(n-1)}{2!}\cdot y_n\cdot 2\right] + [y_{n+1}\cdot x + ny_n\cdot 1] = -\,n^2y_n$$

or $$x^2y_{n+2} + (2n+1)xy_{n+1} + (n^2 - n + n + n^2)y_n = 0$$

or $$x^2y_{n+2} + (2n+1)xy_{n+1} + 2n^2y_n = 0.$$

Example 14. *If* $y = x\cos(\log x)$*, then show that*

$$x^2y_{n+2} + (2n-1)xy_{n+1} + (n^2 - 2n + 2)y_n = 0.$$

Sol. Given $$y = x\cos(\log x) \qquad \text{...}(1)$$

Differentiating, $y_1 = \cos(\log x) - x\sin(\log x)\cdot\frac{1}{x}$

$$\Rightarrow \quad xy_1 = x\cos(\log x) - x\sin(\log x)$$

$$\Rightarrow \quad xy_1 = y - x\sin(\log x) \qquad \text{...}(2)$$

Differentiating again,

$$xy_2 + y_1 = y_1 - \sin(\log x) - x\cos(\log x)\cdot\frac{1}{x}$$

$$\Rightarrow \quad xy_2 = -\sin(\log x) - \cos(\log x)$$

$$\Rightarrow \quad x^2y_2 = -\,x\sin(\log x) - x\cos(\log x)$$

$$\Rightarrow \quad x^2y_2 = (xy_1 - y) - y \qquad [\because \text{ Using (1) and (2)}]$$

$$\Rightarrow \quad x^2y_2 - xy_1 + 2y = 0$$

Differentiating n times by Leibnitz's Theorem,

$$\left[y_{n+2} \cdot x^2 + ny_{n+1} \cdot 2x + \frac{n(n-1)}{2!} \cdot y_n \cdot 2\right] - [y_{n+1} \cdot x + ny_n \cdot 1] + 2y_n = 0$$

$$\Rightarrow \qquad x^2 y_{n+2} + (2n-1)xy_{n+1} + (n^2 - 2n + 2)y_n = 0$$

Example 15. *If $y = x^{n-1} \log x$, prove that $xy_n = (n-1)!$.*

Sol. $\qquad y = x^{n-1} \log x \qquad$ ...(1)

Differentiating $y_1 = x^{n-1} \cdot \dfrac{1}{x} + (n-1)x^{n-2} \log x$

$\Rightarrow \qquad xy_1 = x^{n-1} + (n-1)x^{n-1} \log x$

$\Rightarrow \qquad xy_1 = x^{n-1} + (n-1)y \qquad$ [Using (1)]

$\Rightarrow \quad xy_1 - (n-1)\, y = x^{n-1}$

Differentiating $(n-1)$ times by Leibnitz's Theorem,

$$[y_n \cdot x + (n-1)y_{n-1} \cdot 1] - (n-1)y_{n-1} = (n-1)! \qquad \left[\because \frac{d^n}{dx^n}(x^n) = n!\right]$$

$\Rightarrow \qquad xy_n = (n-1)!.$

4.2. DETERMINATION OF THE VALUE OF THE nth DERIVATIVE OF A FUNCTION FOR x = 0

Sometimes, it is required to find the nth derivative of a function for $x = 0$. This is illustrated by the following solved examples.

Example 1. *Find the value of the nth derivative of $e^{m \sin^{-1} x}$ for $x = 0$.*

(M.D.U. Dec. 2010)

Sol. Let $\qquad y = e^{m \sin^{-1} x} \qquad$ ...(*i*)

$$y_1 = e^{m \sin^{-1} x} \cdot \frac{m}{\sqrt{1-x^2}} = \frac{my}{\sqrt{1-x^2}} \qquad ...(ii)$$

Squaring and cross-multiplying, $(1 - x^2)\, {y_1}^2 = m^2 y^2$

Differentiating again, $(1 - x^2)\, 2y_1 y_2 + y_1^2 (-2x) = m^2 \cdot 2yy_1$

Dividing both sides by $2y_1$, $(1 - x^2)\, y_2 - xy_1 = m^2 y \qquad$ ...(*iii*)

Differentiating every term n times by Leibnitz's Theorem,

$$\left[y_{n+2} \cdot (1-x^2) + ny_{n+1}(-2x) + \frac{n(n-1)}{2!} \cdot y_n(-2)\right]$$

$$- [y_{n+1} \cdot x + ny_n \cdot 1] = m^2 y_n$$

or $\qquad (1 - x^2)y_{n+2} - (2n+1)xy_{n+1} = (m^2 + n^2)y_n \qquad$ ...(*iv*)

Putting $\; x = 0$ in (*i*), (*ii*), (*iii*) and (*iv*)

$$y(0) = e^{m \sin^{-1} 0} = e^0 = 1$$

$$y_1(0) = my(0) = m \cdot 1 = m$$

$$y_2(0) = m^2 y(0) = m^2$$

$$y_{n+2}(0) = (m^2 + n^2)\, y_n(0) \qquad ...(v)$$

Putting $n = 1, 2, 3, 4, \ldots$ in (*v*)

$$y_3(0) = (m^2 + 1^2)\, y_1(0) = (m^2 + 1^2)\, m$$
$$y_4(0) = (m^2 + 2^2)\, y_2(0) = (m^2 + 2^2)\, m^2$$
$$y_5(0) = (m^2 + 3^2)\, y_3(0) = (m^2 + 3^2)\,(m^2 + 1^2)\, m$$
$$y_6(0) = (m^2 + 4^2)\, y_4(0) = (m^2 + 4^2)\,(m^2 + 2^2)\, m^2$$

....

....

Hence when n is odd,

$$y_n(0) = m[m^2 + 1^2]\,[m^2 + 3^2] \ldots [m^2 + (n-2)^2]$$

and when n is even,

$$y_n(0) = m^2[m^2 + 2^2]\,[m^2 + 4^2] \ldots [m^2 + (n-2)^2].$$

Example 2. *If $y = (\sin^{-1} x)^2$, prove that*

$$(1 - x^2)\, y_{n+2} - (2n + 1)\, xy_{n+1} - n^2 y_n = 0.$$

Hence find the value of y_n at $x = 0$. (K.U.K. 2013; U.P.T.U. 2009)

Sol. $$y = (\sin^{-1} x)^2 \qquad \ldots(i)$$

$$y_1 = 2 \sin^{-1} x \cdot \frac{1}{\sqrt{1-x^2}} \qquad \ldots(ii)$$

Squaring and cross-multiplying,

$$(1 - x^2)\, y_1^2 = 4(\sin^{-1} x)^2 = 4y$$

Differentiating again, $(1 - x^2)\, 2y_1 y_2 + y_1^2(-2x) = 4y_1$

Dividing by $2y_1$, $$(1 - x^2)\, y_2 - xy_1 = 2 \qquad \ldots(iii)$$

Differentiating every term n times by Leibnitz's Theorem,

$$\left[y_{n+2}\,(1-x^2) + ny_{n+1}\,(-2x) + \frac{n(n-1)}{2!} \cdot y_n\,(-2)\right] - [y_{n+1}\, x + n \cdot y_n \cdot 1] = 0$$

or $$(1 - x^2) y_{n+2} - (2n + 1) xy_{n+1} - n^2 y_n = 0 \qquad \ldots(iv)$$

Putting $x = 0$ in (*i*), (*ii*), (*iii*) and (*iv*),

$$y(0) = 0,\; y_1(0) = 0,\; y_2(0) = 2,\; y_{n+2}(0) = n^2 y_n(0) \qquad \ldots(v)$$

Putting $n = 1, 2, 3, 4, \ldots$ in (*v*)

$$y_3(0) = 1^2 \cdot y_1(0) = 1^2 \cdot (0) = 0$$
$$y_4(0) = 2^2 \cdot y_2(0) = 2^2 \cdot 2$$
$$y_5(0) = 3^2 \cdot y_3(0) = 3^2 \cdot 0 = 0$$
$$y_6(0) = 4^2 \cdot y_4(0) = 4^2 \cdot 2^2 \cdot 2$$

....

....

Hence, when n is odd, $y_n(0) = 0$

When n is even, $\quad y_n(0) = (n-2)^2 \ldots 4^2 \cdot 2^2 \cdot 2.$

MISCELLANEOUS EXAMPLES

Example 1. *If* $y = x \log \left(\dfrac{x-1}{x+1}\right)$, *prove that*

$$y_m = (-1)^{m-2}(m-2)!\left[\frac{x-m}{(x-1)^m} - \frac{x+m}{(x+1)^m}\right].$$

Sol. $y = x \log \left(\dfrac{x-1}{x+1}\right) = x[\log(x-1) - \log(x+1)]$

or $y = x\log(x-1) - x\log(x+1)$

Differentiating m times by Leibnitz's Theorem, we get

$$y_m = \left[\frac{d^m}{dx^m}\log(x-1)\right]\cdot x + m\left[\frac{d^{m-1}}{dx^{m-1}}\log(x-1)\right]\cdot 1 - \left[\frac{d^m}{dx^m}\log(x+1)\right].\, x - m\left[\frac{d^{m-1}}{dx^{m-1}}\log(x+1)\right]\cdot 1$$

$$= x\cdot\frac{(-1)^{m-1}(m-1)!}{(x-1)^m} + m\cdot\frac{(-1)^{m-2}(m-2)!}{(x-1)^{m-1}} - x\cdot\frac{(-1)^{m-1}(m-1)!}{(x+1)^m} - m\cdot\frac{(-1)^{m-2}(m-2)}{(x+1)^{m-1}}$$

$$= (-1)^{m-2}(m-2)!\left[x\cdot\frac{(-1)(m-1)}{(x-1)^m} + \frac{m}{(x-1)^{m-1}} - x\cdot\frac{(-1)(m-1)}{(x+1)^m} - \frac{m}{(x+1)^{m-1}}\right]$$

$$= (-1)^{m-2}(m-2)!\left[\frac{-mx + x + m(x-1)}{(x-1)^m} - \frac{-mx + x + m(x+1)}{(x+1)^m}\right]$$

$$= (-1)^{m-2}(m-2)!\left[\frac{x-m}{(x-1)^m} - \frac{x+m}{(x+1)^m}\right].$$

Example 2. *If* $u = \dfrac{x^5}{Ax^2 + 2Bx + C}$, *show that*

$$(Ax^2 + 2Bx + C)u_n + 2n(Ax + B)u_{n-1} + n(n-1)A\,u_{n-2} = 0, \text{ if } n > 5.$$

Sol. $(Ax^2 + 2Bx + C)u = x^5$

Differentiating n times by Leibnitz's Theorem, we have

$$u_n\cdot(Ax^2 + 2Bx + C) + nu_{n-1}\cdot(2Ax + 2B) + \frac{n(n-1)}{2!}u_{n-2}\cdot 2A = 0$$

$[\because$ if $n > m$, $y = x^m \Rightarrow y_n = 0]$

$$\Rightarrow (Ax^2 + 2Bx + C)u_n + 2n(Ax + B)u_{n-1} + n(n-1)Au_{n-2} = 0.$$

Example 3. *If $f(x) = \tan x$, show that*

$$f^n(0) - {}^nC_2 f^{n-2}(0) + {}^nC_4 f^{n-4}(0) - \cdots = \sin \frac{n\pi}{2}.$$

Sol. $f(x) = \tan x = \dfrac{\sin x}{\cos x}$

$\Rightarrow \quad f(x) \cdot \cos x = \sin x$

Differentiating n times by Leibnitz's Theorem,

$$f^n(x) \cdot \cos x + {}^nC_1 f^{n-1}(x) \cdot (-\sin x) + {}^nC_2 f^{n-2}(x) \cdot (-\cos x) + {}^nC_3 f^{n-3}(x) \cdot (\sin x)$$
$$+ {}^nC_4 f^{n-4}(x) \cdot \cos x + \cdots = \sin\left(x + \frac{n\pi}{2}\right)$$

Putting $x = 0$, we get

$$f^n(0) - {}^nC_2 f^{n-2}(0) + {}^nC_4 f^{n-4}(0) - \cdots = \sin \frac{n\pi}{2}$$

since $\sin 0 = 0$ and $\cos 0 = 1$.

Example 4. *Find y_n if $y = \dfrac{x^n}{1+x}$.*

Sol. Given $y = \dfrac{x^n}{1+x} = x^n \cdot \dfrac{1}{1+x} = uv$, where

$$u = x^n \quad \text{and} \quad v = \frac{1}{1+x}$$

$$\therefore \quad u_r = \frac{n!}{(n-r)!} x^{n-r} \text{ and } v_r = \frac{(-1)^r \cdot r!}{(1+x)^{r+1}}$$

$\Rightarrow$

$u_n = n!$	$v_1 = \dfrac{-1}{(1+x)^2}$
$u_{n-1} = n! \cdot x$	$v_2 = \dfrac{2!}{(1+x)^3}$
$u_{n-2} = \dfrac{n!}{2!} x^2$	....
....	....
....	
$u_1 = nx^{n-1}$	$v_n = \dfrac{(-1)^n \cdot n!}{(1+x)^{n+1}}$

By Leibnitz's Theorem, we have

$$y_n = \frac{d^n}{dx^n}(uv) = {}^nC_0 u_n v + {}^nC_1 u_{n-1} v_1 + {}^nC_2 u_{n-2} v_2 + \cdots + {}^nC_n u v_n$$

$$= {}^nC_0 \cdot n! \cdot \frac{1}{1+x} + {}^nC_1 \cdot n! \, x \cdot \frac{-1}{(1+x)^2} + {}^nC_2 \cdot \frac{n!}{2!} x^2 \cdot \frac{2!}{(1+x)^3} + \cdots + {}^nC_n \cdot x^n \cdot \frac{(-1)^n \, n!}{(1+x)^{n+1}}$$

$$= \frac{n!}{1+x}\left[{}^nC_0 - {}^nC_1\left(\frac{x}{1+x}\right) + {}^nC_2\left(\frac{x}{1+x}\right)^2 + \cdots + {}^nC_n\left(\frac{x}{1+x}\right)^n\right]$$

$$= \frac{n!}{1+x}\left[1 - \frac{x}{1+x}\right]^n = \frac{n!}{1+x}\left(\frac{1}{1+x}\right)^n = \frac{n!}{(1+x)^{n+1}}.$$

EXERCISE 4

1. Find the nth derivative of
 (i) $x^3 \cos x$ (ii) $x^2 e^x \cos x$ (iii) $x^3 \log x$

2. If $y = x^2 e^x$, show by Leibnitz's theorem that $\dfrac{d^8y}{dx^8} = 28\dfrac{d^2y}{dx^2} - 48\dfrac{dy}{dx} + 21y$.

3. If $y = \sin^{-1} x$, prove that $(1 - x^2)y_{n+2} - (2n + 1)xy_{n+1} - n^2y_n = 0$. *(Kerala 2009)*

4. If $y = e^{m \cos^{-1} x}$, prove that $(1 - x^2)y_{n+2} - (2n + 1)xy_{n+1} - (n^2 + m^2)y_n = 0$. *(M.D.U. Dec. 2012)*

5. If $y = [\log (x + \sqrt{1+x^2})]^2$, prove that $(1 + x^2)y_{n+2} + (2n + 1)xy_{n+1} + n^2y_n = 0$.

6. If $y = \dfrac{\sin^{-1} x}{\sqrt{1-x^2}}$, show that $(1 - x^2)y_{n+2} - (2n + 3)xy_{n+1} - (n + 1)^2 y_n = 0$.

7. If $y = e^{\tan^{-1} x}$, prove that $(1 + x^2)y_{n+2} + [2(n + 1)x - 1]y_{n+1} + n(n + 1)y_n = 0$.

8. If $y = \cos (m \log x)$, prove that $x^2y_{n+2} + (2n + 1)xy_{n+1} + (m^2 + n^2)y_n = 0$. *(M.D.U. Dec. 2011)*

9. If $y = e^{m \cos^{-1} x}$, find $y_n (0)$.

10. If $y = \sin^{-1} x$, prove that $(1 - x^2)y_{n+2} - (2n + 1)xy_{n+1} - n^2y_n = 0$.
 Hence find the value of y_n when $x = 0$.

Answers

1. (i) $x^3 \cos\left(x + \dfrac{n\pi}{2}\right) + 3nx^2 \cos\left(x + (n-1)\dfrac{\pi}{2}\right) + 3n(n-1)x \cos\left(x + (n-2)\dfrac{\pi}{2}\right)$
 $+ n(n-1)(n-2) \cos\left(x + (n-3)\dfrac{\pi}{2}\right)$

 (ii) $2^{n/2} e^x \left[x^2 \cos\left(x + n\dfrac{\pi}{4}\right) + \sqrt{2}\, nx \cos\left(x + (n-1)\dfrac{\pi}{4}\right) + \dfrac{n(n-1)}{2} \cos\left(x + (n-2)\dfrac{\pi}{4}\right)\right]$

 (iii) $\dfrac{(-1)^{n-1}\, n!}{x^{n-3}}\left[\dfrac{1}{n} - \dfrac{3}{n-1} + \dfrac{3}{n-2} - \dfrac{1}{n-3}\right]$

9. When n is odd,
 $y_n(0) = -m\,(m^2 + 1^2)(m^2 + 3^2)(m^2 + 5^2) \ldots [m^2 + (n-2)^2]\, e^{\frac{m\pi}{2}}$
 and when n is even,
 $y_n(0) = m^2\,(m^2 + 2^2)\,(m^2 + 4^2)\,(m^2 + 6^2) \ldots [m^2 + (n-2)^2]\, e^{\frac{m\pi}{2}}$

10. When n is odd,
 $y_n(0) = (n-2)^2\,(n-4)^2 \ldots 3^2 \cdot 1^2 \cdot 1$ and when n is even, $y_n(0) = 0$.

5

Taylor's and Maclaurin's Series

5.1. TAYLOR'S INFINITE SERIES

If a function $f(x)$ possesses **derivatives of all orders** in the interval $(a, a + h)$, then for every integer n, however large, there corresponds a Taylor's development with Lagrange's form of remainder.

We know that $$f(a + h) = f(a) + hf'(a) + \frac{h^2}{2!} f''(a) + \cdots + \frac{h^{n-1}}{(n-1)!} f^{n-1}(a) + \frac{h^n}{n!} f^n(a + \theta h) \quad \text{where } 0 < \theta < 1$$

or $$f(a + h) = S_n + R_n$$

where $$S_n = f(a) + hf(a) + \frac{h^2}{2!} f''(a) + \cdots + \frac{h^{n-1}}{(n-1)!} f^{n-1}(a)$$

$$R_n = \frac{h^n}{n!} f^n(a + \theta h)$$

Now, let $R_n \to 0$ as $n \to \infty$, then $\lim\limits_{n \to \infty} S_n = f(a + h)$

$$= f(a) + hf'(a) + \frac{h^2}{2!} f''(a) + \cdots + \frac{h^{n-1}}{(n-1)!} f^{n-1}(a) + \cdots \text{ ad. inf.}$$

converges to $f(a + h)$.

Thus, if (*i*) $f(x)$ possesses derivatives of all orders in $(a, a + h)$

(*ii*) $R_n = \dfrac{h^n}{n!} f^n(a + \theta h)$, where $0 < \theta < 1$ tends to zero as $n \to \infty$

then $$f(a + h) = f(a) + hf'(a) + \frac{h^2}{2!} f''(a) + \cdots + \frac{h^n}{n!} f^n(a) + \cdots \text{ ad. inf.}$$

Note. To express $f(x)$ in ascending integral powers of $(x - a)$

$$f(x) = f[a + (x - a)] \qquad | \text{ here } h = x - a$$

$$= f(a) + (x - a) f'(a) + \frac{(x-a)^2}{2!} f''(a) + \cdots + \frac{(x-a)^n}{n!} f^n(a) + \cdots \text{ ad. inf.}$$

Application of Taylor's Infinite Series to expand f(x + h)

Method 1. Let $f(x + h)$ = given function

2. Put $h = 0$ and find $f(x)$

3. Differentiate $f(x)$ a number of times and obtain $f'(x), f''(x), f'''(x), \ldots$

4. Substitute the values of $f(x), f'(x), f''(x), \ldots$ in

$$f(x+h) = f(x) + hf'(x) + \frac{h^2}{2!} f''(x) + \cdots$$

ILLUSTRATIVE EXAMPLES

Example 1. *Use Taylor's series to prove that:*

$$\log \sin (x+h) = \log \sin x + h \cot x - \frac{h^2}{2} \operatorname{cosec}^2 x + \frac{h^3}{3} \cot x \operatorname{cosec}^2 x + \cdots$$

Sol. Let $f(x+h) = \log \sin (x+h)$

Putting $h = 0$, $f(x) = \log \sin x$

$$\therefore \quad f'(x) = \frac{\cos x}{\sin x} = \cot x, \; f''(x) = -\operatorname{cosec}^2 x$$

$$f'''(x) = -2 \operatorname{cosec} x (-\operatorname{cosec} x \cot x) = 2 \operatorname{cosec}^2 x \cot x$$

....

$$\therefore \quad \log \sin (x+h) = f(x+h) = f(x) + hf'(x) + \frac{h^2}{2!} f''(x) + \frac{h^3}{3!} f'''(x) + \cdots$$

$$= \log \sin x + h \cot x - \frac{h^2}{2} \operatorname{cosec}^2 x + \frac{h^3}{3} \operatorname{cosec}^2 x \cot x + \cdots$$

Example 2. *Use Taylor's series to prove that:*

$$\sec^{-1}(x+h) = \sec^{-1} x + \frac{h}{x\sqrt{x^2-1}} - \frac{h^2}{2!} \cdot \frac{2x^2-1}{x^2(x^2-1)^{3/2}} + \cdots$$

Sol. Let $f(x+h) = \sec^{-1}(x+h)$

Putting $h = 0$, $f(x) = \sec^{-1} x$

$$f'(x) = \frac{1}{x\sqrt{x^2-1}},$$

$$f''(x) = \frac{d}{dx} [x^{-1}(x^2-1)^{-1/2}] = -x^{-2}(x^2-1)^{-1/2} + x^{-1}. \left(-\tfrac{1}{2}\right)(x^2-1)^{-3/2}(2x)$$

$$= -\frac{1}{x^2(x^2-1)^{1/2}} - \frac{1}{(x^2-1)^{3/2}} = -\frac{x^2-1+x^2}{x^2(x^2-1)^{3/2}} = -\frac{2x^2-1}{x^2(x^2-1)^{3/2}}$$

....

$$\therefore \quad \sec^{-1}(x+h) = f(x+h) = f(x) + hf'(x) + \frac{h^2}{2!} f''(x) + \cdots$$

$$= \sec^{-1} x + \frac{h}{x\sqrt{x^2-1}} - \frac{h^2}{2!} \cdot \frac{2x^2-1}{x^2(x^2-1)^{3/2}} + \cdots$$

Example 3. *Apply Taylor's series to prove that:*

$$\cos\left(\frac{\pi}{4}+h\right) = \frac{1}{\sqrt{2}}\left(1 - h - \frac{h^2}{2!} + \frac{h^3}{3!} \cdots\right).$$

Sol. Let $f(x + h) = \cos (x + h)$

Putting $h = 0$, $f(x) = \cos x$

$$f'(x) = -\sin x, \qquad f''(x) = -\cos x$$
$$f'''(x) = -\sin x, \qquad \ldots\ldots\ldots\ldots$$

....

$$\therefore \quad \cos (x + h) = f(x + h)$$
$$= f(x) + hf'(x) + \frac{h^2}{2!} f''(x) + \frac{h^3}{3!} f'''(x) + \cdots$$
$$= \cos x - h \sin x - \frac{h^2}{2!} \cos x + \frac{h^3}{3!} \sin x \cdots$$

Putting $x = \frac{\pi}{4}$, we have $\cos\left(\frac{\pi}{4} + h\right) = \frac{1}{\sqrt{2}}\left(1 - h - \frac{h^2}{2!} + \frac{h^3}{3!} \cdots\right)$ $\left[\because \cos\frac{\pi}{4} = \sin\frac{\pi}{4} = \frac{1}{\sqrt{2}}\right]$

Example 4. *Expand log sin x in powers of (x – 3).*

Sol. Let $f(x) = \log \sin x$

$$\log \sin x = f(x) = f[3 + (x - 3)] = f(3 + h) \qquad \text{where } h = x - 3$$
$$= f(3) + hf'(3) + \frac{h^2}{2!} f''(3) + \frac{h^3}{3!} f'''(3) + \cdots \qquad \ldots(1)$$

Now $f(x) = \log \sin x, \qquad f(3) = \log \sin 3$

$$f'(x) = \frac{\cos x}{\sin x} = \cot x, \qquad f'(3) = \cot 3$$
$$f''(x) = -\text{cosec}^2 x, \qquad f''(3) = -\text{cosec}^2 3$$
$$f'''(x) = -2 \,\text{cosec}\, x\, (-\text{cosec}\, x \cot x) = 2\, \text{cosec}^2 x \cot x$$
$$f'''(3) = 2\, \text{cosec}^2 3 \cot 3$$

....

$$\therefore \text{ From (1), } \log \sin x = \log \sin 3 + (x - 3) \cot 3 - \frac{(x-3)^2}{2} \text{cosec}^2 3 + \frac{(x-3)^3}{3} \text{cosec}^2 3 \cot 3 + \cdots$$

Example 5. *Use Taylor's Theorem to express the polynomial $2x^3 + 7x^2 + x - 6$ in powers of (x – 2).*

Sol. Let $f(x) = 2x^3 + 7x^2 + x - 6$

$$\therefore \quad 2x^3 + 7x^2 + x - 6 = f(x) = f[2 + (x - 2)] = f(2 + h) \qquad \text{where } h = x - 2$$
$$= f(2) + hf'(2) + \frac{h^2}{2!} f''(2) + \frac{h^3}{3!} f'''(2) \qquad \ldots(1)$$

[as higher derivatives of $f(x)$ vanish]

Now
$$f(x) = 2x^3 + 7x^2 + x - 6 \qquad f(2) = 16 + 28 + 2 - 6 = 40$$
$$f'(x) = 6x^2 + 14x + 1 \qquad f'(2) = 24 + 28 + 1 = 53$$
$$f''(x) = 12x + 14 \qquad f''(2) = 24 + 14 = 38$$
$$f'''(x) = 12 \qquad f'''(2) = 12$$

$\therefore$ From (1), $2x^3 + 7x^2 + x - 6 = 40 + 53h + \dfrac{38h^2}{2!} + \dfrac{12h^3}{3!}$

$= 40 + 53(x-2) + 19(x-2)^2 + 2(x-2)^3.$

Example 6. *If $f(x) = x^3 + 2x^2 - 5x + 11$, calculate the value of $f\left(\dfrac{9}{10}\right)$ by the application of Taylor's series for $f(x + h)$.*

Sol. By Taylor's series $f(x+h) = f(x) + hf'(x) + \dfrac{h^2}{2!}f''(x) + \cdots$

Put $x = 1$ and $h = -\dfrac{1}{10}$

$$f\left(\frac{9}{10}\right) = f(1) - \frac{1}{10}f'(1) + \frac{1}{2!}\cdot\left(\frac{1}{10}\right)^2 f''(1) - \frac{1}{3!}\cdot\left(\frac{1}{10}\right)^3 . f'''(1) + \cdots \text{(A)}$$

Here

$f(x) = x^3 + 2x^2 - 5x + 11 \quad \therefore \quad f(1) = 9$

$f'(x) = 3x^2 + 4x - 5 \quad \therefore \quad f'(1) = 2$

$f''(x) = 6x + 4 \quad \therefore \quad f''(1) = 10$

$f'''(x) = 6 \quad \therefore \quad f'''(1) = 6$

All higher order differential co-efficients vanish.

$\therefore$ From (A), we get $f\left(\dfrac{9}{10}\right) = 9 - \dfrac{1}{10}\cdot 2 + \dfrac{1}{2}\cdot\dfrac{1}{100}\cdot 10 - \dfrac{1}{6}\cdot\dfrac{1}{1000}\cdot 6$

$= 9 - 0.2 + 0.05 - 0.001 = 8.849.$

Example 7. *Calculate the approximate value of $\sqrt{10}$ to four decimal places by taking the first four terms of an appropriate Taylor's series.* (M.D.U. Dec. 2012)

Sol. Let $f(x+h) = \sqrt{x+h}$

Put $h = 0$ then $f(x) = \sqrt{x}$

$$f'(x) = \frac{1}{2}x^{-\frac{1}{2}} = \frac{1}{2\sqrt{x}}$$

$$f''(x) = -\frac{1}{4}x^{-\frac{3}{2}} = -\frac{1}{4x\sqrt{x}}$$

$$f'''(x) = \frac{3}{8}x^{-\frac{5}{2}} = \frac{3}{8x^2\sqrt{x}}$$

....

$\therefore \quad \sqrt{x+h} = f(x+h)$

$$= f(x) + hf'(x) + \frac{h^2}{2!}f''(x) + \frac{h^3}{3!}f'''(x) + \cdots$$

$$= \sqrt{x} + \frac{h}{2\sqrt{x}} - \frac{h^2}{8x\sqrt{x}} + \frac{h^3}{16x^2\sqrt{x}} + \cdots$$

Put $x = 9$ and $h = 1$

$$\sqrt{10} = 3 + \frac{1}{2(3)} - \frac{1}{8(9)(3)} + \frac{1}{16(81)(3)} + \cdots$$

$= 3 + 0.16666 - 0.00463 + 0.00025 + \cdots = 3.16228$ nearly.

Example 8. *Compute to four decimal places, the value of cos 32° by use of Taylor's series.* (K.U.K. Dec. 2010, Dec. 2013; M.D.U. May 2009)

Sol. Let $f(x + h) = \cos (x + h)$

Put $h = 0$, then $f(x) = \cos x$

$f'(x) = -\sin x, \quad f''(x) = -\cos x, \quad f'''(x) = \sin x$ and so on.

$$\therefore \quad \cos (x + h) = f(x + h)$$

$$= f(x) + hf'(x) + \frac{h^2}{2!} f''(x) + \frac{h^3}{3!} f'''(x) + \cdots$$

$$= \cos x - h \sin x - \frac{h^2}{2} \cos x + \frac{h^3}{6} \sin x + \cdots$$

Putting $x = 30°$ and $h = 2° = \frac{\pi}{90} = 0.03490$, we have

$$\cos 32° = \cos 30° - .03490 \sin 30° - \frac{(0.03490)^2}{2} \cos 30° + \frac{(0.03490)^3}{6} \sin 30° + \cdots$$

$$= \frac{\sqrt{3}}{2} - 0.03490 \times \frac{1}{2} - 0.00060 \times \frac{\sqrt{3}}{2} + 0.00000 + \cdots$$

$$= 0.86602 - 0.01745 - 0.00060 \times 0.86602$$

$$= 0.84857 - 0.00051 = 0.84806$$

$= 0.8481$ rounded off to four decimal places.

Example 9. *Given $\log_{10} 4 = 0.6021$, calculate approximately $\log_{10} 404$.* (K.U. 2013)

Sol. Let $f(x + h) = \log_e(x + h)$

Put $h = 0$, then $f(x) = \log_e x$

$$f'(x) = \frac{1}{x}, \qquad f''(x) = -\frac{1}{x^2}$$

$$f'''(x) = \frac{2}{x^3} \quad \text{and so on.}$$

$$\therefore \quad \log_e(x + h) = f(x + h)$$

$$= f(x) + hf'(x) + \frac{h^2}{2!} f''(x) + \frac{h^3}{3!} f'''(x) + \cdots$$

$$= \log_e x + \frac{h}{x} - \frac{h^2}{2x^2} + \frac{2h^3}{6x^3} + \cdots$$

Multiplying by $\log_{10} e$, we have

$$\log_{10}(x + h) = \log_{10} x + \log_{10} e \left(\frac{h}{x} - \frac{h^2}{2x^2} + \frac{h^3}{3x^3} \cdots \right)$$

Putting $x = 400$ and $h = 4$, we have

$$\log_{10} 404 = \log_{10} 400 + 0.4343 \left[\frac{4}{400} - \frac{1}{2} \left(\frac{4}{400} \right)^2 + \frac{1}{3} \left(\frac{4}{400} \right)^3 \cdots \right]$$

$$= \log_{10} (4 \times 10^2) + 0.4343 \left[0.01 - \frac{1}{2} (0.0001) + \frac{1}{3} (0.000001) \ldots \right]$$

$$= \log_{10} 4 + 2 \log_{10} 10 + 0.4343 \, [0.01 - 0.00005 + 0.00000 \ldots]$$

$$= 0.6021 + 2 + 0.4343 \times 0.00995$$
$$= 2.6021 + 0.00432 = 2.6064$$

Example 10. *Use Taylor's series to prove that*

$$\tan^{-1}(x+h) = \tan^{-1} x + (h \sin z) \cdot \frac{\sin z}{1} - (h \sin z)^2 \cdot \frac{\sin 2z}{2} + (h \sin z)^3 \cdot \frac{\sin 3z}{3} - \cdots$$

where $z = \cot^{-1} x$.

Sol. Given $z = \cot^{-1} x \Rightarrow \cot z = x$

$$\therefore \quad \frac{dz}{dx} = \frac{-1}{1+x^2} = \frac{-1}{1+\cot^2 z} = \frac{-1}{\operatorname{cosec}^2 z} = -\sin^2 z$$

Now let $f(x+h) = \tan^{-1}(x+h)$

Putting $h = 0$, $f(x) = \tan^{-1} x$

$$f'(x) = \frac{1}{1+x^2} = \frac{1}{1+\cot^2 z} = \frac{1}{\operatorname{cosec}^2 z} = \sin^2 z$$

$$f''(x) = 2 \sin z \cos z \,.\, \frac{dz}{dx} = \sin 2z\,(-\sin^2 z) = -\sin^2 z \sin 2z$$

$$f'''(x) = -\,[2 \cos 2z \sin^2 z + \sin 2z \cdot 2 \sin z \cos z] \cdot \frac{dz}{dx}$$
$$= -\,2 \sin z\,(\sin z \cos 2z + \cos z \sin 2z) \cdot (-\sin^2 z)$$
$$= 2 \sin^3 z \sin (z + 2z) = 2 \sin^3 z \sin 3z \quad \text{and so on.}$$

$$\therefore \quad \tan^{-1}(x+h) = f(x+h)$$
$$= f(x) + hf'(x) + \frac{h^2}{2!} f''(x) + \frac{h^3}{3!} f'''(x) + \cdots$$
$$= \tan^{-1} x + h \sin^2 z + \frac{h^2}{2}(-\sin^2 z \sin 2z) + \frac{h^3}{6} \cdot 2 \sin^3 z \sin 3z + \cdots$$
$$= \tan^{-1} x + (h \sin z) \cdot \frac{\sin z}{1} - (h \sin z)^2 \cdot \frac{\sin 2z}{2} + (h \sin z)^3 \cdot \frac{\sin 3z}{3} + \cdots$$

EXERCISE 5.1

1. Using Taylor's series, prove that

(*i*) $e^{x+h} = e^x \left(1 + h + \frac{h^2}{2!} + \frac{h^3}{3!} + \cdots\right)$

(*ii*) $a^{x+h} = a^x \left[1 + h \log a + \frac{h^2}{2!}(\log a)^2 + \frac{h^3}{3!}(\log a)^3 + \cdots\right]$

(*iii*) $\frac{1}{x+h} = \frac{1}{x}\left(1 - \frac{h}{x} + \frac{h^2}{x^2} - \frac{h^3}{x^3} + \cdots\right)$

(*iv*) $\log(x+h) = \log x + \frac{h}{x} - \frac{h^2}{2x^2} + \frac{h^3}{3x^3} - \cdots$

(*v*) $\log \cos(x+h) = \log \cos x - h \tan x - \frac{h^2}{2} \sec^2 x - \frac{h^3}{3} \sec^2 x \tan x + \cdots$

(*vi*) $\sin^{-1}(x+h) = \sin^{-1} x + \frac{h}{\sqrt{1-x^2}} + \frac{x}{(1-x^2)^{3/2}} \cdot \frac{h^2}{2!} + \cdots$

(*vii*) $\tan^{-1}(x+h) = \tan^{-1}x + \dfrac{h}{1+x^2} - \dfrac{xh^2}{(1+x^2)^2} + \cdots$

(*viii*) $\tan(x+h) = \tan x + h\sec^2 x + h^2 \sec^2 x \tan x + \dfrac{h^3}{3}\sec^2 x\,(1+3\tan^2 x) + \cdots$

2. Prove that $\dfrac{f(x+h)+f(x-h)}{2} = f(x) + \dfrac{h^2}{2!}f''(x) + \dfrac{h^4}{4!}f^{iv}(x) + \cdots$

3. Expand $\sin(x+h)$ in powers of h and deduce that $\sin(x+h) = \sin x\cos h + \cos x \sin h$.

4. Using Taylor's series, expand

(*i*) e^x in powers of $(x-2)$.

(*ii*) $\sin x$ in powers of $\left(x-\dfrac{\pi}{2}\right)$. Hence find the value of sin 91° correct to 4 decimal places. (*M.D.U. Dec. 2008*)

(*iii*) $\log_e x$ in powers of $(x-1)$. Hence evaluate $\log_e 1.1$ correct to 4 decimal places. (*M.D.U. Dec. 2011*)

(*iv*) $\tan x$ in powers of $\left(x-\dfrac{\pi}{4}\right)$. (*M.D.U. Dec. 2012*)

(*v*) $\tan^{-1}x$ in powers of $(x-1)$.

(*vi*) $2 + x^2 - 3x^5 + 7x^6$ in powers of $(x-1)$.

(*vii*) $\log\cos x$ about the point $\dfrac{\pi}{3}$.

5. Using Taylor's series, compute the value of sin 31° to four decimal places. (*M.D.U. Dec. 2009*)

6. By using Taylor's series prove that

$$\sin(x+y) = \sin x\cos y + \cos x \sin y$$

Hence find the value of sin 46° correct to four places of decimals. (*M.D.U. Dec. 2013*)

7. If $f(x) = x^3 + 8x^2 + 15x - 24$, calculate the value of $f\left(\dfrac{11}{10}\right)$ by using Taylor's series.

8. Calculate the approximate value of $\sqrt{17}$ to four decimal places by taking the first four terms of an appropriate Taylor's series.

9. Prove that:

(*i*) $f(ax) = f(x) + (a-1)x f'(x) + \dfrac{(a-1)^2x^2}{2!}f''(x) + \cdots$

[**Hint.** $f(ax) = f(x + ax - x) = f[x + (a-1)x] = f(x+h)$]

(*ii*) $f\left(\dfrac{x^2}{1+x}\right) = f(x) - \dfrac{x}{1+x}f'(x) + \left(\dfrac{x}{1+x}\right)^2 \cdot \dfrac{f''(x)}{2!} - \left(\dfrac{x}{1+x}\right)^3 \cdot \dfrac{f'''(x)}{3!} + \cdots$ (*M.T.U. 2013*)

$\left[\textbf{Hint. } \dfrac{x^2}{1+x} = x - \dfrac{x}{1+x} = x + h, \text{ where } h = -\dfrac{x}{1+x}\right]$

Answers

4. (*i*) $e^2\left[1 + (x-2) + \dfrac{(x-2)^2}{2!} + \dfrac{(x-2)^3}{3!} + \cdots\right]$

(*ii*) $1 - \dfrac{\left(x-\dfrac{\pi}{2}\right)^2}{2!} + \dfrac{\left(x-\dfrac{\pi}{2}\right)^4}{4!} - \cdots$; 0.9998 (*iii*) $(x-1) - \frac{(x-1)^2}{2} + \frac{(x-1)^3}{3} - \frac{(x-1)^4}{4} + \cdots$; 0.0953

(*iv*) $1 + 2\left(x - \frac{\pi}{4}\right) + 2\left(x - \frac{\pi}{4}\right)^2 + \frac{8}{3}\left(x - \frac{\pi}{4}\right)^3 + \cdots$

(*v*) $\frac{\pi}{4} + \frac{1}{2}(x-1) - \frac{1}{4}(x-1)^2 + \frac{1}{12}(x-1)^3 + \cdots$

(*vi*) $7 + 29(x-1) + 76(x-1)^2 + 110(x-1)^3 + 90(x-1)^4 + 39(x-1)^5 + 7(x-1)^6$.

(*vii*) $\log(0.5) - \sqrt{3}\left(x - \frac{\pi}{3}\right) - 2\left(x - \frac{\pi}{3}\right)^2 - \frac{4\sqrt{3}}{3}\left(x - \frac{\pi}{3}\right)^3 + \cdots$

5. 0.5151 **6.** 0.7193 **7.** 3.511 **8.** 4.123.

5.2. MACLAURIN'S INFINITE SERIES

Putting $a = 0$ and $h = x$ in the Taylor's infinite series, we get Maclaurin's infinite series.

Thus if (*i*) $f(x)$ possesses derivatives of all orders in the interval $(0, x)$ and

(*ii*) Maclaurin's remainder $R_n = \frac{x^n}{n!} f^n(\theta x)$ tends to zero as $n \to \infty$, then

$$\mathbf{f(x) = f(0) + xf'(0) + \frac{x^2}{2!} f''(0) + \frac{x^3}{3!} f'''(0) + \cdots \text{ ad. inf.}}$$

Note. If the function $f(x)$ is denoted by y, then the expansion may be written in the form

$$y = y(0) + xy_1(0) + \frac{x^2}{2!} y_2(0) + \frac{x^3}{3!} y_3(0) + \cdots \text{ ad. inf.}$$

where $y(0), y_1(0), y_2(0), y_3(0), \ldots$ denote the values of $y, y_1, y_2, y_3, \ldots$ respectively for $x = 0$.

ILLUSTRATIVE EXAMPLES

Example 1. *Using Maclaurin's series, expand tan x upto the term containing x^5.*

Sol. Let $f(x) = \tan x \qquad \therefore \quad f(0) = 0$

$$f'(x) = \sec^2 x = 1 + \tan^2 x \qquad \therefore \quad f'(0) = 1$$

$$f''(x) = 2\tan x \sec^2 x = 2\tan x(1 + \tan^2 x)$$

$$= 2\tan x + 2\tan^3 x \qquad \therefore \quad f''(0) = 0$$

$$f'''(x) = 2\sec^2 x + 6\tan^2 x \sec^2 x$$

$$= 2(1 + \tan^2 x) + 6\tan^2 x(1 + \tan^2 x)$$

$$= 2 + 8\tan^2 x + 6\tan^4 x \qquad \therefore \quad f'''(0) = 2$$

$$f^{iv}(x) = 16\tan x \sec^2 x + 24\tan^3 x \sec^2 x$$

$$= 16\tan x(1 + \tan^2 x) + 24\tan^3 x(1 + \tan^2 x)$$

$$= 16\tan x + 40\tan^3 x + 24\tan^5 x \qquad \therefore \quad f^{iv}(0) = 0$$

$$f^{v}(x) = 16\sec^2 x + 120\tan^2 x \sec^2 x + 120\tan^4 x \sec^2 x \quad \therefore \quad f^{v}(0) = 16$$

and so on.

Putting these values in the Maclaurin's series

$$f(x) = f(0) + xf'(0) + \frac{x^2}{2!} f''(0) + \frac{x^3}{3!} f'''(0) + \frac{x^4}{4!} f^{iv}(0) + \cdots, \text{ we get}$$

$$\tan x = 0 + x \cdot 1 + \frac{x^2}{2!} \cdot 0 + \frac{x^3}{3!} \cdot 2 + \frac{x^4}{4!} \cdot 0 + \frac{x^5}{5!} \cdot 16 + \cdots$$

$$= x + \frac{x^3}{3} + \frac{2x^5}{15} + \cdots$$

Second Method

Let $y = \tan x \quad \therefore \quad \mathbf{y(0) = 0}$

$y_1 = \sec^2 x = 1 + \tan^2 x = 1 + y^2 \quad \therefore \quad \mathbf{y_1(0) = 1 + 0 = 1}$

$y_2 = 2yy_1 \quad \therefore \quad \mathbf{y_2(0) = 0}$

$y_3 = 2y_1y_1 + 2yy_2 = 2y_1^2 + 2yy_2 \quad \therefore \quad \mathbf{y_3(0) = 2 \times 1 + 0 = 2}$

$y_4 = 4y_1y_2 + 2y_1y_2 + 2yy_2$

$= 6y_1y_2 + 2yy_3 \quad \therefore \quad \mathbf{y_4(0) = 0 + 0 = 0}$

$y_5 = 6y_2{}^5 + 6y_1y_3 + 2y_1y_3 + 2yy_4$

$= 8y_1y_3 + 6y_2{}^2 + 2yy_4 \quad \therefore \quad \mathbf{y_5(0) = 8 \times 1 \times 2 + 0 + 0 = 16}$

....

$$\therefore \quad \tan x = y(0) + xy_1(0) + \frac{x^2}{2!} y_2(0) + \frac{x^3}{3!} y_3(0) + \frac{x^4}{4!} y_4(0) + \frac{x^5}{5!} y_5(0) + \cdots$$

$$= x + \frac{x^3}{3!} \cdot 2 + \frac{x^5}{5!} \cdot 16 + \cdots = x + \frac{x^3}{3} + \frac{2x^5}{15} + \cdots$$

Example 2. *Using Maclaurin's series, expand log (1 + x). Hence deduce that*

$$\log \sqrt{\frac{1+x}{1-x}} = x + \frac{x^3}{3} + \frac{x^5}{5} + \cdots$$ (M.D.U. Dec. 2011)

Sol. Let $f(x) = \log (1 + x) \quad \therefore \quad f(0) = \log 1 = 0$

$f'(x) = \dfrac{1}{1+x} \quad \therefore \quad f'(0) = 1$

$f''(x) = -\dfrac{1}{(1+x)^2} \quad \therefore \quad f''(0) = -1$

$f'''(x) = \dfrac{2}{(1+x)^3} \quad \therefore \quad f'''(0) = 2$

$f^{iv}(x) = -\dfrac{6}{(1+x)^4} \quad \therefore \quad f^{iv}(0) = -6$ and so on.

$\therefore \quad \log (1 + x) = f(x)$

$$= f(0) + xf'(0) + \frac{x^2}{2!} f''(0) + \frac{x^3}{3!} f'''(0) + \frac{x^4}{4!} f^{iv}(0) + \cdots$$

$$= x - \frac{x^2}{2!} + \frac{2x^3}{3!} - \frac{6x^4}{4!} + \cdots = x - \frac{x^2}{2} + \frac{x^3}{3} - \frac{x^4}{4} + \cdots$$

Changing x to $-x$,

$$\log (1 - x) = -x - \frac{x^2}{2} - \frac{x^3}{3} - \frac{x^4}{4} - \cdots$$

$$\therefore \quad \log\sqrt{\frac{1+x}{1-x}} = \frac{1}{2}[\log(1+x) - \log(1-x)]$$

$$= \frac{1}{2}\left[\left(x - \frac{x^2}{2} + \frac{x^3}{3} - \frac{x^4}{4} + \cdots\right) - \left(-x - \frac{x^2}{2} - \frac{x^3}{3} - \frac{x^4}{4} - \cdots\right)\right]$$

$$= \frac{1}{2}\left[2\left(x + \frac{x^3}{3} + \frac{x^5}{5} + \cdots\right)\right] = x + \frac{x^3}{3} + \frac{x^4}{5} + \cdots$$

Example 3. *Prove that*

(*i*) $\log(\sec x + \tan x) = x + \dfrac{x^3}{6} + \dfrac{x^5}{24} + \cdots$

(*ii*) $\sin^{-1} x = x + \dfrac{1}{2}\cdot\dfrac{x^3}{3} + \dfrac{1\cdot 3}{2\cdot 4}\cdot\dfrac{x^5}{5} + \dfrac{1\cdot 3\cdot 5}{2\cdot 4\cdot 6}\cdot\dfrac{x^7}{7} + \cdots$ (M.D.U. Dec. 2010)

Sol. (*i*) Let $y = \log(\sec x + \tan x)$ $\quad \therefore\ y(0) = \log 1 = 0$

$$y_1 = \frac{1}{\sec x + \tan x}\cdot(\sec x \tan x + \sec^2 x) = \sec x \qquad \therefore\ y_1(0) = 1$$

$$y_2 = \sec x \tan x = y_1 \tan x \qquad \therefore\ y_2(0) = y_1(0) \times 0 = 0$$

$$y_3 = y_2 \tan x + y_1 \sec^2 x = y_2 \tan x + y_1 \cdot y_1^2$$
$$= y_2 \tan x + y_1^3 \qquad \therefore\ y_3(0) = 0 + 1 = 1$$

$$y_4 = y_3 \tan x + y_2 \sec^2 x + 3y_1^2 y_2$$
$$= y_3 \tan x + y_2 \cdot y_1^2 + 3y_1^2 y_2$$
$$= y_3 \tan x + 4y_1^2 y_2 \qquad \therefore\ y_4(0) = 0 + 4 \times 1 \times 0 = 0$$

$$y_5 = y_4 \tan x + y_3 \sec^2 x + 8y_1 y_2^2 + 4y_1^2 y_3$$
$$= y_4 \tan x + y_3 \cdot y_1^2 + 8y_1 y_2^2 + 4y_1^2 y_3$$
$$= y_4 \tan x + 8y_1 y_2^2 + 5y_1^2 y_3 \qquad \therefore\ y_5(0) = 0 + 0 + 5 \times 1^2 \times 1 = 5$$

and so on.

$\therefore\ \log(\sec x + \tan x) = y$

$$= y(0) + xy_1(0) + \frac{x^2}{2!}y_2(0) + \frac{x^3}{3!}y_3(0) + \cdots$$

$$= 0 + x\cdot 1 + \frac{x^2}{2!}\cdot 0 + \frac{x^3}{3!}\cdot 1 + \frac{x^4}{4!}\cdot 0 + \frac{x^5}{5!}.5 + \cdots$$

$$= x + \frac{x^3}{6} + \frac{x^5}{24} + \cdots$$

(*ii*) Let $\quad y = \sin^{-1} x \qquad \therefore\ y(0) = \sin^{-1} 0 = 0$

$$y_1 = \frac{1}{\sqrt{1-x^2}} \qquad \therefore\ y_1(0) = 1$$

$$= (1 - x^2)^{-1/2}$$

$$= 1 + \tfrac{1}{2}x^2 + \frac{\frac{1}{2}(\frac{1}{2}+1)}{2!}x^4 + \frac{\frac{1}{2}(\frac{1}{2}+1)(\frac{1}{2}+2)}{3!}x^6 + \cdots$$

| By Binomial Theorem

$$= 1 + \frac{1}{2}x^2 + \frac{3}{8}x^4 + \frac{5}{16}x^6 + \cdots$$

$$y_2 = x + \frac{3}{2}x^3 + \frac{15}{8}x^5 + \cdots \qquad \therefore \; y_2(0) = 0$$

$$y_3 = 1 + \frac{9}{2}x^2 + \frac{75}{8}x^4 + \cdots \qquad \therefore \; y_3(0) = 1$$

$$y_4 = 9x + \frac{75}{2}x^3 + \cdots \qquad \therefore \; y_4(0) = 0$$

$$y_5 = 9 + \frac{25}{2}x^2 + \cdots \qquad \therefore \; y_5(0) = 9$$

$$y_6 = 225x + \cdots \qquad \therefore \; y_6(0) = 0$$

$$y_7 = 225 + \text{terms containing powers of } x \qquad \therefore \; y_7(0) = 225$$

$$\therefore \qquad \sin^{-1} x = y(0) + xy_1(0) + \frac{x^2}{2!}y_2(0) + \cdots$$

$$= x + \frac{x^3}{3!} + \frac{9x^5}{5!} + \frac{225x^7}{7!} + \cdots$$

$$= x + \frac{1}{2}\cdot\frac{x^3}{3} + \frac{1\cdot 3}{2\cdot 4}\cdot\frac{x^5}{5} + \frac{1\cdot 3\cdot 5}{2\cdot 4\cdot 6}\cdot\frac{x^7}{7} + \cdots$$

Note. Putting $x = \frac{1}{2}$ on both sides, we get

$$\sin^{-1}\frac{1}{2} = \frac{1}{2} + \frac{1}{6}\left(\frac{1}{2}\right)^3 + \frac{3}{40}\left(\frac{1}{2}\right)^5 + \frac{5}{112}\left(\frac{1}{2}\right)^7 + \cdots$$

$$\Rightarrow \qquad \frac{\pi}{6} = 0.5 + 0.0208 + 0.0023 + 0.0003 + \cdots \approx 0.5234$$

$$\Rightarrow \qquad \pi \approx 3.1404 = 3.14 \quad \text{approximately.}$$

5.3. USE OF KNOWN SERIES

When it is very tedius to find the successive derivatives of a function, the use of the following well-known series is often convenient.

(*i*) $e^x = 1 + x + \frac{x^2}{2!} + \frac{x^3}{3!} + \frac{x^4}{4!} + \cdots$ (*ii*) $\log(1 + x) = x - \frac{x^2}{2} + \frac{x^3}{3} - \frac{x^4}{4} + \cdots$

(*iii*) $\sin x = x - \frac{x^3}{3!} + \frac{x^5}{5!} - \frac{x^7}{7!} + \cdots$ (*iv*) $\sinh x = x + \frac{x^3}{3!} + \frac{x^5}{5!} + \frac{x^7}{7!} + \cdots$

(*v*) $\cos x = 1 - \frac{x^2}{2!} + \frac{x^4}{4!} - \frac{x^6}{6!} + \cdots$ (*vi*) $\cosh x = 1 + \frac{x^2}{2!} + \frac{x^4}{4!} + \frac{x^6}{6!} + \cdots$

(*vii*) $\tan x = x + \frac{x^3}{3} + \frac{2x^5}{15} + \cdots$ (*viii*) $\tan^{-1} x = x - \frac{x^3}{3} + \frac{x^5}{5} - \frac{x^7}{7} + \cdots$

Example 4. *Prove that* $e^x \log(1 + x) = x + \frac{x^2}{2} + \frac{x^3}{3} + \frac{3x^5}{40} + \cdots$

Sol. We know that $e^x = 1 + x + \frac{x^2}{2!} + \frac{x^3}{3!} + \frac{x^4}{4!} + \frac{x^5}{5!} + \cdots$

$$= 1 + x + \frac{x^2}{2} + \frac{x^3}{6} + \frac{x^4}{24} + \frac{x^5}{120} + \cdots$$

and $$\log(1 + x) = x - \frac{x^2}{2} + \frac{x^3}{3} - \frac{x^4}{4} + \frac{x^5}{5} - \cdots$$

Multiplying $e^x \log(1 + x) = x + x^2 + \frac{x^3}{2} + \frac{x^4}{6} + \frac{x^5}{24} + \cdots$

$$- \frac{x^2}{2} - \frac{x^3}{2} - \frac{x^4}{4} - \frac{x^5}{12} - \cdots$$

$$+ \frac{x^3}{3} + \frac{x^4}{3} + \frac{x^5}{6} + \cdots$$

$$- \frac{x^4}{4} - \frac{x^5}{4} - \cdots$$

$$+ \frac{x^5}{5} + \cdots$$

....

$$= x + \frac{x^2}{2} + \frac{x^3}{3} + \frac{3x^5}{40} + \cdots$$

Example 5. *Find the first four terms in the expansion of log (1 + tan x).*

Sol. Since $\log(1 + z) = z - \frac{z^2}{2} + \frac{z^3}{3} - \frac{z^4}{4} + \cdots$

$$\therefore \quad \log(1 + \tan x) = \tan x - \frac{1}{2}\tan^2 x + \frac{1}{3}\tan^3 x - \frac{1}{4}\tan^4 x + \cdots$$

$$= \left(x + \frac{x^3}{3} + \frac{2x^5}{15} + \cdots\right) - \frac{1}{2}\left(x + \frac{x^3}{3} + \frac{2x^5}{15} + \cdots\right)^2$$

$$+ \frac{1}{3}\left(x + \frac{x^3}{3} + \frac{2x^5}{15} + \cdots\right)^3 - \frac{1}{4}\left(x + \frac{x^3}{3} + \frac{2x^5}{15} + \cdots\right)^4 + \cdots$$

$$\left[\because \tan x = x + \frac{x^3}{3} + \frac{2x^5}{15} + \cdots\right]$$

$$= \left(x + \frac{x^3}{3} + \frac{2x^5}{15} + \cdots\right) - \frac{1}{2}\left[x + \left(\frac{x^3}{3} + \frac{2x^5}{15} + \cdots\right)\right]^2$$

$$+ \frac{1}{3}\left[x + \left(\frac{x^3}{3} + \frac{2x^5}{15} + \cdots\right)\right]^3 - \frac{1}{4}\left[x + \left(\frac{x^3}{3} + \frac{2x^5}{15} + \cdots\right)\right]^4 + \cdots$$

$$= \left(x + \frac{x^3}{3} + \cdots\right) - \frac{1}{2}\left[x^2 + 2x\left(\frac{x^3}{3} + \cdots\right)\right] + \frac{1}{3}[x^3 + \cdots] - \frac{1}{4}[x^4 + \cdots] + \cdots$$

$$= \left(x + \frac{x^3}{3} + \cdots\right) - \frac{1}{2}\left(x^2 + \frac{2x^4}{3} + \cdots\right) + \frac{1}{3}(x^3 + \ldots) - \frac{1}{4}(x^4 + \ldots) + \cdots$$

$$= x - \frac{1}{2}x^2 + \frac{2}{3}x^3 - \frac{7}{12}x^4 + \cdots$$

Example 6. *Expand log (1 + sin² x) in powers of x as far as the term in x^6.*

Sol. Since $\sin^2 x = \left(x - \frac{x^3}{3!} + \frac{x^5}{5!} - \cdots\right)^2$

$$= \left[x - \left(\frac{x^3}{6} - \frac{x^5}{120} + \ldots\right)\right]^2$$

$$= x^2 - 2x\left(\frac{x^3}{6} - \frac{x^5}{120} + \cdots\right) + \left(\frac{x^3}{6} - \frac{x^5}{120} + \cdots\right)^2.$$

$$= x^2 - \frac{x^4}{3} + \frac{x^6}{60} + \frac{x^6}{36} + \cdots = x^2 - \frac{x^4}{3} + \frac{2x^6}{45} + \cdots = z \text{ (say)}$$

$$\therefore \quad \log(1 + \sin^2 x) = \log(1 + z) = z - \frac{z^2}{2} + \frac{z^3}{3} - \cdots$$

$$= \left(x^2 - \frac{x^4}{3} + \frac{2x^6}{45} - \cdots\right) - \frac{1}{2}\left(x^2 - \frac{x^4}{3} + \cdots\right)^2 + \frac{1}{3}(x^2 - \ldots)^3 - \cdots$$

$$= \left(x^2 - \frac{x^4}{3} + \frac{2x^6}{45} - \cdots\right) - \frac{1}{2}\left(x^4 - \frac{2x^6}{3} - \cdots\right) + \frac{1}{3}(x^2 - \ldots)^3$$

$$= x^2 - \frac{x^4}{3} - \frac{x^4}{2} + \frac{2x^6}{45} + \frac{x^6}{3} + \frac{x^6}{3} \cdots$$

$$= x^2 - \frac{5}{6}x^4 + \frac{32}{45}x^6 - \cdots$$

Example 7. *Show that*

(*i*) $\tan^{-1} \frac{2x}{1 - x^2} = 2\left(x - \frac{x^3}{3} + \frac{x^5}{5} \cdots\right)$ (*ii*) $\tan^{-1} \frac{\sqrt{1 + x^2} - \sqrt{1 - x^2}}{\sqrt{1 + x^2} + \sqrt{1 - x^2}} = \frac{x^2}{2} + \frac{x^6}{12} + \cdots$

Sol. (*i*) Let $y = \tan^{-1} \frac{2x}{1 - x^2}$ Put $x = \tan\theta$ so that $\theta = \tan^{-1} x$

Then $y = \tan^{-1}\left(\frac{2\tan\theta}{1 - \tan^2\theta}\right) = \tan^{-1}(\tan 2\theta) = 2\theta = 2\tan^{-1} x$ $\qquad y(0) = 0$

$$y_1 = \frac{2}{1 + x^2} = 2(1 + x^2)^{-1} = 2(1 - x^2 + x^4 - x^6 + \cdots) \qquad y_1(0) = 2$$

$$y_2 = 2(-2x + 4x^3 - 6x^5 + \cdots) \qquad y_2(0) = 0$$

$$y_3 = 2(-2 + 12x^2 - 30x^4 + \cdots) \qquad y_3(0) = -4$$

$$y_4 = 2(24x - 120x^3 + \cdots) \qquad y_4(0) = 0$$

$$y_5 = 2(24 - 360x^2 + \cdots) \qquad y_5(0) = 48$$

....

$$\therefore \quad \tan^{-1}\left(\frac{2x}{1-x^2}\right) = y(0) + xy_1(0) + \frac{x^2}{2!}\, y_2(0) + \frac{x^3}{3!}\, y_3(0) + \cdots$$

$$= 2x - \frac{4x^3}{3!} + \frac{48x^5}{5!} \ldots = 2\left(x - \frac{x^3}{3} + \frac{x^5}{5} \ldots\right).$$

(*ii*) Let $\qquad y = \tan^{-1} \dfrac{\sqrt{1+x^2} - \sqrt{1-x^2}}{\sqrt{1+x^2} + \sqrt{1-x^2}}$

Put $x^2 = \cos\theta$ so that $\theta = \cos^{-1} x^2$

Then $\qquad y = \tan^{-1} \dfrac{\sqrt{2\cos^2\frac{\theta}{2}} - \sqrt{2\sin^2\frac{\theta}{2}}}{\sqrt{2\cos^2\frac{\theta}{2}} + \sqrt{2\sin^2\frac{\theta}{2}}}$

$$= \tan^{-1}\left(\frac{\cos\frac{\theta}{2} - \sin\frac{\theta}{2}}{\cos\frac{\theta}{2} + \sin\frac{\theta}{2}}\right) = \tan^{-1}\left(\frac{1-\tan\frac{\theta}{2}}{1+\tan\frac{\theta}{2}}\right)$$

$$= \tan^{-1}\left[\tan\left(\frac{\pi}{4} - \frac{\theta}{2}\right)\right] = \frac{\pi}{4} - \frac{\theta}{2} = \frac{\pi}{4} - \frac{1}{2}\cos^{-1} x^2$$

$$\therefore \quad y_1 = 0 - \frac{1}{2}\cdot\frac{-1}{\sqrt{1-x^4}} \cdot 2x = x(1-x^4)^{-\frac{1}{2}}$$

$$= x\left[1 + \frac{1}{2}x^4 + \frac{\frac{1}{2}\cdot\frac{3}{2}}{2!}x^8 + \cdots\right] = x + \frac{1}{2}x^5 + \frac{3}{8}x^9 + \cdots$$

$$y_2 = 1 + \frac{5}{2}x^4 + \frac{27}{8}x^8 + \cdots$$

$$y_3 = 10x^3 + 27x^7 + \cdots$$

$$y_4 = 30x^2 + 189x^6 + \cdots$$

$$y_5 = 60x + 1134x^5 + \cdots$$

$$y_6 = 60 + 5670x^4 + \cdots$$

....

$\therefore \quad y(0) = 0,\ y_1(0) = 0,\ y_2(0) = 1,\ y_3(0) = y_4(0) = y_5(0) = 0,$
$y_6(0) = 60, \ldots$

$$\therefore \quad \tan^{-1}\frac{\sqrt{1+x^2} - \sqrt{1-x^2}}{\sqrt{1+x^2} + \sqrt{1-x^2}} = y(0) + xy_1(0) + \frac{x^2}{2!}\, y_2(0) + \frac{x^3}{3!}\, y_3(0)$$

$$+ \frac{x^4}{4!}\, y_4(0) + \frac{x^5}{5!}\, y_5(0) + \frac{x^6}{6!}\, y_6(0) + \cdots$$

$$= \frac{x^2}{2} + \frac{x^6}{12} + \cdots$$

5.4. NOTE ON LEIBNITZ'S THEOREM FOR SUCCESSIVE DIFFERENTIATION

Differentiating $(1-x^2)y_2$ *i.e.*, $y_2(1-x^2)$, n times by Leibnitz's Theorem, we get

$$y_{n+2}(1-x^2) + ny_{n+1}(-2x) + \frac{n(n-1)}{2!}y_n(-2)$$

Differentiating xy_1 *i.e.*, y_1x, n times by Leibnitz's Theorem, we have

$$y_{n+1}(x) + ny_n \cdot 1.$$

ILLUSTRATIVE EXAMPLES

Example 1. *Expand $e^{a\sin^{-1}x}$ by Maclaurin's Theorem.*

Hence show that $\quad e^{\theta} = 1 + \sin\theta + \frac{1}{2!}\sin^2\theta + \frac{2}{3!}\sin^3\theta + \cdots$

Sol. Let $\quad y = e^{a\sin^{-1}x} \qquad \therefore \quad y(0) = e^0 = 1$

Diff. $\quad y_1 = e^{a\sin^{-1}x} \cdot \frac{a}{\sqrt{1-x^2}} = \frac{ay}{\sqrt{1-x^2}} \qquad \therefore \quad \mathbf{y_1(0) = a}$

Squaring and cross multiplying

$$(1-x^2)y_1^2 = a^2y^2$$

Diff. $\quad (1-x^2)\,2y_1y_2 - 2xy_1^2 = 2a^2yy_1$

or $\quad (1-x^2)y_2 - xy_1 = a^2y \qquad \therefore \quad \mathbf{y_2(0) = a^2}$

Differentiating n times by Leibnitz's Theorem, we get

$$y_{n+2}(1-x^2) + ny_{n+1}(-2x) + \frac{n(n-1)}{2!}y_n(-2) - [y_{n+1}x + ny_n] = a^2y_n$$

or $$(1-x^2)y_{n+2} - (2n+1)xy_{n+1} - (n^2+a^2)y_n = 0$$

Putting $\quad x = 0,\ y_{n+2}(0) = (n^2+a^2)y_n(0) \qquad \ldots(i)$

Putting $\quad n = 1, 2, 3, \ldots$ in (i)

$$y_3(0) = (1^2+a^2)y_1(0) = a(1^2+a^2)$$

$$y_4(0) = (2^2+a^2)y_2(0) = a^2(2^2+a^2)$$

$$y_5(0) = (3^2+a^2)y_3(0) = a(1^2+a^2)(3^2+a^2)$$

....

$$\therefore \quad e^{a\sin^{-1}x} = y(0) + xy_1(0) + \frac{x^2}{2!}y_2(0) + \cdots$$

$$= 1 + ax + \frac{a^2x^2}{2!} + \frac{a(1^2+a^2)}{3!}x^3 + \frac{a^2(2^2+a^2)}{4!}x^4 + \cdots$$

Second Part: Putting $\sin^{-1}x = \theta$ and $a = 1$, we have

$$e^{\theta} = 1 + \sin\theta + \frac{1}{2!}\sin^2\theta + \frac{2}{3!}\sin^3\theta + \frac{5}{4!}\sin^4\theta + \cdots$$

Example 2. *Expand sin (m sin^{-1} x) by Maclaurin's Theorem, as far as x^5.*

Sol. Let $y = \sin(m \sin^{-1} x)$ $\therefore$ $\mathbf{y(0) = \sin 0 = 0}$

Diff. $y_1 = \cos(m \sin^{-1} x) \cdot \dfrac{m}{\sqrt{1-x^2}}$ $\therefore$ $\mathbf{y_1(0) = m}$

Squaring and cross-multiplying

$$(1 - x^2)y_1^2 = m^2 \cos^2 (m \sin^{-1} x) = m^2 [1 - \sin^2 (m \sin^{-1} x)]$$

$$\therefore \quad (1 - x^2)y_1^2 = m^2(1 - y^2)$$

Differentiating,

$$-2x \cdot y_1^2 + (1 - x^2) \cdot 2y_1y_2 = m^2(-2yy_1)$$

or $(1 - x^2)y_2 - xy_1 + m^2y = 0$ $\therefore$ $y_2(0) = 0$

Differentiating n times by Leibnitz's Theorem, we get

$$y_{n+2}(1 - x^2) + ny_{n+1}(-2x) + \frac{n(n-1)}{2!}y_n(-2) - [y_{n+1}x + ny_n \cdot 1] + m^2y_n = 0$$

or $(1 - x^2)y_{n+2} - (2n + 1)xy_{n+1} - (n^2 - m^2)y_n = 0$

Putting $x = 0$, $y_{n+2}(0) = (n^2 - m^2)\, y_n(0)$

Putting $n = 1, 2, 3, \ldots$

$$y_3(0) = (1^2 - m^2)y_1(0) = m\,(1^2 - m^2)$$

$$y_4(0) = (2^2 - m^2)y_2(0) = 0$$

$$y_5(0) = (3^2 - m^2)y_3(0) = m\,(1^2 - m^2)\,(3^2 - m^2)$$

....

$$\therefore \quad \sin(m \sin^{-1} x) = y(0) + xy_1(0) + \frac{x^2}{2!}y_2(0) + \cdots$$

$$= mx + \frac{m(1^2 - m^2)}{3!}x^3 + \frac{m(1^2 - m^2)(3^2 - m^2)}{5!}x^5 + \cdots$$

EXERCISE 5.2

1. Expand the following functions, using Maclaurin's series:

(*i*) e^x (*ii*) $a^x, a > 0$ (*M.D.U. Dec. 2013*)

(*iii*) $\sin x$ (*iv*) $\cos x$

(*v*) $\sec x$ (*vi*) $\sinh x$

(*vii*) $\cosh x$ (*viii*) $\log (1 - x)$.

2. Prove that:

(*i*) $e^{\sin x} = 1 + x + \dfrac{x^2}{2} - \dfrac{x^4}{8} + \cdots$ (*ii*) $e^{x \sin x} = 1 + x^2 + \dfrac{x^4}{3} + \dfrac{x^6}{120} + \cdots$ (*K.U.K. 2009*)

(*iii*) $e^{x \cos x} = 1 + x + \dfrac{x^2}{2} - \dfrac{x^3}{3} + \cdots$ (*iv*) $\dfrac{x}{\sin x} = 1 + \dfrac{x^2}{6} + \dfrac{7x^4}{360} + \cdots$ (*K.U.K. 2012*)

(*v*) $\dfrac{e^x}{\cos x} = 1 + x + \dfrac{2x^2}{2!} + \dfrac{4x^3}{3!} + \cdots$ (*vi*) $\log \sec x = \dfrac{x^2}{2} + \dfrac{x^4}{12} + \dfrac{x^6}{45} + \cdots$

(vii) $\cos^2 x = 1 - x^2 + \frac{x^4}{3} - \cdots$ (viii) $\tan^{-1} x = x - \frac{x^3}{3} + \frac{x^5}{5} - \frac{x^7}{7} + \cdots$

(ix) $\tan^{-1}(1 + x) = \frac{\pi}{4} + \frac{x}{2} - \frac{x^2}{4} + \frac{x^3}{12} - \cdots$

3. Prove that:

(i) $\log(1 + \sin x) = x - \frac{x^2}{2} + \frac{x^3}{6} - \frac{x^4}{12} + \cdots$

(ii) $\log(1 + e^x) = \log 2 + \frac{x}{2} + \frac{x^2}{8} - \frac{x^4}{192} + \cdots$

(iii) $\frac{e^x}{1+e^x} = \frac{1}{2} + \frac{x}{4} - \frac{x^3}{48} + \cdots$ (iv) $e^{ax} \sin bx = bx + abx^2 + \frac{b(3a^2 - b^2)}{3!} x^3 + \cdots$

(v) $e^x \cos x = 1 + x - \frac{2x^3}{3!} - \frac{2^2 x^4}{4!} \cdots$

(vi) $\sqrt{1 + \sin x} = 1 + \frac{x}{2} - \frac{x^2}{8} - \frac{x^3}{48} + \frac{x^4}{384} - \cdots$

4. Assuming the possibility of expansion, show that

(i) $\sin(e^x - 1) = x + \frac{x^2}{2} - \frac{5x^4}{24} + \cdots$ (ii) $\log(1 - x + x^2) = -x + \frac{x^2}{2} + \frac{2x^3}{3} + \frac{x^4}{4} - \cdots$

(iii) $\log[1 - \log(1 - x)] = x + \frac{x^3}{6} + \cdots$ (iv) $(1 + x)^{1+x} = 1 + x + x^2 + \frac{x^3}{2} + \cdots$

(v) $\log \frac{\tan x}{x} = \frac{x^2}{3} + \frac{7x^4}{90} + \cdots$ (vi) $\log \frac{\sin x}{x} = -\left(\frac{x^2}{6} + \frac{x^4}{180} + \frac{x^6}{2835} + \cdots\right)$

(vii) $\sin x \cosh x = x + \frac{x^3}{3} - \frac{x^5}{30} + \cdots$ (viii) $\frac{x}{2}\left(\frac{e^x + 1}{e^x - 1}\right) = 1 + \frac{1}{6} \cdot \frac{x^2}{2!} - \frac{1}{30} \cdot \frac{x^4}{4!} + \cdots$.

5. Prove that:

(i) $\sin^{-1}\left(\frac{2x}{1+x^2}\right) = 2\left(x - \frac{x^3}{3} + \frac{x^5}{5} - \frac{x^7}{7} + \cdots\right)$

(ii) $\tan^{-1} \frac{\sqrt{1+x^2} - 1}{x} = \frac{1}{2}\left(x - \frac{x^3}{3} + \frac{x^5}{5} - \frac{x^7}{7} + \cdots\right)$

(iii) $\sin^{-1}\left(\frac{2x^2}{1+x^4}\right) = 2\left(x^2 - \frac{x^6}{3} + \frac{x^{10}}{5} - \cdots\right)$

(iv) $\tan^{-1} \frac{x}{\sqrt{1-x^2}} = x + \frac{1^2}{3!} x^3 + \frac{1^2 . 3^2}{5!} x^5 + \cdots$

(v) $\sin^{-1}(3x - 4x^3) = 3\left(x + \frac{x^3}{6} + \frac{3x^5}{40} + \cdots\right)$.

6. By forming a differential equation, prove that:

(*i*) $(\sin^{-1} x)^2 = 2 \cdot \dfrac{x^2}{2!} + 2 \cdot 2^2 \cdot \dfrac{x^4}{4!} + 2 \cdot 2^2 \cdot 4^2 \cdot \dfrac{x^6}{6!} + \cdots$

(*ii*) $\log (x + \sqrt{1+x^2}) = x - \dfrac{1}{2} \cdot \dfrac{x^3}{3} + \dfrac{1 \cdot 3}{2 \cdot 4} \cdot \dfrac{x^5}{5} - \cdots$

(*iii*) $\cos (m \sin^{-1} x) = 1 - \dfrac{m^2}{2!} x^2 - \dfrac{m^2(2^2 - m^2)}{4!} x^4 - \dfrac{m^2(2^2 - m^2)(4^2 - m^2)}{6!} x^6 \cdots$

(*iv*) $(x + \sqrt{1+x^2})^n = 1 + nx + \dfrac{n^2x^2}{2!} + \dfrac{n^2(n^2 - 1^2)}{3!} x^3 + \dfrac{n^2(n^2 - 2^2)}{4!} x^4 + \cdots$

Answers

1. (*i*) $1 + x + \dfrac{x^2}{2!} + \dfrac{x^3}{3!} + \dfrac{x^4}{4!} + \cdots$

(*ii*) $1 + x \log a + \dfrac{x^2}{2!} (\log a)^2 + \dfrac{x^3}{3!} (\log a)^3 + \cdots$

(*iii*) $x - \dfrac{x^3}{3!} + \dfrac{x^5}{5!} + \dfrac{x^7}{7!} + \cdots$

(*iv*) $1 - \dfrac{x^2}{2!} + \dfrac{x^4}{4!} - \dfrac{x^6}{6!} + \cdots$

(*v*) $1 + \dfrac{x^2}{2!} + \dfrac{5x^4}{4!} + \cdots$

(*vi*) $x + \dfrac{x^3}{3!} + \dfrac{x^5}{5!} + \dfrac{x^7}{7!} + \cdots$

(*vii*) $1 + \dfrac{x^2}{2!} + \dfrac{x^4}{4!} + \dfrac{x^6}{6!} + \cdots$

(*viii*) $-x - \dfrac{x^2}{2} - \dfrac{x^3}{3} - \dfrac{x^4}{4} - \cdots$.

6
Curvature

6.1. DEFINITIONS

Let P, Q be two neighbouring points on a curve AB. Let arc AP = s, arc AQ = $s + \delta s$ so that arc PQ = δs, A being a fixed point on the curve, from which arcs are measured. Let the tangents at P and Q make angles ψ and $\psi + \delta\psi$ respectively with a fixed line say x-axis. Then angle $\delta\psi$ through which the tangent turns as its point of contact travels along the arc PQ is called the **total curvature** of arc PQ.

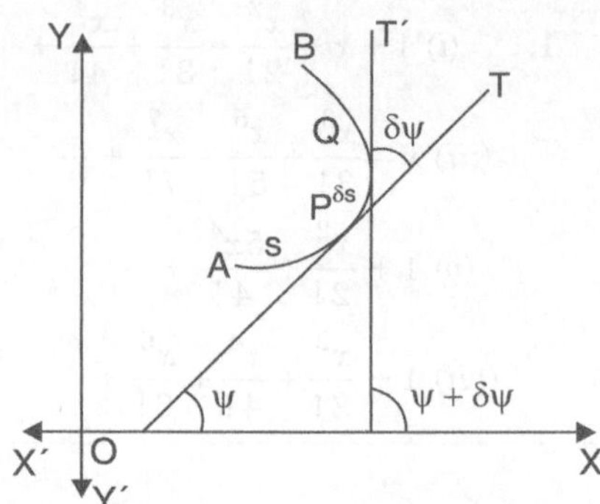

$\dfrac{\delta\psi}{\delta s}$ is called the **average curvature** of arc PQ.

The limiting value of the average curvature when Q → P is called the **curvature of the curve at the point P.** (K.U.K. 2012)

Thus, the curvature k at point $\text{P} = \lim\limits_{Q \to P} \dfrac{\delta\psi}{\delta s} = \lim\limits_{\delta s \to 0} \dfrac{\delta\psi}{\delta s} = \dfrac{d\psi}{ds}$

The reciprocal of the curvature of the curve at P, provided this curvature is not zero, is called the **radius of curvature of the curve at P** and is usually denoted by ρ. Thus $\rho = \dfrac{1}{k} = \dfrac{ds}{d\psi}$.

If PC is normal at P and PC = ρ, then C is called the **centre of curvature of the curve at P.**

The circle with centre C and radius = PC = ρ is called the **circle of curvature of the curve at P.**

Any chord of the circle of curvature at P, **drawn through P** is called a **chord of curvature.**

Note. The relation between s and ψ is called the intrinsic equation of the curve.

The expression $\dfrac{ds}{d\psi}$ for radius of curvature is suitable only for those curves whose instrinsic equation is given.

Example. *Find the radius of curvature at the point* (s, ψ) *on the curve*

$$s = a \log \tan\left(\frac{\pi}{4} + \frac{\psi}{2}\right).$$

Sol. Here $s = a \log \tan\left(\frac{\pi}{4} + \frac{\psi}{2}\right)$

$$\therefore \quad \rho = \frac{ds}{d\psi} = a \cdot \frac{1}{\tan\left(\frac{\pi}{4} + \frac{\psi}{2}\right)} \cdot \sec^2\left(\frac{\pi}{4} + \frac{\psi}{2}\right) \cdot \frac{1}{2}$$

$$= a \cdot \frac{\cos\left(\frac{\pi}{4} + \frac{\psi}{2}\right)}{\sin\left(\frac{\pi}{4} + \frac{\psi}{2}\right)} \cdot \frac{1}{\cos^2\left(\frac{\pi}{4} + \frac{\psi}{2}\right)} \cdot \frac{1}{2} = \frac{a}{2\sin\left(\frac{\pi}{4} + \frac{\psi}{2}\right)\cos\left(\frac{\pi}{4} + \frac{\psi}{2}\right)}$$

$$= \frac{a}{\sin 2\left(\frac{\pi}{4} + \frac{\psi}{2}\right)} = \frac{a}{\sin\left(\frac{\pi}{2} + \psi\right)} = \frac{a}{\cos\psi} = a \sec\psi.$$

6.2. AN IMPORTANT NOTE

The following figures are a useful device to remember the various t-ratios of ψ and ϕ.

(*i*) $\tan\psi = \frac{dy}{dx} \quad \therefore \quad \sin\psi = \frac{dy}{ds}$,

$\cos\psi = \frac{dx}{ds}$ [Fig. (*i*)]

(*ii*) $\tan\psi = \frac{dy}{dx} = y_1 \quad \therefore \quad \sin\psi = \frac{y_1}{\sqrt{1 + y_1^2}}$,

$\cos\psi = \frac{1}{\sqrt{1 + y_1^2}}$. [Fig. (*ii*)]

(*iii*) $\tan\phi = r\frac{d\theta}{dr}, \quad \therefore \sin\phi = r\frac{d\theta}{ds}$,

$\cos\phi = \frac{dr}{ds}$ [Fig. (*iii*)]

(*iv*) $\tan\phi = r\frac{d\theta}{dr} = \frac{r}{\frac{dr}{d\theta}} = \frac{r}{r_1}$

$\therefore \quad \sin\phi = \frac{r}{\sqrt{r^2 + r_1^2}}, \cos\phi = \frac{r_1}{\sqrt{r^2 + r_1^2}}$ [Fig. (*iv*)]

Fig. (*i*)

Fig. (*ii*)

Fig. (*iii*)

Fig. (*iv*)

6.3. RADIUS OF CURVATURE FOR CARTESIAN CURVES

I. Explicit Equation: y = f(x)

Find the radius of curvature at any point of the curve y = f(x).

Equation of curve is $y = f(x)$

Let ψ be the angle which the tangent at any point (x, y) makes with the x-axis. Then

$$\tan\psi = \frac{dy}{dx}.$$

Differentiating both sides w.r.t. s, we have

$$\sec^2\psi\cdot\frac{d\psi}{ds}=\frac{d}{ds}\left(\frac{dy}{dx}\right)\quad\text{or}\quad \sec^2\psi\cdot\frac{1}{\rho}=\frac{d}{dx}\left(\frac{dy}{dx}\right)\cdot\frac{dx}{ds}$$

or $$\sec^2\psi\cdot\frac{1}{\rho}=\frac{d^2y}{dx^2}\cdot\cos\psi \qquad \left[\because \cos\psi=\frac{dx}{ds}\right]$$

$\therefore$ $$\rho=\frac{\sec^3\psi}{\dfrac{d^2y}{dx^2}}=\frac{(1+\tan^2\psi)^{3/2}}{\dfrac{d^2y}{dx^2}}$$

Hence $$\rho=\frac{\left[1+\left(\dfrac{dy}{dx}\right)^2\right]^{3/2}}{\dfrac{d^2y}{dx^2}}=\frac{(1+{y_1}^2)^{3/2}}{y_2}$$

where $y_1=\dfrac{dy}{dx},\ y_2=\dfrac{d^2y}{dx^2}$.

Note. This formula does not hold good when the tangent at (x, y) is parallel to y-axis *i.e.*, perpendicular to x-axis, for then $\dfrac{dy}{dx}$ (= slope of tangent) is infinite.

Since the value of ρ is independent of the choice of axes, therefore, interchanging the axes of x and y, we get

$$\rho=\frac{\left[1+\left(\dfrac{dx}{dy}\right)^2\right]^{3/2}}{d^2x/dy^2}\quad\text{or}\quad\frac{(1+{x_1}^2)^{3/2}}{x_2}$$

II. Implicit Equation: $f(x, y) = 0$.

We know that $$\frac{dy}{dx}=-\frac{f_x}{f_y}\qquad\text{provided } f_y\neq 0$$

and $$\frac{d^2y}{dx^2}=-\frac{f_{xx}(f_y)^2-2f_xf_yf_{xy}+f_{yy}(f_x)^2}{(f_y)^3}$$

Since $$\rho=\frac{\left[1+\left(\dfrac{dy}{dx}\right)^2\right]^{3/2}}{\dfrac{d^2y}{dx^2}}$$

$\therefore$ $$\rho=-\frac{\left[1+\left(-\dfrac{f_x}{f_y}\right)^2\right]^{3/2}\cdot(f_y)^3}{f_{xx}{f_y}^2-2f_xf_yf_{xy}+f_{yy}(f_x)^2}=\frac{[(f_x)^2+(f_y)^2]^{3/2}}{f_{xx}(f_y)^2-2f_xf_yf_{xy}+f_{yy}(f_x)^2}$$

(in magnitude)

III. Parametric Equations: $x=f(t),\ y=\phi(t)$

$$\frac{dy}{dx}=\frac{dy}{dt}\bigg/\frac{dx}{dt}=\frac{y'}{x'}$$

where dashes denote differentiation w.r.t. parameter t.

Also $$\frac{d^2y}{dx^2}=\frac{d}{dx}\left(\frac{y'}{x'}\right)=\frac{d}{dt}\left(\frac{y'}{x'}\right)\cdot\frac{dt}{dx}=\frac{x'y''-y'x''}{x'^2}\cdot\frac{1}{x'}=\frac{x'y''-y'x''}{x'^3}$$

$$\therefore \qquad \rho=\frac{\left[1+\left(\frac{dy}{dx}\right)^2\right]^{3/2}}{\frac{d^2y}{dx^2}}=\frac{\left[1+\left(\frac{y'}{x'}\right)^2\right]^{3/2}\cdot x'^3}{x'y''-y'x''}=\frac{(x'^2+y'^2)^{3/2}}{x'y''-y'x''}.$$

ILLUSTRATIVE EXAMPLES

Example 1. *Find the radius of curvature at the given point of the following curves:*

(*i*) $\sqrt{x}+\sqrt{y}=\sqrt{a}$ *at* $\left(\frac{a}{4},\frac{a}{4}\right)$

(*ii*) $x^3+y^3=3axy$ *at* $\left(\frac{3a}{2},\frac{3a}{2}\right)$. (K.U.K. 2009)

Sol. (*i*) Equation of curve is $\sqrt{x}+\sqrt{y}=\sqrt{a}$ ...(1)

Differentiating w.r.t. x, we have

$$\frac{1}{2}x^{-1/2}+\frac{1}{2}y^{-1/2}\cdot\frac{dy}{dx}=0 \quad \text{or} \quad \frac{dy}{dx}=-\frac{x^{-1/2}}{y^{-1/2}}=-\sqrt{\frac{y}{x}}$$

Differentiating again w.r.t. x, we have

$$\frac{d^2y}{dx^2}=-\frac{x^{1/2}\cdot\frac{1}{2}y^{-1/2}\frac{dy}{dx}-y^{1/2}\cdot\frac{1}{2}x^{-1/2}}{x}$$

$$=-\frac{\sqrt{\frac{x}{y}}\left(-\sqrt{\frac{y}{x}}\right)-\sqrt{\frac{y}{x}}}{2x}=\frac{1+\sqrt{\frac{y}{x}}}{2x}$$

$$=\frac{\sqrt{x}+\sqrt{y}}{2x\sqrt{x}}=\frac{\sqrt{a}}{2x\sqrt{x}} \qquad \text{[from (1)]}$$

At $\left(\frac{a}{4},\frac{a}{4}\right)$, $$\frac{dy}{dx}=-1$$

$$\frac{d^2y}{dx^2}=\frac{\sqrt{a}}{2\cdot\frac{a}{4}\cdot\frac{\sqrt{a}}{2}}=\frac{4}{a}$$

$$\therefore \qquad \rho \text{ at } \left(\frac{a}{4},\frac{a}{4}\right)=\frac{\left[1+\left(\frac{dy}{dx}\right)^2\right]^{3/2}}{\frac{d^2y}{dx^2}}=\frac{(1+1)^{3/2}}{4/a}=\frac{a}{\sqrt{2}}.$$

(*ii*) Equation of curve is $x^3+y^3=3axy$

Differentiating w.r.t. x, we have

$$3x^2 + 3y^2 \frac{dy}{dx} = 3a\left(x\frac{dy}{dx} + y\right) \quad \text{or} \quad (y^2 - ax)\frac{dy}{dx} = ay - x^2 \qquad ...(1)$$

At $\left(\frac{3a}{2}, \frac{3a}{2}\right)$, $\left(\frac{9a^2}{4} - \frac{3a^2}{2}\right)\frac{dy}{dx} = \frac{3a^2}{2} - \frac{9a^2}{4}$

$$\therefore \qquad \frac{dy}{dx} = -1$$

Differentiating (1) w.r.t. x, we have

$$(y^2 - ax)\frac{d^2y}{dx^2} + \left(2y\frac{dy}{dx} - a\right)\frac{dy}{dx} = a\frac{dy}{dx} - 2x$$

At $\left(\frac{3a}{2}, \frac{3a}{2}\right)$, we have

$$\left(\frac{9a^2}{4} - \frac{3a^2}{2}\right)\frac{d^2y}{dx^2} + [3a(-1) - a](-1) = a(-1) - 3a$$

or $$\frac{3a^2}{4}\cdot\frac{d^2y}{dx^2} = -8a \qquad \therefore \qquad \frac{d^2y}{dx^2} = -\frac{32}{3a}$$

Hence $$\rho \text{ at } \left(\frac{3a}{2}, \frac{3a}{2}\right) = \frac{\left[1 + \left(\frac{dy}{dx}\right)^2\right]^{3/2}}{\frac{d^2y}{dx^2}} = \frac{(1+1)^{3/2}}{-\frac{32}{3a}} = \frac{3a}{8\sqrt{2}} \quad \text{(in magnitude).}$$

Second Method

Equation of curve is $\quad x^3 + y^3 = 3axy \quad$ or $\quad f(x, y) = x^3 + y^3 - 3axy = 0$

$$\therefore \qquad f_x = 3x^2 - 3ay, \qquad f_y = 3y^2 - 3ax$$

$$f_{xx} = 6x, \qquad f_{yy} = 6y, \qquad f_{xy} = -3a$$

At the point $\left(\frac{3a}{2}, \frac{3a}{2}\right)$

$$f_x = \frac{27a^2}{4} - \frac{9a^2}{2} = \frac{9a^2}{4}$$

$$f_y = \frac{27a^2}{4} - \frac{9a^2}{2} = \frac{9a^2}{4}$$

$$f_{xx} = 9a, \quad f_{yy} = 9a, \quad f_{xy} = -3a$$

$$\therefore \qquad \rho \text{ at } \left(\frac{3a}{2}, \frac{3a}{2}\right) = -\frac{[(f_x)^2 + (f_y)^2]^{3/2}}{f_{xx}(f_y)^2 - 2f_xf_yf_{xy} + f_{yy}(f_x)^2}$$

$$= -\frac{\left(\frac{81a^4}{16} + \frac{81a^4}{16}\right)^{3/2}}{9a\cdot\frac{81a^4}{16} - 2\cdot\frac{9a^2}{4}\cdot\frac{9a^2}{4}\cdot(-3a) + 9a\cdot\frac{81a^4}{16}}$$

$$= -\frac{\left(\frac{81a^4}{8}\right)^{3/2}}{\frac{81a^5}{16}(9+6+9)} = -\frac{\frac{81a^4}{8}\left(\frac{81a^4}{8}\right)^{1/2}}{\frac{81a^5}{16}(24)} = \frac{1}{12a}\left(\frac{9a^2}{2\sqrt{2}}\right) \quad \text{(In magnitude)}$$

$$= \frac{3a}{8\sqrt{2}}.$$

Example 2. *Show that for the parabola $y^2 = 4ax$, ρ^2 varies as $(SP)^3$, where ρ is the radius of curvature at any point P of the parabola and S is the focus of the parabola.*

(M.D.U. May 2009)

Sol. Equation of parabola is $y^2 = 4ax$ or $x = at^2$, $y = 2at$ [Parametric Equations]

Any point P on it is $(at^2, 2at)$

If dashes denote differentiation w.r.t. parameter t, then

$$x' = 2at, \qquad y' = 2a$$

$$x'' = 2a, \qquad y'' = 0$$

$$\therefore \quad \rho \text{ at P } `t' = \frac{(x'^2 + y'^2)^{3/2}}{x'y'' - y'x''} = \frac{(4a^2t^2 + 4a^2)^{3/2}}{0 - 4a^2}$$

$$= \frac{(4a^2)^{3/2}(t^2+1)^{3/2}}{-4a^2} = -\frac{8a^3(1+t^2)^{3/2}}{4a^2} = 2a(1+t^2)^{3/2} \quad \text{[Numerically]}$$

Also $\quad$ SP = distance between S$(a, 0)$ and P$(at^2, 2at)$

$$= \sqrt{(at^2 - a)^2 + (2at - 0)^2} = \sqrt{a^2(t^2-1)^2 + 4a^2t^2}$$

$$= a\sqrt{(t^2+1)^2} = a(t^2+1)$$

$$\therefore \quad \rho^2 = 4a^2(1+t^2)^3 = \frac{4}{a}\cdot a^3(1+t^2)^3 = \frac{4}{a}(\text{SP})^3 \propto (\text{SP})^3.$$

Example 3. *The tangents at two points P, Q on the cycloid $x = a(\theta - \sin\theta)$, $y = a(1 - \cos\theta)$ are at right angles ; show that if ρ_1, ρ_2 be the radii of curvature at these points, then $\rho_1^2 + \rho_2^2 = 16a^2$.*

Sol. Here $\quad x = a(\theta - \sin\theta), \qquad y = a(1 - \cos\theta)$

$$\Rightarrow \quad x' = a(1 - \cos\theta), \qquad y' = a\sin\theta$$

$$x'' = a\sin\theta, \qquad y'' = a\cos\theta$$

$$\therefore \quad x'^2 + y'^2 = a^2(1 - 2\cos\theta + \cos^2\theta + \sin^2\theta)$$

$$= 2a^2(1 - \cos\theta) = 4a^2\sin^2\theta/2$$

$$x'y'' - y'x'' = a^2\cos\theta(1 - \cos\theta) - a^2\sin^2\theta$$

$$= a^2(\cos\theta - \cos^2\theta - \sin^2\theta)$$

$$= a^2(\cos\theta - 1) = -a^2(1 - \cos\theta) = -2a^2\sin^2\theta/2$$

$$\rho \text{ at } `\theta' = \frac{(x'^2 + y'^2)^{3/2}}{x'y'' - y'x''} = \frac{(4a^2\sin^2\theta/2)^{3/2}}{-2a^2\sin^2\theta/2} = \frac{8a^3\sin^3\theta/2}{-2a^2\sin^2\theta/2}$$

$$= 4a\sin\theta/2 \quad \text{(in magnitude)}$$

Let the points P and Q correspond to

$$\theta = \theta_1 \quad \text{and} \quad \theta = \theta_2, \text{ then}$$

$$\rho_1 = 4a\sin\frac{\theta_1}{2}, \qquad \rho_2 = 4a\sin\frac{\theta_2}{2}$$

Now $$\frac{dy}{dx} = \frac{\frac{dy}{d\theta}}{\frac{dx}{d\theta}} = \frac{y'}{x'} = \frac{\sin\theta}{1-\cos\theta} = \frac{2\sin\theta/2\cos\theta/2}{2\sin^2\theta/2} = \cot\theta/2$$

$\Rightarrow$ Slope of tangent at 'θ' = $\cot\frac{\theta}{2}$

$\therefore$ Slopes of tangents at P and Q are $\cot\frac{\theta_1}{2}$ and $\cot\frac{\theta_2}{2}$.

Since the tangents at P and Q are perpendicular

$$\therefore \quad \cot\frac{\theta_1}{2}\cot\frac{\theta_2}{2} = -1 \qquad (m_1m_2 = -1)$$

$$\Rightarrow \quad \cot\frac{\theta_2}{2} = -\tan\frac{\theta_1}{2} = \cot\left(\frac{\pi}{2}+\frac{\theta_1}{2}\right)$$

$$\Rightarrow \quad \frac{\theta_2}{2} = \frac{\pi}{2}+\frac{\theta_1}{2} \quad \Rightarrow \quad \theta_2 = \pi+\theta_1$$

$$\therefore \quad \rho_2 = 4a\sin\left(\frac{\pi}{2}+\frac{\theta_1}{2}\right) = 4a\cos\frac{\theta_1}{2}$$

Hence $$\rho_1^2+\rho_2^2 = 16a^2\left(\sin^2\frac{\theta_1}{2}+\cos^2\frac{\theta_1}{2}\right) = 16a^2.$$

Example 4. *Prove that the radius of curvature at any point of the astroid $x^{2/3}+y^{2/3} = a^{2/3}$ is three times the length of the perpendicular from the origin to the tangent at that point.*

(K.U.K. 2013; P.T.U. Dec. 2011)

Sol. The Cartesian equation of the curve is $x^{2/3}+y^{2/3} = a^{2/3}$

Its parametric equations are $x = a\cos^3 t,\ y = a\sin^3 t$

$$\Rightarrow \quad x' = -3a\cos^2 t\sin t, \qquad y' = 3a\sin^2 t\cos t$$

$$x'' = -3a(\cos^3 t - 2\cos t\sin^2 t) = 3a\cos t\,(2\sin^2 t-\cos^2 t)$$

$$y'' = 3a(2\sin t\cos^2 t-\sin^3 t) = 3a\sin t\,(2\cos^2 t-\sin^2 t)$$

$$\therefore \quad x'^2+y'^2 = 9a^2(\cos^4 t\sin^2 t+\sin^4 t\cos^2 t) = 9a^2\cos^2 t\sin^2 t$$

$$x'y''-y'x'' = -9a^2\sin^2 t\cos^2 t\,(2\cos^2 t-\sin^2 t) - 9a^2\sin^2 t\cos^2 t\,(2\sin^2 t-\cos^2 t)$$

$$= -9a^2\sin^2 t\cos^2 t\,(\cos^2 t+\sin^2 t) = -9a^2\sin^2 t\cos^2 t$$

$\therefore$ ρ at any point '*t*' *i.e.*, $(a\cos^3 t, a\sin^3 t)$

$$= \frac{(x'^2+y'^2)^{3/2}}{x'y''-y'x''} = \frac{(9a^2\cos^2 t\sin^2 t)^{3/2}}{-9a^2\sin^2 t\cos^2 t}$$

$$= \frac{27a^3\cos^3 t\sin^3 t}{-9a^2\sin^2 t\cos^2 t} = 3a\sin t\cos t \qquad \text{(in magnitude)}$$

Now $$\frac{dy}{dx} = \frac{y'}{x'} = \frac{3a\sin^2 t\cos t}{-3a\cos^2 t\sin t} = -\tan t$$

$\therefore$ Equation of tangent at $(a\cos^3 t, a\sin^3 t)$ is

$$y - a\sin^3 t = -\tan t\,(x-a\cos^3 t)$$

or $$y - a\sin^3 t = -\frac{\sin t}{\cos t}(x-a\cos^3 t)$$

or $$y\cos t - a\sin^3 t\cos t = -x\sin t + a\sin t\cos^3 t$$

or $\quad x \sin t + y \cos t - a \sin t \cos t (\sin^2 t + \cos^2 t) = 0$

or $\quad x \sin t + y \cos t - a \sin t \cos t = 0$

p = length of ⊥ from the origin (0, 0) on the tangent at 't'

$$= \frac{|0 + 0 - a \sin t \cos t|}{\sqrt{\sin^2 t + \cos^2 t}} = a \sin t \cos t$$

Clearly, $\quad \rho = 3p.$

Example 5. *Prove that for the ellipse* $\frac{x^2}{a^2} + \frac{y^2}{b^2} = 1, \quad \rho = \frac{a^2b^2}{p^3}$ *where p is the perpendicular from the centre upon the tangent at (x, y).*

Sol. Any point (x, y) on the ellipse $\frac{x^2}{a^2} + \frac{y^2}{b^2} = 1$ is $(a \cos \theta, b \sin \theta)$ or simply 'θ'.

Parametric equations of the ellipse are

$$x = a \cos \theta, \qquad y = b \sin \theta$$

$$\Rightarrow \quad x' = -a \sin \theta, \qquad y' = b \cos \theta$$

$$x'' = -a \cos \theta, \qquad y'' = -b \sin \theta$$

$$\therefore \quad \rho \text{ at '}\theta\text{'} = \frac{(x'^2 + y'^2)^{3/2}}{x'y'' - y'x''} = \frac{(a^2 \sin^2 \theta + b^2 \cos^2 \theta)^{3/2}}{ab \sin^2 \theta + ab \cos^2 \theta}$$

$$= \frac{(a^2 \sin^2 \theta + b^2 \cos^2 \theta)^{3/2}}{ab} \qquad \text{...(1)}$$

Equation of tangent at 'θ' is $\frac{x}{a} \cos \theta + \frac{y}{b} \sin \theta = 1$

p = length of ⊥ from centre (0, 0) on tangent

$$= \frac{1}{\sqrt{\frac{\cos^2 \theta}{a^2} + \frac{\sin^2 \theta}{b^2}}} = \frac{ab}{\sqrt{a^2 \sin^2 \theta + b^2 \cos^2 \theta}}$$

$$\Rightarrow \quad p^3 = \frac{a^3b^3}{(a^2 \sin^2 \theta + b^2 \cos^2 \theta)^{3/2}} \qquad \text{...(2)}$$

Multiplying (1) and (2), $\rho p^3 = a^2b^2$

Hence $\quad \rho = \frac{a^2b^2}{p^3}$.

EXERCISE 6.1

1. Find the radius of curvature at the point (s, ψ) on the curve $s = a \log (\tan \psi + \sec \psi) + a \tan \psi \sec \psi$.
2. Find the radius of curvature at the given point of the following curves:

   (*i*) $y = 4 \sin x - \sin 2x$ at $x = \frac{\pi}{2}$ $\qquad$ (*ii*) $\sqrt{x} + \sqrt{y} = 1$ at $\left(\frac{1}{4}, \frac{1}{4}\right)$

   (*iii*) $y^2 = \frac{a^2(a - x)}{x}$ at $(a, 0)$ $\qquad$ $\left[\text{Equation of curve is } x = \frac{a^3}{y^2 + a^2}\right]$

   (*iv*) $x^2y = a(x^2 + y^2)$ at $(-2a, 2a)$.

3. Find the radius of curvature of the curve $y = e^x$ at the point where it crosses the y-axis.
4. Find the least value of $|\rho|$ for the curve $y = \log x \cdot (x > 0)$
5. Show that for the rectangular hyperbola $xy = c^2$, $\rho = \dfrac{(x^2 + y^2)^{3/2}}{2c^2}$.
6. (*a*) Find the radius of curvature at any point (o, c) of the catenary $y = c \cosh \dfrac{x}{c}$.

   (*b*) Prove that the radius of curvature for the catenary $y = c \cosh \dfrac{x}{c}$ is equal to the portion of the normal intercepted between the curve and the x-axis and that it varies as the square of the ordinate.

   [**Hint.** Portion of the normal intercepted between the curve and the x-axis is called the *length of the normal* and is given by $y\sqrt{1 + y_1^2}$.]
7. Find the points on the parabola $y^2 = 8x$ at which the radius of curvature is $7\dfrac{13}{16}$.
8. Show that in the parabola $y^2 = 4ax$, the radius of curvature ρ at any point P is twice the part of the normal intercepted between the curve and the directrix.

   [**Hint.** Equation of normal at $P(at^2, 2at)$ is $y + tx = 2at + at^3$.
   It meets the directrix $x = -a$ at $Q(-a, 3at + at^3)$.
   Show that $\rho = 2PQ$.]
9. If ρ_1 and ρ_2 be the radii of curvature at the ends of a focal chord of the parabola $y^2 = 4ax$, then prove that $\rho_1^{-2/3} + \rho_2^{-2/3} = (2a)^{-2/3}$. *(M.D.U. Dec. 2013; P.T.U. 2009)*

   [**Hint.** If P 't_1' and Q 't_2' be the extremities of a focal chord of the parabola $y^2 = 4ax$, then $t_1 t_2 = -1$ or $t_2 = -\dfrac{1}{t_1}$]
10. Find the radius of curvature at any point of the curve:

    (*i*) $x = a \cos^3 \theta$, $y = a \sin^3 \theta$ (*ii*) $x = a(\theta + \sin \theta)$, $y = a(1 - \cos \theta)$ *(M.D.U. May 2008)*

    (*iii*) $x = 3a \cos t - a \cos 3t$, $y = 3a \sin t - a \sin 3t$

    (*iv*) $x = a(\cos t + t \sin t)$, $y = a(\sin t - t \cos t)$

    (*v*) $x = a \sin 2\theta(1 + \cos 2\theta)$, $y = a \cos 2\theta(1 - \cos 2\theta)$.
11. Show that for the curve $x = a \cos \theta\ (1 + \sin \theta)$, $y = a \sin \theta\ (1 + \cos \theta)$ the radius of curvature at $\theta = -\dfrac{\pi}{4}$ is a.
12. Find the radius of curvature at the origin of the two branches of the curve given by $x = 1 - t^2$, $y = t - t^3$.

    [**Hint.** At the origin $x = y = 0 \Rightarrow t = \pm 1$]
13. Show that the radius of curvature at the point $(a \cos^3 \theta, a \sin^3 \theta)$ on the curve $x^{2/3} + y^{2/3} = a^{2/3}$ is $3a \sin \theta \cos \theta$. *(K.U.K. Dec. 2010)*
14. Show that the radius of curvature ρ at any point (x, y) of the curve $x^{2/3} + y^{2/3} = a^{2/3}$ satisfies $\rho^3 = 27axy$.

15. Show that the radius of curvature at any point of the curve $x = a\left(\cos t + \log \tan \frac{t}{2}\right)$, $y = a \sin t$ is inversely proportional to the length of the normal intercepted between the point on the curve and the x-axis.

16. Show that the radius of curvature at a point on the curve $x = ae^{\theta} (\sin \theta - \cos \theta)$, $y = ae^{\theta} (\sin \theta + \cos \theta)$ is twice the distance of the tangent at the point from the origin.

17. Show that the radius of curvature at an end of the major axis of the ellipse $\frac{x^2}{a^2} + \frac{y^2}{b^2} = 1$ is equal to the semi-latus rectum of the ellipse. *(M.D.U. Dec. 2008)*
[**Hint.** At the ends of major axis, $\theta = 0, \pi$]

18. Show that the radius of curvature ρ at P on an ellipse $\frac{x^2}{a^2} + \frac{y^2}{b^2} = 1$ is given by $\rho = \frac{\text{CD}^3}{ab}$ where CD is the semi-diameter conjugate to CP. *(M.D.U. Dec. 2010)*

[**Hint.** If P be the point $(a \cos \theta, b \sin \theta)$, then changing θ to $\frac{\pi}{2} + \theta$, the coordinates of D are $\left(a \cos\left(\frac{\pi}{2} + \theta\right), b \sin\left(\frac{\pi}{2} + \theta\right)\right) = (-a \sin \theta, b \cos \theta)$]

19. If ρ_1 and ρ_2 be the radii of curvature at the extremities of two conjugate semi-diameters of the ellipse $\frac{x^2}{a^2} + \frac{y^2}{b^2} = 1$, prove that $(\rho_1^{2/3} + \rho_2^{2/3})(ab)^{2/3} = a^2 + b^2$. *(P.T.U. Dec. 2011)*

20. Find the radius of curvature for the curve $x = c \log \left[s + \sqrt{s^2 + c^2}\right]$, $y = \sqrt{s^2 + c^2}$.

$$\left[\textbf{Hint.}\ \frac{dy}{dx} = \frac{\frac{dy}{ds}}{\frac{dx}{ds}} = \frac{s}{c} \Rightarrow \tan \psi = \frac{s}{c} \quad \therefore \quad s = c \tan \psi\right]$$

Answers

1. $2a \sec^3 \psi$ 2. (*i*) $\frac{5\sqrt{5}}{4}$ (*ii*) $\frac{1}{\sqrt{2}}$ (*iii*) $\frac{a}{2}$

(*iv*) $2a$ 3. $2\sqrt{2}$ 4. $\frac{3\sqrt{3}}{2}$ 6. (*a*) c

7. $\left(\frac{9}{8}, 3\right)$ and $\left(\frac{9}{8}, -3\right)$

10. (*i*) $3a \sin \theta \cos \theta$ (*ii*) $4a \cos \frac{\theta}{2}$ (*iii*) $3a \sin t$ (*iv*) at

(*v*) $4a \cos 3\theta$ 12. $2\sqrt{2}$ 20. $\frac{c^2 + s^2}{c}$.

6.4. RADIUS OF CURVATURE FOR PEDAL EQUATIONS

If ρ be the radius of curvature for the curve $p = f(r)$, show that $\rho = r\,\frac{dr}{dp}$.

Let the tangent at any point to the curve make an angle ψ with the initial line OX.

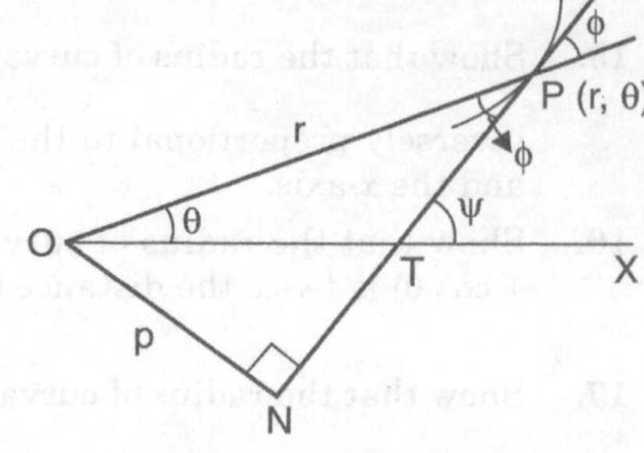

From the figure $\psi = \theta + \phi$

$$\therefore \quad \frac{d\psi}{ds} = \frac{d\theta}{ds} + \frac{d\phi}{ds} \qquad ...(i)$$

Also, we know that $p = r \sin \phi$

Differentiating w.r.t. r,

$$\frac{dp}{dr} = \sin \phi + r \cos \phi \cdot \frac{d\phi}{dr}$$

$$= r\frac{d\theta}{ds} + r \cdot \frac{dr}{ds} \cdot \frac{d\phi}{dr} \qquad \left[\because \sin \phi = r\frac{d\theta}{ds}, \cos \phi = \frac{dr}{ds}\right]$$

$$= r\left[\frac{d\theta}{ds} + \frac{d\phi}{ds}\right] = r\frac{d\psi}{ds} \qquad | \because \text{ of } (i)$$

$$= r \cdot \frac{1}{\rho}. \qquad \text{Hence } \rho = r\frac{dr}{dp}.$$

6.5. RADIUS OF CURVATURE FOR POLAR EQUATIONS

If ρ be the radius of curvature for the curve r = f(θ) or f(r, θ) = 0, prove that

$$\rho = \frac{(r^2 + r_1^2)^{3/2}}{r^2 - rr_2 + 2r_1^2}, \text{ where } r_1 = \frac{dr}{d\theta}; r_2 = \frac{d^2r}{d\theta^2}$$

Let the tangent at any point P(r, θ) to the curve make an angle ψ with the initial line OX.

From the figure $\psi = \theta + \phi$

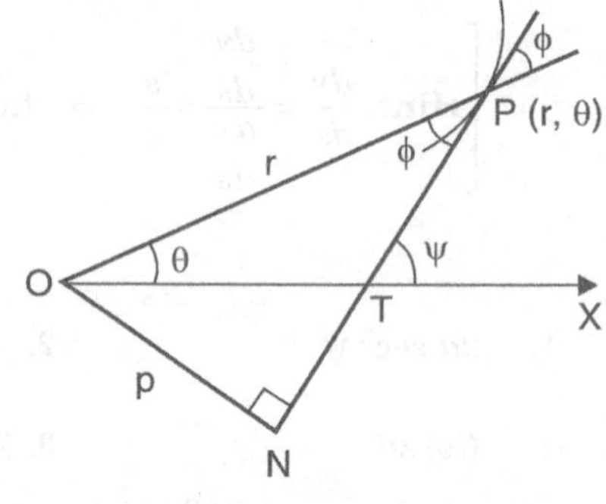

$$\therefore \quad \frac{d\psi}{ds} = \frac{d\theta}{ds} + \frac{d\phi}{ds}$$

$$= \frac{d\theta}{ds} + \frac{d\phi}{d\theta} \cdot \frac{d\theta}{ds} \qquad | \text{ Note}$$

or
$$\frac{1}{\rho} = \frac{d\theta}{ds}\left[1 + \frac{d\phi}{d\theta}\right] \qquad ...(1)$$

Now
$$\tan \phi = r\frac{d\theta}{dr} = \frac{r}{\frac{dr}{d\theta}} = \frac{r}{r_1}$$

Differentiating w.r.t. θ, we have

$$\sec^2 \phi \cdot \frac{d\phi}{d\theta} = \frac{r_1 \cdot r_1 - r \cdot r_2}{r_1^2} = \frac{r_1^2 - rr_2}{r_1^2}$$

$$\therefore \quad \frac{d\phi}{d\theta} = \frac{r_1^2 - rr_2}{r_1^2 \sec^2 \phi} = \frac{r_1^2 - rr_2}{r_1^2 (1 + \tan^2 \phi)} = \frac{r_1^2 - rr_2}{r_1^2\left(1 + \frac{r^2}{r_1^2}\right)} \qquad \left| \because \tan \phi = \frac{r}{r_1}\right.$$

$$= \frac{r_1^2 - rr_2}{r_1^2 + r^2}$$

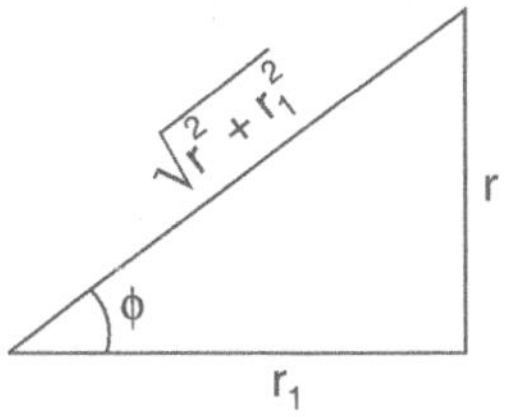

Also $\quad \sin\phi = r\dfrac{d\theta}{ds}$

$$\therefore \quad \frac{d\theta}{ds} = \frac{\sin\phi}{r} = \frac{1}{r}\cdot\frac{r}{\sqrt{r^2+r_1^2}} = \frac{1}{\sqrt{r^2+r_1^2}}$$

$$\therefore \text{ From (1), } \quad \frac{1}{\rho} = \frac{1}{\sqrt{r^2+r_1^2}}\left[1+\frac{r_1^2-rr_2}{r_1^2+r^2}\right] = \frac{r^2-rr_2+2r_1^2}{(r^2+r_1^2)^{3/2}}$$

Hence
$$\rho = \frac{(r^2+r_1^2)^{3/2}}{r^2-rr_2+2r_1^2}$$

Cor. If the equation of curve is given in the form $u = f(\theta)$ where $u = \dfrac{1}{r}$.

Then
$$u = \frac{1}{r} \qquad \therefore \qquad r = \frac{1}{u}$$

$$r_1 = -\frac{1}{u^2}\frac{du}{d\theta} = -\frac{u_1}{u^2} \qquad \text{where} \qquad u_1 = \frac{du}{d\theta}$$

$$r_2 = -\frac{u^2u_2 - u_1\cdot 2uu_1}{u^4} = \frac{2u_1^2-uu_2}{u^2} \qquad \text{where} \qquad u_2 = \frac{d^2u}{d\theta^2}$$

$$\therefore \qquad \rho = \frac{(r^2+r_1^2)^{3/2}}{r^2-rr_2+2r_1^2} = \frac{\left[\dfrac{1}{u^2}+\dfrac{u_1^2}{u^4}\right]^{3/2}}{\dfrac{1}{u^2}-\dfrac{2u_1^2-uu_2}{u^4}+\dfrac{2u_1^2}{u^4}}$$

$$= \frac{(u^2+u_1^2)^{3/2}}{u^6}\cdot\frac{u^4}{u^2-2u_1^2+uu_2+2u_1^2} = \frac{(u^2+u_1^2)^{3/2}}{u^2(u^2+uu_2)} = \frac{(u^2+u_1^2)^{3/2}}{u^3(u+u_2)}$$

ILLUSTRATIVE EXAMPLES

Example 1. *Show that the radius of curvature at any point of the cardioide $r = a(1 + \cos\theta)$ is $\dfrac{2}{3}\sqrt{2ar}$ and prove that $\dfrac{\rho^2}{r}$ is constant.*

Sol. Equation of cardioide is $\quad r = a(1+\cos\theta)$

$$r_1 = -a\sin\theta, \quad r_2 = -a\cos\theta$$

$$\rho = \frac{(r^2+r_1^2)^{3/2}}{r^2-rr_2+2r_1^2} = \frac{[a^2(1+\cos\theta)^2+a^2\sin^2\theta]^{3/2}}{a^2(1+\cos\theta)^2-a(1+\cos\theta)(-a\cos\theta)+2a^2\sin^2\theta}$$

$$= \frac{a^3[(1+\cos\theta)^2+(1-\cos^2\theta)]^{3/2}}{a^2[(1+\cos\theta)^2+\cos\theta(1+\cos\theta)+2(1-\cos^2\theta)]}$$

$$= \frac{a[(1+\cos\theta)(1+\cos\theta+1-\cos\theta)]^{3/2}}{(1+\cos\theta)(1+\cos\theta+\cos\theta+2-2\cos\theta)}$$

$$= \frac{a(1+\cos\theta)^{3/2}\cdot 2^{3/2}}{3(1+\cos\theta)} = \frac{a(1+\cos\theta)^{1/2}\cdot 2\sqrt{2}}{3}$$

$$= \frac{a}{3}\sqrt{\frac{r}{a}} \cdot 2\sqrt{2} \qquad \left| \because \ 1 + \cos\theta = \frac{r}{a} \right.$$

$$= \frac{2}{3}\sqrt{2ar}$$

$$\rho^2 = \frac{8ar}{9} \quad \therefore \quad \frac{\rho^2}{r} = \frac{8a}{9} \text{ which is constant.}$$

Note. Here $\rho = \frac{2}{3}\sqrt{2ar} = \frac{2\sqrt{2a}}{3} \cdot \sqrt{r}$

$\therefore$ ρ varies as $\sqrt{r}$.

Example 2. *Find the radius of curvature at any point of the curve $r^n = a^n \cos n\theta$.* (M.D.U. Dec. 2010)

Sol. Equation of curve is $r^n = a^n \cos n\theta$ ...(1)

Taking logarithms of both sides, we have

$$n \log r = n \log a + \log \cos n\theta$$

Differentiating w.r.t. θ, $\frac{n}{r} \cdot \frac{dr}{d\theta} = 0 + \frac{1}{\cos n\theta}(-n \sin n\theta)$ or $r_1 = \frac{dr}{d\theta} = -r \tan n\theta$

$$\therefore \quad \tan\phi = r\frac{d\theta}{dr} = -\cot n\theta = \tan\left(\frac{\pi}{2} + n\theta\right)$$

$$\phi = \frac{\pi}{2} + n\theta$$

$$p = r \sin\phi = r \sin\left(\frac{\pi}{2} + n\theta\right) = r \cos n\theta = r \cdot \frac{r^n}{a^n} \qquad [\because \text{ Using (1)}]$$

$$\therefore \quad a^n p = r^{n+1} \qquad \text{(Pedal equation)}$$

Differentiating w.r.t. r, $a^n \cdot \frac{dp}{dr} = (n+1) r^n$

$$\therefore \quad \rho = r\frac{dr}{dp} = r \cdot \frac{a^n}{(n+1) r^n} = \frac{a^n}{(n+1) r^{n-1}}.$$

Second Method

Equation of curve is $r^n = a^n \cos n\theta$

Taking logarithms of both sides, we have

$$n \log r = n \log a + \log \cos n\theta$$

Differentiating w.r.t. θ, we have

$$\frac{n}{r} \cdot \frac{dr}{d\theta} = 0 + \frac{1}{\cos n\theta}(-n \sin n\theta)$$

or $$\frac{dr}{d\theta} = -r \tan n\theta \quad \therefore \quad r_1 = -r \tan n\theta$$

Differentiating again w.r.t. θ, we have

$$r_2 = -r(n \sec^2 n\theta) - r_1 \tan n\theta = -nr \sec^2 n\theta - (-r \tan n\theta) \tan n\theta$$

$$= -nr \sec^2 n\theta + r \tan^2 n\theta$$

$$\therefore \quad \rho = \frac{(r^2 + r_1^2)^{3/2}}{r^2 - rr_2 + 2r_1^2} = \frac{(r^2 + r^2 \tan^2 n\theta)^{3/2}}{r^2 - r(-nr \sec^2 n\theta + r \tan^2 n\theta) + 2r^2 \tan^2 n\theta}$$

$$= \frac{(r^2 \sec^2 n\theta)^{3/2}}{r^2 [1 + n \sec^2 n\theta - \tan^2 n\theta + 2 \tan^2 n\theta]}$$

$$= \frac{r^3 \sec^3 n\theta}{r^2 [1+\tan^2 n\theta + n \sec^2 n\theta]} = \frac{r \sec^3 n\theta}{\sec^2 n\theta + n \sec^2 n\theta}$$

$$= \frac{r \sec n\theta}{n+1} = \frac{r}{n+1} \cdot \frac{1}{\cos n\theta} = \frac{r}{n+1} \cdot \frac{a^n}{r^n} = \frac{a^n}{(n+1)\, r^{n-1}}.$$

Example 3. *Find the radius of curvature at any point of the curve* $\theta = \dfrac{\sqrt{r^2 - a^2}}{a} - \cos^{-1} \dfrac{a}{r}$.

Sol. Equation of curve is $\theta = \dfrac{\sqrt{r^2 - a^2}}{a} - \cos^{-1} \dfrac{a}{r}$.

$$\therefore \quad \frac{d\theta}{dr} = \frac{1}{a} \cdot \frac{2r}{2\sqrt{r^2 - a^2}} + \frac{1}{\sqrt{1 - \dfrac{a^2}{r^2}}} \left(-\frac{a}{r^2} \right)$$

$$= \frac{r}{a\sqrt{r^2 - a^2}} - \frac{a}{r\sqrt{r^2 - a^2}} = \frac{r^2 - a^2}{ar\sqrt{r^2 - a^2}} = \frac{\sqrt{r^2 - a^2}}{ar}$$

Since
$$\frac{1}{p^2} = \frac{1}{r^2} + \frac{1}{r^4} \cdot \left(\frac{dr}{d\theta} \right)^2$$

$$\therefore \quad \frac{1}{p^2} = \frac{1}{r^2} + \frac{1}{r^4} \cdot \frac{a^2 r^2}{r^2 - a^2} = \frac{1}{r^2} + \frac{a^2}{r^2(r^2 - a^2)} = \frac{r^2 - a^2 + a^2}{r^2(r^2 - a^2)} = \frac{1}{r^2 - a^2}$$

$$\therefore \quad p^2 = r^2 - a^2 \qquad \text{| Pedal equation}$$

Diff. w.r.t. p,
$$2p = 2r \cdot \frac{dr}{dp}$$

$$\therefore \quad \rho = r\frac{dr}{dp} = p = \sqrt{r^2 - a^2}.$$

Example 4. *Find the radius of curvature at any point (r, θ) on the curve* $\dfrac{l}{r} = 1 + e \cos \theta$.

Sol. $\dfrac{l}{r} = 1 + e \cos \theta$ or $u = \dfrac{1}{l}(1 + e \cos \theta)$ where $u = \dfrac{1}{r}$

$$u_1 = \frac{du}{d\theta} = -\frac{e \sin \theta}{l}; \; u_2 = \frac{d^2u}{d\theta^2} = -\frac{e \cos \theta}{l}$$

$$\therefore \quad \rho = \frac{(u^2 + u_1^2)^{3/2}}{u^3(u + u_2)} = \frac{\left[\dfrac{(1 + e \cos \theta)^2}{l^2} + \dfrac{e^2 \sin^2 \theta}{l^2} \right]^{3/2}}{\dfrac{(1 + e \cos \theta)^3}{l^3} \left[\dfrac{1 + e \cos \theta}{l} - \dfrac{e \cos \theta}{l} \right]}$$

$$= \frac{(1 + e^2 + 2e \cos \theta)^{3/2}}{l^3} \cdot \frac{l^4}{(1 + e \cos \theta)^3} = \frac{l(1 + 2e \cos \theta + e^2)^{3/2}}{(1 + e \cos \theta)^3}.$$

Example 5. *If ϕ be the angle which the radius vector of the curve $r = f(\theta)$ makes with the tangent, prove that* $\rho = \dfrac{r \operatorname{cosec} \phi}{\left(1 + \dfrac{d\phi}{d\theta} \right)}$. *Apply this result to show that* $\rho = \dfrac{a}{2}$ *for the circle* $r = a \cos \theta$.

Sol. We know that $\psi = \theta + \phi$

Differentiating w.r.t. s

$$\frac{d\psi}{ds} = \frac{d\theta}{ds} + \frac{d\phi}{ds} = \frac{d\theta}{ds} + \frac{d\phi}{d\theta} \cdot \frac{d\theta}{ds} \qquad | \text{ Note}$$

$$= \frac{d\theta}{ds}\left[1 + \frac{d\phi}{d\theta}\right]$$

$$\frac{1}{\rho} = \frac{\sin\phi}{r}\left[1 + \frac{d\phi}{d\theta}\right] \qquad \left| \because \sin\phi = r\frac{d\theta}{ds}\right.$$

$$\rho = \frac{r \operatorname{cosec}\phi}{\left[1 + \frac{d\phi}{d\theta}\right]} \qquad \ldots(i)$$

Second Part. Equation of circle is $r = a\cos\theta$

To find ϕ: $\quad \frac{dr}{d\theta} = -a\sin\theta$

$$\tan\phi = r\frac{d\theta}{dr} = -\frac{a\cos\theta}{a\sin\theta} = -\cot\theta = \tan\left(\frac{\pi}{2} + \theta\right)$$

$$\phi = \frac{\pi}{2} + \theta \quad \therefore \quad \frac{d\phi}{d\theta} = 1$$

$\therefore$ From (i), $\quad \rho = \frac{a\cos\theta \cdot \operatorname{cosec}\left(\frac{\pi}{2} + \theta\right)}{1+1} = \frac{a}{2}\cos\theta\sec\theta = \frac{a}{2}$.

EXERCISE 6.2

1. Find the radius of curvature at any point (p, r) on the following curves:
   (i) $p^2 = ar$ (ii) $2ap^2 = r^3$
   (iii) $pa^2 = r^3$ (iv) $\frac{1}{p^2} = \frac{1}{a^2} + \frac{1}{b^2} - \frac{r^2}{a^2b^2}$.
2. In the curve $pa^n = r^{n+1}$, show that the radius of curvature varies inversely as the $(n-1)^{\text{th}}$ power of the radius vector.
3. Show that the radius of curvature at any point of the cardioid $r = a(1 - \cos\theta)$ varies as $\sqrt{r}$.
4. Show that for the curve $r = a(1 + \cos\theta)$, the radius of curvature $\rho = \frac{4a}{3}\cos\frac{\theta}{2}$.
5. If ρ_1, ρ_2 be the radii of curvature at the extremities of any chord of the cardioid $r = a(1 + \cos\theta)$ which passes through the pole, show that ${\rho_1}^2 + {\rho_2}^2 = \frac{16a^2}{9}$. *(P.T.U. May 2009)*
6. Find the radius of curvature at any point (r, θ) of the following curves:
   (i) $r^2\cos 2\theta = a^2$ (ii) $r = ae^{\theta\cot\alpha}$
   (iii) $\frac{2a}{r} = 1 + \cos\theta$ *(K.U.K. Dec. 2013)* (iv) $r^m = a^m \sin m\theta$.
7. For any curve $r = f(\theta)$, prove that $\frac{r}{\rho} = \sin\phi\left(1 + \frac{d\phi}{d\theta}\right)$.

Answers

1. (*i*) $\frac{2r^{3/2}}{\sqrt{a}}$ (*ii*) $\frac{2}{3}\sqrt{2ar}$ (*iii*) $\frac{a^2}{3r}$ (*iv*) $\frac{a^2b^2}{p^3}$

6. (*i*) $\frac{r^3}{a^2}$ (*ii*) r cosec α (*iii*) $\frac{2r^{3/2}}{\sqrt{a}}$ (*iv*) $\frac{a^m}{(m+1)r^{m-1}}$.

6.6. RADIUS OF CURVATURE AT THE ORIGIN

(*a*) Newton's Method

Statement. *If a curve passes through the origin, and axis of x is the tangent at the origin, then*

$$\rho \text{ (at the origin)} = \lim_{\substack{x \to 0 \\ y \to 0}} \left(\frac{x^2}{2y} \right)$$

Proof. Slope of tangent at origin = $\frac{dy}{dx}$ at (0, 0) = $y_1(0)$

Since the tangent at the origin is the axis of x whose slope is 0

$$\therefore \quad y_1(0) = 0$$

$$\therefore \quad \rho \text{ at origin} = \frac{[1 + y_1^2(0)]^{3/2}}{y_2(0)} = \frac{1}{y_2(0)} \quad \ldots(1)$$

Also $\lim_{\substack{x \to 0 \\ y \to 0}} \frac{x^2}{2y}$ | Form $\frac{0}{0}$

$$= \lim_{\substack{x \to 0 \\ y \to 0}} \frac{2x}{2y_1} = \lim_{\substack{x \to 0 \\ y \to 0}} \frac{x}{y_1} \qquad \left| \text{Form } \frac{0}{0}, \ \because \ y_1(0) = 0 \right.$$

$$= \lim_{\substack{x \to 0 \\ y \to 0}} \frac{1}{y_2} = \frac{1}{y_2(0)} \quad \ldots(2)$$

From (1) and (2), ρ at origin = $\lim_{\substack{x \to 0 \\ y \to 0}} \frac{x^2}{2y}$

Note. If the curve passes through the origin and y-axis is the tangent at the origin, then

$$\rho \text{ (at the origin)} = \lim_{\substack{x \to 0 \\ y \to 0}} \left[\frac{y^2}{2x} \right]$$

Cor. Curvature at the origin when polar equation of the curve is given.

If the initial line is the tangent at the pole (origin) then ρ (at the origin or pole) = $\lim_{\substack{x \to 0 \\ y \to 0}} \left[\frac{x^2}{2y} \right]$

$$= \lim_{\substack{r \to 0 \\ \theta \to 0}} \left[\frac{r^2 \cos^2 \theta}{2r \sin \theta} \right] = \lim_{\substack{r \to 0 \\ \theta \to 0}} \left[\frac{r}{2\theta} \cdot \frac{\theta}{\sin \theta} \cdot \cos^2 \theta \right]$$

$$= \lim_{\substack{r \to 0 \\ \theta \to 0}} \left[\frac{r}{2\theta} \right] \qquad \left| \text{Form } \frac{0}{0} \right.$$

$$= \lim_{\substack{r\to 0\\ \theta\to 0}} \frac{\frac{dr}{d\theta}}{2} \qquad \text{| L'Hospital's Rule}$$

$$= \lim_{\substack{r\to 0\\ \theta\to 0}} \left(\frac{1}{2}\frac{dr}{d\theta}\right)$$

Note. Newton's Method for finding the radius of curvature at the origin should be used when x-axis or y-axis is the tangent at the origin.

Tangents at the origin to an algebraic curve are obtained **by equating to zero the lowest degree terms.**

ILLUSTRATIVE EXAMPLES

Example 1. *Find the radius of curvature at the origin for the curves:*

(i) $x^3 - 2x^2y + 3xy^2 - 4y^3 + 5x^2 - 6xy + 7y^2 - 8y = 0$

(ii) $2x^4 + 4x^3y + xy^2 + 6y^3 - 3x^2 - 2xy + y^2 - 4x = 0.$

Sol. *(i)* Equation of curve is $x^3 - 2x^2y + 3xy^2 - 4y^3 + 5x^2 - 6xy + 7y^2 - 8x = 0$ ...(1)

The curve passes through the origin.

Equating to zero the terms of lowest degree, we have y = 0, i.e., x-axis is the tangent at the origin.

$$\therefore \quad \rho \text{ (at the origin)} = \lim_{\substack{x\to 0\\ y\to 0}} \left[\frac{x^2}{2y}\right]$$

Dividing (1) by $2y$, we have $x \cdot \frac{x^2}{2y} - x^2 + \frac{3}{2}xy - 2y^2 + 5\cdot\frac{x^2}{2y} - 3x + \frac{7}{2}y - 4 = 0$

Taking limits as $x \to 0$, $y \to 0$, we have

$$5 \lim_{\substack{x\to 0\\ y\to 0}} \left[\frac{x^2}{2y}\right] - 4 = 0 \qquad \text{(Other terms vanish)}$$

or $\quad 5\rho - 4 = 0 \quad \therefore \quad \rho = \frac{4}{5}.$

(ii) Equation of curve is $2x^4 + 4x^3y + xy^2 + 6y^3 - 3x^2 - 2xy + y^2 - 4x = 0$ ...(1)

The curve passes through the origin.

Equating to zero the terms of lowest degree, we have $x = 0$, *i.e.*, y-axis is the tangent at the origin.

$$\therefore \quad \rho \text{ (at the origin)} = \lim_{\substack{x\to 0\\ y\to 0}} \left[\frac{y^2}{2x}\right]$$

Dividing (1) by $2x$, we have

$$x^3 + 2x^2y + \frac{1}{2}y^2 + 6y\cdot\frac{y^2}{2x} - \frac{3}{2}x - y + \frac{y^2}{2x} - 2 = 0$$

Taking limits as $x \to 0$, $y \to 0$, we have

$$\lim_{\substack{x\to 0\\ y\to 0}} \left[\frac{y^2}{2x}\right] - 2 = 0 \qquad \text{(Other terms vanish)}$$

or $\quad \rho - 2 = 0 \quad \therefore \quad \rho = 2.$

Example 2. *Apply Newton's method to find the radius of curvature at the origin for the cycloid* $x = a(\theta + \sin\theta), \quad y = a(1 - \cos\theta)$.

Sol. $x = a(\theta + \sin\theta), \quad y = a(1 - \cos\theta)$

$$\frac{dx}{d\theta} = a(1 + \cos\theta), \quad \frac{dy}{d\theta} = a\sin\theta$$

$$\frac{dy}{dx} = \frac{\frac{dy}{d\theta}}{\frac{dx}{d\theta}} = \frac{a\sin\theta}{a(1+\cos\theta)} = \frac{2\sin\frac{\theta}{2}\cos\frac{\theta}{2}}{2\cos^2\frac{\theta}{2}} = \tan\frac{\theta}{2}$$

When $\theta = 0$, we have $x = 0$ and $y = 0$

Also $\frac{dy}{dx} = 0$.

$\therefore$ The curve passes through the origin and x-axis is the tangent at the origin.

$$\therefore \quad \rho \text{ (at the origin)} = \lim_{\substack{x\to 0\\ y\to 0}} \left[\frac{x^2}{2y}\right] = \lim_{\theta\to 0} \frac{a^2(\theta + \sin\theta)^2}{2a(1-\cos\theta)} \qquad \left|\ \text{Form } \frac{0}{0}\right.$$

$$= \lim_{\theta\to 0} \frac{a}{2}\cdot\frac{2(\theta + \sin\theta)(1+\cos\theta)}{\sin\theta} \qquad \left|\ \text{Form } \frac{0}{0}\right.$$

$$= \lim_{\theta\to 0} a\cdot\frac{(\theta+\sin\theta)(-\sin\theta) + (1+\cos\theta)(1+\cos\theta)}{\cos\theta}$$

$$= a\cdot\frac{0 + (1+1)(1+1)}{1} = 4a.$$

Example 3. *Find the radius of the curvature of the curve* $r = a\sin n\theta$ *at the pole.*

Sol. Equation of the curve is $r = a\sin n\theta$ ...(*i*)

$$\frac{dr}{d\theta} = na\cos n\theta$$

$$\therefore \quad \tan\phi = r\frac{d\theta}{dr} = \frac{a\sin n\theta}{na\cos n\theta} = \frac{1}{n}\tan n\theta \qquad \text{...(ii)}$$

when $\theta = 0$, from (*i*) $r = 0$ $\quad\therefore$ The curve passes through the pole.

Also from (*ii*), $\tan\phi = 0 \quad \therefore \phi = 0$

$\therefore$ Initial line is the tangent to the curve at pole.

By Newton's method, ρ at pole $= \lim_{\substack{r\to 0\\ \theta\to 0}} \left(\frac{1}{2}\cdot\frac{dr}{d\theta}\right) = \lim_{\substack{r\to 0\\ \theta\to 0}} \left(\frac{1}{2}\cdot na\cos n\theta\right) = \frac{na}{2}$.

(*b*) **Method of Expansion.** When the x-axis or the y-axis is not the tangent at the origin, we use the method of expansion.

Let the equation of the curve be $y = f(x)$

Since it passes through the origin (0, 0) $\quad\therefore \quad f(0) = 0$

By Maclaurin's Expansion,

$$y = f(0) + xf'(0) + \frac{x^2}{2!}f''(0) + \frac{x^3}{3!}f'''(0) + \cdots$$

or

$$y = xf'(0) + \frac{x^2}{2!}f''(0) + \frac{x^3}{3!}f'''(0) + \cdots \qquad [\because \ f(0) = 0]$$

or $$y = px + \frac{1}{2!}qx^2 + \frac{1}{3!}rx^3 + \cdots \quad ...(1)$$

where $p = f'(0) = y_1(0); \quad q = f''(0) = y_2(0)$ etc.

Diff. (1) w.r.t. x, $y_1 = p + qx + \frac{1}{2}rx^2 + \cdots$

Diff. again $y_2 = q + rx + \cdots$

$\therefore$ At the origin, $y_1 = p$ and $y_2 = q$

i.e., $y_1(0) = p$ and $y_2(0) = q$

$$\therefore \quad \rho \text{ at the origin} = \frac{[1 + y_1^2(0)]^{3/2}}{y_2(0)} = \frac{(1 + p^2)^{3/2}}{q}$$

Note. To find the values of p and q

Let the equation of curve be $y = px + \frac{qx^2}{2!} + \cdots$

Put this value of y in the given equation of the curve and equate coefficients of like powers of x on the two sides.

Example 4. *Show that the radius of curvature of the curve* $y^2 = x^2 \frac{(a + x)}{(a - x)}$ *at the origin is* $a\sqrt{2}$. (M.D.U. Dec. 2009)

Sol. Equation of curve is $y^2(a - x) = x^2(a + x)$ ...(1)

It passes through the origin. Equating to zero the lowest degree terms, we have

$$a(y^2 - x^2) = 0 \qquad \therefore \quad y = \pm x \text{ are the tangents at origin.}$$

$\therefore$ Newton's Method is not applicable.

Let $$y = px + \frac{qx^2}{2!} + \cdots$$

Putting this value of y in (1), we get

$$\left(px + \frac{qx^2}{2!} + \cdots\right)^2 (a - x) = x^2(a + x) \quad ...(2)$$

Equating co-effs. of x^2 on both sides, $ap^2 = a \qquad \therefore p = \pm 1$

Equating co-effs. of x^3 on both sides, $-p^2 + apq = 1$ ...(3)

when $p = 1$, from (3), $-1 + aq = 1$ or $q = \frac{2}{a}$

$$\therefore \quad \rho \text{ at the origin} = \frac{(1 + p^2)^{3/2}}{q} = \frac{(1 + 1)^{3/2}}{\frac{2}{a}} = 2\sqrt{2} \cdot \frac{a}{2} = \sqrt{2}a$$

when $p = -1$, from (3), $-1 - aq = 1$ or $q = -\frac{2}{a}$

$$\therefore \quad \rho \text{ at the origin} = \frac{(1 + p^2)^{3/2}}{q} = \frac{(1 + 1)^{3/2}}{-\frac{2}{a}} = \sqrt{2}a \qquad \text{(In magnitude)}$$

Hence ρ at the origin is $\sqrt{2}a$.

EXERCISE 6.3

Find the radius of curvature at the origin for the following curves:

1. $x^3 + y^2 - 2x^2 + 6y = 0$.
2. $2x^4 + 3y^4 + 4x^2y + xy - y^2 + 2x = 0$.
3. $2x^3 + 4x^2y + xy^2 + 5y^3 - x^2 - 2xy + y^2 + 4x = 0$.
4. $y^4 + x^3 + a(x^2 + y^2) - a^2y = 0$.
5. $x^3 + y^3 = 3axy$.
6. $y = 6x + 5x^2 + x^3$.
7. $y = x^2 + 2xy + y^2 + x$. *(M.D.U. Dec. 2011)*
8. $a(y^2 - x^2) = x^3$.

Answers

1. $\frac{3}{2}$ **2.** 1 **3.** 2 **4.** $\frac{a}{2}$

5. $\frac{3a}{2}$ **6.** $\frac{37\sqrt{37}}{10}$ **7.** $\frac{1}{2\sqrt{2}}$ **8.** $2\sqrt{2}\,a$.

6.7. CENTRE OF CURVATURE

To find the co-ordinates of the centre of curvature for any point (x, y) of the curve y = f(x).

Let C(X, Y) be the centre of curvature corresponding to any point P(x, y) on the curve, then

$$PC = \rho.$$

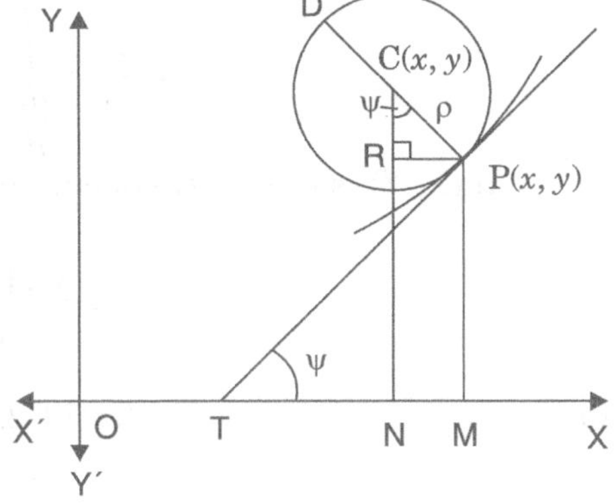

Let the tangent PT make an angle ψ with the +ve direction of x-axis. Draw PM and CN ⊥ s on x-axis and PR ⊥ CN.

$$\angle PCN = 90° - \angle CPR$$
$$= 90° - (90° - \angle RPT)$$
$$= \angle RPT = \angle PTX = \psi$$
$$X = ON = OM - NM = OM - RP$$
$$= x - CP \sin\psi = x - \rho \sin\psi$$
$$Y = NC = NR + RC = MP + RC$$
$$= y + CP\cos\psi = y + \rho\cos\psi$$

Now $$\tan\psi = \frac{dy}{dx} = y_1$$

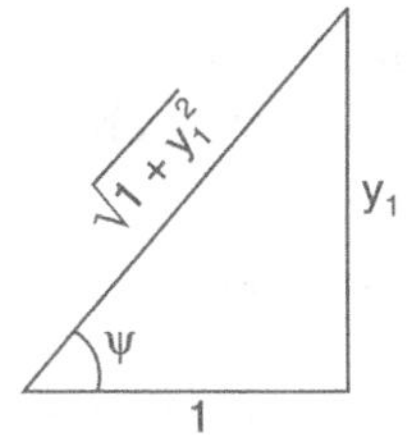

∴ $$\sin\psi = \frac{y_1}{\sqrt{1+y_1^2}} \quad \text{and} \quad \cos\psi = \frac{1}{\sqrt{1+y_1^2}}$$

Also $$\rho = \frac{(1+y_1^2)^{3/2}}{y_2}$$

∴ $$X = x - \frac{y_1(1+y_1^2)}{y_2}, \quad Y = y + \frac{1+y_1^2}{y_2}.$$

Cor. *To find the equation of the circle of curvature at a given point of the curve.*

Let ρ be the radius of curvature and (α, β) the centre of curvature at the given point. Then the equation of the circle of curvature is given by $(x - \alpha)^2 + (y - \beta)^2 = \rho^2$.

6.8. EVOLUTE AND INVOLUTE

The locus of the centres of curvature of the given curve is called the evolute of the curve and the curve itself is called the involute of its evolute.

To find the evolute of a curve:

Rule 1. Let (X, Y) be the co-ordinates of the centre of curvature, then

$$X = x - \frac{y_1(1+y_1^2)}{y_2}, \qquad Y = y + \frac{1+y_1^2}{y_2}$$

2. Eliminate the parameters x and y to find a relation between X and Y.

3. Generalise X and Y to get the required equation of the evolute.

ILLUSTRATIVE EXAMPLES

Example 1. *Find the co-ordinates of the centre of curvature for any point (x, y) on the parabola $y^2 = 4ax$. Also find the equation of the evolute of the parabola.*

Sol. The parametric equations of the parabola $y^2 = 4ax$ are

$$x = at^2, \qquad y = 2at$$

$$\therefore \qquad \frac{dx}{dt} = 2at, \qquad \frac{dy}{dt} = 2a$$

$$y_1 = \frac{dy}{dx} = \frac{\frac{dy}{dt}}{\frac{dx}{dt}} = \frac{2a}{2at} = \frac{1}{t}; \qquad y_2 = -\frac{1}{t^2}\cdot\frac{dt}{dx} = -\frac{1}{t^2}\cdot\frac{1}{2at} = -\frac{1}{2at^3}$$

If (X, Y) is the centre of curvature at 't', then

$$X = x - \frac{y_1(1+y_1^2)}{y_2} = at^2 - \frac{\frac{1}{t}\left(1+\frac{1}{t^2}\right)}{\frac{-1}{2at^3}} = at^2 + 2at^2\left(1+\frac{1}{t^2}\right)$$

or $\qquad X = 3at^2 + 2a \qquad \ldots(1)$

and

$$Y = y + \frac{1+y_1^2}{y_2} = 2at + \frac{1+\frac{1}{t^2}}{\frac{-1}{2at^3}} = 2at - 2at^3\left(1+\frac{1}{t^2}\right) = 2at - 2at^3 - 2at$$

or $\qquad Y = -2at^3 \qquad \ldots(2)$

$\therefore$ The co-ordinates of the centre of curvature at any point (x, y) *i.e.*, at any point 't' are

$$(2a + 3at^2, -2at^3) = \left(2a + 3x, -\frac{2x^{3/2}}{\sqrt{a}}\right). \qquad \left[\because\ t = \sqrt{\frac{x}{a}}\right]$$

Now evolute of a curve is the locus of its centres of curvature.

To eliminate the parameter t between (1) and (2)

From (1), $\qquad t^2 = \dfrac{X - 2a}{3a} \qquad \ldots(3)$

From (2), $\qquad t^3 = -\dfrac{Y}{2a} \qquad \ldots(4)$

Cubing (3), squaring (4) and equating the two values of t^6, we get

$$\frac{(X-2a)^3}{27a^3} = \frac{Y^2}{4a^2} \quad \text{or} \quad 27aY^2 = 4(X-2a)^3$$

Changing X to x and Y to y, the locus of (X, Y) *i.e.*, the evolute of the parabola is

$$27ay^2 = 4(x-2a)^3.$$

Second Method

Equation of parabola is $y^2 = 4ax$

$$\therefore \quad 2yy_1 = 4a \quad \text{or} \quad y_1 = \frac{2a}{y}$$

and
$$y_2 = -\frac{2a}{y^2}\cdot y_1 = -\frac{4a^2}{y^3}$$

If (X, Y) is the centre of curvature at (x, y), then

$$X = x - \frac{y_1(1+y_1^2)}{y_2} = x - \frac{\frac{2a}{y}\left(1+\frac{4a^2}{y^2}\right)}{-\frac{4a^2}{y^3}}$$

$$= x + \frac{y^2}{2a}\left(1+\frac{4a^2}{y^2}\right) = x + \frac{y^2}{2a} + 2a$$

$$= x + \frac{4ax}{2a} + 2a \qquad [\because \; y^2 = 4ax]$$

$$= 3x + 2a \qquad \text{...(1)}$$

and
$$Y = y + \frac{1+y_1^2}{y_2} = y + \frac{1+\frac{4a^2}{y^2}}{-\frac{4a^2}{y^3}}$$

$$= y - \frac{y^3}{4a^2}\left(1+\frac{4a^2}{y^2}\right) = y - \frac{y^3}{4a^2} - y$$

$$= -\frac{(2\sqrt{ax})^3}{4a^2} \qquad [\because \; y = 2\sqrt{ax}]$$

$$= -\frac{2x^{3/2}}{\sqrt{a}} \qquad \text{...(2)}$$

$\therefore$ The centre of curvature at (x, y) is $\left(3x+2a, -\dfrac{2x^{3/2}}{\sqrt{a}}\right)$

To find the evolute, we have to eliminate x from (1) and (2).

Now
$$Y^2 = \frac{4x^3}{a} = \frac{4}{a}\left(\frac{X-2a}{3}\right)^3$$

or
$$27aY^2 = 4(X-2a)^3$$

$\therefore$ The locus of (X, Y), *i.e.*, the evolute of parabola is

$$27ay^2 = 4(x - 2a)^3$$

obtained by changing X to x and Y to y.

Example 2. *Find the co-ordinates of the centre of curvature of the ellipse $x = a \cos \theta$, $y = b \sin \theta$. Hence show that the equation of its evolute is $(ax)^{2/3} + (by)^{2/3} = (a^2 - b^2)^{2/3}$.*

Or

Find the co-ordinates of the centre of curvature at any point (x, y) on the ellipse $\frac{x^2}{a^2} + \frac{y^2}{b^2} = 1$. Also find the evolute of the ellipse.

Sol. Equations of ellipse are $x = a \cos \theta$, $y = b \sin \theta$

$$\therefore \quad \frac{dx}{d\theta} = -a \sin \theta, \quad \frac{dy}{d\theta} = b \cos \theta$$

$$y_1 = \frac{dy}{dx} = \frac{\frac{dy}{d\theta}}{\frac{dx}{d\theta}} = \frac{b \cos \theta}{-a \sin \theta} = -\frac{b}{a} \cot \theta$$

$$y_2 = \frac{b}{a} \operatorname{cosec}^2 \theta \cdot \frac{d\theta}{dx} = \frac{b}{a} \operatorname{cosec}^2 \theta \cdot \frac{1}{-a \sin \theta} = -\frac{b}{a^2} \operatorname{cosec}^3 \theta$$

Let (X, Y) be the co-ordinates of the centre of curvature. Then

$$X = x - \frac{y_1(1 + y_1^2)}{y_2} = a \cos \theta - \frac{-\frac{b}{a} \cot \theta \left(1 + \frac{b^2}{a^2} \cot^2 \theta\right)}{-\frac{b}{a^2} \operatorname{cosec}^3 \theta}$$

$$= a \cos \theta - a \cot \theta \sin^3 \theta \left(1 + \frac{b^2}{a^2} \cot^2 \theta\right)$$

$$= a \cos \theta - a \sin^2 \theta \cos \theta - \frac{b^2}{a} \cos^3 \theta$$

$$= a \cos \theta (1 - \sin^2 \theta) - \frac{b^2}{a} \cos^3 \theta = a \cos^3 \theta - \frac{b^2}{a} \cos^3 \theta$$

$$= \frac{a^2 - b^2}{a} \cos^3 \theta \qquad \text{...}(i)$$

$$Y = y + \frac{1 + y_1^2}{y_2} = b \sin \theta + \frac{1 + \frac{b^2}{a^2} \cot^2 \theta}{-\frac{b}{a^2} \operatorname{cosec}^3 \theta}$$

$$= b \sin \theta - \frac{a^2}{b} \sin^3 \theta \left(1 + \frac{b^2}{a^2} \cot^2 \theta\right) = b \sin \theta - \frac{a^2}{b} \sin^3 \theta - b \sin \theta \cos^2 \theta$$

$$= b \sin \theta [1 - \cos^2 \theta] - \frac{a^2}{b} \sin^3 \theta = b \sin^3 \theta - \frac{a^2}{b} \sin^3 \theta$$

$$= \frac{b^2 - a^2}{b} \sin^3 \theta = -\frac{a^2 - b^2}{b} \sin^3 \theta \qquad ...(ii)$$

Equations (*i*) and (*ii*) give the centre of curvature.

To find the equation of the evolute, we have to eliminate θ between (*i*) and (*ii*)

From (*i*) $\quad a\text{X} = (a^2 - b^2) \cos^3 \theta \quad \therefore \quad (a\text{X})^{2/3} = (a^2 - b^2)^{2/3} \cos^2 \theta \qquad ...(iii)$

From (*ii*) $\quad b\text{Y} = -(a^2 - b^2) \sin^3 \theta \therefore \quad (b\text{Y})^{2/3} = (a^2 - b^2)^{2/3} \sin^2 \theta \qquad ...(iv)$

Adding (*iii*) and (*iv*) $\quad (a\text{X})^{2/3} + (b\text{Y})^{2/3} = (a^2 - b^2)^{2/3}$

∴ Locus of (X, Y) is $\quad (ax)^{2/3} + (by)^{2/3} = (a^2 - b^2)^{2/3}$

which is the required evolute.

Example 3. *Show that equation of the evolute of the curve* $x^{2/3} + y^{2/3} = a^{2/3}$ *is*

$$(x + y)^{2/3} + (x - y)^{2/3} = 2a^{2/3}.$$

Sol. The parametric equations of the curve are

$$x = a \cos^3 \theta, \quad y = a \sin^3 \theta$$

$$\frac{dx}{d\theta} = -3a \cos^2 \theta \sin \theta, \quad \frac{dy}{d\theta} = 3a \sin^2 \theta \cos \theta$$

$$\therefore \quad \frac{dy}{dx} = \frac{\frac{dy}{d\theta}}{\frac{dx}{d\theta}} = \frac{3a \sin^2 \theta \cos \theta}{-3a \cos^2 \theta \sin \theta} = -\tan \theta$$

$$\frac{d^2y}{dx^2} = -\sec^2 \theta \cdot \frac{d\theta}{dx} = -\frac{1}{\cos^2 \theta} \cdot \frac{1}{-3a \cos^2 \theta \sin \theta} = \frac{1}{3a \sin \theta \cos^4 \theta}$$

Let (X, Y) be the co-ordinates of the centre of curvature, then

$$\text{X} = x - \frac{y_1(1 + y_1^2)}{y_2} = a \cos^3 \theta - \frac{-\tan \theta \,(1 + \tan^2 \theta)}{\frac{1}{3a \sin \theta \cos^4 \theta}}$$

$$= a \cos^3 \theta + 3a \sin \theta \cos^4 \theta \cdot \tan \theta \cdot \sec^2 \theta = a \cos^3 \theta + 3a \sin^2 \theta \cos \theta$$

$$= a \cos \theta \,(\cos^2 \theta + 3 \sin^2 \theta) \qquad ...(i)$$

$$\text{Y} = y + \frac{1 + y_1^2}{y_2} = a \sin^3 \theta + \frac{1 + \tan^2 \theta}{\frac{1}{3a \sin \theta \cos^4 \theta}}$$

$$= a \sin^3 \theta + 3a \sin \theta \cos^4 \theta \cdot \sec^2 \theta = a \sin^3 \theta + 3a \sin \theta \cos^2 \theta$$

$$= a \sin \theta \,(\sin^2 \theta + 3 \cos^2 \theta) \qquad ...(ii)$$

(*i*) and (*ii*) give the co-ordinates of the centre of curvature.

To find the equation of the evolute, we have to eliminate θ between (*i*) and (*ii*)

$$\text{X} + \text{Y} = a(\cos^3 \theta + \sin^3 \theta) + 3a \sin \theta \cos \theta \,(\sin \theta + \cos \theta)$$

$$= a[(\cos \theta + \sin \theta)^3 - 3 \cos \theta \sin \theta \,(\cos \theta + \sin \theta)] + 3a \sin \theta \cos \theta \,(\sin \theta + \cos \theta)$$

$$= a[\cos \theta + \sin \theta]^3$$

$$\therefore \quad (\text{X} + \text{Y})^{2/3} = a^{2/3} [\cos \theta + \sin \theta]^2 \qquad ...(iii)$$

Also

$$\text{X} - \text{Y} = a[(\cos^3 \theta - \sin^3 \theta) - 3 \sin \theta \cos \theta \,(\cos \theta - \sin \theta)]$$

$$= a[(\cos \theta - \sin \theta)^3 + 3 \cos \theta \sin \theta \,(\cos \theta - \sin \theta)] - 3 \sin \theta \cos \theta \,(\cos \theta - \sin \theta)]$$

$$= a(\cos\theta - \sin\theta)^3$$

$\therefore \quad (X - Y)^{2/3} = a^{2/3}(\cos\theta - \sin\theta)^2$ ...(iv)

Adding (*iii*) and (*iv*)

$$(X + Y)^{2/3} + (X - Y)^{2/3} = a^{2/3}[(\cos\theta + \sin\theta)^2 + (\cos\theta - \sin\theta)^2]$$

$$= a^{2/3} \cdot 2(\cos^2\theta + \sin^2\theta) = 2a^{2/3}$$

$\therefore$ Locus of (X, Y) is $(x + y)^{2/3} + (x - y)^{2/3} = 2a^{2/3}$, which is the required evolute.

Example 4. *Show that the equation of the circle of curvature at the point* $\left(\frac{a}{4}, \frac{a}{4}\right)$ *on the curve* $\sqrt{x} + \sqrt{y} = \sqrt{a}$ *is* $\left(x - \frac{3}{4}a\right)^2 + \left(y - \frac{3}{4}a\right)^2 = \frac{1}{2}a^2.$

Sol. Equation of curve is $\quad \sqrt{x} + \sqrt{y} = \sqrt{a}$ ...(i)

Diff. w.r.t. x, $\quad \frac{1}{2}x^{-1/2} + \frac{1}{2}y^{-1/2} \cdot y_1 = 0.$ ...(ii)

Diff. again, $\quad -\frac{1}{4}x^{-3/2} - \frac{1}{4}y^{-3/2} \cdot y_1 \cdot y_1 + \frac{1}{2}y^{-1/2}y_2 = 0$ ...(iii)

At the point $\left[\frac{a}{4}, \frac{a}{4}\right]$

From (*ii*), $\quad \frac{1}{2} \cdot \frac{2}{\sqrt{a}} + \frac{1}{2}\frac{2}{\sqrt{a}}y_1 = 0 \qquad \therefore \quad y_1 = -1$

from (*iii*), $\quad -\frac{1}{4} \cdot \frac{4}{a} \cdot \frac{2}{\sqrt{a}} - \frac{1}{4} \cdot \frac{4}{a} \cdot \frac{2}{\sqrt{a}} \cdot (-1)^2 + \frac{1}{2}\frac{2}{\sqrt{a}}y_2 = 0$

or $\quad -\frac{4}{a\sqrt{a}} + \frac{1}{\sqrt{a}}y_2 = 0 \qquad \therefore \quad y_2 = \frac{4}{a}$

$$\rho \text{ (at the given point)} = \frac{(1 + y_1^2)^{3/2}}{y_2} = \frac{(1+1)^{3/2}}{4/a} = 2\sqrt{2}\,\frac{a}{4} = \frac{a}{\sqrt{2}}$$

Let (α, β) be the centre of curvature at $\left[\frac{a}{4}, \frac{a}{4}\right]$. Then

$$\alpha = x - \frac{y_1(1 + y_1^2)}{y_2} = \frac{a}{4} - \frac{-(1+1)}{4/a} = \frac{a}{4} + \frac{a}{2} = \frac{3}{4}a$$

$$\beta = y + \frac{1 + y_1^2}{y_2} = \frac{a}{4} + \frac{1+1}{\frac{4}{a}} = \frac{a}{4} + \frac{a}{2} = \frac{3}{4}a$$

Equation of circle of curvature is

$$(x - \alpha)^2 + (y - \beta)^2 = \rho^2 \quad i.e., \quad \left[x - \frac{3}{4}a\right]^2 + \left[y - \frac{3}{4}a\right]^2 = \frac{1}{2}a^2.$$

Example 5. *Show that the evolute of the tractrix* $x = c\cos t + c\log\tan\frac{t}{2}$, $y = c\sin t$ *is the catenary* $y = c\cosh\frac{x}{c}$.

Sol. Equation of curve is $\quad x = c\cos t + c\log\tan\frac{t}{2}, \; y = c\sin t$

$$\frac{dx}{dt} = -c \sin t + \frac{c}{\tan \frac{t}{2}} \cdot \frac{1}{2} \sec^2 \frac{t}{2} = -c \sin t + \frac{c \cos \frac{t}{2}}{\sin \frac{t}{2}} \cdot \frac{1}{2 \cos^2 \frac{t}{2}}$$

$$= -c \sin t + \frac{c}{2 \sin \frac{t}{2} \cos \frac{t}{2}} = -c \sin t + \frac{c}{\sin t} = \frac{c(1-\sin^2 t)}{\sin t} = \frac{c \cos^2 t}{\sin t}.$$

$$\frac{dy}{dt} = c \cos t$$

$$\therefore \quad y_1 = \frac{dy}{dx} = \frac{\frac{dy}{dt}}{\frac{dx}{dt}} = c \cos t \cdot \frac{\sin t}{c \cos^2 t} = \tan t$$

$$y_2 = \sec^2 t \cdot \frac{dt}{dx} = \frac{1}{\cos^2 t} \cdot \frac{\sin t}{c \cos^2 t} = \frac{\sin t}{c \cos^4 t}$$

If (X, Y) is the centre of curvature at any point on the curve, then

$$X = x - \frac{y_1(1+y_1^2)}{y_2} = c \cos t + c \log \tan \frac{t}{2} - \frac{\tan t(1+\tan^2 t)}{\frac{\sin t}{c \cos^4 t}}$$

$$= c \cos t + c \log \tan \frac{t}{2} - \frac{c \cos^4 t}{\sin t} \cdot \frac{\sin t}{\cos t} \cdot \frac{1}{\cos^2 t}$$

$$= c \cos t + c \log \tan \frac{t}{2} - c \cos t$$

or $$X = c \log \tan \frac{t}{2} \quad \text{...(1)}$$

and $$Y = y + \frac{1+y_1^2}{y_2} = c \sin t + \frac{1+\tan^2 t}{\frac{\sin t}{c \cos^4 t}}$$

$$= c \sin t + \frac{c \cos^4 t}{\sin t} \cdot \sec^2 t = c \sin t + \frac{c \cos^2 t}{\sin t} = \frac{c(\sin^2 t + \cos^2 t)}{\sin t}$$

or $$Y = \frac{c}{\sin t} \quad \text{...(2)}$$

Evolute of the given curve is the locus of (X, Y). Let us eliminate t between (1) and (2)

From (1), $\log \tan \frac{t}{2} = \frac{X}{c} \Rightarrow \tan \frac{t}{2} = e^{\frac{X}{c}}$

From (2), $$\frac{Y}{c} = \frac{1}{\sin t} = \frac{1+\tan^2 \frac{t}{2}}{2 \tan \frac{t}{2}}$$

$$\Rightarrow \quad \frac{Y}{c} = \frac{1}{2}\left(\frac{1}{\tan \frac{t}{2}} + \tan \frac{t}{2}\right) = \frac{1}{2}\left(e^{-\frac{X}{c}} + e^{\frac{X}{c}}\right) \Rightarrow Y = c \cosh \frac{X}{c}$$

Changing X to x, Y to y, the locus of (X, Y) is $y = c \cosh \dfrac{x}{c}$ which is the required equation of evolute.

6.9. CHORD OF CURVATURE

Let PQ be any chord through P making an angle α with CP, the radius of *curvature* at P.

Then from rt. ∠d. ΔPQD

Chord of curvature = PQ = PD cos α

$= 2\rho \cos \alpha$

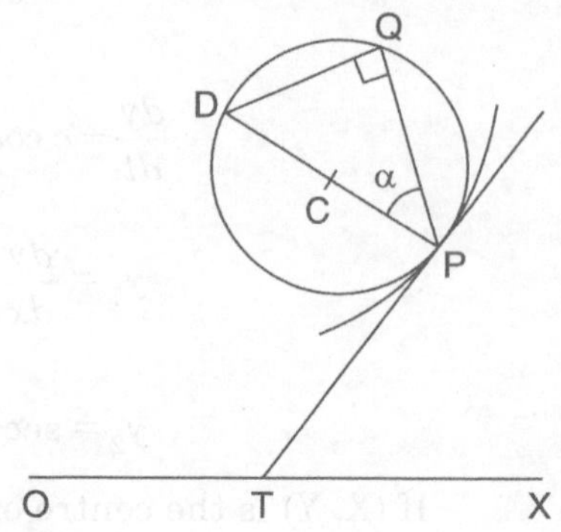

6.10. CHORDS OF CURVATURE PARALLEL TO THE AXES

Let the tangent at P make an angle ψ with the x-axis. Then chord of curvature PA parallel to x-axis makes angle 90° – ψ with CP, the radius of curvature at P and the chord of curvature PB, parallel to y-axis, makes angle ψ with CP.

C_x = length of the chord of curvature PA, parallel to x-axis

$$= \text{PD} \cos (90° - \psi) = 2\rho \sin \psi$$

$$= 2\, \frac{(1+y_1^2)^{3/2}}{y_2} \cdot \frac{y_1}{\sqrt{1+y_1^2}}$$

$\because \tan \psi = y_1$

$\therefore \sin \psi = \dfrac{y_1}{\sqrt{1+y_1^2}}$

$$= \frac{2y_1(1+y_1^2)}{y_2}$$

C_y = length of the chord of curvature PB, parallel to y-axis

$$= \text{PD} \cos \psi = 2\rho \cos \psi = 2 \cdot \frac{(1+y_1^2)^{3/2}}{y_2} \cdot \frac{1}{\sqrt{1+y_1^2}}$$

$$= \frac{2(1+y_1^2)}{y_2}.$$

Example. *If C_x and C_y be the chords of curvature parallel to the axis of x and y respectively at any point of the curve $y = ae^{x/a}$, prove that* $\dfrac{1}{C_x^2} + \dfrac{1}{C_y^2} = \dfrac{1}{2aC_x}$.

Sol. Equation of curve is $y = ae^{x/a}$ ...(*i*)

$$y_1 = ae^{x/a} \cdot \frac{1}{a} = e^{x/a} = \frac{y}{a}, \quad y_2 = \frac{1}{a} \cdot y_1 = \frac{y}{a^2}$$

$$\rho = \frac{(1+y_1^2)^{3/2}}{y_2} = \frac{\left[1+\frac{y^2}{a^2}\right]^{3/2}}{\frac{y}{a^2}} = \frac{(a^2+y^2)^{3/2}}{a^3} \cdot \frac{a^2}{y} = \frac{(a^2+y^2)^{3/2}}{ay}$$

$$\tan\psi = y_1 = \frac{y}{a}$$

$$\therefore \quad \sin\psi = \frac{y}{\sqrt{a^2+y^2}}$$

$$\cos\psi = \frac{a}{\sqrt{a^2+y^2}}$$

$$C_x = 2\rho \sin\psi = 2 \cdot \frac{(a^2+y^2)^{3/2}}{ay} \cdot \frac{y}{\sqrt{a^2+y^2}} = \frac{2(a^2+y^2)}{a}$$

$$C_y = 2\rho \cos\psi = 2 \cdot \frac{(a^2+y^2)^{3/2}}{ay} \cdot \frac{a}{\sqrt{a^2+y^2}} = \frac{2(a^2+y^2)}{y}$$

$$\therefore \quad \frac{1}{C_x^2} + \frac{1}{C_y^2} = \frac{a^2}{4(a^2+y^2)^2} + \frac{y^2}{4(a^2+y^2)^2} = \frac{a^2+y^2}{4(a^2+y^2)^2}$$

$$= \frac{1}{4(a^2+y^2)} = \frac{1}{2aC_x} \qquad \left|\ \because\ C_x = \frac{2(a^2+y^2)}{a}\right.$$

6.11. CHORD OF CURVATURE: POLAR CO-ORDINATES

To find the length of chord of curvature through the pole and perpendicular to the radius vector.

Let the tangent at P make an angle ϕ with the radius vector OP.

PL, the chord of curvature through pole O makes an angle $90° - \phi$ with CP, the radius of curvature and PM, the chord of curvature ⊥ to the radius vector OP makes angle ϕ with CP.

C_0 = length of chord of curvature through the pole

$= \text{PL} = \text{PD} \cos(90° - \phi)$

$= 2\rho \sin\phi$

$$= 2 \cdot \frac{(r^2+r_1^2)^{3/2}}{r^2 - rr_2 + 2r_1^2} \cdot \frac{r}{\sqrt{r^2+r_1^2}}$$

$$\left|\ \because\ \tan\phi = \frac{r}{r_1} \quad \therefore\ \sin\phi = \frac{r}{\sqrt{r^2+r_1^2}}\right.$$

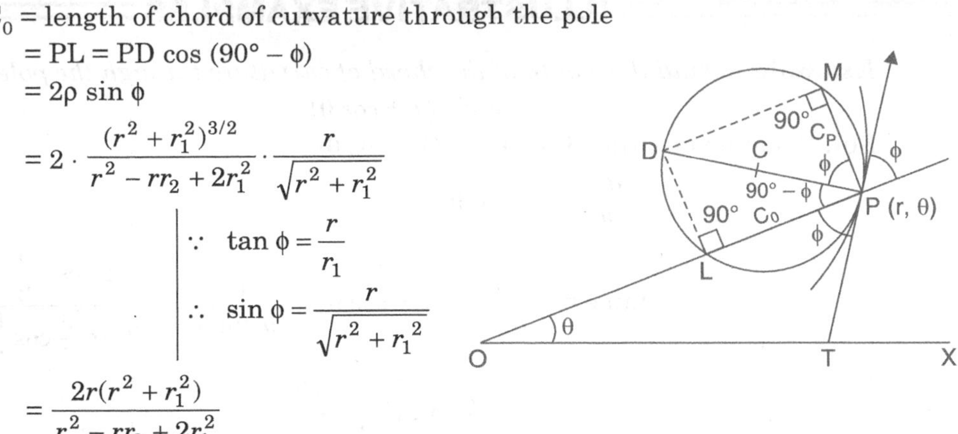

$$= \frac{2r(r^2+r_1^2)}{r^2 - rr_2 + 2r_1^2}$$

C_p = length of chord of *curvature* ⊥ to radius vector

$$= PM = PD \cos \phi = 2\rho \cos \phi$$

$$= 2 \cdot \frac{(r^2 + r_1^2)^{3/2}}{r^2 - rr_2 + 2r_1^2} \cdot \frac{r_1}{\sqrt{r^2 + r_1^2}} = \frac{2r_1(r^2 + r_1^2)}{r^2 - rr_2 + 2r_1^2}.$$

6.12. PEDAL FORM

Show that the chord of curvature through the pole for the curve p = f(r) is given by

$$\mathbf{C_0 = 2p \frac{dr}{dp} = \frac{2\,f(r)}{f'(r)}}$$

Proof. $C_0 = 2\rho \sin \phi$ ...(*i*)

But $\rho = r \dfrac{dr}{dp}$ and $\sin \phi = \dfrac{p}{r}$

∴ From (*i*), $C_0 = 2r \dfrac{dr}{dp} \cdot \dfrac{p}{r} = 2p \dfrac{dr}{dp}$ ...(*ii*)

Also ∵ $p = f(r)$

∴ $\dfrac{dp}{dr} = f'(r), \quad \sin \phi = \dfrac{p}{r} = \dfrac{f(r)}{r}$

From (*i*), $C_0 = 2\rho \sin \phi = 2r \cdot \dfrac{dr}{dp} \sin \phi = 2r \cdot \dfrac{1}{f'(r)} \cdot \dfrac{f(r)}{r} = \dfrac{2f(r)}{f'(r)}$ ...(*iii*)

From (*i*), (*ii*) and (*iii*), $C_0 = 2p \cdot \dfrac{dr}{dp} = \dfrac{2f(r)}{f'(r)}$.

Note. $C_p = 2\rho \cos \phi = 2r \cdot \dfrac{dr}{dp} \cdot \sqrt{1 - \sin^2 \phi}$

$$= 2r \sqrt{1 - \frac{p^2}{r^2}} \cdot \frac{dr}{dp} \qquad \left| \because \sin \phi = \frac{p}{r} \right.$$

$$= 2\sqrt{r^2 - p^2} \cdot \frac{dr}{dp}.$$

ILLUSTRATIVE EXAMPLES

Example 1. *Find the length of the chord of curvature through the pole for the cardioid $r = a\,(1 + \cos \theta)$.*

Sol. Equation of cardioid is $r = a\,(1 + \cos \theta)$ ...(*i*)

∴ $\dfrac{dr}{d\theta} = -\,a \sin \theta$

$$\tan \phi = r \frac{d\theta}{dr} = a(1 + \cos \theta) \cdot \frac{1}{-a \sin \theta} = -\frac{2 \cos^2 \frac{\theta}{2}}{2 \sin \frac{\theta}{2} \cos \frac{\theta}{2}}$$

$$= -\cot \frac{\theta}{2} = \tan \left[\frac{\pi}{2} + \frac{\theta}{2}\right] \qquad \left| \because \phi = \frac{\pi}{2} + \frac{\theta}{2}. \right.$$

$$p = r \sin \phi = r \sin \left[\frac{\pi}{2} + \frac{\theta}{2}\right] = r \cos \frac{\theta}{2}$$

$$p^2 = r^2 \cos^2 \frac{\theta}{2} = r^2 \cdot \frac{1 + \cos \theta}{2} = \frac{r^2}{2} \cdot \frac{r}{a}$$

$\therefore \quad 2ap^2 = r^3 \quad$ (Pedal equation)

$$p = \frac{r^{3/2}}{\sqrt{2a}} \qquad \therefore \quad \frac{dp}{dr} = \frac{3}{2} \cdot \frac{r^{1/2}}{\sqrt{2a}}$$

$$\therefore \quad C_0 = 2p \frac{dr}{dp} = 2 \cdot \frac{r^{3/2}}{\sqrt{2a}} \cdot \frac{2}{3} \cdot \frac{\sqrt{2a}}{r^{1/2}} = \frac{4}{3} r.$$

Example 2. *Prove that the points on the curve $r = f(\theta)$, the circle of curvature at which passes through the pole, are given by the equation $f(\theta) + f''(\theta) = 0$.*

Sol. Since the circle of curvature passes through the pole, the chord of curvature through the pole is OP = r.

But the chord of curvature through the pole.

$$= \frac{2r(r^2 + r_1^2)}{r^2 - rr_2 + 2r_1^2}$$

where the suffixes denote the order of differentiation w.r.t. θ.

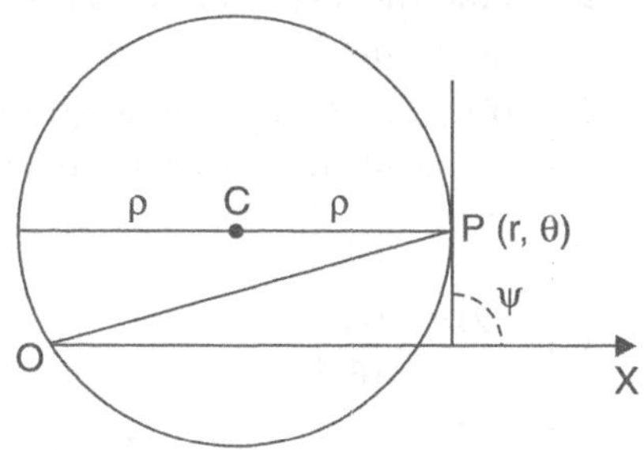

$$\therefore \quad \frac{2r(r^2 + r_1^2)}{r^2 - rr_2 + 2r_1^2} = r$$

$$\Rightarrow \quad 2(r^2 + r_1^2) = r^2 - rr_2 + 2r_1^2$$

$$\Rightarrow \quad r^2 + rr_2 = 0 \quad \Rightarrow \quad r + r_2 = 0$$

$$\Rightarrow \quad f(\theta) + f''(\theta) = 0.$$

EXERCISE 6.4

1. Find the centre of curvature for the point $\left(\frac{3a}{2}, \frac{3a}{2}\right)$ on the folium of Descartes $x^3 + y^3 = 3axy$.
2. Find the coordinates of the centre of curvature at $(at^2, 2at)$ on the parabola $y^2 = 4ax$.
3. Show that the equation of the evolute of the parabola $x^2 = 4ay$ is $4(y - 2a)^3 = 27ax^2$.
4. (*a*) If the centre of curvature of the ellipse $\frac{x^2}{a^2} + \frac{y^2}{b^2} = 1$ at one end of the minor axis lies at the other end, then show that the eccentricity of the ellipse is $\frac{1}{\sqrt{2}}$.

   (*b*) Find the coordinates of the centre of curvature for any point on the hyperbola $\frac{x^2}{a^2} - \frac{y^2}{b^2} = 1$. Hence find its evolute.

   [**Hint.** Take the parametric equations of the hyperbola as $x = a \sec \theta$, $y = b \tan \theta$]

5. Show that the evolute of the rectangular hyperbola $xy = c^2 \left(i.e.,\ x = ct,\ y = \dfrac{c}{t}\right)$ is the curve

$$(x + y)^{2/3} - (x - y)^{2/3} = (4c)^{2/3}.$$

6. Show that the evolute of the cycloid $x = a(\theta - \sin\theta)$, $y = a(1 - \cos\theta)$ is another equal cycloid.

7. Show that the evolute of the curve $x = a(\cos\theta + \theta\sin\theta)$, $y = a(\sin\theta - \theta\cos\theta)$ is $x^2 + y^2 = a^2$.

8. Find the equation of the circle of curvature for the curve $x^3 + y^3 = 3xy$ at the point $\left(\dfrac{3}{2}, \dfrac{3}{2}\right)$ on it.

9. Show that the circle of curvature at the origin of the parabola $y = mx + \dfrac{x^2}{a}$ is $x^2 + y^2 = a(1 + m^2)$ $(y - mx)$.

10. Prove that the chord of curvature parallel to y-axis for the curve $y = a \log \sec\left(\dfrac{x}{a}\right)$ is of constant length.

11. If C_x, C_y be the chords of curvature parallel to the axis of x and y respectively at any point of the curve $y = c\cosh\dfrac{x}{c}$, prove that $4c^2\,(C_x^2 + C_y^2) = C_y^4$.

12. Find the length of the chord of curvature through the pole for the curves:
(*i*) $r = ae^{m\theta}$ (*ii*) $r^n = a^n \cos n\theta$.

13. If C_0 and C_p denote lengths of chords of curvature of the cardioide $r = a(1 + \cos\theta)$, along and perpendicular to the radius vector through any point, show that $3(C_0^2 + C_p^2) = 8aC_0$.

Answers

1. $\left(\dfrac{21a}{16}, \dfrac{21a}{16}\right)$ **2.** $(2a + 3at^2,\ -2at^3)$

4. (*b*) $\left(\dfrac{a^2 + b^2}{a}\sec^3\theta,\ -\dfrac{a^2 + b^2}{b}\tan^3\theta\right)$; $(ax)^{2/3} - (by)^{2/3} = (a^2 + b^2)^{2/3}$

8. $8(x^2 + y^2) - 21(x + y) + 27 = 0$ **12.** (*i*) $2r$ (*ii*) $\dfrac{2r}{n + 1}$.

7

Asymptotes

7.1. (a) BRANCHES OF A CURVE

If in an equation, y has two or more values for every value of x, then we can suppose that in such a case we are given two or more distinct functions. But generally we regard the curves corresponding to these functions, not as different curves but as *different* **branches** of one curve.

Thus if $x^2 + y^2 = a^2$, then $\qquad y = \pm \sqrt{a^2 - x^2}$

So, we say that $y = \sqrt{a^2 - x^2}$ and $y = -\sqrt{a^2 - x^2}$ are the two branches of the curve (circle) $x^2 + y^2 = a^2$, which we call the upper half and the lower half of the circle.

Moreover, we know that the above circle is bounded by a square whose sides are $x = \pm a$, $y = \pm a$. Thus we say that the two branches of the circle are **finite.**

Again consider the hyperbola $\dfrac{x^2}{a^2} - \dfrac{y^2}{b^2} = 1$.

Solving for y, we get $\qquad y = \dfrac{b}{a}\sqrt{x^2 - a^2}$ and $y = -\dfrac{b}{a}\sqrt{x^2 - a^2}$

As $x \to \pm \infty$, y also tends to $\pm \infty$. Here the two branches are **infinite.**

7.1. (b) DEFINITION OF P → ∞, WHERE P IS A POINT ON AN INFINITE BRANCH OF A CURVE

A point P(x, y) on an infinite branch of a curve is said to tend to ∞ *along the curve*, when one at least of x and $y \to +\infty$ or $-\infty$ as P moves along the branch of the curve.

7.2. ASYMPTOTE (Definition) *(K.U.K. 2012)*

I. *A straight line at a finite distance from the origin, is said to be an* **asymptote** *of an infinite branch of a curve, if the perpendicular distance of a point P on that branch from the straight line tends to zero as P tends to infinity along the branch of the curve.*

II. *A straight line at a finite distance from the origin towards which the tangent line to an infinite branch of a curve at a point P on it approaches as a limit, when P tends to infinity, is called an asymptote of the curve.*

ASYMPTOTES PARALLEL TO CO-ORDINATE AXES

7.3. TO FIND THE ASYMPTOTE PARALLEL TO Y-AXIS OF THE CURVE y = f(x)

Let $\qquad x = c \qquad$ ...(1)

be an asymptote of the curve $y = f(x)$.

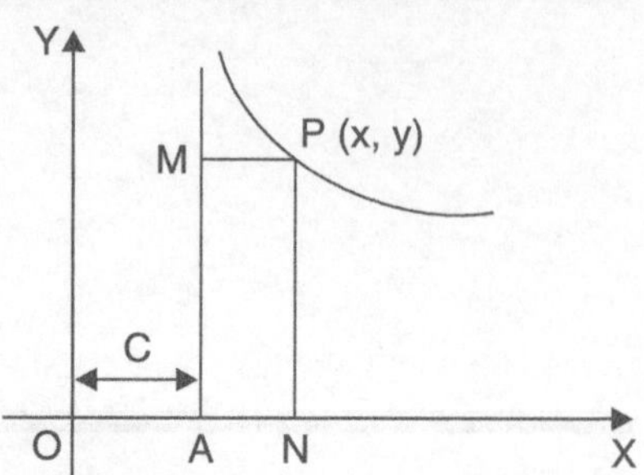

It is required to determine c.

Let δ be the distance of a point $P(x, y)$ on the curve from the line $x = c$.

$$\therefore \qquad \delta = | x - c |$$

Then by definition of an asymptote; if (1) is an asymptote of the curve, then $\delta \to 0$ as $P \to \infty$ along the curve.

As $P \to \infty$ along the curve

$$\delta = | x - c | \to 0 \quad \text{or} \quad x \to c.$$

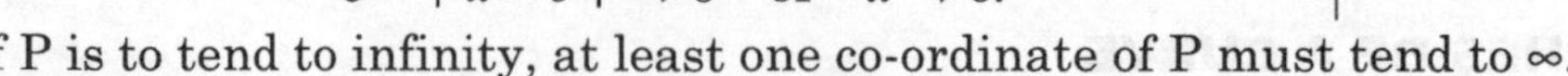

If P is to tend to infinity, at least one co-ordinate of P must tend to ∞.

$\because \; x \to c, \quad \therefore$ it cannot tend to infinity.

$\therefore$ For P to tend to ∞, y must tend to ∞ (*i.e.*, $+\infty$ or $-\infty$).

$\therefore \lim\limits_{y \to \infty} x = c$. Hence c is determined.

Rule. *From the given equation find the definite values c_1, c_2 ... to which x tends, as y tends to $+\infty$ or $-\infty$. Then $x = c_1$, $x = c_2$, ... are the asymptotes of the curve parallel to Y-axis.*

Cor. *Similarly to find the asymptotes || to X-axis, we find from the equation of the curve the definite values d_1, d_2, ... to which y tends as $x \to +\infty$ or $-\infty$. Then $y = d_1$, $y = d_2$, ... are the asymptotes.*

7.4. TO FIND THE ASYMPTOTES PARALLEL TO THE AXES FOR AN ALGEBRAIC CURVE f(x, y) = 0

Let the given equation $f(x, y) = 0$, after arranging in descending powers of y, be

$$y^p \phi_0(x) + y^{p-1}\phi_1(x) + y^{p-2}\phi_2(x) + \cdots + \phi_p(x) = 0 \qquad \text{...(1)}$$

where $\phi_0(x), \phi_1(x), \phi_2(x), \ldots \phi_p(x)$ are polynomials in x.

Let $x = c$ be an asymptote of the curve || to Y-axis.

Dividing (1) throughout by y^p, we have

$$\phi_0(x) + \frac{1}{y}\phi_1(x) + \frac{1}{y^2}\phi_2(x) + \cdots + \frac{1}{y^p}\phi_p(x) = 0 \qquad \text{...(2)}$$

Since $x = c$ is an asymptote of (1) || to Y-axis. $\therefore \lim\limits_{y \to \infty} x = c$.

Then (2) gives $\phi_0(c) = 0$, so that c is a root of the equation

$$\phi_0(x) = 0.$$

If $c_1, c_2, c_3, \ldots$ be the **real** roots of $\phi_0(x) = 0$, then $x = c_1$, $x = c_2$, ... are the asymptotes || to Y-axis.

But from Algebra, we know that if $c_1, c_2, c_3, \ldots$ are the roots of $\phi_0(x) = 0$, then $x - c_1$, $x - c_2$, $x - c_3$, ... are the factors of $\phi_0(x)$, the co-efficient of the highest power of y in (1). Thus we have the following rule:

Rule. *To find the asymptotes parallel to the Y-axis, equate to zero the real linear factors in the co-efficient of the highest power of y in the given equation of the curve.*

If the co-efficient of the highest power of y is a constant or if its linear factors are all imaginary, there will be **no** asymptote || to Y-axis.

Similarly,

Rule to find the asymptotes || to X-axis, *is to equate to zero the real linear factors in the co-efficient of the highest power of x in the equation of the curve.*

If the co-efficient of the highest power of x is a constant or if its linear factors are all imaginary, there will be **no** asymptote parallel to the x-axis.

Example. *Find the asymptotes, parallel to the axes, for the following curves:*

(*a*) $\frac{a^2}{x^2}+\frac{b^2}{y^2}=1$ (*b*) $xy^3 - x^3 = a(x^2 + y^2)$.

Sol. (*a*) The given curve is $\frac{a^2}{x^2}+\frac{b^2}{y^2}=1$ or $a^2y^2 + b^2x^2 = x^2y^2$

or $$x^2y^2 - b^2x^2 - a^2y^2 = 0 \quad ...(1)$$

Equating to zero the co-efficient of the highest power of x *i.e.*, x^2 in (1), we get

$$y^2 - b^2 = 0 \quad \text{or} \quad y = \pm b$$

which are the asymptotes || to x-axis.

Again equating to zero the co-efficient of the highest power of y *i.e.*, y^2 in (1), we get

$$x^2 - a^2 = 0 \quad \text{or} \quad x = \pm a$$

which are the required asymptotes || to y-axis.

(*b*) The equation of the curve is $xy^3 - x^3 = a(x^2 + y^2)$

Here the highest power of x is x^3, having a constant co-efficient. Thus there is no asymptote || to x-axis.

Again equating to zero the co-efficient of y^3, the highest power of y, we get $x = 0$ which is the asymptote || to y-axis, *i.e.*, y-axis itself is the asymptote.

EXERCISE 7.1

Find the asymptotes, parallel to axes, of the following curves (1–6):

1. $\frac{a^2}{x^2}-\frac{b^2}{y^2}=1$ **2.** $x^2y^2 - y^2 - 2 = 0$ **3.** $y^3 - xy^2 = x^2 + 1$

4. $x^2y^2 - a^2(x^2 + y^2) - a^3(x + y) + a^4 = 0$

5. $(x^2 - a^2)(y^2 - b^2) = a^2b^2$

6. $x^2y^2 - x^2y - xy^2 + x + y + 1 = 0$ (*M.D.U. Dec. 2013*)

7. Show that the asymptotes of the curve $x^2y^2 = a^2(x^2 + y^2)$ form a square of side $2a$. (*M.D.U. May 2009*)

Answers

1. $x = \pm a$ **2.** $y = 0, x = \pm 1$ **3.** None

4. $x = \pm a, y = \pm a$ **5.** $x = \pm a, y = \pm b$ **6.** $x = 0, x = 1, y = 0, y = 1$.

OBLIQUE ASYMPTOTES

Def. *An asymptote which is not parallel to the Y-axis is called an* **oblique asymptote,** whose equation is of the form $y = mx + c$, where m and c are finite constants.

7.5. IF y = mx + c IS AN ASYMPTOTE TO THE CURVE f(x, y) = 0, PROVE THAT

$$m = \lim_{x\to\infty}\left(\frac{y}{x}\right) \quad \text{AND} \quad c = \lim_{x\to\infty}(y - mx)$$

Let P(x, y) be **any** point on the infinite branch of the curve $f(x, y) = 0$. Let p = PM be the ⊥ distance of the point P from the line $y = mx + c$. Then

$$p = \frac{y - mx - c}{\sqrt{1+m^2}} \quad \text{...(1)}$$

In this case, the abscissa x tends to infinity as $P(x, y) \to \infty$ along the infinite branch of the curve.

Now from (1), $y - mx = c + p\sqrt{1+m^2}$ or $\frac{y}{x} = m + \frac{c}{x} + \frac{p}{x}\sqrt{1+m^2}$

Since $y = mx + c$ is an asymptote, it follows from the definition that $p \to 0$ as $x \to \infty$.

$$\therefore \quad \lim_{x \to \infty}\left(\frac{y}{x}\right) = \lim_{x \to \infty}\left(m + \frac{c}{x} + \frac{p}{x}\sqrt{1+m^2}\right) = m$$

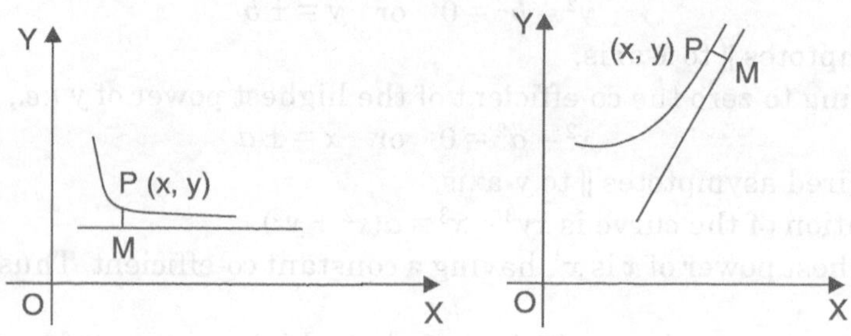

Hence $\lim_{x \to \infty}\left(\frac{y}{x}\right) = m$

Again from (1), $y - mx = c + p\sqrt{1+m^2}$

$$\therefore \quad \lim_{x \to \infty}(y - mx) = \lim_{x \to \infty}(c + p\sqrt{1+m^2}) = c + \sqrt{1+m^2} \cdot \lim_{x \to \infty} p = c + 0 = c$$

Hence $\lim_{x \to \infty}(y - mx) = c$.

7.6. TO FIND THE OBLIQUE ASYMPTOTES OF THE GENERAL ALGEBRAIC CURVE

Let the general algebraic equation of the curve of nth degree be

$$(a_0x^n + a_1x^{n-1}y + a_2x^{n-2}y^2 + \cdots + a_ny^n) + (b_0x^{n-1} + b_1x^{n-2}y + b_2x^{n-3}y^2 + \cdots + b_{n-1}y^{n-1})$$
$$+ (c_0x^{n-2} + c_1x^{n-3}y + \cdots + c_{n-2}y^{n-2}) + \cdots + (k_0x + k_1y) + K = 0.$$

The above equation can be written in the form

$$x^n\phi_n\left(\frac{y}{x}\right) + x^{n-1}\phi_{n-1}\left(\frac{y}{x}\right) + x^{n-2}\phi_{n-2}\left(\frac{y}{x}\right) + \cdots + x\phi_1\left(\frac{y}{x}\right) + \phi_0\left(\frac{y}{x}\right) = 0 \quad \text{...(1)}$$

where $\phi_n\left(\frac{y}{x}\right), \phi_{n-1}\left(\frac{y}{x}\right), \phi_{n-2}\left(\frac{y}{x}\right), \ldots, \phi_0\left(\frac{y}{x}\right)$ are the polynomials in $\frac{y}{x}$ of degree n, $n-1$, $n-2$, ..., 0, at the most.

Let $\quad y = mx + c \quad$ ...(2)

be an asymptote of the curve, where m and c are finite.

[To find m]. Divide both sides of (1) by x^n, we get

$$\phi_n\left(\frac{y}{x}\right) + \frac{1}{x}\phi_{n-1}\left(\frac{y}{x}\right) + \frac{1}{x^2}\phi_{n-2}\left(\frac{y}{x}\right) + \cdots = 0$$

Proceeding to limits as $x \to \infty$, so that $\lim_{x \to \infty}\frac{y}{x} = m$, we get

$$\boldsymbol{\phi_n (m) = 0} \qquad ...(3)$$

This gives the slopes of the asymptotes.

[To find c corresponding to a value of m.]

Let $y - mx = p$, so that as $x \to \infty, p \to c$ $\quad | \because \lim_{x \to \infty} (y - mx) = c$

Now $\quad y = mx + p \quad \therefore \quad \frac{y}{x} = m + \frac{p}{x}$.

Putting this value of $\frac{y}{x}$ in (1), we get

$$x^n \phi_n \left(m + \frac{p}{x}\right) + x^{n-1} \phi_{n-1} \left(m + \frac{p}{x}\right) + x^{n-2} \phi_{n-2} \left(m + \frac{p}{x}\right) + \cdots = 0$$

Expanding each term by Taylor's Theorem, we get

$$x^n \left[\phi_n (m) + \frac{p}{x} \phi_n' (m) + \frac{p^2}{2!.x^2} \phi_n'' (m) + \cdots\right] + x^{n-1} \left[\phi_{n-1} (m) + \frac{p}{x} \phi'_{n-1} (m) + \cdots\right]$$
$$+ x^{n-2} \left[\phi_{n-2} (m) + \frac{p}{x} \phi'_{n-2} (m) + \cdots\right] + \cdots = 0.$$

Putting $\phi_n (m) = 0$ from (3) and arranging the remaining terms in descending powers of x, we get

$$x^{n-1} [p\phi_n'(m) + \phi_{n-1}(m)] + x^{n-2} \left[\frac{p^2}{2!} \phi_n''(m) + p\phi'_{n-1}(m) + \phi_{n-2}(m)\right] + \cdots = 0 \qquad ...(4)$$

Dividing throughout by x^{n-1}, we have

$$[p\phi_n'(m) + \phi_{n-1}(m)] + \frac{1}{x}\left[\frac{p^2}{2!} \phi_n''(m) + p\phi'_{n-1}(m) + \phi_{n-2}(m)\right] + \cdots = 0 \qquad ...(5)$$

Proceeding to limits as $x \to \infty$, so that $p \to c$, we have

$$c\phi_n'(m) + \phi_{n-1} (m) = 0 \qquad ...(6)$$

$\therefore$ If $\phi'_n (m) \neq 0$, which will be the case if $\phi_n(m) = 0$ has no repeated roots, then we get

$$\mathbf{c = -\frac{\phi_{n-1}(m)}{\phi_n'(m)}} \qquad ...(7)$$

Now if $m_1, m_2,...$ are the distinct values of m given by $\phi_n (m) = 0$ and $c_1, c_2, c_3 ,...$ are the corresponding values of c determined by (7), then the asymptotes are

$$y = m_1x + c_1, y = m_2x + c_2, ...$$

Let $\phi_n'(m) = 0$. When $\phi_n'(m) = 0$ and $\phi_{n-1}(m) \neq 0$, then from (7), the value of c is infinite. Thus *there is* **no** *asymptote in this case.*

Case of parallel asymptotes

Suppose $\phi_n'(m) = 0$ and $\phi_{n-1}(m) = 0$. Then (6) becomes an identity and with these values of $\phi_n'(m)$ and $\phi_{n-1}(m)$, the equation (4) reduces to

$$x^{n-2} \left[\frac{p^2}{2!} \phi_n''(m) + p\phi'_{n-1}(m) + \phi_{n-2}(m)\right] + x^{n-3} [...] + \cdots = 0$$

Dividing both sides by x^{n-2}, we get

$$\left[\frac{p^2}{2!} \phi_n''(m) + p\phi'_{n-1}(m) + \phi_{n-2}(m)\right] + \frac{1}{x} [...] ... = 0$$

Proceeding to limits as $x \to \infty$, $p \to c$, we get

$$\frac{c^2}{2!}\phi_n''(m) + c\phi'_{n-1}(m) + \phi_{n-2}(m) = 0$$

which determines two values of c, say c_1 and c_2 provided $\phi_n''(m) \neq 0$.

Then corresponding to a given value of m, there are two values of c *i.e.*, c_1 and c_2. Thus there are two asymptotes $y = mx + c_1$ and $y = mx + c_2$ corresponding to a given slope m. These are clearly parallel.

Working rule for finding oblique asymptotes of an algebraic curve of nth degree.

1. *Find $\phi_n(m)$. This can be obtained by putting $x = 1$, $y = m$ in the highest degree terms of the given equation of the curve.*

2. *Equate $\phi_n(m)$ to zero and solve for m.*

Let its roots be $m_1, m_2, m_3, \ldots$

3. *Find $\phi_{n-1}(m)$ by putting $x = 1$, $y = m$ in the next lower degree terms of the equation. Similarly $\phi_{n-2}(m)$ may be found out by putting $x = 1$ and $y = m$ in the next lower degree terms in the curve and so on.*

4. *Find the values of $c_1, c_2, c_3, \ldots$ corresponding to the values $m_1, m_2, m_3, \ldots$ by using the equation*

$$c = -\frac{\phi_{n-1}(m)}{\phi_n'(m)}.$$

5. *Then the required asymptotes are $y = m_1x + c_1$, $y = m_2x + c_2, \ldots$*

6. *If $\phi_n'(m) = 0$ for some value of m and $\phi_{n-1}(m) \neq 0$ corresponding to that value, then there will be* **no** *asymptote corresponding to that value of m.*

7. *If $\phi_n'(m) = 0$ and $\phi_{n-1}(m) = 0$ for some value of m, the values of c are determined by*

$$\frac{c^2}{2!}\phi_n''(m) + c\phi'_{n-1}(m) + \phi_{n-2}(m) = 0,$$

and this will determine two values of c and thus we shall have two parallel asymptotes corresponding to this value of m.

Important Note:

Before finding the oblique asymptotes, first find the asymptotes parallel to the axes if any. If the number of asymptotes parallel to the axes be equal to the degree of the equation, do not search for oblique asymptotes, as *the total number of asymptotes of a curve cannot exceed the degree of a curve.*

ILLUSTRATIVE EXAMPLES

Example 1. *Find all the asymptotes of the curve*

(*i*) $x^3 + y^3 - 3axy = 0$. (M.D.U. Dec. 2010)

(*ii*) $x^3 - 6x^2y + 11xy^2 - 6y^3 + x + y = 0$ (M.D.U. Dec. 2011)

Sol. (*i*) The equation of the curve is $x^3 + y^3 - 3axy = 0$ ...(1)

Since the co-efficients of x^3 and y^3, the highest degree terms in x and y are constants, the curve has no asymptotes parallel to the x-axis or y-axis. To find the oblique asymptotes, putting $x = 1$, $y = m$ in the two highest degree terms of (1), we have

$$\phi_3(m) = 1 + m^3 \quad \text{and} \quad \phi_2(m) = -3am$$

Now $\phi_3(m) = 0$ gives $1 + m^3 = 0$ or $(1 + m)(1 - m + m^2) = 0$

The only real value is $m = -1$.

$$\therefore \quad c = -\frac{\phi_{n-1}(m)}{\phi_n'(m)} = -\frac{\phi_2(m)}{\phi_3'(m)} = -\frac{-3am}{3m^2} = \frac{a}{m} \qquad | \because \phi_3'(m) = 3m^2$$

When $m = -1, c = \dfrac{a}{-1} = -a$.

$\therefore$ The required asymptote is $(y = mx + c)$ or $y = -1 \cdot x - a$ or $x + y + a = 0$.

(*ii*) The equation of the curve is

$$x^3 - 6x^2y + 11xy^2 - 6y^3 + x + y = 0 \qquad ...(1)$$

Since the co-efficients of x^3 and y^3, the highest degree terms in x and y, are constants, there are no asymptotes parallel to the x-axis or y-axis.

To find oblique asymptotes, putting $x = 1, y = m$ in the third and second degree terms in (1)

$$\phi_3(m) = 1 - 6m + 11m^2 - 6m^3, \qquad \phi_2(m) = 0$$

The slopes of asymptotes are the roots of $\phi_3(m) = 0$

$$\phi_3(m) = 0 \Rightarrow 1 - 6m + 11m^2 - 6m^3 = 0$$

$m = 1$ satisfies it $\therefore$ $m - 1$ is a factor of $\phi_3(m)$.

$\therefore$ $(m - 1)(-6m^2 + 5m - 1) = 0$

$\Rightarrow$ $-(m - 1)(6m^2 - 5m + 1) = 0 \Rightarrow -(m - 1)(2m - 1)(3m - 1) = 0$

$\therefore$ $m = 1, \dfrac{1}{2}, \dfrac{1}{3}$ (all distinct)

Also $\phi_3'(m) = -6 + 22m - 18m^2$

$\therefore$ c is given by $c = -\dfrac{\phi_2(m)}{\phi_3'(m)} = 0$

When $m = 1$, $c = 0$; when $m = \dfrac{1}{2}$, $c = 0$; when $m = \dfrac{1}{3}$, $c = 0$

Putting the corresponding values of m and c in $y = mx + c$, the asymptotes are

$$y = x + 0, \; y = \frac{1}{2}x + 0 \quad \text{and} \quad y = \frac{1}{3}x + 0$$

i.e., $x - y = 0, \; x - 2y = 0$ and $x - 3y = 0$

Example 2. *Find all the asymptotes of the following curves:*

(*i*) $x^3 + 2x^2y - xy^2 - 2y^3 + 3xy + 3y^2 + x + 1 = 0$ (M.D.U. Dec. 2012)

(*ii*) $(x - y)^2(x + 2y - 1) = 3x + y - 7$ (M.D.U. Dec. 2010)

(*iii*) $(x + y)^2(x + y + 2) = x + 9y - 2$ (K.U.K. Dec. 2010, 2013)

(*iv*) $(y - x)(y - 2x)^2 + (y + 3x)(y - 2x) + 2x + 2y - 1 = 0.$

Sol. (*i*) The equation of the curve is

$$x^3 + 2x^2y - xy^2 - 2y^3 + 3xy + 3y^2 + x + 1 = 0 \qquad ...(1)$$

Since the co-efficients of x^3 and y^3, the highest degree terms in x and y, are constants, there are no asymptotes parallel to the x-axis or y-axis.

To find oblique asymptotes, putting $x = 1, y = m$ in the third and second degree terms in (1)

$$\phi_3(m) = 1 + 2m - m^2 - 2m^3, \quad \phi_2(m) = 3m + 3m^2$$

The slopes of asymptotes are the roots of $\phi_3(m) = 0$

$$\phi_3(m) = 0 \Rightarrow (1 + 2m) - m^2(1 + 2m) = 0$$

$\Rightarrow$ $(1 + 2m)(1 - m^2) = 0 \quad \therefore \quad m = -\dfrac{1}{2}, \pm 1$ (all distinct)

Also $\phi_3'(m) = 2 - 2m - 6m^2$

$\therefore$ c is given by $c = -\dfrac{\phi_2(m)}{\phi_3'(m)} = -\dfrac{3m + 3m^2}{2 - 2m - 6m^2} = \dfrac{3m^2 + 3m}{6m^2 + 2m - 2}$

when $m = -\dfrac{1}{2}$, $c = \dfrac{\frac{3}{4} - \frac{3}{2}}{\frac{3}{2} + 1 - 2} = \dfrac{-\frac{3}{4}}{-\frac{3}{2}} = \dfrac{1}{2}$, when $m = 1$, $c = \dfrac{3+3}{6+2-2} = \dfrac{6}{6} = 1$

when $m = -1$, $c = \dfrac{3-3}{6-2-2} = 0$

Putting the corresponding values of m and c in $y = mx + c$, the asymptotes are

$$y = -\frac{1}{2}x + \frac{1}{2}, \quad y = x + 1, \quad \text{and} \quad y = -x + 0$$

i.e., $x + 2y - 1 = 0, \quad x - y + 1 = 0 \quad \text{and} \quad x + y = 0$

(*ii*) The equation of curve is $(x - y)^2(x + 2y - 1) = 3x + y - 7$

or $(x - y)^2(x + 2y) - (x - y)^2 - 3x - y + 7 = 0$ ...(1)

Since the co-efficients of x^3 and y^3, the highest degree terms in x and y, are constants, there are no asymptotes parallel to the x-axis or y-axis.

To find oblique asymptotes, putting $x = 1$, $y = m$ in the third, second and first degree terms in (1)

$$\phi_3(m) = (1 - m)^2(1 + 2m) = (1 - 2m + m^2)(1 + 2m) = 2m^3 - 3m^2 + 1$$

$$\phi_2(m) = -(1 - m)^2 = -m^2 + 2m - 1, \quad \phi_1(m) = -3 - m$$

The slopes of asymptotes are the roots of $\phi_3(m) = 0$

$$\phi_3(m) = 0 \Rightarrow (1 - m^2)(1 + 2m) = 0 \quad \therefore \quad m = 1, 1, -\frac{1}{2}$$

Also $\phi_3'(m) = 6m^2 - 6m, \quad \therefore \quad \phi_3''(m) = 12m - 6, \quad \phi_2'(m) = -2m + 2$

For the non-repeated value $m = -\dfrac{1}{2}$, c is given by

$$c = -\frac{\phi_2(m)}{\phi_3'(m)} \quad \text{or} \quad c = -\frac{-m^2 + 2m - 1}{6m^2 - 6m} = \frac{m^2 - 2m + 1}{6m^2 - 6m}$$

When $m = -\dfrac{1}{2}$, $c = \dfrac{\frac{1}{4} + 1 + 1}{\frac{3}{2} + 3} = \dfrac{9}{4} \times \dfrac{2}{9} = \dfrac{1}{2}$

Corresponding asymptote is $y = mx + c$ *i.e.*, $y = -\dfrac{1}{2}x + \dfrac{1}{2}$ or $x + 2y - 1 = 0$

For the twice repeated value $m = 1, 1$, c is given by

$$\frac{c^2}{2!}\phi_3''(m) + c\phi_2'(m) + \phi_1(m) = 0$$

or $$\frac{c^2}{2}(12m - 6) + c(-2m + 2) + (-3 - m) = 0$$

or $$c^2(6m - 3) + c(-2m + 2) + (-3 - m) = 0$$

When $m = 1$, we get $3c^2 - 4 = 0$ or $c = \pm \dfrac{2}{\sqrt{3}}$

Corresponding asymptotes are $y = mx + c$ *i.e.*, $y = x \pm \dfrac{2}{\sqrt{3}}$

Hence the asymptotes are $x + 2y - 1 = 0$ and $y = x \pm \dfrac{2}{\sqrt{3}}$.

(*iii*) The equation of curve is

$$(x + y)^2(x + y + 2) = x + 9y - 2$$

or $$(x + y)^3 + 2(x + y)^2 - x - 9y + 2 = 0 \quad ...(1)$$

Since the co-efficients of x^3 and y^3, the highest degree terms in x and y, are constants, there are no asymptotes parallel to the x-axis or y-axis.

To find oblique asymptotes, putting $x = 1$, $y = m$ in the third, second and first degree terms in (1)

$$\phi_3(m) = (1 + m)^3, \quad \phi_2(m) = 2(1 + m)^2, \quad \phi_1(m) = -1 - 9m$$

Also $$\phi_0(m) = 2$$

The slopes of asymptotes are the roots of $\phi_3(m) = 0$

$$\phi_3(m) = 0 \Rightarrow (1 + m)^3 = 0 \quad \therefore \quad m = -1, -1, -1$$

Also $$\phi_3'(m) = 3(1 + m)^2, \quad \phi_3''(m) = 6(1 + m), \phi_3'''(m) = 6$$

$$\phi_2'(m) = 4(1 + m), \quad \phi_2''(m) = 4, \quad \phi_1'(m) = -9$$

For the thrice repeated value $m = -1, -1, -1$, c is given by

$$\frac{c^3}{3!}\phi_3'''(m) + \frac{c^2}{2!}\phi_2''(m) + c\,\phi_1'(m) + \phi_0(m) = 0$$

or $$\frac{c^3}{6}(6) + \frac{c^2}{2}(4) + c(-9) + 2 = 0 \quad \text{or} \quad c^3 + 2c^2 - 9c + 2 = 0$$

By inspection $c = 2$ is a root. Using synthetic division

2	1	2	−9	2
		2	8	−2
	1	4	−1	0

The depressed equation is $c^2 + 4c - 1 = 0$

$$\therefore \quad c = \frac{-4 \pm \sqrt{16 + 4}}{2} = -2 \pm \sqrt{5}$$

Hence the asymptotes are $y = mx + c$

i.e., $$y = -x + 2 \quad \text{and} \quad y = -x - 2 \pm \sqrt{5}$$

i.e., $$x + y - 2 = 0 \quad \text{and} \quad x + y + 2 \pm \sqrt{5} = 0.$$

(*iv*) The equation of curve is

$$(y - x)(y - 2x)^2 + (y + 3x)(y - 2x) + 2x + 2y - 1 = 0 \quad ...(1)$$

Since the co-efficients of x^3 and y^3, the highest degree terms in x and y, are constants, there are no asymptotes parallel to the x-axis or y-axis.

To find oblique asymptotes, putting $x = 1$, $y = m$ in the third, second and first degree terms in (1)

$$\phi_3(m) = (m - 1)(m - 2)^2 = (m - 1)(m^2 - 4m + 4) = m^3 - 5m^2 + 8m - 4$$

$$\phi_2(m) = (m + 3)(m - 2) = m^2 + m - 6, \quad \phi_1(m) = 2 + 2m$$

The slopes of the asymptotes are the roots of $\phi_3(m) = 0$

$$\phi_3(m) = 0 \Rightarrow (m - 1)(m - 2)^2 = 0 \quad \therefore \quad m = 1, 2, 2$$

Also $$\phi_3'(m) = 3m^2 - 10m + 8, \phi_3''(m) = 6m - 10$$

$$\phi_2'(m) = 2m + 1$$

For the non-repeated value $m = 1$, c is given by

$$c = -\frac{\phi_2(m)}{\phi_3'(m)} = -\frac{m^2 + m - 6}{3m^2 - 10m + 8}$$

When $m = 1$, $\qquad c = -\dfrac{1+1-6}{3-10+8} = 4$

The corresponding asymptote is $y = mx + c$ *i.e.,* $y = x + 4$ or $x - y + 4 = 0$

For the twice repeated value $m = 2, 2$, c is given by

$$\frac{c^2}{2!}\phi_3''(m) + c\phi_2'(m) + \phi_1(m) = 0$$

or $$\frac{c^2}{2}(6m - 10) + c(2m + 1) + 2 + 2m = 0$$

When $m = 2$, we have $c^2 + 5c + 6 = 0$ or $(c + 2)(c + 3) = 0$ $\therefore$ $c = -2, -3$

The corresponding asymptotes are $y = mx + c$

i.e., $\qquad y = 2x - 2$ and $y = 2x - 3$

i.e., $\qquad 2x - y - 2 = 0$ and $2x - y - 3 = 0$

Hence all the asymptotes of the curve are

$$x - y + 4 = 0,\ 2x - y - 2 = 0 \text{ and } 2x - y - 3 = 0.$$

Example 3. *Find all the asymptotes of the curve*

$$(x - y)^2(x^2 + y^2) - 10(x - y)x^2 + 12y^2 + 2x + y = 0$$

Sol. The equation of the curve is

$$(x - y)^2(x^2 + y^2) - 10(x - y)x^2 + 12y^2 + 2x + y = 0 \qquad ...(1)$$

Since the co-efficients of x^4 and y^4, the highest degree terms in x and y, are constants, there are no asymptotes parallel to the x-axis or y-axis.

To find oblique asymptotes, putting $x = 1$, $y = m$ is the fourth, third, second and first degree terms in (1)

$$\phi_4(m) = (1 - m)^2(1 + m^2) = (1 + m^2 - 2m)(1 + m^2)$$
$$= m^4 - 2m^3 + 2m^2 - 2m + 1$$

$$\phi_3(m) = -10(1 - m) = 10(m - 1), \quad \phi_2(m) = 12\,m^2, \quad \phi_1(m) = 2 + m$$

The slopes of asymptotes are the roots of $\phi_4(m) = 0$

or $\qquad (1 - m)^2(1 + m^2) = 0 \quad \therefore \quad m = 1, 1$

Also $\qquad \phi_4'(m) = 4m^3 - 6m^2 + 4m - 2, \quad \phi_4''(m) = 12m^2 - 12m + 4$

$$\phi_3'(m) = 10$$

For the twice repeated value of m, c is given by

$$\frac{c^2}{2!}\phi_4''(m) + c\phi_3'(m) + \phi_2(m) = 0$$

or $$\frac{c^2}{2}(12m^2 - 12m + 4) + c(10) + 12m^2 = 0$$

Putting $m = 1$, we have $2c^2 + 10c + 12 = 0$

or $\qquad c^2 + 5c + 6 = 0$ or $(c + 2)(c + 3) = 0$

$\therefore \qquad c = -2, -3$

Hence the asymptotes are $y = mx + c$

i.e., $\qquad y = x - 2$ and $y = x - 3$.

EXERCISE 7.2

Find all the asymptotes of the following curves (1–15):

1. $y^3 = x^3 + ax^2$

2. $x^2y + xy^2 + xy + y^2 + 3x = 0$

3. (*a*) $y^2(x - 2a) = x^3 - a^3$ (*b*) $6x^2 + xy - 2y^2 + x + 2y + 1 = 0$

4. $y^3 - 2xy^2 - x^2y + 2x^3 + 3y^2 - 7xy + 2x^2 + 2y + 2x + 1 = 0$

5. $3x^3 + 2x^2y - 7xy^2 + 2y^3 - 14xy + 7y^2 + 4x + 5y = 0$

6. $y^3 - 3x^2y + xy^2 - 3x^3 + 2y^2 + 2xy + 4x + 5y + 6 = 0$

7. $x^3 + 3x^2y - xy^2 - 3y^3 + x^2 - 2xy + 3y^2 + 4x + 7 = 0$

8. $y^3 - xy^2 - x^2y + x^3 + x^2 - y^2 = 0$ (*M.D.U. Dec. 2008*)

9. $4x^2(y - x) + y(y - 2)(x - y) = 4x + 4y - 7$

10. $x^3 + 3x^2y - 4y^3 - x + y + 3 = 0$ (*M.D.U. May 2008*)

11. $x^3 - x^2y - xy^2 + y^3 + 2x^2 - 4y^2 + 2xy + x + y + 1 = 0$

12. $y^3 - 5xy^2 + 8x^2y - 4x^3 - 3y^2 + 9xy - 6x^2 + 2y - 2x + 1 = 0$

13. $x^2(x - y)^2 - a^2(x^2 + y^2) = 0$

14. $x^2y^2(x^2 - y^2)^2 - (x^2 + y^2)^3 = 0$

15. $(x + y + 1)(x^2 + y^2 - xy) - 3xy + x^2 + y^2 + 2x - 3y + 5 = 0.$

16. Show that the parabola $y^2 = 4ax$ has no asymptotes.

17. Show that the asymptotes of the curve $x^2y^2 - a^2(x^2 + y^2) - a^3(x + y) + a^4 = 0$ form a square through two of whose vertices the curve passes.

Answers

1. $3y = 3x + a$

2. $y = 0,\ x = -1,\ x + y = 0$

3. (*a*) $x = 2a,\ x + y + a = 0,\ x - y + a = 0$ (*b*) $14x - 7y + 5 = 0,\ 21x + 14y = 4$

4. $2x - y = 0,\ x - y - 1 = 0,\ x + y + 2 = 0$

5. $6x - 6y - 7 = 0,\ 6x - 2y - 3 = 0,\ 3x + 6y + 5 = 0$

6. $x + y = 0,\ \sqrt{3}\,x - y - 1 = 0,\ \sqrt{3}\,x + y + 1 = 0$

7. $4x - 4y + 1 = 0,\ 2x + 2y - 3 = 0,\ 4x + 12y + 9 = 0$

8. $y = \pm x,\ y = x + 1$

9. $x - y = 0,\ 2x - y + 1 = 0,\ 2x + y - 1 = 0$

10. $x - y = 0,\ x + 2y - 1 = 0,\ x + 2y + 1 = 0$

11. $x + y - 1 = 0,\ 2x - 2y + 3 + \sqrt{5} = 0,\ 2x - 2y + 3 - \sqrt{5} = 0$

12. $x - y = 0,\ 2x - y + 1 = 0,\ 2x - y + 2 = 0$

13. $x = \pm a,\quad x - y = \pm\sqrt{2}a$

14. $x = \pm 1,\ y = \pm 1,\ x - y \pm \sqrt{2} = 0,\ x + y \pm \sqrt{2} = 0$

15. $3x + 3y + 8 = 0.$

7.7. ASYMPTOTES BY INSPECTION

If the equation of a curve of nth degree can be put in the form

$$\mathbf{F_n + F_{n-2} = 0}$$

where F_n is of degree n (i.e., contains terms of degree n and may also contain terms of lower degree) and F_{n-2} is of degree (n – 2) at the most, then **every linear factor of F_n equated to zero will give an asymptote,** *provided no two linear factors of F_n are coincident or differ by a constant.*

Example. *Find all the asymptotes of the following curves:*

(*a*) $\dfrac{x^2}{a^2} - \dfrac{y^2}{b^2} = 1$ (*b*) $xy(x^2 - y^2) + 3x^2 - 4y^2 + 1 = 0.$

Sol. (*a*) The equation of the given curve is $\left(\dfrac{x^2}{a^2} - \dfrac{y^2}{b^2}\right) - 1 = 0.$

The equation is of the form $F_n + F_{n-2} = 0$, [Here $n = 2$], where F_n can be resolved into real linear non-repeated factors.

Thus the asymptotes are given by $F_n = 0$ or $\frac{x^2}{a^2} - \frac{y^2}{b^2} = 0$ or $y^2 = \frac{b^2}{a^2}x^2$ or $y = \pm\frac{b}{a}x$.

(*b*) The equation of the given curve is $[xy(x^2 - y^2)] + (3x^2 - 4y^2 + 1) = 0$

The equation is of the form $F_n + F_{n-2} = 0$, [Here $n = 4$], where F_n can be split up into linear non-repeated factors. Thus the asymptotes are given by $F_n = 0$.

or $\qquad xy(x^2 - y^2) = 0 \quad \text{or} \quad xy(x - y)(x + y) = 0$

$\therefore \qquad x = 0, y = 0, x - y = 0 \quad \text{and} \quad x + y = 0$

are the required asymptotes.

7.8. INTERSECTION OF A CURVE AND ITS ASYMPTOTES

Prove that any asymptote of an algebraic curve of nth degree cuts the curve in (n – 2) points.

Let
$$y = mx + c \qquad ...(1)$$

be an asymptote of the curve
$$x^n\phi_n\left(\frac{y}{x}\right) + x^{n-1}\phi_{n-1}\left(\frac{y}{x}\right) + x^{n-2}\phi_{n-2}\left(\frac{y}{x}\right) + \cdots = 0 \qquad ...(2)$$

Eliminating *y* from (1) and (2), the abscissae of the points of intersection are given by

$$x^n\phi_n\left(m + \frac{c}{x}\right) + x^{n-1}\phi_{n-1}\left(m + \frac{c}{x}\right) + x^{n-2}\phi_{n-2}\left(m + \frac{c}{x}\right) + \cdots = 0$$

Expanding by Taylor's Theorem, we have

$$x^n\left[\phi_n(m) + \frac{c}{x}\cdot\phi_n'(m) + \frac{c^2}{2!x^2}\phi_n''(m) + \cdots\right] + x^{n-1}\left[\phi_{n-1}(m) + \frac{c}{x}\cdot\phi'_{n-1}(m) + \cdots\right]$$
$$+ x^{n-2}[\phi_{n-2}(m) + \cdots] + \cdots = 0$$

or
$$x^n\phi_n(m) + x^{n-1}[c\phi_n'(m) + \phi_{n-1}(m)]$$
$$+ x^{n-2}\left[\frac{c^2}{2!}\phi_n''(m) + c\phi'_{n-1}(m) + \phi_{n-2}(m)\right] + \ldots = 0 \qquad ...(3)$$

Since (1) is an asymptote of (2), $\therefore$ *m* and *c* are given by the equations

$$\phi_n(m) = 0 \quad \text{and} \quad c\phi_n'(m) + \phi_{n-1}(m) = 0$$

Putting these values in (3), we get

$$x^{n-2}\left[\frac{c^2}{2!}\phi_n''(m) + c\phi'_{n-1}(m) + \phi_{n-2}(m)\right] + x^{n-3}[\ldots] \ldots = 0$$

which is of degree $(n - 2)$ and this gives $(n - 2)$ values of x.

Hence (1) cuts (2) in $(n - 2)$ points.

Cor. 1. *If a curve of nth degree has n asymptotes, then they all intersect the curve in* **n(n – 2)** *points.*

Cor. 2. *If the equation of a curve of nth degree can be put in the form $F_n + F_{n-2} = 0$, where F_{n-2} is of degree $(n - 2)$ at the most and F_n consists of n non-repeated linear factors, then the $n(n - 2)$ points of intersection of the curve $[F_n + F_{n-2} = 0]$ and its asymptotes $[F_n = 0]$ lie on the curve $F_{n-2} = 0$.*

Illustrations:

(*i*) For the cubic, $n = 3$ and therefore the asymptotes cut the curve in $3(3 - 2) = 3$ points which lie on a curve of degree $(3 - 2) = 1$ *i.e.*, on a straight line.

(*ii*) For a curve of fourth degree, $n = 4$ and therefore the asymptotes cut the curve in $4(4 - 2) = 8$ points which lie on a curve of degree $4 - 2 = 2$ *i.e.*, on a conic.

Rule to find the equation of the curve on which lie the points of intersection of the curve and its asymptotes.

1. *Find all the asymptotes of the given curve and hence the joint equation of all the asymptotes.*

2. *Write this equation in such a way that the highest degree term in x is the same as in the equation of the curve. Denote this by* $F_n = 0$.

3. *Re-write the given equation of the curve in the form* $F_n + F_{n-2} = 0$, *where* F_n *is of step 2 above.*

4. *Then the equation* $F_{n-2} = 0$ *gives the curve on which lie the points of intersection of the curve and its asymptotes.*

ILLUSTRATIVE EXAMPLES

Example 1. *Show that the points of intersection of the curve*

$$2y^3 - 2x^2y - 4xy^2 + 4x^3 - 14xy + 6y^2 + 4x^2 + 6y + 1 = 0$$

and its asymptotes lie on the straight line $8x + 2y + 1 = 0$.

Sol. The equation of the given curve is

$$2y^3 - 2x^2y - 4xy^2 + 4x^3 - 14xy + 6y^2 + 4x^2 + 6y + 1 = 0 \quad ...(1)$$

To find the asymptotes of (1)

The curve has no asymptotes parallel to x-axis or y-axis.

($\because$ the co-efficients of x^3 and y^3, the highest degree terms in x and y are constants)

Putting $x = 1$, $y = m$ in the two highest (*i.e.*, third and second) degree terms, we get

$$\phi_n(m) = \phi_3(m) = 2m^3 - 2m - 4m^2 + 4$$

and $$\phi_{n-1}(m) = \phi_2(m) = -14m + 6m^2 + 4$$

The slopes of the asymptotes are given by $\phi_n(m) = 0$

or $$2m^3 - 2m - 4m^2 + 4 = 0 \quad \text{or} \quad 2m(m^2 - 1) - 4(m^2 - 1) = 0$$

or $$(2m - 4)(m^2 - 1) = 0$$

$\therefore$ $$m = 2, 1, -1.$$

Also $\phi_n'(m) = \phi_3'(m) = 6m^2 - 2 - 8m$

$$c = -\frac{\phi_{n-1}(m)}{\phi_n'(m)} = -\frac{\phi_2(m)}{\phi_3'(m)} = -\frac{-14m + 6m^2 + 4}{6m^2 - 2 - 8m} = \frac{7m - 3m^2 - 2}{3m^2 - 4m - 1}$$

When $m = 2$, $c = \dfrac{14 - 12 - 2}{12 - 8 - 1} = 0$, When $m = 1$, $c = \dfrac{7 - 3 - 2}{3 - 4 - 1} = \dfrac{2}{-2} = -1$.

When $m = -1$, $c = \dfrac{-7 - 3 - 2}{3 + 4 - 1} = -2$.

Putting these values of m and c in $y = mx + c$, the asymptotes are

$$y = 2x + 0, \ y = x - 1, \quad \text{and} \quad y = -x - 2$$

or $$2x - y = 0, \ x - y - 1 = 0 \quad \text{and} \quad x + y + 2 = 0.$$

Thus the three asymptotes will cut the curve again in 3(3 – 2) = 3 points.

The joint equation of the asymptotes is

$$(2x - y)(x - y - 1)(x + y + 2) = 0$$

or
$$y^3 - x^2y - 2xy^2 + 2x^3 - 7xy + 3y^2 + 2x^2 + 2y - 4x = 0$$

Multiplying by 2

$$2y^3 - 2x^2y - 4xy^2 + 4x^3 - 14xy + 6y^2 + 4x^2 + 4y - 8x = 0.$$

Now the equation of the curve is

$$2y^3 - 2x^2y - 4xy^2 + 4x^3 - 14xy + 6y^2 + 4x^2 + 6y + 1 = 0$$

which can be re-written as

$$[2y^3 - 2x^2y - 4xy^2 + 4x^3 - 14xy + 6y^2 + 4x^2 + 4y - 8x] + [2y + 8x + 1] = 0$$

which is of the form $F_n + F_{n-2} = 0$

Hence the points of intersection lie on the curve

$$F_{n-2} = 0 \quad \text{or} \quad 2y + 8x + 1 = 0 \quad \text{or} \quad 8x + 2y + 1 = 0$$

which is the required straight line.

Example 2. *Show that the asymptotes of the curve*

$$(x^2 - y^2)(y^2 - 4x^2) + 6x^3 - 5x^2y - 3xy^2 + 2y^3 - x^2 + 3xy - 1 = 0$$

cut the curve again in eight points which lie on a circle of radius unity.

Sol. The given equation of the curve is

$$(x^2 - y^2)(y^2 - 4x^2) + 6x^3 - 5x^2y - 3xy^2 + 2y^3 - x^2 + 3xy - 1 = 0 \quad ...(1)$$

The curve has no asymptotes parallel to x-axis or y-axis, since the co-efficients of x^4 and y^4 are constants.

Putting $x = 1$, $y = m$ in the two highest degree terms in (1),

we have
$$\phi_n(m) = \phi_4(m) = (1 - m^2)(m^2 - 4)$$

and
$$\phi_{n-1}(m) = \phi_3(m) = 6 - 5m - 3m^2 + 2m^3$$

The slopes of the asymptotes are given by $\phi_n(m) = 0$ or $(1 - m^2)(m^2 - 4) = 0$ $\therefore$ $\mathbf{m = \pm 1, \pm 2}$.

Also $\phi_4'(m) = (1 - m^2)\, 2m + (m^2 - 4)(-2m) = 2m(1 - m^2 - m^2 + 4) = -2m(2m^2 - 5)$

Now
$$c = -\frac{\phi_{n-1}(m)}{\phi_n'(m)} = -\frac{\phi_3(m)}{\phi_4'(m)} = -\frac{6 - 5m - 3m^2 + 2m^3}{-2m(2m^2 - 5)} = \frac{6 - 5m - 3m^2 + 2m^3}{2m(2m^2 - 5)}$$

When $m = 1$, $c = \dfrac{6 - 5 - 3 + 2}{2(2 - 5)} = 0$, when $m = -1$, $c = \dfrac{6 + 5 - 3 + 2}{-2(2 - 5)} = 1$,

when $m = 2$, $c = \dfrac{6 - 10 - 12 + 16}{4(8 - 5)} = 0$ and when $m = -2$, $c = \dfrac{6 + 10 - 12 - 16}{-4(8 - 5)} = 1$.

Putting these values of m and c in $y = mx + c$, the required asymptotes are

$$y = x + 0,\; y = -x + 1,\; y = 2x + 0 \quad \text{and} \quad y = -2x + 1$$

or
$$x - y = 0,\; x + y - 1 = 0,\; y - 2x = 0 \quad \text{and} \quad y + 2x - 1 = 0.$$

The four asymptotes will cut the curve in 4(4 – 2) = 8 points.

The joint equation of all the asymptotes is

$$(x - y)(x + y - 1)(y - 2x)(y + 2x - 1) = 0$$

or
$$[(x^2 - y^2) - x + y]\,[(y^2 - 4x^2) - y + 2x] = 0$$

or
$$(x^2 - y^2)(y^2 - 4x^2) + 6x^3 - 3xy^2 + 2y^3 - 2x^2 - y^2 + 3xy = 0 \quad ...(2)$$

The equation of the curve (1) can be written as

$$[(x^2 - y^2)(y^2 - 4x^2) + 6x^3 - 3xy^2 + 2y^3 - 2x^2 - y^2 + 3xy] + (x^2 + y^2 - 1) = 0$$

which is of the form $F_n + F_{n-2} = 0$ [Here $n = 4$]

Hence the points of intersection lie on the curve $F_{n-2} = 0$ or $x^2 + y^2 - 1 = 0$ which is a circle of radius unity.

7.9. EQUATION OF THE CURVE WITH GIVEN ASYMPTOTES AND SATISFYING SOME OTHER CONDITION

Example 1. *Find the equation of the hyperbola having $x + y - 1 = 0$ and $x - y + 2 = 0$ as its asymptotes and passing through the origin.*

Sol. The joint equation of the asymptotes is

$$F_2 = (x + y - 1)(x - y + 2) = 0 \quad ...(1)$$

Since the equation of the curve which has asymptotes given by $F_n = 0$ is of the form $F_n + F_{n-2} = 0$.

$\therefore$ The equation of the curve which has asymptotes given by (1) is of the form

$$F_2 + F_0 = 0$$

or $$(x + y - 1)(x - y + 2) + k = 0 \quad ...(2)$$ | $\because$ F_0 is of zero degree and a constant.

This passes through the origin (0, 0), if $(-1)(2) + k = 0 \quad \therefore \quad k = 2$.

Putting this value of k in (2), the required equation is $(x + y - 1)(x - y + 2) + 2 = 0$

or $$x^2 - y^2 + x + 3y = 0.$$

Example 2. *Find the equation of the cubic which has the same asymptotes as the curve $x^3 - 6x^2y + 11xy^2 - 6y^3 + x + y + 1 = 0$ and which passes through the points (0, 0), (1, 0) and (0, 1).*

Sol. The given equation of the curve is

$$x^3 - 6x^2y + 11xy^2 - 6y^3 + x + y + 1 = 0$$

or $$(x - y)(x - 2y)(x - 3y) + (x + y + 1) = 0$$

which is of the form $F_n + F_{n-2} = 0$ (where $n = 3$) and F_n consists of linear non-repeated factors.

Thus by inspection method, the asymptotes are given by $F_n = 0$

or $$(x - y)(x - 2y)(x - 3y) = 0 \quad i.e., \quad x - y = 0, x - 2y = 0 \quad \text{and} \quad x - 3y = 0$$

The joint equation of the asymptotes is

$$F_3 \equiv (x - y)(x - 2y)(x - 3y) = 0 \quad ...(1)$$

Since the equation of any curve which has asymptotes $F_n = 0$ is of the form

$$F_n + F_{n-2} = 0.$$

$\therefore$ The equation of any cubic having the asymptotes given by (1) is of the form

$$F_3 + F_1 = 0$$

or $$F_3 + (ax + by + c) = 0$$

or $$(x - y)(x - 2y)(x - 3y) + ax + by + c = 0 \quad ...(2)$$

where $ax + by + c$ is a general linear expression.

Since the curve (1) passes through (0, 0), (1, 0) and (0, 1).

$\therefore \quad c = 0, 1 + a + c = 0$ and $-6 + b + c = 0$ which give $c = 0, a = -1, b = 6$.

Putting these values in (2), the required cubic is

$$(x - y)(x - 2y)(x - 3y) - x + 6y = 0$$

or $$x^3 - 6x^2y + 11xy^2 - 6y^3 - x + 6y = 0.$$

EXERCISE 7.3

1. Find the asymptotes of the curve $x^2y - xy^2 + xy + y^2 + x - y = 0$ and show that they cut the curve in three points which lie on the straight line $x + y = 0$.
2. Find the equation of the straight line on which lie the three points of intersection of the curve $x^3 + 2x^2y - xy^2 - 2y^3 + 4y^2 + 2xy + y - 1 = 0$ and its asymptotes.
3. Show that the asymptotes of the curve $x^3 - 2y^3 + xy(2x - y) + y(x - y) + 1 = 0$ cut the curve in three points which lie on the straight line $x - y + 1 = 0$.
4. Find the asymptotes of the curve $xy(x^2 - y^2) + x^2 + y^2 = a^2$ and show that the eight points of intersection of the curve and its asymptotes lie on a circle whose centre is at the origin.
5. Show that the eight points of intersection of the curve $x^4 - 5x^2y^2 + 4y^4 + x^2 - y^2 + x + y + 1 = 0$ and its asymptotes lie on a rectangular hyperbola.
6. Show that the four asymptotes of the curve $xy(x^2 - y^2) + 9x^2 + 25y^2 = 114$ cut the curve in eight points which lie on an ellipse of eccentricity $\frac{4}{5}$.
7. Find the equation of the cubic which has the same asymptotes as the curve $x^3 - 6x^2y + 11xy^2 - 6y^3 + 4x + 5y + 7 = 0$ and which passes through the points (0, 0), (0, 2) and (2, 0).
8. Find the equation of the curve which has the same asymptotes as the curve $x^3 - 6x^2y + 11xy^2 - 6y^3 + x + y + 1 = 0$ and which touches the axis of y at the origin and passes through the point (3, 2). *(K.U.K. 2012)*
[**Hint.** Equation of tangent at the origin is obtained by equating to zero the lowest degree terms in the equation of the curve.]

Answers

2. $x + 3y - 1 = 0$
7. $x^3 - 6x^2y + 11xy^2 - 6y^3 - 4x + 24y = 0$
8. $x^3 - 6x^2y + 11xy^2 - 6y^3 - x = 0.$

7.10. ASYMPTOTES OF POLAR CURVES

If α is a root of the equation $f(\theta) = 0$, then $r \sin(\theta - \alpha) = \frac{1}{f'(\alpha)}$ is an asymptote of the polar curve $\frac{1}{r} = f(\theta)$. *(K.U.K. 2013)*

Working rule for finding the asymptotes of polar curves.

1. Write down the given equation as $\frac{1}{r} = f(\theta)$.
2. Equate $f(\theta)$ to zero and solve for θ. Let the roots be $\theta = \theta_1, \theta_2, \theta_3, \ldots$
3. Find $f'(\theta)$ and calculate $f'(\theta)$ at $\theta = \theta_1, \theta_2, \theta_3, \ldots$
4. Then write asymptotes as $r \sin(\theta - \theta_1) = \frac{1}{f'(\theta_1)}$, $r \sin(\theta - \theta_2) = \frac{1}{f'(\theta_2)}, \ldots$

Note 1. *If* $\frac{1}{f'(\theta_1)}$ *is infinite, the curve has no asymptote corresponding to the direction* $\theta = \theta_1$.

Note 2. *The following results will be helpful.*

$$\left.\begin{array}{ll} (i) & \text{If } \sin\theta = 0, \text{ then } \theta = n\pi \\ (ii) & \text{If } \cos\theta = 0, \text{ then } \theta = (2n+1)\dfrac{\pi}{2} \\ (iii) & \text{If } \sin\theta = \sin\alpha, \text{ then } \theta = n\pi + (-1)^n\alpha \\ (iv) & \text{If } \cos\theta = \cos\alpha, \text{ then } \theta = 2n\pi \pm \alpha \\ (v) & \text{If } \tan\theta = \tan\alpha, \text{ then } \theta = n\pi + \alpha \\ (vi) & \sin(n\pi + \theta) = (-1)^n \sin\theta \\ (vii) & \cos(n\pi + \theta) = (-1)^n \cos\theta \\ (viii) & \tan(n\pi + \theta) = \tan\theta \end{array}\right\} \quad n \in I$$

ILLUSTRATIVE EXAMPLES

Example 1. *Find the asymptotes of the curve* $r(\theta^2 - \pi^2) = 2a\theta$.

Sol. The equation of the curve is

$$r(\theta^2 - \pi^2) = 2a\theta \quad \text{or} \quad \frac{1}{r} = \frac{\theta^2 - \pi^2}{2a\theta} = f(\theta), \text{ say.}$$

$$\therefore \qquad f(\theta) = 0 \text{ gives } \frac{\theta^2 - \pi^2}{2a\theta} = 0 \;\Rightarrow\; \theta^2 - \pi^2 = 0 \quad \text{or} \quad \theta = \pm\pi$$

$$\text{Now} \qquad f'(\theta) = \frac{1}{2a}\,\frac{\theta \cdot 2\theta - (\theta^2 - \pi^2)\cdot 1}{\theta^2} = \frac{\theta^2 + \pi^2}{2a\theta^2}.$$

$$\therefore \qquad f'(\pi) = \frac{\pi^2 + \pi^2}{2a\pi^2} = \frac{1}{a} \quad \text{and} \quad f'(-\pi) = \frac{\pi^2 + \pi^2}{2a\pi^2} = \frac{1}{a}$$

$\therefore$ The asymptote corresponding to $\theta = \pi$ is

$$r\sin(\theta - \pi) = \frac{1}{f'(\pi)} \quad \text{or} \quad -r\sin(\pi - \theta) = a$$

$$\text{or} \qquad -r\sin\theta = a \quad \text{or} \quad r\sin\theta + a = 0 \qquad \text{...(1)}$$

Again the asymptote corresponding to $\theta = -\pi$ is

$$r\sin(\theta + \pi) = \frac{1}{f'(-\pi)} \quad \text{or} \quad -r\sin\theta = a$$

or $\qquad r\sin\theta + a = 0$, which is the same as (1).

Hence there is only one asymptote $r\sin\theta + a = 0$.

Example 2. *Find the asymptotes of the curve* $r = \dfrac{2a}{1 - 2\cos\theta}$.

Sol. The equation of the curve is

$$r = \frac{2a}{1 - 2\cos\theta} \quad \text{or} \quad \frac{1}{r} = \frac{1 - 2\cos\theta}{2a} = f(\theta), \text{ say.}$$

Now $f(\theta) = 0$ gives $1 - 2\cos\theta = 0$ or $\cos\theta = \dfrac{1}{2} = \cos\dfrac{\pi}{3}$. $\therefore$ $\theta = 2n\pi \pm \dfrac{\pi}{3}$

$$\text{Also} \qquad f'(\theta) = \frac{1}{2a}[0 + 2\sin\theta] = \frac{1}{a}\sin\theta.$$

$\therefore \quad f'\left(2n\pi \pm \frac{\pi}{3}\right) = \frac{1}{a} \sin\left(2n\pi \pm \frac{\pi}{3}\right) = \frac{1}{a} \sin\left(\pm \frac{\pi}{3}\right)$ $\quad | \because \quad \sin(2n\pi + \alpha) = \sin \alpha$

$\therefore$ Required asymptotes are given by

$$r \sin\left[\theta - \left(2n\pi \pm \frac{\pi}{3}\right)\right] = \frac{1}{f'\left(2n\pi \pm \frac{\pi}{3}\right)} \quad \text{or} \quad -r \sin\left[2n\pi \pm \frac{\pi}{3} - \theta\right] = \frac{a}{\sin\left(\pm \frac{\pi}{3}\right)}$$

or $$-r \sin\left(\pm \frac{\pi}{3} - \theta\right) = \frac{a}{\sin\left(\pm \frac{\pi}{3}\right)}$$

Taking upper signs, $-r \sin\left(\frac{\pi}{3} - \theta\right) = \dfrac{a}{\sin\left(\frac{\pi}{3}\right)}$ or $r \sin\left(\frac{\pi}{3} - \theta\right) = \dfrac{-2a}{\sqrt{3}}$.

Taking lower signs, $-r \sin\left(-\frac{\pi}{3} - \theta\right) = \dfrac{a}{\sin\left(-\frac{\pi}{3}\right)}$ or $r \sin\left(\frac{\pi}{3} + \theta\right) = -\dfrac{a}{\sin \frac{\pi}{3}} = -\dfrac{2a}{\sqrt{3}}$.

Hence the required asymptotes are

$$r \sin\left(\frac{\pi}{3} \pm \theta\right) + \frac{2a}{\sqrt{3}} = 0.$$

Example 3. *Find the asymptotes of the curve $r = a \operatorname{cosec} \theta + b$.*

Sol. The equation of the given curve is

$$r = a \operatorname{cosec} \theta + b = \frac{a}{\sin \theta} + b = \frac{a + b \sin \theta}{\sin \theta} \quad \text{or} \quad \frac{1}{r} = \frac{\sin \theta}{a + b \sin \theta} = f(\theta), \text{ say.}$$

$f(\theta) = 0$ gives $\sin \theta = 0 \quad \therefore \quad \theta = n\pi.$

$$\therefore \quad f'(\theta) = \frac{(a + b \sin \theta) \cos \theta - \sin \theta \cdot b \cos \theta}{(a + b \sin \theta)^2} = \frac{a \cos \theta}{(a + b \sin \theta)^2}$$

$$\therefore \quad f'(n\pi) = \frac{a \cos(n\pi)}{(a + b \sin n\pi)^2} = \frac{a(-1)^n}{(a + b \cdot 0)^2} = \frac{1}{a}(-1)^n,$$

$\therefore$ The asymptotes are given by

$$r \sin(\theta - n\pi) = \frac{1}{f(n\pi)} \quad \text{or} \quad -r \sin(n\pi - \theta) = \frac{a}{(-1)^n}$$

or $$-(-1)^n r \sin(-\theta) = \frac{a}{(-1)^n} \qquad | \because \quad \sin(n\pi + \alpha) = (-1)^n \sin \alpha$$

or $$(-1)^n r \sin \theta = \frac{a}{(-1)^n} \quad \text{or} \quad (-1)^{2n} r \sin \theta = a \quad \text{or} \quad r \sin \theta = a.$$

Hence there is only one asymptote $r \sin \theta = a$.

Example 4. *Find the asymptotes of the curve* $r = a \sec\theta + b \tan\theta$.

Sol. The equation of the curve is

$$r = a \sec\theta + b\tan\theta = \frac{a + b\sin\theta}{\cos\theta}$$

which can be re-written as $\dfrac{1}{r} = \dfrac{\cos\theta}{a + b\sin\theta} = f(\theta)$, say.

Now $f(\theta) = 0$ gives $\dfrac{\cos\theta}{a + b\sin\theta} = 0$ or $\cos\theta = 0$ $\therefore$ $\theta = (2n+1)\dfrac{\pi}{2}$

$$f'(\theta) = \frac{(a + b\sin\theta)(-\sin\theta) - \cos\theta \cdot b\cos\theta}{(a + b\sin\theta)^2}$$

$$= \frac{-a\sin\theta - b(\sin^2\theta + \cos^2\theta)}{(a + b\sin\theta)^2} = \frac{-(a\sin\theta + b)}{(a + b\sin\theta)^2}$$

$$\therefore \quad f'\left(2n + 1\frac{\pi}{2}\right) = -\frac{a\sin(2n+1)\dfrac{\pi}{2} + b}{\left[a + b\sin(2n+1)\dfrac{\pi}{2}\right]^2}$$

$$= -\frac{a(-1)^n + b}{[a + b(-1)^n]^2}$$

$\because \sin(2n+1)\dfrac{\pi}{2} = \sin\left(n\pi + \dfrac{\pi}{2}\right) = (-1)^n \sin\dfrac{\pi}{2} = (-1)^n$

Thus the asymptotes are given by

$$r\sin\left[\theta - (2n+1)\frac{\pi}{2}\right] = \frac{1}{f'\left[(2n+1)\dfrac{\pi}{2}\right]}$$

or $$-r\sin\left(n\pi + \frac{\pi}{2} - \theta\right) = -\frac{[a + b(-1)^n]^2}{b + a(-1)^n}$$

or $$(-1)^n r\sin\left(\frac{\pi}{2} - \theta\right) = \frac{[a + b(-1)^n]^2}{b + a(-1)^n}$$

or $$r\cos\theta = \frac{[a + b(-1)^n]^2}{(-1)^n[b + a(-1)^n]} = \frac{[a + b(-1)^n]^2}{b(-1)^n + a} = a + b(-1)^n$$

Putting $n = 0, 1, 2, 3, \ldots$, we get the asymptotes as $r\cos\theta = a + b$ and $r\cos\theta = a - b$.

Hence there are only two asymptotes of the curve given by $r\cos\theta = a \pm b$.

7.11. CIRCULAR ASYMPTOTES

Definition. If the equation of the curve be $r = f(\theta)$ and if $\lim\limits_{\theta \to \infty} f(\theta) = a$, then the circle $r = a$ is called the **circular asymptote** of the curve $r = f(\theta)$.

It is clear that in such a case, the curve approaches the circle as $\theta \to \infty$.

Example. *Find the circular asymptote of the curve $r(\theta + \sin\theta) = 2\theta + \cos\theta$.*

Sol. The equation of the given curve is

$$r(\theta + \sin\theta) = 2\theta + \cos\theta \quad \text{or} \quad r = \frac{2\theta + \cos\theta}{\theta + \sin\theta} = f(\theta), \text{ say.}$$

$$\lim_{\theta\to\infty} = \lim_{\theta\to\infty} \frac{2\theta + \cos\theta}{\theta + \sin\theta} \qquad | \text{ Divide num. and den. by } \theta$$

$$= \lim_{\theta\to\infty} \frac{2 + \frac{1}{\theta}\cos\theta}{1 + \frac{1}{\theta}\sin\theta} = \frac{2 + 0 \times \text{a finite value between } -1 \text{ and } 1}{1 + 0 \times \text{a finite value between } -1 \text{ and } 1} = \frac{2+0}{1+0} = 2.$$

Hence $r = 2$ is the required circular asymptote.

EXERCISE 7.4

Find the asymptotes of the following polar curves (1–10):

1. $r\theta = a$
2. $r = \dfrac{a\theta}{\theta - 1}$
3. $r(1 - e^{\theta}) = a$
4. $r \log\theta = a$
5. $r = a\tan\theta$
6. $r \sin n\theta = a$
7. $r\sin\theta = ae^{\theta}$
8. $r = a(\sec\theta + \tan\theta)$
9. $r\sin\theta = 2\cos 2\theta$ *(K.U.K. 2009, Dec. 2013)*
10. $r = \dfrac{3a\sin\theta\cos\theta}{\sin^3\theta + \cos^3\theta}$
11. Prove that the curve $r = \dfrac{a}{1 - \cos\theta}$ has no asymptotes.

Find the circular asymptotes of the following curves:

12. $r(e^{\theta} - 1) = a(e^{\theta} + 1)$
13. $r(2\theta^2 - 3\theta + 7) = 6\theta^2 + 5\theta - 1.$

Answers

1. $r\sin\theta = a$
2. $r\sin(\theta - 1) = a$
3. $r\sin\theta + a = 0$
4. $r\sin(\theta - 1) = a$
5. $r\cos\theta = \pm a$
6. $r\sin\left(\theta - \dfrac{m\pi}{n}\right) = \dfrac{a(-1)^m}{n}$
7. $r\sin\theta = ae^{n\pi}$
8. $r\cos\theta = 0,\ r\cos\theta = 2a$
9. $r\sin\theta = 2$
10. $r(\sin\theta + \cos\theta) + a = 0$
12. $r = a$
13. $r = 3.$

8

Curve Tracing

8.1. PROCEDURE FOR TRACING OF CARTESIAN CURVES

I. Symmetry. Apply the following rules to ascertain whether the given curve is symmetrical about any line or not.

(*i*) *Symmetry about the x-axis.* If the equation of a curve remains unaltered when y is changed to $-y$ *i.e.,* if only **even** powers of y occur in the equation, then the curve is symmetrical about x-axis. Thus the parabola $y^2 = 4ax$ is symmetrical about the x-axis.

(*ii*) *Symmetry about the y-axis.* If only **even** powers of x occur in the equation of the curve, the curve is symmetrical about the y-axis. Thus the curve $x^2 = 4y$ is symmetrical about the y-axis.

(*iii*) *Symmetry in opposite quadrants.* If the equation of the curve remains unaltered when **both** x and y are changed to $-x$ and $-y$ respectively, the curve is symmetrical in opposite quadrants. Thus the curve $xy = c^2$ is symmetrical in opposite quadrants.

(*iv*) *Symmetry about the line* $y = x$. If the equation of the curve is unchanged when x and y are interchanged (*i.e.,* x is changed to y and y to x), the curve is symmetrical about the line $y = x$. Thus the curve $x^3 + y^3 = 3axy$ is symmetrical about the line $y = x$.

II. Origin. (*i*) Find whether the curve passes through the origin or not. It will pass through the origin if the equation of the curve is free from the constant term.

(*ii*) If the curve passes through the origin, find the equations of the tangents at the origin by equating to zero the lowest degree terms in the equation of the curve.

(*iii*) If there is only one tangent, find whether the curve lies below or above the tangent in the neighbourhood of the origin. If there are two tangents, then origin is a double point. Find the nature of the double point.

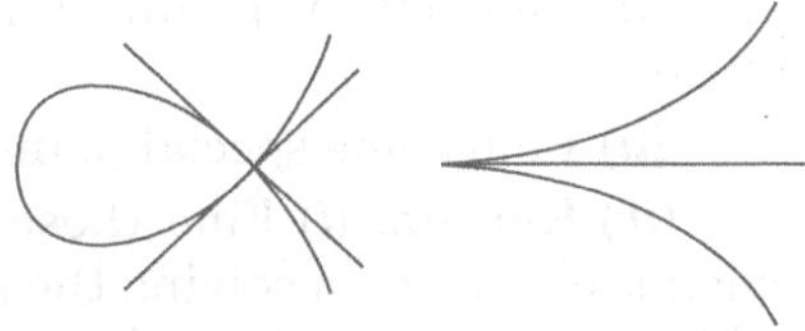

If the two tangents are real and distinct, then origin is a node.

If the two tangents are real and coincident, then origin is a cusp.

If the two tangents are imaginary, then origin is a conjugate point (or an isolated point).

III. Axes Intersection. Find the points where the curve cuts the co-ordinate axes. Find also the tangents at these points if necessary and the position of the curve relative to these tangents.

If $y = x$ or $y = -x$ is a line of symmetry, find the point of intersection of the curve and the line and also the tangents at that point.

IV. Asymptotes. (*a*) Find all the asymptotes (parallel to the axes and oblique) of the curve if any.

(*i*) *Asymptotes* || *to x-axis*. Equate to zero the co-efficient of the highest power of x, present in the given equation of the curve. Resolve L.H.S. into real linear factors. Each factor, equated to zero, gives an asymptotes || to x-axis.

(*ii*) *Asymptotes* || *to y-axis*. [Replace x by y in (*i*)].

(*iii*) *Oblique Asymptotes*. Putting $x = 1$, $y = m$ in the highest degree terms of the given equation, find $\phi_n(m)$.

Putting $x = 1$, $y = m$ in the next lower degree terms of the given equation, find $\phi_{n-1}(m)$. Similarly find $\phi_n(m)$.

Solve $\phi_n(m) = 0$ for m. These are the slopes of the asymptotes.

For distinct, real, values of m, find c from the equation

$$c\phi_n'(m) + \phi_{n-1}(m) = 0$$

For two real, equal, values of m, find c from the equation

$$\frac{c^2}{2!}\phi_n''(m) + c\phi'_{n-1}(m) + \phi_{n-2}(m) = 0.$$

Putting the values of m and corresponding values of c in $y = mx + c$, we get the required asymptotes.

(*b*) Find the position of the curve w.r.t. oblique asymptotes.

If the equation of a curve or one of its branches is of the form

$$y = mx + c + \frac{A}{x} + \frac{B}{x^2} + \cdots, \text{ then } \quad y = mx + c \text{ is an asymptote to the curve.}$$

If y_c and y_a stand for the ordinates of a point on the curve and a point on the asymptote (both having same abscissa), then $y_c - y_a = \frac{A}{x} + \frac{B}{x^2} + \cdots$

If $y_c - y_a > 0$, the curve lies above the asymptote.

If $y_c - y_a < 0$, the curve lies below the asymptote.

V. Special Points. (*i*) Find the points on the curve where $\frac{dy}{dx} = 0$ or ∞ *i.e.*, the points where the tangent is || or ⊥ to the x-axis. At such points, the ordinates or abscissae generally change the character from increasing to decreasing or *vice versa*.

(*ii*) Find out the points of inflexion, if they exist and the multiple points if any and their nature.

(*iii*) Find some special points on the curve if necessary.

(*iv*) **Region.** (*i*) Find those regions of xy-plane where the curve does not exist. Such a region is obtained on solving the given equation for one variable in terms of the other variable and finding out the set of values of one variable which make the other imaginary or undefined.

Illustration. Consider the curve $y^2(2a - x) = x^3$.

Here $$y = \pm x\sqrt{\frac{x}{2a - x}}$$

When x is +ve and $> 2a$, y is imaginary ($\because \; 2a - x$ is –ve) and so no part of the curve lies to the right of the line $x = 2a$.

Again when $x < 0$, y is again imaginary, so the curve does not lie to the left of the line $x = 0$ *i.e.*, y-axis.

(*ii*) Solve the equation for y (say) in terms of x. Consider $x = 0$ and observe how y varies as x increases and finally tends to $+\infty$, paying special attention to those values of x for which $y = 0$ or $y \to \infty$.

If as $x \to a, y \to 0$, then the special point is $(a, 0)$. Shift the origin to the point $(a, 0)$ and study the curve at the new origin.

Again if as $x \to b, y \to \infty$, then $x = b$ is an asymptote || to y-axis and we draw the shape of the curve near the line $x = b$.

(*iii*) Similarly observe the variations of y as x decreases from 0 to $-\infty$. But if the curve is symmetrical about y-axis, only +ve values of x need be considered and the curve for –ve values of x is traced by symmetry.

(*iv*) For limitations of the curve, we sometimes change from the Cartesian to polar or parametric form.

Important Note. In the examples, inconvenient steps may be omitted without any disadvantage.

ILLUSTRATIVE EXAMPLES

Example 1. *Trace the curve* $y = x^3 - 3ax^2$.

Sol. *The equation of the curve is* $\mathbf{y = x^3 - 3ax^2}$ ...(1)

1. Symmetry. The curve is not symmetrical about any line.

2. Origin. (*i*) The curve passes through the origin.

(*ii*) Equating to zero the lowest degree terms, the tangent at the origin is given by $y = 0$, *i.e.*, x-axis.

3. Axes Intersection. The curve meets x-axis, where putting $y = 0$ in (1), we get

$$x^3 - 3ax^2 = 0 \quad \text{or} \quad x^2(x - 3a) = 0. \quad \therefore \quad x = 0, 3a.$$

Hence the curve meets x-axis in the points (0, 0) and $(3a, 0)$. The curve meets y-axis only at the origin.

4. Asymptotes. The curve has no asymptote.

5. Special points. From (1), $\dfrac{dy}{dx} = 3x^2 - 6ax = 3x(x - 2a)$.

$\therefore \quad \dfrac{dy}{dx} = 0$ when $3x(x - 2a) = 0$ *i.e.*, if $x = 0$ or $2a$.

From (1), when $x = 0$, $y = 0$ and when $x = 2a$, $y = 8a^3 - 12a^3 = -4a^3$.

Thus the tangent at $(2a, -4a^3)$ is || to x-axis and the tangent at (0, 0) coincides with the x-axis. $\dfrac{dy}{dx}$ is not infinite for any real value of x.

6. Region. The equation of the curve can be written as $y = x^2(x - 3a)$

For values of $x < 3a$, y is –ve and for values of $x > 3a$, y is +ve.

When $x = 0$, $y = 0$. As x increases from 0 to $2a$, y decreases from 0 to $-4a^3$. As x increases from $2a$ to $3a$, y increases from $-4a^3$ to 0 and as x further increases from $3a$ to ∞, y also increases from 0 to ∞.

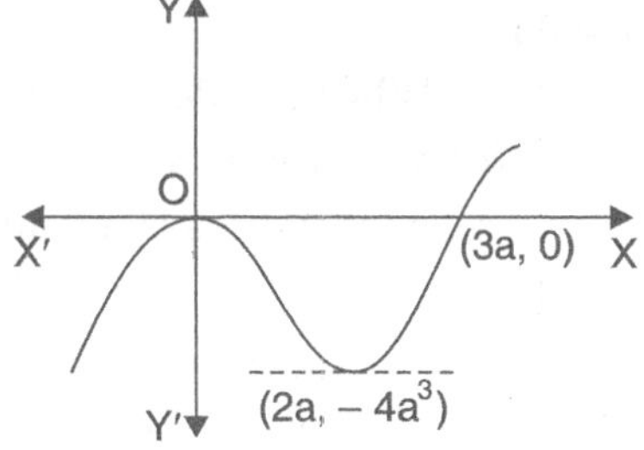

As x decreases from 0 to $-\infty$, y also decreases from 0 to $-\infty$.

Hence the shape of the curve is as shown in the adjoining figure.

Example 2. *Trace the curve* $y^2(a + x) = x^2(3a - x)$.

Sol. *The equation of the curve is*

$$\mathbf{y^2(a + x) = x^2(3a - x)} \qquad ...(1)$$

1. Symmetry. Since the equation (1) contains only even powers of y, $\therefore$ The curve is symmetrical about x-axis.

The curve is *not* symmetrical about y-axis.

2. Origin. (*i*) The curve passes through the origin.

(*ii*) Equating to zero, the lowest degree terms, the tangents at the origin are given by $ay^2 - 3ax^2 = 0$ or $y = \pm\sqrt{3}x$.

Since the two tangents are real and different,

$\therefore$ Origin is a node.

(*iii*) From (1), we have $y^2 = x^2 \dfrac{(3a - x)}{a + x}$.

Since $\dfrac{3a - x}{a + x} < \dfrac{3a}{a}$ *i.e.,* < 3 for small +ve x. $\therefore$ y is less than $\sqrt{3}x$.

Hence the curve lies below the tangent $y = \sqrt{3}x$ for small +ve values of x.

3. Axes Intersection. The curve meets x-axis, where putting $y = 0$, we have

$$x^2(3a - x) = 0. \quad \therefore \quad x = 0 \quad \text{or} \quad 3a$$

i.e., in the points (0, 0) and A(3a, 0).

Similarly it meets y-axis in the origin only.

Shifting the origin to (3a, 0), the equation (1) transforms to

$$Y^2(4a + X) = (3a + X)^2(-X).$$

Equating to zero the lowest degree terms, the tangent at the new origin is given by $-9a^2X = 0$ or $X = 0$. Thus the tangent at A(3a, 0) is $\parallel$ to y-axis.

4. Asymptotes. Equating to zero the co-efficient of y^2, the asymptote $\parallel$ to y-axis is $x + a = 0$.

There is no other asymptote of the curve.

5. Region. Solving for y and considering the +ve values ($\because$ the curve is symmetrical about the x-axis), we have

$$y = x\sqrt{\frac{3a - x}{a + x}} \qquad ...(2)$$

(*i*) When $x = 0$, $y = 0$. When x is +ve and small, y is real. As x goes on increasing, y also increases upto the point for which $x = \sqrt{3}a$ and then decreases and finally becomes zero at $x = 3a$.

(*ii*) When $x > 3a$, y is imaginary. Hence no part of the curve lies beyond the line $x = 3a$.

(*iii*) When x is –ve and numerically small, from (2), y is real.

Also for values of x under consideration,

$$3a - x > 3(a + x)$$

$\therefore$ y is numerically $> -\sqrt{3}x$.

$\therefore$ The curve lies above the tangent in the second quadrant.

(*iv*) When $x < -a$, from (2), y is imaginary. $\therefore$ The curve does not lie beyond $x = -a$.

Again when x is –ve and numerically increases from 0 to a, y is –ve and decreases from 0 to $-\infty$. Hence the shape of the curve is as shown in the figure.

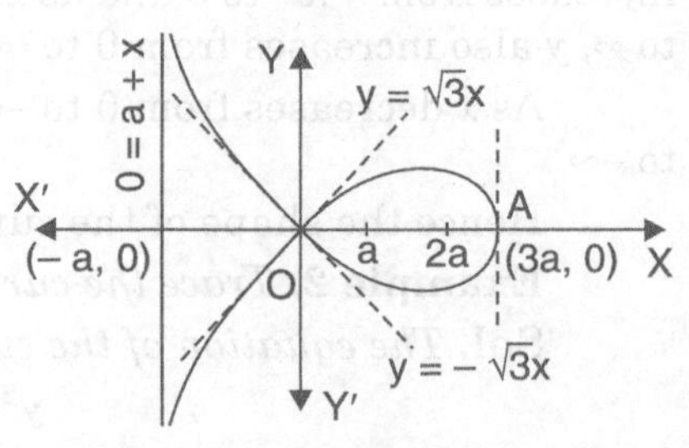

Example 3. *Trace the curve $y^2(2a - x) = x^3$. [Cissoid]*

Sol. *The equation of the curve is* $\mathbf{y^2(2a - x) = x^3}$ ...(1)

1. Symmetry. Since (1) contains even powers of y, the curve is symmetrical about x-axis.

2. Origin. (*i*) The curve passes through the origin.

(*ii*) The tangents at the origin are given by $y^2 = 0$. Since the two tangents are real and coincident, ∴ Origin is a cusp.

3. Axes Intersection. The curve meets x-axis and y-axis at the origin only.

4. Asymptotes. Equating to zero the coefficient of y^2, the highest degree term in y, the asymptote || to y-axis is $x - 2a = 0$. There is no other asymptote of the curve.

5. Special points. From (1),

$$y = \frac{x^{3/2}}{\sqrt{2a - x}} = x^{3/2}(2a - x)^{-1/2} \quad ...(2)$$

$$\therefore \quad \frac{dy}{dx} = \frac{3}{2}x^{1/2}(2a - x)^{-1/2} - \frac{1}{2}(2a - x)^{-3/2}\cdot(-1)\cdot x^{3/2}$$

$$= \frac{1}{2}(2a - x)^{-3/2}\cdot x^{1/2}[3(2a - x) + x] = \frac{\sqrt{x}(3a - x)}{(2a - x)^{3/2}}$$

$$\therefore \quad \frac{dy}{dx} = 0 \text{ when } \sqrt{x}(3a - x) = 0 \quad \text{or} \quad x = 0, 3a.$$

Rejecting $x = 3a$ ∵ when $x = 3a$, from (2), y is imaginary.

∴ The tangent at $x = 0$ is || to x-axis.

Again $\frac{dy}{dx} \to \infty$ when $x \to 2a$. From (2), when $x \to 2a$, $y \to \infty$.

Thus $x = 2a$ is an asymptote.

6. Region. From (1), $y = x\sqrt{\frac{x}{2a - x}}$ | Taking +ve root.

When $x < 0$ or when $x > 2a$, y is imaginary.

Thus the curve does not lie to the left of y-axis and to the right of the line $x = 2a$.

When $x = 0$, $y = 0$. As x increases from 0 to $2a$, y also increases from 0 to ∞.

Hence the shape of the curve is as shown in the adjoining figure.

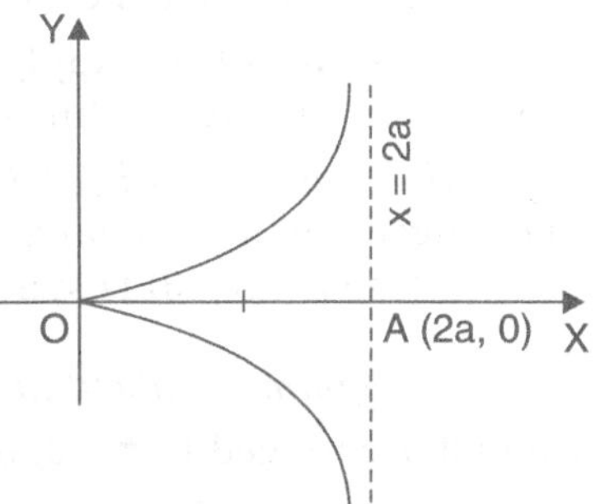

Example 4. *Trace the curve*

$$y^2(a^2 + x^2) = x^2(a^2 - x^2) \quad or \quad x^2(x^2 + y^2) = a^2(x^2 - y^2).$$

Sol. *The equation of the curve is*

$$\mathbf{y^2(a^2 + x^2) = x^2(a^2 - x^2)} \quad ...(1)$$

1. Symmetry. Since in (1), powers of both x and y are even, the curve is symmetrical about both axes.

2. Origin. The curve passes through the origin and the tangents at the origin are $y^2 = x^2$ or $y = \pm x$, which are real and distinct showing that origin is a node.

3. Axes Intersection. The curve meets x-axis at A(a, 0) and A′($-a$, 0). Transferring the origin to (a, 0) the equation (1) transforms to $Y^2[a^2 + (X + a)^2] = (X + a)^2[a^2 - (X + a)^2]$

or $$Y^2[2a^2 + 2aX + X^2] = (X + a)^2[-2aX - X^2]$$

Equating to zero the lowest degree terms, the tangent at the new origin is $X = 0$ *i.e.*, new y-axis. Thus the tangent at $A(a, 0)$ is $\parallel$ to y-axis. Similarly the tangent at $A'(-a, 0)$ is also parallel to y-axis (by symmetry).

4. Asymptotes. The curve has no asymptotes.

5. Region. (*i*) Solving (1) for y, we have $y = \pm x\sqrt{\dfrac{a^2 - x^2}{a^2 + x^2}}$

Now y is real if and only if $a^2 - x^2 \geq 0$ or if $x^2 \leq a^2$ or $|x| \leq a$.

$\therefore$ The whole curve lies between the lines $x = \pm a$.

(*ii*) For the branch in the first quadrant.

$$y = x\sqrt{\frac{a^2 - x^2}{a^2 + x^2}}$$

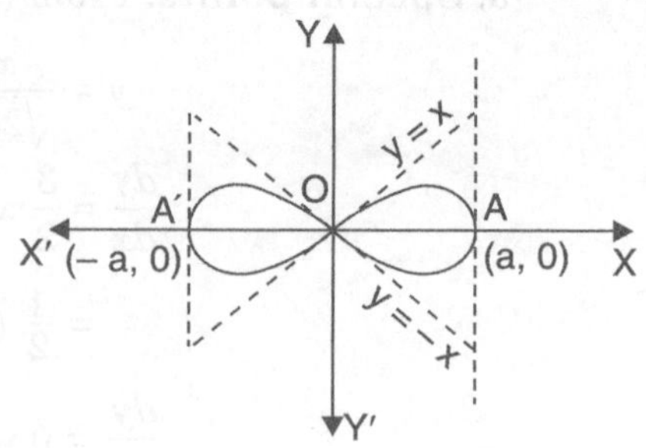

When $x = 0$, $y = 0$. As x increases, y also increases and goes on increasing until $x = a\sqrt{\sqrt{2} - 1}$, where $\dfrac{dy}{dx} = 0$ *i.e.*, the tangent is parallel to x-axis. As x increases from $a\sqrt{\sqrt{2} - 1}$ to a, y decreases and ultimately becomes zero when $x = a$.

The shape of the curve is as shown in the figure.

8.2. PROCEDURE FOR TRACING POLAR CURVES

Let the equation of the curve be $f(r, \theta) = 0$.

I. Symmetry. Apply the following rules to ascertain whether the given curve is symmetrical about any line or not:

(*i*) *Symmetry about the initial line or x-axis.* If the equation of the curve remains unchanged when θ is changed to $-\theta$, the curve is symmetrical about the initial line. Thus the curve $r = a(1 + \cos\theta)$ is symmetrical about the initial line.

(*ii*) *Symmetry about the line $\theta = \dfrac{\pi}{2}$ or y-axis.* If the equation of the curve remains unaltered when θ is changed to $\pi - \theta$, or when θ is changed to $-\theta$ and r to $-r$, the curve is symmetrical about the line $\theta = \dfrac{\pi}{2}$. Thus the curves $r = a(1 + \sin\theta)$, $r = a\sin 3\theta$ are symmetrical about the line $\theta = \dfrac{\pi}{2}$ or y-axis.

(*iii*) *Symmetry about the line $\theta = \dfrac{\pi}{4}$ or the line $y = x$.* If the equation of the curve remains unchanged when θ is changed to $\dfrac{\pi}{2} - \theta$, the curve is said to be symmetrical about the line $\theta = \dfrac{\pi}{4}$. Thus the curve $r = a\sin 2\theta$ is symmetrical about the line $\theta = \dfrac{\pi}{4}$ or the line $y = x$.

(*iv*) *Symmetry about the line* $\theta = \frac{3\pi}{4}$ *or the line* $y = -x$. If the equation of the curve remains unchanged when θ is changed to $\frac{3\pi}{2} - \theta$, the curve is symmetrical about the line $\theta = \frac{3\pi}{4}$. Thus the curve $r = a \sin 2\theta$ is symmetrical about the line $\theta = \frac{3\pi}{4}$.

(*v*) *Symmetry about the pole.* If the equation of the curve remains unchanged when r is changed to $-r$, the curve is said to be symmetrical about the pole. Thus the curve $r^2 = a^2 \cos 2\theta$ is symmetrical about the pole.

II. Origin or Pole. (*i*) *Find whether the curve passes through the pole or not.* This can be done by putting $r = 0$ in the equation and then finding some real value of θ.

If it is not possible to find a real value of θ for which $r = 0$, then the curve does not pass through the pole.

(*ii*) *Find the tangents at the pole.* Putting $r = 0$, the real values of θ give the tangents at the pole.

For instance, consider the curve $r = a(1 - \sin \theta)$

Putting $r = 0$, we get $\sin \theta = 1$ or $\theta = \frac{\pi}{2}$.

Hence the curve passes through the pole and $\theta = \frac{\pi}{2}$ is the tangent at the pole.

(*iii*) Find the points where the curve meets the initial line and the line $\theta = \frac{\pi}{2}$.

III. Value of ϕ. (*i*) Find $\tan \phi = r\frac{d\theta}{dr}$ and hence ϕ.

(*ii*) Find the points, where $\phi = 0$ or $\frac{\pi}{2}$.

IV. Asymptotes. If $r \to \infty$ and $\theta \to \theta_1$, then there is an asymptote. Find it by usual method. [θ_1 is a fixed number].

V. Special Points. Find some points on the curve for convenient values of θ.

VI. Region. Solve the given equation for r or θ. Find the regions in which the curve does not lie. This can be done in the following manner.

(*i*) If r is imaginary, when $\alpha < \theta < \beta$, then there is no part of the curve between the lines $\theta = \alpha$ and $\theta = \beta$. Consider the curve $r^2 = a^2 \sin 2\theta$. When $\frac{\pi}{2} < \theta < \pi$, $\sin 2\theta$ is –ve,

$\therefore$ r^2 is –ve and so r is imaginary.

Hence the curve does not lie between the lines $\theta = \frac{\pi}{2}$ and $\theta = \pi$.

(*ii*) If the greatest numerical value of r be a, the curve lies entirely within the circle $r = a$. Also, if the least numerical value of r be b, the curve lies outside the circle $r = b$.

(*iii*) Trace the variations of r when θ varies in the intervals $(0, \infty)$ and $(-\infty, 0)$ marking values of θ for which $r = 0$ or attains a minimum or maximum value. Plot the points so obtained.

[**Note.** When r is a periodic function of θ, the –ve values of θ need not be taken into account. We may consider values from $\theta = 0$ to those values of θ, where the values begin to repeat].

VII. Conversion into Cartesians. Transform the equation to Cartesian co-ordinates whenever required.

ILLUSTRATIVE EXAMPLES

Example 1. *Trace the curve* $r = a(1 + \cos\theta)$ *(Cardioid).* (P.T.U. Dec. 2011)

Sol. *The equation of the curve is* $\mathbf{r = a(1 + \cos\theta)}$ ...(1)

1. Symmetry. The equation of curve remains unchanged when θ is changed to $-\theta$.

$\therefore$ The curve is symmetrical about initial line.

2. Origin or Pole. (*i*) When $\theta = \pi$, $r = 0$, $\therefore$ Pole lies on the curve.

(*ii*) The tangent at the pole is $\theta = \pi$.

(*iii*) The curve cuts the initial line $\theta = 0$ at $(2a, 0)$ and the lines $\theta = \pm\dfrac{\pi}{2}$ at $\left(a, \dfrac{\pi}{2}\right), \left(a, -\dfrac{\pi}{2}\right)$.

3. Value of ϕ.

$$\tan\phi = r\frac{d\theta}{dr} = -\frac{(1+\cos\theta)}{\sin\theta} = -\cot\frac{\theta}{2} = \tan\left(\frac{\pi}{2}+\frac{\theta}{2}\right)$$

$$\Rightarrow \qquad \phi = \frac{\pi}{2}+\frac{\theta}{2} \quad \therefore \quad \phi = \frac{\pi}{2} \text{ when } \theta = 0 \text{ and } r = 2a,$$

i.e., at $(2a, 0)$ the tangent is perpendicular to initial line.

4. Asymptotes. There are no asymptotes, because for any finite value of θ, r does not tend to infinity.

5. Special Points.

The corresponding values of θ and r are given below

$\theta = 0$	$\dfrac{\pi}{4}$	$\dfrac{\pi}{2}$	π
$r = 2a$	$a\left(1+\dfrac{1}{\sqrt{2}}\right)$	a	0

6. Region. (*i*) r is never greater than $2a$

$\therefore$ No portion of the curve lies to the right of the tangent at $(2a, 0)$.

(*ii*) Since $|\cos\theta| \le 1$, $\therefore$ $r \le 2a$

$\therefore$ Curve lies entirely within the circle $r = 2a$.

(*iii*) When θ increases from 0 to π, r remains positive and decreases from $2a$ to 0.

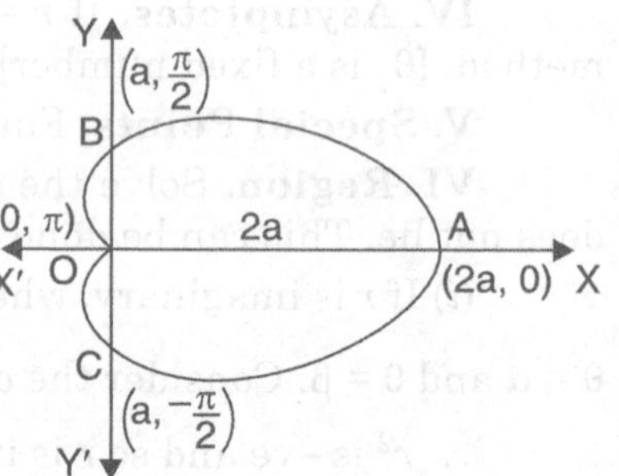

When θ increases from π to 2π, r remains positive and increases from 0 to $2a$.

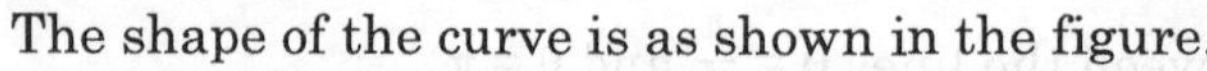

The shape of the curve is as shown in the figure.

Example 2. *Trace the curve* $r = 2 + 3\cos\theta$. (K.U.K. Dec. 2010)

Sol. *The equation of the curve is* $\mathbf{r = 2 + 3\cos\theta}$ ...(1)

1. Symmetry. Equation of curve remains unchanged when θ is changed to $-\theta$,

$\therefore$ Curve is symmetrical about initial line.

2. Origin or Pole. When $r = 0$, $\cos\theta = -\frac{2}{3} < 1$ numerically.

$\therefore$ The curve passes through the pole.

Now $\cos\theta = -\frac{2}{3} = \cos(\pi - \alpha)$ [say]

$\therefore$ $\theta = (\pi - \alpha)$ is tangent to the curve at the pole where α is given by $\cos(\pi - \alpha) = -\frac{2}{3}$.

3. Value of ϕ. From (1), $$\frac{dr}{d\theta} = -3 \sin \theta$$

$$\tan \phi = r\frac{d\theta}{dr} = \frac{2 + 3\cos\theta}{-3\sin\theta} = 0$$

When $2 + 3 \cos \theta = 0$ or $\cos \theta = -\frac{2}{3}$ or $\theta = (\pi - \alpha)$

$\therefore$ At $(0, \pi - \alpha)$ the tangent to curve is parallel to initial line.

$\tan \phi = \infty$ when $\sin \theta = 0$, *i.e.*, when $\theta = 0$ or π.

$\therefore$ At (5, 0) and $(-1, \pi)$, the tangent is perpendicular to the initial line.

4. Asymptotes. Curve has no asymptotes because for any real value of θ, r is never infinite.

5. Special Points. The values of r for corresponding values of θ, are as follows:

$\theta = 0$	$\frac{\pi}{2}$	π
$r = 5$	2	-1

6. Region. (*i*) Since $|\cos \theta| \le 1$, $\therefore$ $r \le 5$.

$\therefore$ Entire curve lies within the circle $r = 5$.

(*ii*) As θ increases from 0 to $\frac{\pi}{2}$, r remains positive but decreases from 5 to 2.

When θ increases from $\frac{\pi}{2}$ to π, r decreases from 2 to 0 and then from 0 to -1.

Thus the shape of the curve is as shown in the figure.

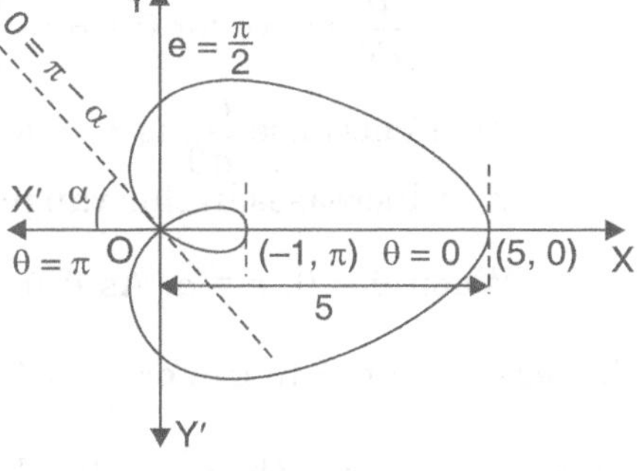

Example 3. *Trace the curve $r^2 = a^2 \cos 2\theta$.* (U.P.T.U. 2009; M.T.U. 2011)

Sol. *The equation of the curve is* $\mathbf{r^2 = a^2 \cos 2\theta}$ ...(1)

1. Symmetry (*i*) Since (1) remains unaltered when θ is changed to $-\theta$, the curve is symmetrical about the initial line.

(*ii*) Since the equation remains unchanged when θ is changed to $\pi - \theta$, the curve is symmetrical about the line $\theta = \frac{\pi}{2}$.

(*iii*) Since (1) remains unaltered when r is changed to $-r$, the curve is symmetrical about the pole.

2. Origin or Pole. (*i*) Putting $r = 0$, we get $\cos 2\theta = 0$, or $2\theta = \pm\frac{\pi}{2}$ or $\theta = \pm\frac{\pi}{4}$. Since it is possible to find real values of θ for which $r = 0$, therefore the curve passes through the pole and $\theta = \pm\frac{\pi}{4}$ are the tangents at the pole.

(*ii*) When $\theta = 0$, $r^2 = a^2$ or $r = \pm a$. Thus the curve meets the initial line ($\theta = 0$) in the points $(\pm a, 0)$.

3. Asymptotes. Since there is no finite value of θ for which $r \to \infty$, therefore the curve has no asymptote.

4. Value of ϕ. From (1), $2r\dfrac{dr}{d\theta} = -2a^2 \sin 2\theta$

$$\therefore \quad \tan\phi = r\frac{d\theta}{dr} = \frac{r}{\dfrac{dr}{d\theta}} = \frac{r}{\dfrac{-a^2 \sin 2\theta}{r}} = \frac{a^2 \cos 2\theta}{-a^2 \sin 2\theta} = -\cot 2\theta = \tan\left(\frac{\pi}{2} + 2\theta\right)$$

$$\therefore \quad \phi = \frac{\pi}{2} + 2\theta$$

when $\theta = 0$, $\phi = \dfrac{\pi}{2}$. Thus at the points $(\pm a, 0)$, the tangents are $\perp$ to the initial line.

5. Special Points and Region.

From (1), $\quad r = a\sqrt{\cos 2\theta}$ | *Taking +ve root,* $\because$ *the curve is symmetrical about the pole.*

$$\therefore \quad \frac{dr}{d\theta} = -\frac{a \sin 2\theta}{\sqrt{\cos 2\theta}}$$

$\because \quad \dfrac{dr}{d\theta}$ is –ve for $0 < \theta < \dfrac{\pi}{4}$, $\quad \therefore \quad r$ decreases in this range.

Again because $\dfrac{dr}{d\theta}$ is + ve for $\dfrac{3\pi}{4} < \theta < \pi$,

$\therefore \quad r$ increases in this range.

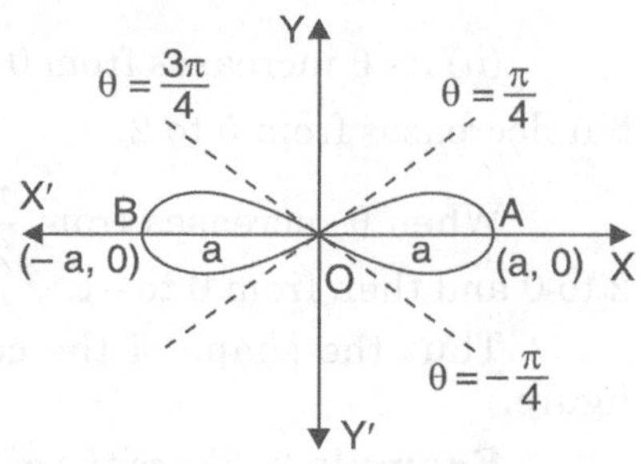

When $\theta = 0$, $r = a$. As θ increases from 0 to $\dfrac{\pi}{4}$, r decreases from a to 0. For $\dfrac{\pi}{4} < \theta < \dfrac{3\pi}{4}$, r is imaginary.

Thus no portion of the curve lies between the lines $\theta = \dfrac{\pi}{4}$ and $\theta = \dfrac{3\pi}{4}$.

Again as θ increases from $\dfrac{3\pi}{4}$ to π, r is +ve and increases from 0 to a. Thus we trace the part of the curve above the initial line. The part of the curve below the initial line can be traced by symmetry. Hence the shape of the curve is as shown in the figure.

Example 4. *Trace the curve* $r = a \sin 3\theta$ [*Three leaved rose*].

Sol. *The equation of the curve is* $\mathbf{r = a \sin 3\theta}$ ...(1)

1. Symmetry. Since (1) remains unaltered when θ is changed to $\pi - \theta$, therefore the curve is symmetrical about the line $\theta = \dfrac{\pi}{2}$.

2. Pole. Putting $r = 0$, we get $\sin 3\theta = 0$, which gives

$$3\theta = 0, \quad \pi, \quad 2\pi, \quad 3\pi, \ldots$$

$$\text{or} \quad \theta = 0, \quad \frac{\pi}{3}, \quad \frac{2\pi}{3}, \quad \pi, \ldots$$

$\therefore$ The curve passes through the pole and the tangents at the pole are $\theta = 0$, $\theta = \dfrac{\pi}{3}$ and $\theta = \dfrac{2\pi}{3}$, the other values of θ give the same tangents.

3. Asymptotes. Since for no values of θ, $r \to \infty$, therefore the curve has no asymptote.

4. Special Points and Region. (*i*) From (1), $r = a \sin 3\theta$.

$\because \quad |\sin 3\theta| \le 1$.

$\therefore \quad r$ is never greater than a. Thus the curve lies entirely within the circle $r = a$.

(*ii*) Some special points are

$\theta = 0$	$\frac{\pi}{6}$	$\frac{\pi}{3}$	$\frac{\pi}{2}$	$\frac{2\pi}{3}$	$\frac{5\pi}{6}$	π
$r = 0$	a	0	$-a$	0	a	0

When θ increases from 0 to $\frac{\pi}{6}$, r is +ve and increases from 0 to a.

When θ increases from $\frac{\pi}{6}$ to $\frac{\pi}{3}$, r is +ve and decreases from a to 0. Thus we get a loop between the lines $\theta = 0$ and $\theta = \frac{\pi}{3}$.

As θ increases from $\frac{\pi}{6}$ to $\frac{\pi}{2}$, r is –ve and numerically increases from 0 to a. As θ increases from $\frac{\pi}{2}$ to $\frac{2\pi}{3}$, r is –ve and numerically decreases from a to 0. Thus we get a second loop between the lines $\theta = \frac{\pi}{3}$ and $\theta = \frac{2\pi}{3}$ in the opposite direction [$\because$ r is –ve]. As θ increases from $\frac{2\pi}{3}$ to $\frac{5\pi}{6}$, r remains +ve and increases from 0 to a. As θ increases from $\frac{5\pi}{6}$ to π, r remains +ve and decreases from a to 0. Thus we have the third loop between the lines $\theta = \frac{2\pi}{3}$ and $\theta = \pi$.

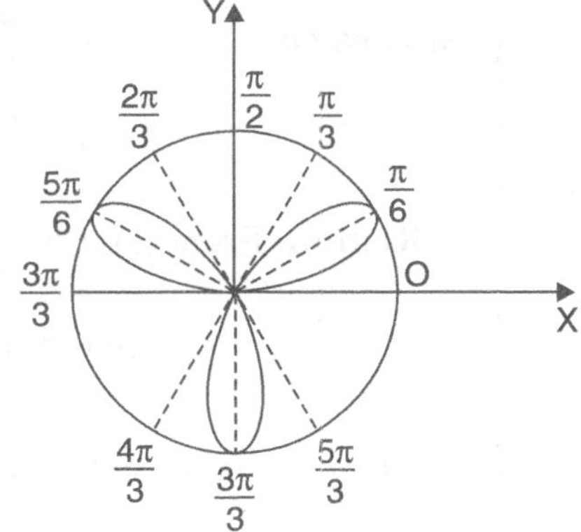

When θ varies from π to 2π, the same loops are repeated and we do not get any new loop. As r is periodic, therefore values of θ outside [0, 2π] need not be considered.

Thus the shape of the curve is as shown in the figure.

Note. The curve $r = a \sin n\theta$ or $r = a \cos n\theta$ consists of n or $2n$ loops according as n is odd or even.

To trace the curve $r = a \sin n\theta$,

(*i*) Put $r = 0$, thus getting $\sin n\theta = 0$

or $$n\theta = 0, \pi, 2\pi, \ldots \quad \text{or} \quad \theta = 0, \frac{\pi}{n}, \frac{2\pi}{n}, \ldots$$

Draw the lines $\theta = 0, \theta = \frac{\pi}{n}, \theta = \frac{2\pi}{n}, \ldots$

These lines divide the circle $r = a$ into $2n$ equal parts.

(*ii*) If n is odd, place the n loops in alternate divisions.

(*iii*) If n is even, place the $2n$ loops, one in each division.

Similarly for $r = a \cos n\theta$.

Example 5. *Trace the curve* $r\theta = a$. [*Reciprocal spiral*] (K.U.K. 2012)

Sol. *The equation of the curve is* $\mathbf{r\theta = a}$. ...(1)

Let us consider positive values of θ, *i.e.*, values of θ in (0, ∞) first.

1. Symmetry. Since (1) remains unaltered when θ is changed to $-\theta$ and r to $-r$, therefore, the curve is symmetrical about the line $\theta = \dfrac{\pi}{2}$, *i.e.*, y-axis.

2. Pole. Since r does not become zero for any finite value of θ, therefore, the curve does not pass through the pole.

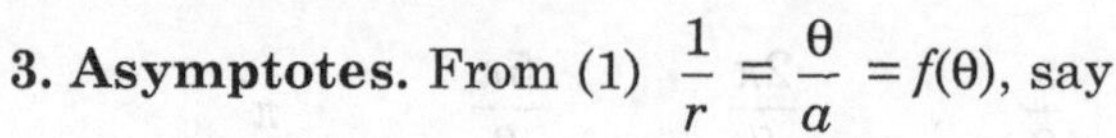

3. Asymptotes. From (1) $\dfrac{1}{r} = \dfrac{\theta}{a} = f(\theta)$, say

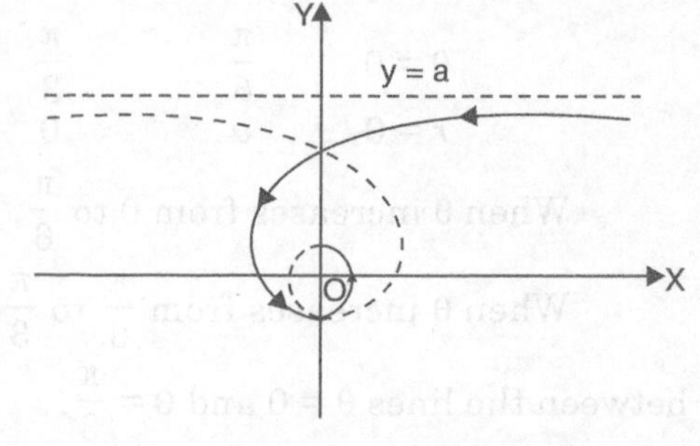

$$\therefore \quad f(\theta) = 0 \;\Rightarrow\; \frac{\theta}{a} = 0 \quad \text{or} \quad \theta = 0$$

Also $f'(\theta) = \dfrac{1}{a}$ so that $f'(0) = \dfrac{1}{a}$

The asymptote corresponding to $\theta = 0$ is

$$r \sin(\theta - 0) = \frac{1}{f'(0)} \quad \text{or } r \sin\theta = a$$

or $\qquad y = a \qquad (\because\; r\sin\theta = y)$

4. Special points. From (1), $r \to \infty$ when $\theta \to 0^+$

Also, when

$$\theta = \frac{\pi}{6} \quad \frac{\pi}{3} \quad \frac{\pi}{2} \quad \frac{2\pi}{3} \quad \pi \quad \frac{3\pi}{2} \quad 2\pi \ldots$$

$$r = \frac{6a}{\pi} \quad \frac{3a}{\pi} \quad \frac{2a}{\pi} \quad \frac{3a}{2\pi} \quad \frac{a}{\pi} \quad \frac{2a}{3\pi} \quad \frac{a}{2\pi} \ldots$$

5. Region. From (1) $r = \dfrac{a}{\theta}$

$$\Rightarrow \quad r\sin\theta = a\frac{\sin\theta}{\theta} < a \qquad \left(\because\; \frac{\sin\theta}{\theta} < 1\right)$$

$$\Rightarrow \quad y < a \qquad (\because\; r\sin\theta = y)$$

$\therefore$ The curve lies entirely below the asymptote $y = a$

Also

$$r = \frac{a}{\theta} \;\Rightarrow\; \frac{dr}{d\theta} = -\frac{a}{\theta^2} < 0$$

$\therefore$ As θ increases from 0 to ∞, r is positive and decreases from ∞ to 0.

Hence for positive values of θ, the shape of the curve is shown thick in the figure. The part of the curve for negative values of θ is shown dotted in the figure.

Example 6. *Trace the curve* $x^3 + y^3 = 3axy$. [*Folium of Descarte's*] (K.U.K. 2012)

Sol. *The equation of the given curve is* $\mathbf{x^3 + y^3 = 3axy}$ ...(1)

1. Symmetry. Since (1) remains unaltered when x and y are interchanged,

$\therefore$ The curve is symmetrical about the line $y = x$.

2. Origin. The curve passes through the origin and equating to zero the lowest degree terms, the tangents at the origin are given by $xy = 0$ or $x = 0$, $y = 0$.

These tangents being real and distinct, origin is a node.

3. Axes Intersection. (*i*) The curve meets the axes only at the origin (0, 0).

(*ii*) The line $y = x$ meets the curve in the point $\left(\dfrac{3a}{2}, \dfrac{3a}{2}\right)$ besides the origin, and the tangent at the point $\left(\dfrac{3a}{2}, \dfrac{3a}{2}\right)$ is $x + y = 3a$.

4. Asymptotes. (*i*) The curve has no asymptotes || to axes.

(*ii*) There is an oblique asymptote which is determined as below.

Putting $x = 1$, $y = m$ in the highest (*i.e.*, third) degree terms and equating to zero, we get

$$\phi_3(m) = 1 + m^3 = 0 \qquad \text{which gives} \quad m = -1$$

Now $\phi_2(m) = -3am$ | *Putting x = 1, y = m in the second degree terms*

$$\therefore \quad c = -\frac{\phi_2(m)}{\phi_3'(m)} = -\frac{-3am}{3m^2} = \frac{a}{m} = -a \qquad | \because m = -1$$

$\therefore$ The asymptote is $y = mx + c$ or $y = -x - a$ or $x + y + a = 0$.

5. Region. (*i*) x and y cannot both be –ve,

$\because$ if x and y are both –ve, then LHS of (1) is –ve and RHS of (1) is +ve, which is impossible.

$\therefore$ No portion of the curve lies in the third quadrant.

(*ii*) Transforming (1) to polars [by putting $x = r\cos\theta$, $y = r\sin\theta$],

we get $r^3(\cos^3\theta + \sin^3\theta) = 3ar^2 \sin\theta\cos\theta$

or
$$r = \frac{3a\sin\theta\cos\theta}{\cos^3\theta + \sin^3\theta} \qquad ...(2)$$

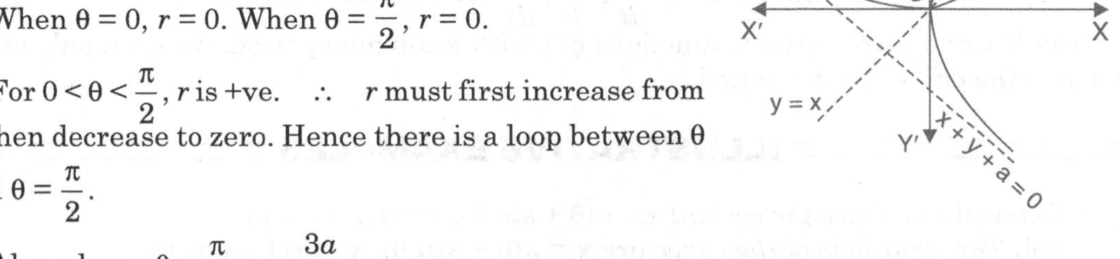

When $\theta = 0$, $r = 0$. When $\theta = \frac{\pi}{2}$, $r = 0$.

For $0 < \theta < \frac{\pi}{2}$, r is +ve. $\therefore$ r must first increase from 0 and then decrease to zero. Hence there is a loop between $\theta = 0$ and $\theta = \frac{\pi}{2}$.

Also when $\theta = \frac{\pi}{4}$, $r = \frac{3a}{\sqrt{2}}$.

Again as θ increases from $\frac{\pi}{2}$ to $\frac{3\pi}{4}$, $\sin\theta$ remains +ve and decreases while $\cos\theta$ becomes –ve and numerically increases. $\therefore$ r is –ve in this range and numerically increases from 0 to ∞. Thus we have the portion of the curve in the fourth quadrant.

As θ increases from $\frac{3\pi}{4}$ to π, r remains +ve and decreases from ∞ to 0.

Thus we have the portion of the curve in the second quadrant. As θ increases from π to 2π, we do not get any new point on the curve.

Hence the shape of the curve is as shown in the figure.

8.3. PROCEDURE FOR TRACING PARAMETRIC CURVES

Method I. *Eliminate the parameter if possible and obtain the corresponding Cartesian equation of the curve which can be traced as explained earlier.*

Method II. This method is used when the parameter cannot be easily eliminated from the given parametric equations.

In this case we proceed as follows:

Let $x = f(t)$, $y = \phi(t)$ be the given parametric equations of the curve, where t is the parameter.

1. Limits. If possible find the greatest and the least values of x and y and therefore the lines || to the axes between which the curve lies or does not lie.

2. Symmetry. (*i*) If $x = f(t)$ be an even function of t and $y = \phi(t)$ be an odd function of t, then the curve is *symmetrical about x-axis*.

(*ii*) If x be an odd function of t and y, an even function of t, then the curve is *symmetrical about the y-axis.*

(*iii*) If x and y are both odd functions of t, then the curve is *symmetrical in opposite quadrants.*

3. Origin. If on putting $x = 0$, we get a real value of t, which also makes $y = 0$, then the curve passes through the origin.

4. Axes Intersection. Find the points of intersection of the curve and the co-ordinate axes.

5. Asymptotes. Find the asymptotes if any.

6. Special Points. Find points on the curve, where $\frac{dy}{dx} = 0$ or ∞.

7. Region

(*i*) Find the regions in which the curve does not exist.

(*ii*) Consider the signs of $\frac{dx}{dt}$ and $\frac{dy}{dt}$ in suitable intervals of t.

(*iii*) Consider the values of x, y, $\frac{dx}{dt}, \frac{dy}{dt}, \frac{dy}{dx}$ for suitable values of t.

(*iv*) If x and y are periodic functions of t with a common period, we need only study the portion of the curve for one period.

ILLUSTRATIVE EXAMPLES

Example 1. *Trace the cycloid* $x = a(\theta + \sin \theta)$, $y = a(1 - \cos \theta)$.

Sol. *The equations of the curve are* $\mathbf{x = a(\theta + \sin \theta), y = a(1 - \cos \theta)}$ ...(*i*)

1. Limits. The greatest value of y is $2a$ and the least value is 0. Hence the curve lies between the lines $y = 0$ and $y = 2a$.

2. Symmetry. Since $x = a(\theta + \sin \theta)$ is an odd function of θ $[\because f(-\theta) = -f(\theta)]$ and $y = a(1 - \cos \theta)$ is an even function of θ $[\because f(-\theta) = f(\theta)]$.

$\therefore$ The curve is symmetrical about the y-axis.

3. Origin and Axes Intersection

(*i*) When $x = 0$, $\theta + \sin \theta = 0$ or $\theta = 0$ and then $y = 0$

$\therefore$ The curve passes through the origin.

(*ii*) The curve cuts the x-axis (putting $y = 0$) where $1 - \cos \theta = 0$

or $\cos \theta = 1$ or $\theta = 0$ and then $x = 0$ *i.e.,* at (0, 0).

The curve cuts y-axis (putting $x = 0$) where $\theta + \sin \theta = 0$

or $\theta = 0$ and then $y = 0$, *i.e.,* at (0, 0).

4. Asymptotes. When $\theta \to \infty$, $x \to \infty$ then y does not tend to a limit.

$\therefore$ The curve has no asymptote $\parallel$ to x-axis.

Also y cannot tend to ∞ $[\because |\cos \theta| \le 1]$, $\therefore$ The curve has no asymptote $\parallel$ to y-axis.

5. Special Points. $\frac{dx}{d\theta} = a(1 + \cos \theta)$, $\frac{dy}{d\theta} = a \sin \theta$

$$\therefore \quad \frac{dy}{dx} = \frac{a \sin \theta}{a(1 + \cos \theta)} = \frac{2 \sin \theta/2 \cos \theta/2}{2 \cos^2 \theta/2} = \tan \theta/2.$$

$$\therefore \quad \frac{dy}{dx} = 0 \text{ when } \theta/2 = 0 \quad i.e., \quad \text{when } \theta = 0.$$

But when $\theta = 0$, $x = 0$, $y = 0$. $\therefore$ The tangent at (0, 0) is $\parallel$ to x-axis *i.e.,* x-axis itself.

Again $\frac{dy}{dx} = \infty$ when $\tan \frac{\theta}{2} = \infty$ or $\frac{\theta}{2} = \frac{\pi}{2}$ or $\theta = \pi$.

$\theta = \pi$ gives $x = a\pi$, $y = 2a$. Thus at $(a\pi, 2a)$ the tangent is $\perp$ to x-axis.

6. Region. (*i*) $y = a\,(1 - \cos\theta) = 2a\sin^2\dfrac{\theta}{2}$ $\quad\therefore\quad \sin\dfrac{\theta}{2} = \sqrt{\dfrac{y}{2a}}$.

When y is – ve, $\sin\dfrac{\theta}{2}$ is imaginary. $\therefore$ No portion of the curve lies below the x-axis.

(*ii*) We have the following table:

$\theta = 0$	$\dfrac{\pi}{2}$	π	2π	$-\dfrac{\pi}{2}$	$-\pi$
$x = 0$	$a\left(\dfrac{\pi}{2}+1\right)$	$a\pi$	$2a\pi$	$-a\left(\dfrac{\pi}{2}+1\right)$	$-a\pi$
$y = 0$	a	$2a$	0	0	$2a$
$\dfrac{dy}{dx} = 0$	1	∞	0	-1	$-\infty$

(*iii*) As x increases from 0 to $a\pi$, y also increases from 0 to $2a$.

But as x further increases to $2a\pi$, y decreases to zero again.

$\therefore$ For +ve values of θ from 0 to π, we can draw the portion OAC of the curve in the first quadrant.

In fact as x further increases, y again increases to $2a$ and then decreases to 0 again and so on.

The curve for –ve values of x can be traced by symmetry [$\because$ the curve is symmetrical about y-axis].

The curve consists of congruent arches extending to ∞ on both sides of y-axis. The shape of the curve is as shown in the figure.

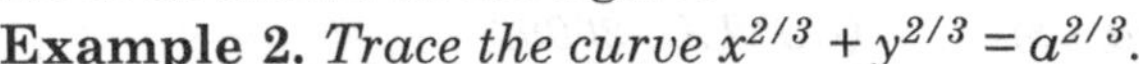

Example 2. *Trace the curve* $x^{2/3} + y^{2/3} = a^{2/3}$. (P.T.U. 2010)

Sol. The parametric equations of the curve are $x = a\cos^3 t$, $y = a\sin^3 t$

(*i*) **Limits.** $|\,x\,| \le a$ and $|\,y\,| \le a$

$\Rightarrow$ the curve lies entirely within the square bounded by the lines $x = \pm a$, $y = \pm a$.

(*ii*) **Symmetry.** x is an even function of t and y is an odd function of t.

$\Rightarrow$ the curve is symmetrical about the x-axis.

(*iii*) **Origin and Axes Intersection.**

When $x = 0$, $\cos^3 t = 0 \;\Rightarrow\; t = \dfrac{\pi}{2}$ and then $y = a \ne 0$

$\therefore$ The curve does not pass through the origin.

The curve meets x-axis (putting $y = 0$) where $\sin^3 t = 0$ or $t = 0, \pi, 2\pi, \ldots$

$\therefore$ The curve meets x-axis at $(a, 0)$ and $(-a, 0)$.

Similarly the curve meets y-axis at $(0, a)$ and $(0, -a)$.

(*iv*) **Asymptotes.** The curve has no asymptotes.

(*v*) **Special Points.** $\dfrac{dx}{dt} = -\,3a\cos^2 t\sin t,$

$$\frac{dy}{dt} = 3a\sin^2 t\cos t$$

$$\Rightarrow \quad \frac{dy}{dx} = \frac{\dfrac{dy}{dt}}{\dfrac{dx}{dt}} = -\tan t$$

$$\therefore \quad \frac{dy}{dx} = 0 \text{ when } t = 0 \quad\text{or}\quad \pi$$

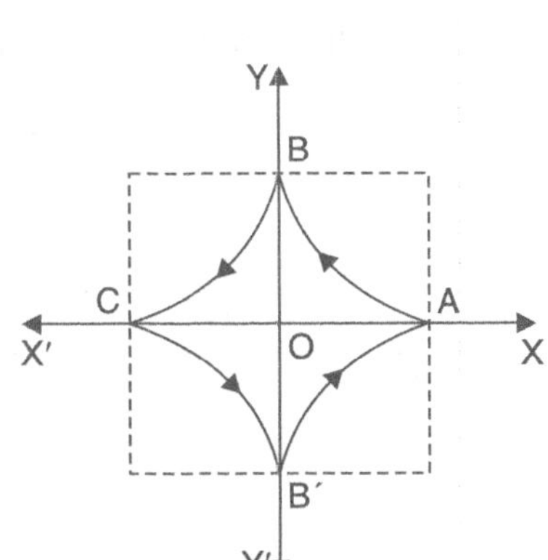

and $\frac{dy}{dx} \to \infty$ when $t = \frac{\pi}{2}$.

As t increases from 0 to $\frac{\pi}{2}$

x is positive and decreases from a to 0

y is positive and increases from 0 to a

Also $\frac{dy}{dx}$ varies from 0 to ∞.

Thus the portion AB is traced.

As t increases from $\frac{\pi}{2}$ to π

x is negative and increases numerically from 0 to $-a$.

y is positive and decreases from a to 0.

Also $\frac{dy}{dx}$ varies from ∞ to 0.

Thus the portion BC is traced.

As t increases from π to 2π. We get the curve CB′A which is the reflection of the curve CBA in the x-axis. Since x and y are periodic functions with period 2π, no new points are obtained.

Hence the shape of the curve is as shown in the figure.

EXERCISE 8

Trace the following curves:

1. $y = (x - 2)(x + 1)^2$
2. $y^2(a - x) = x^2(a + x)$ *(K.U.K. Dec. 2013; M.T.U. 2013)*
3. $x(x^2 + y^2) = a(x^2 - y^2)$ or $y^2(a + x) = x^2(a - x)$ *(U.P.T.U. 2009)*
4. $ay^2 = x(x - a)^2$
5. $xy^2 = a^2(a - x)$
6. $y(x^2 + 4a^2) = 8a^3$
7. $a^2y^2 = x^3(2a - x)$
8. $a^4y^2 = a^2x^4 - x^6$
9. $a^2y^2 = x^2(a^2 - x^2)$ *(P.T.U. May 2009)*
10. $y^2(1 - x^2) = x^2(1 + x^2)$
11. $(x^2 - 1)y = x^2 + 1$ *(K.U.K. 2009)*
12. $r = a(1 - \cos\theta)$ *(P.T.U. 2010)*
13. $r = a(1 + \sin\theta)$ *(P.T.U. Dec. 2011)*
14. $r = a + b\cos\theta$ $(a > b)$ [Limicon]
15. $r = a\cos 3\theta$
16. $r = a\sin 2\theta$
17. $r = a\cos 2\theta$
18. $r = \dfrac{a\sin^2\theta}{\cos\theta}$ (Cissoid)

    [**Hint.** Change to Cartesian Coordinates]
19. $x = a(\theta + \sin\theta)$, $y = a(1 + \cos\theta)$
20. $x = a(\theta - \sin\theta)$, $y = a(1 - \cos\theta)$
21. $x = a(\theta - \sin\theta)$, $y = a(1 + \cos\theta)$
22. $x = a\left(\cos t + \log\tan\frac{t}{2}\right)$, $y = a\sin t$ [Tractrix]

Answers

1.

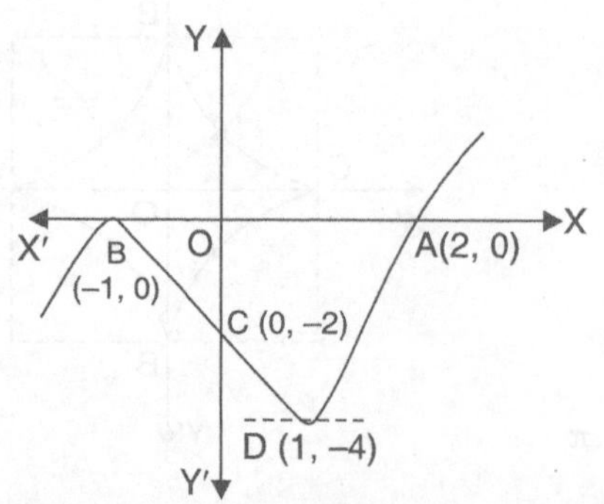

2.

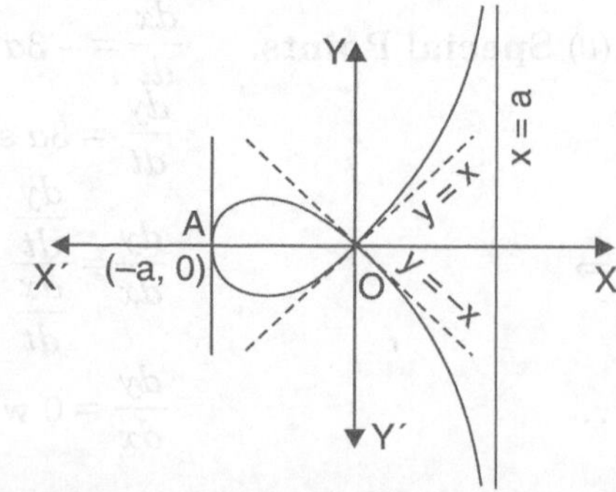

3.

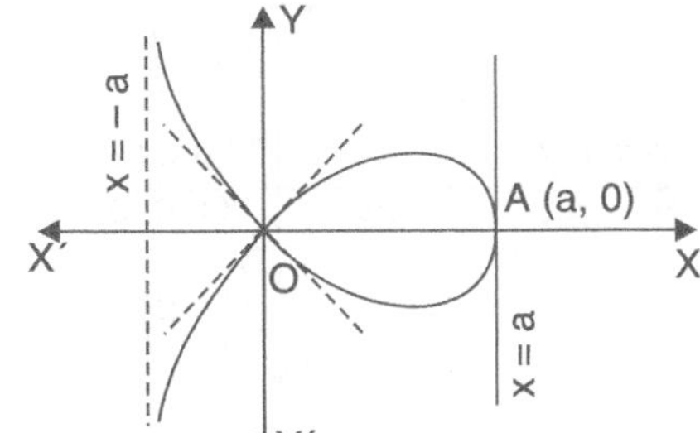
Y
x = – a
A (a, 0)
X′
O
X
x = a
Y′

4.

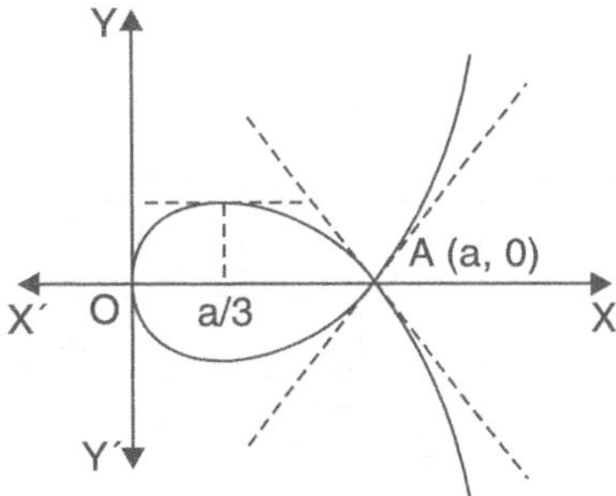
Y
A (a, 0)
X′
O
a/3
X
Y′

5.

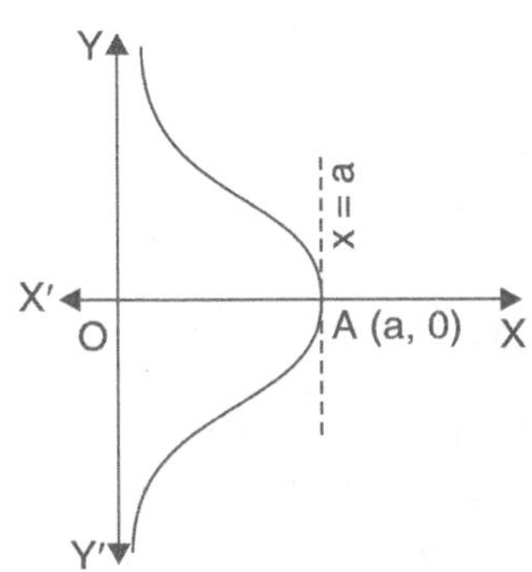
Y
x = a
X′
O
A (a, 0)
X
Y′

6.

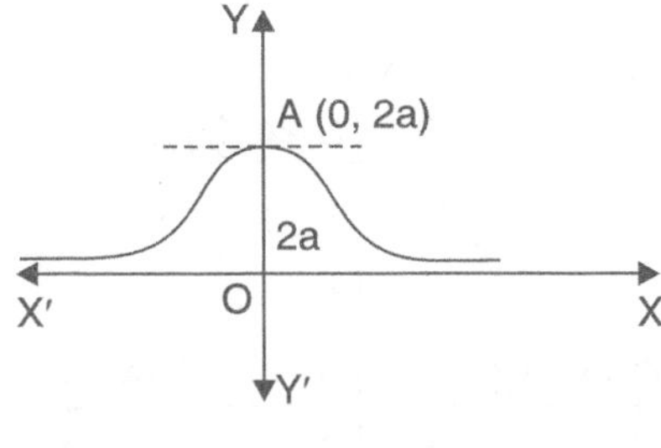
Y
A (0, 2a)
2a
X′
O
X
Y′

7.

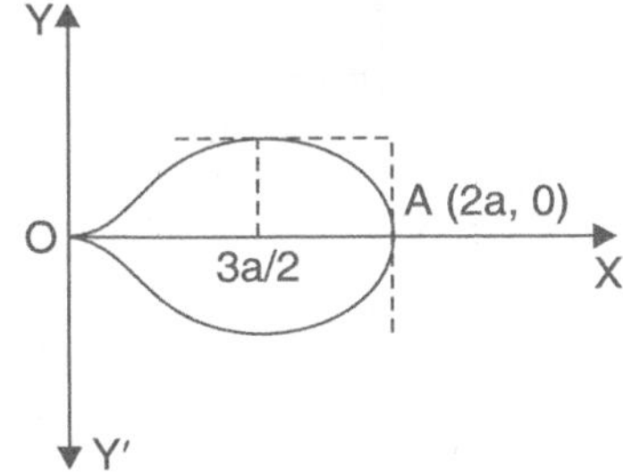
Y
O
A (2a, 0)
3a/2
X
Y′

8.

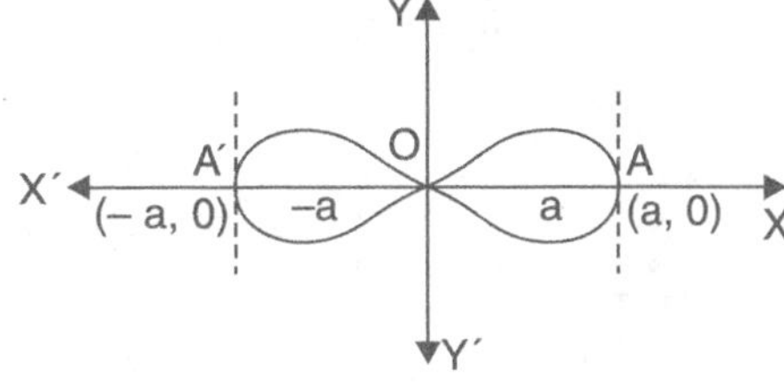
Y
A′
O
A
X′
(– a, 0)
–a
a
(a, 0)
X
Y′

9.

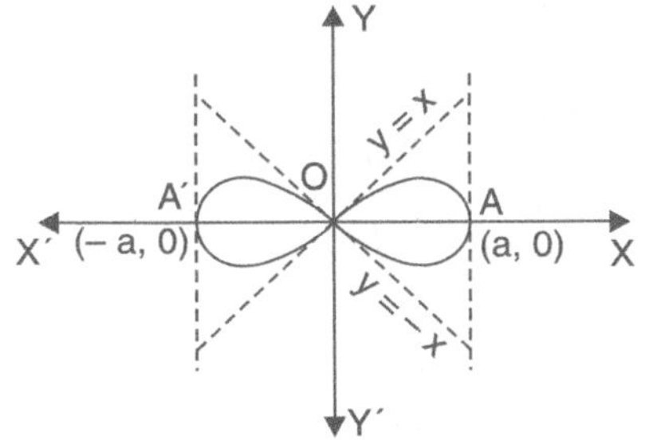
Y
O
y = x
A′
A
X′
(– a, 0)
(a, 0)
X
y = – x
Y′

10.

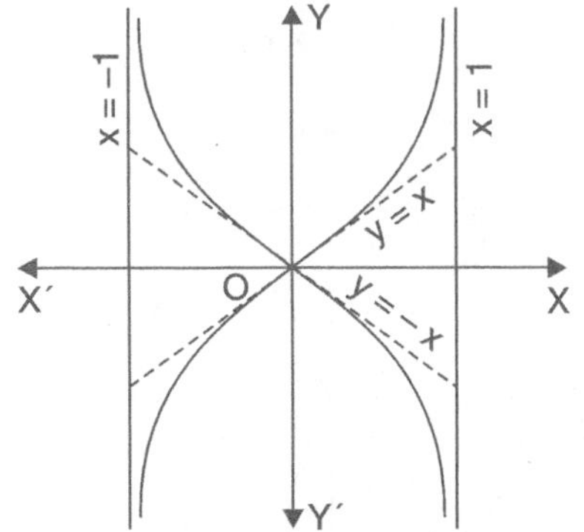
Y
x = –1
x = 1
y = x
X′
O
y = – x
X
Y′

11.

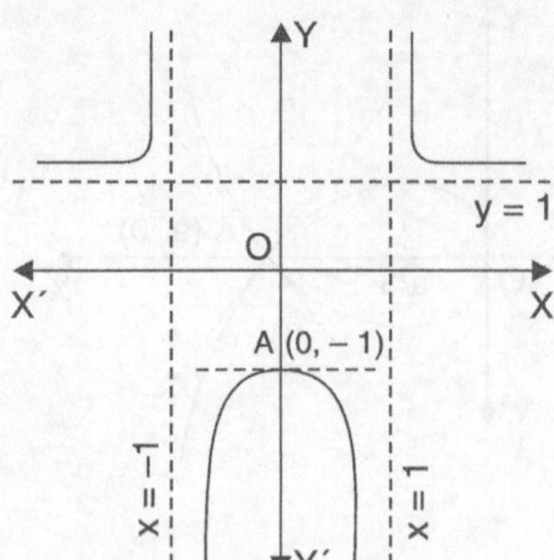

12.

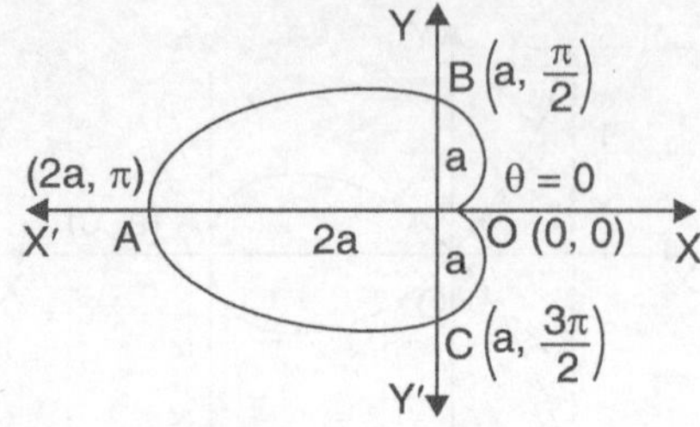

13.

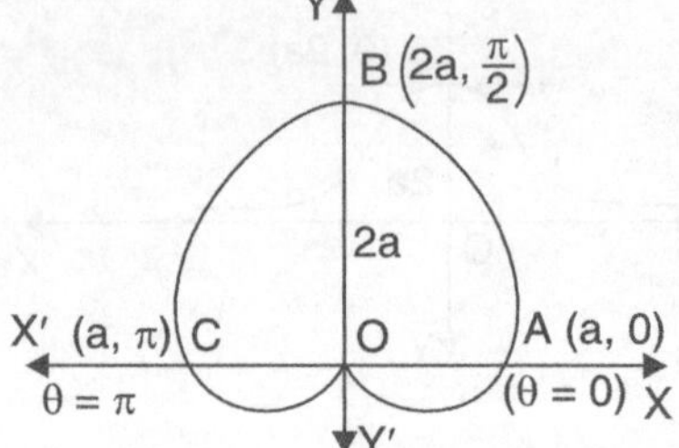

14.

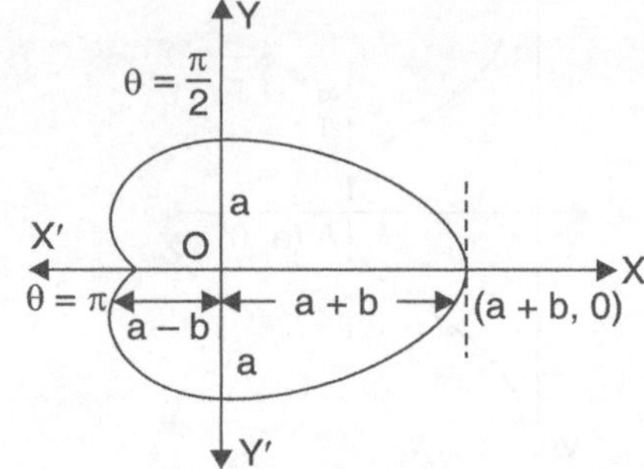

15.

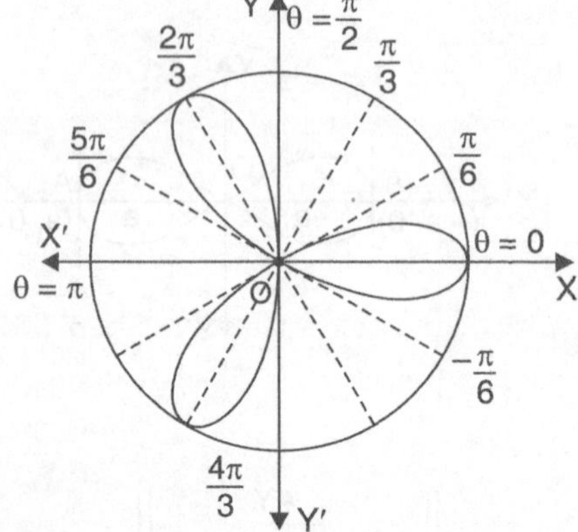

16.

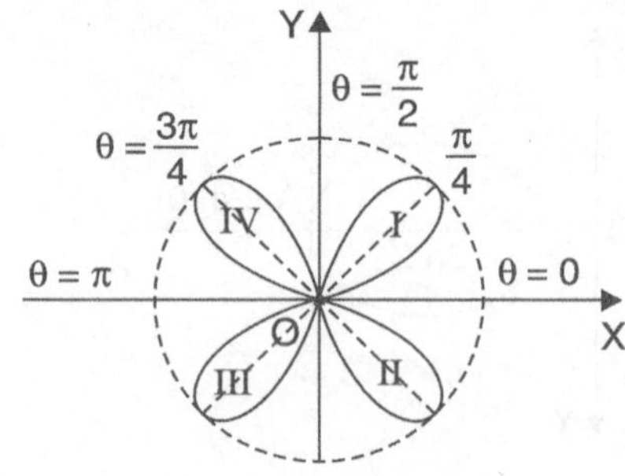

17.

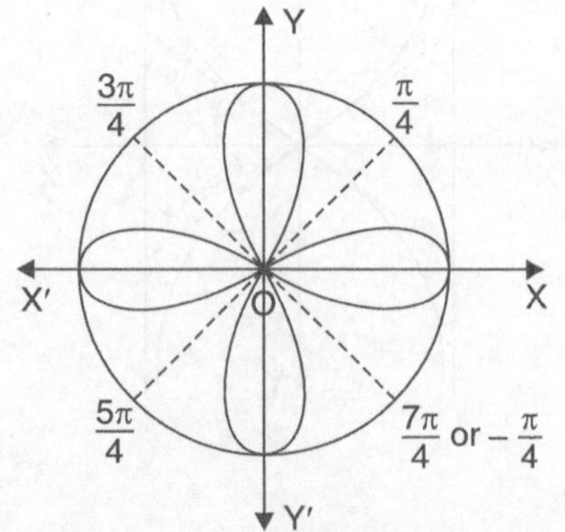

18.

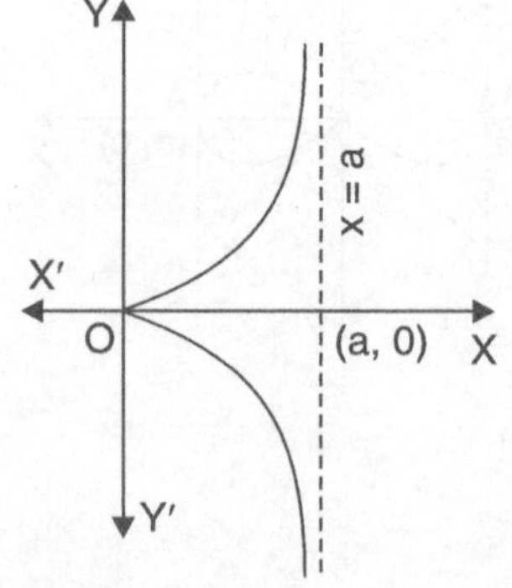

19.

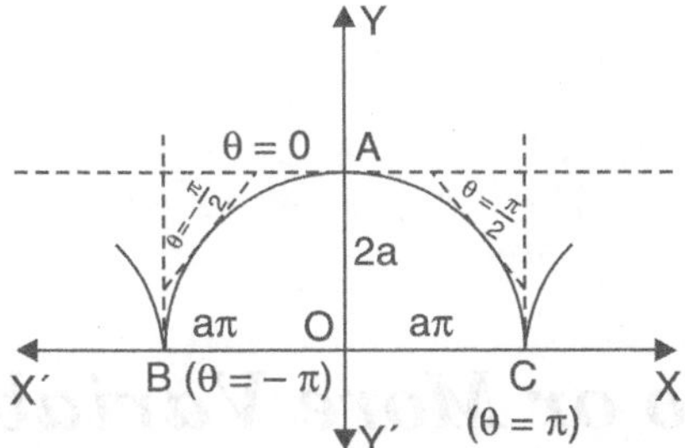
Y
θ = 0
A
θ = −π/2
θ = π/2
2a
aπ
O
aπ
X′
B (θ = − π)
C
(θ = π)
X
Y′

20.

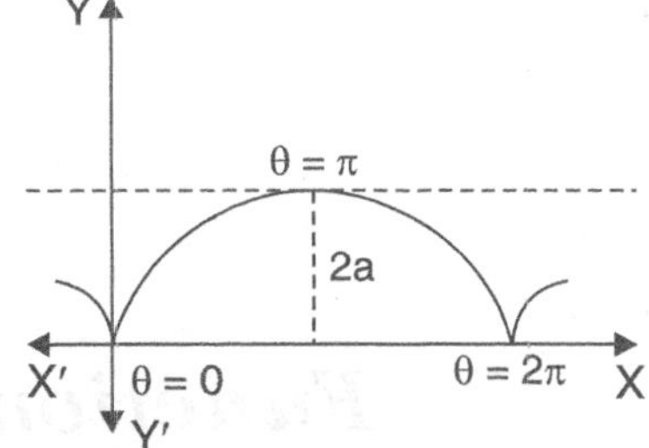
Y
θ = π
2a
X′
θ = 0
θ = 2π
X
Y′

21.

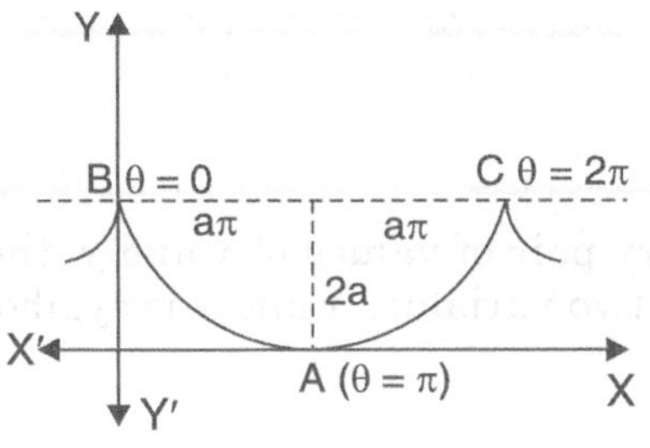
Y
B θ = 0
C θ = 2π
aπ
aπ
2a
X′
A (θ = π)
X
Y′

22.

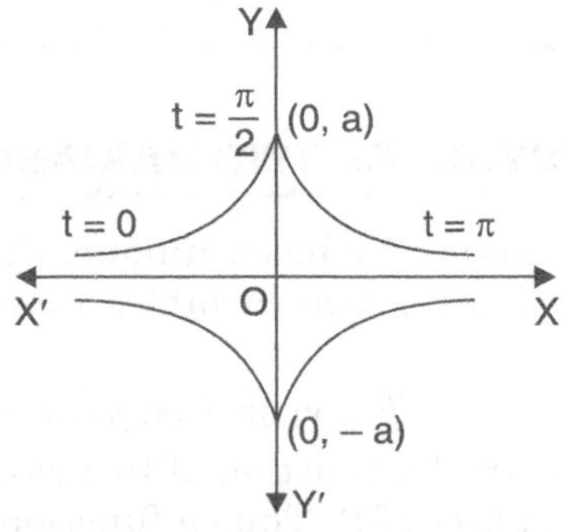
Y
t = π/2
(0, a)
t = 0
t = π
X′
O
X
(0, – a)
Y′

9

Functions of Two or More Variables

(Limit and Continuity)

9.1. FUNCTION OF TWO VARIABLES

If a quantity z has a unique, finite value for every pair of values of x and y, then z is called a function of two variables x and y. A function of two variables x and y is symbolically written as

$$f(x, y) \text{ or } F(x, y) \text{ or } \phi(x, y).$$

Domain of a function of two variables is a subset of $R^2 = R \times R = \{(x, y) : x, y \in R\}$ and range is a subset of R. Thus a function f of two variables is denoted as

$$f : S \to R \text{ where } S \subset R^2.$$

Similarly, a function f of three variables is denoted as $F : S \to R$ where $S \subset R^3$.

9.2. NEIGHBOURHOOD OF A POINT (a, b)

Every point (a, b) in R^2 has two types of neighbourhoods:

(*i*) **Square Neighbourhood**

The interior of the square with centre at (a, b), sides parallel to the coordinate axes and each side = 2δ is called a square neighbourhood of the point (a, b). For every positive value of δ, we get a square neighbourhood of (a, b).

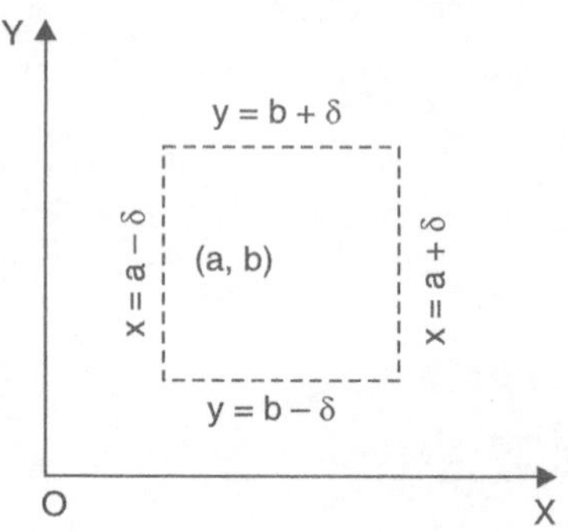

Thus a square neighbourhood of (a, b) is

$$\{(x, y) : a - \delta < x < a + \delta,\ b - \delta < y < b + \delta\}$$
$$= \{(x, y) : |x - a| < \delta,\ |y - b| < \delta\}$$

Similarly a neighbourhood of (a, b, c) in the form of a cube is

$$\{(x, y, z) : a - \delta < x < a + \delta,\ b - \delta < y < b + \delta,\ c - \delta < z < c + \delta\}$$
$$= \{(x, y, z) : |x - a| < \delta,\ |y - b| < \delta,\ |z - c| < \delta\}$$

(*ii*) **Circular Neighbourhood**

The interior of the circle with centre at (a, b) and radius δ is called a circular neighbourhood of the point (a, b). For every positive value of δ, we get a circular nbd of (a, b).

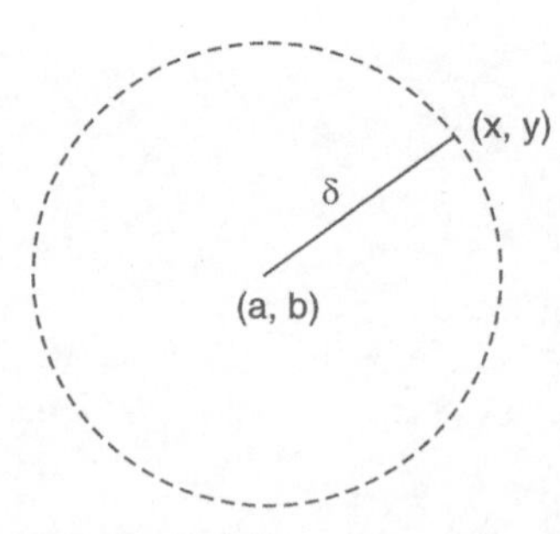

Thus a circular neighbourhood of (a, b) is

$\{(x, y) : |(x, y) - (a, b)| < \delta\}$ where $|(x, y) - (a, b)|$ stands for the distance between the points (x, y) and (a, b)

i.e., $|(x, y) - (a, b)| = \sqrt{(x-a)^2 + (y-b)^2}$

Similarly, a spherical neighbourhood of (a, b, c) is

$\{(x, y, z) : |(x, y, z) - (a, b, c)| < \delta\}$

where $|(x, y, z) - (a, b, c)| = \sqrt{(x-a)^2 + (y-b)^2 + (z-c)^2}$.

9.3. LIMIT OF A FUNCTION OF TWO VARIABLES

A function $f(x, y)$ is said to tend to a limit l as the point (x, y) tends to the point (a, b) if corresponding to any pre-assigned positive number ε, however small, we can find a positive number δ (depending on ε) such that

Def. 1. $|f(x, y) - l| < \varepsilon$

for all points (x, y) other than (a, b) for which $|x - a| < \delta$ and $|y - b| < \delta$

This definition of limit is based on square neighbourhood of a point.

Def. 2. $|f(x, y) - l| < \varepsilon$

for all points (x, y) other than (a, b) for which

$$|(x, y) - (a, b)| < \delta$$

This definition of limit is based on square neighbourhood of a point.

Note 1. A function $f(x, y)$ tends to a limit l as the point (x, y) tends to point (a, b) is symbolically written as

$$\underset{(x,y)\to(a,b)}{\text{Lt}} f(x, y) = l$$

Note 2. $\underset{(x,y)\to(a,b)}{\text{Lt}} f(x, y)$ if it exists, is unique

Note 3. We know that if f is a function of single variable x, then $\underset{x\to a}{\text{Lt}} f(x)$ exists iff

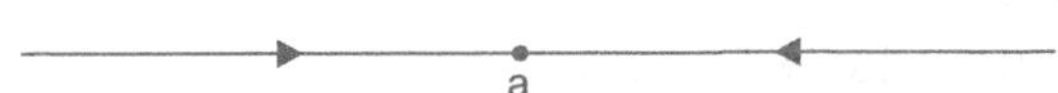

$\underset{x\to a^-}{\text{Lt}} f(x) = \underset{x\to a^+}{\text{Lt}} f(x)$, *i.e.,* the limit is independent of the path along which we approach the point 'a'.

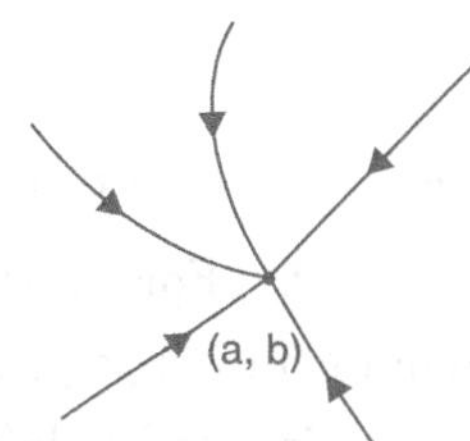

Similarly, if f is a function of two variables x and y, then $\underset{(x,y)\to(a,b)}{\text{Lt}} f(x, y)$ exists if this limit is independent of the path along which we approach the point (a, b).

9.4. CONTINUITY OF A FUNCTION OF TWO VARIABLES

A function $f(x, y)$ is said to be continuous at the point (a, b) if

$$\underset{(x,y)\to(a,b)}{\text{Lt}} f(x, y) \text{ exists and } = f(a, b)$$

Thus $f(x, y)$ is said to be continuous at the point (a, b) if given $\varepsilon > 0$, there exists a real number $\delta > 0$ such that

$$|f(x, y) - f(a, b)| < \varepsilon \text{ for } |(x, y) - (a, b)| < \delta$$

9.5. CONTINUITY OF A FUNCTION OF THREE VARIABLES

A function $f(x, y, z)$ is said to be continuous at the point (a, b, c) if

$$\underset{(x, y, z)\to(a, b, c)}{\text{Lt}} f(x, y, z) \text{ exists and } = f(a, b, c).$$

Thus $f(x, y, z)$ is said to be continuous at the point (a, b, c) if given $\varepsilon > 0$, there exists a real number $\delta > 0$ such that $|f(x, y, z) - f(a, b, c)| < \varepsilon$ for $|(x, y, z) - (a, b, c)| < \delta$.

ILLUSTRATIVE EXAMPLES

Example 1. *Let $f : R^2 \to R$ be defined as $f(x, y) = x^2 + y^2$.*

Show that $\underset{(x, y)\to(0,0)}{Lt} f(x, y) = 0$

Sol. Let $\varepsilon > 0$ be given

$$|f(x, y) - 0| = |x^2 + y^2| = x^2 + y^2 < \varepsilon$$

whenever $\sqrt{x^2 + y^2} < \sqrt{\varepsilon}$

i.e., whenever $|(x, y) - (0, 0)| < \delta$ where $\delta = \sqrt{\varepsilon}$

$\therefore$ For every $\varepsilon > 0$, there exists $\delta (= \sqrt{\varepsilon}) > 0$ such that

$$|f(x, y) - 0| < \varepsilon \text{ whenever } |(x, y) - (0, 0)| < \delta$$

Hence by definition of limit, $\underset{(x, y)\to(0, 0)}{\text{Lt}} f(x, y) = 0$

Example 2. *Let $f : R^3 \to R$ be defined by $f(x, y, z) = x^2 + y^2 + z^2$*

Show that $\underset{(x, y, z)\to(0,0,0)}{Lt} f(x, y, z) = 0.$

Sol. Let $\varepsilon > 0$ be given

$$|f(x, y, z) - 0| = |x^2 + y^2 + z^2|$$

$$= x^2 + y^2 + z^2 < \varepsilon \text{ whenever } \sqrt{x^2 + y^2 + z^2} < \sqrt{\varepsilon}$$

i.e., whenever $|(x, y, z) - (0, 0, 0)| < \delta$ where $\delta = \sqrt{\varepsilon}$

$\therefore$ For every $\varepsilon > 0$, there exists $\delta (= \sqrt{\varepsilon}) > 0$ such that $|f(x, y, z) - 0| < \varepsilon$

whenever $|(x, y, z) - (0, 0, 0)| < \delta$

Hence by definition of limit, $\underset{(x, y, z)\to(0, 0, 0)}{\text{Lt}} f(x, y, z) = 0.$

Example 3. *Let $A = \{(x, y) : 0 < x < 1, 0 < y < 1, x, y \in R\}$. Let $f : A \to R$ defined by $f(x, y) = x + y$. Show that*

$$\underset{\substack{(x, y)\to\left(0, \frac{1}{2}\right) \\ (x, y)\in A}}{Lt} f(x, y) = \frac{1}{2}.$$

Sol. Let $\varepsilon > 0$ be given.

$$\left| f(x, y) - \frac{1}{2} \right| = \left| x + y - \frac{1}{2} \right|$$

$$= \left| x + \left(y - \frac{1}{2} \right) \right| \le |x| + \left| y - \frac{1}{2} \right| < \frac{\varepsilon}{2} + \frac{\varepsilon}{2} = \varepsilon$$

whenever $|x| < \frac{\varepsilon}{2}$ and $\left| y - \frac{1}{2} \right| < \frac{\varepsilon}{2}$

i.e., whenever $|x - 0| < \delta$ and $\left| y - \frac{1}{2} \right| < \delta$ where $\delta = \frac{\varepsilon}{2}$

$\therefore$ For every $\varepsilon > 0$, there exists $\delta \left(= \frac{\varepsilon}{2} \right) > 0$ such that

$$\left| f(x, y) - \frac{1}{2} \right| < \varepsilon \text{ whenever } |x - 0| < \delta \text{ and } \left| y - \frac{1}{2} \right| < \delta$$

Hence by definition of limit, $\underset{(x,y)\to\left(0,\frac{1}{2}\right)}{\text{Lt}} f(x, y) = \frac{1}{2}$.

Example 4. *Let $f(x, y) = x + y$. Show that $f(x, y)$ is continuous at $\left(\frac{1}{2}, \frac{1}{3} \right)$.*

Sol. $f\left(\frac{1}{2}, \frac{1}{3} \right) = \frac{1}{2} + \frac{1}{3} = \frac{5}{6}$

Let $\varepsilon > 0$ be given.

$$\left| f(x, y) - f\left(\frac{1}{2}, \frac{1}{3} \right) \right| = \left| (x + y) - \left(\frac{1}{2} + \frac{1}{3} \right) \right|$$

$$= \left| \left(x - \frac{1}{2} \right) + \left(y - \frac{1}{3} \right) \right| \le \left| x - \frac{1}{2} \right| + \left| y - \frac{1}{3} \right| < \frac{\varepsilon}{2} + \frac{\varepsilon}{2} = \varepsilon$$

whenever $\left| x - \frac{1}{2} \right| < \frac{\varepsilon}{2}$ and $\left| y - \frac{1}{3} \right| < \frac{\varepsilon}{2}$

i.e., whenever $\left| x - \frac{1}{2} \right| < \delta$ and $\left| y - \frac{1}{3} \right| < \delta$ where $\delta = \frac{\varepsilon}{2}$

$\therefore$ For every $\varepsilon > 0$, there exists $\delta \left(= \frac{\varepsilon}{2} \right) > 0$ such that

$$\left| f(x, y) - f\left(\frac{1}{2}, \frac{1}{3} \right) \right| < \varepsilon \text{ whenever } \left| x - \frac{1}{2} \right| < \delta \text{ and } \left| y - \frac{1}{3} \right| < \delta$$

Hence by definition of continuity, $f(x, y)$ is continuous at $\left(\frac{1}{2}, \frac{1}{3} \right)$.

Example 5. *Let $f : R^2 \to R$ be defined by*

$$f(x, y) = \begin{cases} \dfrac{xy}{x^2 + y^2}, & (x, y) \neq (0, 0) \\ 0 \quad , & (x, y) = (0, 0) \end{cases}$$

Prove that $\underset{(x, y)\to(0,0)}{\text{Lt}} f(x, y)$ *does not exist.*

Sol. We know that if $\underset{(x, y)\to(a, b)}{\text{Lt}} f(x, y)$ exists, then this limit is independent of the path along which (x, y) approaches the point (a, b).

Here, let $(x, y) \to (0, 0)$ along the path $y = mx$ where m is any real number.

As $x \to 0$, from $y = mx$, we have $y \to 0$.

Now $\underset{(x,y)\to(0,0)}{\text{Lt}} f(x, y) = \underset{(x,y)\to(0,0)}{\text{Lt}} \dfrac{xy}{x^2 + y^2}$ (Putting $y = mx$)

$$= \underset{x\to 0}{\text{Lt}} \frac{x \,.\, mx}{x^2 + m^2x^2} = \underset{x\to 0}{\text{Lt}} \frac{mx^2}{x^2\,(1 + m^2)}$$

$$= \underset{x\to 0}{\text{Lt}} \frac{m}{1 + m^2} = \frac{m}{1 + m^2}$$

which is different for different values of m.

$\therefore$ $\underset{(x,y)\to(0,0)}{\text{Lt}} f(x)$ does not exist.

Example 6. *Prove that* $\underset{(x,y)\to(0,0)}{\text{Lt}} f(x, y)$ *does not exist, where*

$$f(x, y) = \frac{x^2 - y^2}{x^2 + y^2},\ (x, y) \neq (0, 0).$$

Sol. We know that if $\underset{(x,y)\to(a,b)}{\text{Lt}} f(x, y)$ exists, then this limit is independent of the path along which (x, y) approaches the point (a, b).

Here, let $(x, y) \to (0, 0)$ along the path $y = mx$ where m is any real number.

As $x \to 0$, from $y = mx$, we have $y \to 0$

Now $\underset{(x,y)\to(0,0)}{\text{Lt}} f(x, y) = \underset{(x,y)\to(0,0)}{\text{Lt}} \dfrac{x^2 - y^2}{x^2 + y^2}$ (Putting $y = mx$)

$$= \underset{x\to 0}{\text{Lt}} \frac{x^2 - m^2x^2}{x^2 + m^2x^2} = \underset{x\to 0}{\text{Lt}} \frac{x^2\,(1 - m^2)}{x^2\,(1 + m^2)}$$

$$= \underset{x\to 0}{\text{Lt}} \frac{1 - m^2}{1 + m^2} = \frac{1 - m^2}{1 + m^2}$$

which is different for different values of m.

$\therefore$ $\underset{(x,y)\to(0,0)}{\text{Lt}} f(x, y)$ does not exist.

Example 7. *Prove that* $\underset{(x,y)\to(0,0)}{Lt} f(x, y)$ *does not exist, where*

$$f(x, y) = \frac{xy^2}{x^2 + y^4}, \ (x, y) \neq (0, 0).$$

Sol. Let $(x, y) \to (0, 0)$ along the path $y = m\sqrt{x}$.

As $x \to 0$, from $y = m\sqrt{x}$, we have $y \to 0$.

$$\underset{(x,y)\to(0,0)}{\text{Lt}} f(x, y) = \underset{(x,y)\to(0,0)}{\text{Lt}} \frac{xy^2}{x^2 + y^4} \qquad \text{(Putting } y = m\sqrt{x}\text{)}$$

$$= \underset{x\to 0}{\text{Lt}} \frac{x \cdot m^2 x}{x^2 + m^4 x^2} = \underset{x\to 0}{\text{Lt}} \frac{m^2 x^2}{x^2(1+m^4)}$$

$$= \underset{x\to 0}{\text{Lt}} \frac{m^2}{1+m^4} = \frac{m^2}{1+m^4}$$

which is different for different values of m.

$\therefore \quad \underset{(x,y)\to(0,0)}{\text{Lt}} f(x, y)$ does not exist.

Example 8. *Prove that* $\underset{(x,y)\to(0,0)}{Lt} \dfrac{x^2 y}{x^4 + y^2}$ *does not exist.*

Sol. Let $(x, y) \to (0, 0)$ along the path $x = m\sqrt{y}$

As $y \to 0$, from $x = m\sqrt{y}$, we have $x \to 0$.

$$\therefore \quad \underset{(x,y)\to(0,0)}{\text{Lt}} \frac{x^2 y}{x^4 + y^2} = \underset{y\to 0}{\text{Lt}} \frac{m^2 y \,.\, y}{m^4 y^2 + y^2} = \underset{y\to 0}{\text{Lt}} \frac{m^2 y^2}{y^2 (m^4 + 1)}$$

$$= \underset{y\to 0}{\text{Lt}} \frac{m^2}{m^4 + 1} = \frac{m^2}{m^4 + 1}$$

which is different for different values of m.

$\therefore \quad \underset{(x,y)\to(0,0)}{\text{Lt}} \dfrac{x^2 y}{x^4 + y^2}$ does not exist.

Example 9. *Let* $f(x, y) = y \sin \dfrac{1}{x} + x \sin \dfrac{1}{y}$, *where* $x \neq 0, y \neq 0$. *Prove that* $f(x, y) \to 0$ *as* $(x, y) \to (0, 0)$.

Sol. Let $\varepsilon > 0$ be given.

$$| f(x, y) - 0 | = \left| y \sin \frac{1}{x} + x \sin \frac{1}{y} \right| \leq \left| y \sin \frac{1}{x} \right| + \left| x \sin \frac{1}{y} \right|$$

$$= | y | \left| \sin \frac{1}{x} \right| + | x | \left| \sin \frac{1}{y} \right|$$

$$\leq |y| + |x| \qquad \left[\because \left|\sin\frac{1}{x}\right| \leq 1 \text{ and } \left|\sin\frac{1}{y}\right| \leq 1\right]$$

$$< \frac{\varepsilon}{2} + \frac{\varepsilon}{2} = \varepsilon$$

whenever $|x| < \frac{\varepsilon}{2}$ and $|y| < \frac{\varepsilon}{2}$

i.e., whenever $|x| < \delta$ and $|y| < \delta$ where $\delta = \frac{\varepsilon}{2}$

$\therefore$ For every $\varepsilon > 0$, there exists $\delta\left(= \frac{\varepsilon}{2}\right) > 0$ such that $|f(x, y) - 0| < \varepsilon$ whenever $|x - 0| < \delta$ and $|y - 0| < \delta$. Hence by definition of limit,

$$\underset{(x,y)\to(0,0)}{\text{Lt}} f(x, y) = 0.$$

Example 10. *Let $A = \{(x, y) : 0 < x < 1, 0 < y < 1\}$ and $f : A \to R$ be defined by $f(x, y) = x + y$. Prove that f is continuous at every point of the domain A.*

Sol. Let (α, β) be any point of A.

Let us prove that $f(x, y)$ is continuous at (α, β)

i.e.,
$$\underset{(x,y)\to(\alpha,\beta)}{\text{Lt}} f(x, y) = f(\alpha, \beta)$$

Let $\varepsilon > 0$ be given

$$|f(x, y) - f(\alpha, \beta)| = |(x + y) - (\alpha + \beta)| = |(x - \alpha) + (y - \beta)|$$

$$\leq |x - \alpha| + |y - \beta|$$

$$< \frac{\varepsilon}{2} + \frac{\varepsilon}{2} = \varepsilon$$

whenever $|x - \alpha| < \frac{\varepsilon}{2}$ and $|y - \beta| < \frac{\varepsilon}{2}$

i.e., whenever $|x - \alpha| < \delta$ and $|y - \beta| < \delta$ where $\delta = \frac{\varepsilon}{2}$

$\therefore$ For every $\varepsilon > 0$, there exists $\delta\left(= \frac{\varepsilon}{2}\right) > 0$ such that

$|f(x, y) - f(\alpha, \beta)| < \varepsilon$ whenever $|x - \alpha| < \delta$ and $|y - \beta| < \delta$.

Hence by definition of continuity, $f(x, y)$ is continuous at (α, β). Since (α, β) is any point of A, therefore, f is continuous at every point of A.

Example 11. *Show that the function $f : R^2 \to R$ defined by*

$$f(x, y) = \begin{cases} xy\left(\dfrac{x^2 - y^2}{x^2 + y^2}\right), & (x, y) \neq (0, 0) \\ 0, & \text{Otherwise} \end{cases}$$

is continuous at (0, 0).

Sol. Let $\varepsilon > 0$ be given.

$$|f(x, y) - f(0, 0)| = \left| xy\left(\frac{x^2 - y^2}{x^2 + y^2}\right) - 0 \right|$$

$$= |xy| \left| \frac{x^2 - y^2}{x^2 + y^2} \right|$$

$$\leq |xy| \qquad \left[\because \ |x^2 - y^2| \leq |x^2 + y^2| \ \therefore \ \left|\frac{x^2 - y^2}{x^2 + y^2}\right| \leq 1\right]$$

or $\qquad [f(x, y) - f(0, 0)] \leq |x|\,|y|$

$$< \sqrt{\varepsilon} \times \sqrt{\varepsilon} = \varepsilon$$

whenever $\quad |x| < \sqrt{\varepsilon}$ and $|y| < \sqrt{\varepsilon}$

i.e., whenever $|x| < \delta$ and $|y| < \delta$ where $\delta = \sqrt{\varepsilon}$

$\therefore$ For every $\varepsilon > 0$, there exists $\delta\ (= \sqrt{\varepsilon}) > 0$ such that $|f(x, y) - f(0, 0)| < \varepsilon$ whenever $|x - 0| < \delta$ and $|y - 0| < \delta$.

Hence by definition of continuity, $f(x, y)$ is continuous at (0, 0).

Example 12. *Let $f(x, y) = \sqrt{|xy|}$. Show that $f(x, y)$ is continuous at the origin.*

Sol. Let $\varepsilon > 0$ be given

$$|f(x, y) - f(0, 0)| = |\sqrt{|xy|} - 0| = \sqrt{|xy|} = \sqrt{|x||y|} = \sqrt{|x|}\,.\,\sqrt{|y|} < \sqrt{\varepsilon}\,.\,\sqrt{\varepsilon} = \varepsilon$$

whenever $\qquad \sqrt{|x|} < \sqrt{\varepsilon}$ and $\sqrt{|y|} < \sqrt{\varepsilon}$

i.e., whenever $|x| < \varepsilon$ and $|y| < \varepsilon$

i.e., whenever $|x - 0| < \delta$ and $|y - 0| < \delta$ where $\delta = \varepsilon$

$\therefore$ For every $\varepsilon > 0$, there exists $\delta\ (= \varepsilon) > 0$ such that

$|f(x, y) - f(0, 0)| < \varepsilon$ whenever $|x - 0| < \delta$ and $|y - 0| < \delta$.

Hence by definition of continuity, $f(x, y)$ is continuous at (0, 0) the origin.

Example 13. *Let* $\phi(y, z) = \dfrac{yz}{\sqrt{y^2 + z^2}}$, $(y, z) \neq (0, 0)$

$\qquad = 0$, *when* $(y, z) = (0, 0)$.

Show that $\phi(y, z)$ is continuous at (0, 0).

Sol. Let $\varepsilon > 0$ be given.

$$|\phi(y, z) - \phi(0, 0)| = \left| \frac{yz}{\sqrt{y^2 + z^2}} - 0 \right| = \left| \frac{yz}{\sqrt{y^2 + z^2}} \right|$$

(Put $y = r\cos\theta$ and $z = r\sin\theta$ so that $y^2 + z^2 = r^2$)

$$= \left| \frac{r\cos\theta \cdot r\sin\theta}{r} \right| = \left| \frac{1}{2} r \sin 2\theta \right| = \frac{1}{2} r\,|\sin 2\theta| \leq \frac{1}{2} r < r$$

or $\qquad |\phi(y, z) - \phi(0, 0)| < \sqrt{y^2 + z^2} < \varepsilon$

whenever $\qquad \sqrt{y^2 + z^2} < \varepsilon$

i.e., whenever $|(y, z) - (0, 0)| < \delta$ where $\delta = \varepsilon$

$\therefore$ For ever y $\varepsilon > 0$, there exists $\delta\ (= \varepsilon) > 0$ such that

$|\phi(y, z) - \phi(0, 0)| < \varepsilon$ whenever $|(y, z) - (0, 0)| < \delta$.

Hence by definition of continuity, $\phi(y, z)$ is continuous at (0, 0).

EXERCISE 9

1. Let $f(x, y) = x^2 + y^2$. Show that $f(x, y)$ is continuous at the origin.
2. Let $f(x, y, z) = x^2 + y^2 + z^2$. Show that $f(x, y, z)$ is continuous at the origin.
3. Let $f : R^2 \to R$ be defined as

$$f(x, y) = \begin{cases} \dfrac{xy}{x^2 + y^2}, & \text{if } (x, y) \neq (0, 0) \\ 0 \quad , & \text{if } (x, y) = (0, 0) \end{cases}$$

Show that f is not continuous at (0, 0).

4. Show that $f(x, y) = \begin{cases} \dfrac{x^2 - y^2}{x^2 + y^2}, & \text{if } (x, y) \neq (0, 0) \\ 0 \quad , & \text{if } (x, y) = (0, 0) \end{cases}$

is not continuous at (0, 0).

5. Prove that $\lim\limits_{(x,y)\to(0,0)} f(x, y)$ does not exist, where

$$f(x, y) = \frac{2xy}{x^2 + y^2}, \ (x, y) \neq (0, 0)$$

6. Show that the following functions are discontinuous at the origin.

(i) $f(x, y) = \begin{cases} \dfrac{x^4 - y^4}{x^4 + y^4}, & \text{if } (x, y) \neq (0, 0) \\ 0 \quad , & \text{if } (x, y) = (0, 0) \end{cases}$

(ii) $f(x, y) = \begin{cases} \dfrac{x^2 y^2}{x^4 + y^4}, & \text{if } (x, y) \neq (0, 0) \\ 0 \quad , & \text{if } (x, y) = (0, 0) \end{cases}$

10

Partial Differentiation

10.1. FUNCTIONS OF TWO VARIABLES

If three variables x, y, z are so related that the value of z depends upon the values of x and y, then z is called a function of two variables x and y, and this is denoted by $z = f(x, y)$.

z is called the dependent variable while x and y are called independent variables.

For example, the area of a triangle is determined when its base and altitude are known. Thus, area of a triangle is a function of two variables, base and altitude.

(In a similar way, a function of more than two variables can be defined).

Geometrically. Let $z = f(x, y)$ be a function of two independent variables x and y defined for all pairs of values of x and y which belong to an area A of the xy-plane. Then to each point (x, y) of this area corresponds a value of z given by the relation $z = f(x, y)$. Representing all these values (x, y, z) by points in space, we get a surface.

Hence the function z = f(x, y) represents a surface.

10.2. PARTIAL DERIVATIVES OF FIRST ORDER

Let $z = f(x, y)$ be a function of two independent variables x and y. If y is kept constant and x alone is allowed to vary, then z becomes a function of x only. The derivative of z with respect to x, treating y as constant, is called partial derivative of z w.r.t. x and is denoted by

$$\frac{\partial z}{\partial x} \quad \text{or} \quad \frac{\partial f}{\partial x} \quad \text{or} \quad f_x.$$

Thus,
$$\frac{\partial z}{\partial x} = \lim_{h \to 0} \frac{f(x+h, y) - f(x, y)}{h}$$

Similarly, the derivative of z with respect to y, treating x as constant, is called partial derivative of z w.r.t. y and is denoted by $\frac{\partial z}{\partial y}$ or $\frac{\partial f}{\partial y}$ or f_y.

Thus,
$$\frac{\partial z}{\partial y} = \lim_{k \to 0} \frac{f(x, y+k) - f(x, y)}{k}$$

$\frac{\partial z}{\partial x}$ and $\frac{\partial z}{\partial y}$ are called **first order partial derivatives of z.**

[**In general,** if z is a function of two or more independent variables, then the **partial derivative of z** w.r.t. **any one** of the independent variables is the **ordinary derivative of z** w.r.t. that variable, **treating all other variables as constant.**]

Geometrically. Let $z = f(x, y)$ be a function of two variables x and y. Then by Art. 10.1, it represents a surface S. If $y = k$, a constant, then $y = k$ represents a plane parallel to the zx-plane.

$\therefore$ $z = f(x, y)$ and $y = k$ represent a plane curve C which is the section of S by $y = k$.

$\frac{\partial z}{\partial x}$ represents the slope of tangent to C at (x, k, z).

Thus, $\frac{\partial z}{\partial x}$ gives the slope of the tangent drawn to the curve of intersection of the surface $z = f(x, y)$ and a plane parallel to zx-plane.

Similarly, $\frac{\partial z}{\partial y}$ gives the slope of the tangent drawn to the curve of intersection of the surface $z = f(x, y)$ and a plane parallel to yz-plane.

10.3. PARTIAL DERIVATIVES OF HIGHER ORDER

Since the first order partial derivatives $\frac{\partial z}{\partial x}$ and $\frac{\partial z}{\partial y}$ are themselves functions of x and y, they can be further differentiated partially w.r.t. x as well as y. These are called second order partial derivatives of z. The usual notations for these second order partial derivatives are:

$$\frac{\partial}{\partial x}\left(\frac{\partial z}{\partial x}\right) = \frac{\partial^2 z}{\partial x^2} \text{ or } f_{xx}; \qquad \frac{\partial}{\partial y}\left(\frac{\partial z}{\partial y}\right) = \frac{\partial^2 z}{\partial y^2} \text{ or } f_{yy}$$

$$\frac{\partial}{\partial x}\left(\frac{\partial z}{\partial y}\right) = \frac{\partial^2 z}{\partial x \partial y} \text{ or } f_{xy}; \qquad \frac{\partial}{\partial y}\left(\frac{\partial z}{\partial x}\right) = \frac{\partial^2 z}{\partial y \partial x} \text{ or } f_{yx}$$

In general, $$\frac{\partial^2 z}{\partial x \partial y} = \frac{\partial^2 z}{\partial y \partial x} \text{ or } f_{xy} = f_{yx}.$$

Note 1. If $z = f(x)$, a function of single independent variable x, we get $\frac{dz}{dx}$.

If $z = f(x, y)$, a function of two independent variables x and y, we get $\frac{\partial z}{\partial x}$ and $\frac{\partial z}{\partial y}$.

Similarly, for a function of more than two independent variables $x_1, x_2, \ldots, x_n$, we get $\frac{\partial z}{\partial x_1}, \frac{\partial z}{\partial x_2}, \ldots, \frac{\partial z}{\partial x_n}$.

Note 2. (*i*) If $z = u + v$, where $u = f(x, y)$, $v = \phi(x, y)$ then z is a function of x and y.

$$\therefore \quad \frac{\partial z}{\partial x} = \frac{\partial u}{\partial x} + \frac{\partial v}{\partial x}; \quad \frac{\partial z}{\partial y} = \frac{\partial u}{\partial y} + \frac{\partial v}{\partial y}.$$

(*ii*) If $z = uv$, where $u = f(x, y)$, $v = \phi(x, y)$ then $\frac{\partial z}{\partial x} = \frac{\partial}{\partial x}(uv) = u\frac{\partial v}{\partial x} + v\frac{\partial u}{\partial x}$

$$\frac{\partial z}{\partial y} = \frac{\partial}{\partial y}(uv) = u\frac{\partial v}{\partial y} + v\frac{\partial u}{\partial y}$$

(*iii*) If $z = \dfrac{u}{v}$, where $u = f(x, y)$, $v = \phi\,(x, y)$ then $\dfrac{\partial z}{\partial x} = \dfrac{\partial}{\partial x}\left(\dfrac{u}{v}\right) = \dfrac{v\dfrac{\partial u}{\partial x} - u\dfrac{\partial v}{\partial x}}{v^2}$

$$\frac{\partial z}{\partial y} = \frac{\partial}{\partial y}\left(\frac{u}{v}\right) = \frac{v\dfrac{\partial u}{\partial y} - u\dfrac{\partial v}{\partial y}}{v^2}$$

(*iv*) If $z = f(u)$, where $u = \phi\,(x, y)$ then $\dfrac{\partial z}{\partial x} = \dfrac{dz}{du}\cdot\dfrac{\partial u}{\partial x}$; $\dfrac{\partial z}{\partial y} = \dfrac{dz}{du}\cdot\dfrac{\partial u}{\partial y}$.

ILLUSTRATIVE EXAMPLES

Example 1. *Find the first order partial derivatives of the following:*

(*i*) $u = tan^{-1}\dfrac{x^2 + y^2}{x + y}$ (*ii*) $u = cos^{-1}\left(\dfrac{x}{y}\right)$.

Sol. (*i*) $u = \tan^{-1}\dfrac{x^2 + y^2}{x + y}$

$$\frac{\partial u}{\partial x} = \frac{1}{1 + \left(\dfrac{x^2 + y^2}{x + y}\right)^2}\cdot\frac{\partial}{\partial x}\left(\frac{x^2 + y^2}{x + y}\right)$$

$$= \frac{(x + y)^2}{(x + y^2) + (x^2 + y^2)^2}\cdot\frac{(x + y)\dfrac{\partial}{\partial x}(x^2 + y^2) - (x^2 + y^2)\dfrac{\partial}{\partial x}(x + y)}{(x + y)^2}$$

$$= \frac{(x + y)\cdot 2x - (x^2 + y^2)\cdot 1}{(x + y)^2 + (x^2 + y^2)^2}$$

$$\Rightarrow \quad \frac{\partial u}{\partial x} = \frac{x^2 + 2xy - y^2}{(x + y)^2 + (x^2 + y^2)^2} \qquad \ldots(1)$$

[Since u remains the same if we interchange x and y, u is symmetrical w.r.t. x and y. Interchanging x and y in (1), we have]

Similarly, $\dfrac{\partial u}{\partial y} = \dfrac{y^2 + 2xy - x^2}{(x + y)^2 + (x^2 + y^2)^2}$

(*ii*) $u = \cos^{-1}\left(\dfrac{x}{y}\right)$

$$\frac{\partial u}{\partial x} = \frac{-1}{\sqrt{1 - \left(\dfrac{x}{y}\right)^2}}\cdot\frac{\partial}{\partial x}\left(\frac{x}{y}\right) = \frac{-y}{\sqrt{y^2 - x^2}}\cdot\frac{1}{y} = \frac{-1}{\sqrt{y^2 - x^2}}$$

$$\frac{\partial u}{\partial y} = \frac{-1}{\sqrt{1 - \left(\dfrac{x}{y}\right)^2}}\cdot\frac{\partial}{\partial y}\left(\frac{x}{y}\right) = \frac{-y}{\sqrt{y^2 - x^2}}\left(-\frac{x}{y^2}\right) = \frac{x}{y\sqrt{y^2 - x^2}}.$$

Example 2. *If $z(x + y) = x^2 + y^2$, show that* $\left(\frac{\partial z}{\partial x} - \frac{\partial z}{\partial y}\right)^2 = 4\left(1 - \frac{\partial z}{\partial x} - \frac{\partial z}{\partial y}\right)$.

Sol. $$z = \frac{x^2 + y^2}{x + y}$$ [z is symmetrical w.r.t. x and y]

$$\frac{\partial z}{\partial x} = \frac{(x+y)\frac{\partial}{\partial x}(x^2+y^2) - (x^2+y^2)\frac{\partial}{\partial x}(x+y)}{(x+y)^2}$$

$$= \frac{(x+y)\cdot 2x - (x^2+y^2)\cdot 1}{(x+y)^2} = \frac{x^2+2xy-y^2}{(x+y)^2}$$

Similarly, $$\frac{\partial z}{\partial y} = \frac{y^2+2xy-x^2}{(x+y)^2}$$

Now $$\left(\frac{\partial z}{\partial x} - \frac{\partial z}{\partial y}\right)^2 = \left[\frac{2x^2-2y^2}{(x+y)^2}\right]^2 = \frac{4(x+y)^2(x-y)^2}{(x+y)^4} = \frac{4(x-y)^2}{(x+y)^2}$$

$$4\left(1 - \frac{\partial z}{\partial x} - \frac{\partial z}{\partial y}\right) = 4\left[1 - \frac{x^2+2xy-y^2}{(x+y)^2} - \frac{y^2+2xy-x^2}{(x+y)^2}\right]$$

$$= 4\left[\frac{x^2+2xy+y^2-x^2-2xy+y^2-y^2-2xy+x^2}{(x+y)^2}\right]$$

$$= \frac{4(x^2-2xy+y^2)}{(x+y)^2} = \frac{4(x-y)^2}{(x+y)^2}$$

$$\therefore \quad \left(\frac{\partial z}{\partial x} - \frac{\partial z}{\partial y}\right)^2 = 4\left(1 - \frac{\partial z}{\partial x} - \frac{\partial z}{\partial y}\right).$$

Example 3. *Prove that if* $f(x, y) = \frac{1}{\sqrt{y}} \cdot e^{-\frac{(x-a)^2}{4y}}$, *then* $f_{xy} = f_{yx}$.

Sol. $$f(x, y) = \frac{1}{\sqrt{y}} \cdot e^{-\frac{(x-a)^2}{4y}} = y^{-\frac{1}{2}} e^{-\frac{(x-a)^2}{4y}}$$

$$f_x = \frac{\partial f}{\partial x} = y^{-\frac{1}{2}} \cdot e^{-\frac{(x-a)^2}{4y}} \cdot \frac{\partial}{\partial x}\left[-\frac{(x-a)^2}{4y}\right]$$

$$= y^{-\frac{1}{2}} \cdot e^{-\frac{(x-a)^2}{4y}} \cdot \left[-\frac{2(x-a)}{4y}\right] = -\frac{1}{2} y^{-\frac{3}{2}} (x-a) e^{-\frac{(x-a)^2}{4y}}$$

$$f_y = \frac{\partial f}{\partial y} = -\frac{1}{2} y^{-\frac{3}{2}} \cdot e^{-\frac{(x-a)^2}{4y}} + y^{-\frac{1}{2}} \cdot e^{-\frac{(x-a)^2}{4y}} \cdot \frac{\partial}{\partial y}\left[-\frac{(x-a)^2}{4y}\right]$$

$$= e^{-\frac{(x-a)^2}{4y}} \left[-\frac{1}{2} y^{-\frac{3}{2}} + y^{-\frac{1}{2}} \cdot \frac{(x-a)^2}{4y^2}\right] = \frac{1}{4} y^{-\frac{3}{2}} e^{-\frac{(x-a)^2}{4y}} \quad [-2 + y^{-1}(x-a)^2]$$

$$f_{xy} = \frac{\partial}{\partial x}\left(\frac{\partial f}{\partial y}\right)$$

$$= \frac{1}{4} y^{-\frac{3}{2}} \left\{ e^{-\frac{(x-a)^2}{4y}} \cdot \frac{\partial}{\partial x}\left[-\frac{(x-a)^2}{4y}\right] \cdot [-2 + y^{-1}(x-a)^2] + e^{-\frac{(x-a)^2}{4y}} \cdot 2y^{-1}(x-a) \right\}$$

$$= \frac{1}{4} y^{-\frac{3}{2}} \cdot e^{-\frac{(x-a)^2}{4y}} \left\{ -\frac{2(x-a)}{4y} [-2 + y^{-1}(x-a)^2] + 2y^{-1}(x-a) \right\}$$

$$= \frac{1}{4} y^{-\frac{3}{2}} \cdot e^{-\frac{(x-a)^2}{y}} \cdot \frac{x-a}{y} \left\{ -\frac{1}{2} [-2 + y^{-1}(x-a)^2] + 2 \right\}$$

$$= \frac{1}{4} y^{-\frac{5}{2}} (x-a)\, e^{-\frac{(x-a)^2}{4y}} \left[3 - \frac{(x-a)^2}{2y} \right]$$

$$f_{yx} = \frac{\partial}{\partial y}\left(\frac{\partial f}{\partial x}\right) = -\frac{1}{2}(x-a)\left[-\frac{3}{2} y^{-\frac{5}{2}} \cdot e^{-\frac{(x-a)^2}{4y}} + y^{-\frac{3}{2}} \cdot e^{-\frac{(x-a)^2}{4y}} \cdot \frac{(x-a)^2}{4y^2} \right]$$

$$= -\frac{1}{4}(x-a)\, y^{-\frac{5}{2}} \cdot e^{-\frac{(x-a)^2}{4y}} \left[-3 + \frac{(x-a)^2}{2y} \right]$$

$$= \frac{1}{4} y^{-\frac{5}{2}} (x-a)\, e^{-\frac{(x-a)^2}{4y}} \left[3 - \frac{(x-a)^2}{2y} \right]$$

$\therefore \quad f_{xy} = f_{yx}.$

Example 4. *If* $u = x^y$, *show that* $\dfrac{\partial^3 u}{\partial x^2 \partial y} = \dfrac{\partial^3 u}{\partial x \partial y \partial x}$.

Sol.

$$u = x^y$$

$$\frac{\partial u}{\partial y} = x^y \log x$$

$$\frac{\partial^2 u}{\partial x \partial y} = \frac{\partial}{\partial x}\left(\frac{\partial u}{\partial y}\right) = yx^{y-1} \log x + x^y \cdot \frac{1}{x} = x^{y-1}(y \log x + 1)$$

$$\frac{\partial^3 u}{\partial x^2 \partial y} = \frac{\partial}{\partial x}\left(\frac{\partial^2 u}{\partial x \partial y}\right) = \frac{\partial}{\partial x}[x^{y-1}(y \log x + 1)] \quad ...(1)$$

$$\frac{\partial u}{\partial x} = yx^{y-1}$$

$$\frac{\partial^2 u}{\partial y \partial x} = \frac{\partial}{\partial y}\left(\frac{\partial u}{\partial x}\right) = x^{y-1} + yx^{y-1} \log x = x^{y-1}(y \log x + 1)$$

$$\frac{\partial^3 u}{\partial x \partial y \partial x} = \frac{\partial}{\partial x}\left(\frac{\partial^2 u}{\partial y \partial x}\right) = \frac{\partial}{\partial x}[x^{y-1}(y \log x + 1)] \quad ...(2)$$

From (1) and (2), $\dfrac{\partial^3 u}{\partial x^2 \partial y} = \dfrac{\partial^3 u}{\partial x \partial y \partial x}$.

Example 5. *If* $\theta = t^n e^{-\frac{r^2}{4t}}$, *find the value of n which will make* $\frac{1}{r^2}\frac{\partial}{\partial r}\left(r^2 \frac{\partial \theta}{\partial r}\right) = \frac{\partial \theta}{\partial t}$. (P.T.U. 2011)

Sol. We have $\theta = t^n e^{-\frac{r^2}{4t}}$

$$\frac{\partial \theta}{\partial r} = t^n \cdot e^{-\frac{r^2}{4t}} \cdot \left(-\frac{2r}{4t}\right) = -\frac{1}{2} r t^{n-1} e^{-\frac{r^2}{4t}}$$

$\therefore$
$$r^2 \frac{\partial \theta}{\partial r} = -\frac{1}{2} r^3 \cdot t^{n-1} e^{-\frac{r^2}{4t}}$$

$$\frac{\partial}{\partial r}\left(r^2 \frac{\partial \theta}{\partial r}\right) = -\frac{1}{2} t^{n-1} \left[3r^2 e^{-\frac{r^2}{4t}} + r^3 e^{-\frac{r^2}{4t}}\left(-\frac{2r}{4t}\right)\right] = -\frac{1}{2} t^{n-1} r^2 e^{-\frac{r^2}{4t}}\left[3 - \frac{r^2}{2t}\right]$$

$\therefore$
$$\frac{1}{r^2}\frac{\partial}{\partial r}\left(r^2 \frac{\partial \theta}{\partial r}\right) = \frac{1}{2} t^{n-1} e^{-\frac{r^2}{4t}}\left(\frac{r^2}{2t} - 3\right)$$

Also,
$$\frac{\partial \theta}{\partial t} = n t^{n-1} e^{-\frac{r^2}{4t}} + t^n e^{-\frac{r^2}{4t}} \cdot \left(\frac{r^2}{4t^2}\right) = t^{n-1} e^{-\frac{r^2}{4t}}\left(n + \frac{r^2}{4t}\right)$$

Since,
$$\frac{1}{r^2}\frac{\partial}{\partial r}\left(r^2 \frac{\partial \theta}{\partial r}\right) = \frac{\partial \theta}{\partial t} \qquad \text{[Given]}$$

$\therefore$
$$\frac{1}{2} t^{n-1} e^{-\frac{r^2}{4t}}\left(\frac{r^2}{2t} - 3\right) = t^{n-1} e^{-\frac{r^2}{4t}}\left(n + \frac{r^2}{4t}\right)$$

$\Rightarrow$
$$\frac{r^2}{4t} - \frac{3}{2} = n + \frac{r^2}{4t} \quad \therefore \quad n = -\frac{3}{2}.$$

Example 6. *If* $u = (1 - 2xy + y^2)^{-1/2}$, *prove that* $\frac{\partial}{\partial x}\left\{(1 - x^2)\frac{\partial u}{\partial x}\right\} + \frac{\partial}{\partial y}\left\{y^2 \frac{\partial u}{\partial y}\right\} = 0.$

Sol.
$$u = (1 - 2xy + y^2)^{-1/2} = V^{-1/2}, \quad \text{where } V = 1 - 2xy + y^2$$

$$\frac{\partial u}{\partial x} = -\frac{1}{2} V^{-3/2} \cdot \frac{\partial V}{\partial x} = -\frac{1}{2} V^{-3/2}(-2y) = yV^{-3/2}$$

$$\frac{\partial^2 u}{\partial x^2} = y \cdot \frac{\partial}{\partial x}(V^{-3/2}) = y \cdot \left(-\frac{3}{2}\right) V^{-5/2} \cdot \frac{\partial V}{\partial x} = -\frac{3}{2} yV^{-5/2}(-2y) = 3y^2 V^{-5/2}$$

$\therefore$
$$\frac{\partial}{\partial x}\left\{(1 - x^2)\frac{\partial u}{\partial x}\right\} = (1 - x^2) \cdot \frac{\partial^2 u}{\partial x^2} + \frac{\partial u}{\partial x} \cdot \frac{\partial}{\partial x}(1 - x^2)$$
$$= (1 - x^2) \cdot 3y^2 V^{-5/2} + yV^{-3/2}(-2x) = yV^{-3/2}[3yV^{-1}(1 - x^2) - 2x] \qquad \text{...(1)}$$

Also, $\dfrac{\partial u}{\partial y} = -\dfrac{1}{2}V^{-3/2}\dfrac{\partial V}{\partial y} = -\dfrac{1}{2}V^{-3/2}.(-2x+2y) = V^{-3/2}.(x-y)$

$$\frac{\partial^2 u}{\partial y^2} = V^{-3/2}\cdot\frac{\partial}{\partial y}(x-y) + (x-y)\cdot\frac{\partial}{\partial y}(V^{-3/2})$$
$$= V^{-3/2}\cdot(-1) + (x-y)\cdot\left(-\frac{3}{2}V^{-5/3}\right)\cdot\frac{\partial V}{\partial y}$$
$$= -V^{-3/2} - \frac{3}{2}(x-y)V^{-5/2}\cdot(-2x+2y) = -V^{-3/2} + 3(x-y)^2V^{-5/2}$$

$$\therefore \quad \frac{\partial}{\partial y}\left\{y^2\frac{\partial u}{\partial y}\right\} = y^2\cdot\frac{\partial^2 u}{\partial y^2} + \frac{\partial u}{\partial y}\cdot\frac{\partial}{\partial y}(y^2)$$
$$= y^2[-V^{-3/2} + 3(x-y)^2V^{-5/2}] + V^{-3/2}(x-y)\cdot 2y$$
$$= y\,V^{-3/2}[-y + 3y(x-y)^2V^{-1} + 2(x-y)]$$
$$= y\,V^{-3/2}[3y(x-y)^2V^{-1} + (2x-3y)] \qquad ...(2)$$

Adding (1) and (2), we have

$$\frac{\partial}{\partial x}\left\{(1-x^2)\frac{\partial u}{\partial x}\right\} + \frac{\partial}{\partial y}\left\{y^2\frac{\partial u}{\partial y}\right\} = yV^{-3/2}[3yV^{-1}(1-x^2) - 2x + 3y(x-y)^2V^{-1} + 2x - 3y]$$
$$= y\,V^{-3/2}[3yV^{-1}(1-x^2+x^2-2xy+y^2) - 3y]$$
$$= yV^{-2/3}[3yV^{-1}(1-2xy+y^2) - 3y]$$
$$= yV^{-3/2}[3y-3y] \qquad |\because \quad V = 1-2xy+y^2$$
$$= 0.$$

Example 7. *If* $u = \log(x^3+y^3+z^3-3xyz)$, *show that*

(*i*) $\left(\dfrac{\partial}{\partial x}+\dfrac{\partial}{\partial y}+\dfrac{\partial}{\partial z}\right)^2 u = -\dfrac{9}{(x+y+z)^2}$ (U.K.T.U. 2011; P.T.U. Jan. 2009)

(*ii*) $\dfrac{\partial^2 u}{\partial x^2}+\dfrac{\partial^2 u}{\partial y^2}+\dfrac{\partial^2 u}{\partial z^2}+2\dfrac{\partial^2 u}{\partial y\partial z}+2\dfrac{\partial^2 u}{\partial z\partial x}+2\dfrac{\partial^2 u}{\partial x\partial y} = \dfrac{-9}{(x+y+z)^2}$.

Sol. (*i*) $u = \log(x^3+y^3+z^3-3xyz)$

$$\frac{\partial u}{\partial x} = \frac{3x^2-3yz}{x^3+y^3+z^3-3xyz}; \quad \frac{\partial u}{\partial y} = \frac{3y^2-3zx}{x^3+y^3+z^3-3xyz}$$
$$\frac{\partial u}{\partial z} = \frac{3z^2-3xy}{x^3+y^3+z^3-3xyz}$$

Adding, $\dfrac{\partial u}{\partial x}+\dfrac{\partial u}{\partial y}+\dfrac{\partial u}{\partial z} = \dfrac{3(x^2+y^2+z^2-xy-yz-zx)}{x^3+y^3+z^3-3xyz} = \dfrac{3}{x+y+z}$

$[\because \quad x^3+y^3+z^3-3xyz = (x+y+z)(x^2+y^2+z^2-xy-yz-zx)]$

Now $\left(\dfrac{\partial}{\partial x}+\dfrac{\partial}{\partial y}+\dfrac{\partial}{\partial z}\right)^2 u = \left(\dfrac{\partial}{\partial x}+\dfrac{\partial}{\partial y}+\dfrac{\partial}{\partial z}\right)\left(\dfrac{\partial}{\partial x}+\dfrac{\partial}{\partial y}+\dfrac{\partial}{\partial z}\right)u$

$$= \left(\frac{\partial}{\partial x}+\frac{\partial}{\partial y}+\frac{\partial}{\partial z}\right)\left(\frac{\partial u}{\partial x}+\frac{\partial u}{\partial y}+\frac{\partial u}{\partial z}\right) = \left(\frac{\partial}{\partial x}+\frac{\partial}{\partial y}+\frac{\partial}{\partial z}\right)\left(\frac{3}{x+y+z}\right)$$
$$= -\frac{3}{(x+y+z)^2} - \frac{3}{(x+y+z)^2} - \frac{3}{(x+y+z)^2} = -\frac{9}{(x+y+z)^2}$$

...(1)

(ii) $$\left(\frac{\partial}{\partial x}+\frac{\partial}{\partial y}+\frac{\partial}{\partial z}\right)^2 u=\left(\frac{\partial}{\partial x}+\frac{\partial}{\partial y}+\frac{\partial}{\partial z}\right)\left(\frac{\partial}{\partial x}+\frac{\partial}{\partial y}+\frac{\partial}{\partial z}\right) u=\left(\frac{\partial}{\partial x}+\frac{\partial}{\partial y}+\frac{\partial}{\partial z}\right)\left(\frac{\partial u}{\partial x}+\frac{\partial u}{\partial y}+\frac{\partial u}{\partial z}\right)$$

$$=\frac{\partial}{\partial x}\left(\frac{\partial u}{\partial x}+\frac{\partial u}{\partial y}+\frac{\partial u}{\partial z}\right)+\frac{\partial}{\partial y}\left(\frac{\partial u}{\partial x}+\frac{\partial u}{\partial y}+\frac{\partial u}{\partial z}\right)+\frac{\partial}{\partial z}\left(\frac{\partial u}{\partial x}+\frac{\partial u}{\partial y}+\frac{\partial u}{\partial z}\right)$$

$$=\frac{\partial^2 u}{\partial x^2}+\frac{\partial^2 u}{\partial x\partial y}+\frac{\partial^2 u}{\partial x\partial z}+\frac{\partial^2 u}{\partial y\partial x}+\frac{\partial^2 u}{\partial y^2}+\frac{\partial^2 u}{\partial y\partial z}+\frac{\partial^2 u}{\partial z\partial x}+\frac{\partial^2 u}{\partial z\partial y}+\frac{\partial^2 u}{\partial z^2}$$

$$=\frac{\partial^2 u}{\partial x^2}+\frac{\partial^2 u}{\partial y^2}+\frac{\partial^2 u}{\partial z^2}+2\frac{\partial^2 u}{\partial y\partial z}+2\frac{\partial^2 u}{\partial z\partial x}+2\frac{\partial^2 u}{\partial x\partial y}$$

$$\left[\because\ \frac{\partial^2 u}{\partial y\partial x}=\frac{\partial^2 u}{\partial x\partial y},\ \frac{\partial^2 u}{\partial z\partial y}=\frac{\partial^2 u}{\partial y\partial z},\ \frac{\partial^2 u}{\partial x\partial z}=\frac{\partial^2 u}{\partial z\partial x}\right]$$

$$\therefore\quad \frac{\partial^2 u}{\partial x^2}+\frac{\partial^2 u}{\partial y^2}+\frac{\partial^2 u}{\partial z^2}+2\frac{\partial^2 u}{\partial y\partial z}+2\frac{\partial^2 u}{\partial z\partial x}+2\frac{\partial^2 u}{\partial x\partial y}=\left(\frac{\partial}{\partial x}+\frac{\partial}{\partial y}+\frac{\partial}{\partial z}\right)^2 u=\frac{-9}{(x+y+z)^2}$$

[from (1)]

Example 8. *If $x^x y^y z^z = c$, show that at $x = y = z$, $\frac{\partial^2 z}{\partial x\partial y} = -(x \log ex)^{-1}$.*

(M.T.U. 2013; K.U.K. 2012; Kerala 2008)

Sol. $x^x y^y z^z = c$ defines z as a function of x and y.

Taking logs, $x \log x + y \log y + z \log z = \log c$

Differentiating partially w.r.t. y, we have

$$y\cdot\frac{1}{y}+1\cdot\log y+z\cdot\frac{1}{z}\cdot\frac{\partial z}{\partial y}+1\cdot\log z\cdot\frac{\partial z}{\partial y}=0$$

or $$1+\log y+(1+\log z)\frac{\partial z}{\partial y}=0 \qquad ...(1)$$

or $$\frac{\partial z}{\partial y}=-\frac{1+\log y}{1+\log z}$$

Similarly, $$\frac{\partial z}{\partial x}=-\frac{1+\log x}{1+\log z} \qquad ...(2)$$

Differentiating (1) partially w.r.t. x, we have

$$\left(\frac{1}{z}\frac{\partial z}{\partial x}\right)\frac{\partial z}{\partial y}+(1+\log z)\frac{\partial^2 z}{\partial x\partial y}=0 \quad\text{or}\quad \frac{\partial^2 z}{\partial x\partial y}=-\frac{1}{z(1+\log z)}\frac{\partial z}{\partial x}\cdot\frac{\partial z}{\partial y} \qquad ...(3)$$

When $x = y = z$

From (2), $\frac{\partial z}{\partial y}=-1,\ \frac{\partial z}{\partial x}=-1$

From (3), $$\frac{\partial^2 z}{\partial x\partial y}=-\frac{1}{x(1+\log x)}(-1)(-1)=-\frac{1}{x(\log e+\log x)}=-\frac{1}{x(\log ex)}=-(x\log ex)^{-1}.$$

Example 9. *If $u = f(r)$ and $x = r \cos \theta$, $y = r \sin \theta$, prove that $\frac{\partial^2 u}{\partial x^2}+\frac{\partial^2 u}{\partial y^2}=f''(r)+\frac{1}{r}f'(r)$.*

Sol. Given $x = r\cos\theta, \quad y = r\sin\theta$

$\Rightarrow \quad x^2 + y^2 = r^2(\cos^2\theta + \sin\theta)$

$$\Rightarrow \qquad r^2 = x^2 + y^2 \qquad ...(1)$$

Differentiating partially w.r.t. x, we get $2r\dfrac{\partial r}{\partial x} = 2x$ or $\dfrac{\partial r}{\partial x} = \dfrac{x}{r}$

Similarly, $\dfrac{\partial r}{\partial y} = \dfrac{y}{r}$

Now $u = f(r)$

$$\therefore \qquad \frac{\partial u}{\partial x} = f'(r)\cdot\frac{\partial r}{\partial x} = \frac{x}{r}f'(r)$$

Differentiating again w.r.t. x, we get

$$\frac{\partial^2 u}{\partial x^2} = \frac{1}{r}f'(r) + x\cdot\left(-\frac{1}{r^2}\frac{\partial r}{\partial x}\right)f'(r) + \frac{x}{r}f''(r)\cdot\frac{\partial r}{\partial x}$$

$$\left[\because \frac{\partial}{\partial x}(uvw) = vw\frac{\partial}{\partial x}(u) + uw\frac{\partial}{\partial x}(v) + uv\frac{\partial}{\partial x}(w)\right]$$

$$= \frac{1}{r}f'(r) - \frac{x}{r^2}\cdot\frac{x}{r}f'(r) + \frac{x}{r}\cdot f''(r)\cdot\frac{x}{r} = \frac{1}{r}f'(r) - \frac{x^2}{r^3}f'(r) + \frac{x^2}{r^2}f''(r)$$

$$= \frac{r^2 - x^2}{r^3}f'(r) + \frac{x^2}{r^2}f''(r) = \frac{y^2}{r^3}f'(r) + \frac{x^2}{r^2}f''(r) \qquad | \text{ Using (1)}$$

Similarly, $\dfrac{\partial^2 u}{\partial y^2} = \dfrac{x^2}{r^3}f'(r) + \dfrac{y^2}{r^2}f''(r)$

$$\therefore \qquad \frac{\partial^2 u}{\partial x^2} + \frac{\partial^2 u}{\partial y^2} = \frac{x^2 + y^2}{r^3}f'(r) + \frac{x^2 + y^2}{r^2}f''(r) = \frac{r^2}{r^3}f'(r) + \frac{r^2}{r^2}f''(r) = f''(r) + \frac{1}{r}f'(r).$$

Example 10. *If $x = r\cos\theta$, $y = r\sin\theta$, prove that*

(*i*) $\dfrac{\partial r}{\partial x} = \dfrac{\partial x}{\partial r}$ (*ii*) $\dfrac{1}{r}\cdot\dfrac{\partial x}{\partial \theta} = r\dfrac{\partial \theta}{\partial x}$ (*iii*) $\dfrac{\partial^2 \theta}{\partial x^2} + \dfrac{\partial^2 \theta}{\partial y^2} = 0.$

Sol. (*i*) $\dfrac{\partial r}{\partial x}$ means $\left(\dfrac{\partial r}{\partial x}\right)_y$ = The partial derivative of r w.r.t. x, treating y as constant.

$\therefore$ We express r in terms of x and y.

Squaring and adding the given relations, $r^2 = x^2 + y^2$

Differentiating partially w.r.t. x, we get $2r\dfrac{\partial r}{\partial x} = 2x$ or $\dfrac{\partial r}{\partial x} = \dfrac{x}{r}$

$\dfrac{\partial x}{\partial r}$ means $\left(\dfrac{\partial x}{\partial r}\right)_\theta$ = The partial derivative of x w.r.t. r treating θ as constant.

$\therefore$ We express x in terms of r and θ.

Thus, $x = r\cos\theta$ (given)

$$\frac{\partial x}{\partial r} = \cos\theta = \frac{x}{r} \qquad \left(\because \cos\theta = \frac{x}{r}\right)$$

$$\therefore \qquad \frac{\partial r}{\partial x} = \frac{\partial x}{\partial r}$$

(*ii*) Expressing x in terms of r and θ, we have $x = r\cos\theta$

$$\Rightarrow \qquad \frac{\partial x}{\partial \theta} = -r\sin\theta = -y \quad \Rightarrow \quad \frac{1}{r}\frac{\partial x}{\partial \theta} = -\frac{y}{r}$$

Expressing θ in terms of x and y, we have $\tan\theta = \frac{y}{x}$ or $\theta = \tan^{-1}\frac{y}{x}$

$$\frac{\partial \theta}{\partial x} = \frac{1}{1+\frac{y^2}{x^2}}\cdot\left(-\frac{y}{x^2}\right) = -\frac{y}{x^2+y^2} = \frac{-y}{r^2(\cos^2\theta+\sin^2\theta)} = -\frac{y}{r^2}$$

$$\Rightarrow \qquad r\frac{\partial \theta}{\partial x} = -\frac{y}{r} \qquad \therefore \quad \frac{1}{r}\cdot\frac{\partial x}{\partial \theta} = r\frac{\partial \theta}{\partial x}.$$

(*iii*) Expressing θ in terms of x and y, we have $\tan\theta = \frac{y}{x}$ or $\theta = \tan^{-1}\frac{y}{x}$

$$\frac{\partial \theta}{\partial x} = \frac{1}{1+\frac{y^2}{x^2}}\cdot\left(-\frac{y}{x^2}\right) = -\frac{y}{x^2+y^2} = -y(x^2+y^2)^{-1}$$

$$\frac{\partial^2 \theta}{\partial x^2} = y(x^2+y^2)^{-2}\cdot 2x = \frac{2xy}{(x^2+y^2)^2}$$

$$\frac{\partial \theta}{\partial y} = \frac{1}{1+\frac{y^2}{x^2}}\cdot\frac{1}{x} = \frac{x}{x^2+y^2} = x(x^2+y^2)^{-1}$$

$$\frac{\partial^2 \theta}{\partial x^2} = -x(x^2+y^2)^{-2}\cdot 2y = -\frac{2xy}{(x^2+y^2)^2}$$

$$\therefore \qquad \frac{\partial^2 \theta}{\partial x^2} + \frac{\partial^2 \theta}{\partial y^2} = 0.$$

Example 11. *If* $x = e^{r\cos\theta}$ *cos* (r *sin* θ) *and* $y = e^{r\cos\theta}$ *sin* (r *sin* θ), *prove that*

$$\frac{\partial x}{\partial r} = \frac{1}{r}\cdot\frac{\partial y}{\partial \theta},\ \frac{\partial y}{\partial r} = -\frac{1}{r}\cdot\frac{\partial x}{\partial \theta}.$$

Hence deduce that $\frac{\partial^2 x}{\partial r^2} + \frac{1}{r}\frac{\partial x}{\partial r} + \frac{1}{r^2}\frac{\partial^2 x}{\partial \theta^2} = 0.$

Sol. $\qquad x = e^{r\cos\theta}\cos(r\sin\theta)$

$$\therefore \qquad \frac{\partial x}{\partial r} = e^{r\cos\theta}\cdot\cos\theta\cdot\cos(r\sin\theta) - e^{r\cos\theta}\sin(r\sin\theta)\cdot\sin\theta$$

$$= e^{r\cos\theta}\,[\cos\theta\cos(r\sin\theta) - \sin\theta\sin(r\sin\theta)]$$

$$= e^{r\cos\theta}\cos(\theta + r\sin\theta) \qquad ...(1)$$

$$\frac{\partial x}{\partial \theta} = e^{r\cos\theta}\cdot(-r\sin\theta)\cdot\cos(r\sin\theta) - e^{r\cos\theta}\sin(r\sin\theta)\cdot r\cos\theta$$

$$= -re^{r\cos\theta}\,[\sin\theta\cos(r\sin\theta) + \cos\theta\sin(r\sin\theta)]$$

$$= -re^{r\cos\theta}\sin(\theta + r\sin\theta) \qquad ...(2)$$

Also, $\qquad y = e^{r\cos\theta}\sin(r\sin\theta)$

$$\therefore \quad \frac{\partial y}{\partial r} = e^{r\cos\theta}\cdot\cos\theta\cdot\sin(r\sin\theta) + e^{r\cos\theta}\cdot\cos(r\sin\theta)\sin\theta$$

$$= e^{r\cos\theta}[\sin\theta\cos(r\sin\theta) + \cos\theta\sin(r\sin\theta)]$$

$$= e^{r\cos\theta}\sin(\theta + r\sin\theta) \quad ...(3)$$

$$\frac{\partial y}{\partial \theta} = e^{r\cos\theta}(-r\sin\theta)\sin(r\sin\theta) + e^{r\cos\theta}\cos(r\sin\theta)\times r\cos\theta$$

$$= re^{r\cos\theta}[\cos\theta\cos(r\sin\theta) - \sin\theta\sin(r\sin\theta)]$$

$$= re^{r\cos\theta}\cos(\theta + r\sin\theta) \quad ...(4)$$

From (1) and (4), $$\frac{\partial x}{\partial r} = \frac{1}{r}\cdot\frac{\partial y}{\partial \theta} \quad ...(5)$$

From (2) and (3), $$\frac{\partial y}{\partial r} = -\frac{1}{r}\cdot\frac{\partial x}{\partial \theta} \quad ...(6)$$

From (5), $$\frac{\partial^2 x}{\partial r^2} = -\frac{1}{r^2}\cdot\frac{\partial y}{\partial \theta} + \frac{1}{r}\cdot\frac{\partial^2 y}{\partial r\partial \theta}$$

From (6), $$\frac{\partial x}{\partial \theta} = -r\frac{\partial y}{\partial r}$$

$$\therefore \quad \frac{\partial^2 x}{\partial \theta^2} = -r\frac{\partial^2 y}{\partial \theta\partial r} = -r\frac{\partial^2 y}{\partial r\partial \theta}$$

$$\therefore \quad \frac{\partial^2 x}{\partial r^2} + \frac{1}{r}\cdot\frac{\partial x}{\partial r} + \frac{1}{r^2}\cdot\frac{\partial^2 x}{\partial \theta^2} = -\frac{1}{r^2}\cdot\frac{\partial y}{\partial \theta} + \frac{1}{r}\cdot\frac{\partial^2 y}{\partial r\partial \theta} + \frac{1}{r^2}\cdot\frac{\partial y}{\partial \theta} - \frac{1}{r}\cdot\frac{\partial^2 y}{\partial r\partial \theta} = 0.$$

Example 12. *If* $\frac{x^2}{a^2+u} + \frac{y^2}{b^2+u} + \frac{z^2}{c^2+u} = 1$*, prove that*

$$\left(\frac{\partial u}{\partial x}\right)^2 + \left(\frac{\partial u}{\partial y}\right)^2 + \left(\frac{\partial u}{\partial z}\right)^2 = 2\left(x\frac{\partial u}{\partial x} + y\frac{\partial u}{\partial y} + z\frac{\partial u}{\partial z}\right).$$

Sol. Given $$\frac{x^2}{a^2+u} + \frac{y^2}{b^2+u} + \frac{z^2}{c^2+u} = 1 \quad ...(1)$$

or $$x^2(a^2+u)^{-1} + y^2(b^2+u)^{-1} + z^2(c^2+u)^{-1} = 1$$

Differentiating partially w.r.t. x, we have

$$2x(a^2+u)^{-1} - x^2(a^2+u)^{-2}\cdot\frac{\partial u}{\partial x} - y^2(b^2+u)^{-2}\cdot\frac{\partial u}{\partial x} - z^2(c^2+u)^{-2}\cdot\frac{\partial u}{\partial x} = 0$$

or $$\frac{2x}{a^2+u} = \left[\frac{x^2}{(a^2+u)^2} + \frac{y^2}{(b^2+u)^2} + \frac{z^2}{(c^2+u)^2}\right]\frac{\partial u}{\partial x}$$

or $$\frac{2x}{a^2+u} = V\frac{\partial u}{\partial x} \text{ where } V = \frac{x^2}{(a^2+u)^2} + \frac{y^2}{(b^2+u)^2} + \frac{z^2}{(c^2+u)^2}$$

or $$\frac{\partial u}{\partial x} = \frac{2x}{V(a^2+u)}$$

Similarly, $$\frac{\partial u}{\partial y} = \frac{2y}{V(b^2+u)} \quad \text{and} \quad \frac{\partial u}{\partial z} = \frac{2z}{V(c^2+u)}$$

$$\therefore \quad \left(\frac{\partial u}{\partial x}\right)^2+\left(\frac{\partial u}{\partial y}\right)^2+\left(\frac{\partial u}{\partial z}\right)^2 = \frac{4}{V^2}\left[\frac{x^2}{(a^2+u)^2}+\frac{y^2}{(b^2+u)^2}+\frac{z^2}{(c^2+u)^2}\right]$$

$$= \frac{4}{V^2}(V) = \frac{4}{V} \qquad ...(2)$$

Now, $$2\left(x\frac{\partial u}{\partial x}+y\frac{\partial u}{\partial y}+z\frac{\partial u}{\partial z}\right) = 2\left[\frac{2x^2}{V(a^2+u)}+\frac{2y^2}{V(b^2+u)}+\frac{2z^2}{V(c^2+u)}\right]$$

$$= \frac{4}{V}\left[\frac{x^2}{a^2+u}+\frac{y^2}{b^2+u}+\frac{z^2}{c^2+u}\right]$$

$$= \frac{4}{V}(1) \qquad \text{[Using (1)]}$$

$$= \left(\frac{\partial u}{\partial x}\right)^2+\left(\frac{\partial u}{\partial y}\right)^2+\left(\frac{\partial u}{\partial z}\right)^2 \qquad \text{[Using (2)]}$$

Example 13. *If $u = lx + my$, $v = mx - ly$, show that*

$$\left(\frac{\partial u}{\partial x}\right)_y\left(\frac{\partial x}{\partial u}\right)_v = \frac{l^2}{l^2+m^2}, \quad \left(\frac{\partial y}{\partial v}\right)_x\left(\frac{\partial v}{\partial y}\right)_u = \frac{l^2+m^2}{l^2}.$$

Sol. Given

$$u = lx + my \qquad ...(1)$$

$$v = mx - ly \qquad ...(2)$$

(*i*) $\left(\frac{\partial u}{\partial x}\right)_y$ = The partial derivative of u w.r.t. x keeping y constant.

∴ We need a relation expressing u as a function of x and y.

From (1), $$\left(\frac{\partial u}{\partial x}\right)_y = l$$

$\left(\frac{\partial x}{\partial u}\right)_v$ = The partial derivative of x w.r.t. u keeping v constant.

∴ We need a relation expressing x as a function of u and v.

Eliminating y between (1) and (2) by multiplying (1) by l, (2) by m and adding the products, we have

$$lu + mv = (l^2 + m^2)x \quad \text{or} \quad x = \frac{lu+mv}{l^2+m^2}$$

$$\therefore \quad \left(\frac{\partial x}{\partial u}\right)_v = \frac{l}{l^2+m^2}$$

Hence, $$\left(\frac{\partial u}{\partial x}\right)_y\left(\frac{\partial x}{\partial u}\right)_v = \frac{l^2}{l^2+m^2}$$

(*ii*) $\left(\frac{\partial y}{\partial v}\right)_x$ = The partial derivative of y w.r.t. v keeping x constant.

∴ We need a relation expressing y as a function of v and x.

From (2), $$y = \frac{mx-v}{l} \qquad \therefore \quad \left(\frac{\partial y}{\partial v}\right)_x = -\frac{1}{l}$$

Also $\left(\dfrac{\partial v}{\partial y}\right)_u$ = Partial derivative of v w.r.t. y keeping u constant

$\therefore$ We need a relation expressing v as a function of y and u.

Eliminating x between (1) and (2), we have $v = \dfrac{mu - (l^2 + m^2)y}{l}$

$$\therefore \qquad \left(\frac{\partial v}{\partial y}\right)_u = -\frac{l^2 + m^2}{l}$$

Hence $$\left(\frac{\partial y}{\partial v}\right)_x \left(\frac{\partial v}{\partial y}\right)_u = \left(-\frac{1}{l}\right)\left(-\frac{l^2 + m^2}{l}\right) = \frac{l^2 + m^2}{l^2}.$$

EXERCISE 10.1

1. Find the first order partial derivatives of the following functions:
   (i) $u = y^x$ (ii) $u = \log (x^2 + y^2)$
   (iii) $u = x^2 \sin \dfrac{y}{x}$ (iv) $u = \dfrac{x}{y} \tan^{-1}\left(\dfrac{y}{x}\right)$.
2. If $u = x^2 + y^2 + z^2$, prove that $xu_x + yu_y + zu_z = 2u$.
3. If $z = \log (x^2 + xy + y^2)$, prove that $x\dfrac{\partial z}{\partial x} + y\dfrac{\partial z}{\partial y} = 2$.
4. If $u = x^2y + y^2z + z^2x$, prove that $u_x + u_y + u_z = (x + y + z)^2$.
5. If $u = \dfrac{y}{z} + \dfrac{z}{x} + \dfrac{x}{y}$, show that $x\dfrac{\partial u}{\partial x} + y\dfrac{\partial u}{\partial y} + z\dfrac{\partial u}{\partial z} = 0$.
6. If $f(x, y) = x^3y - xy^3$, find $\left[\dfrac{1}{\dfrac{\partial f}{\partial x}} + \dfrac{1}{\dfrac{\partial f}{\partial y}}\right]_{\substack{x=1\\y=2}}$.
7. (i) If $u = \log (\tan x + \tan y)$, prove that $\sin 2x\dfrac{\partial u}{\partial x} + \sin 2y\dfrac{\partial u}{\partial y} = 2$.
   (ii) If $u(x, y, z) = \log (\tan x + \tan y + \tan z)$, show that $\sin 2x\dfrac{\partial u}{\partial x} + \sin 2y\dfrac{\partial u}{\partial y} + \sin 2z\dfrac{\partial u}{\partial z} = 2$. *(M.T.U. 2011)*
8. If $f(x, y, z) = \begin{vmatrix} x^2 & y^2 & z^2 \\ x & y & z \\ 1 & 1 & 1 \end{vmatrix}$, prove that $f_x + f_y + f_z = 0$.
9. Verify that $\dfrac{\partial^2 u}{\partial x \partial y} = \dfrac{\partial^2 u}{\partial y \partial x}$ for the following functions:
   (i) $u = ax^2 + 2hxy + by^2$ (ii) $u = \tan^{-1}\left(\dfrac{x}{y}\right)$ (iii) $u = \log\left(\dfrac{x^2 + y^2}{xy}\right)$ *(U.P.T.U. 2008)*
   (iv) $u = e^{ax} \sin by$ (v) $u = \log (x \sin y + y \sin x)$.
10. If $z = \log (e^x + e^y)$, show that $r\,t - s^2 = 0$; where $r = \dfrac{\partial^2 z}{\partial x^2}$, $t = \dfrac{\partial^2 z}{\partial y^2}$, $s = \dfrac{\partial^2 z}{\partial x \partial y}$.
11. If $u = \tan^{-1} \dfrac{xy}{\sqrt{1 + x^2 + y^2}}$, show that $\dfrac{\partial^2 u}{\partial x \partial y} = (1 + x^2 + y^2)^{-3/2}$.

12. If $u = e^{xyz}$, prove that $\dfrac{\partial^3 u}{\partial x \partial y \partial z} = (1 + 3xyz + x^2y^2z^2)e^{xyz}$. (*M.D.U. Dec. 2012; U.P.T.U. 2009*)

13. If $u = \log(x^2 + y^2) + \tan^{-1}\dfrac{y}{x}$, prove that $\dfrac{\partial^2 u}{\partial x^2} + \dfrac{\partial^2 u}{\partial y^2} = 0$.

14. If $u = \tan^{-1}\left(\dfrac{2xy}{x^2 - y^2}\right)$, prove that $\dfrac{\partial^2 u}{\partial x^2} + \dfrac{\partial^2 u}{\partial y^2} = 0$.

15. Verify that $f_{xy} = f_{yx}$ when f is equal to

(*i*) $\sin^{-1}\left(\dfrac{y}{x}\right)$ (*ii*) $\log x \tan^{-1}(x^2 + y^2)$.

16. Find the value of n so that the equation $V = r^n(3\cos^2\theta - 1)$ satisfies the relation

$$\frac{\partial}{\partial r}\left(r^2 \frac{\partial V}{\partial r}\right) + \frac{1}{\sin\theta}\frac{\partial}{\partial\theta}\left(\sin\theta \frac{\partial V}{\partial\theta}\right) = 0.$$

17. If $z = \tan(y + ax) - (y - ax)^{3/2}$, show that $\dfrac{\partial^2 z}{\partial x^2} = a^2 \dfrac{\partial^2 z}{\partial y^2}$.

18. If $V = (x^2 + y^2 + z^2)^{-1/2}$, prove that $\dfrac{\partial^2 V}{\partial x^2} + \dfrac{\partial^2 V}{\partial y^2} + \dfrac{\partial^2 V}{\partial z^2} = 0$.

19. If $V = r^m$ where $r^2 = x^2 + y^2 + z^2$, show that $V_{xx} + V_{yy} + V_{zz} = m(m + 1)\, r^{m-2}$. (*M.D.U. Dec. 2008; P.T.U. 2010*)

20. If $u = \log(x^2 + y^2 + z^2)$, prove that $x\dfrac{\partial^2 u}{\partial y \partial z} = y\dfrac{\partial^2 u}{\partial z \partial x} = z\dfrac{\partial^2 u}{\partial x \partial y}$.

21. If $u = \log\sqrt{x^2 + y^2 + z^2}$, prove that $(x^2 + y^2 + z^2)\left(\dfrac{\partial^2 u}{\partial x^2} + \dfrac{\partial^2 u}{\partial y^2} + \dfrac{\partial^2 u}{\partial z^2}\right) = 1$. (*U.K.T.U. 2011*)

22. If $u = x^2\tan^{-1}\dfrac{y}{x} - y^2\tan^{-1}\dfrac{x}{y}$, find the value of $\dfrac{\partial^2 u}{\partial x \partial y}$. (*U.K.T.U. 2012*)

23. If $x^2 + y^2 + z^2 = \dfrac{1}{u^2}$, prove that $\dfrac{\partial^2 u}{\partial x^2} + \dfrac{\partial^2 u}{\partial y^2} + \dfrac{\partial^2 u}{\partial z^2} = 0$.

24. If $u = \sqrt{x^2 + y^2 + z^2}$, show that

(*i*) $\left(\dfrac{\partial u}{\partial x}\right)^2 + \left(\dfrac{\partial u}{\partial y}\right)^2 + \left(\dfrac{\partial u}{\partial z}\right)^2 = 1$ (*ii*) $\dfrac{\partial^2 u}{\partial x^2} + \dfrac{\partial^2 u}{\partial y^2} + \dfrac{\partial^2 u}{\partial z^2} = \dfrac{2}{u}$.

25. If $u = e^{x-at}\cos(x - at)$, show that $\dfrac{\partial^2 u}{\partial t^2} = a^2\dfrac{\partial^2 u}{\partial x^2}$.

26. If $v = \dfrac{1}{\sqrt{t}} \cdot e^{\frac{-x^2}{4a^2 t}}$, prove that $\dfrac{\partial v}{\partial t} = a^2\dfrac{\partial^2 v}{\partial x^2}$. (*K.U.K. Dec. 2010*)

27. If $u = (1 - 2xy + y^2)^{-1/2}$, prove that $x\dfrac{\partial u}{\partial x} - y\dfrac{\partial u}{\partial y} = y^2u^3$.

28. If $u = e^x(x\cos y - y\sin y)$, show that $\dfrac{\partial^2 u}{\partial x^2} + \dfrac{\partial^2 u}{\partial y^2} = 0$.

29. If $x = r\cos\theta$, $y = r\sin\theta$, prove that

(*i*) $\dfrac{\partial^2 r}{\partial x^2} \cdot \dfrac{\partial^2 r}{\partial y^2} = \left(\dfrac{\partial^2 r}{\partial x \partial y}\right)^2$ (*ii*) $\left(\dfrac{\partial r}{\partial x}\right)^2 + \left(\dfrac{\partial r}{\partial y}\right)^2 = 1$ (*iii*) $\dfrac{\partial^2 r}{\partial x^2} + \dfrac{\partial^2 r}{\partial y^2} = \dfrac{1}{r}\left[\left(\dfrac{\partial r}{\partial x}\right)^2 + \left(\dfrac{\partial r}{\partial y}\right)^2\right]$.

30. If $x^2 = au + bv$, $y^2 = au - bv$, prove that $\left(\frac{\partial u}{\partial x}\right)_y \cdot \left(\frac{\partial x}{\partial u}\right)_v = \frac{1}{2} = \left(\frac{\partial v}{\partial y}\right)_x \cdot \left(\frac{\partial y}{\partial v}\right)_u$. *(G.B.T.U. 2013)*

31. Show that $\frac{\partial^2 z}{\partial x^2} - 2\frac{\partial^2 z}{\partial x \partial y} + \frac{\partial^2 z}{\partial y^2} = 0$, where $z = xf(x + y) + yg(x + y)$.

32. If $u = f(ax^2 + 2hxy + by^2)$, $v = \phi(ax^2 + 2hxy + by^2)$, prove that $\frac{\partial}{\partial y}\left(u\frac{\partial v}{\partial x}\right) = \frac{\partial}{\partial x}\left(u\frac{\partial v}{\partial y}\right)$.

[**Hint.** Given $u = f(z)$, $v = \phi(z)$, where $z = ax^2 + 2hxy + by^2$.

We have to prove that $\frac{\partial u}{\partial y}\cdot\frac{\partial v}{\partial x} + u\frac{\partial^2 v}{\partial y \partial x} = \frac{\partial u}{\partial x}\cdot\frac{\partial v}{\partial y} + u\frac{\partial^2 v}{\partial x \partial y}$ or $\frac{\partial u}{\partial y}\cdot\frac{\partial v}{\partial x} = \frac{\partial u}{\partial x}\cdot\frac{\partial v}{\partial y}$ $\left(\because \frac{\partial^2 v}{\partial y \partial x} = \frac{\partial^2 v}{\partial x \partial y}\right)$

or $$\frac{\partial u}{\partial x} : \frac{\partial u}{\partial y} = \frac{\partial v}{\partial x} : \frac{\partial v}{\partial y}\Bigg]$$

33. If $u = \log(x^3 + y^3 - x^2y - xy^2)$, show that $\frac{\partial^2 u}{\partial x^2} + 2\frac{\partial^2 u}{\partial x \partial y} + \frac{\partial^2 u}{\partial y^2} = -4(x + y)^{-2}$.

[**Hint.** $u = \log\{x^2(x - y) - y^2(x - y)\} = \log(x - y)(x^2 - y^2) = \log(x - y)^2(x + y)$
$= 2\log(x - y) + \log(x + y)$]

34. (*a*) If $u = f(r)$ where $r^2 = x^2 + y^2$, prove that

$$\frac{\partial^2 u}{\partial x^2} + \frac{\partial^2 u}{\partial y^2} = f''(r) + \frac{1}{r}f'(r)$$

(*b*) If $V = f(r)$ and $r^2 = x^2 + y^2 + z^2$, prove that $V_{xx} + V_{yy} + V_{zz} = f''(r) + \frac{2}{r}f'(r)$. *(M.D.U. May 2008)*

35. If $z = f(x + ay) + \phi(x - ay)$, prove that $\frac{\partial^2 z}{\partial y^2} = a^2\frac{\partial^2 z}{\partial x^2}$.

36. Find p and q, if $x = \sqrt{a}(\sin u + \cos v)$, $y = \sqrt{a}(\cos u - \sin v)$, $z = 1 + \sin(u - v)$ where p and q mean $\frac{\partial z}{\partial x}$ and $\frac{\partial z}{\partial y}$ respectively.

$$\left[\textbf{Hint. } x^2 + y^2 = 2az, \therefore z = \frac{x^2 + y^2}{2a}\right]$$

37. The equation $\frac{\partial u}{\partial t} = \mu\frac{\partial^2 u}{\partial x^2}$ refers to the conduction of heat along a bar without radiation. Show that if $u = Ae^{-gx}\sin(nt - gx)$, where A, g, n are positive constants then $g = \sqrt{\frac{n}{2\mu}}$.

Answers

1. (*i*) $y^x \log y, xy^{x-1}$ (*ii*) $\frac{2x}{x^2 + y^2}, \frac{2y}{x^2 + y^2}$ (*iii*) $2x\sin\frac{y}{x} - y\cos\frac{y}{x}, x\cos\frac{y}{x}$

(*iv*) $\frac{-x}{x^2 + y^2} + \frac{1}{y}\tan^{-1}\frac{y}{x}, \frac{x^2}{y(x^2 + y^2)} - \frac{x}{y^2}\tan^{-1}\frac{y}{x}$ **6.** $-\frac{13}{22}$ **16.** 2, –3

22. $\frac{x^2 - y^2}{x^2 + y^2}$ **36.** $p = \frac{x}{a}, q = \frac{y}{a}$.

10.4. HOMOGENEOUS FUNCTIONS

(K.U.K. Dec. 2013; P.T.U. May 2009)

A function $f(x, y)$ is said to be homogeneous of degree (or order) n in the variables x and y if it can be expressed in the form $x^n \phi\left(\frac{y}{x}\right)$ or $y^n \phi\left(\frac{x}{y}\right)$.

An alternative test for a function $f(x, y)$ to be homogeneous of degree (or order) n is that

$$f(tx, ty) = t^n f(x, y).$$

For example, if $f(x, y) = \dfrac{x+y}{\sqrt{x}+\sqrt{y}}$, then

$$(i)\ f(x, y) = \frac{x\left(1+\dfrac{y}{x}\right)}{\sqrt{x}\left(1+\sqrt{\dfrac{y}{x}}\right)} = x^{1/2}\,\phi\left(\frac{y}{x}\right)$$

$\Rightarrow$ $f(x, y)$ is a homogeneous function of degree $\frac{1}{2}$ in x and y.

$$(ii)\ f(x, y) = \frac{y\left(\dfrac{x}{y}+1\right)}{\sqrt{y}\left(\sqrt{\dfrac{x}{y}}+1\right)} = y^{1/2}\,\phi\left(\frac{x}{y}\right)$$

$\Rightarrow$ $f(x, y)$ is a homogeneous function of degree $\frac{1}{2}$ in x and y.

$$(iii)\ f(tx, ty) = \frac{tx+ty}{\sqrt{tx}+\sqrt{ty}} = \frac{t(x+y)}{\sqrt{t}\,(\sqrt{x}+\sqrt{y})} = t^{1/2} f(x, y)$$

$\Rightarrow$ $f(x, y)$ is a homogeneous function of degree $\frac{1}{2}$ in x and y.

Similarly, a function $f(x, y, z)$ is said to be homogeneous of degree (or order) n in the variables x, y, z if

$$f(x, y, z) = x^n\,\phi\left(\frac{y}{x}, \frac{z}{x}\right) \quad \text{or} \quad y^n\phi\left(\frac{x}{y}, \frac{z}{y}\right) \quad \text{or} \quad z^n\,\phi\left(\frac{x}{z}, \frac{y}{z}\right).$$

Alternative test is $f(tx, ty, tz) = t^n f(x, y, z)$.

10.5. EULER'S THEOREM ON HOMOGENEOUS FUNCTIONS

(G.B.T.U. 2013; K.U.K. Dec. 2013; U.K.T.U. 2011; P.T.U. 2009, 2010)

If u is a homogeneous function of degree n in x and y, then $\mathbf{x\dfrac{\partial u}{\partial x} + y\dfrac{\partial u}{\partial y} = nu}$

Since u is a homogeneous function of degree n in x and y, it can be expressed as $u = x^n f\left(\frac{y}{x}\right)$

$$\therefore \qquad \frac{\partial u}{\partial x} = nx^{n-1} f\left(\frac{y}{x}\right) = x^n f'\left(\frac{y}{x}\right)\cdot\left(-\frac{y}{x^2}\right)$$

$$\Rightarrow \qquad x\frac{\partial u}{\partial x} = nx^n f\left(\frac{y}{x}\right) - x^{n-1} yf'\left(\frac{y}{x}\right) \qquad ...(1)$$

Also $$\frac{\partial u}{\partial y} = x^n f'\left(\frac{y}{x}\right) \cdot \frac{1}{x} = x^{n-1} f'\left(\frac{y}{x}\right)$$

$$\Rightarrow \quad y\frac{\partial u}{\partial y} = x^{n-1} yf'\left(\frac{y}{x}\right) \qquad ...(2)$$

Adding (1) and (2), we get $x\frac{\partial u}{\partial x} + y\frac{\partial u}{\partial y} = nx^n f\left(\frac{y}{x}\right) = nu.$

Note. Euler's theorem can be extended to a homogeneous function of any number of variables. Thus, if u is a homogeneous function of degree n in x, y and z, then $x\frac{\partial u}{\partial x} + y\frac{\partial u}{\partial y} + z\frac{\partial u}{\partial z} = nu.$

10.6. IF u IS A HOMOGENEOUS FUNCTION OF DEGREE n IN x AND y, THEN $\mathbf{x^2 \frac{\partial^2 u}{\partial x^2} + 2xy \frac{\partial^2 u}{\partial x \partial y} + y^2 \frac{\partial^2 u}{\partial y^2} = n(n-1)u}$

Since u is a homogeneous function of degree n in x and y

$\therefore$ By Euler's Theorem, we have $$x\frac{\partial u}{\partial x} + y\frac{\partial u}{\partial y} = nu \qquad ...(1)$$

Differentiating (1) partially w.r.t. x, we have $$1 \cdot \frac{\partial u}{\partial x} + x\frac{\partial^2 u}{\partial x^2} + y \cdot \frac{\partial^2 u}{\partial x \partial y} = n\frac{\partial u}{\partial x} \qquad ...(2)$$

Differentiating (1) partially, w.r.t. y, we have $$x\frac{\partial^2 u}{\partial y \partial x} + 1 \cdot \frac{\partial u}{\partial y} + y \cdot \frac{\partial^2 u}{\partial y^2} = n \cdot \frac{\partial u}{\partial y}$$

But $$\frac{\partial^2 u}{\partial y \partial x} = \frac{\partial^2 u}{\partial x \partial y}$$

$$\therefore \quad x\frac{\partial^2 u}{\partial x \partial y} + \frac{\partial u}{\partial y} + y\frac{\partial^2 u}{\partial y^2} = n\frac{\partial u}{\partial y} \qquad ...(3)$$

Multiplying (2) by x, (3) by y and adding

$$x^2\frac{\partial^2 u}{\partial x^2} + 2xy\frac{\partial^2 u}{\partial x \partial y} + y^2\frac{\partial^2 u}{\partial y^2} + \left(x\frac{\partial u}{\partial x} + y\frac{\partial u}{\partial y}\right) = n\left(x\frac{\partial u}{\partial x} + y\frac{\partial u}{\partial y}\right)$$

or $$x^2\frac{\partial^2 u}{\partial x^2} + 2xy\frac{\partial^2 u}{\partial x \partial y} + y^2\frac{\partial^2 u}{\partial y^2} + nu = n \cdot nu \qquad \text{[Using (1)]}$$

$$\Rightarrow \quad x^2\frac{\partial^2 u}{\partial x^2} + 2xy\frac{\partial^2 u}{\partial x \partial y} + y^2\frac{\partial^2 u}{\partial y^2} = n^2 u - nu = n(n-1)u.$$

ILLUSTRATIVE EXAMPLES

Example 1. *Verify Euler's theorem for the functions:*

(*i*) $u = (x^{1/2} + y^{1/2})(x^n + y^n)$ (*ii*) $u = \sin^{-1}\frac{x}{y} + \tan^{-1}\frac{y}{x}$.

Sol. (*i*) $$u = (x^{1/2} + y^{1/2})(x^n + y^n) \qquad ...(1)$$

$$= x^{1/2}\left(1 + \frac{y^{1/2}}{x^{1/2}}\right)x^n\left(1 + \frac{y^n}{x^n}\right) = x^{n+1/2}\left[1 + \left(\frac{y}{x}\right)^{1/2}\right]\left[1 + \left(\frac{y}{x}\right)^n\right] = x^{n+1/2} f\left(\frac{y}{x}\right)$$

[OR $f(tx, ty) = t^{n+1/2} f(x, y)$]

$\Rightarrow$ u is a homogeneous function of degree $\left(n+\frac{1}{2}\right)$ in x and y

$\therefore$ By Euler's theorem, we have $x\frac{\partial u}{\partial x}+y\frac{\partial u}{\partial y}=\left(n+\frac{1}{2}\right)u$ ...(2)

From (1), $$\frac{\partial u}{\partial x}=\frac{1}{2}x^{-1/2}(x^n+y^n)+nx^{n-1}(x^{1/2}+y^{1/2})$$

$$x\frac{\partial u}{\partial x}=\frac{1}{2}x^{1/2}(x^n+y^n)+nx^n(x^{1/2}+y^{1/2})$$

$$\frac{\partial u}{\partial y}=\frac{1}{2}y^{-1/2}(x^n+y^n)+ny^{n-1}(x^{1/2}+y^{1/2})$$

$$y\frac{\partial u}{\partial y}=\frac{1}{2}y^{1/2}(x^n+y^n)+ny^n(x^{1/2}+y^{1/2})$$

$$\therefore \quad x\frac{\partial u}{\partial x}+y\frac{\partial u}{\partial y}=\frac{1}{2}(x^{1/2}+y^{1/2})(x^n+y^n)+n(x^n+y^n)(x^{1/2}+y^{1/2})$$

$=\frac{1}{2}u+nu=\left(n+\frac{1}{2}\right)u$ which is the same as (2). Hence the verification.

(*ii*) $$u=\sin^{-1}\frac{x}{y}+\tan^{-1}\frac{y}{x} \quad \text{...(1)}$$

$$=\operatorname{cosec}^{-1}\frac{y}{x}+\tan^{-1}\frac{y}{x}=x^0 f\left(\frac{y}{x}\right)$$

[OR $f(tx, ty)=f(x, y)=t^0 f(x, y)$]

$\Rightarrow$ u is a homogeneous function of degree 0 in x and y.

$\therefore$ By Euler's theorem, we have $x\frac{\partial u}{\partial x}+y\frac{\partial u}{\partial y}=0\times u=0$ ...(2)

From (1), $$\frac{\partial u}{\partial x}=\frac{1}{\sqrt{1-\frac{x^2}{y^2}}}\cdot\frac{1}{y}+\frac{1}{1+\frac{y^2}{x^2}}\left(-\frac{y}{x^2}\right)=\frac{1}{\sqrt{y^2-x^2}}-\frac{y}{\sqrt{x^2+y^2}}$$

$$x\frac{\partial u}{\partial x}=\frac{x}{\sqrt{y^2-x^2}}-\frac{xy}{x^2+y^2}$$

$$\frac{\partial u}{\partial y}=\frac{1}{\sqrt{1-\frac{x^2}{y^2}}}\left(-\frac{x}{y^2}\right)+\frac{1}{1+\frac{y^2}{x^2}}\cdot\frac{1}{x}=-\frac{x}{y\sqrt{y^2-x^2}}+\frac{x}{x^2+y^2}$$

$$y\frac{\partial u}{\partial y}=-\frac{x}{\sqrt{y^2-x^2}}+\frac{xy}{x^2+y^2}$$

$\therefore$ $x\frac{\partial u}{\partial x}+y\frac{\partial u}{\partial y}=0$ which is the same as (2). Hence the verification.

Example 2. *If* $u=\tan^{-1}\frac{x^3+y^3}{x-y}$, *prove that* $x\frac{\partial u}{\partial x}+y\frac{\partial u}{\partial y}=\sin 2u$. (M.D.U. Dec. 2013)

Sol. Here u is not a homogeneous function but $\tan u = \dfrac{x^3+y^3}{x-y} = \dfrac{x^3\left[1+\left(\dfrac{y}{x}\right)^3\right]}{x\left[1-\dfrac{y}{x}\right]} = x^2 f\left(\dfrac{y}{x}\right)$

is a homogeneous function of degree 2 in x and y.

$\therefore$ By Euler's theorem, we have

$$x\frac{\partial}{\partial x}(\tan u) + y\frac{\partial}{\partial y}(\tan u) = 2\tan u \quad \text{or} \quad x\sec^2 u\frac{\partial u}{\partial x} + y\sec^2 u\frac{\partial u}{\partial y} = 2\tan u$$

or $$x\frac{\partial u}{\partial x} + y\frac{\partial u}{\partial y} = \frac{2\sin u}{\cos u}\,.\cos^2 u = 2\sin u\cos u = \sin 2u.$$

Example 3. *If* $u = \sin^{-1}\left(\dfrac{x+2y+3z}{\sqrt{x^8+y^8+z^8}}\right)$, *show that* $x\dfrac{\partial u}{\partial x} + y\dfrac{\partial u}{\partial y} + z\dfrac{\partial u}{\partial z} + 3\tan u = 0.$

(M.D.U. Dec. 2009; U.K.T.U. 2010)

Sol. Here u is not a homogeneous function.

$$\sin u = f(x, y, z) = \frac{x+2y+3z}{\sqrt{x^8+y^8+z^8}}$$

$$f(tx, ty, tz) = \frac{t\,(x+2y+3z)}{t^4\sqrt{x^8+y^8+z^8}} = t^{-3} f(x, y, z)$$

$\Rightarrow$ $\sin u$ is a homogeneous function of degree -3 in x, y, z.

$\therefore$ By Euler's theorem, we have

$$x\frac{\partial}{\partial x}(\sin u) + y\frac{\partial}{\partial y}(\sin u) + z\frac{\partial}{\partial z}(\sin u) = -3\sin u$$

or $$x\cos u\frac{\partial u}{\partial x} + y\cos u\frac{\partial u}{\partial y} + z\cos u\frac{\partial u}{\partial z} + 3\sin u = 0$$

or $$x\frac{\partial u}{\partial x} + y\frac{\partial u}{\partial y} + z\frac{\partial u}{\partial z} + 3\tan u = 0.$$

Example 4. *If* $u = \log\dfrac{x^4+y^4}{x+y}$, *show that* $x\dfrac{\partial u}{\partial x} + y\dfrac{\partial u}{\partial y} = 3.$

(U.P.T.U. 2009; U.K.T.U. 2012)

Sol. Here u is not a homogeneous function

$$u = \log\frac{x^4+y^4}{x+y} \quad\Rightarrow\quad u = \log_e\left(\frac{x^4+y^4}{x+y}\right) \quad\Rightarrow\quad e^u = \frac{x^4+y^4}{x+y}$$

which is a homogeneous function of degree 3 in x, y.

$\therefore$ By Euler's theorem, we have $x\dfrac{\partial}{\partial x}(e^u) + y\dfrac{\partial}{\partial y}(e^u) = 3\times e^u$

or $$xe^u\frac{\partial u}{\partial x} + ye^u\frac{\partial u}{\partial y} = 3e^u \quad \text{or} \quad x\frac{\partial u}{\partial x} + y\frac{\partial u}{\partial y} = 3.$$

Example 5. *If* $u = \dfrac{x^2 y}{x+y}$, *show that* $x\dfrac{\partial^2 u}{\partial x^2} + y\dfrac{\partial^2 u}{\partial y\partial x} = \dfrac{\partial u}{\partial x}.$

Sol. Here $u = \dfrac{x^2 y}{x + y}$ is a homogeneous function of degree 2 in x and y.

$\therefore$ By Euler's theorem, we have

$$x \frac{\partial u}{\partial x} + y \frac{\partial u}{\partial y} = 2u \qquad \text{...(1)}$$

Differentiating (1) partially w.r.t. x,

$$x \frac{\partial^2 u}{\partial x^2} + 1 \frac{\partial u}{\partial x} + y \frac{\partial^2 u}{\partial x \partial y} = 2 \frac{\partial u}{\partial x}$$

$$\Rightarrow \quad x \frac{\partial^2 u}{\partial x^2} + y \frac{\partial^2 u}{\partial y \partial x} = \frac{\partial u}{\partial x} \qquad \left[\because \frac{\partial^2 u}{\partial x \partial y} = \frac{\partial^2 u}{\partial y \partial x}\right]$$

Example 6. *If* $u = \sin^{-1} \dfrac{x + y}{\sqrt{x} + \sqrt{y}}$, *prove that* $x^2 \dfrac{\partial^2 u}{\partial x^2} + 2xy \dfrac{\partial^2 u}{\partial x \partial y} + y^2 \dfrac{\partial^2 u}{\partial y^2} = -\dfrac{\sin u \cos 2u}{4 \cos^3 u}$.

(M.D.U. Dec. 2012, Dec. 2013)

Sol. $u = \sin^{-1} \dfrac{x + y}{\sqrt{x} + \sqrt{y}}$ is not a homogeneous function but

$\sin u = \dfrac{x + y}{\sqrt{x} + \sqrt{y}}$ is a homogeneous function of degree $\dfrac{1}{2}$ in x and y.

$\therefore$ By Euler's theorem, we have

$$x \frac{\partial}{\partial x} (\sin u) + y \frac{\partial}{\partial y} (\sin u) = \frac{1}{2} \sin u$$

or $$x \cos u \frac{\partial u}{\partial x} + y \cos u \frac{\partial u}{\partial y} = \frac{1}{2} \sin u \quad \text{or} \quad x \frac{\partial u}{\partial x} + y \frac{\partial u}{\partial y} = \frac{1}{2} \tan u \qquad \text{...(1)}$$

Differentiating (1) partially w.r.t. x,

$$x \frac{\partial^2 u}{\partial x^2} + 1 . \frac{\partial u}{\partial x} + y \frac{\partial^2 u}{\partial x \partial y} = \frac{1}{2} \sec^2 u \frac{\partial u}{\partial x}$$

or $$x \frac{\partial^2 u}{\partial x^2} + y \frac{\partial^2 u}{\partial x \partial y} = \left(\frac{1}{2} \sec^2 u - 1\right) \frac{\partial u}{\partial x} \qquad \text{...(2)}$$

Differentiating (1) partially w.r.t. y,

$$x \frac{\partial^2 u}{\partial y \partial x} + y \frac{\partial^2 u}{\partial y^2} + 1 . \frac{\partial u}{\partial y} = \frac{1}{2} \sec^2 u \frac{\partial u}{\partial y}$$

or $$x \frac{\partial^2 u}{\partial x \partial y} + y \frac{\partial^2 u}{\partial y^2} = \left(\frac{1}{2} \sec^2 u - 1\right) \frac{\partial u}{\partial y} \qquad \text{...(3)} \qquad \left[\because \frac{\partial^2 u}{\partial y \partial x} = \frac{\partial^2 u}{\partial x \partial y}\right]$$

Multiplying (2) by x, (3) by y and adding

$$x^2 \frac{\partial^2 u}{\partial x^2} + 2xy \frac{\partial^2 u}{\partial x \partial y} + y^2 \frac{\partial^2 u}{\partial y^2} = \left(\frac{1}{2} \sec^2 u - 1\right) \left(x \frac{\partial u}{\partial x} + y \frac{\partial u}{\partial y}\right)$$

$$= \left(\frac{1}{2\cos^2 u} - 1\right) \cdot \frac{1}{2} \tan u \qquad \text{[Using (1)]}$$

$$= -\frac{2\cos^2 u - 1}{2\cos^2 u} \cdot \frac{1}{2} \frac{\sin u}{\cos u} = -\frac{\sin u \cos 2u}{4\cos^3 u} \qquad [\because \; 2\cos^2 u - 1 = \cos 2u]$$

Example 7. *If* $z = xf\left(\frac{y}{x}\right) + g\left(\frac{y}{x}\right)$, *show that* $x^2 \frac{\partial^2 z}{\partial x^2} + 2xy \frac{\partial^2 z}{\partial x \partial y} + y^2 \frac{\partial^2 z}{\partial y^2} = 0.$

(Kerala 2008)

Sol. Let $u = xf\left(\frac{y}{x}\right)$ *and* $v = g\left(\frac{y}{x}\right) = x^0 g\left(\frac{y}{x}\right)$

so that $z = u + v$ ...(1)

Since u is a homogeneous function of degree $n = 1$ in x, y

$$\therefore \quad x^2 \frac{\partial^2 u}{\partial x^2} + 2xy \frac{\partial^2 u}{\partial x \partial y} + y^2 \frac{\partial^2 u}{\partial y^2} = n(n-1)u = 0 \qquad ...(2)$$

Since v is a homogeneous function of degree $n = 0$ in x, y

$$\therefore \quad x^2 \frac{\partial^2 v}{\partial x^2} + 2xy \frac{\partial^2 v}{\partial x \partial y} + y^2 \frac{\partial^2 v}{\partial y^2} = n(n-1)v = 0 \qquad ...(3)$$

Adding (2) and (3), we have

$$x^2 \frac{\partial^2}{\partial x^2}(u+v) + 2xy \frac{\partial^2}{\partial x \partial y}(u+v) + y^2 \frac{\partial^2}{\partial y^2}(u+v) = 0$$

or $$x^2 \frac{\partial^2 z}{\partial x^2} + 2xy \frac{\partial^2 z}{\partial x \partial y} + y^2 \frac{\partial^2 z}{\partial y^2} = 0. \qquad \text{[Using (1)]}$$

EXERCISE 10.2

1. Verify Euler's theorem for the functions:

   (*i*) $f(x, y) = ax^2 + 2hxy + by^2$ (*ii*) $u = \dfrac{x^{1/4} + y^{1/4}}{x^{1/5} + y^{1/5}}$

   (*iii*) $f(x, y) = \dfrac{x^2 (x^2 - y^2)^3}{(x^2 + y^2)^3}$ (*iv*) $f(x, y, z) = 3x^2yz + 5xy^2z + 4z^4$

   (*v*) $u = \log\left(\dfrac{x^2 + y^2}{xy}\right)$. (*U.K.T.U. 2011*)

2. (*i*) If $u = f\left(\frac{y}{x}\right)$, show that $x\frac{\partial u}{\partial x} + y\frac{\partial u}{\partial y} = 0$ (*ii*) If $u = xf\left(\frac{y}{x}\right)$, prove that $x\frac{\partial u}{\partial x} + y\frac{\partial u}{\partial y} = u$.

   (*iii*) If $z = xyf\left(\frac{x}{y}\right)$, prove that $x\frac{\partial z}{\partial x} + y\frac{\partial z}{\partial y} = 2z$ (*M.T.U. 2013*)

3. If $V = \dfrac{x^3y^3}{x^3 + y^3}$, show that $x\frac{\partial V}{\partial x} + y\frac{\partial V}{\partial y} = 3V$.

4. If $u = \sin^{-1}\frac{x}{y} + \tan^{-1}\frac{y}{x}$, then find the value of $x\frac{\partial u}{\partial x} + y\frac{\partial u}{\partial y}$.

   (*M.D.U. Dec. 2010, Dec. 2011; P.T.U. Dec. 2010*)

5. If $f(x, y) = \sqrt{x^2 - y^2}\,\sin^{-1}\frac{y}{x}$, prove that $x\frac{\partial f}{\partial x} + y\frac{\partial f}{\partial y} = f(x, y)$.

6. If $f(x, y) = \sqrt{y^2 - x^2}\,\sin^{-1}\frac{x}{y} + \dfrac{x^2 - y^2}{\sqrt{x^2 + y^2}}$, show that $x\frac{\partial f}{\partial x} + y\frac{\partial f}{\partial y} - f(x, y) = 0$.

7. If $f(x, y) = \frac{1}{x^2} + \frac{1}{xy} + \frac{\log x - \log y}{x^2 + y^2}$, show that $x\frac{\partial f}{\partial x} + y\frac{\partial f}{\partial y} + 2f(x, y) = 0$.

8. If $u = \cos\left(\frac{xy + yz + zx}{x^2 + y^2 + z^2}\right)$, prove that $x\frac{\partial u}{\partial x} + y\frac{\partial u}{\partial y} + z\frac{\partial u}{\partial z} = 0$.

9. If $u = \cos^{-1}\frac{x + y}{\sqrt{x} + \sqrt{y}}$, show that $x\frac{\partial u}{\partial x} + y\frac{\partial u}{\partial y} + \frac{1}{2}\cot u = 0$.

10. If $u = \sin^{-1}\left(\frac{x^2 + y^2}{x + y}\right)$, prove that $x\frac{\partial u}{\partial x} + y\frac{\partial u}{\partial y} = \tan u$.

11. (*i*) If $u = \sin^{-1}\frac{\sqrt{x} - \sqrt{y}}{\sqrt{x} + \sqrt{y}}$, show that $\frac{\partial u}{\partial x} = -\frac{y}{x}\cdot\frac{\partial u}{\partial y}$.

(*ii*) If $\sin u = \frac{x^2y^2}{x + y}$, show that $x\frac{\partial u}{\partial x} + y\frac{\partial u}{\partial y} = 3\tan u$.

(*iii*) Show that $x\frac{\partial u}{\partial x} + y\frac{\partial u}{\partial y} = 2u\log u$ where $\log u = \frac{x^3 + y^3}{3x + 4y}$.

12. (*i*) If $u = \log\left(\frac{x^5 + y^5 + z^5}{x^2 + y^2 + z^2}\right)$, show that $x\frac{\partial u}{\partial x} + y\frac{\partial u}{\partial y} + z\frac{\partial u}{\partial z} = 3$.

(*ii*) Show that $xu_x + yu_y + zu_z = 2\tan u$, where $u = \sin^{-1}\left(\frac{x^3 + y^3 + z^3}{ax + by + cz}\right)$. (*P.T.U. 2008*)

(*iii*) If $u = \frac{x}{y + z} + \frac{y}{z + x} + \frac{z}{x + y}$, show that $x\frac{\partial u}{\partial x} + y\frac{\partial u}{\partial y} + z\frac{\partial u}{\partial z} = 0$.

13. If $u = \frac{x^2y^2}{x + y}$, show that

(*i*) $x\frac{\partial^2 u}{\partial x^2} + y\frac{\partial^2 u}{\partial x\partial y} = 2\frac{\partial u}{\partial x}$ (*ii*) $x\frac{\partial^2 u}{\partial x\partial y} + y\frac{\partial^2 u}{\partial y^2} = 2\frac{\partial u}{\partial y}$.

14. Given $z = x^n f_1\left(\frac{y}{x}\right) + y^{-n} f_2\left(\frac{x}{y}\right)$, prove that $x^2\frac{\partial^2 z}{\partial x^2} + 2xy\frac{\partial^2 z}{\partial x\partial y} + y^2\frac{\partial^2 z}{\partial y^2} + x\frac{\partial z}{\partial x} + y\frac{\partial z}{\partial y} = n^2 z$.

(*K.U.K. 2009, 2012*)

15. If $u = (x^2 + y^2)^{1/3}$, show that $x^2\frac{\partial^2 u}{\partial x^2} + 2xy\frac{\partial^2 u}{\partial x\partial y} + y^2\frac{\partial^2 u}{\partial y^2} = -\frac{2u}{9}$.

16. If $u = \tan^{-1}\frac{x^3 + y^3}{x - y}$, prove that $x^2\frac{\partial^2 u}{\partial x^2} + 2xy\frac{\partial^2 u}{\partial x\partial y} + y^2\frac{\partial^2 u}{\partial y^2} = \sin 4u - \sin 2u = 2\cos 3u\sin u$.

(*M.D.U. Dec. 2008; P.T.U. 2010, 2011*)

17. If $u = \tan^{-1}\left(\frac{y^2}{x}\right)$, show that $x^2\frac{\partial^2 u}{\partial x^2} + 2xy\frac{\partial^2 u}{\partial x\partial y} + y^2\frac{\partial^2 u}{\partial y^2} = -\sin 2u\sin^2 u$. (*P.T.U. May 2011*)

18. If $u = x^2\tan^{-1}\left(\frac{y}{x}\right) - y^2\tan^{-1}\left(\frac{x}{y}\right)$, then evaluate $x^2\frac{\partial^2 u}{\partial x^2} + 2xy\frac{\partial^2 u}{\partial x\partial y} + y^2\frac{\partial^2 u}{\partial y^2}$.

(*U.K.T.U. 2010; M.D.U. May 2009*)

19. If $u = \operatorname{cosec}^{-1}\left(\frac{x^{1/2} + y^{1/2}}{x^{1/3} + y^{1/3}}\right)^{1/2}$, prove that $x^2\frac{\partial^2 u}{\partial x^2} + 2xy\frac{\partial^2 u}{\partial x\partial y} + y^2\frac{\partial^2 u}{\partial y^2} = \frac{\tan u}{144}(13 + \tan^2 u)$.

(*M.D.U. Dec. 2010*)

Answers

4. 0 18. $2u$.

10.7. COMPOSITE FUNCTIONS

(*i*) If $u = f(x, y)$ where $x = \phi(t)$, $y = \psi(t)$

then u is called a composite function of (the **single variable**) t and we can find $\frac{du}{dt}$.

(*ii*) If $z = f(x, y)$ where $x = \phi(u, v)$, $y = \psi(u, v)$

then z is called a composite function of (**two variables**) u and v so that we can find $\frac{\partial z}{\partial u}$ and $\frac{\partial z}{\partial v}$.

10.8. DIFFERENTIATION OF COMPOSITE FUNCTIONS

If u is composite function of t, defined by the relations u = f(x, y); x = ϕ(t), y = ψ(t), then

$$\frac{du}{dt} = \frac{\partial u}{\partial x} \cdot \frac{dx}{dt} + \frac{\partial u}{\partial y} \cdot \frac{dy}{dt}$$

Proof. Here $u = f(x, y)$ ...(1)

Let δt be an increment in t and δx, δy, δu the corresponding increments in x, y and u respectively. Then, we have

$$u + \delta u = f(x + \delta x, y + \delta y) \quad ...(2)$$

Subtracting (1) from (2), we get

$$\delta u = f(x + \delta x, y + \delta y) - f(x, y)$$

$$= f(x + \delta x, y + \delta y) - f(x, y + \delta y) + f(x, y + \delta y) - f(x, y)$$

$$\frac{\delta u}{\delta t} = \frac{f(x + \delta x, y + \delta y) - f(x, y + \delta y)}{\delta t} + \frac{f(x, y + \delta y) - f(x, y)}{\delta t}$$

$$= \frac{f(x + \delta x, y + \delta y) - f(x, y + \delta y)}{\delta x} \cdot \frac{\delta x}{\delta t} + \frac{f(x, y + \delta y) - f(x, y)}{\delta y} \cdot \frac{\delta x}{\delta t} \quad ...(3)$$

As $\delta t \to 0$, δx and δy both $\to 0$, so that

$$\lim_{\delta t \to 0} \frac{\delta u}{\delta t} = \frac{du}{dt}, \quad \lim_{\delta t \to 0} \frac{\delta x}{\delta t} = \frac{dx}{dt}, \quad \lim_{\delta t \to 0} \frac{\delta y}{\delta t} = \frac{dy}{dt}$$

$$\lim_{\delta x \to 0} \frac{f(x + \delta x, y + \delta y) - f(x, y + \delta y)}{\delta x} = \frac{\partial f}{\partial x} = \frac{\partial u}{\partial x}$$

and
$$\lim_{\delta y \to 0} \frac{f(x, y + \delta y) - f(x, y)}{\delta y} = \frac{\partial f}{\partial y} = \frac{\partial u}{\partial y}$$

∴ From (1),
$$\frac{du}{dt} = \frac{\partial u}{\partial x} \cdot \frac{dx}{dt} + \frac{\partial u}{\partial y} \cdot \frac{dy}{dt}$$

$\frac{du}{dt}$ is called the **total derivative** of u to distinguish it from the partial derivatives $\frac{\partial u}{\partial x}$ and $\frac{\partial u}{\partial y}$.

Cor. 1. If $u = f(x, y, z)$ and x, y, z are function of t, then y is a composite function of t and

$$\frac{du}{dt} = \frac{\partial u}{\partial x}\cdot\frac{dx}{dt} + \frac{\partial u}{\partial y}\cdot\frac{dy}{dt} + \frac{\partial u}{\partial z}\cdot\frac{dz}{dt}.$$

Cor. 2. If $z = f(x, y)$ and x, y are functions of u and v, then

$$\frac{\partial z}{\partial u} = \frac{\partial z}{\partial x}\cdot\frac{\partial x}{\partial u} + \frac{\partial z}{\partial y}\cdot\frac{\partial y}{\partial u};\quad \frac{\partial z}{\partial v} = \frac{\partial z}{\partial x}\cdot\frac{\partial x}{\partial v} + \frac{\partial z}{\partial y}\cdot\frac{\partial y}{\partial v}.$$

Cor. 3. If $u = f(x, y)$ where $y = \phi(x)$ then since $x = \psi(x)$, u is a composite function of x.

$$\therefore \quad \frac{du}{dx} = \frac{\partial u}{\partial x}\cdot\frac{dx}{dx} + \frac{\partial u}{\partial y}\cdot\frac{dy}{dx} \quad\Rightarrow\quad \frac{du}{dx} = \frac{\partial u}{\partial x} + \frac{\partial u}{\partial y}\cdot\frac{dy}{dx}.$$

Cor. 4. If we are given an **implicit function f(x, y) = c,** then $u = f(x, y)$ where $u = c$

Using Cor. 3, we have $\dfrac{du}{dx} = \dfrac{\partial u}{\partial x} + \dfrac{\partial u}{\partial y}\cdot\dfrac{dy}{dx}$

But $\dfrac{du}{dx} = 0 \quad \therefore \quad \dfrac{\partial u}{\partial x} + \dfrac{\partial u}{\partial y}\cdot\dfrac{dy}{dx} = 0$ or $\dfrac{\partial f}{\partial x} + \dfrac{\partial f}{\partial y}\cdot\dfrac{dy}{dx} = 0$

$$\therefore \quad \frac{dy}{dx} = -\frac{\dfrac{\partial f}{\partial x}}{\dfrac{\partial f}{\partial y}} = -\frac{f_x}{f_y}$$

Hence the differential coefficient of $f(x, y)$ w.r.t. x is $\dfrac{df}{dx} = \dfrac{\partial f}{\partial x} + \dfrac{\partial f}{\partial y}\cdot\dfrac{dy}{dx}$.

Cor. 5. If $f(x, y) = c$, then by Cor. 4, we have $\dfrac{dy}{dx} = -\dfrac{f_x}{f_y}$

Differentiating again w.r.t. x, we get

$$\frac{d^2u}{dx^2} = -\frac{f_y\dfrac{d}{dx}(f_x) - f_x\dfrac{d}{dx}(f_y)}{f_y^2} = -\frac{f_y\left[\dfrac{\partial f_x}{\partial x} + \dfrac{\partial f_x}{\partial y}\cdot\dfrac{dy}{dx}\right] - f_x\left[\dfrac{\partial f_y}{\partial x} + \dfrac{\partial f_y}{\partial y}\cdot\dfrac{dy}{dx}\right]}{f_y^2}$$

$$= -\frac{f_y\left[f_{xx} - f_{yx}\cdot\dfrac{f_x}{f_y}\right] - f_x\left[f_{xy} - f_{yy}\cdot\dfrac{f_x}{f_y}\right]}{f_y^2} = -\frac{f_{xx}f_y^2 - f_xf_yf_{xy} - f_xf_yf_{xy} - f_{yy}f_x^2}{f_y^3}$$

Hence $\dfrac{\mathbf{d^2y}}{\mathbf{dx^2}} = -\dfrac{\mathbf{f_{xx}f_y^2 - 2f_xf_yf_{xy} + f_{yy}f_x^2}}{\mathbf{f_y^3}}$.

ILLUSTRATIVE EXAMPLES

Example 1. *If* $u = \sin^{-1}(x - y)$, $x = 3t$, $y = 4t^3$, *show that* $\dfrac{du}{dt} = \dfrac{3}{\sqrt{1 - t^2}}$.

Sol. The given equations define u as a composite function of t.

$$\therefore \quad \frac{du}{dt} = \frac{\partial u}{\partial x} \cdot \frac{dx}{dt} + \frac{\partial u}{\partial y} \cdot \frac{dy}{dt}$$

$$= \frac{1}{\sqrt{1-(x-y)^2}} . 3 + \frac{1}{\sqrt{1-(x-y)^2}} (-1) . 12 t^2$$

$$= \frac{3(1-4t^2)}{\sqrt{1-(x-y)^2}} = \frac{3(1-4t^2)}{\sqrt{1-(3t-4t^3)^2}} = \frac{3(1-4t^2)}{\sqrt{1-9t^2+24t^4-16t^6}}$$

$$= \frac{3(1-4t^2)}{\sqrt{(1-t^2)(1-8t^2+16t^4)}} = \frac{3(1-4t^2)}{\sqrt{(1-t^2)(1-4t^2)^2}} = \frac{3}{\sqrt{1-t^2}}.$$

Example 2. *If $z = 2xy^2 - 3x^2y$ and if x increases at the rate of 2 cm per second when it passes through the value $x = 3$ cm, show that if y is passing through the value $y = 1$ cm, y must be decreasing at the rate of $2\frac{2}{15}$ cm per second, in order that z shall remain constant.*

Sol. Given : $z = 2xy^2 - 3x^2y$ and $\frac{dx}{dt} = 2$ cm/sec when $x = 3$ cm, we have to find $\frac{dy}{dt}$ when $y = 1$ cm.

Now $\quad \frac{dz}{dt} = \frac{\partial z}{\partial x} \cdot \frac{dx}{dt} + \frac{\partial z}{\partial y} \cdot \frac{dy}{dt} = (2y^2 - 6xy) \frac{dx}{dt} + (4xy - 3x^2) \frac{dy}{dt}$

Since z remains constant, $\frac{dz}{dt} = 0$

$$\therefore \quad 0 = (2y^2 - 6 \times 3 \times y) \times 2 + (4 \times 3 \times y - 3 \times 3^2) \frac{dy}{dt}$$

$$\Rightarrow \quad 0 = (4y^2 - 36y) + (12y - 27) \frac{dy}{dt}$$

When $y = 1$ cm, we have

$$0 = (4 - 36) + (12 - 27) \frac{dy}{dt}$$

$$\Rightarrow \quad \frac{dy}{dt} = -\frac{32}{15} \text{ which is negative}$$

$\therefore$ When $y = 1$ cm, y is decreasing at the rate of $2\frac{2}{15}$ cm/sec.

Example 3. *If z is a function of x and y, where $x = e^u + e^{-v}$ and $y = e^{-u} - e^v$, show that*

$$\frac{\partial z}{\partial u} - \frac{\partial z}{\partial v} = x\frac{\partial z}{\partial x} - y\frac{\partial z}{\partial y}.$$

Sol. Here z is a composite function of u and v.

$$\therefore \quad \frac{\partial z}{\partial u} = \frac{\partial z}{\partial x} \cdot \frac{\partial x}{\partial u} + \frac{\partial z}{\partial y} \cdot \frac{\partial y}{\partial u} = \frac{\partial z}{\partial x} \cdot e^u + \frac{\partial z}{\partial y}(-e^{-u})$$

and

$$\frac{\partial z}{\partial v} = \frac{\partial z}{\partial x} \cdot \frac{\partial x}{\partial v} + \frac{\partial z}{\partial y} \cdot \frac{\partial y}{\partial v} = \frac{\partial z}{\partial x}(-e^{-v}) + \frac{\partial z}{\partial y}(-e^v)$$

Subtracting, $\quad \frac{\partial z}{\partial u} - \frac{\partial z}{\partial v} = (e^u + e^{-v})\frac{\partial z}{\partial x} - (e^{-u} - e^v)\frac{\partial z}{\partial y} = x\frac{\partial z}{\partial x} - y\frac{\partial z}{\partial y}.$

Example 4. *If* $u = f(y - z, z - x, x - y)$, *prove that* $\frac{\partial u}{\partial x} + \frac{\partial u}{\partial y} + \frac{\partial u}{\partial z} = 0.$

(U.K.T.U. 2010; Kerala 2010; P.T.U. Dec. 2011)

Sol. Here $u = f(X, Y, Z)$ where $X = y - z, Y = z - x, Z = x - y$

$\therefore$ u is a composite function of x, y and z.

$$\frac{\partial u}{\partial x} = \frac{\partial u}{\partial X} \cdot \frac{\partial X}{\partial x} + \frac{\partial u}{\partial Y} \cdot \frac{\partial Y}{\partial x} + \frac{\partial u}{\partial Z} \cdot \frac{\partial Z}{\partial x} = \frac{\partial u}{\partial X}(0) + \frac{\partial u}{\partial Y}(-1) + \frac{\partial u}{\partial Z}(1)$$

$$\frac{\partial u}{\partial y} = \frac{\partial u}{\partial X} \cdot \frac{\partial X}{\partial y} + \frac{\partial u}{\partial Y} \cdot \frac{\partial Y}{\partial y} + \frac{\partial u}{\partial Z} \cdot \frac{\partial Z}{\partial y} = \frac{\partial u}{\partial X}(1) + \frac{\partial u}{\partial Y}(0) + \frac{\partial u}{\partial Z}(-1)$$

$$\frac{\partial u}{\partial z} = \frac{\partial u}{\partial X} \cdot \frac{\partial X}{\partial z} + \frac{\partial u}{\partial Y} \cdot \frac{\partial Y}{\partial z} + \frac{\partial u}{\partial Z} \cdot \frac{\partial Z}{\partial z} = \frac{\partial u}{\partial X}(-1) + \frac{\partial u}{\partial Y}(1) + \frac{\partial u}{\partial Z}(0)$$

Adding, $\frac{\partial u}{\partial x} + \frac{\partial u}{\partial y} + \frac{\partial u}{\partial z} = 0.$

Example 5. *If* $w = f(x, y)$, $x = r \cos \theta$, $y = r \sin \theta$, *show that*

$$\left(\frac{\partial w}{\partial r}\right)^2 + \frac{1}{r^2}\left(\frac{\partial w}{\partial \theta}\right)^2 = \left(\frac{\partial f}{\partial x}\right)^2 + \left(\frac{\partial f}{\partial y}\right)^2.$$

Sol. The given equations define w as a composite function of r and θ.

$$\frac{\partial w}{\partial r} = \frac{\partial w}{\partial x} \cdot \frac{\partial x}{\partial r} + \frac{\partial w}{\partial y} \cdot \frac{\partial y}{\partial r} = \frac{\partial w}{\partial x} \cdot \cos \theta + \frac{\partial w}{\partial y} \cdot \sin \theta$$

or $$\frac{\partial w}{\partial r} = \frac{\partial f}{\partial x} \cos \theta + \frac{\partial f}{\partial y} \sin \theta \quad \ldots(1) \quad [\because \ w = f(x, y)]$$

Also $$\frac{\partial w}{\partial \theta} = \frac{\partial w}{\partial x} \cdot \frac{\partial x}{\partial \theta} + \frac{\partial w}{\partial y} \cdot \frac{\partial y}{\partial \theta} = \frac{\partial w}{\partial x}(-r \sin \theta) + \frac{\partial w}{\partial y}(r \cos \theta)$$

or $$\frac{1}{r}\frac{\partial w}{\partial \theta} = -\frac{\partial f}{\partial x} \sin \theta + \frac{\partial f}{\partial y} \cos \theta \quad \ldots(2)$$

Squaring and adding (1) and (2), we get

$$\left(\frac{\partial w}{\partial r}\right)^2 + \frac{1}{r^2}\left(\frac{\partial w}{\partial \theta}\right)^2 = \left(\frac{\partial f}{\partial x}\right)^2 + \left(\frac{\partial f}{\partial y}\right)^2.$$

Example 6. *If u is a homogeneous function of nth degree in x, y, z, prove that*

$$x\frac{\partial u}{\partial x} + y\frac{\partial u}{\partial y} + z\frac{\partial u}{\partial z} = nu.$$

Sol. Since u is a homogeneous function of degree n in x, y, z, let

$$u = x^n f\left(\frac{y}{x}, \frac{z}{x}\right)$$

or $$u = x^n f(t, s) \text{ where } t = \frac{y}{x}, \quad s = \frac{z}{x}$$

Here f is a composite function of x, y, z.

$$\therefore \quad \frac{\partial u}{\partial x} = nx^{n-1} f(t, s) + x^n \left(\frac{\partial f}{\partial t} \cdot \frac{\partial t}{\partial x} + \frac{\partial f}{\partial s} \cdot \frac{\partial s}{\partial x}\right)$$

$$= nx^{n-1} f(t, s) + x^n \left[\frac{\partial f}{\partial t} \cdot \left(-\frac{y}{x^2}\right) + \frac{\partial f}{\partial s}\left(-\frac{z}{x^2}\right)\right]$$

$$\Rightarrow \qquad x\frac{\partial u}{\partial x} = nx^n f(t,s) - yx^{n-1}\frac{\partial f}{\partial t} - zx^{n-1}\frac{\partial f}{\partial s} \qquad ...(1)$$

$$\frac{\partial u}{\partial y} = x^n\left(\frac{\partial f}{\partial t}\cdot\frac{\partial t}{\partial y} + \frac{\partial f}{\partial s}\cdot\frac{\partial s}{\partial y}\right) = x^n\left[\frac{\partial f}{\partial t}\cdot\frac{1}{x} + \frac{\partial f}{\partial s}\cdot 0\right]$$

$$\Rightarrow \qquad y\frac{\partial u}{\partial y} = yx^{n-1}\frac{\partial f}{\partial t} \qquad ...(2)$$

$$\frac{\partial u}{\partial z} = x^n\left[\frac{\partial f}{\partial t}\cdot\frac{\partial t}{\partial z} + \frac{\partial f}{\partial s}\cdot\frac{\partial s}{\partial z}\right] = x^n\left[\frac{\partial f}{\partial t}\cdot 0 + \frac{\partial f}{\partial s}\cdot\frac{1}{x}\right]$$

$$\Rightarrow \qquad z\frac{\partial u}{\partial z} = zx^{n-1}\frac{\partial f}{\partial s} \qquad ...(3)$$

Adding (1), (2) and (3), we have

$$x\frac{\partial u}{\partial x} + y\frac{\partial u}{\partial y} + z\frac{\partial u}{\partial z} = nx^n f(t, s) = nu$$

Example 7. *If by the substitution $u = x^2 - y^2$, $v = 2xy$, $f(x, y) = \theta(u, v)$, show that* $\dfrac{\partial^2 f}{\partial x^2} + \dfrac{\partial^2 f}{\partial y^2} = 4(x^2 + y^2)\left(\dfrac{\partial^2\theta}{\partial u^2} + \dfrac{\partial^2\theta}{\partial v^2}\right)$.

Sol. Here $\quad f(x, y) = \theta(u, v) \quad$ and $\quad u = x^2 - y^2,\ v = 2xy$

$\Rightarrow$ f is a function of u, v and u, v are functions of x, y

$\Rightarrow$ f is a composite function of x, y.

$$\therefore \qquad \frac{\partial f}{\partial x} = \frac{\partial f}{\partial u}\cdot\frac{\partial u}{\partial x} + \frac{\partial f}{\partial v}\cdot\frac{\partial v}{\partial x} = \frac{\partial\theta}{\partial u}\cdot 2x + \frac{\partial\theta}{\partial v}\cdot 2y \qquad [\because\ f = \theta \text{ (given)}]$$

$$= 2\left(x\frac{\partial\theta}{\partial u} + y\frac{\partial\theta}{\partial v}\right) = 2\left(x\frac{\partial}{\partial u} + y\frac{\partial}{\partial v}\right)\theta$$

$$\Rightarrow \qquad \frac{\partial}{\partial x} = 2\left(x\frac{\partial}{\partial u} + y\frac{\partial}{\partial v}\right) \qquad ...(1)$$

$$\text{Also} \qquad \frac{\partial f}{\partial y} = \frac{\partial f}{\partial u}\cdot\frac{\partial u}{\partial y} + \frac{\partial f}{\partial v}\cdot\frac{\partial v}{\partial y} = \frac{\partial\theta}{\partial u}\cdot(-2y) + \frac{\partial\theta}{\partial v}\cdot 2x \qquad [\because\ f = \theta \text{ (given)}]$$

$$= 2\left(-y\frac{\partial\theta}{\partial u} + x\frac{\partial\theta}{\partial v}\right) = 2\left(-y\frac{\partial}{\partial u} + x\frac{\partial}{\partial v}\right)\theta$$

$$\Rightarrow \qquad \frac{\partial}{\partial y} = 2\left(-y\frac{\partial}{\partial u} + x\frac{\partial}{\partial v}\right) \qquad ...(2)$$

$$\text{Now} \qquad \frac{\partial^2 f}{\partial x^2} = \frac{\partial}{\partial x}\left(\frac{\partial f}{\partial x}\right) = 2\left(x\frac{\partial}{\partial u} + y\frac{\partial}{\partial v}\right)\left(2x\frac{\partial\theta}{\partial u} + 2y\frac{\partial\theta}{\partial v}\right) \qquad [\text{Using (1)}]$$

$$= 4\left(x^2\frac{\partial^2\theta}{\partial u^2} + xy\frac{\partial^2\theta}{\partial u\partial v} + yx\frac{\partial^2\theta}{\partial v\partial u} + y^2\frac{\partial^2\theta}{\partial v^2}\right)$$

$$\Rightarrow \qquad \frac{\partial^2 f}{\partial x^2} = 4\left(x^2\frac{\partial^2\theta}{\partial u^2} + 2xy\frac{\partial^2\theta}{\partial u\partial v} + y^2\frac{\partial^2\theta}{\partial v^2}\right) \qquad ...(3) \quad \left[\because\ \frac{\partial^2\theta}{\partial u\partial v} = \frac{\partial^2\theta}{\partial v\partial u}\right]$$

and $$\frac{\partial^2 f}{\partial y^2} = \frac{\partial}{\partial y}\left(\frac{\partial f}{\partial y}\right) = 2\left(-y\frac{\partial}{\partial u} + x\frac{\partial}{\partial v}\right)\left(-2y\frac{\partial \theta}{\partial u} + 2x\frac{\partial \theta}{\partial v}\right) \quad \text{[Using (2)]}$$

$$= 4\left(y^2\frac{\partial^2 \theta}{\partial u^2} - yx\frac{\partial^2 \theta}{\partial u \partial v} - xy\frac{\partial^2 \theta}{\partial v \partial u} + x^2\frac{\partial^2 \theta}{\partial v^2}\right)$$

$$\Rightarrow \quad \frac{\partial^2 f}{\partial y^2} = 4\left(y^2\frac{\partial^2 \theta}{\partial u^2} - 2xy\frac{\partial^2 \theta}{\partial u \partial v} + x^2\frac{\partial^2 \theta}{\partial v^2}\right) \quad \ldots(4) \quad \left[\because \frac{\partial^2 \theta}{\partial u \partial v} = \frac{\partial^2 \theta}{\partial v \partial u}\right]$$

Adding (3) and (4), we get

$$\frac{\partial^2 f}{\partial x^2} + \frac{\partial^2 f}{\partial y^2} = 4\left[(x^2 + y^2)\frac{\partial^2 \theta}{\partial u^2} + (y^2 + x^2)\frac{\partial^2 \theta}{\partial v^2}\right]$$

$$= 4\,(x^2 + y^2)\left(\frac{\partial^2 \theta}{\partial u^2} + \frac{\partial^2 \theta}{\partial v^2}\right).$$

Example 8. *If $x + y = 2e^{\theta}\cos\phi$ and $x - y = 2ie^{\theta}\sin\phi$, show that $\frac{\partial^2 u}{\partial \theta^2} + \frac{\partial^2 u}{\partial \phi^2} = 4xy\frac{\partial^2 u}{\partial x \partial y}$, $\left(i = \sqrt{-1}\right)$.*

Sol. Here $\quad x + y = 2e^{\theta}\cos\phi$ and $x - y = 2ie^{\theta}\sin\phi$

Adding $\quad 2x = 2e^{\theta}(\cos\phi + i\sin\phi)$

$\Rightarrow \quad x = e^{\theta}\cdot e^{i\phi} = e^{\theta + i\phi}$ (By Euler's Theorem)

Subtracting $\quad 2y = 2e^{\theta}(\cos\phi - i\sin\phi)$

$\Rightarrow \quad y = e^{\theta}\cdot e^{-i\phi} = e^{\theta - i\phi}$

Now u is a function of x, y and x, y are functions of θ, ϕ

$\Rightarrow$ u is a composite function of θ, ϕ.

$$\therefore \quad \frac{\partial u}{\partial \theta} = \frac{\partial u}{\partial x}\cdot\frac{\partial x}{\partial \theta} + \frac{\partial u}{\partial y}\cdot\frac{\partial y}{\partial \theta} = \frac{\partial u}{\partial x}\cdot e^{\theta + i\phi} + \frac{\partial u}{\partial y}\cdot e^{\theta - i\phi}$$

$$= x\frac{\partial u}{\partial x} + y\frac{\partial u}{\partial y} = \left(x\frac{\partial}{\partial x} + y\frac{\partial}{\partial y}\right)u$$

$$\Rightarrow \quad \frac{\partial}{\partial \theta} = x\frac{\partial}{\partial x} + y\frac{\partial}{\partial y} \quad \ldots(1)$$

Also, $$\frac{\partial u}{\partial \phi} = \frac{\partial u}{\partial x}\cdot\frac{\partial x}{\partial \phi} + \frac{\partial u}{\partial y}\cdot\frac{\partial y}{\partial \phi} = \frac{\partial u}{\partial x}\cdot ie^{\theta + i\phi} + \frac{\partial u}{\partial y}\cdot(-ie^{\theta - i\phi})$$

$$= ix\frac{\partial u}{\partial x} - iy\frac{\partial u}{\partial y} = \left(ix\frac{\partial}{\partial x} - iy\frac{\partial}{\partial y}\right)u$$

$$\Rightarrow \quad \frac{\partial}{\partial \phi} = ix\frac{\partial}{\partial x} - iy\frac{\partial}{\partial y} \quad \ldots(2)$$

$$\therefore \quad \frac{\partial^2 u}{\partial \theta^2} = \frac{\partial}{\partial \theta}\left(\frac{\partial u}{\partial \theta}\right) = \left(x\frac{\partial}{\partial x} + y\frac{\partial}{\partial y}\right)\left(x\frac{\partial u}{\partial x} + y\frac{\partial u}{\partial y}\right) \quad \text{[Using (1)]}$$

$$= x^2 \frac{\partial^2 u}{\partial x^2} + xy \frac{\partial^2 u}{\partial x \partial y} + yx \frac{\partial^2 u}{\partial y \partial x} + y^2 \frac{\partial^2 u}{\partial y^2}$$

$$\Rightarrow \quad \frac{\partial^2 u}{\partial \theta^2} = x^2 \frac{\partial^2 u}{\partial x^2} + 2xy \frac{\partial^2 u}{\partial x \partial y} + y^2 \frac{\partial^2 u}{\partial y^2} \quad \ldots(3) \left[\because \frac{\partial^2 u}{\partial x \partial y} = \frac{\partial^2 u}{\partial y \partial x}\right]$$

and
$$\frac{\partial^2 u}{\partial \phi^2} = \frac{\partial}{\partial \phi}\left(\frac{\partial u}{\partial \phi}\right)$$

$$= \left(ix \frac{\partial}{\partial x} - iy \frac{\partial}{\partial y}\right)\left(ix \frac{\partial u}{\partial x} - iy \frac{\partial u}{\partial y}\right) \quad \text{[Using (2)]}$$

$$= i^2 \left(x^2 \frac{\partial^2 u}{\partial x^2} - xy \frac{\partial^2 u}{\partial x \partial y} - yx \frac{\partial^2 u}{\partial y \partial x} + y^2 \frac{\partial^2 u}{\partial y^2}\right)$$

$$= -\left(x^2 \frac{\partial^2 u}{\partial x^2} - 2xy \frac{\partial^2 u}{\partial x \partial y} + y^2 \frac{\partial^2 u}{\partial y^2}\right) \quad \ldots(4) \left[\because \frac{\partial^2 u}{\partial x \partial y} = \frac{\partial^2 u}{\partial y \partial x}\right]$$

Adding (3) and (4), we get

$$\frac{\partial^2 u}{\partial \theta^2} + \frac{\partial^2 u}{\partial \phi^2} = 4xy \frac{\partial^2 u}{\partial x \partial y}.$$

Example 9. *If* $z = \sqrt{x^2 + y^2}$ *and* $x^3 + y^3 + 3axy = 5a^2$, *find the value of* $\frac{dz}{dx}$, *when* $x = y = a$. (P.T.U. May 2009)

Sol. The given equations are of the form $z = f(x, y)$ and $\phi(x, y) = c$

$\therefore$ z is composite function of x.

$$\Rightarrow \quad \frac{dz}{dx} = \frac{\partial z}{\partial x} \cdot \frac{dx}{dx} + \frac{\partial z}{\partial y} \cdot \frac{dy}{dx} = \frac{\partial z}{\partial x} + \frac{\partial z}{\partial y} \cdot \frac{dy}{dx} \quad \ldots(1)$$

Now
$$\frac{\partial z}{\partial x} = \frac{1}{2} (x^2 + y^2)^{-1/2} \cdot 2x = \frac{x}{\sqrt{x^2 + y^2}}$$

Similarly,
$$\frac{\partial z}{\partial y} = \frac{y}{\sqrt{x^2 + y^2}}$$

Also, differentiating $x^3 + y^3 + 3axy = 5a^2$ w.r.t. x, we have

$$3x^2 + 3y^2 \cdot \frac{dy}{dx} + 3ay + 3ax \cdot \frac{dy}{dx} = 0 \quad \text{or} \quad (y^2 + ax) \frac{dy}{dx} = -(x^2 + ay)$$

$$\therefore \quad \frac{dy}{dx} = -\frac{x^2 + ay}{y^2 + ax}$$

$\therefore$ From (1),
$$\frac{dz}{dx} = \frac{x}{\sqrt{x^2 + y^2}} + \frac{y}{\sqrt{x^2 + y^2}}\left(-\frac{x^2 + ay}{y^2 + ax}\right)$$

$$\left[\frac{dz}{dx}\right]_{\substack{x = a \\ y = a}} = \frac{a}{\sqrt{a^2 + a^2}} + \frac{a}{\sqrt{a^2 + a^2}} \cdot \frac{a^2 + a^2}{a^2 + a^2} = 0.$$

Example 10. *If* $u = xe^y z$, *where* $y = \sqrt{a^2 - x^2}$, $z = \sin^2 x$, *find* $\frac{du}{dx}$.

Sol. Here u is a function of x, y and z while y and z are functions of x.

$$\therefore \quad \frac{du}{dx} = \frac{\partial u}{\partial x}\cdot\frac{dx}{dx} + \frac{\partial u}{\partial y}\cdot\frac{dy}{dx} + \frac{\partial u}{\partial z}\cdot\frac{dz}{dx}$$

$$= e^y z \cdot 1 + xe^y z \cdot \frac{1}{2}(a^2 - x^2)^{-1/2}(-2x) + xe^y \cdot 2\sin x\cos x$$

$$= e^y\left[z - \frac{x^2 z}{\sqrt{a^2 - x^2}} + x\sin 2x\right].$$

Example 11. *Find* $\frac{du}{dx}$ *if* $u = \sin(x^2 + y^2)$, *where* $a^2x^2 + b^2y^2 = c^2$.

Sol. The given equations are the form $u = f(x, y)$ and $\phi(x, y) = k$

$\therefore$ u is a composite function of x.

$$\frac{du}{dx} = \frac{\partial u}{\partial x}\cdot\frac{dx}{dx} + \frac{\partial u}{\partial y}\cdot\frac{dy}{dx} = \frac{\partial u}{\partial x} + \frac{\partial u}{\partial y}\cdot\frac{dy}{dx} \quad ...(1)$$

Now $\frac{\partial u}{\partial x} = 2x\cos(x^2 + y^2)$, $\frac{\partial u}{\partial y} = 2y\cos(x^2 + y^2)$

Also, differentiating $a^2x^2 + b^2y^2 = c$ w.r.t. x, we have

$$2a^2x + 2b^2y\cdot\frac{dy}{dx} = 0 \quad \text{or} \quad \frac{dy}{dx} = -\frac{a^2x}{b^2y}$$

$\therefore$ From (1),

$$\frac{du}{dx} = 2x\cos(x^2 + y^2) + 2y\cos(x^2 + y^2)\cdot\left[-\frac{a^2x}{b^2y}\right]$$

$$= 2\left[x - \frac{a^2x}{b^2}\right]\cos(x^2 + y^2) = \frac{2(b^2 - a^2)x}{b^2}\cdot\cos(x^2 + y^2).$$

Example 12. *Find* $\frac{dy}{dx}$, *when*

(*i*) $x^y + y^x = c$ (*ii*) $(\cos x)^y = (\sin y)^x$.

Sol. (*i*) Let $f(x, y) = x^y + y^x$, then $f(x, y) = c$

[Using Cor. 4] $\therefore$ $\frac{dy}{dx} = -\frac{f_x}{f_y} \Rightarrow \frac{dy}{dx} = -\frac{yx^{y-1} + y^x\log y}{x^y\log x + xy^{x-1}}.$

(*ii*) Let $f(x, y) = (\cos x)^y - (\sin y)^x = 0$

$$\frac{dy}{dx} = -\frac{f_x}{f_y} = -\frac{y(\cos x)^{y-1}\cdot(-\sin x) - (\sin y)^x\log(\sin y)}{(\cos x)^y\log(\cos x) - x(\sin y)^{x-1}\cdot\cos y}$$

$$= \frac{y(\cos x)^{y-1}\sin x + (\cos x)^y\log(\sin y)}{(\cos x)^y\log(\cos x) - x(\cos x)^y(\sin y)^{-1}\cos y}$$

$[\because \quad (\sin y)^x = (\cos x)^y]$

$$= \frac{(\cos x)^y\left[y\cdot\frac{\sin x}{\cos x} + \log\sin y\right]}{(\cos x)^y[\log\cos x - x\cot y]} = \frac{y\tan x + \log\sin y}{\log\cos x - x\cot y}.$$

Example 13. *If $f(x, y) = 0$, $\phi(y, z) = 0$, show that* $\dfrac{\partial f}{\partial y} \cdot \dfrac{\partial \phi}{\partial z} \cdot \dfrac{dz}{dx} = \dfrac{\partial f}{\partial x} \cdot \dfrac{\partial \phi}{\partial y}$.

Sol.
$$\frac{dz}{dx} = \frac{dz}{dy} \cdot \frac{dy}{dx} \qquad ...(1)$$

$f(x, y) = 0$ gives $\dfrac{dy}{dx} = -\dfrac{\frac{\partial f}{\partial x}}{\frac{\partial f}{\partial y}}$; $\phi(y, z) = 0$ gives $\dfrac{dz}{dy} = -\dfrac{\frac{\partial \phi}{\partial y}}{\frac{\partial \phi}{\partial z}}$

$\therefore$ From (1),
$$\frac{dz}{dx} = \frac{\frac{\partial f}{\partial x} \cdot \frac{\partial \phi}{\partial y}}{\frac{\partial f}{\partial y} \cdot \frac{\partial \phi}{\partial z}} \Rightarrow \frac{\partial f}{\partial y} \cdot \frac{\partial \phi}{\partial z} \cdot \frac{dz}{dx} = \frac{\partial f}{\partial x} \cdot \frac{\partial \phi}{\partial y}.$$

Example 14. *If $\phi(x, y, z) = 0$, show that* $\left(\dfrac{\partial y}{\partial z}\right)_x \left(\dfrac{\partial z}{\partial x}\right)_y \left(\dfrac{\partial x}{\partial y}\right)_z = -1$. (P.T.U. Dec. 2011)

Sol. The given relation defines y as a function of x and z. Treating x as constant
$$\left(\frac{\partial y}{\partial z}\right)_x = -\frac{\frac{\partial \phi}{\partial z}}{\frac{\partial \phi}{\partial y}}$$

The given relation defines z as a function of x and y. Treating y as constant $\left(\dfrac{\partial z}{\partial x}\right)_y = -\dfrac{\frac{\partial \phi}{\partial x}}{\frac{\partial \phi}{\partial z}}$

Similarly,
$$\left(\frac{\partial x}{\partial y}\right)_z = -\frac{\frac{\partial \phi}{\partial y}}{\frac{\partial \phi}{\partial x}}$$

Multiplying, we get the desired result.

Example 15. *Prove that* $\dfrac{\partial^2 z}{\partial x^2} + \dfrac{\partial^2 z}{\partial y^2} = \dfrac{\partial^2 z}{\partial u^2} + \dfrac{\partial^2 z}{\partial v^2}$ (P.T.U. Dec. 2008; P.T.U. 2011)

where $x = u \cos \alpha - v \sin \alpha$, $y = u \sin \alpha + v \cos \alpha$

Or

By changing the independent variables u and v to x and y by means of the relations $x = u \cos \alpha - v \sin \alpha$, $y = u \sin \alpha + v \cos \alpha$, show that $\dfrac{\partial^2 z}{\partial u^2} + \dfrac{\partial^2 z}{\partial v^2}$ *transforms into* $\dfrac{\partial^2 z}{\partial x^2} + \dfrac{\partial^2 z}{\partial y^2}$.

Sol. Here z is a composite function of u and v
$$\frac{\partial z}{\partial u} = \frac{\partial z}{\partial x} \cdot \frac{\partial x}{\partial u} + \frac{\partial z}{\partial y} \cdot \frac{\partial y}{\partial u} = \cos \alpha \frac{\partial z}{\partial x} + \sin \alpha \frac{\partial z}{\partial y}$$

or
$$\frac{\partial}{\partial u}(z) = \left(\cos \alpha \frac{\partial}{\partial x} + \sin \alpha \frac{\partial}{\partial y}\right) z \Rightarrow \frac{\partial}{\partial u} \equiv \cos \alpha \frac{\partial}{\partial x} + \sin \alpha \frac{\partial}{\partial y} \qquad ...(1)$$

Also
$$\frac{\partial z}{\partial v} = \frac{\partial z}{\partial x} \cdot \frac{\partial x}{\partial v} + \frac{\partial z}{\partial y} \cdot \frac{\partial y}{\partial v} = -\sin \alpha \frac{\partial z}{\partial x} + \cos \alpha \frac{\partial z}{\partial y}$$

or $$\frac{\partial}{\partial v}(z) = \left(-\sin\alpha\frac{\partial}{\partial x} + \cos\alpha\frac{\partial}{\partial y}\right) z$$

$$\Rightarrow \quad \frac{\partial}{\partial v} \equiv -\sin\alpha\frac{\partial}{\partial x} + \cos\alpha\frac{\partial}{\partial y} \quad ...(2)$$

Now we shall make use of the equivalence of operators as given by (1) and (2).

$$\frac{\partial^2 z}{\partial u^2} = \frac{\partial}{\partial u}\left(\frac{\partial z}{\partial u}\right) = \left(\cos\alpha\frac{\partial}{\partial x} + \sin\alpha\frac{\partial}{\partial y}\right)\left(\cos\alpha\frac{\partial z}{\partial x} + \sin\alpha\frac{\partial z}{\partial y}\right)$$

$$= \cos^2\alpha\frac{\partial^2 z}{\partial x^2} + \cos\alpha\sin\alpha\frac{\partial^2 z}{\partial x\partial y} + \sin\alpha\cos\alpha\frac{\partial^2 z}{\partial y\partial x} + \sin^2\alpha\frac{\partial^2 z}{\partial y^2}$$

$$= \cos^2\alpha\frac{\partial^2 z}{\partial x^2} + 2\cos\alpha\sin\alpha\frac{\partial^2 z}{\partial x\partial y} + \sin^2\alpha\frac{\partial^2 z}{\partial y^2} \quad ...(3)$$

$$\frac{\partial^2 z}{\partial v^2} = \frac{\partial}{\partial v}\left(\frac{\partial z}{\partial v}\right) = \left(-\sin\alpha\frac{\partial}{\partial x} + \cos\alpha\frac{\partial}{\partial y}\right)\left(-\sin\alpha\frac{\partial z}{\partial x} + \cos\alpha\frac{\partial z}{\partial y}\right)$$

$$= \sin^2\alpha\frac{\partial^2 z}{\partial x^2} - \sin\alpha\cos\alpha\frac{\partial^2 z}{\partial x\partial y} - \cos\alpha\sin\alpha\frac{\partial^2 z}{\partial y\partial x} + \cos^2\alpha\frac{\partial^2 z}{\partial y^2}$$

$$= \sin^2\alpha\frac{\partial^2 z}{\partial x^2} - 2\cos\alpha\sin\alpha\frac{\partial^2 z}{\partial x\partial y} + \cos^2\alpha\frac{\partial^2 z}{\partial y^2} \quad ...(4)$$

Adding (3) and (4), $\dfrac{\partial^2 z}{\partial u^2} + \dfrac{\partial^2 z}{\partial v^2} = \dfrac{\partial^2 z}{\partial x^2} + \dfrac{\partial^2 z}{\partial y^2}$.

Example 16. *Transform the equation* $\dfrac{\partial^2 u}{\partial x^2} + \dfrac{\partial^2 u}{\partial y^2} = 0$ *into polar co-ordinates.*

(K.U.K. Dec. 2010; M.D.U. Dec. 2009)

Sol. The relations connecting cartesian co-ordinates, (x, y) with polar co-ordinates (r, θ) are

$$x = r\cos\theta,\ y = r\sin\theta$$

Squaring and adding, $r^2 = x^2 + y^2$

Dividing, $\tan\theta = \dfrac{y}{x}$

$$r = \sqrt{x^2 + y^2},\ \theta = \tan^{-1}\left(\frac{y}{x}\right)$$

$$\frac{\partial r}{\partial x} = \frac{x}{\sqrt{x^2+y^2}} = \frac{r\cos\theta}{r} = \cos\theta$$

$$\frac{\partial r}{\partial y} = \frac{y}{\sqrt{x^2+y^2}} = \frac{r\sin\theta}{r} = \sin\theta$$

$$\frac{\partial \theta}{\partial x} = \frac{1}{1+\frac{y^2}{x^2}}\cdot\left(-\frac{y}{x^2}\right) = -\frac{y}{x^2+y^2} = -\frac{r\sin\theta}{r^2} = -\frac{\sin\theta}{r}$$

$$\frac{\partial \theta}{\partial y} = \frac{1}{1+\frac{y^2}{x^2}}\cdot\frac{1}{x} = \frac{x}{x^2+y^2} = \frac{r\cos\theta}{r^2} = \frac{\cos\theta}{r}$$

Here u is a composite function of x and y

$$\frac{\partial u}{\partial x} = \frac{\partial u}{\partial r}\cdot\frac{\partial r}{\partial x} + \frac{\partial u}{\partial \theta}\cdot\frac{\partial \theta}{\partial x} = \cos\theta\,\frac{\partial u}{\partial r} - \frac{\sin\theta}{r}\cdot\frac{\partial u}{\partial \theta}$$

or $$\frac{\partial}{\partial x}(u) = \left(\cos\theta\,\frac{\partial}{\partial r} - \frac{\sin\theta}{r}\cdot\frac{\partial}{\partial \theta}\right)u$$

$\Rightarrow$ $$\frac{\partial}{\partial x} \equiv \cos\theta\,\frac{\partial}{\partial r} - \frac{\sin\theta}{r}\cdot\frac{\partial}{\partial \theta} \qquad ...(1)$$

Also, $$\frac{\partial u}{\partial y} = \frac{\partial u}{\partial r}\cdot\frac{\partial r}{\partial y} + \frac{\partial u}{\partial \theta}\cdot\frac{\partial \theta}{\partial y} = \sin\theta\,\frac{\partial u}{\partial r} + \frac{\cos\theta}{r}\cdot\frac{\partial u}{\partial \theta}$$

or $$\frac{\partial}{\partial y}(u) = \left(\sin\theta\,\frac{\partial}{\partial r} + \frac{\cos\theta}{r}\cdot\frac{\partial}{\partial \theta}\right)u \;\Rightarrow\; \frac{\partial}{\partial y} \equiv \sin\theta\,\frac{\partial}{\partial r} + \frac{\cos\theta}{r}\cdot\frac{\partial}{\partial \theta} \qquad ...(2)$$

Now we shall make use of the equivalence of cartesian and polar operators as given by (1) and (2),

$$\frac{\partial^2 u}{\partial x^2} = \frac{\partial}{\partial x}\left(\frac{\partial u}{\partial x}\right) = \left(\cos\theta\,\frac{\partial}{\partial r} - \frac{\sin\theta}{r}\cdot\frac{\partial}{\partial \theta}\right)\left(\cos\theta\,\frac{\partial u}{\partial r} - \frac{\sin\theta}{r}\cdot\frac{\partial u}{\partial \theta}\right)$$

$$= \cos\theta\,\frac{\partial}{\partial r}\left(\cos\theta\,\frac{\partial u}{\partial r} - \frac{\sin\theta}{r}\cdot\frac{\partial u}{\partial \theta}\right) - \frac{\sin\theta}{r}\cdot\frac{\partial}{\partial \theta}\left(\cos\theta\,\frac{\partial u}{\partial r} - \frac{\sin\theta}{r}\cdot\frac{\partial u}{\partial \theta}\right)$$

$$= \cos\theta\left[\cos\theta\,\frac{\partial^2 u}{\partial r^2} - \sin\theta\,\frac{\partial u}{\partial \theta}\left(-\frac{1}{r^2}\right) - \frac{\sin\theta}{r}\cdot\frac{\partial^2 u}{\partial r\,\partial \theta}\right]$$

$$- \frac{\sin\theta}{r}\left[-\sin\theta\,\frac{\partial u}{\partial r} + \cos\theta\,\frac{\partial^2 u}{\partial \theta\,\partial r} - \frac{\cos\theta}{r}\,\frac{\partial u}{\partial \theta} - \frac{\sin\theta}{r}\cdot\frac{\partial^2 u}{\partial \theta^2}\right]$$

$$= \cos^2\theta\,\frac{\partial^2 u}{\partial r^2} + \frac{2\cos\theta\sin\theta}{r^2}\cdot\frac{\partial u}{\partial \theta} + \frac{\sin^2\theta}{r}\cdot\frac{\partial u}{\partial r} - \frac{2\cos\theta\sin\theta}{r}\cdot\frac{\partial^2 u}{\partial r\,\partial \theta} + \frac{\sin^2\theta}{r^2}\cdot\frac{\partial^2 u}{\partial \theta^2} \qquad ...(3)$$

$$\frac{\partial^2 u}{\partial y^2} = \frac{\partial}{\partial y}\left(\frac{\partial u}{\partial y}\right) = \left(\sin\theta\,\frac{\partial}{\partial r} + \frac{\cos\theta}{r}\cdot\frac{\partial}{\partial \theta}\right)\left(\sin\theta\,\frac{\partial u}{\partial r} + \frac{\cos\theta}{r}\cdot\frac{\partial u}{\partial \theta}\right)$$

$$= \sin\theta\,\frac{\partial}{\partial r}\left(\sin\theta\,\frac{\partial u}{\partial r} + \frac{\cos\theta}{r}\cdot\frac{\partial u}{\partial \theta}\right) + \frac{\cos\theta}{r}\cdot\frac{\partial}{\partial \theta}\left(\sin\theta\,\frac{\partial u}{\partial r} + \frac{\cos\theta}{r}\cdot\frac{\partial u}{\partial \theta}\right)$$

$$= \sin\theta\left[\sin\theta\,\frac{\partial^2 u}{\partial r^2} - \frac{\cos\theta}{r^2}\cdot\frac{\partial u}{\partial \theta} + \frac{\cos\theta}{r}\cdot\frac{\partial^2 u}{\partial r\,\partial \theta}\right]$$

$$+ \frac{\cos\theta}{r}\left[\cos\theta\,\frac{\partial u}{\partial r} + \sin\theta\,\frac{\partial^2 u}{\partial \theta\,\partial r} - \frac{\sin\theta}{r}\cdot\frac{\partial u}{\partial \theta} + \frac{\cos\theta}{r}\cdot\frac{\partial^2 u}{\partial \theta^2}\right]$$

$$= \sin^2\theta\,\frac{\partial^2 u}{\partial r^2} - \frac{2\cos\theta\sin\theta}{r^2}\cdot\frac{\partial u}{\partial \theta} + \frac{\cos^2\theta}{r}\cdot\frac{\partial u}{\partial r} + \frac{2\cos\theta\sin\theta}{r}\cdot\frac{\partial^2 u}{\partial r\,\partial \theta} + \frac{\cos^2\theta}{r^2}\cdot\frac{\partial^2 u}{\partial \theta^2} \qquad ...(4)$$

Adding (3) and (4), $$\frac{\partial^2 u}{\partial x^2} + \frac{\partial^2 u}{\partial y^2} = \frac{\partial^2 u}{\partial r^2} + \frac{1}{r}\,\frac{\partial u}{\partial r} + \frac{1}{r^2}\cdot\frac{\partial^2 u}{\partial \theta^2}$$

$\therefore$ $\dfrac{\partial^2 u}{\partial x^2} + \dfrac{\partial^2 u}{\partial y^2} = 0$ transforms into $\dfrac{\partial^2 u}{\partial r^2} + \dfrac{1}{r}\cdot\dfrac{\partial u}{\partial r} + \dfrac{1}{r^2}\cdot\dfrac{\partial^2 u}{\partial \theta^2} = 0.$

EXERCISE 10.3

1. (*a*) Find $\dfrac{du}{dt}$ when $u = x^2 + y^2$, $x = at^2$, $y = 2at$. Also verify by direct substitution.

(*b*) If $u = x^2 + y^2 + z^2$ and $x = e^{2t}$, $y = e^{2t} \cos 3t$, $z = e^{2t} \sin 3t$; find $\dfrac{du}{dt}$ as a total derivative and verify the result by direct substitution.

2. If $u = \sin \dfrac{x}{y}$, $x = e^t$, $y = t^2$, find $\dfrac{du}{dt}$ as a function of t. Verify your result by direct substitution.

3. If $u = x^3 + y^3$, where $x = a \cos t$, $y = b \sin t$, find $\dfrac{du}{dt}$ and verify the result.

4. (*a*) At a given instant the sides of a rectangle are 4 m and 3 m respectively and they are increasing at the rate of 1.5 m/sec and 0.5 m/sec respectively. Find the rate at which the area is increasing at that instant.

(*b*) At a given instant the major and minor axes of an ellipse are 60 cm and 40 cm respectively and they are increasing at the rate of 2 cm/sec and 1.5 cm/sec respectively. Find the rate at which the area is increasing at that instant.

(*c*) If $z = 2xy - 3x^2y$ and if x increases at the rate of 3 cm/sec at the instant when $x = 4$ cm and $y = 2$ cm, at what rate must y be changing in order that z shall be neither increasing nor decreasing?

5. (*a*) If $z = u^2 + v^2$, $u = r \cos \theta$, $v = r \sin \theta$, find $\dfrac{\partial z}{\partial r}$ and $\dfrac{\partial z}{\partial \theta}$.

(*b*) If $z = \log (u^2 + v)$, $u = e^{x^2 + y^2}$, $v = x^2 + y$, find $\dfrac{\partial z}{\partial x}$ and $\dfrac{\partial z}{\partial y}$.

(*c*) If $u = \tan^{-1} \left(\dfrac{y}{x}\right)$ where $x = e^t - e^{-t}$ and $y = e^t + e^{-t}$, find $\dfrac{du}{dt}$.

6. If $u = f(r, s)$, $r = x + y$, $s = x - y$, show that $\dfrac{\partial u}{\partial x} + \dfrac{\partial u}{\partial y} = 2\dfrac{\partial u}{\partial r}$.

7. If $z = e^{ax+by} f(ax - by)$, prove that $b\dfrac{\partial z}{\partial x} + a\dfrac{\partial z}{\partial y} = 2abz$.

8. (*i*) If $x = u + v$, $y = uv$ and z is a function of x, y ; show that $u\dfrac{\partial z}{\partial u} + v\dfrac{\partial z}{\partial v} = x\dfrac{\partial z}{\partial x} + 2y\dfrac{\partial z}{\partial y}$.

(*ii*) If $u = f(r, s)$, $r = x + at$, $s = y + bt$, show that $\dfrac{\partial u}{\partial t} = a\dfrac{\partial u}{\partial x} + b\dfrac{\partial u}{\partial y}$.

9. If $u = x \log (xy)$, where $x^3 + y^3 + 3xy = 1$, find $\dfrac{du}{dx}$.

10. (*a*) If $u = f(r, s, t)$ and $r = \dfrac{x}{y}$, $s = \dfrac{y}{z}$, $t = \dfrac{z}{x}$, prove that $x\dfrac{\partial u}{\partial x} + y\dfrac{\partial u}{\partial y} + z\dfrac{\partial u}{\partial z} = 0$.

(*M.T.U. 2013; U.K.T.U. 2011, 2012*)

(*b*) If $x = u + v + w$, $y = vw + wu + uv$, $z = uvw$ and F is a function of x, y, z, show that

$$u\frac{\partial F}{\partial u} + v\frac{\partial F}{\partial v} + w\frac{\partial F}{\partial w} = x\frac{\partial F}{\partial x} + 2y\frac{\partial F}{\partial y} + 3z\frac{\partial F}{\partial z}.$$

(*c*) If $u = f(2x - 3y, 3y - 4z, 4z - 2x)$, prove that $\dfrac{1}{2}\dfrac{\partial u}{\partial x} + \dfrac{1}{3}\dfrac{\partial u}{\partial y} + \dfrac{1}{4}\dfrac{\partial u}{\partial z} = 0$.

(*Kerala 2008; P.T.U. 2011*)

(*d*) If $u = f(x^2 + 2yz, y^2 + 2zx)$, prove that $(y^2 - zx)\dfrac{\partial u}{\partial x} + (x^2 - yz)\dfrac{\partial u}{\partial y} + (z^2 - xy)\dfrac{\partial u}{\partial z} = 0$.

11. If $z = x^2y$ and $x^2 + xy + y^2 = 1$, show that $\dfrac{dz}{dx} = 2xy - \dfrac{x^2(2x + y)}{x + 2y}$.

12. If $x^3 + 3x^2y + 6xy^2 + y^3 = 1$, find $\frac{dy}{dx}$.

13. If $x^y = y^x$, show that $\frac{dy}{dx} = \frac{y(y - x \log y)}{x(x - y \log x)}$ using partial derivative method.

14. If the curves $f(x, y) = 0$ and $\phi(x, y) = 0$ touch, show that at the point of contact, $\frac{\partial f}{\partial x} \cdot \frac{\partial \phi}{\partial y} = \frac{\partial f}{\partial y} \cdot \frac{\partial \phi}{\partial x}$.

15. (*a*) If $f(x, y) = 0$, show that $\frac{d^2y}{dx^2} = -\frac{q^2r - 2pqs + p^2t}{q^3}$.

(*b*) If $f(x, y) = 0$, show that $\left(\frac{\partial f}{\partial y}\right)^3 \cdot \frac{d^2y}{dx^2} = 2\left(\frac{\partial f}{\partial x}\right)\left(\frac{\partial f}{\partial y}\right)\left(\frac{\partial^2 f}{\partial x \partial y}\right) - \left(\frac{\partial f}{\partial y}\right)^2\left(\frac{\partial^2 f}{\partial x^2}\right) - \left(\frac{\partial f}{\partial x}\right)^2\left(\frac{\partial^2 f}{\partial y^2}\right)$.

16. (*a*) If $x^3 + y^3 - 3axy = 0$, find $\frac{dy}{dx}$ and $\frac{d^2y}{dx^2}$ using partial derivative method.

(*b*) Prove that if $y^3 - 3ax^2 + x^3 = 0$, then $\frac{d^2y}{dx^2} + \frac{2a^2x^2}{y^5} = 0$.

17. If $z = xyf\left(\frac{y}{x}\right)$ and z is a constant, show that $\frac{f'\left(\frac{y}{x}\right)}{f\left(\frac{y}{x}\right)} = \frac{x\left[y + x\frac{dy}{dx}\right]}{y\left[y - x\frac{dy}{dx}\right]}$.

18. By changing the independent variables x and y to u and v by means of the relations $u = x - ay$, $v = x + ay$, show that $a^2 \frac{\partial^2 z}{\partial x^2} - \frac{\partial^2 z}{\partial y^2}$ transforms into $4a^2 \frac{\partial^2 z}{\partial u \partial v}$.

19. If z is a function of x and y, and u and v be two other variables such that $u = lx + my$, $v = ly - mx$, show that $\frac{\partial^2 z}{\partial x^2} + \frac{\partial^2 z}{\partial y^2} = (l^2 + m^2)\left(\frac{\partial^2 z}{\partial u^2} + \frac{\partial^2 z}{\partial v^2}\right)$.

20. If $x = e^r \cos \theta$, $y = e^r \sin \theta$, show that $\frac{\partial^2 u}{\partial x^2} + \frac{\partial^2 u}{\partial y^2} = e^{-2r}\left(\frac{\partial^2 u}{\partial r^2} + \frac{\partial^2 u}{\partial \theta^2}\right)$. (*G.B.T.U. 2013*)

21. If $x = \rho \cos \phi$, $y = \rho \sin \phi$, show that $\left(\frac{\partial v}{\partial x}\right)^2 + \left(\frac{\partial v}{\partial y}\right)^2 = \left(\frac{\partial v}{\partial \rho}\right)^2 + \frac{1}{\rho^2}\left(\frac{\partial v}{\partial \phi}\right)^2$.

22. Prove that $\frac{\partial^2 u}{\partial x^2} + \frac{\partial^2 u}{\partial y^2} = \frac{\partial^2 u}{\partial s^2} + \frac{\partial^2 u}{\partial t^2}$, where $x = s \cos \alpha - t \sin \alpha$ and $y = s \sin \alpha + t \cos \alpha$.

Answers

1. (*a*) $4a^2t(t^2 + 2)$ (*b*) $8e^{4t}$

2. $\frac{e^t(t-2)}{t^3} \cos\left(\frac{e^t}{t^2}\right)$

3. $-3a^3 \cos^2 t \sin t + 3b^3 \sin^2 t \cos t$

4. (*a*) 6.5 m^2/sec (*b*) 85π cm^2/sec
(*c*) Decreasing at the rate of 3.3 cm/sec

5. (*a*) $2r$, 0 (*b*) $\frac{2x(2u^2 + 1)}{u^2 + v}, \frac{4yu^2 + 1}{u^2 + v}$ (*c*) $\frac{-2}{e^{2t} + e^{-2t}}$

9. $1 + \log(xy) - \frac{x(x^2 + y)}{y(y^2 + x)}$

12. $-\frac{x^2 + 2xy + 2y^2}{x^2 + 4xy + y^2}$

16. (*a*) $\frac{x^2 - ay}{ax - y^2}, \frac{2a^3xy}{(ax - y^2)^3}$.

10.9. JACOBIANS

If u and v are functions of two independent variables x and y, then the determinant $\begin{vmatrix} \frac{\partial u}{\partial x} & \frac{\partial u}{\partial y} \\ \frac{\partial v}{\partial x} & \frac{\partial v}{\partial y} \end{vmatrix}$ is called Jacobian of u, v with respect to x, y and is denoted by the symbol $J\left(\frac{u, v}{x, y}\right)$ or $\frac{\partial(u, v)}{\partial(x, y)}$.

Similarly, if u, v, w be functions of x, y, z, then the Jacobian of u, v, w with respect to x, y, z is

$$J\left(\frac{u, v, w}{x, y, z}\right) \quad \text{or} \quad \frac{\partial(u, v, w)}{\partial(x, y, z)} = \begin{vmatrix} \frac{\partial u}{\partial x} & \frac{\partial u}{\partial y} & \frac{\partial u}{\partial z} \\ \frac{\partial v}{\partial x} & \frac{\partial v}{\partial y} & \frac{\partial v}{\partial z} \\ \frac{\partial w}{\partial x} & \frac{\partial w}{\partial y} & \frac{\partial w}{\partial z} \end{vmatrix}.$$

10.10. PROPERTIES OF JACOBIANS

I. If u, v are functions of r, s where r, s are functions of x, y, then

$$\frac{\partial(\mathbf{u}, \mathbf{v})}{\partial(\mathbf{x}, \mathbf{y})} = \frac{\partial(\mathbf{u}, \mathbf{v})}{\partial(\mathbf{r}, \mathbf{s})} \times \frac{\partial(\mathbf{r}, \mathbf{s})}{\partial(\mathbf{x}, \mathbf{y})}. \qquad \text{[Chain Rule for Jacobians]}$$

Proof. Since u, v are composite functions of x, y

$$\therefore \quad \left.\begin{aligned} \frac{\partial u}{\partial x} &= \frac{\partial u}{\partial r}\cdot\frac{\partial r}{\partial x} + \frac{\partial u}{\partial s}\cdot\frac{\partial s}{\partial x} = u_r r_x + u_s s_x \\ \frac{\partial u}{\partial y} &= \frac{\partial u}{\partial r}\cdot\frac{\partial r}{\partial y} + \frac{\partial u}{\partial s}\cdot\frac{\partial s}{\partial y} = u_r r_y + u_s s_y \\ \frac{\partial v}{\partial x} &= \frac{\partial v}{\partial r}\cdot\frac{\partial r}{\partial x} + \frac{\partial v}{\partial s}\cdot\frac{\partial s}{\partial x} = v_r r_x + v_s s_x \\ \frac{\partial v}{\partial y} &= \frac{\partial v}{\partial r}\cdot\frac{\partial r}{\partial y} + \frac{\partial v}{\partial s}\cdot\frac{\partial s}{\partial y} = v_r r_y + v_s s_y \end{aligned}\right] \quad \text{...(A)}$$

Now $\quad \frac{\partial(u, v)}{\partial(r, s)} \times \frac{\partial(r, s)}{\partial(x, y)} = \begin{vmatrix} u_r & u_s \\ v_r & v_s \end{vmatrix} \cdot \begin{vmatrix} r_x & r_y \\ s_x & s_y \end{vmatrix}$

Interchanging rows and columns in the second determinant

$$= \begin{vmatrix} u_r & u_s \\ v_r & v_s \end{vmatrix} \cdot \begin{vmatrix} r_x & s_x \\ r_y & s_y \end{vmatrix} = \begin{vmatrix} u_r r_x + u_s s_x & u_r r_y + u_s s_y \\ v_r r_x + v_s s_x & v_r r_y + v_x s_y \end{vmatrix} = \begin{vmatrix} \frac{\partial u}{\partial x} & \frac{\partial u}{\partial y} \\ \frac{\partial v}{\partial x} & \frac{\partial v}{\partial y} \end{vmatrix} \quad \text{[Using (A)]}$$

$$= \frac{\partial(u, v)}{\partial(x, y)}.$$

II. If J_1 is the Jacobian of u, v, with respect to x, y and J_2 is the Jacobian of x, y, with respect to u, v, then $J_1 J_2 = 1$ i.e., $\frac{\partial(\mathbf{u}, \mathbf{v})}{\partial(\mathbf{x}, \mathbf{y})} \cdot \frac{\partial(\mathbf{x}, \mathbf{y})}{\partial(\mathbf{u}, \mathbf{v})} = 1.$

Proof. Let $u = u(x, y)$ and $v = v(x, y)$, so that u and v are functions of x, y. Differentiating partially w.r.t. u and v, we get

$$\left.\begin{aligned} 1 &= \frac{\partial u}{\partial x}\cdot\frac{\partial x}{\partial u}+\frac{\partial u}{\partial y}\cdot\frac{\partial y}{\partial u} = u_x x_u + u_y y_u \\ 0 &= \frac{\partial u}{\partial x}\cdot\frac{\partial x}{\partial v}+\frac{\partial u}{\partial y}\cdot\frac{\partial y}{\partial v} = u_x x_v + u_y y_v \\ 0 &= \frac{\partial v}{\partial x}\cdot\frac{\partial x}{\partial u}+\frac{\partial v}{\partial y}\cdot\frac{\partial y}{\partial u} = v_x x_u + v_y y_u \\ 1 &= \frac{\partial v}{\partial x}\cdot\frac{\partial x}{\partial v}+\frac{\partial v}{\partial y}\cdot\frac{\partial y}{\partial v} = v_x x_v + v_y y_v \end{aligned}\right] \quad \text{...(A)}$$

Now $$\frac{\partial(u, v)}{\partial(x, y)}\cdot\frac{\partial(x, y)}{\partial(u, v)} = \begin{vmatrix} u_x & u_y \\ v_x & v_y \end{vmatrix}\begin{vmatrix} x_u & x_v \\ y_u & y_v \end{vmatrix}$$

Interchanging rows and columns in the second determinant

$$= \begin{vmatrix} u_x & u_y \\ v_x & v_y \end{vmatrix}\begin{vmatrix} x_u & y_u \\ x_v & y_v \end{vmatrix} = \begin{vmatrix} u_x x_u + u_y y_u & u_x x_v + u_y y_v \\ v_x x_u + v_y y_u & v_x x_v + v_y y_v \end{vmatrix}$$

$$= \begin{vmatrix} 1 & 0 \\ 0 & 1 \end{vmatrix} = 1. \quad \text{[Using (A)]}$$

III. Jacobian of Implicit Functions

If $f_1(x, y, u, v) = 0, \quad f_2(x, y, u, v) = 0$, then

$$\frac{\partial(u, v)}{\partial(x, y)} = (-1)^2 \frac{\dfrac{\partial(f_1, f_2)}{\partial(x, y)}}{\dfrac{\partial(f_1, f_2)}{\partial(u, v)}}$$

In general,

$$\frac{\partial(u_1, u_2, \ldots, u_n)}{\partial(x_1, x_2, \ldots, x_n)} = (-1)^n \frac{\partial(f_1, f_2, \ldots, f_n)/\partial(x_1, x_2, \ldots, x_n)}{\partial(f_1, f_2, \ldots, f_n)/\partial(u_1, u_2, \ldots, u_n)}$$

ILLUSTRATIVE EXAMPLES

Example 1. *If* $r = \sqrt{x^2 + y^2}$, $\theta = \tan^{-1}\dfrac{y}{x}$, *evaluate* $\dfrac{\partial(r, \theta)}{\partial(x, y)}$.

Sol. $$r = \sqrt{x^2 + y^2}, \quad \theta = \tan^{-1}\frac{y}{x}$$

$$\frac{\partial r}{\partial x} = \frac{x}{\sqrt{x^2 + y^2}}, \qquad \frac{\partial r}{\partial y} = \frac{y}{\sqrt{x^2 + y^2}}$$

$$\frac{\partial \theta}{\partial x} = \frac{1}{1+\dfrac{y^2}{x^2}}\left(-\frac{y}{x^2}\right) = -\frac{y}{x^2 + y^2}, \qquad \frac{\partial \theta}{\partial y} = \frac{1}{1+\dfrac{y^2}{x^2}}\left(\frac{1}{x}\right) = \frac{x}{x^2 + y^2}$$

$$\frac{\partial(r, \theta)}{\partial(x, y)} = \begin{vmatrix} \dfrac{\partial r}{\partial x} & \dfrac{\partial r}{\partial y} \\ \dfrac{\partial \theta}{\partial x} & \dfrac{\partial \theta}{\partial y} \end{vmatrix} = \begin{vmatrix} \dfrac{x}{\sqrt{x^2 + y^2}} & \dfrac{y}{\sqrt{x^2 + y^2}} \\ -\dfrac{y}{x^2 + y^2} & \dfrac{x}{x^2 + y^2} \end{vmatrix}$$

$$= \frac{x^2}{(x^2+y^2)^{3/2}} + \frac{y^2}{(x^2+y^2)^{3/2}} = \frac{x^2+y^2}{(x^2+y^2)^{3/2}} = \frac{1}{\sqrt{x^2+y^2}}.$$

Example 2. *If $x = r \sin\theta \cos\phi$, $y = r \sin\theta \sin\phi$, $z = r\cos\theta$, show that* $\frac{\partial(x, y, z)}{\partial(r, \theta, \phi)} = r^2 \sin\theta$.

Sol.
$$\frac{\partial(x, y, z)}{\partial(r, \theta, \phi)} = \begin{vmatrix} \frac{\partial x}{\partial r} & \frac{\partial x}{\partial \theta} & \frac{\partial x}{\partial \phi} \\ \frac{\partial y}{\partial r} & \frac{\partial y}{\partial \theta} & \frac{\partial y}{\partial \phi} \\ \frac{\partial z}{\partial r} & \frac{\partial z}{\partial \theta} & \frac{\partial z}{\partial \phi} \end{vmatrix} = \begin{vmatrix} \sin\theta\cos\phi & r\cos\theta\cos\phi & -r\sin\theta\sin\phi \\ \sin\theta\sin\phi & r\cos\theta\sin\phi & r\sin\theta\cos\phi \\ \cos\theta & -r\sin\theta & 0 \end{vmatrix}$$

Taking out common factors (r from second column and $r \sin\theta$ from third column)

$$= r^2 \sin\theta \begin{vmatrix} \sin\theta\cos\phi & \cos\theta\cos\phi & -\sin\phi \\ \sin\theta\sin\phi & \cos\theta\sin\phi & \cos\phi \\ \cos\theta & -\sin\theta & 0 \end{vmatrix}$$

Expanding by third row

$$= r^2 \sin\theta \left\{ \cos\theta \begin{vmatrix} \cos\theta\cos\phi & -\sin\phi \\ \cos\theta\sin\phi & \cos\phi \end{vmatrix} + \sin\theta \begin{vmatrix} \sin\theta\cos\phi & -\sin\phi \\ \sin\theta\sin\phi & \cos\phi \end{vmatrix} \right\}$$

$$= r^2 \sin\theta\,[\cos\theta\,(\cos\theta\cos^2\phi + \cos\theta\sin^2\phi) + \sin\theta\,(\sin\theta\cos^2\phi + \sin\theta\sin^2\phi)]$$

$$= r^2 \sin\theta\,(\cos^2\theta + \sin^2\theta) = r^2\sin\theta.$$

Note 1. Here (x, y, z) and (r, θ, ϕ) are respectively the Cartesian and spherical polar coordinates of a point.

Note 2. $\frac{\partial(r,\theta,\phi)}{\partial(x,y,z)} = \frac{1}{\frac{\partial(x,y,z)}{\partial(r,\theta,\phi)}} = \frac{1}{r^2\sin\theta}$ *(M.D.U. Dec. 2011)*

Example 3. *If $x = e^u \cos v$, $y = e^u \sin v$, find* $J = \frac{\partial(u, v)}{\partial(x, y)}$, $J' = \frac{\partial(x, y)}{\partial(u, v)}$ *and hence show that $JJ' = 1$.* (K.U.K. Dec. 2013)

Sol. Given $x = e^u \cos v$, $y = e^u \sin v$

$$\therefore \quad J' = \frac{\partial(x, y)}{\partial(u, v)} = \begin{vmatrix} \frac{\partial x}{\partial u} & \frac{\partial x}{\partial v} \\ \frac{\partial y}{\partial u} & \frac{\partial y}{\partial v} \end{vmatrix} = \begin{vmatrix} e^u\cos v & -e^u\sin v \\ e^u\sin v & e^u\cos v \end{vmatrix}$$

$$= e^{2u}(\cos^2 v + \sin^2 v) = e^{2u}$$

Also, $\quad x^2 + y^2 = e^{2u}(\cos^2 v + \sin^2 v) = e^{2u}$ and $\frac{y}{x} = \tan v$

$\Rightarrow \quad \log(x^2 + y^2) = 2u$ and $\frac{y}{x} = \tan v$

$\Rightarrow \quad u = \frac{1}{2}\log(x^2 + y^2)$, $v = \tan^{-1}\frac{y}{x}$

$$\frac{\partial u}{\partial x} = \frac{1}{2}\cdot\frac{2x}{x^2+y^2} = \frac{x}{x^2+y^2}, \quad \frac{\partial u}{\partial y} = \frac{y}{x^2+y^2}$$

$$\frac{\partial v}{\partial x} = \frac{1}{1+\frac{y^2}{x^2}}\left(-\frac{y}{x^2}\right) = \frac{-y}{x^2+y^2}, \frac{\partial v}{2y} = \frac{1}{1+\frac{y^2}{x^2}} \cdot \frac{1}{x} = \frac{x}{x^2+y^2}$$

$$\therefore \qquad J = \frac{\partial(u,v)}{\partial(x,y)} = \begin{vmatrix} \frac{\partial u}{\partial x} & \frac{\partial u}{\partial y} \\ \frac{\partial v}{\partial x} & \frac{\partial v}{\partial y} \end{vmatrix} = \begin{vmatrix} \frac{x}{x^2+y^2} & \frac{y}{x^2+y^2} \\ \frac{-y}{x^2+y^2} & \frac{x}{x^2+y^2} \end{vmatrix} = \frac{x^2+y^2}{(x^2+y^2)^2}$$

$$= \frac{1}{x^2+y^2} = \frac{1}{e^{2u}} = e^{-2u}$$

Hence, $\qquad JJ' = e^{-2u} \times e^{2u} = e^0 = 1.$

Example 4. *If* $u = \frac{x}{y-z}, v = \frac{y}{z-x}, w = \frac{z}{x-y}$, *show that* $\frac{\partial(u,v,w)}{\partial(x,y,z)} = 0.$

(M.D.U. Dec. 2010)

Sol. Here $\qquad u = \frac{x}{y-z}, v = \frac{y}{z-x}, w = \frac{z}{x-y}$

$$\Rightarrow \qquad \log u = \log x - \log (y-z) \qquad ...(1)$$

$$\log v = \log y - \log (z-x) \qquad ...(2)$$

$$\log w = \log z - \log (x-y) \qquad ...(3)$$

Differentiating (1) partially w.r.t. x

$$\frac{1}{u} \cdot \frac{\partial u}{\partial x} = \frac{1}{x} \quad \Rightarrow \quad \frac{\partial u}{\partial x} = \frac{u}{x}$$

Differentiating (1) partially w.r.t. y

$$\frac{1}{u} \cdot \frac{\partial u}{\partial y} = -\frac{1}{y-z} \quad \Rightarrow \quad \frac{\partial u}{\partial y} = \frac{-u}{y-z}$$

Differentiating (1) partially w.r.t. z

$$\frac{1}{u} \cdot \frac{\partial u}{\partial z} = -\frac{1}{y-z}(-1) \quad \Rightarrow \quad \frac{\partial u}{\partial z} = \frac{u}{y-z}$$

Similarly from (2) and (3), we have

$$\frac{\partial v}{\partial x} = \frac{v}{z-x}, \quad \frac{\partial v}{\partial y} = \frac{v}{y}, \quad \frac{\partial v}{\partial z} = \frac{-v}{z-x}$$

$$\frac{\partial w}{\partial x} = \frac{-w}{x-y}, \quad \frac{\partial w}{\partial y} = \frac{w}{x-y}, \quad \frac{\partial w}{\partial z} = \frac{w}{z}$$

$$\therefore \frac{\partial(u,v,w)}{\partial(x,y,z)} = \begin{vmatrix} \frac{\partial u}{\partial x} & \frac{\partial u}{\partial y} & \frac{\partial u}{\partial z} \\ \frac{\partial v}{\partial x} & \frac{\partial v}{\partial y} & \frac{\partial v}{\partial z} \\ \frac{\partial w}{\partial x} & \frac{\partial w}{\partial y} & \frac{\partial w}{\partial z} \end{vmatrix}$$

$$= \begin{vmatrix} \frac{u}{x} & \frac{-u}{y-z} & \frac{u}{y-z} \\ \frac{v}{z-x} & \frac{v}{y} & \frac{-v}{z-x} \\ \frac{-w}{x-y} & \frac{w}{x-y} & \frac{w}{z} \end{vmatrix}$$

Taking out u, v, w from R_1, R_2, R_3 respectively

$$= uvw \begin{vmatrix} \frac{1}{x} & \frac{-1}{y-z} & \frac{1}{y-z} \\ \frac{1}{z-x} & \frac{1}{y} & \frac{-1}{z-x} \\ \frac{-1}{x-y} & \frac{1}{x-y} & \frac{1}{z} \end{vmatrix}$$ Multiplying R_1, R_2, R_3 by $y-z, z-x, x-y$ respectively

$$= \frac{uvw}{(y-z)(z-x)(x-y)} \begin{vmatrix} \frac{y-z}{x} & -1 & 1 \\ 1 & \frac{z-x}{y} & -1 \\ -1 & 1 & \frac{x-y}{z} \end{vmatrix}$$ Multiplying C_1, C_2, C_3 by x, y, z respectively

$$= \frac{uvw}{xyz(y-z)(z-x)(x-y)} \begin{vmatrix} y-z & -y & z \\ x & z-x & -z \\ -x & y & x-y \end{vmatrix}$$ Operating $C_1 \to C_1 + C_2 + C_3$

$$= \frac{1}{(y-z)^2 (z-x)^2 (x-y)^2} \begin{vmatrix} 0 & -y & z \\ 0 & z-x & -z \\ 0 & y & x-y \end{vmatrix} = 0$$

$$\left[\because \frac{u}{x} = \frac{1}{y-z}, \frac{v}{y} = \frac{1}{z-x}, \frac{w}{z} = \frac{1}{x-y}\right]$$

EXERCISE 10.4

1. If $u = x^2 - 2y$, $v = x + y$, prove that $\frac{\partial(u, v)}{\partial(x, y)} = 2x + 2$.

2. (*a*) If $u = x(1 - y)$, $v = xy$; prove that $JJ' = 1$.

   (*b*) Prove that $JJ' = 1$ if $x = uv$, $y = \frac{u}{v}$

   [**Hint.** $u = vy \Rightarrow x = v^2 y \therefore v = \sqrt{\frac{x}{y}}$ and $u = \sqrt{xy}$]

3. (*a*) In polar coordinates, $x = r \cos \theta$, $y = r \sin \theta$, show that $\frac{\partial (x, y)}{\partial (r, \theta)} = r$.

   (*b*) If $x = r \cos \theta$, $y = r \sin \theta$, verify that $\frac{\partial(x, y)}{\partial(r, \theta)} \cdot \frac{\partial(r, \theta)}{\partial(x, y)} = 1$. (*P.T.U. 2011*)

   $\left[\textbf{Hint. } r = \sqrt{x^2 + y^2}, \theta = \tan^{-1} \frac{y}{x}.\right]$

4. If $x = r \cos \theta$, $y = r \sin \theta$, $z = z$, evaluate $\frac{\partial(x, y, z)}{\partial(r, \theta, z)}$.

   Note. Here (x, y, z) and (r, θ, z) are respectively the Cartesian and cylindrical coordinates of a point.

5. (*a*) If $y_1 = \frac{x_2 x_3}{x_1}$, $y_2 = \frac{x_3 x_1}{x_2}$, $y_3 = \frac{x_1 x_2}{x_3}$, show that $\frac{\partial (y_1, y_2, y_3)}{\partial (x_1, x_2, x_3)} = 4$. (*U.K.T.U. 2011*)

   (*b*) If $u = \frac{yz}{x}$, $v = \frac{zx}{y}$, $w = \frac{xy}{z}$, show that $\frac{\partial (x, y, z)}{\partial (u, v, w)} = \frac{1}{4}$

6. If $u = \dfrac{y^2}{2x}, v = \dfrac{x^2 + y^2}{2x}$, find $\dfrac{\partial(u, v)}{\partial(x, y)}$.

7. If $x = a \cosh \xi \cos \eta, y = a \sinh \xi \sin \eta$, show that $\dfrac{\partial(x, y)}{\partial(\xi, \eta)} = \dfrac{a^2}{2}(\cosh 2\xi - \cos 2\eta)$.

$\left[\textbf{Hint. } \cosh^2 x = \dfrac{\cosh 2x + 1}{2}, \sinh^2 x = \dfrac{\cosh 2x - 1}{2}\right]$

8. If $F = xu + v - y, G = u^2 + vy + w, H = zu - v + vw$, compute $\dfrac{\partial(F, G, H)}{\partial(u, w, v)}$.

9. If $u = \dfrac{x + y}{1 - xy}$ and $v = \tan^{-1} x + \tan^{-1} y$, find $\dfrac{\partial(u, v)}{\partial(x, y)}$.

10. If $u = x^2 - 2y, v = x + y + z, w = x - 2y + 3z$, find $\dfrac{\partial(u, v, w)}{\partial(x, y, z)}$.

11. If $u = x + y + z, uv = y + z, uvw = z$, show that $\dfrac{\partial(x, y, z)}{\partial(u, v, w)} = u^2 v$.

(G.B.T.U. 2013; K.U.K. 2009; P.T.U. Jan. 2010, Dec. 2011)

12. If $u = xyz, v = xy + yz + zx, w = x + y + z$, show that $\dfrac{\partial(u, v, w)}{\partial(x, y, z)} = (x - y)(y - z)(z - x)$.

(P.T.U. 2008)

13. If $y_1 = x_1^3, y_2 = e^{x_2}$ and $y_3 = x_1 + \sin x_3$, find $\dfrac{\partial(y_1, y_2, y_3)}{\partial(x_1, x_2, x_3)}$.

14. If $u = 2xy, v = x^2 - y^2, x = r \cos \theta, y = r \sin \theta$, evaluate $\dfrac{\partial(u, v)}{\partial(r, \theta)}$.

$\left[\textbf{Hint. } \text{Use } \dfrac{\partial(u, v)}{\partial(r, \theta)} = \dfrac{\partial(u, v)}{\partial(x, y)} \times \dfrac{\partial(x, y)}{\partial(r, \theta)} \quad \text{OR} \quad \text{Eliminate } x \text{ and } y\right]$

15. If $u = \sin^{-1} x + \sin^{-1} y, v = x\sqrt{1 - y^2} + y\sqrt{1 - x^2}$, find $\dfrac{\partial(u, v)}{\partial(x, y)}$.

16. If $x = uv$ and $y = \dfrac{u + v}{u - v}$, find $\dfrac{\partial(u, v)}{\partial(x, y)}$.

17. If $u = xy + yz + zx, v = x^2 + y^2 + z^2, w = x + y + z$, find $\dfrac{\partial(u, v, w)}{\partial(x, y, z)}$.

18. If $y_1 = 1 - x_1, y_2 = x_1(1 - x_2), y_3 = x_1 x_2 (1 - x_3)$, find $\dfrac{\partial(y_1, y_2, y_3)}{\partial(x_1, x_2, x_3)}$.

19. If $x = \sqrt{vw}, y = \sqrt{wu}, z = \sqrt{uv}$ and $u = r \sin \theta \cos \phi, v = r \sin \theta \sin \phi, w = r \cos \theta$, then evaluate the Jacobian $\dfrac{\partial(x, y, z)}{\partial(r, \theta, \phi)}$.

20. If $u = xyz, v = x^2 + y^2 + z^2, w = x + y + z$, find the Jacobian $\dfrac{\partial(x, y, z)}{\partial(u, v, w)}$. *(U.K.T.U 2012)*

21. If $x^2 + y^2 + u^2 - v^2 = 0$ and $uv + xy = 0$, prove that $\dfrac{\partial(u, v)}{\partial(x, y)} = \dfrac{x^2 - y^2}{u^2 + v^2}$.

[**Hint.** Let $f_1 = x^2 + y^2 + u^2 - v^2, f_2 = uv + xy$

Find $\dfrac{\partial(f_1, f_2)}{\partial(x, y)}$ and $\dfrac{\partial(f_1, f_2)}{\partial(u, v)}$, then $\dfrac{\partial(u, v)}{\partial(x, y)} = (-1)^2 \dfrac{\partial(f_1, f_2)/\partial(x, y)}{\partial(f_1, f_2)/\partial(u, v)}$]

Answers

2. x	4. r	6. $-\dfrac{y}{2x}$
8. $xw - x - xyv + 2uv - z$	9. 0	10. $10x + 4$
13. $3x_1^2 e^{x_2} \cos x_3$	14. $-4r^3$	15. 0
16. $\dfrac{(u-v)^2}{4uv}$	17. 0	18. $-x_1^2 x_2$
19. $\dfrac{1}{4} r^2 \sin\theta$	20. $\dfrac{-1}{2(x-y)(y-z)(z-x)}$.	

10.11. TAYLOR'S THEOREM FOR A FUNCTION OF TWO VARIABLES

We know that by Taylor's theorem for a function $f(x)$ of single variable x,

$$f(x+h) = f(x) + hf'(x) + \frac{h^2}{2!} f''(x) + \frac{h^3}{3!} f'''(x) + \cdots$$

Now let $f(x, y)$ be a function of two independent variables x and y. **If y is kept constant,** then by Taylor's theorem for a function of a single variable x, we have

$$f(x+h, \mathbf{y}+\mathbf{k}) = f(x, \mathbf{y}+\mathbf{k}) + h\frac{\partial}{\partial x} f(x, \mathbf{y}+\mathbf{k}) + \frac{h^2}{2!}\cdot\frac{\partial^2}{\partial x^2} f(x, \mathbf{y}+\mathbf{k}) + \frac{h^3}{3!}\cdot\frac{\partial^3}{\partial x^3} f(x, \mathbf{y}+\mathbf{k}) + \cdots \quad \text{...(1)}$$

Now **keeping x constant** and applying Taylor's theorem for a function of a single variable y, we have

$$f(\mathbf{x}, y+k) = f(\mathbf{x}, y) + k\frac{\partial}{\partial y} f(\mathbf{x}, y) + \frac{k^2}{2!}\cdot\frac{\partial^2}{\partial y^2} f(\mathbf{x}, \mathbf{y}) + \frac{k^3}{3!}\cdot\frac{\partial^3}{\partial y^3} f(\mathbf{x}, \mathbf{y}) + \cdots \quad \text{...(2)}$$

Using (2), we can write (1) as

$$f(x+h, y+k) = \left[f(x, y) + k\frac{\partial}{\partial y} f(x, y) + \frac{k^2}{2!}\cdot\frac{\partial^2}{\partial y^2} f(x, y) + \frac{k^3}{3!}\cdot\frac{\partial^3}{\partial y^3} f(x, y) + \cdots\right]$$

$$+ h\frac{\partial}{\partial x}\left[f(x, y) + k\frac{\partial}{\partial y} f(x, y) + \frac{k^2}{2!}\cdot\frac{\partial^2}{\partial y^2} f(x, y) + \cdots\right]$$

$$+ \frac{h^2}{2!}\cdot\frac{\partial^2}{\partial x^2}\left[f(x, y) + k\frac{\partial}{\partial y} f(x, y) + \cdots\right] + \frac{h^3}{3!}\cdot\frac{\partial^3}{\partial x^3}\,[f(x, y) + \cdots] + \cdots$$

$$= \left[f(x, y) + k\frac{\partial f}{\partial y} + \frac{k^2}{2!}\cdot\frac{\partial^2 f}{\partial y^2} + \frac{k^3}{3!}\cdot\frac{\partial^3 f}{\partial y^3} + \cdots\right]$$

$$+ \left[h\frac{\partial f}{\partial x} + hk\frac{\partial^2 f}{\partial x \partial y} + \frac{hk^2}{2!}\cdot\frac{\partial^3 f}{\partial x \partial y^2} + \cdots\right]$$

$$+ \left[\frac{h^2}{2!}\cdot\frac{\partial^2 f}{\partial x^2} + \frac{h^2 k}{2!}\cdot\frac{\partial^3 f}{\partial x^2 \partial y} + \cdots\right]$$

$$+ \frac{h^3}{3!}\frac{\partial^3 f}{\partial x^3} + \cdots$$

$$= f(x, y) + \left(h\frac{\partial f}{\partial x} + k\frac{\partial f}{\partial y}\right) + \left(\frac{h^2}{2!}\cdot\frac{\partial^2 f}{\partial x^2} + hk\frac{\partial^2 f}{\partial x \partial y} + \frac{k^2}{2!}\cdot\frac{\partial^2 f}{\partial y^2}\right)$$

$$+ \left(\frac{h^3}{3!}\cdot\frac{\partial^3 f}{\partial x^3} + \frac{h^2 k}{2!}\cdot\frac{\partial^3 f}{\partial x^2 \partial y} + \frac{hk^2}{2!}\cdot\frac{\partial^3 f}{\partial x \partial y^2} + \frac{k^3}{3!}\cdot\frac{\partial^3 f}{\partial y^3}\right) + \cdots$$

$$= f(x, y) + \left(h\frac{\partial f}{\partial x} + k\frac{\partial f}{\partial y}\right) + \frac{1}{2!}\left(h^2\frac{\partial^2 f}{\partial x^2} + 2hk\frac{\partial^2 f}{\partial x \partial y} + k^2\frac{\partial^2 f}{\partial y^2}\right)$$

$$+ \frac{1}{3!}\left(h^3\frac{\partial^3 f}{\partial x^3} + 3h^2 k\frac{\partial^3 f}{\partial x^2 \partial y} + 3hk^2\frac{\partial^3 f}{\partial x \partial y^2} + k^3\frac{\partial^3 f}{\partial y^3}\right) + \cdots$$

$$= f(x, y) + \left(h\frac{\partial}{\partial x} + k\frac{\partial}{\partial y}\right)f + \frac{1}{2!}\left(h\frac{\partial}{\partial x} + k\frac{\partial}{\partial y}\right)^2 f + \frac{1}{3!}\left(h\frac{\partial}{\partial x} + k\frac{\partial}{\partial y}\right)^3 f + \cdots$$

Cor. 1. Putting $x = a$ and $y = b$, we have

$$f(a + h, b + k) + f(a, b) + [hf_x(a, b) + kf_y(a, b)] + \frac{1}{2!}[h^2 f_{xx}(a, b) + 2hkf_{xy}(a, b) + k^2 f_{yy}(a, b)]$$

$$+ \frac{1}{3!}[h^3 f_{xxx}(a, b) + 3h^2 k f_{xxy}(a, b) + 3hk^2 f_{xyy}(a, b) + k^3 f_{yyy}(a, b)] + \cdots$$

Cor. 2. In Cor. 1, putting $a + h = x$ and $b + k = y$ so that $h = x - a$ and $k = y - b$, we have

$$f(x, y) = f(a, b) + [(x - a) f_x(a, b) + (y - b) f_y(a, b)] + \frac{1}{2!}[(x - a)^2 f_{xx}(a, b) + 2(x - a)(y - b)$$

$$f_{xy}(a, b) + (y - b)^2 f_{yy}(a, b)] + \cdots$$

Cor. 3. Putting $a = 0$, $b = 0$ in Cor. 2, we have

$$f(x, y) = f(0, 0) + [xf_x(0, 0) + yf_y(0, 0)] + \frac{1}{2!}[x^2 f_{xx}(0, 0) + 2xyf_{xy}(0, 0) + y^2 f_{yy}(0, 0)] + \cdots$$

This is called Maclaurin's theorem for two variables.

Note. Cor. 3 is used to expand $f(x, y)$ in powers of x and y [or to expand $f(x, y)$ in the neighbourhood of origin (0, 0)].

Cor. 2 is used to expand $f(x, y)$ in the neighbourhood of (a, b).

ILLUSTRATIVE EXAMPLES

Example 1. *Expand $e^x \sin y$ in powers of x and y as far as terms of the third degree.*

Sol. Here

$f(x, y) = e^x \sin y,$	$f(0, 0) = 0$
$f_x(x, y) = e^x \sin y,$	$f_x(0, 0) = 0$
$f_y(x, y) = e^x \cos y,$	$f_y(0, 0) = 1$
$f_{xx}(x, y) = e^x \sin y,$	$f_{xx}(0, 0) = 0$
$f_{xy}(x, y) = e^x \cos y,$	$f_{xy}(0, 0) = 1$
$f_{yy}(x, y) = -e^x \sin y,$	$f_{yy}(0, 0) = 0$
$f_{xxx}(x, y) = e^x \sin y,$	$f_{xxx}(0, 0) = 0$
$f_{xxy}(x, y) = e^x \cos y,$	$f_{xxy}(0, 0) = 1$

$$f_{xyy}(x, y) = -e^x \sin y, \qquad f_{xyy}(0, 0) = 0$$
$$f_{yyy}(x, y) = -e^x \cos y, \qquad f_{yyy}(0, 0) = -1$$

....

$\therefore \quad e^x \sin y = f(x, y)$

$$= f(0, 0) + [xf_x(0, 0) + y f_y(0, 0)] + \frac{1}{2!}[x^2 f_{xx}(0, 0) + 2xy f_{xy}(0, 0) + y^2 f_{yy}(0, 0)]$$
$$+ \frac{1}{3!}[x^3 f_{xxx}(0, 0) + 3x^2y f_{xxy}(0, 0) + 3xy^2 f_{xyy}(0, 0) + y^3 f_{yyy}(0, 0)] + \cdots$$
$$= 0 + [x \cdot 0 + y \cdot 1] + \frac{1}{2!}[x^2 \cdot 0 + 2xy \cdot 1 + y^2 \cdot 0] + \frac{1}{3!}[x^3 \cdot 0 + 3x^2y \cdot 1 + 3xy^2$$
$$+ 0 + y^3(-1)] + \cdots$$
$$= y + xy + \frac{1}{2}x^2y - \frac{1}{6}y^3 + \cdots$$

Example 2. *Expand $f(x, y) = \tan^{-1} \frac{y}{x}$ in powers of $(x - 1)$ and $(y - 1)$ upto third degree terms. Hence compute $f(1.1, 0.9)$ approximately.*

(U.K.T.U. 2011, 2012; P.T.U. 2008; Kerala 2009)

Sol. Here $\quad f(x, y) = \tan^{-1} \frac{y}{x}, f(1, 1) = \tan^{-1} 1 = \frac{\pi}{4}$

$$f_x(x, y) = \frac{1}{1 + \frac{y^2}{x^2}} \cdot \left(-\frac{y}{x^2}\right) = -\frac{y}{x^2 + y^2}, \; f_x(1, 1) = -\frac{1}{2}$$

$$f_y(x, y) = \frac{1}{1 + \frac{y^2}{x^2}} \cdot \frac{1}{x} = \frac{x}{x^2 + y^2}, \; f_y(1, 1) = \frac{1}{2}$$

$$f_{xx}(x, y) = -y(-1)(x^2 + y^2)^{-2} \cdot 2x = \frac{2xy}{(x^2 + y^2)^2}, \qquad f_{xx}(1, 1) = \frac{1}{2}$$

$$f_{xy}(x, y) = \frac{(x^2 + y^2) \cdot 1 - x \cdot 2x}{(x^2 + y^2)^2} = \frac{y^2 - x^2}{(x^2 + y^2)^2}, \qquad f_{xy}(1, 1) = 0$$

$$f_{yy}(x, y) = x(-1)(x^2 + y^2)^{-2} \cdot 2y = -\frac{2xy}{(x^2 + y^2)^2}, \qquad f_{yy}(1, 1) = -\frac{1}{2}$$

$$f_{xxx}(x, y) = 2y \cdot \frac{(x^2 + y^2)^2 \cdot 1 - x \cdot 2(x^2 + y^2) \cdot 2x}{(x^2 + y^2)^4} = 2y \cdot \frac{(x^2 + y^2) - 4x^2}{(x^2 + y^2)^3}$$

$$= \frac{2y(y^2 - 3x^2)}{(x^2 + y^2)^3}, \quad f_{xxx}(1, 1) = -\frac{1}{2}$$

$$f_{xxy}(x, y) = \frac{(x^2 + y^2)^2 (-2x) - (y^2 - x^2) \cdot 2(x^2 + y^2) \cdot 2x}{(x^2 + y^2)^4}$$

$$= \frac{(x^2+y^2)(-2x) - 4x(y^2-x^2)}{(x^2+y^2)^3} = \frac{2x(x^2-3y^2)}{(x^2+y^2)^3}, \quad f_{xxy}(1, 1) = -\frac{1}{2}$$

$$f_{xyy}(x, y) = -2y \cdot \frac{(x^2+y^2)^2 \cdot 1 - x \cdot 2(x^2+y^2) \cdot 2x}{(x^2+y^2)^4} = \frac{-2y(x^2+y^2-4x^2)}{(x^2+y^2)^3}$$

$$= \frac{2y(3x^2-y^2)}{(x^2+y^2)^3}, \quad f_{xyy}(1, 1) = \frac{1}{2}$$

$$f_{yyy}(x, y) = -2x \cdot \frac{(x^2+y^2)^2 \cdot 1 - y \cdot 2(x^2+y^2) \cdot 2y}{(x^2+y^2)^4} = \frac{-2x(x^2+y^2-4y^2)}{(x^2+y^2)^3}$$

$$= \frac{2x(3y^2-x^2)}{(x^2+y^2)^3}, \quad f_{yyy}(1, 1) = \frac{1}{2}$$

$$\therefore \quad \tan^{-1}\frac{y}{x} = f(x, y) = f(1, 1) + [(x-1) f_x(1, 1) + (y-1) f_y(1, 1)]$$

$$+ \frac{1}{2!}[(x-1)^2 f_{xx}(1, 1) + 2(x-1)(y-1) f_{xy}(1, 1) + (y-1)^2 f_{yy}(1, 1)]$$

$$+ \frac{1}{3!}[(x-1)^3 f_{xxx}(1, 1) + 3(x-1)^2(y-1) f_{xxy}(1, 1)$$

$$+ 3(x-1)(y-1)^2 f_{xyy}(1, 1) + (y-1)^3 f_{yyy}(1, 1)] + \cdots$$

$$= \frac{\pi}{4} + \left[(x-1)\left(-\frac{1}{2}\right) + (y-1)\left(\frac{1}{2}\right)\right] + \frac{1}{2}\left[(x-1)^2\left(\frac{1}{2}\right) + 2(x-1)(y-1)(0) + (y-1)^2\left(-\frac{1}{2}\right)\right]$$

$$+ \frac{1}{6}\left[(x-1)^3\left(-\frac{1}{2}\right) + 3(x-1)^2(y-1)\left(-\frac{1}{2}\right) + 3(x-1)(y-1)^2\left(\frac{1}{2}\right) + (y-1)^3\left(\frac{1}{2}\right)\right] + \cdots$$

$$= \frac{\pi}{4} - \frac{1}{2}[(x-1) - (y-1)] + \frac{1}{4}[(x-1)^2 - (y-1)^2] - \frac{1}{12}[(x-1)^3$$

$$+ 3(x-1)^2(y-1) - 3(x-1)(y-1)^2 - (y-1)^3] + \cdots$$

Putting $x = 1.1$ and $y = 0.9$, we get

$$f(1.1, 0.9) = \frac{\pi}{4} - \frac{1}{2}(0.2) + \frac{1}{4}(0) - \frac{1}{12}[(0.1)^3 + 3(0.1)^2(-0.1) - 3(0.1)(-0.1)^2 - (-0.1)^3]$$

$$= 0.7854 - 0.1 + 0.0003 = 0.6857.$$

Example 3. *Expand $e^x \log(1+y)$ in powers of x and y upto terms of third degree.*

(M.D.U. May 2008; P.T.U. 2010; Kerala 2008)

Sol. Here

$$f(x, y) = e^x \log(1+y), \qquad f(0, 0) = 0$$

$$f_x(x, y) = e^x \log(1+y), \qquad f_x(0, 0) = 0$$

$$f_y(x, y) = \frac{e^x}{1+y}, \qquad f_y(0, 0) = 1$$

$$f_{xx}(x, y) = e^x \log(1+y), \qquad f_{xx}(0, 0) = 0$$

$$f_{xy}(x, y) = \frac{e^x}{1+y}, \qquad f_{xy}(0, 0) = 1$$

$$f_{yy}(x, y) = -\frac{e^x}{(1+y)^2}, \qquad f_{yy}(0, 0) = -1$$

$$f_{xxx}(x, y) = e^x \log (1 + y), \qquad f_{xxx}(0, 0) = 0$$

$$f_{xxy}(x, y) = \frac{e^x}{1+y}, \qquad f_{xxy}(0, 0) = 1$$

$$f_{xyy}(x, y) = -\frac{e^x}{(1+y)^2}, \qquad f_{xyy}(0, 0) = -1$$

$$f_{yyy}(x, y) = \frac{2e^x}{(1+y)^3}, \qquad f_{yyy}(0, 0) = 2$$

....

$\therefore \quad e^x \log (1 + y) = f(x, y)$

$$= f(0, 0) + [xf_x (0, 0) + yf_y(0, 0)] + \frac{1}{2!} [x^2 f_{xx}(0, 0) + 2xyf_{xy} (0, 0) + y^2 f_{yy}(0, 0)]$$

$$+ \frac{1}{3!} [x^3 f_{xxx}(0, 0) + 3x^2y f_{xxy}(0, 0) + 3xy^2 f_{xyy}(0, 0) + y^3 f_{yyy}(0, 0)] + \cdots$$

$$= 0 + [x \cdot 0 + y \cdot 1] + \frac{1}{2}[x^2 \cdot 0 + 2xy \cdot 1 + y^2(-1)] + \frac{1}{6} [x^3 \cdot 0 + 3x^2y \cdot 1 + 3xy^2 \cdot (-1) + y^3 \cdot 2] + \cdots$$

$$= y + xy - \frac{1}{2}y^2 + \frac{1}{2}x^2y - \frac{1}{2}xy^2 + \frac{1}{3}y^3 \ldots$$

Example 4. *Expand* $\frac{(x+h)(y+k)}{x+h+y+k}$ *in powers of h, k upto and inclusive of the second degree terms.*

Sol. Here $\quad f(x + h, y + k) = \frac{(x+h)(y+k)}{x+h+y+k}$

Putting $h = k = 0$, we have $f(x, y) = \frac{xy}{x+y}$

$$f_x = \frac{(x+y) \cdot y - xy \cdot 1}{(x+y)^2} = \frac{y^2}{(x+y)^2}; \qquad f_y = \frac{x^2}{(x+y)^2}, \text{ by symmetry}$$

$$f_{xx} = -\frac{2y^2}{(x+y)^3}, f_{yy} = -\frac{2x^2}{(x+y)^3}$$

$$f_{xy} = \frac{(x+y)^2 \cdot 2x - x^2 \cdot 2(x+y)}{(x+y)^4} = \frac{2x\,(x+y) - 2x^2}{(x+y)^3} = \frac{2xy}{(x+y)^3}$$

$$\therefore \quad \frac{(x+h)(y+k)}{x+h+y+k} = f(x + h, y + k) = f(x, y) + [hf_x + kf_y] + \frac{1}{2!} [h^2 f_{xx} + 2hk f_{xy} + k^2 f_{yy}] + \cdots$$

$$= \frac{xy}{x+y} + \left[h \cdot \frac{y^2}{(x+y)^2} + k \cdot \frac{x^2}{(x+y)^2}\right]$$

$$+ \frac{1}{2}\left[h^2 \cdot \frac{-2y^2}{(x+y)^3} + 2hk \frac{2xy}{(x+y)^3} + k^2 \cdot \frac{-2x^2}{(x+y)^3}\right] + \cdots$$

$$= \frac{xy}{x+y} + \frac{y^2}{(x+y)^2} \cdot h + \frac{x^2}{(x+y)^2} \cdot k - \frac{y^2}{(x+y)^3} \cdot h^2$$

$$+ \frac{2xy}{(x+y)^3} \cdot hk - \frac{x^2}{(x+y)^3} \cdot k^2 + \cdots$$

Example 5. *Expand $x^2y + 3y - 2$ in powers of $(x - 1)$ and $(y + 2)$ using Taylor's Theorem.* (P.T.U. 2009, 2011)

Sol. Expansion of $f(x, y)$ in powers of $(x - a)$ and $(y - b)$ is given by

$$f(x, y) = f(a, b) + [(x - a)f_x(a, b) + (y - b)f_y(a, b)] + \frac{1}{2!}[(x - a)^2 f_{xx}(a, b)$$
$$+ 2(x - a)(y - b) f_{xy}(a, b) + (y - b)^2 f_{yy}(a, b)] + \frac{1}{3!}[(x - a)^3 f_{xxx}(a, b)$$
$$+ 3(x - a)^2(y - b)f_{xxy}(a, b) + 3(x - a)(y - b)^2 f_{xyy}(a, b) + (y - b)^3 f_{yyy}(a, b)]$$
$$+ \cdots \qquad \ldots(1)$$

Here $f(x, y) = x^2y + 3y - 2,\ a = 1,\ b = -2$

$f(1, -2) = 1^2 \times (-2) + 3(-2) - 2 = -10$

$f_x = 2xy, \quad f_x(1, -2) = 2(1)(-2) = -4; \quad f_y = x^2 + 3, \quad f_y(1, -2) = 1^2 + 3 = 4$

$f_{xx} = 2y, \quad f_{xx}(1, -2) = 2(-2) = -4; \quad f_{xy} = 2x, \quad f_{xy}(1, -2) = 2(1) = 2$

$f_{yy} = 0, \quad f_{yy}(1, -2) = 0; \quad f_{xxx} = 0, \quad f_{xxx}(1, -2) = 0$

$f_{xxy} = 2, \quad f_{xxy}(1, -2) = 2; \quad f_{xyy} = 0, \quad f_{xyy}(1, -2) = 0$

$f_{yyy} = 0, \quad f_{yyy}(1, -2) = 0$

All higher order partial derivatives vanish.

$\therefore$ From (1), we have

$x^2y + 3y - 1 = f(x, y)$

$$= -10 + [(x - 1)(-4) + (y + 2)(4)] + \frac{1}{2}[(x - 1)^2(-4) + 2(x - 1)(y + 2)(2) + (y + 2)^2(0)]$$
$$+ \frac{1}{6}[(x - 1)^3(0) + 3(x - 1)^2(y + 2)(2) + 3(x - 1)(y + 2)^2(0) + (y + 2)^3(0)]$$
$$= -10 - 4(x - 1) + 4(y + 2) - 2(x - 1)^2 + 2(x - 1)(y + 2) + (x - 1)^2(y + 2).$$

EXERCISE 10.5

1. Show that $e^y \log(1 + x) = x + xy - \dfrac{x^2}{2}$ approximately.

   [**Hint.** Find the expansion at (0, 0)]
2. Expand $e^x \cos y$ in powers of x and y as far as the terms of third degree.
3. Expand $e^{ax} \sin by$ in powers of x and y as far as the terms of third degree.
4. Expand e^{xy} at (1, 1) in powers of $(x - 1)$ and $(y - 1)$.
5. Expand $e^x \cos y$ about $\left(1, \dfrac{\pi}{4}\right)$. (*U.P.T.U. 2008; P.T.U. Dec. 2011*)
6. Expand $(1 + x + y^2)^{1/2}$ about (1, 0).
7. Expand $\sin(x + h)(y + k)$ by Taylor's Theorem.
8. Expand $\sin xy$ in powers of $(x - 1)$ and $\left(y - \dfrac{\pi}{2}\right)$ upto the second degree terms.

   (*U.K.T.U. 2011; U.P.T.U. 2008; P.T.U. 2010*)
9. Expand $f(x, y) = xy^2 + \cos xy$ in powers of $(x - 1)$ and $\left(y - \dfrac{\pi}{2}\right)$ upto the second degree terms.

   (*K.U.K. Dec. 2010*)
10. If $f(x, y) = \tan^{-1} xy$, compute $f(0.9, -1.2)$ approximately. (*M.D.U. May 2009*)

    [**Hint.** Expand $f(x, y)$ in powers of $(x - 1)$ and $(y + 1)$, *i.e.*, in the neighbourhood of $(1, -1)$.]

Answers

2. $1 + x + \frac{1}{2!}(x^2 - y^2) + \frac{1}{3!}(x^3 - 3xy^2) + \cdots$ 3. $by + abxy + \frac{1}{3!}(3a^2 bx^2y - b^3y^3) + \cdots$

4. $e\left\{1 + (x-1) + (y-1) + \frac{1}{2!}\left((x-1)^2 + 2(x-1)(y-1) + (y-1)^2\right) + \cdots\right\}$

5. $\frac{e}{\sqrt{2}}\left\{1 + (x-1) - \left(y - \frac{\pi}{4}\right) + \frac{(x-1)^2}{2} - (x-1)\left(y - \frac{\pi}{4}\right) - \frac{1}{2}\left(y - \frac{\pi}{4}\right)^2 + \cdots\right\}$

6. $\sqrt{2}\left\{1 + \frac{x-1}{4} - \frac{(x-1)^2}{32} + \frac{y^2}{4} + \cdots\right\}$

7. $\sin xy + (hy + kx)\cos xy + hk\cos xy - \frac{1}{2}(hy + kx)^2 \sin xy + \cdots$

8. $1 - \frac{1}{8}\pi^2(x-1)^2 - \frac{1}{2}\pi(x-1)\left(y - \frac{\pi}{2}\right) - \frac{1}{2}\left(y - \frac{\pi}{2}\right)^2$

9. $\frac{\pi^2}{4} + (x-1)\left(\frac{\pi^2}{4} - \frac{\pi}{2}\right) + \left(y - \frac{\pi}{2}\right)(\pi - 1) + (x-1)\left(y - \frac{\pi}{2}\right)(\pi - 1) + \left(y - \frac{\pi}{2}\right)^2$

10. -0.8294.

10.12. MAXIMA AND MINIMA OF FUNCTIONS OF TWO VARIABLES

A function $f(x, y)$ is said to have a **maximum value at x = a, y = b if f(a, b) > f(a + h, b + k),** for small and independent values of h and k, positive or negative.

A function $f(x, y)$ is said to have a **minimum value at x = a, y = b** if $f(a, b) < f(a + h, b + k)$, for small and independent values of h and k, positive or negative.

Thus $f(x, y)$ has a maximum or minimum value at a point (a, b) according as

$$\Delta f = f(a + h, b + k) - f(a, b) < \text{or} > 0.$$

Geometrically, the surface $z = f(x, y)$ has a maximum at the point (a, b) if $f(a, b)$ is the greatest in the small neighbourhood of the point (a, b). As such, the surface descends in all directions at this point. The point of maxima may well be compared with the highest point of a dome. The surface has a minimum at the point (a, b) if $f(a, b)$ is the smallest in the small neighbourhood of the point (a, b). As such, the surface ascends in all directions at this point. The point of minima may well be compared with the lowest point of a bowl. However, sometimes, at the point (a, b), the tangent plane is horizontal and the surface descends in certain directions and ascends in other directions. Such a point is called a *saddle point*.

A maximum or a minimum value of a function is called its *extreme value*.

10.13. CONDITIONS FOR f(x, y) TO BE MAXIMUM OR MINIMUM

By Taylor's theorem, we have

$$\Delta f = f(a + h, b + k) - f(a, b)$$

$$= [hf_x(a, b) + kf_y(a, b)] + \frac{1}{2!}[h^2 f_{xx}(a, b) + 2hkf_{xy}(a, b) + k^2 f_{yy}(a, b)] + \cdots \quad \text{...(1)}$$

For small values of h and k, the second and higher order terms are still smaller and may be neglected. Thus, sign of Δf = sign of $[hf_x(a, b) + kf_y(a, b)]$

Taking $h = 0$, the sign of Δf changes with the sign of k. Similarly taking $k = 0$, the sign of Δf changes with the sign of h. Since Δf changes sign with h and k, $f(x, y)$ cannot have a maximum or a minimum value at (a, b) unless $f_x(a, b) = 0 = f_y(a, b)$.

Hence the necessary conditions for $f(x, y)$ *to have a maximum or a minimum value at* (a, b) *are*

$$f_x(a, b) = 0, f_y(a, b) = 0$$

If these conditions are satisfied, then for small values of h and k, we have, from (1),

$$\Delta f = \frac{1}{2!}[h^2 f_{xx}(a, b) + 2hk f_{xy}(a, b) + k^2 f_{yy}(a, b)] + \cdots$$

$$= \frac{1}{2!}[h^2 r + 2hks + k^2 t] + \cdots \text{ where } r = f_{xx}(a, b),\ s = f_{xy}(a, b),\ t = f_{yy}(a, b)$$

or

$$\Delta f = \frac{1}{2r}[h^2 r^2 + 2hkrs + k^2 rt] + \cdots = \frac{1}{2r}[(h^2 r^2 + 2hkrs + k^2 s^2) + k^2 rt - k^2 s^2)] + \cdots$$

$$= \frac{1}{2r}[(hr + ks)^2 + k^2(rt - s^2)] + \cdots \qquad \ldots(2)$$

Now $(hr + ks)^2$ is always positive and $k^2(rt - s^2)$ will be positive if $rt - s^2 > 0$. In this case, Δf will have the same sign as that of r for all values of h and k.

Hence if $rt - s^2 > 0$, then $f(x, y)$ has a maximum or a minimum at (a, b) according as $r < 0$ or $r > 0$.

If $rt - s^2 < 0$, then Δf changes sign with h and k. Hence there is neither a maximum nor a minimum value at (a, b). The point (a, b) is a saddle point in this case.

If $rt - s^2 = 0$, no conclusion can be drawn about a maximum or a minimum value at (a, b) and hence further investigation is required.

Note. The point (a, b) is called a *stationary point* if $f_x(a, b) = 0$, $f_y(a, b) = 0$. The value $f(a, b)$ is called a *stationary value*. Thus every extreme value is a stationary value but the converse may not be true.

10.14. RULE TO FIND THE EXTREME VALUES OF A FUNCTION z = f(x, y)

(*i*) Find $\frac{\partial z}{\partial x}$ and $\frac{\partial z}{\partial y}$.

(*ii*) Solve $\frac{\partial z}{\partial x} = 0$ and $\frac{\partial z}{\partial y} = 0$ simultaneously.

Let (a, b); (c, d) ··· be the solutions of these equations.

(*iii*) For each solution in step (*ii*), find $r = \frac{\partial^2 f}{\partial x^2}$, $s = \frac{\partial^2 f}{\partial x \partial y}$, $t = \frac{\partial^2 f}{\partial y^2}$.

(*iv*) (*a*) If $rt - s^2 > 0$ and $r < 0$ for a particular solution (a, b) of step (*ii*), then z has a maximum value at (a, b).

(*b*) If $rt - s^2 > 0$ and $r > 0$ for a particular solution (a, b) of step (*ii*), then z has a minimum value at (a, b).

(*c*) If $rt - s^2 < 0$ for a particular solution (a, b) of step (*ii*), then z has no extreme value at (a, b).

(*d*) If $rt - s^2 = 0$, the case is doubtful and requires further investigation.

ILLUSTRATIVE EXAMPLES

Example 1. *Examine the function* $x^3 + y^3 - 3axy$ *for maxima and minima.*

(U.P.T.U. 2009; G.B.T.U. 2012)

Sol. Here $f(x, y) = x^3 + y^3 - 3axy$

$$f_x = 3x^2 - 3ay, \quad f_y = 3y^2 - 3ax, \quad r = f_{xx} = 6x, \ s = f_{xy} = -3a, \ t = f_{yy} = 6y$$

Now for extreme values $f_x = 0$ and $f_y = 0$

$\Rightarrow \quad x^2 - ay = 0 \quad ...(1) \qquad y^2 - ax = 0 \quad ...(2)$

From (1), $\quad y = \dfrac{x^2}{a}$

$\therefore$ From (2), $\quad \dfrac{x^4}{a^2} - ax = 0$ or $x(x^3 - a^3) = 0$ or $x = 0, a$

when $x = 0,\ y = 0$; when $x = a,\ y = a$

$\therefore$ There are two stationary points (0, 0) and (a, a).

Now $\quad rt - s^2 = 36xy - 9a^2$

At (0, 0), $rt - s^2 = -9a^2 < 0$

$\Rightarrow$ there is no extreme value at (0, 0).

At (a, a), $rt - s^2 = 36a^2 - 9a^2 = 27a^2 > 0$

$\Rightarrow$ $f(x, y)$ has extreme value at (a, a)

Now $\quad r = 6a$

If $a > 0$, $r > 0$ so that $f(x, y)$ has a minimum value at (a, a).

Minimum value $= f(a, a) = a^3 + a^3 - 3a^3 = -a^3$

If $a < 0$, $r < 0$ so that $f(x, y)$ has a maximum value at (a, a).

Maximum value $= f(a, a) = a^3 + a^3 - 3a^3 = -a^3$

Example 2. *Discuss the maxima and minima of* $x^3y^2(1 - x - y)$. (P.T.U. May 2009)

Sol. Here $\quad f(x, y) = x^3y^2(1 - x - y) = x^3y^2 - x^4y^2 - x^3y^3$

$$f_x = 3x^2y^2 - 4x^3y^2 - 3x^2y^3,\ f_y = 2x^3y - 2x^4y - 3x^3y^2$$

$$r = f_{xx} = 6xy^2 - 12x^2y^2 - 6xy^3;\ s = f_{xy} = 6x^2y - 8x^3y - 9x^2y^2;\ t = f_{yy} = 2x^3 - 2x^4 - 6x^3y$$

Now for extreme values $\quad f_x = 0$ and $f_y = 0$

$\Rightarrow \quad x^2y^2(3 - 4x - 3y) = 0 \quad ...(1) \qquad x^3y(2 - 2x - 3y) = 0 \quad ...(2)$

Solving (1) and (2), the stationary points are (0, 0) and $(\frac{1}{2}, \frac{1}{3})$.

Now $\quad rt - s^2 = 6xy^2(1 - 2x - y)\ 2x^3(1 - x - 3y) - [x^2y(6 - 8x - 9y)]^2$

$$= 12x^4y^2(1 - 2x - y)(1 - x - 3y) - x^4y^2(6 - 8x - 9y)^2$$

At (0, 0), $\quad rt - s^2 = 0$

$\therefore$ At (0, 0) further investigation is needed.

Consider $f(a + h, b + k) - f(a, b) = f(h, k) - f(0, 0) = h^3k^2(1 - h - k)$

$= h^3k^2$ (neglecting $h^4k^2 + h^3k^3$ which is very small as compared to h^3k^2)

which is > 0 if $h > 0$ and < 0 if $h < 0$

[For example, $f(.1, .2) > 0$ and $f(-.1, .2) < 0$]

$\therefore$ $f(a + h, b + k) - f(a, b)$ does not keep same sign for all small values of h and k (positive or negative).

$\therefore$ There is no extreme value at (0, 0).

At $\left(\dfrac{1}{2}, \dfrac{1}{3}\right)$, $\quad rt - s^2 = 12 \cdot \dfrac{1}{16} \cdot \dfrac{1}{9}\left(-\dfrac{1}{3}\right)\left(-\dfrac{1}{2}\right) - \dfrac{1}{16} \cdot \dfrac{1}{9}(1) = \dfrac{1}{144}(2 - 1) = \dfrac{1}{14} > 0.$

Also $\quad r = 6 \cdot \dfrac{1}{2} \cdot \dfrac{1}{9} - 12 \cdot \dfrac{1}{4} \cdot \dfrac{1}{9} - 6 \cdot \dfrac{1}{2} \cdot \dfrac{1}{8} = \dfrac{1}{3} - \dfrac{1}{3} - \dfrac{3}{8} = -\dfrac{3}{8} < 0.$

$\therefore$ $f(x, y)$ has a maximum value at $(\frac{1}{2}, \frac{1}{3})$

Maximum value $= f\left(\dfrac{1}{2}, \dfrac{1}{3}\right) = \dfrac{1}{8} \cdot \dfrac{1}{9}\left(1 - \dfrac{1}{2} - \dfrac{1}{3}\right) = \dfrac{1}{432}.$

Example 3. *Locate the stationary points of* $x^4 + y^4 - 2x^2 + 4xy - 2y^2$ *and determine their nature.* (M.T.U. 2013; P.T.U. Dec. 2010)

Sol. Here $f(x, y) = x^4 + y^4 - 2x^2 + 4xy - 2y^2$

$$f_x = 4x^3 - 4x + 4y, \quad f_y = 4y^3 + 4x - 4y$$

$$r = f_{xx} = 12x^2 - 4, \quad s = f_{xy} = 4, \; t = f_{yy} = 12y^2 - 4$$

Now $f_x = 0$ and $f_y = 0$

$\Rightarrow$ $x^3 - x + y = 0$ ...(1) $y^3 + x - y = 0$ ...(2)

Adding (1) and (2), $x^3 + y^3 = 0 \Rightarrow y = -x$.

Putting $y = -x$ in (1), we get $x^3 - 2x = 0$ or $x(x^2 - 2) = 0 \quad \therefore \quad x = 0, \pm\sqrt{2}$.

Since $y = -x$, the stationary points are (0, 0), $(\sqrt{2}, -\sqrt{2})$ and $(-\sqrt{2}, \sqrt{2})$.

Now $rt - s^2 = (12x^2 - 4)(12y^2 - 4) - 16$

At (0, 0), $rt - s^2 = 16 - 16 = 0$.

$\therefore$ At (0, 0), further investigation is needed.

Now $f(0, 0) = 0$

$$f(x, y) = x^4 + y^4 - 2(x^2 - 2xy + y^2) = x^4 + y^4 - 2(x - y)^2$$

When h and k are small, $f(h, k) = 2h^4 > 0$ for $h = k$

and $f(h, k) = -2(h - k)^2 < 0$ for $h \neq k$

(neglecting $h^4 + k^4$ which is very small)

$\Rightarrow$ $f(0, 0) < f(h, k)$ for $h = k$

$f(0, 0) > f(h, k)$ for $h \neq k$

$\Rightarrow$ There is no extreme value at (0, 0).

At $(\sqrt{2}, -\sqrt{2})$, $rt - s^2 = (12 \times 2 - 4)^2 - 16 = 384 > 0$

$$r = 12 \times 2 - 4 = 20 > 0$$

$\therefore$ $f(x, y)$ has a minimum value at $(\sqrt{2}, -\sqrt{2})$.

$$\text{Minimum Value} = (\sqrt{2})^4 + (-\sqrt{2})^4 - 2(\sqrt{2})^2 + 4\sqrt{2}(-\sqrt{2}) - 2(-\sqrt{2})^2$$

$$= 4 + 4 - 4 - 8 - 4 = -8.$$

Similarly, $f(x, y)$ has a minimum value -8 at $(-\sqrt{2}, \sqrt{2})$.

Example 4. *Examine for minimum and maximum values:* $\sin x + \sin y + \sin (x + y)$. (M.D.U. Dec. 2013)

Sol. Here $f(x, y) = \sin x + \sin y + \sin (x + y)$

$$f_x = \cos x + \cos (x + y), \; f_y = \cos y + \cos (x + y)$$

$$r = f_{xx} = -\sin x - \sin (x + y), \; s = f_{xy} = -\sin (x + y),$$

$$t = f_{yy} = -\sin y - \sin (x + y)$$

Now $f_x = 0$ and $f_y = 0$

$\Rightarrow$ $\cos x + \cos (x + y) = 0$ ...(1) $\cos y + \cos (x + y) = 0$ ...(2)

Subtracting (2) from (1), $\cos x - \cos y = 0$ or $\cos x = \cos y \quad \therefore \quad x = y$

From (1), $\cos x + \cos 2x = 0$ or $\cos 2x = -\cos x = \cos (\pi - x)$ or $2x = \pi - x \quad \therefore \quad x = \dfrac{\pi}{3}$

$\therefore$ $x = y = \dfrac{\pi}{3}$ is a stationary point.

At $\left(\dfrac{\pi}{3}, \dfrac{\pi}{3}\right)$, $r = -\dfrac{\sqrt{3}}{2} - \dfrac{\sqrt{3}}{2} = -\sqrt{3} < 0$, $s = \dfrac{\sqrt{3}}{2}$, $t = -\dfrac{\sqrt{3}}{2} - \dfrac{\sqrt{3}}{2} = -\sqrt{3}$

or $rt - s^2 = 3 - \dfrac{3}{4} = \dfrac{9}{4} > 0$

Also $r < 0$

$\therefore$ $f(x, y)$ has a maximum value at $\left(\dfrac{\pi}{3}, \dfrac{\pi}{3}\right)$.

$$\text{Maximum value} = f\left(\frac{\pi}{3}, \frac{\pi}{3}\right) = \sin\frac{\pi}{3} + \sin\frac{\pi}{3} + \sin\frac{2\pi}{3} = \frac{\sqrt{3}}{2} + \frac{\sqrt{3}}{2} + \frac{\sqrt{3}}{2} = \frac{3\sqrt{3}}{2}.$$

Example 5. *A rectangular box, open at the top, is to have a given capacity. Find the dimensions of the box requiring least material for its construction.* (G.B.T.U. 2010)

Sol. Let x, y and z be the length, breadth and height respectively. Let V be the given capacity and S, the surface.

(V is given $\Rightarrow$ V is constant)

$$V = xyz \quad \text{or} \quad z = \frac{V}{xy} \qquad \text{...(1)}$$

$$S = xy + 2xz + 2yz = xy + \frac{2V}{y} + \frac{2V}{x} = f(x, y) \qquad \text{[Using (1)]}$$

$$f_x = y - \frac{2V}{x^2}, \; f_y = x - \frac{2V}{y^2}$$

$$r = f_{xx} = \frac{4V}{x^3}, \; s = f_{xy} = 1, \; t = f_{yy} = \frac{4V}{y^3}$$

Now $\qquad f_x = 0 \quad \text{and} \quad f_y = 0$

$$\Rightarrow \qquad y - \frac{2V}{x^2} = 0 \qquad \text{...(1)} \qquad x - \frac{2V}{y^2} = 0 \qquad \text{...(2)}$$

From (1), $\qquad y = \dfrac{2V}{x^2}$

$$\therefore \text{ From (2), } x - 2V \cdot \frac{x^4}{4V^2} = 0 \text{ or } x\left(1 - \frac{x^3}{2V}\right) = 0 \text{ or } x = (2V)^{1/3} \qquad | \because \; x \neq 0$$

and $\qquad y = \dfrac{2V}{x^2} = \dfrac{2V}{(2V)^{3/2}} = (2V)^{1/3}$

$\therefore$ $x = y = (2V)^{1/3}$ is a stationary point.

At this point, $\qquad r = \dfrac{4V}{2V} = 2 > 0, \; s = 1, \; t = \dfrac{4V}{2V} = 2$

so that $\qquad rt - s^2 = 4 - 1 = 3 > 0$ and $r > 0$

$\Rightarrow$ S is minimum when $x = y = (2V)^{1/3}$

Also $\qquad z = \dfrac{V}{xy} = \dfrac{V}{(2V)^{2/3}} = \dfrac{\frac{1}{2}(2V)}{(2V)^{2/3}} = \dfrac{1}{2}(2V)^{1/3} = \dfrac{y}{2}$

Hence S is minimum when $x = y = 2z = (2V)^{1/3}$.

Example 6. *Prove that if the perimeter of a triangle is constant, then its area is maximum when the triangle is equilateral.* (M.D.U. Dec. 2010, Dec. 2011)

Sol. Let a, b, c be the sides of a triangle whose perimeter $2s$ is constant.

Then $\qquad 2s = a + b + c$ or $c = 2s - a - b \qquad$ ...(1)

$$\Delta = \sqrt{s(s-a)(s-b)(s-c)} = \sqrt{s(s-a)(s-b)(a+b-s)} \qquad | \text{ Using (1)}$$

Let $\qquad z = \Delta^2 = s(s-a)(s-b)(a+b-s) = f(a, b)$

$$f_a = s(s-b)\frac{\partial}{\partial a}[(s-a)(a+b-s)]$$

$$= s(s-b)[-(a+b-s) + (s-a)] = s(s-b)(2s - 2a - b)$$

$$f_b = s(s-a)\frac{\partial}{\partial b}[(s-b)(a+b-s)]$$
$$= s(s-a)[-(a+b-s)+(s-b)] = s(s-a)(2s-a-2b)$$
$$r = f_{aa} = -2s(s-b),\; s = f_{ab} = s[-(2s-a-2b)-(s-a)] = s(2a+2b-3s)$$
$$t = f_{bb} = -2s(s-a)$$

Now $\quad f_a = 0$ and $f_b = 0$

$\Rightarrow \quad s(s-b)(2s-2a-b) = 0$ and $s(s-a)(2s-a-2b) = 0$

$\Rightarrow \quad (s-b)(2s-2a-b) = 0$ ...(1)

$(s-a)(2s-a-2b) = 0$ ...(2) $\quad | \because \; s \neq 0$

From (1), $\quad s = b$ or $2s = 2a + b$

When $s = b$, from (2), $\quad (b-a)(-a) = 0$ or $b = a$ $\quad | \because \; a \neq 0$

when $2s = 2a + b$, from (2), $\quad \frac{b}{2}(a-b) = 0$ or $a = b$ $\quad | \because \; b \neq 0$

If we express z as a function of b and c, we similarly get $b = c$

$$\therefore \quad a = b = c = \frac{2s}{3}$$
$$r = -2s\left(\frac{s}{3}\right) = -\frac{2s^2}{3} < 0$$
$$\text{'}s\text{'} = s\left(\frac{4s}{3}+\frac{4s}{3}-3s\right) = s\left(-\frac{s}{3}\right) = -\frac{s^2}{3}$$
$$t = -2s\left(\frac{s}{3}\right) = -\frac{2s^2}{3}$$
$$\text{'}rt - s^2\text{'} = \frac{4s^4}{9} - \frac{s^4}{9} = \frac{s^4}{3} > 0. \text{ Also } r < 0$$

$\therefore \quad z$ and hence Δ is maximum when $a = b = c = \frac{2s}{3}$ *i.e.*, when the triangle is equilateral.

Example 7. *Show that the rectangular solid of maximum volume that can be inscribed in a given sphere is a cube.* (K.U.K. Dec. 2013)

Sol. Let x, y, z be the length, breadth and height of the rectangular solid. If V is the volume of the solid, then

$$\text{V} = xyz \qquad \text{...(1)}$$

Since each diagonal of the solid passes through the centre of the sphere

$\therefore \quad$ each diagonal = diameter of sphere = d (say)

$$\Rightarrow \quad \sqrt{x^2+y^2+z^2} = d \text{ or } x^2+y^2+z^2 = d^2 \text{ or } z = \sqrt{d^2-x^2-y^2}$$

$\therefore \quad$ From (1), $\quad \text{V} = xy\sqrt{d^2-x^2-y^2}$

or
$$\text{V}^2 = x^2y^2(d^2-x^2-y^2) = d^2x^2y^2 - x^4y^2 - x^2y^4 = f(x, y)$$
$$f_x = 2d^2xy^2 - 4x^3y^2 - 2xy^4 = 2xy^2(d^2-2x^2-y^2)$$
$$f_y = 2d^2x^2y - 2x^4y - 4x^2y^3 = 2x^2y(d^2-x^2-2y^2)$$
$$r = f_{xx} = 2d^2y^2 - 12x^2y^2 - 2y^4,\; s = f_{xy} = 4d^2xy - 8x^3y - 8xy^3$$
$$t = f_{yy} = 2d^2x^2 - 2x^4 - 12x^2y^2$$

Now $f_x = 0$ and $f_y = 0$

$\Rightarrow$ $d^2 - 2x^2 - y^2 = 0$ ...(1) $[\because\ x \neq 0, y \neq 0]$

$d^2 - x^2 - 2y^2 = 0$ ...(2)

Subtracting (2) from (1), $-x^2 + y^2 = 0 \Rightarrow y = x$

$\therefore$ From (1), $x = y = \dfrac{d}{\sqrt{3}}$

$$z = \sqrt{d^2 - \frac{d^2}{3} - \frac{d^2}{3}} = \frac{d}{\sqrt{3}}$$

At $x = y = \dfrac{d}{\sqrt{3}}$,

$$r = 2d^2 \cdot \frac{d^2}{3} - 12 \cdot \frac{d^2}{3} \cdot \frac{d^2}{3} - 2 \cdot \frac{d^4}{9} = -\frac{8d^4}{9} < 0$$

$$s = 4d^2 \cdot \frac{d^2}{3} - 8 \cdot \frac{d^3}{3\sqrt{3}} \cdot \frac{d}{\sqrt{3}} - 8 \cdot \frac{d}{\sqrt{3}} \cdot \frac{d^3}{3\sqrt{3}} = -\frac{4d^4}{9}$$

$$t = 2d^2 \cdot \frac{d^2}{3} - 2 \cdot \frac{d^4}{9} - 12 \cdot \frac{d^2}{3} \cdot \frac{d^2}{3} = -\frac{8d^4}{9}$$

$$rt - s^2 = \frac{64d^8}{81} - \frac{16d^8}{81} = \frac{16d^8}{27} > 0, \text{ Also } r < 0.$$

$\therefore$ V^2 and hence, V is maximum when $x = y = z$ *i.e.*, when the rectangular solid is a cube.

Note. Try it by using Lagrange's method. Take the sphere as $x^2 + y^2 + z^2 = a^2$. Let the edges of the rectangular solid be $2x$, $2y$, $2z$ which are parallel to the axes. Then its volume $V = 8xyz$ and $F(x, y, z) = 8xyz + \lambda(x^2 + y^2 + z^2 - a^2)$.

For stationary values, $dF = 0$

$\Rightarrow$ $8yz + \lambda(2x) = 0$ ...(1)

$8zx + \lambda(2y) = 0$ ...(2)

$8xy + \lambda(2z) = 0$ ...(3)

Multiplying (1) by x, (2) by y and subtracting, we get

$2\lambda(x^2 - y^2) = 0 \Rightarrow x = y$

Similarly, from (2) and (3), we get $y = z$

Combining $x = y = z$

Thus the rectangular solid is a cube.

☞Read this note after doing Art. 10.15.

Example 8. *Find the volume of the largest rectangular parallelopiped that can be inscribed in the ellipsoid,* $\dfrac{x^2}{a^2} + \dfrac{y^2}{b^2} + \dfrac{z^2}{c^2} = 1$. (M.D.U. May 2008; U.K.T.U. 2010)

Sol. Let (x, y, z) be a vertex of the parallelopiped, then it lies on the ellipsoid

$$\frac{x^2}{a^2} + \frac{y^2}{b^2} + \frac{z^2}{c^2} = 1 \qquad ...(1)$$

Also its dimensions are $2x$, $2y$, $2z$ so that the volume V is given by $V = 2x \cdot 2y \cdot 2z = 8xyz$

$$\Rightarrow \quad V^2 = 64x^2y^2z^2 = 64x^2y^2c^2\left(1 - \frac{x^2}{a^2} - \frac{y^2}{b^2}\right) = 64c^2\left(x^2y^2 - \frac{x^4y^2}{a^2} - \frac{x^2y^4}{b^2}\right) = f(x, y)$$

$$f_x = 64c^2\left(2xy^2 - \frac{4x^3y^2}{a^2} - \frac{2xy^4}{b^2}\right); \qquad f_y = 64c^2\left(2x^2y - \frac{2x^4y}{a^2} - \frac{4x^2y^3}{b^2}\right)$$

$$r = f_{xx} = 64c^2\left(2y^2 - \frac{12x^2y^2}{a^2} - \frac{2y^4}{b^2}\right); \qquad s = f_{xy} = 64c^2\left(4xy - \frac{8x^3y}{a^2} - \frac{8xy^3}{b^2}\right)$$

$$t = f_{yy} = 64c^2\left(2x^2 - \frac{2x^4}{a^2} - \frac{12x^2y^2}{b^2}\right)$$

Now $f_x = 0$ and $f_y = 0$

$$\Rightarrow \quad 128c^2xy^2\left(1 - \frac{2x^2}{a^2} - \frac{y^2}{b^2}\right) = 0 \quad \text{and} \quad 128c^2x^2y\left(1 - \frac{x^2}{a^2} - \frac{2y^2}{b^2}\right) = 0$$

$$\Rightarrow \quad 1 - \frac{2x^2}{a^2} - \frac{y^2}{b^2} = 0 \qquad \text{...(1)} \qquad [\because \; x \neq 0, y \neq 0]$$

$$1 - \frac{x^2}{a^2} - \frac{2y^2}{b^2} = 0 \qquad \text{...(2)}$$

Subtracting (2) from (1), $-\dfrac{x^2}{a^2} + \dfrac{y^2}{b^2} = 0$ or $y = \dfrac{bx}{a}$

$\therefore$ From (1), $\quad 1 - \dfrac{2x^2}{a^2} - \dfrac{x^2}{a^2} = 0$ or $x^2 = \dfrac{a^2}{3}$ or $x = \dfrac{a}{\sqrt{3}}$

$$\therefore \quad y = \frac{b}{a} \cdot \frac{a}{\sqrt{3}} = \frac{b}{\sqrt{3}}$$

and

$$z^2 = c^2\left(1 - \frac{x^2}{a^2} - \frac{y^2}{b^2}\right) = c^2\left(1 - \frac{1}{3} - \frac{1}{3}\right) = \frac{c^2}{3}$$

$$\therefore \quad z = \frac{c}{\sqrt{3}}$$

Thus $x = \dfrac{a}{\sqrt{3}}, y = \dfrac{b}{\sqrt{3}}$ is a stationary point.

At this point, $r = 64c^2\left[\dfrac{2b^2}{3} - \dfrac{12}{a^2} \cdot \dfrac{a^2}{3} \cdot \dfrac{b^2}{3} - \dfrac{2}{b^2} \cdot \dfrac{b^4}{9}\right] = -\dfrac{512}{9}\, b^2c^2 < 0$

$$s = 64c^2\left[4 \cdot \frac{a}{\sqrt{3}} \cdot \frac{b}{\sqrt{3}} - \frac{8}{a^2} \cdot \frac{a^3}{3\sqrt{3}} \cdot \frac{b}{\sqrt{3}} - \frac{8}{b^2} \cdot \frac{a}{\sqrt{3}} \cdot \frac{b^3}{3\sqrt{3}}\right] = -\frac{256}{9}\, abc^2$$

$$t = 64c^2\left[2 \cdot \frac{a^2}{3} - \frac{2}{a^2} \cdot \frac{a^4}{9} - \frac{12}{b^2} \cdot \frac{a^2}{3} \cdot \frac{b^2}{3}\right] = -\frac{512}{9}\, a^2c^2$$

$$rt - s^2 = \left(\frac{512}{9}\right)^2 a^2b^2c^4 - \left(\frac{256}{9}\right)^2 a^2b^2c^4 = \left(\frac{256}{9}\right)^2 a^2b^2c^4\,(4-1) > 0.$$

Also $\quad r < 0$

$\therefore$ V^2 and hence, V is maximum.

when $\quad x = \dfrac{a}{\sqrt{3}}, y = \dfrac{b}{\sqrt{3}}, z = \dfrac{c}{\sqrt{3}}$

$$\text{Maximum volume} = 8 \cdot \frac{a}{\sqrt{3}} \cdot \frac{b}{\sqrt{3}} \cdot \frac{c}{\sqrt{3}} = \frac{8abc}{3\sqrt{3}}.$$

Note. Try it by using Lagrange's Method. Take the edges of the parallelopiped as $2x$, $2y$, $2z$ which are parallel to the axes. Then its volume $V = 8xyz$ and

$$F(x, y, z) = 8xyz + \lambda\left(\frac{x^2}{a^2} + \frac{y^2}{b^2} + \frac{z^2}{c^2} - 1\right)$$

For stationary values, $dF = 0$

$$\Rightarrow \quad 8yz + \lambda\left(\frac{2x}{a^2}\right) = 0 \quad ...(1)$$

$$8zx + \lambda\left(\frac{2y}{b^2}\right) = 0 \quad ...(2)$$

$$8xy + \lambda\left(\frac{2z}{c^2}\right) = 0 \quad ...(3)$$

Multiplying (1) by x, (2) by y and subtracting, we get

$$2\lambda\left(\frac{x^2}{a^2} - \frac{y^2}{b^2}\right) = 0 \quad \Rightarrow \quad \frac{x}{a} = \frac{y}{b}$$

Similarly, from (2) and (3), we get $\frac{y}{b} = \frac{z}{c}$

Combining, $\frac{x}{a} = \frac{y}{b} = \frac{z}{c} = \frac{1}{\sqrt{3}}$, since $\frac{x^2}{a^2} + \frac{y^2}{b^2} + \frac{z^2}{c^2} = 1$

☞Read this note after doing Art. 10.15.

Example 9. *In a plane triangle ABC, find the maximum value of cos A cos B cos C.*

(G.B.T.U. 2010; Kerala 2008, 2010)

Sol. $\cos A \cos B \cos C = \cos A \cos B \cos[\pi - (A + B)]$ $\quad | \because A + B + C = \pi$

$$= -\cos A \cos B \cos(A + B) = f(A, B)$$

$$\frac{\partial f}{\partial A} = -\cos B[-\sin A \cos(A + B) - \cos A \sin(A + B)]$$

$$= \cos B \sin[A + (A + B)] = \cos B \sin(2A + B)$$

$$\frac{\partial f}{\partial B} = -\cos A[-\sin B \cos(A + B) - \cos B \sin(A + B)]$$

$$= \cos A \sin[B + (A + B)] = \cos A \sin(A + 2B)$$

$$r = 2 \cos B \cos(2A + B)$$

$$s = -\sin A \sin(A + 2B) + \cos A \cos(A + 2B)$$

$$= \cos[A + (A + 2B)] = \cos(2A + 2B)$$

$$t = 2 \cos A \cos(A + 2B)$$

Now $\frac{\partial f}{\partial A} = 0$ and $\frac{\partial f}{\partial B} = 0$

$$\Rightarrow \quad \cos B \sin(2A + B) = 0 \quad ...(1)$$

$$\cos A \sin(A + 2B) = 0 \quad ...(2)$$

If $\cos B = 0$, then $B = \frac{\pi}{2}$

From (2), $\cos A \sin(A + \pi) = 0$ or $\cos A(-\sin A) = 0$

$\Rightarrow$ either $\cos A = 0$ *i.e.*, $A = \frac{\pi}{2}$ which is not possible $\quad (\because A + B + C = \pi \Rightarrow C = 0)$

or $\quad \sin A = 0$ *i.e.*, $A = 0$ or π which is not possible.

$\therefore \quad \cos B \neq 0$. Similarly $\cos A \neq 0$

$\therefore$ From (1), $\quad \sin(2A + B) = 0$ or $2A + B = \pi$

From (2), $\quad \sin(A + 2B) = 0$ or $A + 2B = \pi$

Solving these equations, $A = B = \frac{\pi}{3}$

At $A = B = \dfrac{\pi}{3}$

$$r = 2\cos\frac{\pi}{3}\cos\pi = -1, \quad s = \cos\frac{4\pi}{3} = \cos\left(\pi + \frac{\pi}{3}\right) = -\cos\pi = -\frac{1}{2}$$

$$t = 2\cos\frac{\pi}{3}\cos\pi = -1$$

$$rt - s^2 = 1 - \frac{1}{4} = \frac{3}{4} > 0$$

Also $\qquad r = -1 < 0$

$\Rightarrow$ $f(A, B)$ is maximum at $A = B = \dfrac{\pi}{3}$

Maximum value $= f\left(\dfrac{\pi}{3}, \dfrac{\pi}{3}\right) = -\cos\dfrac{\pi}{3}\cos\dfrac{\pi}{3}\cos\dfrac{2\pi}{3} = -\left(\dfrac{1}{2}\right)\left(\dfrac{1}{2}\right)\left(-\dfrac{1}{2}\right) = \dfrac{1}{8}$.

Example 10. *Show that of all triangles inscribed in a circle, the one with maximum area is equilateral.*

Sol. Let ABC be a triangle inscribed in a circle of radius R. Let x, y, z be the angles of ΔABC so that $0 < x < \pi$, $0 < y < \pi$, $0 < z < \pi$ and $x + y + z = \pi$. If S denotes the area of ΔABC, then $S = \dfrac{1}{2} ab \sin z$, where a, b denote the lengths of the sides containing the angle z.

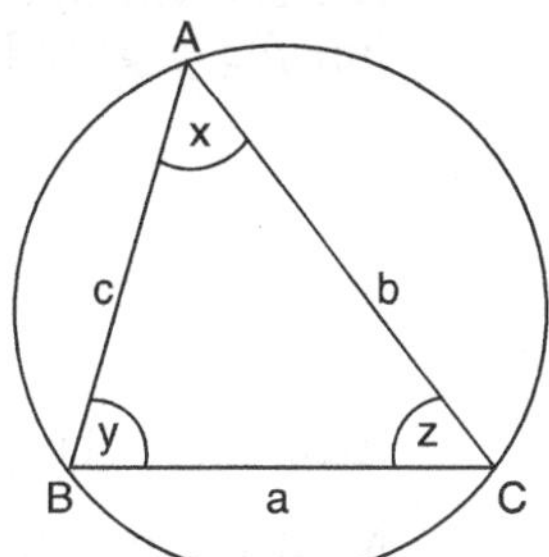

Since $\qquad R = \dfrac{a}{2\sin x} = \dfrac{b}{2\sin y} = \dfrac{c}{2\sin z}$

$\therefore \qquad a = 2R\sin x,\ b = 2R\sin y$

and $\qquad S = \dfrac{1}{2}\cdot 2R\sin x \cdot 2R\sin y \cdot \sin[\pi - (x + y)]$

$$= 2R^2 \sin x \sin y \sin(x + y)$$

We have to maximize S, or equivalently $\dfrac{S}{2R^2}$.

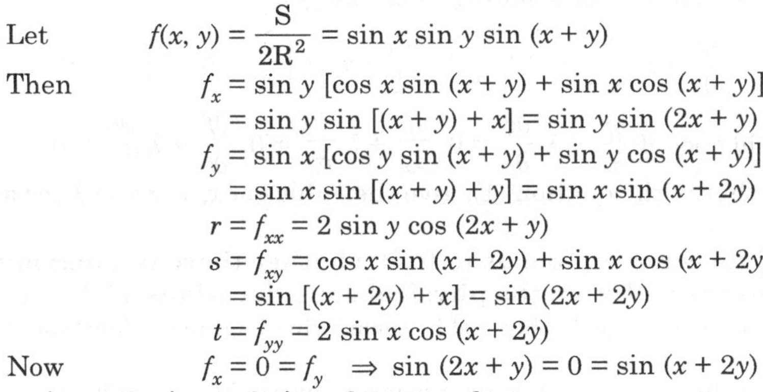

Let $\qquad f(x, y) = \dfrac{S}{2R^2} = \sin x \sin y \sin(x + y)$

Then
$$f_x = \sin y[\cos x \sin(x + y) + \sin x \cos(x + y)]$$
$$= \sin y \sin[(x + y) + x] = \sin y \sin(2x + y)$$
$$f_y = \sin x[\cos y \sin(x + y) + \sin y \cos(x + y)]$$
$$= \sin x \sin[(x + y) + y] = \sin x \sin(x + 2y)$$
$$r = f_{xx} = 2\sin y \cos(2x + y)$$
$$s = f_{xy} = \cos x \sin(x + 2y) + \sin x \cos(x + 2y)$$
$$= \sin[(x + 2y) + x] = \sin(2x + 2y)$$
$$t = f_{yy} = 2\sin x \cos(x + 2y)$$

Now $\qquad f_x = 0 = f_y \Rightarrow \sin(2x + y) = 0 = \sin(x + 2y)$

because $\sin x \neq 0$, $\sin y \neq 0$ since $0 < x < \pi$, $0 < y < \pi$.

$\Rightarrow \qquad 2x + y = \pi$ and $x + 2y = \pi$

so that $\qquad x = y = \dfrac{\pi}{3} \qquad \left(\text{and hence } z = \pi - x - y = \dfrac{\pi}{3}\right)$

At $\qquad x = \dfrac{\pi}{3},\ y = \dfrac{\pi}{3}$

$$r = 2\sin\frac{\pi}{3}\cos\pi = 2 \times \frac{\sqrt{3}}{2} \times (-1) = -\sqrt{3}$$

$$s = \sin \frac{4\pi}{3} = \sin\left(\pi + \frac{\pi}{3}\right) = -\sin\frac{\pi}{3} = -\frac{\sqrt{3}}{2}$$

$$t = 2 \sin \frac{\pi}{3} \cos \pi = 2 \times \frac{\sqrt{3}}{2} \times (-1) = -\sqrt{3}$$

Since $rt - s^2 = 3 - \frac{3}{4} = \frac{9}{4} > 0$ and $r < 0$, therefore, $f(x, y)$ has a maximum at $\left(\frac{\pi}{3}, \frac{\pi}{3}\right)$.

$\Rightarrow$ S is maximum when $x = y = z = \frac{\pi}{3}$

$\Rightarrow$ area of inscribed triangle is maximum when the triangle is equilateral.

10.15. LAGRANGE'S METHOD OF UNDETERMINED MULTIPLIERS

To find the maximum or minimum values of a function of three (or more) variables, when the variables are not independent but are connected by some given relation, we try to convert the given function to the one, having least number of independent variables with the help of the given relation. [See examples with Art. 10.13 and 10.14]

When this procedure is not practicable, we use Lagrange's method.

Let $f(x, y, z)$ be a function of x, y, z which is to be examined for maximum or minimum value.

Let the variables x, y, z be connected by the relation $\phi(x, y, z) = 0$ ...(1)

For $f(x, y, z)$ to have a maximum or minimum value, the necessary condition is

$$\frac{\partial f}{\partial x} = 0, \frac{\partial f}{\partial y} = 0, \frac{\partial f}{\partial z} = 0$$

$$\therefore \quad \frac{\partial f}{\partial x} dx + \frac{\partial f}{\partial y} dy + \frac{\partial f}{\partial z} dz = 0 \quad \text{...(2)}$$

Also, from (1), taking differentials, we get $\frac{\partial \phi}{\partial x} dx + \frac{\partial \phi}{\partial y} dy + \frac{\partial \phi}{\partial z} dz = 0$ ...(3)

Multiplying (3) by a parameter λ and adding to (2), we get

$$\left(\frac{\partial f}{\partial x} + \lambda \frac{\partial \phi}{\partial x}\right) dx + \left(\frac{\partial f}{\partial y} + \lambda \frac{\partial \phi}{\partial y}\right) dy + \left(\frac{\partial f}{\partial z} + \lambda \frac{\partial \phi}{\partial z}\right) dz = 0$$

This equation will hold good if $\frac{\partial f}{\partial x} + \lambda \frac{\partial \phi}{\partial s} = 0, \frac{\partial f}{\partial y} + \lambda \frac{\partial \phi}{\partial y} = 0, \frac{\partial f}{\partial z} + \lambda \frac{\partial \phi}{\partial z} = 0$

These equations together with equation (1), give the values of x, y, z and λ for a maximum or minimum.

Lagrange's method does not enable us to find whether there is a maximum or minimum. This fact is determined from the physical considerations of the problem.

Note. The above equations can be easily obtained by considering Lagrange's function

$$F(x, y, z) = f(x, y, z) + \lambda\phi(x, y, z)$$

and considering the stationary values of $F(x, y, z)$. For stationary values of $F(x, y, z)$, $dF = 0$

$$\Rightarrow \quad \left(\frac{\partial f}{\partial x} + \lambda \frac{\partial \phi}{\partial x}\right) dx + \left(\frac{\partial f}{\partial y} + \lambda \frac{\partial \phi}{\partial y}\right) dy + \left(\frac{\partial f}{\partial z} + \lambda \frac{\partial \phi}{\partial z}\right) dz = 0$$

$$\Rightarrow \quad \frac{\partial f}{\partial x} + \lambda \frac{\partial \phi}{\partial x} = 0, \frac{\partial f}{\partial y} + \lambda \frac{\partial \phi}{\partial y} = 0, \frac{\partial f}{\partial z} + \lambda \frac{\partial \phi}{\partial z} = 0.$$

Example 11. *Find the maximum and minimum distances of the point (3, 4, 12) from the sphere*

$$x^2 + y^2 + z^2 = 1. \quad \text{(U.K.T.U. 2010)}$$

Sol. Let (x, y, z) be any point on the sphere.

Distance of the point A(3, 4, 12) from (x, y, z) is given by $\sqrt{(x-3)^2+(y-4)^2+(z-12)^2}$

If the distance is maximum or minimum, so will be the square of the distance.

Let $f(x, y, z) = (x-3)^2 + (y-4)^2 + (z-12)^2$ ...(1)

subject to the condition that $\phi(x, y, z) = x^2 + y^2 + z^2 - 1 = 0$ ...(2)

Consider Lagrange's function $F(x, y, z) = f(x, y, z) + \lambda\phi(x, y, z)$

$$= (x-3)^2 + (y-4)^2 + (z-12)^2 + \lambda(x^2+y^2+z^2-1)$$

For stationary values, $dF = 0$.

$$\Rightarrow \quad [2(x-3)+2\lambda x]dx + [2(y-4)+2\lambda y]dy + [2(z-12)+2\lambda z]dz = 0$$

$$\Rightarrow \quad 2(x-3) + 2\lambda x = 0 \qquad ...(3)$$

$$2(y-4) + 2\lambda y = 0 \qquad ...(4)$$

$$2(z-12) + 2\lambda z = 0 \qquad ...(5)$$

Multiplying (3) by x, (4) by y, (5) by z and adding, we get

$$2(x^2+y^2+z^2) - 6x - 8y - 24z + 2\lambda(x^2+y^2+z^2) = 0$$

or $$2 - 6x - 8y - 24z + 2\lambda = 0 \qquad | \text{ Using (2)}$$

or $$3x + 4y + 12z = 1 + \lambda \qquad ...(6)$$

From (3), (4) and (5), $x = \dfrac{3}{1+\lambda}, y = \dfrac{4}{1+\lambda}, z = \dfrac{12}{1+\lambda}$.

Putting these values of x, y, z in (6), we have

$$\frac{9}{1+\lambda} + \frac{16}{1+\lambda} + \frac{144}{1+\lambda} = 1 + \lambda \quad \text{or} \quad (1+\lambda)^2 = 169 \text{ or } 1 + \lambda = \pm 13$$

$$\therefore \quad \lambda = 12 \quad \text{or} \quad -14$$

When $\lambda = 12$, $\quad x = \dfrac{3}{13}, \quad y = \dfrac{4}{13}, \quad z = \dfrac{12}{13}$

When $\lambda = -14$, $\quad x = -\dfrac{3}{13}, \quad y = -\dfrac{4}{13}, \quad z = -\dfrac{12}{13}$

Thus, we get two points $P\left(\dfrac{3}{13}, \dfrac{4}{13}, \dfrac{12}{13}\right)$ and $Q\left(-\dfrac{3}{13}, -\dfrac{4}{13}, -\dfrac{12}{13}\right)$

on the sphere which are at a *maximum or minimum* distance from the given point A.

Now $$AP = \sqrt{\left(3-\frac{3}{13}\right)^2 + \left(4-\frac{4}{13}\right)^2 + \left(12-\frac{12}{13}\right)^2} = 12$$

$$AQ = \sqrt{\left(3+\frac{3}{13}\right)^2 + \left(4+\frac{4}{13}\right)^2 + \left(12+\frac{12}{13}\right)^2} = 14$$

$\therefore \quad P\left(\dfrac{3}{13}, \dfrac{4}{13}, \dfrac{12}{13}\right)$ is at a minimum distance from A and the minimum distance = 12.

$Q\left(-\dfrac{3}{13}, -\dfrac{4}{13}, -\dfrac{12}{13}\right)$ is at a maximum distance from A and the maximum distance = 14.

Example 12. *Find the minimum value of* $x^2 + y^2 + z^2$, *given that* $ax + by + cz = p$.

(U.K.T.U. 2012; U.P.T.U. 2009)

Sol. Let $f(x, y, z) = x^2 + y^2 + z^2$ ...(1)

where $\phi(x, y, z) = ax + by + cz - p = 0$ ...(2)

Consider Lagrange's function, $F(x, y, z) = (x^2 + y^2 + z^2) + \lambda(ax + by + cz - p)$

For stationary values, $dF = 0$

$$\Rightarrow \quad (2x + \lambda a)dx + (2y + \lambda b)dy + (2z + \lambda c)dz = 0$$

$\Rightarrow \quad 2x + \lambda a = 0 \quad ...(3) \qquad 2y + \lambda b = 0 \quad ...(4)$

$2z + \lambda c = 0 \quad ...(5)$

Multiplying (3) by x, (4) by y, (5) by z and adding, we get

$$2(x^2 + y^2 + z^2) + \lambda(ax + by + cz) = 0$$

or $$2f + \lambda p = 0$$ | Using (1) and (2)

$$\therefore \quad \lambda = -\frac{2f}{p}$$

From (3), (4) and (5), $x = \frac{af}{p}, \; y = \frac{bf}{p}, \; z = \frac{cf}{p}$

$\therefore$ From (1), $f = \frac{(a^2 + b^2 + c^2) f^2}{p^2}$ or $f = \frac{p^2}{a^2 + b^2 + c^2}$.

This is the **maximum or minimum** value of u. Now u is the square of the distance of any point P(x, y, z) on the plane (2) from the origin. Also, the length of perpendicular from O on the plane is $\frac{p}{\sqrt{a^2 + b^2 + c^2}}$.

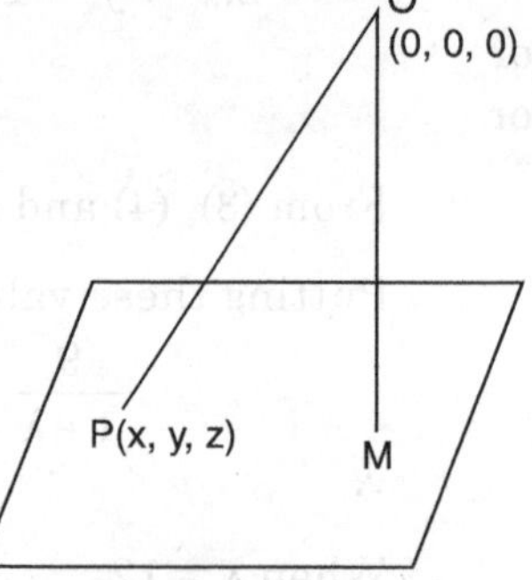

Clearly, OP is least when P coincides with M, the foot of the perpendicular from O on the plane.

Hence the minimum value of $f = \frac{p^2}{a^2 + b^2 + c^2}$.

Note. Try it by eliminating z so that

$$f(x, y) = x^2 + y^2 + \left(\frac{p - ax - by}{c}\right)^2$$

Example 13. *Find the minimum value of $ax + by + cz$ subject to the condition $\frac{a^2}{x} + \frac{b^2}{y} + \frac{c^2}{z} = 1$.*

Sol. Let $$f(x, y, z) = ax + by + cz \quad ...(1)$$

where $$\phi(x, y, z) = \frac{a^2}{x} + \frac{b^2}{y} + \frac{c^2}{z} - 1 = 0 \quad ...(2)$$

Consider Lagrange's function

$$F(x, y, z) = (ax + by + cz) + \lambda\left(\frac{a^2}{x} + \frac{b^2}{y} + \frac{c^2}{z} - 1\right)$$

For stationary values, $dF = 0$

$$\Rightarrow \quad \left(a - \frac{\lambda a^2}{x^2}\right) dx + \left(b - \frac{\lambda b^2}{y^2}\right) dy + \left(c - \frac{\lambda c^2}{z^2}\right) dz = 0$$

$$\Rightarrow \quad a - \frac{\lambda a^2}{x^2} = 0 \quad ...(3)$$

$$b - \frac{\lambda b^2}{y^2} = 0 \quad ...(4)$$

$$c - \frac{\lambda c^2}{z^2} = 0 \quad ...(5)$$

From (3), (4) and (5)

$$x = \sqrt{\lambda a}\,,\; y = \sqrt{\lambda b}\,,\; z = \sqrt{\lambda c}$$

Putting these values in (2), we have

$$\frac{a^2}{\sqrt{\lambda a}} + \frac{b^2}{\sqrt{\lambda b}} + \frac{c^2}{\sqrt{\lambda c}} = 1$$

$$\Rightarrow \quad \sqrt{\lambda} = a^{3/2} + b^{3/2} + c^{3/2} \quad \Rightarrow \quad \lambda = (a^{3/2} + b^{3/2} + c^{3/2})^2$$

$$\therefore \quad x = a^{1/2}\,(a^{3/2} + b^{3/2} + c^{3/2}), \quad y = b^{1/2}\,(a^{3/2} + b^{3/2} + c^{3/2}),$$

$$z = c^{1/2}\,(a^{3/2} + b^{3/2} + c^{3/2})$$

$\Rightarrow$ the stationary point is

$$(a^{1/2}\,(a^{3/2} + b^{3/2} + c^{3/2}),\quad b^{1/2}\,(a^{3/2} + b^{3/2} + c^{3/2}),\quad c^{1/2}\,(a^{3/2} + b^{3/2} + c^{3/2}))$$

and the minimum value of f is

$$a^{3/2}\,(a^{3/2} + b^{3/2} + c^{3/2}) + b^{3/2}\,(a^{3/2} + b^{3/2} + c^{3/2}) + c^{3/2}\,(a^{3/2} + b^{3/2} + c^{3/2})$$
$$= (a^{3/2} + b^{3/2} + c^{3/2})^2.$$

Example 14. *Given $x + y + z = a$, find the maximum value of $x^m y^n z^p$.*

(M.D.U. Dec. 2010)

Sol. Let $\quad f(x, y, z) = x^m y^n z^p \quad$ ...(1)

where $\quad \phi(x, y, z) = x + y + z - a = 0 \quad$ ...(2)

Consider Lagrange's function

$$F(x, y, z) = x^m y^n z^p + \lambda(x + y + z - a)$$

For stationary values, $dF = 0$

$$\Rightarrow \quad (m\, x^{m-1}\, y^n\, z^p + \lambda)dx + (n\, x^m y^{n-1}\, z^p + \lambda)dy + (p\, x^m y^n\, z^{p-1} + \lambda)dz = 0$$

$$\Rightarrow \quad m\, x^{m-1} y^n\, z^p + \lambda = 0 \qquad ...(3)$$

$$n\, x^m y^{n-1}\, z^p + \lambda = 0 \qquad ...(4)$$

$$p\, x^m y^n\, z^{p-1} + \lambda = 0 \qquad ...(5)$$

Multiplying (3), (4), (5) by x, y, z respectively and adding

$$(m + n + p)x^m y^n z^p + \lambda(x + y + z) = 0$$

$$\Rightarrow \quad (m + n + p)f + \lambda a = 0 \qquad \text{[by (2)]}$$

$$\Rightarrow \quad \lambda = -\frac{(m + n + p)f}{a}$$

From (3), multiplying by x, we have

$$mf + \lambda x = 0 \quad \Rightarrow \quad x = -\frac{mf}{\lambda}$$

$$\therefore \quad x = \frac{ma}{m + n + p}$$

Similarly, $\quad y = \dfrac{na}{m + n + p}, \quad z = \dfrac{pa}{m + n + p}$

$\therefore$ Maximum value of f

$$= \left(\frac{ma}{m + n + p}\right)^m \left(\frac{na}{m + n + p}\right)^n \left(\frac{pa}{m + n + p}\right)^p$$

$$= \frac{m^m n^n p^p\, a^{m+n+p}}{(m + n + p)^{m+n+p}}.$$

Example 15. *Find the dimensions of the rectangular box, open at the top, of maximum capacity whose surface is 432 sq cm.* (M.D.U. Dec. 2008)

Sol. Let x, y, z be the edges of the box.

Capacity of box $= xyz$, and

Surface area $= xy + 2yz + 2zx = 432$

Let $$f(x, y, z) = xyz \quad \text{...(1)}$$

where $$\phi(x, y, z) = xy + 2yz + 2zx - 432 = 0 \quad \text{...(2)}$$

Consider Lagrange's function

$$F(x, y, z) = xyz + \lambda(xy + 2yz + 2zx - 432)$$

For stationary values, $dF = 0$

$$\Rightarrow \quad [yz + \lambda(y + 2z)]dx + [xz + \lambda(x + 2z)]dy + [xy + \lambda(2y + 2x)]dz = 0$$

$$\Rightarrow \quad yz + \lambda(y + 2z) = 0 \quad \text{...(3)}$$

$$xz + \lambda(x + 2z) = 0 \quad \text{...(4)}$$

$$xy + \lambda(2y + 2x) = 0 \quad \text{...(5)}$$

Multiplying (3) by x, (4) by y and subtracting, we get

$$2z\lambda(x - y) = 0 \Rightarrow x = y \quad (\because \lambda \neq 0, z \neq 0)$$

Multiplying (4) by y, (5) by z and subtracting, we get

$$x\lambda(y - 2z) = 0 \Rightarrow y = 2z$$

$$\therefore \quad x = y = 2z$$

$\therefore$ From (2) $x \cdot x + x \cdot x + x \cdot x = 432$

$$\Rightarrow \quad 3x^2 = 432 \Rightarrow x^2 = 144$$

$$\therefore \quad x = 12 \Rightarrow y = 12, \quad z = 6$$

Hence the dimensions of the box are 12 cm, 12 cm, 6 cm.

Example 16. *Prove that the stationary values of* $u = \dfrac{x^2}{a^4} + \dfrac{y^2}{b^4} + \dfrac{z^2}{c^4}$

where $lx + my + nz = 0$ *and* $\dfrac{x^2}{a^2} + \dfrac{y^2}{b^2} + \dfrac{z^2}{c^2} = 1$ *are the roots of the equation*

$$\frac{l^2a^4}{1 - a^2u} + \frac{m^2b^4}{1 - b^2u} + \frac{n^2c^4}{1 - c^2u} = 0.$$

Sol. Consider Lagrange's function,

$$F(x, y, z) = \left(\frac{x^2}{a^4} + \frac{y^2}{b^4} + \frac{z^2}{c^4}\right) + \lambda\,(lx + my + nz) + \mu\left(\frac{x^2}{a^2} + \frac{y^2}{b^2} + \frac{z^2}{c^2} - 1\right)$$

For stationary values, $dF = 0$

$$\Rightarrow \quad \left(\frac{2x}{a^4} + \lambda l + \frac{2\mu x}{a^2}\right)dx + \left(\frac{2y}{b^4} + \lambda m + \frac{2\mu y}{b^2}\right)dy + \left(\frac{2z}{c^4} + \lambda n + \frac{2\mu z}{c^2}\right)dz = 0$$

$$\Rightarrow \quad \frac{2x}{a^4} + \lambda l + \frac{2\mu x}{a^2} = 0 \quad \text{...(1)}$$

$$\Rightarrow \quad \frac{2y}{b^4} + \lambda m + \frac{2\mu y}{b^2} = 0 \quad \text{...(2)}$$

$$\frac{2z}{c^4} + \lambda n + \frac{2\mu z}{c^2} = 0 \quad \text{...(3)}$$

Multiplying (1), (2), (3) by x, y, z respectively and adding, we get

$$2\left(\frac{x^2}{a^4}+\frac{y^2}{b^4}+\frac{z^2}{c^4}\right)+\lambda\,(lx+my+nz)+2\mu\left(\frac{x^2}{a^2}+\frac{y^2}{b^2}+\frac{z^2}{c^2}\right)=0$$

$\Rightarrow$ $$2u+\lambda(0)+2\mu\,(1)=0 \qquad \text{| from given relations}$$

$\Rightarrow$ $$\mu=-u$$

$\therefore$ Equation (1) becomes $\dfrac{2x}{a^4}+\lambda l-\dfrac{2ux}{a^2}=0$ or $\dfrac{2x}{a^4}(1-a^2u)=-\lambda l$ or $x=-\dfrac{\lambda la^4}{2(1-a^2u)}$

Similarly, $$y=-\frac{\lambda mb^4}{2(1-b^2u)},\; z=-\frac{\lambda nc^4}{2(1-c^2u)}$$

To eliminate λ between them, multiply these values of x, y, z by l, m, n respectively and add. Then

$$lx+my+nz=-\frac{\lambda}{2}\left[\frac{l^2a^4}{1-a^2u}+\frac{m^2b^4}{1-b^2u}+\frac{n^2c^4}{1-c^2u}\right]$$

Since $lx+my+nz=0$, we have $\dfrac{l^2a^4}{1-a^2u}+\dfrac{m^2b^4}{1-b^2u}+\dfrac{n^2c^4}{1-c^2u}=0$

which is a quadratic in u and gives two stationary values of u.

EXERCISE 10.6

1. Examine for extreme values:

   (i) $x^2+y^2+6x+12$ (G.B.T.U. 2012) (ii) x^3+y^3+3xy (iii) $2(x^2-y^2)-x^4+y^4$

   (iv) $x^2y^2-5x^2-8xy-5y^2$ (v) $x^3+3xy^2-15x^2-15y^2+72x$

   (vi) $x^2-xy+y^2+3x-2y+1$ (vii) $x^3+y^3-63(x+y)+12xy$ (U.K.T.U. 2011) (viii) $xy+\dfrac{a^3}{x}+\dfrac{a^3}{y}$

   (ix) $\sin x\sin y\sin(x+y)$ $(0<x, y<\pi)$

   (x) $\cos x+\cos y+\cos(x+y)$ $(0<x, y<\pi)$.
2. Find the stationary points of the function $z=x^3y^2(12-x-y)$ satisfying the condition $x>0, y>0$ and examine their nature.
3. Find the minimum values of $x^2+y^2+z^2$ when

   (i) $x+y+z=3a$ (ii) $xyz=a^3$.
4. Find the minimum value of the function $x+y+z$ subject to the condition $\dfrac{a}{x}+\dfrac{b}{y}+\dfrac{c}{z}=1$.
5. Find the minimum value of the function $x^2+y^2+z^2$ subject to the condition

   (i) $ax+by+cz=a+b+c$ (P.T.U. 2011) (ii) $xy+yz+zx=3a^2$.
6. Find the points on the surface $z^2=xy+1$ nearest to the origin.
7. Given $f(x, y, z)=\dfrac{5xyz}{x+2y+4z}$, find the values of x, y, z for which $f(x, y, z)$ is a maximum, subject to the condition $xyz=8$.
8. A rectangular box, open at the top, is to have a volume of 32 c.c. Find the dimensions of the box requiring least material for its construction. (G.B.T.U. 2011; P.T.U. 2008; U.P.T.U. 2008)
9. The sum of three positive numbers in constant. Prove that their product is maximum when they are equal.

10. Find the greatest and the least values that the function $f(x, y) = xy$ takes on the ellipse $\frac{x^2}{8} + \frac{y^2}{2} = 1$.

11. Test the function $f(x, y) = (x^2 + y^2)\, e^{-(x^2+y^2)}$ for maxima and minima for points not on the circle $x^2 + y^2 = 1$.

12. Find the shortest distance between the lines

$$\frac{x-3}{1} = \frac{y-5}{-2} = \frac{z-7}{1} \text{ and } \frac{x+1}{7} = \frac{y+1}{-6} = \frac{z+1}{1}$$

[**Hint.** Any point P on the first line is $(\lambda + 3, -2\lambda + 5, \lambda + 7)$ and any point Q on the second line is $(7\mu - 1, -6\mu - 1, \mu - 1)$.

$$PQ = \sqrt{(\lambda + 3 - 7\mu + 1)^2 + (-2\lambda + 5 + 6\lambda + 1)^2 + (\lambda + 7 - \mu + 1)^2}$$

$$= \sqrt{6\lambda^2 + 86\mu^2 - 40\lambda\mu + 116}$$

Let $u = (PQ)^2 = 6\lambda^2 + 86\mu^2 - 40\lambda\mu + 116$]

13. Find the minimum distance from the point (1, 2, 0) to the cone $z^2 = x^2 + y^2$.

14. Find a point on the paraboloid $z = x^2 + y^2$ nearest to the point (3, – 6, 4).

15. A tent on a square base of side x, has its sides vertical of height y and the top is a regular pyramid of height h. Find x and y in terms of h, if the canvas required for its construction is to be minimum for the tent to have a given capacity.

16. The temperature T at any point (x, y, z) in space is $T = 400\, xyz^2$. Find the highest temperature on the surface of the unit sphere $x^2 + y^2 + z^2 = 1$.

17. Find the shortest and the longest distances from the point (1, 2, –1) to the sphere $x^2 + y^2 + z^2 = 24$ *(G.B.T.U. 2010, 2013; P.T.U. Dec. 2011)*

18. Find the maximum and minimum distances from the origin to the curve $5x^2 + 6xy + 5y^2 - 8 = 0$. *(M.D.U. May 2008)*

19. Prove that of all the rectangular parallelopipeds of the same volume, the cube has the least surface.

20. Prove that of all the rectangulars parallelopipeds of given surface, cube has the maximum volume.

21. Divide 24 into three parts such that the continued product of the first, square of the second and the cube of the third may be maximum. *(K.U.K. Dec. 2010)*

22. Find the stationary values of $x^2 + y^2 + z^2$ subject to $ax^2 + by^2 + cz^2 = 1$ and $lx + my + nz = 0$.

23. If $u = a^3x^2 + b^3y^2 + c^3z^2$ where $x^{-1} + y^{-1} + z^{-1} = 1$, show that the stationary value of u is given by

$$x = \frac{\Sigma a}{a},\ y = \frac{\Sigma a}{b},\ z = \frac{\Sigma a}{c}.$$ *(P.T.U. Jan. 2009)*

Answers

1. *(i)* Min. value = 3 at (–3, 0) *(ii)* Max. value = 1 at (–1, –1)

(iii) Max. value = 1 at (±1, 0); Min. value = –1 at (0, ±1)

(iv) Max. value = 0 at (0, 0) *(v)* Max. value = 112 at (4, 0); Min. value = 108 at (6, 0)

(vi) Min. value $= -\frac{4}{3}$ at $\left(-\frac{4}{3}, \frac{1}{3}\right)$

(vii) Max. value = 784 at (–7, –7) ; Min. value = –216 at (3, 3)

(viii) Min. value = $3a^2$ at (a, a)

(*ix*) Max. value $= \frac{3\sqrt{3}}{8}$ at $\left(\frac{\pi}{3}, \frac{\pi}{3}\right)$; Min. value $= -\frac{3\sqrt{3}}{8}$ at $\left(\frac{2\pi}{3}, \frac{2\pi}{3}\right)$

(*x*) Min. value $= -\frac{3}{2}$ at $\left(\frac{2\pi}{3}, \frac{2\pi}{3}\right)$

2. Maximum at (6, 4) **3.** (*i*) $3a^2$ (*ii*) $3a^2$

4. $(\sqrt{a}+\sqrt{b}+\sqrt{c})^2$ at $(\sqrt{a}(\sqrt{a}+\sqrt{b}+\sqrt{c}), \sqrt{b}(\sqrt{a}+\sqrt{b}+\sqrt{c}), \sqrt{c}(\sqrt{a}+\sqrt{b}+\sqrt{c}))$

5. (*i*) $\frac{(a+b+c)^2}{a^2+b^2+c^2}$ (*ii*) $3a^2$ at (a, a, a) and $(-a, -a, -a)$

6. (0, 0, ± 1) **7.** (4, 2, 1) **8.** 4 cm × 4 cm × 2 cm

10. 2, – 2 **11.** Minimum value 0 at (0, 0) **12.** $2\sqrt{29}$

13. $\sqrt{\frac{5}{2}}$ **14.** (1, – 2, 5)

15. $x = \sqrt{5}h, y = \frac{1}{2}h$ **16.** 50 **17.** $\sqrt{6}, 3\sqrt{6}$

18. 4, 1 **21.** 4, 8, 12

22. Two stationary values of u are roots of $\frac{l^2}{au-1} + \frac{m^2}{bu-1} + \frac{n^2}{cu-1} = 0$.

10.16. ERRORS AND APPROXIMATIONS

Let $z = f(x, y)$ ...(1)

If δx, δy are small increments in x and y, respectively and δz, the corresponding increment in z, then

$$z + \delta z = f(x + \delta x, y + \delta y) \quad ...(2)$$

Subtracting (1) from (2), we get

$$\delta z = f(x+\delta x, y+\delta y) - f(x, y) = f(x, y) + \delta x \frac{\partial f}{\partial x} + \delta y \frac{\partial f}{\partial y} + \cdots - f(x, y)$$

$$= \delta x \frac{\partial f}{\partial x} + \delta y \frac{\partial f}{\partial y} \text{ (approximately)} \mid \text{Neglecting higher powers of } \delta x, \delta y$$

$$\therefore \quad \delta z = \frac{\partial f}{\partial x} \cdot \delta x + \frac{\partial f}{\partial y} \cdot \delta y \text{ (approximately)}$$

⇒ If δx and δy are small changes (or errors) in x and y respectively, then an approximate change (or error) in z is δz.

Replacing δx, δy, δz by dx, dy, dz respectively, we have $\mathbf{dz = \frac{\partial f}{\partial x} dx + \frac{\partial f}{\partial y} dy}$.

Note 1. If δx is the error in x, then **relative error** $= \frac{\delta x}{x}$

percentage error $= \frac{\delta x}{x} \times 100$.

Note 2. $f'(x)\, dx$ is called the **differential of** $f(x)$.

ILLUSTRATIVE EXAMPLES

Example 1. *Find the percentage error in the area of an ellipse when an error of +1 per cent is made in measuring the major and minor axes.*

Sol. If x and y are semi-major and semi-minor axes of the ellipse, then its area A is given by $A = \pi xy$

Taking logs, $\log A = \log \pi + \log x + \log y$

Taking differentials $\frac{dA}{A} = 0 + \frac{dx}{x} + \frac{dy}{y}$

or
$$\frac{dA}{A} \times 100 = \frac{dx}{x} \times 100 + \frac{dy}{y} \times 100 = 1 + 1 = 2.$$

$\therefore$ Error in the area = 2%.

Example 2. *The work that must be done to propel a ship of displacement 'D' for a distance 'S' in time 't' is proportional to* $\frac{S^2 D^{2/3}}{t^2}$. *Find approximately the increase of work necessary when the displacement is increased by 1%, the time diminished by 1% and the distance diminished by 2%.*

Sol. Let the work done be W, then $W \propto \frac{S^2 D^{2/3}}{t^2}$

$$\Rightarrow \quad W = k \cdot \frac{S^2 D^{2/3}}{t^2}, \text{ where } k \text{ is a constant.}$$

Taking logs, $\log W = \log k + 2 \log S + \frac{2}{3} \log D - 2 \log t$

Taking differentials,
$$\frac{dW}{W} = 0 + 2\frac{dS}{S} + \frac{2}{3}\frac{dD}{D} - 2\frac{dt}{t}$$

or
$$\frac{dW}{W} \times 100 = 2\frac{dS}{S} \times 100 + \frac{2}{3}\frac{dD}{D} \times 100 - 2\frac{dt}{t} \times 100$$

$$= 2(-2) + \frac{2}{3}(1) - 2(-1) = -\frac{4}{3}$$

$\therefore$ Approximate increase of work $= -\frac{4}{3}\%$.

Example 3. *The height of a tower is determined by observing the elevations θ and φ of its summit from two points in a direct line with the foot of the tower and at a distance 'a' apart. Show that the error in the calculated height due to small errors dθ and dφ is approximately*

$$a(\sin^2 \theta \, d\phi - \sin^2 \phi \, d\theta) \operatorname{cosec}^2 (\theta - \phi).$$

Sol. Let h be the height of the tower AB and C and D, the two points of observation so that

$$CD = a, \angle ACB = \theta,$$
$$\angle ADB = \phi. \text{ Let } AC = x$$

From rt. $\angle d$. ΔBAC, $\quad x = h\cot\theta \quad$ ...(1)

From rt. $\angle d$. ΔBAD, $\quad x + a = h\cot\phi \quad$ ...(2)

Subtracting (1) from (2),

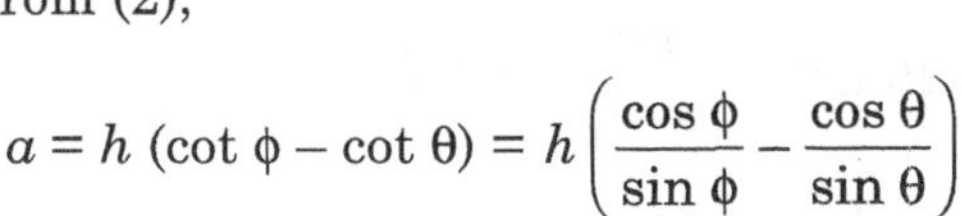

$$a = h\,(\cot\phi - \cot\theta) = h\left(\frac{\cos\phi}{\sin\phi} - \frac{\cos\theta}{\sin\theta}\right)$$

$$= h\left[\frac{\sin\theta\cos\phi - \cos\theta\sin\phi}{\sin\theta\sin\phi}\right] = \frac{h\sin(\theta-\phi)}{\sin\theta\sin\phi}$$

$$\Rightarrow \qquad h = \frac{a\sin\theta\sin\phi}{\sin(\theta-\phi)} = f(\theta,\phi) \qquad \text{...(1)}$$

Taking logs, $\quad \log h = \log a + \log\sin\theta + \log\sin\phi - \log\sin(\theta-\phi)$

Taking differentials

$$\frac{dh}{h} = 0 + \frac{\cos\theta}{\sin\theta}d\theta + \frac{\cos\phi}{\sin\phi}d\phi - \frac{\cos(\theta-\phi)}{\sin(\theta-\phi)}(d\theta - d\phi)$$

or
$$\frac{dh}{h} = \left[\frac{\cos\theta}{\sin\theta} - \frac{\cos(\theta-\phi)}{\sin(\theta-\phi)}\right]d\theta + \left[\frac{\cos\phi}{\sin\phi} + \frac{\cos(\theta-\phi)}{\sin(\theta-\phi)}\right]d\phi$$

$$= \frac{\sin(\theta-\phi)\cos\theta - \cos(\theta-\phi)\sin\theta}{\sin\theta\sin(\theta-\phi)}d\theta + \frac{\sin(\theta-\phi)\cos\phi + \cos(\theta-\phi)\sin\phi}{\sin\phi\sin(\theta-\phi)}d\phi$$

$$= \frac{\sin[(\theta-\phi)-\theta]}{\sin\theta\sin(\theta-\phi)}d\theta + \frac{\sin[(\theta-\phi)+\phi]}{\sin\phi\sin(\theta-\phi)}d\phi$$

$$= \frac{\sin(-\phi)}{\sin\theta\sin(\theta-\phi)}d\theta + \frac{\sin\theta}{\sin\phi\sin(\theta-\phi)}d\phi$$

$$= \frac{\sin^2\theta\,d\phi - \sin^2\phi\,d\theta}{\sin\theta\sin\phi\sin(\theta-\phi)} \qquad [\because \sin(-\phi) = -\sin\phi]$$

$$\therefore \qquad dh = h\cdot\frac{\sin^2\theta\,d\phi - \sin^2\phi\,d\theta}{\sin\theta\sin\phi\sin(\theta-\phi)} = \frac{a\sin\theta\sin\phi}{\sin(\theta-\phi)}\cdot\frac{\sin^2\theta\,d\phi - \sin^2\phi\,d\theta}{\sin\theta\sin\phi\sin(\theta-\phi)}$$

[Using (1)]

$$= a(\sin^2\theta\,d\phi - \sin^2\phi\,d\theta)\,\text{cosec}^2(\theta-\phi).$$

Example 4. *If the sides and angles of a triangle ABC vary in such a way that its circum-radius remains constant, prove that* $\dfrac{da}{\cos A} + \dfrac{db}{\cos B} + \dfrac{dc}{\cos C} = 0.$

Sol. The circum-radius R of a ΔABC is given by

$$R = \frac{a}{2 \sin A} = \frac{b}{2 \sin B} = \frac{c}{2 \sin C}$$

Now $\quad a = 2R \sin A \quad$ (R is constant)

Taking differentials, $da = 2R \cos A \, dA$ or $\frac{da}{\cos A} = 2R \, dA$

Similarly, $\quad \frac{db}{\cos B} = 2R \, dB; \quad \frac{dc}{\cos C} = 2R \, dC$

Adding, $$\frac{da}{\cos A} + \frac{db}{\cos B} + \frac{dc}{\cos C} = 2R\,(dA + dB + dC) \quad ...(1)$$

Also, $\quad A + B + C = \pi$

Taking differentials, $\quad dA + dB + dC = 0$

$\therefore$ From (1), $$\frac{da}{\cos A} + \frac{db}{\cos B} + \frac{dc}{\cos C} = 0.$$

Example 5. *If* $x^2 + y^2 + z^2 - 2xyz = 1$, *show that* $\frac{dx}{\sqrt{1-x^2}} + \frac{dy}{\sqrt{1-y^2}} + \frac{dz}{\sqrt{1-z^2}} = 0$.

Sol. Given $\quad x^2 + y^2 + z^2 - 2xyz = 1 \quad ...(1)$

Taking differentials, we get

$$2xdx + 2ydy + 2zdz - 2yzdx - 2zxdy - 2xydz = 0$$

or $$(x - yz)\,dx + (y - zx)\,dy + (z - xy)\,dz = 0 \quad ...(2)$$

Now $\quad (x - yz)^2 = x^2 - 2xyz + y^2z^2 = 1 - y^2 - z^2 + y^2z^2 \quad$ | Using (1)

$$= (1 - y^2) - z^2(1 - y^2) = (1 - y^2)(1 - z^2)$$

$\Rightarrow$ $$x - yz = \sqrt{(1-y^2)(1-z^2)}$$

Similarly, $\quad y - zx = \sqrt{(1-z^2)(1-x^2)}$ and $z - xy = \sqrt{(1-x^2)(1-y^2)}$

$\therefore$ From (2), $$\sqrt{(1-y^2)(1-z^2)}\,dx + \sqrt{(1-z^2)(1-x^2)}\,dy + \sqrt{(1-x^2)(1-y^2)}\,dz = 0$$

Dividing by $\sqrt{(1-x^2)(1-y^2)(1-z^2)}$, we get $\frac{dx}{\sqrt{1-x^2}} + \frac{dy}{\sqrt{1-y^2}} + \frac{dz}{\sqrt{1-z^2}} = 0.$

EXERCISE 10.7

1. Find the percentage error in the area of a rectangle when an error of +1 per cent is made in measuring its length and breadth.
2. The radius of a circle is found to be 100 cm. Find the relative error in the area of the circle due to an error of 1 mm in measuring the radius.
3. The radius of a sphere is found to be 10 cm with a possible error of 0.02 cm. What is the relative error in computing the volume?
4. What error in the common logarithm of a number will be produced by an error of 1% in the number. *(M.T.U. 2013)*

5. The density of the atmosphere at height z metres is given by $\rho = \rho_0 e^{\frac{-z}{800}}$, where ρ_0 is the density at sea level. Find the approximate increase in height corresponding to a fall of 1% in the density.

6. If the kinetic energy $k = \dfrac{wv^2}{2g}$, find approximately the change in the kinetic energy as w changes from 49 to 49.5 and v changes from 1600 to 1590.

7. The indicated horse-power I of an engine is calculated from the formula $I = \dfrac{PLAN}{33000}$, where $A = \dfrac{\pi}{4} d^2$. Assuming that errors of r per cent may have been made in measuring P, L, N and d, find the greatest possible error in I.

8. The period T of a simple pendulum is $T = 2\pi \sqrt{\dfrac{l}{g}}$. Find the maximum error in T due to possible errors upto 1% in l and 2.5% in g.

9. The quantity Q of water flowing over a notch is given by $Q = \dfrac{8}{15} \times 0.64 \times \sqrt{2}gH^{5/2}$, where H is the head of the notch. What is the percentage error in Q caused by measuring H as 0.198 instead of 0.2?

10. If $pV^2 = k$ and the respective errors in p and V are respectively 0.05 and 0.025, show that the error in k is 10%. *(K.U.K. Dec. 2012)*

11. The range R of a projectile which starts with a velocity v at an elevation α is given by $R = \dfrac{v^2 \sin 2\alpha}{g}$. Find the percentage error in R due to an error of 1% in v and an error of $\dfrac{1}{2}$ % in α.

12. If $ax^2 + by^2 + cz^2 = 1$ and $lx + my + nz = 0$, prove that $\dfrac{dx}{bny - cmz} = \dfrac{dy}{clz - anx} = \dfrac{dz}{amx - bly}$.

Answers

1. 2%	**2.** 0.002 sq cm.	**3.** 0.006 cc	**4.** 0.0043429
5. 8 m	**6.** – 4500 units	**7.** $5r$%	**8.** 1.75%
9. – 2.5%	**11.** $2 + \alpha \cot 2\alpha$.		

10.17. DIFFERENTIATION UNDER INTEGRAL SIGN

If a function $f(x, \alpha)$ of the two variables x and α, α being called a parameter, be integrated w.r.t. x between the limits a and b, $\int_a^b f(x, \alpha)\, dx$ is a function of α. For example,

$$\int_0^{\pi/2} \sin \alpha\, x dx = -\left[\frac{\cos \alpha x}{\alpha}\right]_0^{\pi/2} = -\frac{1}{\alpha}\left(\cos \frac{\pi}{2}\alpha - 1\right) = \frac{1}{\alpha}\left(1 - \cos \frac{\pi}{2}\alpha\right)$$

Thus in general $\int_a^b f(x, \alpha) dx = F(\alpha)$.

10.18. THEOREM (Leibnitz's Rule)

If $f(x, \alpha)$ and $\frac{\partial}{\partial x}$ $[f(x, \alpha)]$ be continuous functions of x and α, then

$$\frac{d}{d\alpha}\left[\int_a^b f(x, \alpha)\, dx\right] = \int_a^b \frac{\partial}{\partial \alpha}[f(x, \alpha)]\, dx$$ **where a and b are constants independent of α.**

Proof. Let $$F(\alpha) = \int_a^b f(x, \alpha)\, dx \quad ...(i)$$

When α changes to α + δα, let F(α) change to F(α + δα)

Then $$F(\alpha + \delta\alpha) = \int_a^b f(x, \alpha + \delta x)\, dx \quad ...(ii)$$

Subtracting *(i)* from *(ii)*

$$F(\alpha + \delta\alpha) - F(\alpha) = \int_a^b f(x, \alpha + \delta\alpha)\, dx - \int_a^b f(x, \alpha)\, dx$$

$$= \int_a^b [f(x, \alpha + \delta\alpha) - f(x, \alpha)]dx \quad ...(iii)$$

Now by the Mean Value Theorem of Differential Calculus, we have

$$f(x, \alpha + \delta\alpha) - f(x, \alpha) = \delta\alpha \frac{\partial}{\partial\alpha} f(x, \alpha + \theta \,.\, \delta\alpha) \text{ where } 0 < \theta < 1$$

$\therefore$ From *(iii)*, $$F(\alpha + \delta\alpha) - F(\alpha) = \delta\alpha \int_a^b \frac{\partial}{\partial\alpha} f(x, \alpha + \theta \,.\, \delta x)\, dx$$

$$\frac{F(\alpha + \delta\alpha) - F(\alpha)}{\delta\alpha} = \int_a^b \frac{\partial}{\partial\alpha} f(x, \alpha + \theta.\, \delta\alpha)\, dx$$

Proceeding to the limit as $\delta\alpha \to 0$, we get

$$\frac{d}{d\alpha}[F(\alpha)] = \int_a^b \frac{\partial}{\partial x}[f(x, \alpha)]\, dx \qquad \left[\because \frac{\partial}{\partial\alpha}(f(x, \alpha)) \text{ is continuous}\right]$$

or $$\frac{d}{d\alpha}\left[\int_a^b f(x, \alpha)\, dx\right] = \int_a^b \frac{\partial}{\partial\alpha}[f(x, \alpha)]\, dx.$$

Note. If the limits of integration *a* and *b* are not independent of α, then

$$\frac{d}{d\alpha}\left[\int_a^b f(x, \alpha)\, dx\right] = \int_a^b \frac{\partial}{\partial\alpha}[f(x, \alpha)]\, dx + \frac{db}{d\alpha} f(b, \alpha) - \frac{da}{d\alpha} f(a, \alpha)$$

[The proof is beyond the scope of this book.]

ILLUSTRATIVE EXAMPLES

Example 1. *Evaluate the integral* $\int_0^1 \frac{x^\alpha - 1}{\log x}\, dx$ *by applying differentiation under the integral sign* ($\alpha \geq 0$).

Sol. Let $$F(\alpha) = \int_0^1 \frac{x^\alpha - 1}{\log x}\, dx \quad ...(i)$$

Differentiating both sides w.r.t. α

$$F'(\alpha) = \int_0^1 \frac{\partial}{\partial\alpha}\left[\frac{x^\alpha - 1}{\log x}\right] dx \qquad \left|\ \frac{d}{dx}(a^x) = a^x \log a\right.$$

$$= \int_0^1 \frac{1}{\log x} x^\alpha \log x \, dx = \int_0^1 x^\alpha \, dx = \left[\frac{x^{\alpha+1}}{\alpha+1}\right]_0^1 \text{ for } \alpha > -1 = \frac{1}{1+\alpha}$$

Integrating both sides w.r.t. α, $F(\alpha) = \log(1+\alpha) + c$ ...(*ii*)

From (*i*), when $\alpha = 0$, $F(0) = \int_0^1 \frac{1-1}{\log x} dx = 0$

$\therefore$ From (*ii*), $0 = \log 1 + c = 0 + c \Rightarrow c = 0$

$\therefore$ $F(\alpha) = \log(1+\alpha)$ where $\alpha > -1$

Hence $\displaystyle\int_0^1 \frac{x^\alpha - 1}{\log x} dx = \log(1+\alpha).$

Example 2. *Evaluate* $\displaystyle\int_0^{\pi/2} \frac{dx}{(a^2 \cos^2 x + b^2 \sin^2 x)^2}$.

Sol. Let $$I = \int_0^{\pi/2} \frac{dx}{(a^2 \cos^2 x + b^2 \sin^2 x)^2} \quad ...(i)$$

Let us first evaluate $\displaystyle\int_0^{\pi/2} \frac{dx}{a^2 \cos^2 x + b^2 \sin^2 x}$

Dividing the numerator and denominator by $\cos^2 x$

$$\int_0^{\pi/2} \frac{dx}{a^2 \cos^2 x + b^2 \sin^2 x} = \int_0^{\pi/2} \frac{\sec^2 x \, dx}{a^2 + b^2 \tan^2 x}$$

$\left[\text{Put } \tan x = t,\ \sec^2 x \, dx = dt, \text{ when } x = 0, t = 0\ ; \text{ when } x = \frac{\pi}{2} \cdot t \to \infty\right]$

$$= \int_0^\infty \frac{dt}{a^2 + b^2 t^2} = \frac{1}{b^2} \int_0^\infty \frac{dt}{t^2 + \frac{a^2}{b^2}} = \frac{1}{b^2} \cdot \frac{1}{\frac{a}{b}} \left[\tan^{-1} \frac{t}{\frac{a}{b}}\right]_0^\infty$$

$$= \frac{1}{ab} [\tan^{-1} \infty - \tan^{-1} 0] = \frac{\pi}{2ab}$$

$$\therefore \quad \int_0^{\pi/2} \frac{dx}{a^2 \cos^2 x + b^2 \sin^2 x} = \frac{\pi}{2ab} \quad ...(ii)$$

Differentiating both sides w.r.t. a

$$\int_0^{\pi/2} \frac{\partial}{\partial a} [(a^2 \cos^2 x + b^2 \sin^2 x)^{-1}] \, dx = \frac{\partial}{\partial a}\left[\frac{\pi}{2ab}\right]$$

or $$\int_0^{\pi/2} - (a^2 \cos^2 x + b^2 \sin^2 x)^{-2} \, 2a \cos^2 x \, dx = \frac{\pi}{2b} \cdot (-a^{-2})$$

i.e., $$\int_0^{\pi/2} \frac{-2a \cos^2 x}{(a^2 \cos^2 x + b^2 \sin^2 x)^2} dx = \frac{\pi}{2a^2 b}$$

Dividing both sides by $-2a$

$$\int_0^{\pi/2} \frac{\cos^2 x}{(a^2 \cos^2 x + b^2 \sin^2 x)^2} dx = \frac{\pi}{4a^3 b} \quad ...(iii)$$

Similarly differentiating (*ii*), partially w.r.t. b

$$\int_0^{\pi/2} \frac{\sin^2 x}{(a^2 \cos^2 x + b^2 \sin^2 x)^2} dx = \frac{\pi}{4ab^3}$$

Adding (*iii*) and (*iv*) $\int_0^{\pi/2} \frac{\cos^2 x + \sin^2 x}{(a^2 \cos^2 x + b^2 \sin^2 x)^2}\, dx = \frac{\pi}{4a^3 b} + \frac{\pi}{4ab^3}$ ...(*iv*)

$$\Rightarrow \quad \int_0^{\pi/2} \frac{dx}{(a^2 \cos^2 x + b^2 \sin^2 x)^2} = \frac{\pi}{4a^3 b^3}(a^2 + b^2)$$

Hence $$\mathrm{I} = \frac{\pi(a^2 + b^2)}{4a^3 b^3}.$$

Example 3. *Evaluate* $\int_0^\infty \frac{\tan^{-1} ax}{x(1+x^2)}\, dx \ (a \ge 0)$ *by applying differentiation under the integral sign.* (M.D.U. May 2008, Dec. 2009, Dec. 2012, Dec. 2013)

Sol. Let $$\mathrm{F}(a) = \int_0^\infty \frac{\tan^{-1} ax}{x(1+x^2)}\, dx \quad ...(i)$$

Differentiating both sides w.r.t. a

$$\mathrm{F}'(a) = \int_0^\infty \frac{\partial}{\partial a}\left[\frac{\tan^{-1} ax}{x(1+x^2)}\right] dx = \int_0^\infty \frac{1}{x(1+x^2)} \cdot \frac{1}{1+a^2x^2} \cdot x\, dx$$

$$= \int_0^\infty \frac{dx}{(1+x^2)(1+a^2x^2)} = \frac{1}{1-a^2}\int_0^\infty \left[\frac{1}{1+x^2} - \frac{a^2}{1+a^2x^2}\right] dx$$

| Partial fractions by putting $x^2 = t$

$$= \frac{1}{1-a^2}\left[\tan^{-1} x\right]_0^\infty - \frac{a^2}{1-a^2}\int_0^\infty \frac{dx}{1+a^2x^2}$$

$$= \frac{1}{1-a^2} \cdot [\tan^{-1} \infty - \tan^{-1} 0] - \frac{a^2}{1-a^2} \cdot \frac{1}{a^2}\int_0^\infty \frac{dx}{x^2 + \frac{1}{a}}$$

$$= \frac{1}{1-a^2} \cdot \frac{\pi}{2} - \frac{1}{1-a^2} \cdot \frac{1}{\frac{1}{a}}\left[\tan^{-1} \frac{x}{\frac{1}{a}}\right]_0^\infty$$

$$= \frac{1}{1-a^2} \cdot \frac{\pi}{2} - \frac{a}{1-a^2}\, [\tan^{-1} \infty - \tan^{-1} 0]$$

$$= \frac{1}{1-a^2} \cdot \left[\frac{\pi}{2} - a \cdot \frac{\pi}{2}\right] = \frac{1}{1-a^2} \cdot \frac{\pi}{2}[1-a] = \frac{\pi}{2(1+a)}$$

Integrating both sides w.r.t. a, $\mathrm{F}(a) = \frac{\pi}{2} \log(1+a) + c$ ...(*ii*)

From (*i*), when $a = 0$, $\mathrm{F}(0) = 0$

$\therefore$ From (*ii*), $$0 = \frac{\pi}{2} \log 1 + c = 0 + c \Rightarrow c = 0$$

$\therefore$ $$\mathrm{F}(a) = \frac{\pi}{2} \log(1+a)$$

Hence $$\int_0^\infty \frac{\tan^{-1} ax}{x(1+x^2)}\, dx = \frac{\pi}{2} \log(1+a).$$

Example 4. *If* $|a| < 1$, *prove that* $\int_0^{\pi} \log(1 + a\cos x)\,dx = \pi \log\left[\frac{1}{2} + \frac{1}{2}\sqrt{1-a^2}\right]$.

(K.U.K. Dec. 2013; M.D.U. Dec. 2010)

Sol. Let $\quad F(a) = \int_0^{\pi} \log(1 + a\cos x)\,dx \qquad \ldots(i)$

Differentiating w.r.t. a, $F'(a) = \int_0^{\pi} \frac{\partial}{\partial a}[\log(1 + a\cos x)]dx$

$$= \int_0^{\pi} \frac{1}{1+a\cos x}\cdot\cos x\,dx = \frac{1}{a}\int_0^{\pi}\frac{a\cos x}{1+a\cos x}dx$$

$$= \frac{1}{a}\int_0^{\pi}\frac{(1+a\cos x)-1}{1+a\cos x}dx$$

$$= \frac{1}{a}\int_0^{\pi}\left[1-\frac{1}{1+a\cos x}\right]dx = \frac{1}{a}\Big[x\Big]_0^{\pi} - \frac{1}{a}\cdot\frac{1}{\sqrt{1-a^2}}\left[\cos^{-1}\frac{a+1\cos x}{1+a\cos x}\right]_0^{\pi}$$

$\mid\because\ |a| < 1$

$$= \frac{\pi}{a} - \frac{1}{a\sqrt{1-a^2}}\left[\cos^{-1}\frac{a-1}{1-a} - \cos^{-1}\frac{a+1}{1+a}\right]$$

$$= \frac{\pi}{a} - \frac{1}{a\sqrt{1-a^2}}[\cos^{-1}(-1) - \cos^{-1}1]$$

$$= \frac{\pi}{a} - \frac{1}{a\sqrt{1-a^2}}[\pi - 0] = \frac{\pi}{a} - \frac{\pi}{a\sqrt{1-a^2}}$$

Integrating w.r.t. a

$$F(a) = \int\frac{\pi}{a}\,da - \pi\int\frac{da}{a\sqrt{1-a^2}} + c \qquad \text{Put } a = \frac{1}{t}$$

$$= \pi\log a - \pi\int\frac{-\frac{1}{t^2}dt}{\frac{1}{t}\sqrt{1-\frac{1}{t^2}}} + c = \pi\log a + \pi\int\frac{dt}{\sqrt{t^2-1}} + c$$

$$= \pi\log a + \pi\log(t + \sqrt{t^2-1}) + c = \pi\left[\log a + \log\left(\frac{1}{a} + \sqrt{\frac{1}{a^2}-1}\right)\right] + c$$

$$= \pi\log\left[a\left(\frac{1}{a} + \sqrt{\frac{1}{a^2}-1}\right)\right] + c = \pi\log(1 + \sqrt{1-a^2}) + c \qquad \ldots(ii)$$

From (i), when $a = 0$, $F(0) = 0$

$\therefore$ From (ii), $0 = \pi\log 2 + c \Rightarrow c = -\pi\log 2$

$$\therefore \quad F(a) = \pi\log(1 + \sqrt{1-a^2}) - \pi\log 2 = \pi\log\frac{1+\sqrt{1-a^2}}{2}$$

$\Big|\because\ \log m - \log n = \log\frac{m}{n}$

Hence, $\quad \int_0^{\pi}\log(1 + a\cos x)\,dx = \pi\log\left(\frac{1}{2} + \frac{1}{2}\sqrt{1-a^2}\right).$

Example 5. *Evaluate the following integral by differentiating under integral sign*

$$\int_0^\infty e^{-\left(x^2+\frac{a^2}{x^2}\right)} dx \qquad \text{(K.U.K. 2012)}$$

Solution. Let $F(a) = \int_0^\infty e^{-\left(x^2+\frac{a^2}{x^2}\right)} dx$ ...(1)

Differentiating w.r.t. 'a', we get

$$F'(a) = \int_0^\infty \frac{\partial}{\partial a}\left[e^{-\left(x^2+\frac{a^2}{x^2}\right)}\right] dx = \int_0^\infty e^{-\left(x^2+\frac{a^2}{x^2}\right)} \cdot \frac{-2a}{x^2} dx$$

Putting $\frac{a}{x} = y$ so that $-\frac{a}{x^2} dx = dy$

or $\frac{a}{x^2} dx = -dy$

When $x \to 0, y \to \infty$ and when $x \to \infty, y \to 0$

$$\therefore \quad F'(a) = \int_\infty^0 e^{-\left(\frac{a^2}{y^2}+y^2\right)} \cdot (-2)(-dy) = 2\int_\infty^0 e^{-\left(y^2+\frac{a^2}{y^2}\right)} dy$$

Interchanging the limits of integration and changing the variable y to x.

$$F'(a) = -2\int_0^\infty e^{-\left(x^2+\frac{a^2}{x^2}\right)} dx = -2F(a) \qquad \text{[Using (1)]}$$

$$\Rightarrow \quad \frac{F'(a)}{F(a)} = -2$$

Integrating w.r.t. 'a'

$$\log F(a) = -2a + \log c$$

$$\Rightarrow \quad \log \frac{F(a)}{c} = -2a$$

$$\Rightarrow \quad \frac{F(a)}{c} = e^{-2a}$$

$$\therefore \quad F(a) = ce^{-2a} \qquad ...(2)$$

When $a = 0$, from (1)

$$F(0) = \int_0^\infty e^{-x^2} dx = \frac{\Gamma\left(\frac{1}{2}\right)}{2} = \frac{\sqrt{\pi}}{2}$$

[See Chapter 12, Gamma Function]

When $a = 0$, from (2)

$$F(0) = c$$

$$\Rightarrow \quad c = \frac{\sqrt{\pi}}{2}$$

$\therefore$ From (2)

$$F(a) = \frac{\sqrt{\pi}}{2} e^{-2a}$$

$$\Rightarrow \quad \int_0^\infty e^{-\left(x^2+\frac{a^2}{x^2}\right)} dx = \frac{\sqrt{\pi}}{2} e^{-2a}$$

Example 6. *Evaluate* $\int_0^a \frac{\log(1+ax)}{1+x^2} dx$ *and hence show that* $\int_0^1 \frac{\log(1+x)}{1+x^2} dx = \frac{\pi}{8} \log 2.$

(M.D.U. May 2009)

Sol. Let $\quad F(a) = \int_0^a \frac{\log(1+ax)}{1+x^2} dx \qquad \text{...}(i)$

[Here the upper limit involves the parameter a]

Differentiating w.r.t. a

$$F'(a) = \int_0^a \frac{\partial}{\partial a}\left[\frac{\log(1+ax)}{1+x^2}\right] dx + \frac{\log(1+a^2)}{1+a^2} \cdot \frac{d}{da}(a) - \frac{\log(1+a\,.\,0)}{1+0} \cdot \frac{d}{da}(0)$$

$$= \int_0^a \frac{1}{1+x^2} \cdot \frac{1}{1+ax} \cdot x\, dx + \frac{\log(1+a^2)}{1+a^2} \qquad \text{[See Note with 10.18]}$$

$$= \int_0^a \frac{x}{(1+ax)(1+x^2)} dx + \frac{\log(1+a^2)}{1+a^2} \qquad \text{...}(ii)$$

To resolve the integrand into partial fractions:

Let $\quad \frac{x}{(1+ax)(1+x^2)} = \frac{A}{1+ax} + \frac{Bx+C}{1+x^2} \quad \Rightarrow \quad x = A(1+x^2) + (Bx+C)(1+ax)$

Putting $1 + ax = 0$ *i.e.*, $x = -\frac{1}{a}$, we get $-\frac{1}{a} = A\left(1+\frac{1}{a^2}\right) \quad \therefore \quad A = \frac{-a}{1+a^2}$

Comparing co-efficients of x^2, we get

$$0 = A + aB \quad \Rightarrow \quad B = -\frac{A}{a} = \frac{1}{1+a^2}$$

Comparing constant terms, we get $0 = A + C \quad \Rightarrow \quad C = -A = \frac{a}{1+a^2}$

$$\therefore \quad \frac{x}{(1+ax)(1+x^2)} = \frac{1}{1+a^2}\left[\frac{-a}{1+ax} + \frac{x+a}{1+x^2}\right]$$

$$\Rightarrow \int_0^a \frac{x}{(1+ax)(1+x^2)} dx = \frac{1}{1+a^2}\left[-\log(1+ax) + \frac{1}{2}\log(1+x^2) + a\tan^{-1} x\right]_0^a$$

$$= \frac{1}{1+a^2}\left[-\log(1+a^2) + \frac{1}{2}\log(1+a^2) + a\tan^{-1} a\right]$$

$$= \frac{1}{1+a^2}\left[-\frac{1}{2}\log(1+a^2) + a\tan^{-1} a\right]$$

$\therefore$ From (*ii*), we get

$$F'(a) = \frac{1}{1+a^2}\left[-\frac{1}{2}\log(1+a^2) + a\tan^{-1} a\right] + \frac{\log(1+a^2)}{1+a^2}$$

$$= \frac{1}{1+a^2}\left[\frac{1}{2}\log(1+a^2) + a\tan^{-1} a\right]$$

Integrating w.r.t. a, $F(a) = \frac{1}{2}\int \log(1+a^2) \cdot \frac{1}{1+a^2}\, da + \int \frac{a\tan^{-1} a}{1+a^2}\, da + c$

[Integrating by parts]

$$= \frac{1}{2}\left[\log(1+a^2)\tan^{-1} a - \int \frac{2a}{1+a^2}\tan^{-1} a\, da\right] + \int \frac{a\tan^{-1} a}{1+a^2}\, da + c$$

$$= \frac{1}{2}\log(1+a^2)\tan^{-1} a + c \qquad ...(iii)$$

Putting $a = 0$ in (*i*), $F(0) = 0$

Putting $a = 0$ in (*iii*), $F(0) = c$ or $c = 0$

$\therefore$ From (*iii*), we have $F(a) = \frac{1}{2}\log(1+a^2)\tan^{-1} a$

Hence $\int_0^a \frac{\log(1+ax)}{1+x^2}\, dx = \frac{1}{2}\log(1+a^2)\tan^{-1} a$

Putting $a = 1$, we get $\int_0^1 \frac{\log(1+x)}{1+x^2}\, dx = \frac{\pi}{8}\log 2.$

Example 7. *By successive differentiation of* $\int_0^1 x^m\, dx = \frac{1}{m+1}$ *w.r.t. m, evaluate* $\int_0^1 x^m (\log x)^n\, dx$.

Sol. Given $\int_0^1 x^m\, dx = \frac{1}{m+1}$

Differentiating w.r.t. m, we have

$$\int_0^1 \frac{\partial}{\partial m}(x^m)\, dx = \frac{d}{dm}\left(\frac{1}{m+1}\right)$$

$$\Rightarrow \int_0^1 x^m \log x\, dx = \frac{-1}{(m+1)^2}$$

Differentiating again w.r.t. m, we have

$$\int_0^1 \frac{\partial}{\partial m}(x^m \log x)\, dx = \frac{d}{dm}\left[\frac{-1}{(m+1)^2}\right]$$

$$\Rightarrow \int_0^1 x^m (\log x)^2\, dx = \frac{(-1)(-2)}{(m+1)^3}$$

Similarly, $\int_0^1 x^m (\log x)^3\, dx = \frac{(-1)(-2)(-3)}{(m+1)^4}$

$$\int_0^1 x^m (\log x)^4 \, dx = \frac{(-1)(-2)(-3)(-4)}{(m+1)^5}$$

....

....

$$\int_0^1 x^m (\log x)^n \, dx = \frac{(-1)(-2)(-3)(-4)\ldots(-n)}{(m+1)^{n+1}} = \frac{(-1)^n \cdot n!}{(m+1)^{n+1}}.$$

Example 8. *If* $y = \int_0^x f(t) \sin [k(x-t)] \, dt$, *prove that y satisfies the differential equation*

$$\frac{d^2 y}{dx^2} + k^2 y = k\, f(x).$$

Sol. Given
$$y = \int_0^x f(t) \sin [k(x-t)] \, dt \qquad \ldots(1)$$

[Here the upper limit involves the parameter x.]

Differentiating w.r.t. x, we have

$$\frac{dy}{dx} = \int_0^x \frac{\partial}{\partial x} [f(t) \sin \{k(x-t)\}] \, dt + f(x) \sin [k(x-x)] \cdot \frac{d}{dx}(x) - f(0) \sin [k(x-0)] \cdot \frac{d}{dx}(0)$$

$$= \int_0^x k f(t) \cos [k(x-t)] \, dt$$

Differentiating again w.r.t. x, we have

$$\frac{d^2 y}{dx^2} = \int_0^x \frac{\partial}{\partial x} [kf(t) \cos \{k(x-t)\}] \, dt + kf(x) \cos [k(x-x)] \cdot \frac{d}{dx}(x) - kf(0) \cos [k(x-0)] \cdot \frac{d}{dx}(0)$$

$$= \int_0^x -k^2 f(t) \sin [k(x-t)] \, dt + kf(x)$$

$$= -k^2 \int_0^x f(t) \sin [k(x-t)] \, dt + kf(x)$$

$$= -k^2 y + kf(x) \qquad \text{[by (1)]}$$

$$\Rightarrow \qquad \frac{d^2 y}{dx^2} + k^2 y = kf(x).$$

EXERCISE 10.8

1. Differentiating $\int_0^x \frac{dx}{x^2 + a^2} = \frac{1}{a} \tan^{-1} \frac{x}{a}$ under the integral sign, evaluate $\int_0^x \frac{dx}{(x^2 + a^2)^2}$.

2. Evaluate the integral $\int_0^\infty e^{-x} \frac{\sin ax}{x} \, dx.$

3. Using the result $\int_0^\pi \frac{dx}{a + b \cos x} = \frac{\pi}{\sqrt{a^2 - b^2}}$, $a > b$ evaluate

   (*i*) $\int_0^\pi \frac{dx}{(a + b \cos x)^2}$ (*ii*) $\int_0^\pi \frac{\cos x \, dx}{(a + b \cos x)^2}$.

4. Prove that $\int_0^{\pi/2} \frac{\log(1+a\sin^2 x)}{\sin^2 x}\,dx = \pi[\sqrt{1+a}-1]$.

5. (i) Prove that for $y > 0$, $\int_0^{\infty} e^{-xy}\frac{\sin x}{x}\,dx = \cot^{-1} y$. Hence evaluate $\int_0^{\infty}\frac{\sin x}{x}\,dx$

$$\left[\textbf{Hint. } \int e^{ax}\sin bx\,dx = \frac{e^{ax}}{a^2+b^2}(a\sin bx - b\cos bx)\right]$$

(ii) $\int_0^{\infty}\frac{e^{-x}}{x}(1-e^{-ax})\,dx = \log(1+a), \ (a \ge 0)$.

6. (a) Using $\int_0^{\infty} e^{-x^2}dx = \frac{\sqrt{\pi}}{2}$, show that $\int_0^{\infty} e^{-x^2}\cos\alpha x\,dx = \frac{\sqrt{\pi}}{2}e^{-\frac{1}{4}\alpha^2}$. *(K.U.K. 2009)*

(b) Prove that $\int_0^{\infty} xe^{-x^2}\sin\alpha x\,dx = \frac{\sqrt{\pi}}{4}\alpha e^{-\frac{\alpha^2}{4}}$.

7. If $|a| < 1$, evaluate $\int_0^{\pi}\frac{\log(1+a\cos x)}{\cos x}\,dx$.

8. (i) Prove that $\int_0^{\pi/2}\frac{\log(1+\cos\alpha\cos x)}{\cos x}\,dx = \frac{1}{2}\left(\frac{\pi^2}{4}-\alpha^2\right)$.

(ii) Prove that $\int_0^{\pi}\frac{\log(1+\sin\alpha\cos x)}{\cos x}\,dx = \pi\alpha$. *(M.D.U. Dec. 2008)*

9. Prove that if $a > b$, $\int_0^{\pi/2}\log\left(\frac{a+b\sin\theta}{a-b\sin\theta}\right)\frac{d\theta}{\sin\theta} = \pi\sin^{-1}\frac{b}{a}$. *(M.D.U. Dec. 2010)*

10. If $\alpha > 0, \beta > 0$, prove that $\int_0^{\pi/2}\log(\alpha\cos^2\theta + \beta\sin^2\theta)\,d\theta = \pi\log\left[\frac{1}{2}(\sqrt{\alpha}+\sqrt{\beta})\right]$.

11. Prove that $\int_{-\pi/2}^{\pi/2}\frac{\log(1+a\sin x)}{\sin x}\,dx = \pi\sin^{-1}a$, where $|a| < 1$.

12. Prove that $\frac{d}{da}\left[\int_0^{a^2}\tan^{-1}\frac{x}{a}\,dx\right] = 2a\tan^{-1}a - \frac{1}{2}\log(a^2+1)$ and verify by direct integration.

13. Prove that $\int_{\frac{\pi}{2}-\alpha}^{\pi/2}\sin\theta\cos^{-1}(\cos\alpha\,\mathrm{cosec}\,\theta)d\theta = \frac{\pi}{2}(1-\cos\alpha)$.

Answers

1. $\frac{x}{2a^2(x^2+a^2)} + \frac{1}{2a^3}\tan^{-1}\frac{x}{a}$ 2. $\tan^{-1} a$

3. (i) $\frac{\pi a}{(a^2-b^2)^{3/2}}$ (ii) $-\frac{\pi b}{(a^2-b^2)^{3/2}}$ 5. (i) $\frac{\pi}{2}$ 7. $\pi\sin^{-1} a$.

SECTION D

Integral Calculus

11

Applications of Single Integration

(Volumes and Surfaces of Solids of Revolution)

11.1. DEFINITIONS

If a plane area is revolved about a fixed straight line lying in its own plane, then the body so generated by the plane area is called the **solid of revolution** and the surface generated by the boundary of the plane area is called the **surface of revolution.** The fixed line about which the plane area rotates is called the **axis of revolution.**

For example; a right-angled triangle when revolved about one of its sides (forming the rt. angle) generates a right circular cone.

Remember. The section of a solid of revolution by a plane perpendicular to the axis of revolution is a circle, having its centre on the axis of revolution.

11.2. VOLUME FORMULAE FOR CARTESIAN EQUATIONS

The volume of the solid generated by the revolution about the x-axis of the area bounded by the curve y = f(x), the x-axis and the ordinates x = a, x = b is

$$\int_a^b \pi y^2 dx$$

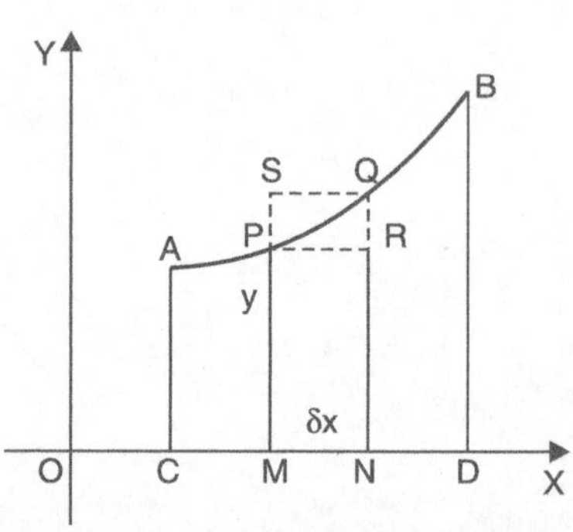

Let AB be the curve $y = f(x)$ and CA, DB be the ordinates $x = a$, $x = b$ respectively.

Let $P(x, y)$ be **any point** on the curve AB.

Draw PM ⊥ OX.

$\therefore \quad OM = x$ and $PM = y$.

Let V denote the volume of the solid generated by the revolution about X-axis of the area ACMP.

As x increases, MP moves towards the right, V also increases.

Thus, the volume V depends on x and is therefore a function of x.

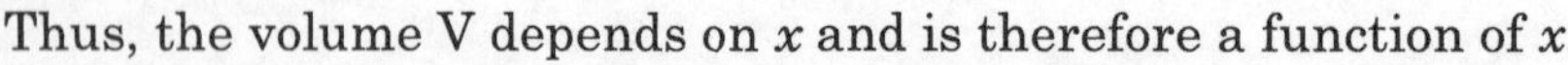

Let $Q(x + \delta x, y + \delta y)$ be a point on the curve in the neighbourhood of P. Then the volume of the solid generated by the revolution about x-axis of the area ACNQ will be $V = \delta V$, so that the volume of the solid generated by the revolution about x-axis of the area PMNQ is δV.

Complete the rectangle PRQS.

Then the volume of the solid generated by the revolution about the x-axis of the area PMNQ lies between the right circular cylinders generated by the rectangles PMNR and SMNQ *i.e.*, δV lies between $\pi y^2\,\delta x$ and $\pi(y+\delta y)^2\,\delta x$

or $\dfrac{\delta V}{\delta x}$ lies between πy^2 and $\pi(y+\delta y)^2$

In the limiting position as $Q \to P$, $\delta x \to 0$ and $\therefore\ \delta y \to 0$

$\therefore$ $\lim\limits_{\delta x \to 0} \dfrac{\delta V}{\delta x}$ lies between πy^2 and $\lim\limits_{\delta y \to 0} \pi(y+\delta y)^2$

or $\dfrac{dV}{dx}$ lies between πy^2 and a quantity which approaches πy^2.

$$\therefore \qquad \frac{dV}{dx} = \pi y^2$$

$$\therefore \qquad \int_a^b \pi y^2\,dx = \int_a^b \frac{dV}{dx}\,dx \left[V\right]_a^b$$

$$= (\text{volume V when } x = b) - (\text{volume V when } x = a)$$

$$= \text{volume generated by the area ACDB} - 0.$$

Hence the volume of the solid generated by the area ACDB about the x-axis is $\int_a^b \pi y^2\,dx$.

Cor. Revolution about y-axis. *The volume of the solid generated by revolution about the y-axis of the area bounded by the curve $x = f(y)$, the y-axis and the abscissae $y = a$, $y = b$ is* $\int_a^b \pi x^2\,dy$.

The result follows immediately on interchanging x and y in the above proposition.

An important advice.

(*i*) If the generating curve is **symmetrical about the x-axis** and it is required to find the volume generated by the *revolution of the area about the-axis,* in such a case we shall revolve only **one** of the symmetrical portions, because the part of the curve above the x-axis generates the same volume as the part of the curve below the x-axis when revolved about the x-axis. Thus the volume generated by the revolution of the area DCOAB about OX = volume generated by the area OABM or OCDM.

(*ii*) If the curve is *symmetrical about y-axis* and the curve be made to revolve *about the x-axis* (the curve lying only on one side of the x-axis), then the volume generated = 2 × volume generated by the portion OAB lying to the right of y-axis.

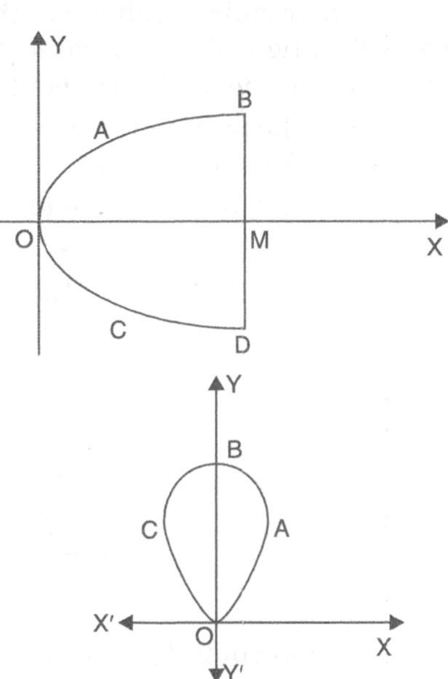

ILLUSTRATIVE EXAMPLES

Example 1. (*a*) *Find the volume of a sphere of radius a.*

(*b*) *Find the volume of a spherical cap of height h cut off from a sphere of radius a.*

Sol. (*a*) The sphere is the solid or revolution generated by the revolution of a semi-circular area about its bounding diameter.

Let the equation of the circle of radius a be

$$x^2 + y^2 = a^2 \qquad ...(1)$$

whose centre is O.

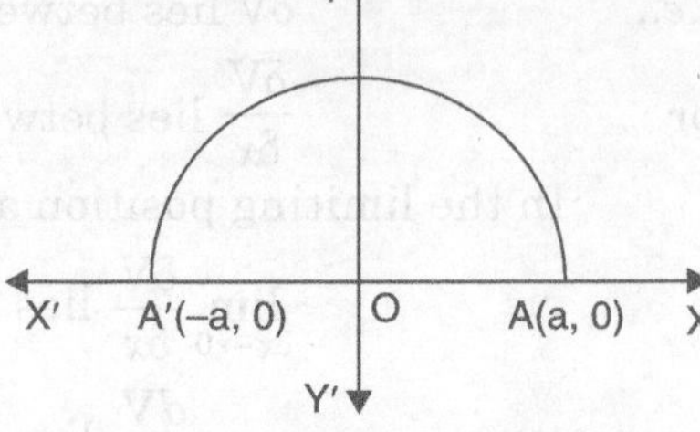

Now for the semi-circle above the x-axis, x varies from $-a$ to a.

$\therefore$ Required volume of the sphere

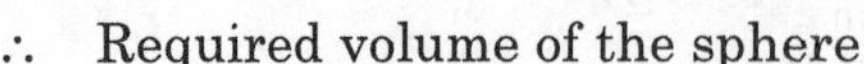

$$= \int_{-a}^{a} \pi y^2\, dx = \pi \int_{-a}^{a} (a^2 - x^2)\, dx \qquad \text{| From (1)}$$

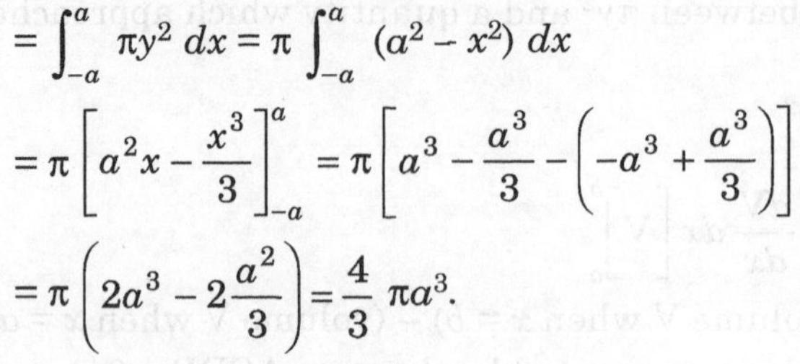

$$= \pi \left[a^2 x - \frac{x^3}{3} \right]_{-a}^{a} = \pi \left[a^3 - \frac{a^3}{3} - \left(-a^3 + \frac{a^3}{3} \right) \right]$$

$$= \pi \left(2a^3 - 2\frac{a^2}{3} \right) = \frac{4}{3} \pi a^3.$$

(*b*) Let the equation of the circle be

$$x^2 + y^2 = a^2 \qquad ...(1)$$

Let the plane PBP′ cut the circle at a distance h from A, so that BA = h and

$\therefore$ OB = $a - h$, as OA = a.

The required volume of the spherical cap (shown shaded in figure) is generated by the revolution of the area PBA of the circle (1) about the x-axis.

For the portion of the curve PA, x varies from OB to OA *i.e.*, from $a - h$ to a.

$\therefore$ Required volume of spherical cap

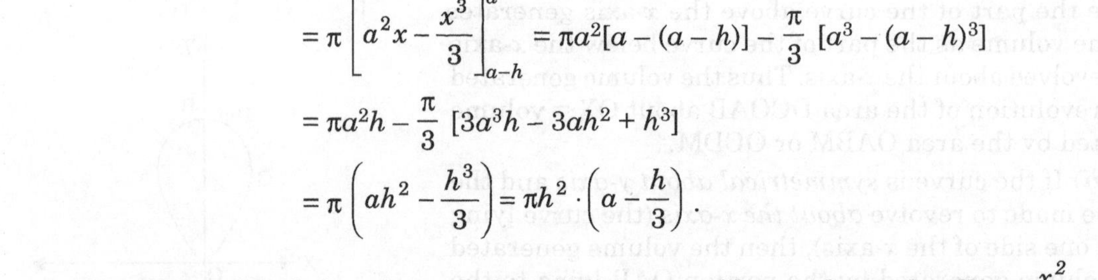

$$= \int_{a-h}^{a} \pi y^2\, dx = \int_{a-h}^{a} \pi (a^2 - x^2)\, dx \qquad \text{| From (1)}$$

$$= \pi \left[a^2 x - \frac{x^3}{3} \right]_{a-h}^{a} = \pi a^2 [a - (a - h)] - \frac{\pi}{3} [a^3 - (a - h)^3]$$

$$= \pi a^2 h - \frac{\pi}{3} [3a^2 h - 3ah^2 + h^3]$$

$$= \pi \left(ah^2 - \frac{h^3}{3} \right) = \pi h^2 \cdot \left(a - \frac{h}{3} \right)$$

Example 2. *Find the volume of the solid generated by revolving the ellipse* $\frac{x^2}{a^2} + \frac{y^2}{b^2} = 1$ *about the major axis.* (P.T.U. 2010)

Sol. The equation of the ellipse is $\frac{x^2}{a^2} + \frac{y^2}{b^2} = 1$

or $$\frac{y^2}{b^2} = 1 - \frac{x^2}{a^2} = \frac{a^2 - x^2}{a^2}$$

$\therefore$ $$y^2 = \frac{b^2}{a^2} (a^2 - x^2) \qquad ...(1)$$

Since the ellipse is symmetrical about y-axis.

$\therefore$ Required volume generated by the ellipse by the revolution about x-axis

= 2 × volume generated by the arc in first quadrant about the x-axis.

Now for the arc in the first quadrant, x varies from 0 to a.

$\therefore$ Required volume

$$= 2\int_0^a \pi y^2\, dx = 2\pi \int_0^a \frac{b^2}{a^2}(a^2 - x^2)\, dx \qquad | \text{ From (1)}$$

$$= 2\pi \frac{b^2}{a^2}\left[a^2 x - \frac{x^3}{3}\right]_0^a = 2\pi \frac{b^2}{a^2}\left[a^3 - \frac{a^3}{3} - 0\right]$$

$$= 2\pi \frac{b^2}{a^2} \cdot \frac{2a^3}{3} = \frac{4}{3}\pi ab^2.$$

Example 3. *Find the volume of the solid formed by the revolution about the x-axis, of the loop of the curve* $y^2 = \dfrac{x^2(a+x)}{(a-x)}$. (K.U.K. 2013)

Sol. The equation of the curve is $y^2 = \dfrac{x^2(a+x)}{(a-x)}$

or
$$y^2(a - x) = x^2(a + x) \qquad \text{...(1)}$$

To trace the curve roughly to find the limits of integration.

(*i*) The curve is symmetrical about the x-axis.

(*ii*) The curve passes through the origin and the tangents at the origin are given by $y^2 = x^2$ or $y = \pm x$.

(*iii*) The curve meets the x-axis at the point $(-a, 0)$.

The curve meets y-axis only at origin.

(*iv*) The curve has an asymptote $a - x = 0$

i.e., $x = a$ || to y-axis.

Thus the shape of the curve is roughly as shown in the figure.

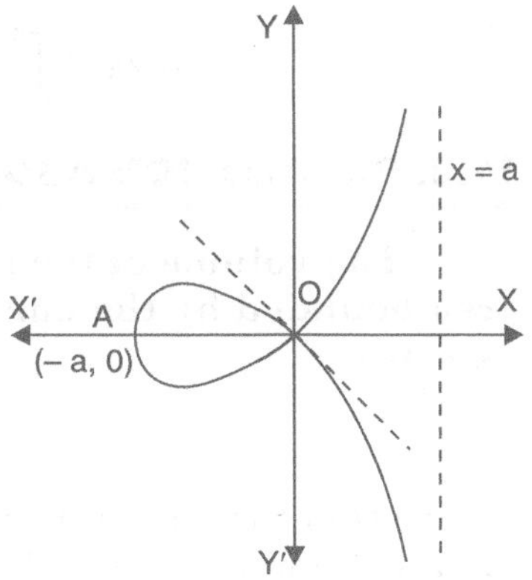

For the upper half of the loop, x varies from $-a$ to 0.

$\therefore$ Required volume

$$= \int_{-a}^0 \pi y^2\, dx$$

$$= \pi \int_{-a}^0 \frac{x^2(a+x)}{a-x}\, dx \qquad | \text{ From (1)}$$

$$= \pi \int_{-a}^0 \frac{ax^2 + x^3}{a-x}\, dx \qquad \text{[Divide the num. by the denominator]}$$

$$= \pi \int_{-a}^0 \left[-x^2 - 2ax - 2a^2 + \frac{2a^3}{a-x}\right] dx$$

$$= \pi\left[-\frac{x^3}{3} - ax^2 - 2a^2x - 2a^3 \log (a - x)\right]_{-a}^{0}$$

$$= \pi\left[-2a^3 \log a - \left(\frac{a^3}{3} - a^3 + 2a^3 - 2a^3 \log 2a\right)\right]$$

$$= \pi\,[-\tfrac{4}{3}a^3 + 2a^3 (\log 2a - \log a)] = 2\pi\, a^3 (\log 2 - \tfrac{2}{3}).$$

Example 4. *Find the volume of the solid generated by the revolution of the curve $y(a^2 + x^2) = a^3$ about its asymptote.*

Sol. The equation of the curve is $y(x^2 + a^2) = a^3$ ...(1)

To trace the curve roughly to find the limits of integration:

(*i*) The curve is symmetrical about y-axis.

(*ii*) The curve does not pass through the origin.

(*iii*) The curve does not meet x-axis. It meets y-axis in the point $(0, a)$.

(*iv*) The asymptote || to x-axis is $y = 0$ *i.e.*, the x-axis itself. The curve has no other asymptote.

The shape of the curve is as shown in the figure.

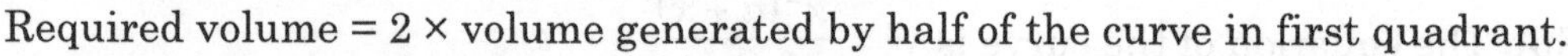

Required volume = 2 × volume generated by half of the curve in first quadrant.

For the arc in first quadrant x varies from 0 to ∞

∴ Required volume

$$= 2\int_0^\infty \pi y^2\, dx = 2\pi \int_0^\infty \left(\frac{a^3}{a^2 + x^2}\right)^2 dx \quad \left[\begin{array}{l}\text{Put } x = a\tan\theta \;\therefore\; dx = a\sec^2\theta\, d\theta \\ \text{when } x = 0, \theta = 0\,;\ \text{when } x = \infty, \theta = \pi/2\end{array}\right]$$

$$= 2\pi a^6 \int_0^{\pi/2} \frac{a \sec^2\theta\, d\theta}{(a^2 + a^2\tan^2\theta)^2} = 2\pi a^6 \int_0^{\pi/2} \frac{a\sec^2\theta\, d\theta}{a^4 \sec^4\theta}$$

$$= 2\pi a^3 \int_0^{\pi/2} \cos^2\theta\, d\theta = 2\pi a^3 \cdot \frac{1}{2}\cdot\frac{\pi}{2} = \frac{1}{2}\pi^2 a^3.$$

11.3. REVOLUTION ABOUT ANY AXIS

The volume of the solid generated by the revolution about any axis CD of the area bounded by the curve AB, the axis CD and the perpendiculars AC, BD on the axis, is

$$\int_{OC}^{OD} \pi (PM)^2\, d\,(OM)$$

where O is a fixed point on the axis CD and PM is perpendicular from any point P of the curve AB on CD.

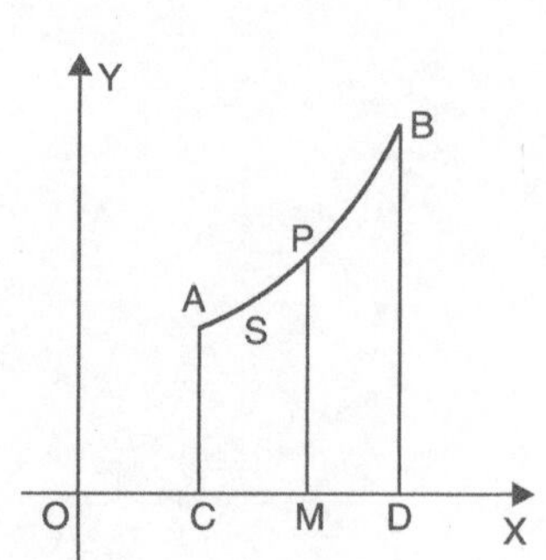

Take the fixed point O on CD as origin and CD, the axis of revolution as x-axis.

Let OY ⊥ to CD be taken as y-axis.

Let P(x, y) be any point on the curve referred to OC and OY as axes.

Draw PM ⊥ OC, so that

$$OM = x \quad \text{and} \quad MP = y.$$

If OC $= a$ and OD $= b$, then the required volume

$$= \int_a^b \pi y^2 \, dx = \pi \int_a^b (\text{PM})^2 \cdot d(\text{OM}).$$

Practical method for questions:

1. Take any point P(x, y) on the curve.
2. Draw PM $\perp$ on the line about which the curve is to be revolved and find the length of PM.
3. Find the distance OM of the foot of $\perp$ from a fixed point O (say) on the line and take its differential.
4. Then use the formula $\int \pi(\text{PM})^2 \, d(\text{OM})$ with proper limits of integration.

ILLUSTRATIVE EXAMPLES

Example 1. *Find the volume of the solid formed by the revolution of the cissoid $y^3(2a - x) = x^3$ about its asymptote.*

Sol. The equation of the curve is

$$y^2 (2a - x) = x^3$$

or
$$y^2 = \frac{x^3}{2a - x} \qquad \ldots(1)$$

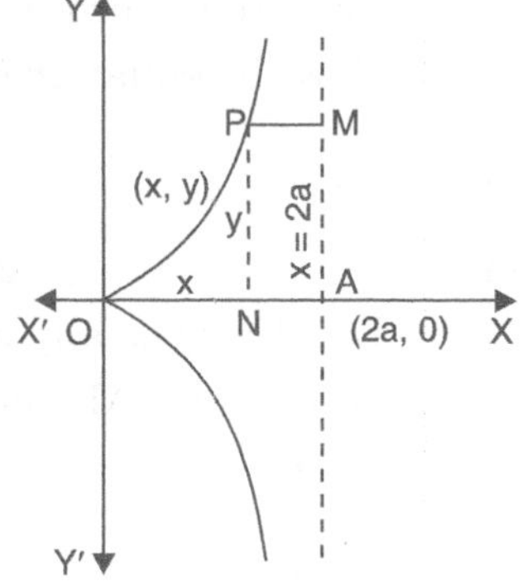

The curve is symmetrical about the x-axis and the asymptote is the line

$$2a - x = 0 \quad \text{or} \quad x = 2a.$$

If P(x, y) be any point on the curve and PM $\perp$ on the asymptote (the axis of revolution), and PN $\perp$ OX, then PM = NA = OA $-$ NO $= 2a - x$ and AM = NP $= y$, where A is the point of intersection of the asymptote and the x-axis.

$\therefore$ Required volume $= 2 \int \pi \, (\text{PM})^2 \cdot d\,(\text{AM}) \qquad \ldots(2)$

Now
$$\text{AM} = y = \frac{x^{3/2}}{\sqrt{2a - x}} \qquad | \text{ From (1)}$$

$\therefore$
$$d(\text{AM}) = dy = \frac{(2a - x)^{1/2} \cdot \frac{3}{2} x^{1/2} - x^{3/2} \cdot \frac{1}{2} (2a - x)^{-1/2} \cdot (-1)}{2a - x} \, dx$$

$$= \frac{3x^{1/2}(2a - x) + x^{3/2}}{2(2a - x)^{3/2}} \, dx = \frac{x^{1/2} (6a - 3x + x)}{2(2a - x)^{3/2}} \, dx$$

$$= \frac{\sqrt{x}\,(3a - x)}{(2a - x)^{3/2}} \, dx$$

$\therefore$ From (2),

Required volume
$$= 2\pi \int_0^{2a} (2a - x)^2 \cdot \frac{\sqrt{x}\,(3a - x)}{(2a - x)^{3/2}} \, dx$$

$$= 2\pi \int_0^{\pi/2} (3a - x)\sqrt{2a - x} \, dx$$

| Put $x = 2a \sin^2 \theta \;\; \therefore \;\; dx = 4a \sin \theta \cos \theta \, d\theta$
when $x = 0, \theta = 0$; when $x = 2a, \theta = \pi/2$

$$= 2\pi \int_0^{\pi/2} (3a - 2a \sin^2 \theta) \cdot \sqrt{2a \sin^2 \theta} \times \sqrt{2a(1 - \sin^2 \theta)} \cdot 4a \sin \theta \cos \theta \, d\theta$$

$$= 16\pi a^3 \int_0^{\pi/2} (3 - 2\sin^2\theta)\cdot \sin^2\theta\cos^2\theta\, d\theta$$

$$= 16\pi a^3 \int_0^{\pi/2} (3\sin^2\theta\cos^2\theta - 2\sin^4\theta\cos^2\theta)\, d\theta$$

$$= 16\pi a^3 \left[3\cdot\frac{1\cdot 1}{4\cdot 2}\cdot\frac{\pi}{2} - 2\cdot\frac{3\cdot 1\cdot 1}{6\cdot 4\cdot 2}\cdot\frac{\pi}{2}\right]$$

$$= 16\pi a^3 \left[\frac{3\pi}{16} - \frac{\pi}{16}\right] = 2\pi^2 a^3.$$

Example 2. *The area cut off from the parabola $y^2 = 4ax$ by the chord joining the vertex to an end of the latus rectum is rotated through four right angles about the chord. Find the volume of the solid generated.*

Sol. The equation of the parabola is $y^2 = 4ax$.

Let LSL′ be the latus rectum and A the vertex of the parabola.

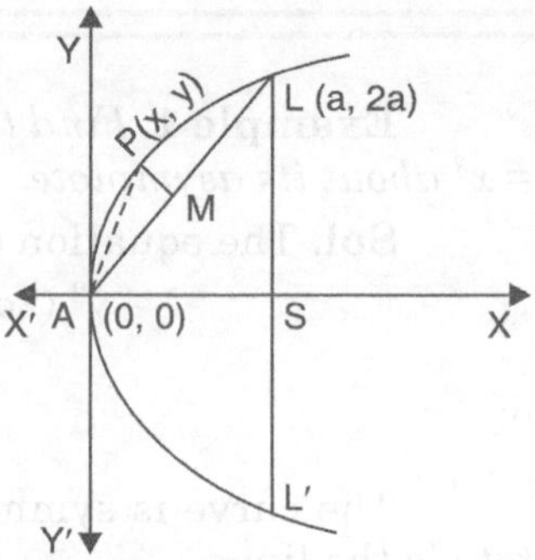

Then the co-ordinates of L are $(a, 2a)$ and of A are $(0, 0)$.

Now the equation of the line AL is

or $$y - 0 = \frac{2a - 0}{a - 0}(x - 0)$$

or $$2x - y = 0$$

Let $P(x, y)$ be any point on the arc AL of the parabola.

Draw PM ⊥ AL and join AP.

Then PM = ⊥ distance of $P(x, y)$ from the line $2x - y = 0$

$$= \frac{2x - y}{\sqrt{4+1}} = \frac{2x - y}{\sqrt{5}} = \frac{2x - 2\sqrt{ax}}{\sqrt{5}} \qquad | \because\ y^2 = 4ax$$

$$= \frac{2\sqrt{x}(\sqrt{x} - \sqrt{a})}{\sqrt{5}}$$

$$\therefore\quad PM^2 = \frac{4x(x + a - 2\sqrt{ax})}{5}$$

Now $$AM^2 = AP^2 - PM^2 = x^2 + y^2 - \frac{4x(x + a - 2\sqrt{ax})}{5}$$

$$= x^2 + 4ax - \tfrac{4}{5}x(x + a - 2\sqrt{ax}) \qquad | \because\ y^2 = 4ax$$

$$= \tfrac{1}{5}[5x^2 + 20ax - 4x^2 - 4ax + 8\sqrt{a}x^{3/2}]$$

$$= \tfrac{1}{5}(x^2 + 16ax + 8\sqrt{a}\,x^{3/2}) = \tfrac{1}{5}(x + 4\sqrt{ax})^2$$

$$\therefore\quad AM = \frac{1}{\sqrt{5}}(x + 4\sqrt{a}\sqrt{x})$$

$$\therefore\ d(AM) = \frac{1}{\sqrt{5}}\left(1 + 4\sqrt{a}\,.\,\frac{1}{2\sqrt{x}}\right)dx = \frac{1}{\sqrt{5}}\left(\frac{\sqrt{x} + 2\sqrt{a}}{\sqrt{x}}\right)dx$$

$\therefore$ Required volume $= \int \pi\,(\text{PM})^2 \cdot d\,(\text{AM})$

$$= \pi \int_0^a \frac{4x(\sqrt{x}-\sqrt{a})^2}{5} \cdot \frac{1}{\sqrt{5}} \left(\frac{\sqrt{x}+2\sqrt{a}}{\sqrt{x}} \right) dx$$

| $\because$ for the arc AL, x varies from 0 to a

$$= \frac{4\pi}{5\sqrt{5}} \int_0^a \sqrt{x}(\sqrt{x}-\sqrt{a})^2 (\sqrt{x}+2\sqrt{a})\, dx$$

| Put $\mathbf{x = at^2}$ $\therefore$ $dx = 2at\, dt$
| when $x = 0, t = 0$; when $x = a, t = 1$

$$= \frac{4\pi}{5\sqrt{5}} \int_0^1 \sqrt{a}\, t(\sqrt{a}\, t - \sqrt{a})^2 \cdot (\sqrt{a}\, t + 2\sqrt{a})\; 2at\, dt$$

$$= \frac{8\pi a^3}{5\sqrt{5}} \int_0^1 t^2\,(t+2)\,(t^2-2t+1)\, dt = \frac{8\pi a^3}{5\sqrt{5}} \int_0^1 t^2\,(t^3-3t+2)\, dt$$

$$= \frac{8\pi a^3}{5\sqrt{5}} \left[\frac{t^6}{6} - 3\frac{t^4}{4} + 2\frac{t^3}{3} \right]_0^1 = \frac{8\pi a^3}{5\sqrt{5}} \left[\frac{1}{6} - \frac{3}{4} + \frac{2}{3} \right]$$

$$= \frac{8\pi a^3}{5\sqrt{5}} \left[\frac{2-9+8}{12} \right] = \frac{2\pi a^3}{15\sqrt{5}}.$$

11.4. VOLUME FORMULAE FOR PARAMETRIC EQUATIONS

(*i*) **The volume of the solid generated by the revolution about the x-axis, of the area bounded by the curve x = f(t), y = φ(t), the x-axis and the ordinates, where t = a, t = b is**

$$\int_a^b \pi y^2 \frac{dx}{dt}\, dt$$

(*ii*) **The volume of the solid generated by the revolution about the y-axis, of the area bounded by the curve x = f(t), y = φ(t), the y-axis and the abscissae at the points, where t = a, t = b is**

$$\int_a^b \pi x^2 \frac{dy}{dt}\, dt.$$

ILLUSTRATIVE EXAMPLES

Example 1. *Find the volume of the solid formed by the revolution of one arch of the cycloid*

(*i*) $x = a(\theta - \sin\theta)$, $y = a(1 - \cos\theta)$ *about its base*

(*ii*) $x = a\,(\theta + \sin\theta)$, $y = a\,(1 - \cos\theta)$ *about the tangent at the vertex.*

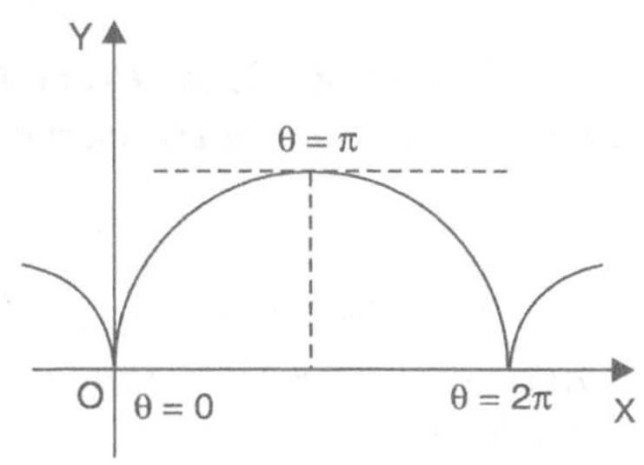

Sol. (*i*) The equations of the cycloid are

$$x = a\,(\theta - \sin\theta),\; y = a(1 - \cos\theta)$$

The cycloid is symmetrical about the line through the point, where $\theta = \pi$, $\perp$ to x-axis.

For the first half of the cycloid, in the first quadrant, θ varies from 0 to π.

The base is x-axis.

$\therefore$ Required volume of the solid generated by revolution about its base (x-axis)

$$= 2 \int_0^{\pi} \pi y^2 \cdot \frac{dx}{d\theta}\, d\theta$$

$$= 2\pi \int_0^{\pi} a^2(1 - \cos\theta)^2 \cdot a(1 - \cos\theta)\, d\theta \qquad | \text{ From (1)}$$

$$= 2\pi a^3 \int_0^{\pi} \left(2\sin^2\frac{\theta}{2}\right)^2 \cdot \left(2\sin^2\frac{\theta}{2}\right) d\theta$$

| Put $\theta/2 = t$ $\therefore$ $d\theta = 2dt$
| when $\theta = 0, t = 0$; when $\theta = \pi, t = \pi/2$

$$= 2\pi a^3 \int_0^{\pi/2} (2\sin^2 t)^3 \cdot 2\, dt = 32\, \pi a^3 \int_0^{\pi/2} \sin^6 t\, dt$$

$$= 32\pi a^3 \cdot \frac{5 \cdot 3 \cdot 1}{6 \cdot 4 \cdot 2} \cdot \frac{\pi}{2} = 5\pi^2 a^3.$$

(*ii*) The equations of the cycloid are

$$x = a(\theta + \sin\theta),$$
$$y = a\,(1 - \cos\theta)$$

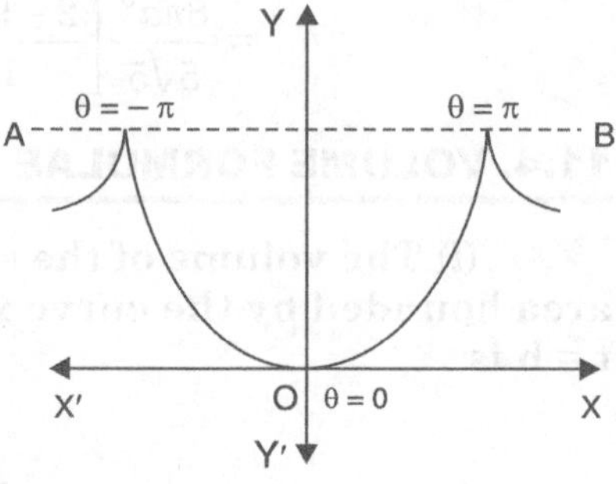

The cycloid is symmetrical about the y-axis and the tangent at the vertex is x-axis.

For the half of the curve, θ varies from 0 to π.

$\therefore$ Required volume

$$= 2 \int_0^{\pi} \pi y^2 \frac{dx}{d\theta}\, d\theta$$

$$= 2\pi \int_0^{\pi} a^2(1 - \cos\theta)^2 \cdot a(1 + \cos\theta)\, d\theta$$

$$= 2\pi a^3 \int_0^{\pi} \left(2\sin^2\frac{\theta}{2}\right)^2 \cdot \left(2\cos^2\frac{\theta}{2}\right) d\theta$$

$$= 2\pi a^3 \int_0^{\pi/2} (2\sin^2 t)^2\, (2\cos^2 t) \cdot 2dt \qquad \Big| \text{ where } \frac{\theta}{2} = t$$

$$= 32\pi a^3 \int_0^{\pi/2} \sin^4 t \cos^2 t\, dt$$

$$= 32\pi a^3 \cdot \frac{3 \cdot 1 \cdot 1}{6 \cdot 4 \cdot 2} \cdot \frac{\pi}{2} = \pi^2 a^3.$$

Example 2. (*a*) *Prove that the volume of the solid generated by the revolution about the x-axis of the loop of the curve*

$$x = t^2,\ y = t - \frac{1}{3}t^3 \text{ is } \frac{3\pi}{4}.$$

(*b*) *Find the volume of the spindle-shaped solid generated by revolving the astroid $x^{2/3} + y^{2/3} = a^{2/3}$ about the x-axis.*

Sol. (*a*) The equations of the curve are

$$x = t^2,\ y = t - \tfrac{1}{3}t^3 \qquad \text{...(1)}$$

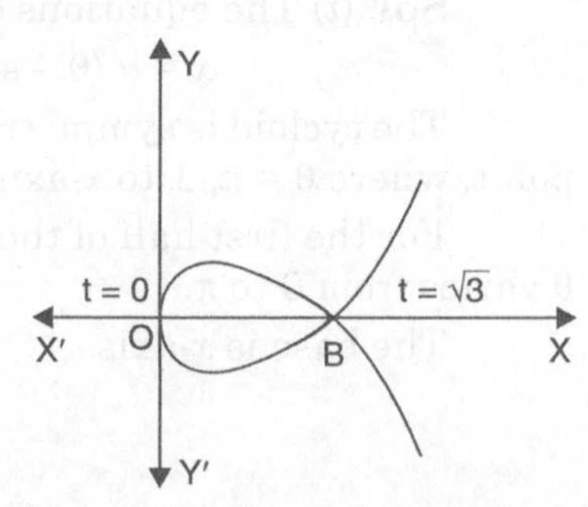

The curve is symmetrical about the x-axis and for the loop, putting $y = 0$, we get

$$t(1 - \tfrac{1}{3} t^2) = 0$$

$$\therefore \qquad t = 0, \pm \sqrt{3}.$$

Thus for the upper half of the loop, t varies from 0 to $\sqrt{3}$.

$$\therefore \quad \text{Required volume} = \int_0^{\sqrt{3}} \pi y^2 \frac{dx}{dt}\, dt = \int_0^{\sqrt{3}} \pi(t - \frac{1}{2} t^3)^2 \cdot 2t\, dt$$

$$= 2\pi \int_0^{\sqrt{3}} t(t^2 + \frac{1}{9} t^6 - \frac{2}{3} t^4)\, dt = 2\pi \int_0^{\sqrt{3}} (t^3 + \frac{1}{9} t^7 - \frac{2}{3} t^5)\, dt$$

$$= 2\pi \left[\frac{t^4}{4} + \frac{1}{9} \cdot \frac{t^8}{8} - \frac{2}{3} \cdot \frac{t^6}{6} \right]_0^{\sqrt{3}} = 2\pi \left[\frac{9}{4} + \frac{1}{9} \cdot \frac{81}{8} - \frac{2}{3} \cdot \frac{27}{6} \right]$$

$$= 2\pi \left[\frac{9}{4} + \frac{9}{8} - 3 \right] = 2\pi \left[\frac{18 + 9 - 24}{8} \right] = \frac{3\pi}{4}.$$

(*b*) The equation of the curve is $x^{2/3} + y^{2/3} = a^{2/3}$.

The parametric equations of the curve are

$$x = a \cos^3 t,\ y = a \sin^3 t \quad ...(1)$$

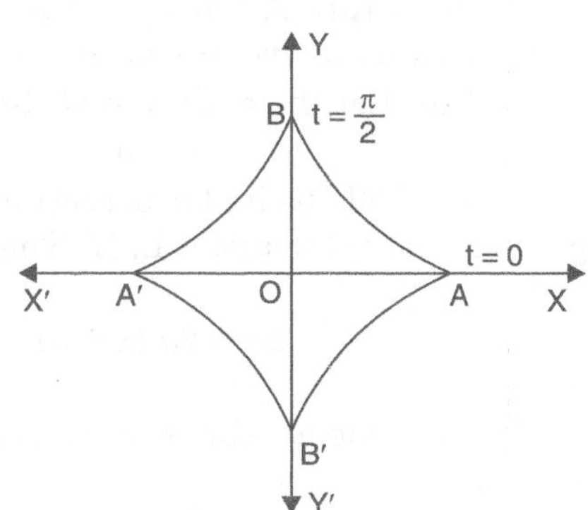

The curve is symmetrical about both the axes and for the portion of the curve in the first quadrant, t varies from 0 to $\frac{\pi}{2}$.

$\therefore$ Required volume

= 2 × volume generated by the arc in the first quadrant

$$= 2 \int_0^{\pi/2} \pi y^2 \frac{dx}{dt}\, dt$$

$$= 2\pi \int_0^{\pi/2} a^2 \sin^6 t \cdot (-3a \cos^2 t \sin t)\, dt$$

$$= -6\pi a^3 \int_0^{\pi/2} \sin^7 t \cdot \cos^2 t\, dt = -6\pi a^3 \cdot \frac{6 \cdot 4 \cdot 2 \cdot 1}{9 \cdot 7 \cdot 5 \cdot 3 \cdot 1}$$

$$= \frac{-32}{105} \pi a^3 = \frac{32}{105} \pi a^3 \text{ (in magnitude).}$$

11.5. VOLUME BETWEEN TWO SOLIDS

The volume of the solid generated by the revolution about the x-axis of the area bounded by the curves $y = f(x)$, $y = \phi(x)$, and the ordinates $x = a$, $x = b$ is

$$\int_a^b \pi(y_1^2 - y_2^2)\, dx$$

where y_1 is the 'y' of the upper curve and y_2 that of the lower curve.

ILLUSTRATIVE EXAMPLES

Example 1. *Find the volume of the solid generated by the revolution of the plane area bounded by $y^2 = 9x$ and $y = 3x$ about the x-axis.*

Sol. The line cuts the parabola at A(1, 3).

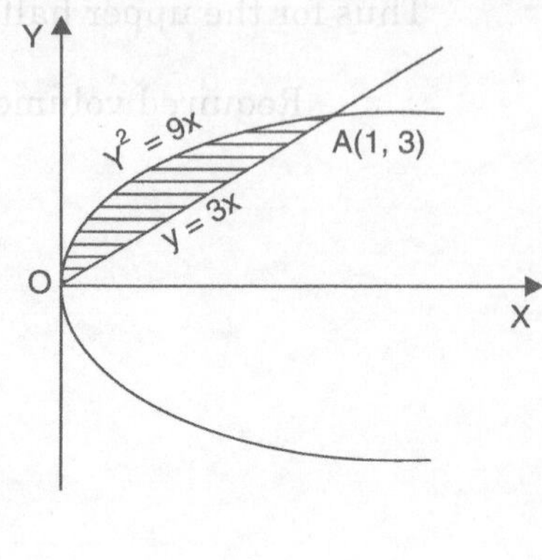

Required volume

$$= \int_0^1 \pi(y_1^2 - y_2^2)\, dx$$

$$= \int_0^1 \pi(\underset{\text{(for parabola)}}{\underset{\downarrow}{9x}} - \underset{\text{(for line)}}{\underset{\downarrow}{9x^2}})\, dx$$

$$= \pi\left[\frac{9x^2}{2} - 3x^3\right]_0^1$$

$$= \pi\left(\frac{9}{2} - 3\right) = \frac{3\pi}{2}.$$

Example 2. *The figure bounded by a parabola and the tangents at the extremities of its latus rectum revolves about the axis of the parabola. Find the volume of the solid thus generated.*

Sol. Let the equation of the parabola be

$$y^2 = 4ax \qquad \text{...(1)}$$

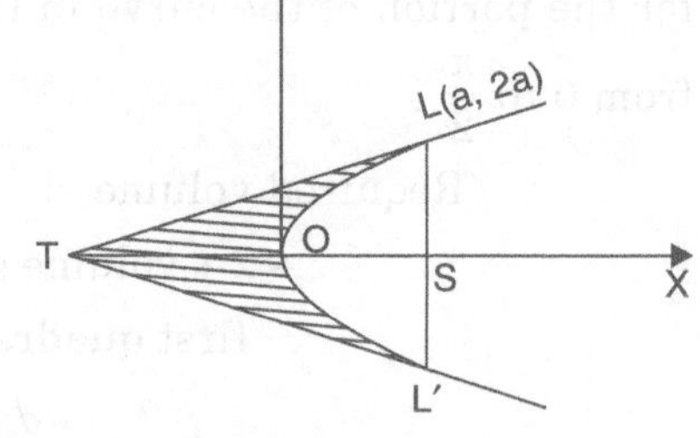

Let LSL′ be its latus rectum and LT, L′T be the tangents at the extremities L, L′. The equation of the tangent at L $(a, 2a)$ is

$$y \cdot 2a = 2a\,(x + a) \quad | \text{ Using } yy_1 = 2a(x + x_1)$$

or

$$y = x + a \qquad \text{...(2)}$$

This meets the x-axis, where putting $y = 0$, we get $x = -\, a$.

Thus the point T is $(-a, 0)$. The curve and the two tangents are symmetrical about the x-axis.

$$\therefore \quad \text{Required volume} = \underset{\text{(for tangent)}}{\int_{-a}^{a} \pi y^2\, dx} - \underset{\text{(for parabola)}}{\int_0^a \pi y^2\, dx}$$

$$= \pi \int_{-a}^{a} (x + a)^2\, dx - \pi \int_0^a 4ax\, dx$$

$$= \pi \left| \frac{(x + a)^3}{3} \right|_{-a}^{a} - 2a\pi \left| x^2 \right|_0^a$$

$$= \frac{\pi}{3}(8a^3 - 0) - 2a\pi(a^2 - 0) = \frac{2}{3}\pi a^3.$$

11.6. THE VOLUME FORMULAE FOR POLAR CURVES

The volume of the solid generated by the revolution of the area bounded by the curve r = f(θ) and the radii vectors θ = α, θ = β

(*i*) about the initial line OX (θ = 0) is

$$\int_\beta^\alpha \frac{2}{3} \pi r^3 \sin\theta\, d\theta$$

(*ii*) **about the line OY** $\left(\theta = \frac{\pi}{2}\right)$ **is**

$$\int_{\alpha}^{\beta} \frac{2}{3} \pi r^3 \cos\theta \, d\theta.$$

ILLUSTRATIVE EXAMPLES

Example 1. *Show that the volume of the solid formed by the revolution of the curve $r = a + b\cos\theta$ $(a > b)$ about the initial line is*

$$\frac{4}{3}\pi a(a^2 + b^2).$$

Hence or otherwise show that the volume of solid generated by the revolution of the cardioid $r = a(1 + \cos\theta)$ about the initial line is

$$\frac{8}{3}\pi a^2.$$

Sol. The equation of the curve is

$$r = a + b\cos\theta \; (a > b) \qquad ...(1)$$

The curve is symmetrical about the initial line and for the upper half of the curve θ varies from 0 to π.

Y
θ = π
θ = 0
B O A X

∴ Required volume

$$= \int_0^{\pi} \frac{2}{3} \pi r^3 \sin\theta \, d\theta$$

$$= \frac{2}{3}\pi \int_0^{\pi} (a + b\cos\theta)^3 \sin\theta \, d\theta$$

$$= -\frac{2}{3}\frac{\pi}{b} \int_0^{\pi} (a + b\cos\theta)^3 (-b\sin\theta \, d\theta)$$

$$= -\frac{2}{3}\frac{\pi}{b}\left[\frac{(a + b\cos\theta)^4}{4}\right]_0^{\pi} = -\frac{2\pi}{3b}\left[\frac{(a-b)^4}{4} - \frac{(a+b)^4}{4}\right]$$

$$= \frac{\pi}{6b}[(a+b)^4 - (a-b)^4]$$

$$= \frac{\pi}{6b}[(a+b)^2 + (a-b)^2][(a+b)^2 - (a-b)^2]$$

$$= \frac{\pi}{6b} \cdot 2(a^2 + b^2) \cdot 4ab = \frac{4}{3}\pi a(a^2 + b^2) \qquad ...(2)$$

Putting $b = a$, in (1), the curve becomes $r = a\,(1 + \cos\theta)$ which is the given cardioid.

∴ Required volume by the revolution of the cardioid about the initial line [Putting $b = a$ in (2)] is

$$\tfrac{4}{3}\pi\, a(a^2 + a^2) = \tfrac{8}{3}\pi a^3.$$

Example 2. *Find the volume of the solid generated by revolving one loop of the lemniscate $r^2 = a^2 \cos 2\theta$ about the line $\theta = \frac{\pi}{2}$.*

Sol. The given curve is

$$r^2 = a^2 \cos 2\theta \qquad \text{...(1)}$$

The curve is symmetrical about the initial line

For a loop, putting $r = 0$ in (1), we get $\cos 2\theta = 0$

$$\therefore \quad 2\theta = \pm \frac{\pi}{2}$$

$$\therefore \quad \theta = \pm \frac{\pi}{4} \quad \textit{[two consecutive values]}$$

Thus as θ varies from $-\frac{\pi}{4}$ to $\frac{\pi}{4}$, we get a loop.

For the upper half of this loop, θ varies from 0 to $\frac{\pi}{4}$.

$\therefore$ Required volume obtained by revolution of the loop about OY $\left(\theta = \frac{\pi}{2}\right)$

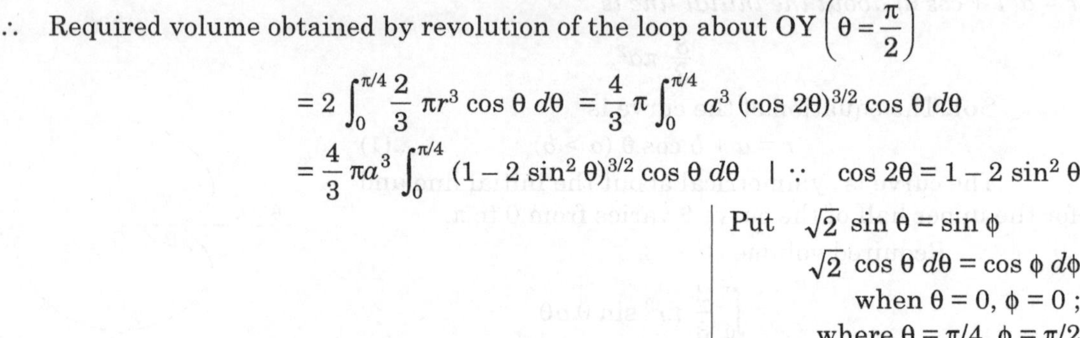

$$= 2\int_0^{\pi/4} \frac{2}{3}\pi r^3 \cos\theta \, d\theta = \frac{4}{3}\pi \int_0^{\pi/4} a^3 (\cos 2\theta)^{3/2} \cos\theta \, d\theta$$

$$= \frac{4}{3}\pi a^3 \int_0^{\pi/4} (1 - 2\sin^2\theta)^{3/2} \cos\theta \, d\theta \quad \Big| \because \quad \cos 2\theta = 1 - 2\sin^2\theta$$

Put $\sqrt{2}\sin\theta = \sin\phi$

$\sqrt{2}\cos\theta \, d\theta = \cos\phi \, d\phi$

when $\theta = 0, \phi = 0$;

where $\theta = \pi/4, \phi = \pi/2$

$$= \frac{4}{3}\pi a^3 \int_0^{\pi/2} (1 - \sin^2\phi)^{3/2} \frac{1}{\sqrt{2}} \cos\phi \, d\phi$$

$$= \frac{4}{3\sqrt{2}}\pi a^3 \int_0^{\pi/2} \cos^4\phi \, d\phi = \frac{4}{3\sqrt{2}}\pi a^3 \cdot \frac{3 \cdot 1}{4 \cdot 2}\frac{\pi}{2} = \frac{\pi^2 a^3}{4\sqrt{2}}.$$

EXERCISE 11.1

1. A basin is formed by the revolution of the curve $x^3 = 64y$, $(y > 0)$ about the axis of y. If the depth of the basin is 8 units, how many cubic units of water will it hold?
2. Show that the volume of the spherical cap of a sphere whose height is h and base has radius c is $\frac{\pi h}{6}(3c^2 + h^2)$.
3. A segment is cut off from a sphere of radius a by a plane at a distance $\frac{a}{2}$ from the centre. Show that the volume of the segment is $\frac{5}{32}$ of the volume of the sphere.
4. Find the volume generated by revolving about $y = 2$ the area in the first quadrant bounded by the parabola $8y = x^2$, the y-axis and the line $y = 2$.
5. Prove that the volume of the solid generated by the revolution of an ellipse round its minor axis is the mean proportional between the volumes of the solids generated by the revolution of the ellipse and the auxiliary circle round the major axis.
6. Find the volume of the solid obtained by revolving $x = a\cos\theta$, $y = b\sin\theta$ about the y-axis.

   $\left[\textbf{Hint}. \text{ Eliminating } \theta, \text{ we have } \frac{x^2}{a^2} + \frac{y^2}{b^2} = 1\right]$ *(M.D.U. Dec. 2012)*

7. Find the volume when the loop of the curve $y^2 = x(2x - 1)^2$ revolves about the x-axis.

8. (*a*) Show that the volume of solid generated by the revolution of the loop of $x(x^2 + y^2) = a(x^2 - y^2)$ about the x-axis is $2\pi a^3 (\log 2 - \frac{2}{3})$.

   (*b*) Find the volume of the solid formed by the revolution about the x-axis, of the following curves:
   (*i*) $y^2(a + x) = x^2(a - x)$ (*ii*) $y^2 = x^2(a - x)$.

   (*c*) The loop of the curve $2ay^2 = x(x - a)^2$ revolves about x-axis, find the volume of the solid so generated.

9. Show that the volume of the solid obtained by the revolution of the curve $a^2y^2 = x^2 (a^2 - x^2)$ about the x-axis is $\frac{4}{15} a^3$.

10. Find the volume of the solid formed by the revolution of the area enclosed by the curve $xy^2 = a^2(a - x)$ and its asymptote about the y-axis.

11. If the curve $(a - x) y^2 = a^2x$ revolves about its asymptote, find the volume so formed.

12. The smaller segment of the ellipse $\frac{x^2}{a^2} + \frac{y^2}{b^2} = 1$ cut off by the chord $x = \frac{a}{2}$ revolves completely about this chord. Find the volume generated.

13. Show that the volume of the solid generated by the revolution of the cycloid $x = a (\theta + \sin \theta)$, $y = a(1 - \cos \theta)$ about y-axis is $\pi a^3 \left(\frac{3}{2}\pi^2 - \frac{8}{3}\right)$.

14. The arc lying between $\theta = -\frac{\pi}{2}$ and $\theta = \frac{\pi}{2}$ of the curve $x = a \cos^3 \theta$, $y = a \sin^3 \theta$ rotates about the axis of x. Find the volume of the solid so generated.

15. Find the volume of the solid generated by the arc of cissoid $x = 2a \sin^2 t$, $y = \frac{2a \sin^3 t}{\cos t}$ about its asymptote.
[**Hint.** Changing to cartesian co-ordinates, the equation of the cissoid is $y^2 (2a - x) = x^3$.]

16. Prove that the volume of reel formed by the revolution of the cycloid $x = a(\theta + \sin \theta)$, $y = a (1 - \cos \theta)$ about the tangent at the vertex is π^2a^3.

17. A loop of the curve $y^2 = a^2 (1 - x^2)$ is rotated about the y-axis. Find the volume generated.

18. (*a*) The segment of the parabola $y^2 = 4ax$ which is cut off by the latus rectum revolves about the directrix. Find the volume of the solid generated.

    (*b*) The area bounded by the parabola $y^2 = 4ax$, its axis and its latus rectum revolves about the latus rectum. Find the volume generated.

    (*c*) Find the volume of the reel-shaped solid formed by the revolution about the y-axis, of the part of the parabola $y^2 = 4ax$ cut off by the latusrectum. (*M.D.U. Dec. 2010*)

    (*d*) Find the volume generated by revolving the portion of the prabola $y^2 = 4ax$ cut off by its latusrectum about the axis of the parabola.

19. (*a*) Find the volume generated by revolving the area bounded by the curve $\sqrt{x} + \sqrt{y} = \sqrt{a}$, $x = 0$, $y = 0$ about the x-axis.

    (*b*) Find the volume formed by the revolution of loop of the curve $y^2(a + x) = x^2(3a - x)$, about the x-axis. (*P.T.U. Dec. 2011*)

    (*c*) Find the volume generated by the revolution of the curve $27ay^2 = 4(x - 2a)^3$ about the x-axis.

    (*d*) Find the volume of the paraboloid generated by the revolution of $y^2 = 4ax$ about x-axis from $x = 0$ to $x = b$.

(*e*) Area bounded by x-axis, $y^2 = 4ax$ and the ordinate $x = 3a$ is revolved about x-axis. Find the volume generated.

20. (*a*) Find the volume generated by revolving about x-axis, the area cut off from the parabola $9y = 4(9 - x^2)$ by the line $4x + 3y = 12$.

(*b*) The area bounded by the parabola $y^2 = 4x$ and the straight line $4x - 3y + 2 = 0$ is rotated about the y-axis. Obtain the volume of the solid formed by this revolution.

$$\left[\textbf{Hint. } \text{Use } \int_a^b \pi\,(x_1^2 - x_2^2)\,dy.\right]$$

21. Find the volume generated by revolving about OX the area bounded by curves $x^2 + y^2 = 25$, $3x - 4y = 0$, $y = 0$ lying in the first quadrant.

22. Show that the volume of the solid generated by revolving the area included between the curves $y^2 = x^3$ and $x^2 = y^3$ about x-axis is $\frac{5\pi}{28}$.

23. The cardioid $r = a(1 - \cos\theta)$ revolves about the initial line. Find the volume of the solid generated. (*P.T.U. 2011*)

24. Show that if the area lying within the cardioid $r = 2a(1 + \cos\theta)$ and without the parabola $r(1 + \cos\theta) = 2a$, revolves about the initial line, the volume generated is $18\,\pi a^3$.

25. (*a*) Find the volume of the solid generated by the revolution of the cardioid $r = a(1 + \cos\theta)$ about the initial line. (*M.D.U. Dec. 2010; Kerala 2010*)

(*b*) Find the volume of the solid formed by revolving a loop of the lemniscate $r^2 = a^2\cos 2\theta$ about the initial line. (*M.D.U. May 2009*)

Answers

1. $\frac{1536}{5}\pi$ cubic units **4.** $\frac{128}{15}\pi$ **6.** $\frac{4}{3}\pi a^2 b$

7. $\frac{\pi}{48}$ **8.** (*b*) (*i*) $2\pi a^3\left(\log 2 - \frac{2}{3}\right)$ (*ii*) $\frac{\pi a^4}{12}$

(*c*) $\frac{\pi a^3}{24}$ **10.** $\frac{1}{2}\pi^2 a^3$ **11.** $\frac{1}{2}\pi^2 a^3$

12. $\pi a^2 b\left(\frac{3\sqrt{3}}{4} - \frac{\pi}{3}\right)$ **14.** $\frac{16\pi a^2}{105}$ **15.** $2\pi^2 a^3$

17. $\frac{\pi^3}{4}$ **18.** (*a*) $\frac{128\,\pi a^3}{15}$ (*b*) $\frac{32}{15}\pi a^3$

(*c*) $\frac{4}{5}\pi a^3$ (*d*) $2\pi a^3$

19. (*a*) $\frac{\pi a^3}{12}$ (*b*) $\pi(8\log 2 - 3)a^3$ (*c*) $48\pi a^3$

(*d*) $2ab^2\pi$ (*e*) $18\pi a^3$ **20.** (*a*) $\frac{48\pi}{5}$

(*b*) $\frac{\pi}{20}$ **21.** $\frac{50\pi}{3}$ **23.** $\frac{8}{3}\pi a^3$

25. (*a*) $\frac{8}{3}\pi a^3$ (*b*) $\frac{\pi a^3}{4}\left[\frac{1}{\sqrt{2}}\log(\sqrt{2} + 1) - \frac{1}{3}\right]$

11.7. THREE PRACTICAL FORMS OF THE SURFACE FORMULA

(*i*) **Surface formula for Cartesian equations.** *The curved surface of the solid generated by the revolution about the x-axis, of the area bounded by the curve v = f(x), the x-axis and the ordinates x = a, x = b is*

$$\int_{x=a}^{x=b} 2\pi y \frac{ds}{dx}\,.\,dx, \quad where \quad \frac{ds}{dx} = \sqrt{1+\left(\frac{dy}{dx}\right)^2}$$

(*ii*) **Surface formula for Parametric equations.** *The curved surface of solid generated by the revolution about the x-axis, of the area bounded by the curve x = f(t), y = φ (t), the x-axis and the ordinates at the points, where t = a, t = b is*

$$\int_{t=a}^{t=b} 2\pi y \frac{ds}{dt}\,.\,dt, \quad where \quad \frac{ds}{dt} = \sqrt{\left(\frac{dx}{dt}\right)^2 + \left(\frac{dy}{dt}\right)^2}$$

(*iii*) **Surface formula for Polar equations.** *The curved surface of the solid generated by the revolution, about the initial line, of the area bounded by the curve r = f(θ) and the radii vectors θ = α, θ = β is*

$$\int_{\theta=\alpha}^{\theta=\beta} 2\pi y \frac{ds}{d\theta}\, d\theta, \quad where \quad \frac{ds}{d\theta} = \sqrt{r^2 + \left(\frac{dr}{d\theta}\right)^2} \quad \text{and} \quad y = r\sin\theta.$$

ILLUSTRATIVE EXAMPLES

Example 1. *Find the surface generated by the revolution of arc of the catenary*

$y = c \cosh \frac{x}{c}$ from x = 0 to any point (x, y) about the axis of x.

Sol. The equation of the curve is $y = c \cosh \frac{x}{c}$

$$\therefore \quad \frac{dy}{dx} = c \cdot \sinh\left(\frac{x}{c}\right) \cdot \frac{1}{c} = \sinh \frac{x}{c}$$

$$\therefore \quad \frac{ds}{dx} = \sqrt{1+\left(\frac{dy}{dx}\right)^2} = \sqrt{1+\sinh^2 \frac{x}{c}} = \cosh \frac{x}{c}$$

If the arc be measured from the vertex (where $x = 0$) to any point (x, y), then

$$\text{the required surface} = \int_{x=a}^{x} 2\pi y \frac{ds}{dx}\, dx = \int_0^x 2\pi c \cdot \cosh \frac{x}{c} \cdot \cosh \frac{x}{c}\, dx = \int_0^x 2\pi c \cosh^2 \frac{x}{c} dx$$

$$= \pi c \int_0^x \left[1 + \cosh\left(\frac{2x}{c}\right)\right] dx \qquad \left|\ \because \ \cosh^2\theta = \frac{1+\cosh 2\theta}{2}\right.$$

$$= \pi c \left[x + \frac{c}{2}\sinh\left(\frac{2x}{c}\right)\right]_0^x = \pi c \left[x + \frac{c}{2}\sinh\left(\frac{2x}{c}\right)\right].$$

Example 2. *Find the surface of the solid generated by the revolution of the ellipse $x^2 + 4y^2 = 16$ about its major axis.*

Sol. The equation of the ellipse is $x^2 + 4y^2 = 16$

or $$4y^2 = 16 - x^2 \quad \text{or} \quad y = \frac{\sqrt{16 - x^2}}{2} \qquad \text{...(1)}$$

$$\therefore \quad \frac{dy}{dx} = \frac{1}{2} \cdot \frac{1}{2} (16 - x^2)^{-1/2} (-2x) = \frac{-x}{2\sqrt{16 - x^2}}$$

$$\frac{ds}{dx} = \sqrt{1 + \left(\frac{dy}{dx}\right)^2} = \sqrt{1 + \frac{x^2}{4(16 - x^2)}} = \sqrt{\frac{64 - 3x^2}{4(16 - x^2)}}$$

The ellipse $x^2 + 4y^2 = 16$ meets x-axis, where putting $y = 0$, we get

$$x^2 = 16 \quad \text{or} \quad x = \pm 4.$$

Thus for the upper half of the ellipse in first quadrant, x varies from 0 to 4.

Since the ellipse is symmetrical about y-axis.

$\therefore$ Required surface = 2 × surface generated by the arc in the first quadrant.

$$= 2\int_0^4 2\pi y \frac{ds}{dx} \cdot dx = 4\pi \int_0^4 \frac{\sqrt{16 - x^2}}{2} \cdot \sqrt{\frac{64 - 3x^2}{4(16 - x^2)}} \, dx$$

$$= \pi \int_0^4 \sqrt{64 - 3x^2} \, dx = \sqrt{3}\,\pi \int_0^4 \sqrt{\frac{64}{3} - x^2 dx}$$

$$= \pi\sqrt{3} \left[\frac{x}{2} \cdot \sqrt{\frac{64}{3} - x^2} + \frac{64}{3} \cdot \frac{1}{2} \sin^{-1}\left(\frac{x}{8/\sqrt{3}}\right) \right]_0^4$$

$$= \sqrt{3}\,\pi \left[2\sqrt{\frac{64}{3} - 16} + \frac{32}{3} \sin^{-1}\left(\frac{\sqrt{3}}{2}\right) \right]$$

$$= \sqrt{3}\,\pi \left[2 \cdot \frac{4}{\sqrt{3}} + \frac{32}{3} \cdot \frac{\pi}{3} \right] = 8\pi \left[1 + \frac{4\pi}{3\sqrt{3}} \right].$$

Example 3. *A parabolic reflector of an automobile head light is 12 cm in diameter and 4 cm deep. Find the cost of plating of the front portion of the reflector if the cost of plating is ₹ 5.00 per cm².*

Sol. Let $y^2 = 4ax$ be the equation of the parabola. Since OA = 4 cm and AP = 6 cm, the coordinates of P arc (4, 6). As P lies on the parabola, we have

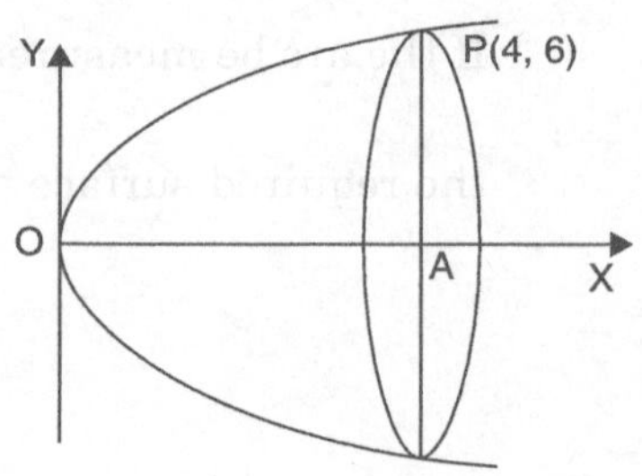

$$6^2 = 4a \times 4 \quad \text{or} \quad a = \frac{9}{4}.$$

$\therefore$ Equation of parabola is

$$y^2 = 9x$$

The surface S of the parabolic reflector is formed by revolving the arc OP of the parabola about x-axis.

For the arc OP,

$$y = 3\sqrt{x} \quad \text{and} \quad \frac{dy}{dx} = \frac{3}{2\sqrt{x}}$$

$$\therefore \quad S = \int_0^4 2\pi y \frac{dS}{dx} dx = 2\pi \int_0^4 3\sqrt{x} \sqrt{1 + \left(\frac{dy}{dx}\right)^2} \, dx$$

$$= 6\pi \int_0^4 \sqrt{x} \sqrt{1+\frac{9}{4x}}\, dx$$

$$= 3\pi \int_0^4 \sqrt{4x+9}\, dx = 3\pi \left[\frac{(4x+9)^{3/2}}{\frac{3}{2}\times 4}\right]_0^4$$

$$= \frac{\pi}{2}(125-27) = 49\pi = 49 \times \frac{22}{7} = 154 \text{ cm}^2.$$

Cost of plating = ₹ (154 × 5) = ₹ 770.

Example 4. *Find the surface of the solid generated by the revolution of the astroid*

$$x^{2/3} + y^{2/3} = a^{2/3} \quad \text{or} \quad x = a\cos^3 t,\ y = a\sin^3 t$$

about the x-axis.

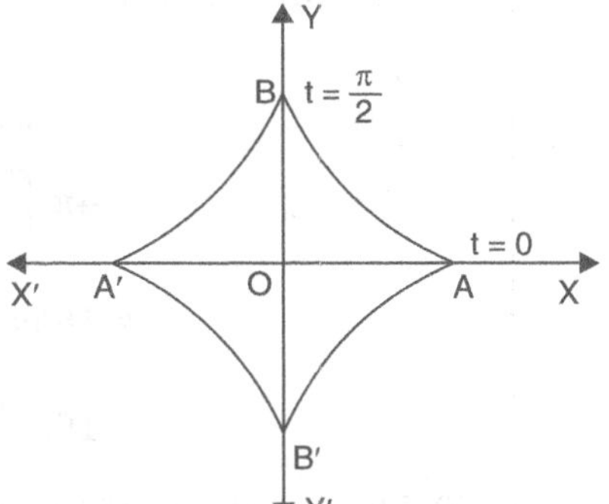

Sol. The parametric equations of the astroid are

$$x = a\cos^3 t,\ y = a\sin^3 t$$

The curve is symmetrical about both the axes and it lies in the square bounded by the lines.

$$x = \pm a,\ y = \pm a.$$

For the portion of the curve in the first quadrant, t varies from 0 to $\frac{\pi}{2}$.

Now $$\frac{dx}{dt} = -3a\cos^2 t \sin t \quad \text{and} \quad \frac{dy}{dt} = 3a\sin^2 t\cos t$$

$\therefore$ $$\frac{ds}{dt} = \sqrt{9a^2\cos^4 t\sin^2 t + 9a^2\sin^4 t\cos^2 t}$$

$$= \sqrt{9a^2\sin^2 t\cos^2 t(\cos^2 t + \sin^2 t)} = 3a\sin t\cos t$$

$\therefore$ Required surface $$= 2\int_{t=0}^{\pi/2} 2\pi y \frac{ds}{dt}\, dt$$

$$= 4\pi \int_0^{\pi/2} a\sin^3 t \cdot 3a\sin t\cos t\, dt$$

$$= 12\pi a^2 \int_0^{\pi/2} \sin^4 t\cos t\, dt = 12\pi a^2 \left[\frac{\sin^5 t}{5}\right]_0^{\pi/2}$$

$$= \frac{12}{5}\pi a^2 (1-0) = \frac{12}{5}\pi a^2.$$

Example 5. *Find the surface area of the solid generated by revolving the cycloid.*

(*i*) $x = a(\theta - \sin\theta),\ y = a(1-\cos\theta)$ *about the x-axis.*

(*ii*) $x = a(\theta + \sin\theta),\ y = a(1-\cos\theta)$ *about the tangent at the vertex.*

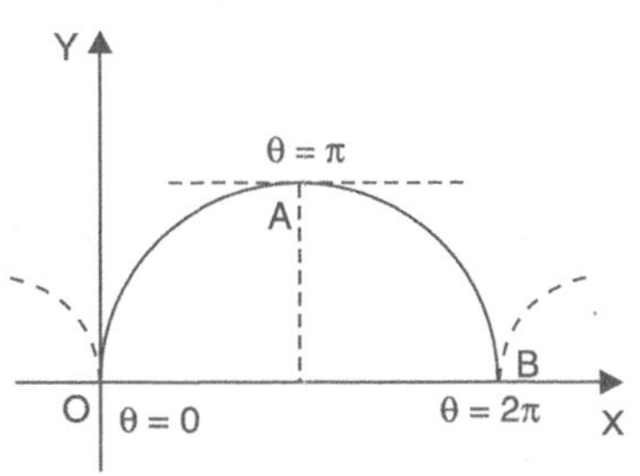

Sol. (*i*) The equations of the cycloid are

$$x = a(\theta - \sin\theta),\ y = a(1-\cos\theta)$$

The cycloid is symmetrical about the line through the point

$$\theta = \pi, \perp \text{ to the } x\text{-axis}.$$

For the first half of the cycloid, θ varies from 0 to π.

Now $$\frac{dx}{d\theta} = a(1-\cos\theta),$$

$$\frac{dy}{d\theta} = a \sin \theta$$

$$\therefore \quad \frac{ds}{d\theta} = \sqrt{\left(\frac{dx}{d\theta}\right)^2 + \left(\frac{dy}{d\theta}\right)^2} = \sqrt{a^2 (1 - \cos \theta)^2 + a^2 \sin^2 \theta}$$

$$= a\sqrt{1 + \cos^2 \theta - 2 \cos \theta + \sin^2 \theta} = a\sqrt{2(1 - \cos \theta)}$$

$$= a\sqrt{2 . 2 \sin^2 \frac{\theta}{2}} = 2a \sin \frac{\theta}{2}.$$

$$\therefore \quad \text{Required surface} = 2 \int_0^{\pi} 2\pi y \frac{ds}{d\theta}\, d\theta$$

$$= 4\pi \int_0^{\pi} a(1 - \cos \theta) \cdot 2a \sin \frac{\theta}{2}\, d\theta$$

$$= 4\pi \int_0^{\pi} a\left(2 \sin^2 \frac{\theta}{2}\right) \cdot 2a \sin \frac{\theta}{2}\, d\theta$$

$$= 16\pi a^2 \int_0^{\pi} \sin^3 \frac{\theta}{2}\, d\theta \qquad \left|\begin{array}{l}\text{Put } \theta/2 = t \\ \therefore \; d\theta = 2dt\end{array}\right.$$

$$= 16\pi a^2 \int_0^{\pi/2} \sin^3 t \cdot 2\, dt = 32\pi a^2 \frac{2}{3 \cdot 1} = \frac{64}{3} \pi a^2.$$

(*ii*) The equations of the cycloid are

$$x = a(\theta + \sin \theta),$$
$$y = a(1 - \cos \theta)$$

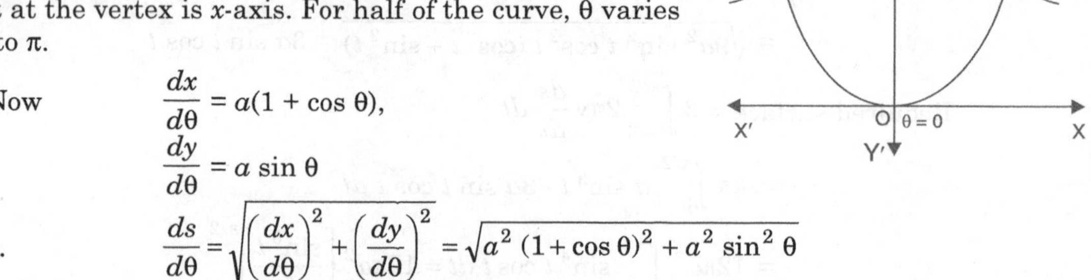

The cycloid is symmetrical about the y-axis and the tangent at the vertex is x-axis. For half of the curve, θ varies from 0 to π.

Now $$\frac{dx}{d\theta} = a(1 + \cos \theta),$$

$$\frac{dy}{d\theta} = a \sin \theta$$

$$\therefore \quad \frac{ds}{d\theta} = \sqrt{\left(\frac{dx}{d\theta}\right)^2 + \left(\frac{dy}{d\theta}\right)^2} = \sqrt{a^2 (1 + \cos \theta)^2 + a^2 \sin^2 \theta}$$

$$= a\sqrt{1 + \cos^2 \theta + 2 \cos \theta + \sin^2 \theta} = a\sqrt{2(1 + \cos \theta)}$$

$$= a\sqrt{2 \cdot 2 \cos^2 \frac{\theta}{2}} = 2a \cos \frac{\theta}{2}$$

$$\therefore \quad \text{Required surface} = 2 \int_0^{\pi} 2\pi y \frac{ds}{d\theta}\, d\theta$$

$$= 4\pi \int_0^{\pi} a(1 - \cos \theta) \cdot 2a \cos \frac{\theta}{2}\, d\theta$$

$$= 4\pi \int_0^{\pi} a\left(2 \sin^2 \frac{\theta}{2}\right) \cdot 2a \cos \frac{\theta}{2}\, d\theta$$

$$= 16\pi a^2 \int_0^{\pi} \sin^2 \frac{\theta}{2} \cos \frac{\theta}{2}\, d\theta$$

$$= 16\pi a^2 \left| \frac{\sin^3 \theta/2}{3 \cdot \frac{1}{2}} \right|_0^{\pi} = \frac{32\,\pi a^2}{3}[1 - 0] = \frac{32}{3} \pi a^2.$$

Example 6. *Determine the area of the surface generated by the revolution of the loops of the curve $r^2 = a^2 \cos 2\theta$ about the initial line.*

Sol. The equation of the curves is $r^2 = a^2 \cos 2\theta$ ...(1)

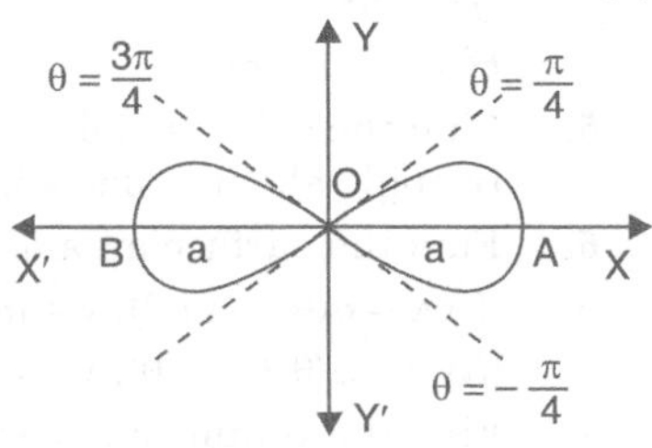

The curve is symmetrical about the initial line and the line

$$\theta = \frac{\pi}{2}.$$

For a loop, putting $r = 0$, we get $\cos 2\theta = 0$

$$2\theta = \pm\frac{\pi}{2}$$

or $$\theta = +\frac{\pi}{4}, -\frac{\pi}{4}$$ [*two consecutive values*]

The curve consists of two equal loops and $\theta = \frac{-\pi}{4}$, $\theta = \frac{\pi}{4}$ are the tangents at the pole. In the first quadrant, for half of the loop, θ varies from 0 to $\frac{\pi}{4}$.

From (1),

$$2r\frac{dr}{d\theta} = -2a^2 \sin 2\theta \quad \text{or} \quad \frac{dr}{d\theta} = \frac{-a^2 \sin 2\theta}{r}$$

$$\therefore \quad \frac{ds}{d\theta} = \sqrt{r^2 + \left(\frac{dr}{d\theta}\right)^2} = \sqrt{a^2 \cos 2\theta + \frac{a^4 \sin^2 2\theta}{r^2}}$$

$$= \sqrt{a^2 \cos 2\theta + \frac{a^4 \sin^2 2\theta}{a^2 \cos 2\theta}} = \sqrt{\frac{a^2 (\cos^2 2\theta + \sin^2 2\theta)}{\cos 2\theta}}$$

$$= \frac{a}{\sqrt{\cos 2\theta}}$$

$$\therefore \quad \text{Required surface} = 2\int_0^{\pi/4} 2\pi y \frac{ds}{d\theta} \cdot d\theta = \pi \int_0^{\pi/4} r \sin^2 \theta \cdot \frac{a}{\sqrt{\cos 2\theta}} \cdot d\theta$$

$$= 4\pi \int_0^{\pi/4} a\sqrt{\cos 2\theta} \cdot \sin\theta \cdot \frac{a}{\sqrt{\cos 2\theta}} \cdot d\theta$$

$$= 4\pi a^2 \int_0^{\pi/4} \sin\theta \, d\theta = 4\pi a^2 \Big[-\cos\theta\Big]_0^{\pi/4}$$

$$= 4\pi a^2 \left[-\frac{1}{\sqrt{2}} + 1\right] = 2\pi a^2 (2 - \sqrt{2}).$$

EXERCISE 11.2

1. The circle $x^2 + y^2 = a^2$ revolves about the x-axis. Find the surface area of the whole volume so generated. (Or Find the surface area of a sphere of radius a.)
2. Find the area of the surface formed by the revolution of the parabola $y^2 = 4ax$ about the x-axis by the arc from the vertex to one end of the latus rectum.
3. Find the area of the surface generated by revolving the arc of the catenary $y = c \cosh \frac{x}{c}$ from $x = 0$ to $x = c$ about x-axis.

4. Prove that the surface and volume of the ellipsoid formed by the revolution of the ellipse $\frac{x^2}{a^2}+\frac{y^2}{b^2}=1$, round its major axis are respectively $2\pi ab = \sqrt{1-e^2}+e^{-1}\sin^{-1} e$ and $\frac{4}{3}\pi ab^2$, where $b^2 = a^2(1-e^2)$.

5. Find the volume and surface of the right circular cone formed by the revolution of a right angled triangle about *a* side which contains the right angle.

6. Find the surface area of the solid generated by revolving the cycloid

(*i*) $x = a(\theta - \sin\theta)$, $y = a(1-\cos\theta)$ about the *x*-axis.

(*ii*) $x = a(\theta + \sin\theta)$, $y = a(1-\cos\theta)$ about the tangent at the vertex. *(M.D.U. Dec. 2009)*

7. Find the volume and surface area of the solid generated by revolving the cycloid $x = a(\theta + \sin\theta)$, $y = a(1+\cos\theta)$ about its base.

8. The portion between two consecutive cusps of the cycloid $x = a(\theta + \sin\theta)$, $y = a(1+\cos\theta)$ is revolved about the *x*-axis. Prove that the area of the surface so formed is to the area of the cycloid as 64 : 9.

9. Prove that the surface area of the solid generated by the revolution, about the *x*-axis, of the loop of the curve $x = t^2$, $y = t - \frac{1}{3}t^3$ is 3π.

10. Find the surface generated by the revolution of the loop of the curve $3ay^2 = x(x-a)^2$ about the *x*-axis.

11. The curve $r = a(1+\cos\theta)$ revolves about the initial line. Find the area of the surface formed. *(M.D.U. Dec. 2008)*

12. (*a*) Find the surface of the solid generated by the revolution of the astroid $x = a\cos^3 t$, $y = a\sin^3 t$, about the *y*-axis.

(*b*) The arc of the curve $x^{2/3} + y^{2/3} = a^{2/3}$ in the first quadrant revolves about *x*-axis. Show that the area of the surface generated is $\frac{6\pi a^2}{5}$.

(*c*) Prove that the surface of the solid generated by the revolution of the tractrix

$$x = a\cos t + \frac{a}{2}\log\tan^2\frac{t}{2},\ y = a\sin t, \text{ about } x\text{-axis is } 4\pi a^2.$$

Answers

1. $4\pi a^2$ **2.** $\frac{8\pi a^2}{3}(2\sqrt{2}-1)$ **3.** $\frac{\pi c^2}{2}(2+\sinh 2)$

5. $\frac{1}{3}\pi r^2 h$, $\pi r l$, where *r* is the base radius, *h* the height and *l* the slant height of the cone.

6. (*i*) $\frac{64\pi a^2}{3}$ (*ii*) $\frac{32\pi a^2}{3}$ **7.** $5\pi^2 a^3$, $\frac{64}{3}\pi a^2$

10. $\frac{\pi a^2}{3}$ **11.** $\frac{32}{5}\pi a^2$ **12.** (*a*) $\frac{12\pi a^2}{5}$.

12

Multiple Integrals

12.1. DOUBLE INTEGRALS

The definite integral $\int_a^b f(x)\, dx$ is defined as the limits of the sum $f(x_1)\delta x_1 + f(x_2)\delta x_2 + \cdots + f(x_n)\delta x_n$ when $n \to \infty$ and each of the lengths $\delta x_1, \delta x_2, \ldots, \delta x_n$ tends to zero. Here $\delta x_1, \delta x_2, \ldots, \delta x_n$ are n sub-intervals into which the range $b - a$ has been divided and $x_1, x_2, \ldots, x_n$ are values of x lying respectively in the first, second, ..., nth sub-interval.

A double integral is its counterpart in two dimensions. Let a single-valued and bounded function $f(x, y)$ of two independent variables x, y be defined in a closed region R of the xy-plane. Divide the region R into sub-regions by drawing lines parallel to co-ordinate axes. Number the rectangles which lie entirely inside the region R, from 1 to n. Let (x_r, y_r) be any point inside the rth rectangle whose area is δA_r.

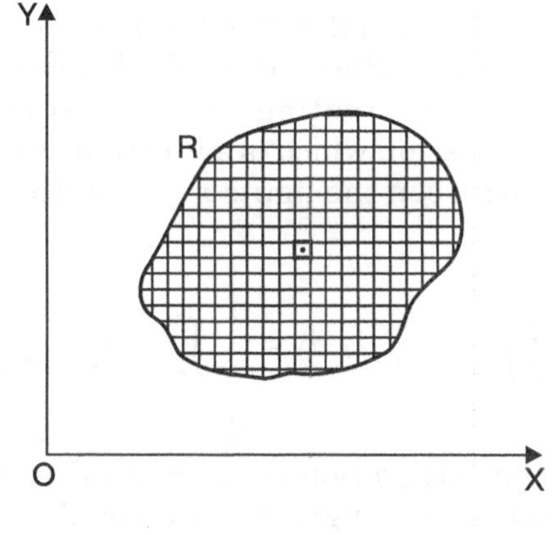

Consider the sum

$$f(x_1, y_1)\delta A_1 + f(x_2, y_2)\delta A_2 + \cdots + f(x_n, y_n)\delta A_n$$

$$= \sum_{r=1}^{n} f(x_r, y_r)\delta A_r \qquad \ldots(1)$$

Let the number of these sub-regions increase indefinitely, such that the largest linear dimension (*i.e.*, diagonal) of δA_r approaches zero. The limit of the sum (1), if it exists, irrespective of the mode of sub-division, is called the *double integral* of $f(x, y)$ over the region R and is denoted by $\iint_R f(x, y)\, dA$

In other words,
$$\lim_{\substack{n \to \infty \\ \delta A_r \to 0}} \sum_{r=1}^{n} f(x_r, y_r)\, \delta A_r = \iint_R f(x, y)\, dA$$

which is also expressed as
$$\iint_R f(x, y)\, dxdy \quad \text{or} \quad \iint_R f(x, y)\, dy\, dx$$

12.2. EVALUATION OF DOUBLE INTEGRALS

The methods of evaluating the double integrals depend upon the nature of the curves bounding the region R. Let the region R be bounded by the curves $x = x_1$, $x = x_2$ and $y = y_1$, $y = y_2$.

(i) When x_1, x_2 are functions of y and y_1, y_2 are constants. Let AB and CD be the curves $x_1 = \phi_1(y)$ and $x_2 = \phi_2(y)$.

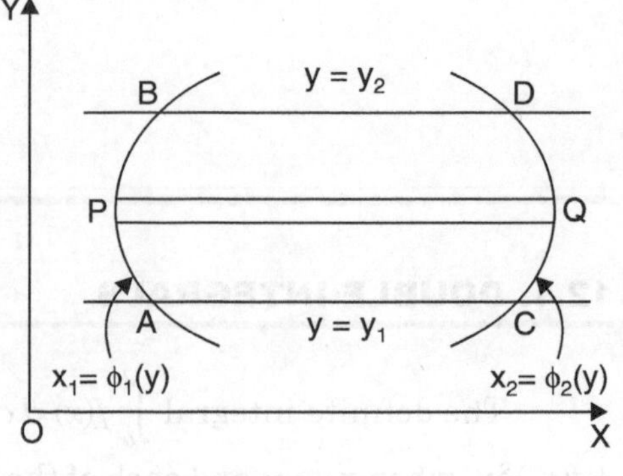

Take a horizontal strip PQ of width δy. Here the double integral is evaluated first w.r.t. x (treating y as a constant). The resulting expression which is a function of y is integrated w.r.t. y between the limits $y = y_1$ and $y = y_2$. Thus

$$\iint_R f(x, y)\, dx\, dy = \boxed{\int_{y_1}^{y_2} \boxed{\int_{x_1 = \phi_1(y)}^{x_2 = \phi_2(y)} f(x, y)\, dx}\, dy}$$

the integration being carried from the inner to the outer rectangle. Geometrically, the integral in the inner rectangle indicates that the integration is performed along the horizontal strip PQ (keeping y constant) while the outer rectangle corresponds to the sliding of the strip PQ from AC to BD thus covering the entire region ABDC of integration.

(ii) When y_1, y_2 are functions of x and x_1, x_2 are constants. Let AB and CD be the curves $y_1 = \phi_1(x)$ and $y_2 = \phi_2(x)$. Take a vertical strip PQ of width δx. Here the double integral is evaluated first w.r.t. y (treating x as constant). The resulting expression which is a function of x is integrated w.r.t. x between the limits $x = x_1$ and $x = x_2$. Thus,

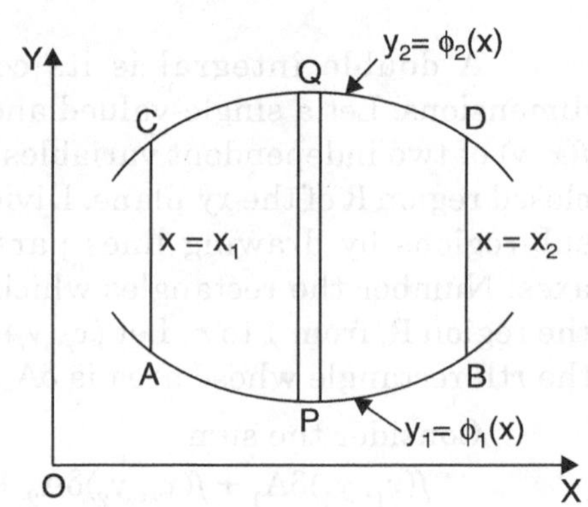

$$\iint_R f(x, y)\, dx\, dy = \boxed{\int_{x_1}^{x_2} \boxed{\int_{y_1 = \phi_1(x)}^{y_2 = \phi_2(x)} f(x, y)\, dy}\, dx}$$

the integration being carried from the inner to the outer rectangle. Geometrically, the integral in the inner rectanlge indicates that the integration is performed along the vertical strip PQ (keeping x constant) while the outer rectangle corresponds to the sliding of the strip PQ from AC to BD thus covering the entire region ABDC of integration.

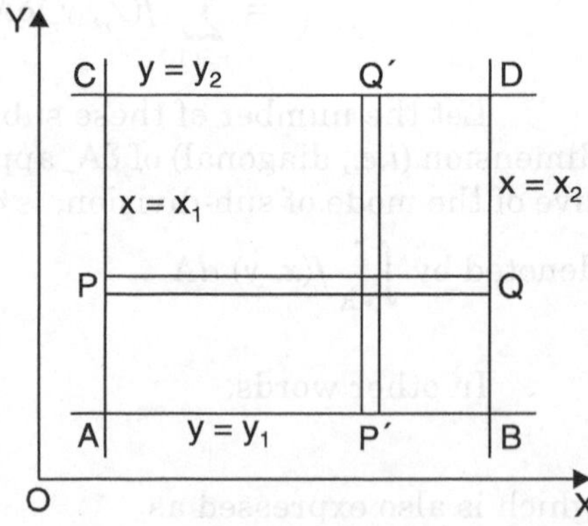

(iii) When x_1, x_2, y_1, y_2 are constants. Here the region of integration R is the rectangle ABDC. It is immaterial whether we integrate first along the horizontal strip PQ and then slide it from AB to CD; or we integrate first along the vertical strip P′Q′ and then slide it from AC to BD. Thus the order of integration is immaterial, provided the limits of integration are changed accordingly.

$$\iint_R f(x, y)\, dx\, dy = \boxed{\int_{y_1}^{y_2} \boxed{\int_{x_1}^{x_2} f(x, y)\, dx}\, dy} = \boxed{\int_{x_1}^{x_2} \boxed{\int_{y_1}^{y_2} f(x, y)\, dy}\, dx}$$

Note 1. From cases (*i*) and (*ii*) above, we observe that *integration is to be performed w.r.t. that variable having variable limits first and then w.r.t. the variable with constant limits.*

Note 2. If $f(x, y)$ has discontinuities within or on the boundary of the region of integration, then the change of the order of integration does not result into the same integrals.

ILLUSTRATIVE EXAMPLES

Example 1. *Prove that* $\int_1^2 \int_3^4 (xy + e^y)\, dy\, dx = \int_3^4 \int_1^2 (xy + e^y)\, dx\, dy.$

Sol. $\int_1^2 \int_3^4 (xy + e^y)\, dy\, dx = \int_1^2 \left[\int_3^4 (xy + e^y)\, dy\right] dx$

$$= \int_1^2 \left[\frac{xy^2}{2} + e^y\right]_3^4 dx = \int_1^2 \left(8x + e^4 - \frac{9}{2}x - e^3\right) dx$$

$$= \int_1^2 \left(\frac{7}{2}x + e^4 - e^3\right) dx = \left[\frac{7x^2}{4} + (e^4 - e^3)x\right]_1^2$$

$$= 7 + 2(e^4 - e^3) - \frac{7}{4} - (e^4 - e^3) = \frac{21}{4} + e^4 - e^3$$

$$\int_3^4 \int_1^2 (xy + e^y)\, dx\, dy = \int_3^4 \left[\int_1^2 (xy + e^y)\, dx\right] dy = \int_3^4 \left[\frac{yx^2}{2} + xe^y\right]_1^2 dy$$

$$= \int_3^4 \left(2y + 2e^y - \frac{y}{2} - e^y\right) dy = \int_3^4 \left(\frac{3y}{2} + e^y\right) dy$$

$$= \left[\frac{3y^2}{4} + e^y\right]_3^4 = 12 + e^4 - \frac{27}{4} - e^3 = \frac{21}{4} + e^4 - e^3$$

Hence the result.

Example 2. *Evaluate* $\int_0^1 \int_0^1 \frac{dx\, dy}{\sqrt{(1 - x^2)(1 - y^2)}}$.

Sol. $\int_0^1 \int_0^1 \frac{dx\, dy}{\sqrt{(1 - x^2)(1 - y^2)}} = \int_0^1 \left[\int_0^1 \frac{dx}{\sqrt{(1 - x^2)(1 - y^2)}}\right] dy = \int_0^1 \frac{1}{\sqrt{1 - y^2}} \left[\sin^{-1} x\right]_0^1 dy$

$$= \int_0^1 \frac{1}{\sqrt{1 - y^2}} \cdot \frac{\pi}{2}\, dy = \frac{\pi}{2} \left[\sin^{-1} y\right]_0^1 = \frac{\pi}{2} \cdot \frac{\pi}{2} = \frac{\pi^2}{4}.$$

Example 3. *Show that* $\int_0^1 dx \int_0^1 \frac{x - y}{(x + y)^3}\, dy \neq \int_0^1 dy \int_0^1 \frac{x - y}{(x + y)^3}\, dx.$

Sol. LHS $= \int_0^1 dx \int_0^1 \frac{2x-(x+y)}{(x+y)^3}\, dy = \int_0^1 dx \int_0^1 \left[\frac{2x}{(x+y)^3} - \frac{1}{(x+y)^2}\right] dy$

$$= \int_0^1 \left[2x \cdot \frac{(x+y)^{-2}}{-2} - \frac{(x+y)^{-1}}{-1}\right]_0^1 dx = \int_0^1 \left[\frac{-x}{(x+y)^2} + \frac{1}{x+y}\right]_0^1 dx$$

$$= \int_0^1 \left[\frac{-x}{(x+1)^2} + \frac{1}{x+1} + \frac{1}{x} - \frac{1}{x}\right] dx = \int_0^1 \frac{-x+x+1}{(x+1)^2}\, dx = \int_0^1 \frac{1}{(x+1)^2}\, dx$$

$$= \left[-\frac{1}{x+1}\right]_0^1 = -\frac{1}{2} + 1 = \frac{1}{2}$$

$$\text{RHS} = \int_0^1 dy \int_0^1 \frac{(x+y)-2y}{(x+y)^3}\, dx = \int_0^1 dy \int_0^1 \left[\frac{1}{(x+y)^2} - \frac{2y}{(x+y)^3}\right] dx$$

$$= \int_0^1 \left[\frac{(x+y)^{-1}}{-1} - 2y \cdot \frac{(x+y)^{-2}}{-2}\right]_0^1 dy = \int_0^1 \left[-\frac{1}{x+y} + \frac{y}{(x+y)^2}\right]_0^1 dy$$

$$= \int_0^1 \left[-\frac{1}{1+y} + \frac{y}{(1+y)^2} + \frac{1}{y} - \frac{1}{y}\right] dy = \int_0^1 \frac{-1-y+y}{(1+y)^2}\, dy = -\int_0^1 \frac{1}{(1+y)^2}\, dy$$

$$= \left[\frac{1}{1+y}\right]_0^1 = \frac{1}{2} - 1 = -\frac{1}{2}$$

$\therefore$ The two integrals are not equal.

Example 4. *Evaluate* $\int_0^1 \int_0^{\sqrt{1+x^2}} \frac{dy\, dx}{1+x^2+y^2}$. (Kerala 2012)

Sol. $I = \int_0^1 \left[\int_0^{\sqrt{1+x^2}} \frac{1}{(1+x^2)+y^2}\, dy\right] dx$

$$= \int_0^1 \frac{1}{\sqrt{1+x^2}} \left[\tan^{-1} \frac{y}{\sqrt{1+x^2}}\right]_0^{\sqrt{1+x^2}} dx$$

$$= \int_0^1 \frac{1}{\sqrt{1+x^2}} [\tan^{-1} 1 - \tan^{-1} 0]\, dx = \frac{\pi}{4} \int_0^1 \frac{dx}{\sqrt{1+x^2}}$$

$$= \frac{\pi}{4} \left[\log (x + \sqrt{1+x^2})\right]_0^1 = \frac{\pi}{4} [\log (1+\sqrt{2}) - \log 1] = \frac{\pi}{4} \log (\sqrt{2}+1).$$

Example 5. *Evaluate* $\int_0^a \int_0^{\sqrt{a^2-y^2}} \sqrt{a^2-x^2-y^2}\, dx\, dy$.

Sol. $I = \int_0^a \left[\int_0^{\sqrt{a^2-y^2}} \sqrt{(a^2-y^2)-x^2}\, dx\right] dy.$

$$= \int_0^a \left[\frac{x\sqrt{a^2 - y^2 - x^2}}{2} + \frac{a^2 - y^2}{2} \sin^{-1} \frac{x}{\sqrt{a^2 - y^2}} \right]_0^{\sqrt{a^2 - y^2}} dy$$

$$\left[\because \int \sqrt{a^2 - x^2}\, dx = \frac{x\sqrt{a^2 - x^2}}{2} + \frac{a^2}{2} \sin^{-1} \frac{x}{a} \right]$$

$$= \int_0^a \frac{a^2 - y^2}{2} \sin^{-1} 1\, dy = \frac{\pi}{4} \int_0^a (a^2 - y^2)\, dy$$

$$= \frac{\pi}{4} \left[a^2 y - \frac{y^3}{3} \right]_0^a = \frac{\pi}{4} \left[a^3 - \frac{a^3}{3} \right] = \frac{\pi a^3}{6}.$$

Example 6. *Evaluate* $\iint e^{2x+3y}\, dx\, dy$ *over the triangle bounded by* $x = 0$, $y = 0$ *and* $x + y = 1$.

Sol. The region R of integration is the triangle OAB. Here x varies from 0 to 1 and y varies from x-axis upto the line $x + y = 1$ *i.e.*, from 0 to $1 - x$.

$\therefore$ The region R can be expressed as

$$0 \le x \le 1,\ 0 \le y \le 1 - x$$

Y, X, B (0,1), $y = 1 - x$, $x = 0$, (1, 0), O, $y = 0$, A

$$\therefore \iint_R e^{2x+3y}\, dx\, dy$$

$$= \int_0^1 \int_0^{1-x} e^{2x+3y}\, dy\, dx$$

$$= \int_0^1 \left[\frac{1}{3} e^{2x+3y} \right]_0^{1-x} dx$$

$$= \frac{1}{3} \int_0^1 (e^{3-x} - e^{2x})\, dx$$

$$= \frac{1}{3} \left[-e^{3-x} - \frac{1}{2} e^{2x} \right]_0^1 = -\frac{1}{3} \left[\left(e^2 + \frac{1}{2} e^2 \right) - \left(e^3 + \frac{1}{2} \right) \right]$$

$$= -\frac{1}{3} \left[-e^2(e - 1) + \frac{1}{2}(e^2 - 1) \right]$$

$$= \frac{1}{6} (e - 1)\, [2e^2 - (e + 1)] = \frac{1}{6} (e - 1)(2e^2 - e - 1)$$

$$= \frac{1}{6} (e - 1)(e - 1)(2e + 1) = \frac{1}{6} (e - 1)^2 (2e + 1).$$

Example 7. *Evaluate* $\iint_R y\, dx\, dy$, *where R is the region bounded by the parabolas* $y^2 = 4x$ *and* $x^2 = 4y$.

Sol. Solving $y^2 = 4x$ and $x^2 = 4y$, we have

$$\left(\frac{x^2}{4}\right)^2 = 4x \quad \text{or} \quad x(x^3 - 64) = 0$$

$\therefore \quad x = 0, 4$

When $\quad x = 4,\ y = 4$

$\therefore$ Co-ordinates of A are (4, 4)

The region R can be expressed as

$$0 \le x \le 4, \frac{x^2}{4} \le y \le 2\sqrt{x}$$

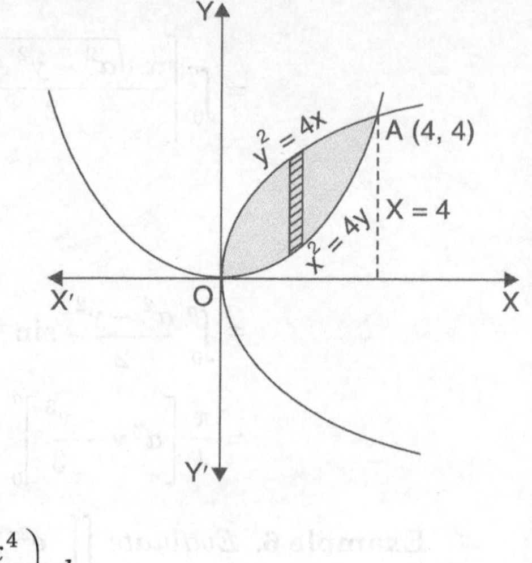

$$\therefore \quad \iint_R y\,dx\,dy = \int_0^4 \int_{x^2/4}^{2\sqrt{x}} y\,dy\,dx$$

$$= \int_0^4 \frac{1}{2}\left[y^2\right]_{x^2/4}^{2\sqrt{x}} dx = \frac{1}{2}\int_0^4 \left(4x - \frac{x^4}{16}\right) dx$$

$$= \frac{1}{2}\left[2x^2 - \frac{x^5}{80}\right]_0^4 = \frac{1}{2}\left[32 - \frac{1024}{80}\right] = \frac{48}{5}.$$

Example 8. *Evaluate* $\iint (x + y)^2\,dx\,dy$ *over the area bounded by the ellipse*

$$\frac{x^2}{a^2} + \frac{y^2}{b^2} = 1. \qquad \text{(K.U.K. 2009; U.K.T.U. 2011)}$$

Sol. For the ellipse

$$\frac{y}{b} = \pm\sqrt{1 - \frac{x^2}{a^2}} \quad \text{or} \quad y = \pm\frac{b}{a}\sqrt{a^2 - x^2}$$

$\therefore$ The region of integration R can be expressed as

$$-a \le x \le a, -\frac{b}{a}\sqrt{a^2 - x^2} \le y \le \frac{b}{a}\sqrt{a^2 - x^2}$$

$$\therefore \quad \iint_R (x + y)^2\,dx\,dy = \iint_R (x^2 + y^2 + 2xy)\,dx\,dy$$

$$= \int_{-a}^{a} \int_{-b/a\sqrt{a^2 - x^2}}^{b/a\sqrt{a^2 - x^2}} (x^2 + y^2 + 2xy)\,dy\,dx$$

$$= \int_{-a}^{a} \int_{-b/a\sqrt{a^2 - x^2}}^{b/a\sqrt{a^2 - x^2}} (x^2 + y^2)\,dy\,dx + \int_{-a}^{a} \int_{-b/a\sqrt{a^2 - x^2}}^{b/a\sqrt{a^2 - x^2}} 2xy\,dy\,dx$$

$$= \int_{-a}^{a} \int_{0}^{b/a\sqrt{a^2 - x^2}} 2(x^2 + y^2)\,dy\,dx + 0$$

[Since $(x^2 + y^2)$ is an even function of y and $2xy$ is an odd function of y]

$$= \int_{-a}^{a} \left[2\left(x^2 y + \frac{y^3}{3}\right)\right]_0^{b/a\sqrt{a^2 - x^2}} dx$$

$$= 2\int_{-a}^{a} \left[x^2 \cdot \frac{b}{a}\sqrt{a^2 - x^2} + \frac{1}{3} \cdot \frac{b^3}{a^3}(a^2 - x^2)^{3/2}\right] dx$$

$$= 4\int_0^a \left[\frac{b}{a}x^2\sqrt{a^2-x^2} + \frac{b^3}{3a^3}(a^2-x^2)^{3/2}\right]dx$$

$$= 4\int_0^{\pi/2}\left(\frac{b}{a}\cdot a^2\sin^2\theta\cdot a\cos\theta + \frac{b^3}{3a^3}\cdot a^3\cos^3\theta\right)\times a\cos\theta\, d\theta$$

on putting $x = a\sin\theta$ and $dx = a\cos\theta\, d\theta$

$$= 4\int_0^{\pi/2}\left(a^3 b\sin^2\theta\cos^2\theta + \frac{ab^3}{3}\cos^4\theta\right)d\theta$$

$$= 4\left[a^3 b\cdot\frac{1\cdot 1}{4\cdot 2}\cdot\frac{\pi}{2} + \frac{ab^3}{3}\cdot\frac{3\cdot 1}{4\cdot 2}\cdot\frac{\pi}{2}\right] = \frac{\pi}{4}(a^3 b + ab^3) = \frac{\pi}{4}ab(a^2+b^2).$$

EXERCISE 12.1

Evaluate the following integrals (1–10):

1. $\int_0^3\int_0^1 (x^2+3y^2)\,dy\,dx.$

2. $\int_0^3\int_1^2 xy\,(1+x+y)\,dy\,dx.$

3. $\int_1^a\int_1^b \frac{dy\,dx}{xy}.$

4. $\int_1^2\int_0^x \frac{dy\,dx}{x^2+y^2}.$ *(P.T.U. May 2011)*

5. $\int_0^1 dx\int_0^x e^{y/x}\,dy.$

6. $\int_0^1\int_{x^2}^x (x^2+3y+2)\,dy\,dx.$

7. $\int_0^1\int_0^{x^2} e^{y/x}\,dy\,dx.$ *(Kerala 2009)*

8. $\int_0^1\int_y^{y^2+1} x^2 y\,dx\,dy.$

9. $\int_0^{2a}\int_0^{\sqrt{2ax-x^2}} (x^2+y^2)\,dy\,dx.$

10. $\int_0^1\int_x^{\sqrt{x}} (x^2+y^2)\,dy\,dx.$

11. Evaluate $\iint (x^2+y^2)\,dx\,dy$ over the region in the positive quadrant for which $x + y \le 1$.

12. Evaluate $\iint x^2y^2\,dx\,dy$ over the circle $x^2+y^2=1$.

13. Evaluate $\iint xy\,dx\,dy$ over the positive quadrant of the circle $x^2+y^2=a^2$.
(U.P.T.U. 2009; M.D.U. Dec. 2011)

14. Compute the value of $\iint_R y\,dx\,dy$, where R is the region in the first quadrant bounded by the ellipse $\frac{x^2}{a^2}+\frac{y^2}{b^2}=1$.

15. Evaluate $\iint_S \sqrt{xy-y^2}\,dx\,dy$, where S is a triangle with vertices (0, 0), (10, 1) and (1, 1).

16. Evaluate $\iint xy\,(x+y)\,dx\,dy$ over the area between $y = x^2$ and $y = x$. *(U.K.T.U. 2011, 2012)*

17. Evaluate $\iint_A xy\,dx\,dy$, where A is the domain bounded by x-axis, ordinate $x = 2a$ and the curve $x^2 = 4ay$. *(M.T.U. 2012; G.B.T.U. 2010)*

18. Evaluate $\iint_R xy\,dx\,dy$, where R is the region inside the square $|x| + |y| = 1$.

Answers

1. 12	2. $30\frac{3}{4}$	3. $\log a \log b$	4. $\frac{\pi}{4} \log 2$
5. $\frac{1}{2}(e-1)$	6. $\frac{7}{12}$	7. $\frac{1}{2}$	8. $\frac{67}{120}$
9. $\frac{3\pi a^4}{4}$	10. $\frac{3}{35}$	11. $\frac{1}{6}$	12. $\frac{\pi}{24}$
13. $\frac{a^4}{8}$	14. $\frac{ab^2}{3}$	15. 6	16. $\frac{3}{56}$
17. $\frac{a^4}{3}$	18. $\frac{1}{6}$.		

12.3. EVALUATION OF DOUBLE INTEGRALS IN POLAR CO-ORDINATES

To evaluate $\int_{\theta_1}^{\theta_2}\int_{r_1}^{r_2} f(r, \theta)\, dr\, d\theta$ over the region bounded by the straight lines $\theta = \theta_1$, $\theta = \theta_2$ and the curves $r = r_1$, $r = r_2$, we first integrate w.r.t. r between the limits $r = r_1$ and $r = r_2$ (treating θ as a constant). The resulting expression is then integrated w.r.t. θ between the limits $\theta = \theta_1$ and $\theta = \theta_2$.

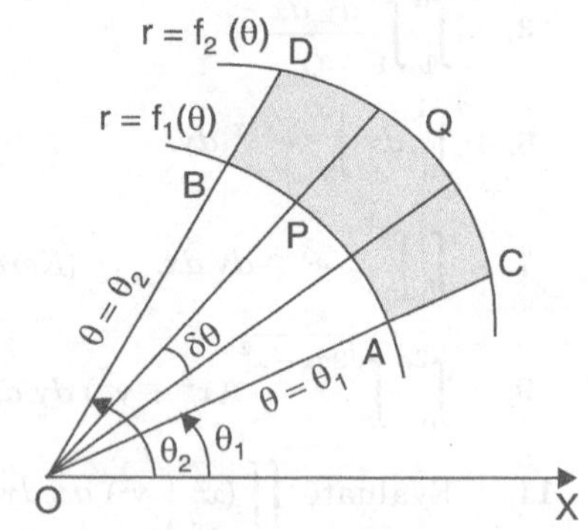

Geometrically, AB and CD are the curves $r = f_1(\theta)$ and $r = f_2(\theta)$ bounded by the lines $\theta = \theta_1$ and $\theta = \theta_2$ so that ACDB is the region of integration. PQ is a wedge of angular thickness $\delta\theta$. Then $\int_{r=r_1}^{r=r_2} f(r,\theta)\, dr$ indicates that the integration is performed along PQ (*i.e.*, r varies, θ is constant) and the integration w.r.t. θ means rotation of this strip PQ from AC to BD. dr indicates that the integration is performed along PQ (*i.e.*, r varies, θ is constant) and the integration w.r.t. θ means rotation of this strip PQ from AC to BD.

ILLUSTRATIVE EXAMPLES

Example 1. *Evaluate* $\int_0^{\pi/2}\left[\int_0^{a\cos\theta} r\sqrt{a^2 - r^2}\, dr\right] d\theta.$

Sol.
$$I = \int_0^{\pi/2}\left[\int_0^{a\cos\theta} -\frac{1}{2}(a^2 - r^2)^{1/2}(-2r)\, dr\right] d\theta$$

$$= \int_0^{\pi/2}\left[-\frac{1}{2}\cdot\frac{(a^2 - r^2)^{3/2}}{3/2}\right]_0^{a\cos\theta} d\theta$$

$$= -\frac{1}{3}\int_0^{\pi/2}(a^3\sin^3\theta - a^3)\, d\theta = -\frac{a^3}{3}\left[\frac{2}{3} - \frac{\pi}{2}\right] = \frac{a^3}{18}(3\pi - 4).$$

Example 2. *Evaluate* $\iint r^3\, dr\, d\theta$, *over the area bounded between the circles* $r = 2\cos\theta$ *and* $r = 4\cos\theta$. (M.D.U. Dec. 2012; Kerala 2009)

Sol. The region of integration R is shown shaded. Here r varies from $2\cos\theta$ to $4\cos\theta$ while θ varies from $-\frac{\pi}{2}$ to $\frac{\pi}{2}$.

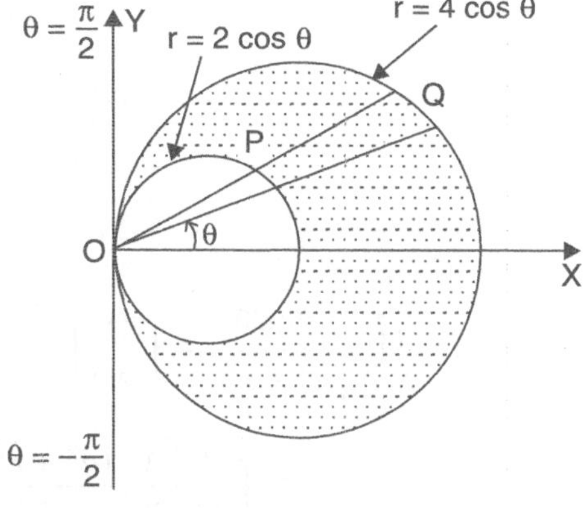

$$\therefore \quad \iint_R r^3\,dr\,d\theta = \int_{-\pi/2}^{\pi/2}\int_{2\cos\theta}^{4\cos\theta} r^3 dr\,d\theta$$

$$= \int_{-\pi/2}^{\pi/2}\left[\frac{r^4}{4}\right]_{2\cos\theta}^{4\cos\theta} d\theta$$

$$= \int_{-\pi/2}^{\pi/2}\frac{1}{4}(256\cos^4\theta - 16\cos^4\theta)\,d\theta$$

$$= 60\int_{-\pi/2}^{\pi/2}\cos^4\theta\,d\theta$$

$$= 120\int_0^{\pi/2}\cos^4\theta\,d\theta \qquad \text{[Since } \cos^4\theta \text{ is an even function of } \theta]$$

$$= 120 \times \frac{3\times 1}{4\times 2}\cdot\frac{\pi}{2} = \frac{45}{2}\pi.$$

Example 3. *Evaluate* $\iint r\sin\theta\,dr\,d\theta$ *over the area of the cardioid* $r = a(1+\cos\theta)$ *above the initial line.*

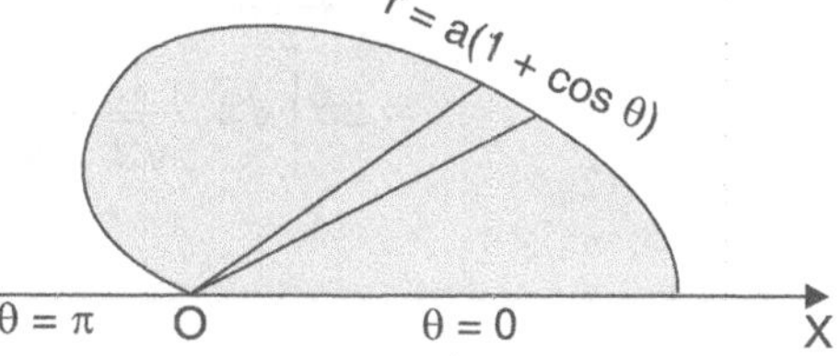

Sol. The region of integration R is covered by radial strips whose ends are $r = 0$ and $r = a(1+\cos\theta)$, the strips starting from $\theta = 0$ and ending at $\theta = \pi$.

$$\therefore \quad \iint_R r\sin\theta\,dr\,d\theta = \int_0^{\pi}\int_0^{a(1+\cos\theta)} r\sin\theta\,dr\,d\theta\cdot$$

$$= \int_0^{\pi}\sin\theta\left[\frac{r^2}{2}\right]_0^{a(1+\cos\theta)} d\theta = \frac{1}{2}\int_0^{\pi}\sin\theta\,.\,a^2(1+\cos\theta)^2\,d\theta$$

$$= \frac{a^2}{2}\int_0^{\pi} 2\sin\frac{\theta}{2}\cos\frac{\theta}{2}\,.\left(2\cos^2\frac{\theta}{2}\right)^2 d\theta = 4a^2\int_0^{\pi}\sin\frac{\theta}{2}\cos^5\frac{\theta}{2}\,d\theta$$

Putting $\cos\frac{\theta}{2} = t$ so that $\sin\frac{\theta}{2}\,d\theta = -2dt$

$$\iint_R r\sin\theta\,dr\,d\theta = 4a^2\int_1^0 t^5(-2dt) = 8a^2\int_0^1 t^5 dt = 8a^2\left[\frac{t^6}{6}\right]_0^1 = 8a^2\left(\frac{1}{6}\right) = \frac{4a^2}{3}$$

Example 4. *Evaluate* $\iint \frac{r\,dr\,d\theta}{\sqrt{a^2+r^2}}$ *over one loop of the lemniscate* $r^2 = a^2\cos 2\theta$.

Sol. The region of integration R is covered by radial strips whose ends are $r = 0$ and $r = a\sqrt{\cos 2\theta}$, the strips starting from $\theta = -\frac{\pi}{4}$ and ending at $\theta = \frac{\pi}{4}$.

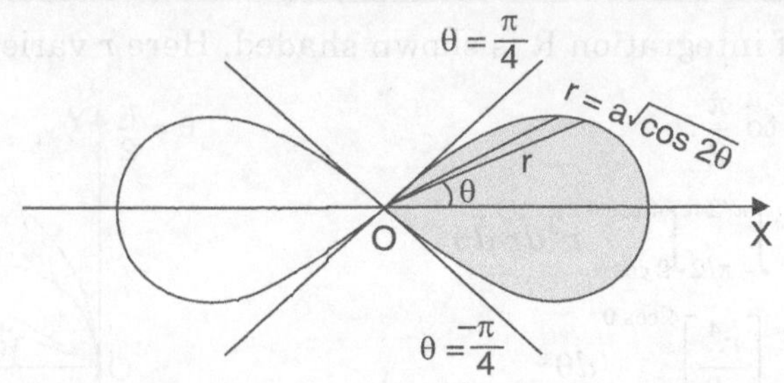

$$\therefore \quad \iint_R \frac{r\,dr\,d\theta}{\sqrt{a^2+r^2}} = \int_{-\pi/4}^{\pi/4}\int_0^{a\sqrt{\cos 2\theta}} \frac{1}{2}(a^2+r^2)^{-1/2} \cdot 2r\,dr\,d\theta$$

$$= \int_{-\pi/4}^{\pi/4}\left[\frac{1}{2}\cdot\frac{(a^2+r^2)^{1/2}}{1/2}\right]_0^{a\sqrt{\cos 2\theta}} d\theta = \int_{-\pi/4}^{\pi/4} [(a^2+a^2\cos 2\theta)^{1/2} - a]\,d\theta$$

$$= a\int_{-\pi/4}^{\pi/4} [(1+\cos 2\theta)^{1/2} - 1]\,d\theta = a\int_{-\pi/4}^{\pi/4} [(2\cos^2\theta)^{1/2} - 1]\,d\theta$$

$$= a\int_{-\pi/4}^{\pi/4} (\sqrt{2}\cos\theta - 1)\,d\theta = 2a\int_0^{\pi/4} (\sqrt{2}\cos\theta - 1)\,d\theta = 2a\left[\sqrt{2}\sin\theta - \theta\right]_0^{\pi/4}$$

$$= 2a\left[\sqrt{2}\cdot\frac{1}{\sqrt{2}} - \frac{\pi}{4}\right] = 2a\left(1-\frac{\pi}{4}\right).$$

EXERCISE 12.2

Evaluate the following integrals (1–4):

1. $\int_0^{\pi}\int_0^{a\sin\theta} r\,dr\,d\theta.$

2. $\int_0^{\pi/2}\int_0^{a\cos\theta} r\sin\theta\,dr\,d\theta.$

3. $\int_0^{\pi/2}\int_a^{a(1+\cos\theta)} r\,dr\,d\theta.$

4. $\int_0^{\pi}\int_0^{a(1+\cos\theta)} r^2\cos\theta\,dr\,d\theta.$

5. Show that $\iint_R r^2\sin\theta\,dr\,d\theta = \frac{2a^3}{3}$, where R is the region bounded by the semi-circle $r = 2a\cos\theta$, above the initial line.
6. Evaluate $\iint r^3\,dr\,d\theta$ over the area included between the circles $r = 2\sin\theta$ and $r = 4\sin\theta$.
7. Evaluate $\iint r\sin\theta\,dr\,d\theta$ over the cardioid $r = a(1-\cos\theta)$ above the initial line. *(Kerala 2010)*
8. Evaluate $\iint r^3\,dr\,d\theta$ over the area included between the circles $r = 2a\cos\theta$, $r = 2b\cos\theta$, $(b < a)$.

Answers

1. $\frac{\pi a^2}{4}$ 2. $\frac{a^2}{6}$ 3. $a^2\left(1+\frac{\pi}{8}\right)$

4. $\frac{5}{8}\pi a^3$ 6. $\frac{45\pi}{2}$ 7. $\frac{4a^2}{3}$

8. $\frac{3\pi}{2}(a^4-b^4)$.

12.4. CHANGE OF ORDER OF INTEGRATION

In a double integral, if the limits of integration are constant, then the order of integration is immaterial, provided the limits of integration are changed accordingly. Thus

$$\int_c^d \int_a^b f(x, y)\, dx\, dy = \int_a^b \int_c^d f(x, y)\, dy\, dx .$$

But if the limits of integration are variable, a change in the order of integration necessitates change in the limits of integration. A rough sketch of the region of integration helps in fixing the new limits of integration.

ILLUSTRATIVE EXAMPLES

Example 1. *Change the order of integration in* $\int_0^a \int_y^a \frac{x\, dx\, dy}{x^2 + y^2}$ *and hence evaluate the same.* (M.D.U. Dec. 2012; Kerala 2008)

Sol. From the limits of integration, it is clear that the region of integration is bounded by $x = y$, $x = a$, $y = 0$ and $y = a$. Thus the region of integration is the ΔOAB and is divided into horizontal strips. For changing the order of integration, we divide the region of integration into vertical strips. The new limits of integration become: y varies from 0 to x and x varies from 0 to a.

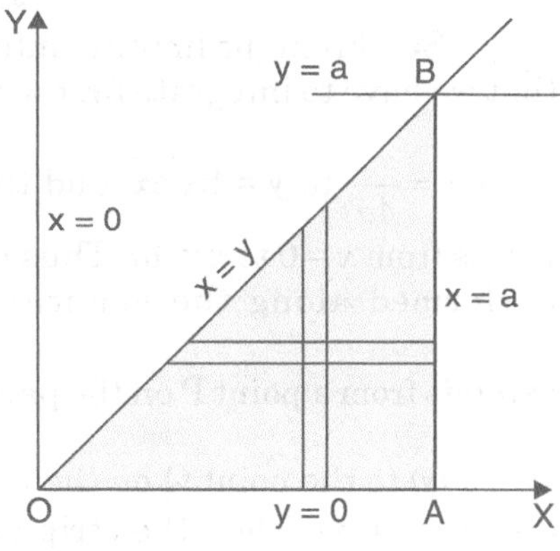

$$\int_0^a \int_y^a \frac{x\, dx\, dy}{x^2 + y^2} = \int_0^a \int_0^x \frac{x\, dy\, dx}{x^2 + y^2}$$

$$= \int_0^a x . \left[\frac{1}{x} \tan^{-1} \frac{y}{x} \right]_0^x dx$$

$$= \int_0^a \frac{\pi}{4}\, dx = \frac{\pi}{4} . \left[x \right]_0^a = \frac{\pi a}{4} .$$

Example 2. *Change the order of integration in* $I = \int_0^1 \int_{x^2}^{2-x} xy\, dy\, dx$ *and hence evaluate the same.* (Kerala 2010, 2012; M.D.U. Dec. 2010; P.T.U. 2009, 2011)

Sol. From the limits of integration, it is clear that we have to integrate first with respect to y which varies from $y = x^2$ to $y = 2 - x$ and then with respect to x which varies from $x = 0$ to $x = 1$. The region of integration (shown shaded) is divided into vertical strips. For changing the order of integration, we divide the region of integration into horizontal strips.

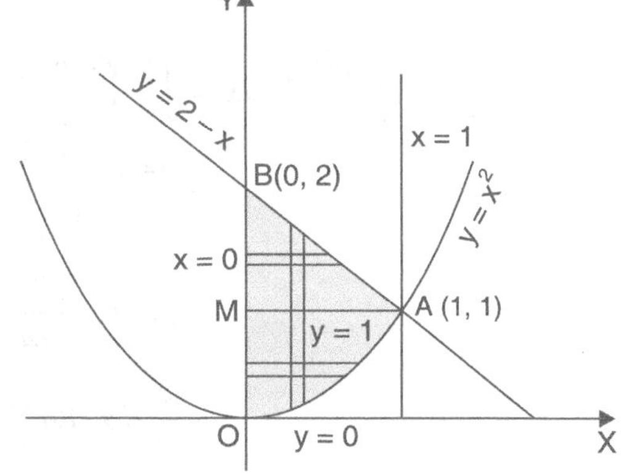

Solving $y = x^2$ and $y = 2 - x$, the co-ordinates of A are (1, 1). Draw AM $\perp$ OY. The region of integration is divided into two parts, OAM and MAB.

For the region OAM, x varies from 0 to $\sqrt{y}$ and y varies from 0 to 1. For the region MAB, x varies from 0 to $2 - y$ and y varies from 1 to 2.

$$\therefore \quad \int_0^1 \int_{x^2}^{2-x} xy\, dx\, dy = \int_0^1 \int_0^{\sqrt{y}} xy\, dx\, dy + \int_1^2 \int_0^{2-y} xy\, dx\, dy$$

$$= \int_0^1 y \cdot \left[\frac{x^2}{2}\right]_0^{\sqrt{y}} dy + \int_1^2 y \cdot \left[\frac{x^2}{2}\right]_0^{2-y} dy = \frac{1}{2}\int_0^1 y^2 dy + \frac{1}{2}\int_1^2 y(2-y)^2\, dy$$

$$= \frac{1}{2} \cdot \left[\frac{y^3}{3}\right]_0^1 + \frac{1}{2}\int_1^2 (4y - 4y^2 + y^3)\, dy = \frac{1}{6} + \frac{1}{2}\left[2y^2 - \frac{4y^3}{3} + \frac{y^4}{4}\right]_1^2$$

$$= \frac{1}{6} + \frac{1}{2}\left[\left(8 - \frac{32}{3} + 4\right) - \left(2 - \frac{4}{3} + \frac{1}{4}\right)\right] = \frac{3}{8}.$$

Example 3. *Change the order of integration in the following integral and evaluate:*

$$\int_0^{4a} \int_{x^2/4a}^{2\sqrt{ax}} dy\, dx.$$ (P.T.U. Jan. 2010; M.D.U. May 2009)

Sol. From the limits of integration, it is clear that we have to integrate first w.r.t. y which varies from $y = \dfrac{x^2}{4a}$ to $y = 2\sqrt{ax}$ and then w.r.t. x which varies from $x = 0$ to $x = 4a$. Thus integration is first performed along the vertical strip PQ which extends from a point P on the parabola $y = \dfrac{x^2}{4a}$ (*i.e.*, $x^2 = 4ay$) to the point Q on the parabola $y = 2\sqrt{ax}$ (*i.e.*, $y^2 = 4ax$). Then the strip slides from O to A $(4a, 4a)$, the point of intersection of the two parabolas.

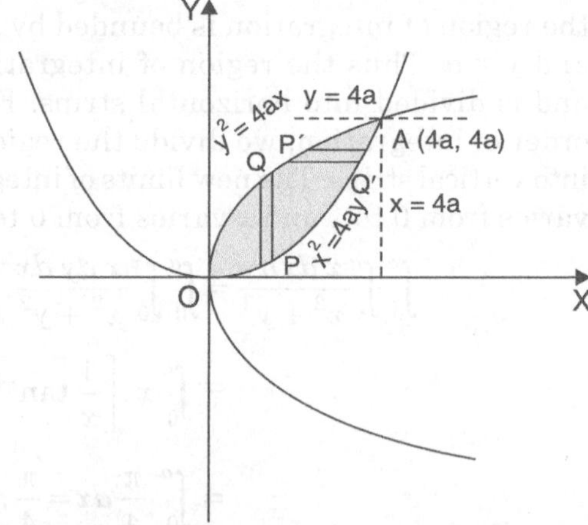

For changing the order of integration, we divide the region of integration OPAQO into horizontal strips P′Q′ which extend from P′ on the parabola $y^2 = 4ax$ *i.e.*, $x = \dfrac{y^2}{4a}$ to Q′ on the parabola $x^2 = 4ay$ *i.e.*, $x = 2\sqrt{ay}$. Then this strip slides from O to A$(4a, 4a)$, *i.e.*, varies from 0 to $4a$.

$$\therefore \quad \int_0^{4a} \int_{x^2/4a}^{2\sqrt{ax}} dy\, dx = \int_0^{4a} \int_{y^2/4a}^{2\sqrt{ay}} dx\, dy$$

$$= \int_0^{4a} \left[x\right]_{y^2/4a}^{2\sqrt{ay}} dy = \int_0^{4a} \left(2\sqrt{ay} - \frac{y^2}{4a}\right) dy$$

$$= \left[2\sqrt{a} \cdot \frac{y^{3/2}}{3/2} - \frac{y^3}{12a}\right]_0^{4a} = \frac{4}{3}\sqrt{a} \cdot (4a)^{3/2} - \frac{64a^3}{12a}$$

$$= \frac{4}{3}\sqrt{a} \cdot 8a^{3/2} - \frac{16a^2}{3} = \frac{32a^2}{3} - \frac{16a^2}{3} = \frac{16a^2}{3}.$$

Example 4. *Change the order of integration in the integral* $\displaystyle\int_{-a}^{a} \int_0^{\sqrt{a^2 - y^2}} f(x, y)\, dx\, dy.$

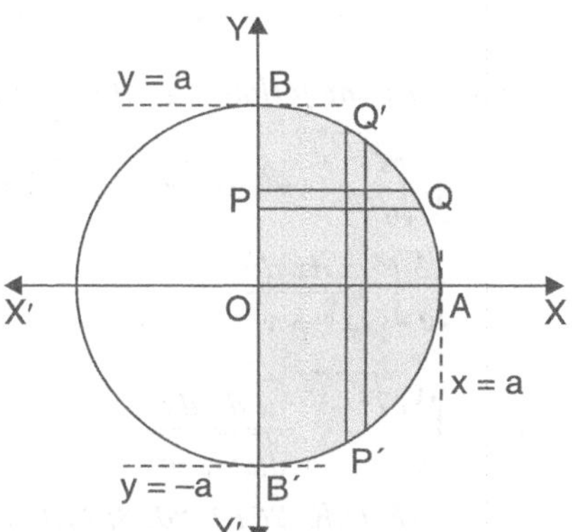

Sol. From the limits of integration, it is clear that we have to integrate first w.r.t. x which varies from $x = 0$ to $x = \sqrt{a^2 - y^2}$ and then w.r.t. y which varies from $y = -a$ to $y = a$. Thus integration is first performed along the horizontal strip PQ which extends from a point P on $x = 0$ (*i.e.*, y-axis) to the point Q on the circle $x = \sqrt{a^2 - y^2}$ (*i.e.*, $x^2 = a^2 - y^2$ or $x^2 + y^2 = a^2$). Then the strip slides from B′ to B.

For changing the order of integration, we divide the region of integration B′AQBPB′ into vertical strips P′Q′ which extend from P′ on the circle $y = -\sqrt{a^2 - x^2}$ to Q′ on the circle $y = \sqrt{a^2 - x^2}$. [$x^2 + y^2 = a^2 \Rightarrow y = \pm\sqrt{a^2 - x^2}$; for points in the 4th quadrant. $y = -\sqrt{a^2 - x^2}$ and for points in the first quadrant, $y = \sqrt{a^2 - x^2}$] . Then this strip slides from y-axis ($x = 0$) to A, where $x = a$.

$$\therefore \quad \int_{-a}^{a}\int_{0}^{\sqrt{a^2 - x^2}} f(x, y)\, dx\, dy = \int_{0}^{a}\int_{-\sqrt{a^2 - x^2}}^{\sqrt{a^2 - x^2}} f(x, y)\, dy\, dx .$$

Example 5. *By changing the order of integration, evaluate*

$$\int_0^a \int_{y^2/a}^{y} \frac{y}{(a - x)\sqrt{ax - y^2}}\, dx\, dy .$$

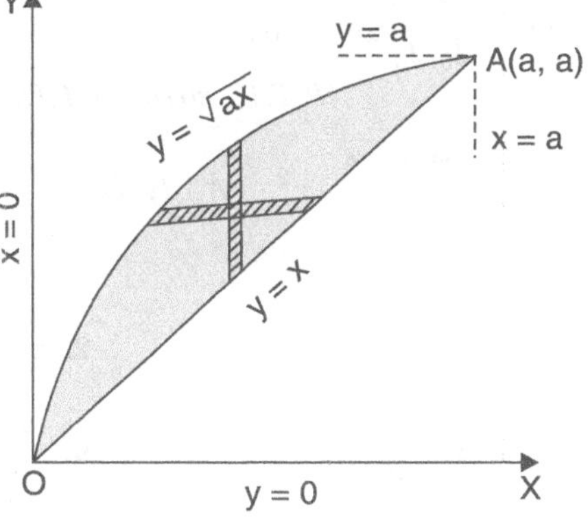

Sol. The given limits show that the area of integration (shown shaded) lies between $x = \dfrac{y^2}{a}$, $x = y$, $y = 0$, $y = a$. We can consider it as lying between

$$y = x,\ y = \sqrt{ax}\ ,\ x = 0,\ x = a.$$

$$\therefore \quad \int_0^a \int_{\frac{y^2}{a}}^{y} \frac{y}{(a - x)\sqrt{ax - y^2}}\, dx\, dy = \int_0^a \int_x^{\sqrt{ax}} \frac{y}{(a - x)\sqrt{ax - y^2}}\, dy\, dx$$

$$= \int_0^a \int_x^{\sqrt{ax}} -\frac{1}{2} \cdot \frac{1}{a - x} \cdot (ax - y^2)^{-1/2} \cdot (-2y)\, dy dx$$

$$= \int_0^a \frac{-1}{2(a - x)} \cdot \left[\frac{(ax - y^2)^{1/2}}{\frac{1}{2}}\right]_x^{\sqrt{ax}} dx = \int_0^a \frac{-1}{a - x}\left[0 - (ax - x^2)^{1/2}\right] dx$$

$$= \int_0^a \frac{(ax - x^2)^{1/2}}{a - x}\, dx = \int_0^a \frac{\left[x\,(a - x)\right]^{1/2}}{a - x}\, dx$$

$$= \int_0^a \left(\frac{x}{a - x}\right)^{1/2} dx \qquad \text{Put } x = a \sin^2 \theta,\ dx = 2a \sin \theta \cos \theta\, d\theta$$

$$= \int_0^{\frac{\pi}{2}} \left(\frac{a \sin^2 \theta}{a \cos^2 \theta}\right)^{1/2} \cdot 2a \sin \theta \cos \theta\, d\theta = 2a \int_0^{\frac{\pi}{2}} \sin^2 \theta\, d\theta = 2a \cdot \frac{1}{2} \cdot \frac{\pi}{2} = \frac{\pi a}{2}.$$

EXERCISE 12.3

Evaluate the following integrals by changing the order of integration:

1. $\int_0^1 \int_0^{\sqrt{1-x^2}} y^2 \, dy \, dx$. (*M.D.U. Dec. 2010*)

2. $\int_0^1 \int_{4y}^4 e^{x^2} \, dx \, dy$. (*Kerala 2012*)

3. $\int_0^4 \int_y^4 \frac{x \, dx \, dy}{x^2 + y^2}$.

4. $\int_0^{a/\sqrt{2}} \int_y^{\sqrt{a^2 - y^2}} x \, dx \, dy$.

5. $\int_0^1 \int_x^{\sqrt{2 - x^2}} \frac{x \, dy \, dx}{\sqrt{x^2 + y^2}}$.
(*K.U.K. 2012, 2013; M.D.U. Dec. 2013*)

6. $\int_0^a \int_{x^2/a}^{2a - x} xy \, dy \, dx$. (*U.P.T.U. 2008*)

7. $\int_0^b \int_0^{a/b \sqrt{b^2 - y^2}} xy \, dx \, dy$.

8. $\int_0^a \int_{\sqrt{ax}}^a \frac{y^2 \, dy \, dx}{\sqrt{y^4 - a^2 x^2}}$.

9. $\int_0^2 \int_{\sqrt{2y}}^2 \frac{x^2 \, dx \, dy}{\sqrt{x^4 - 4y^2}}$. (*M.D.U. Dec. 2011*)

10. $\int_0^\infty \int_x^\infty \frac{e^{-y}}{y} \, dy \, dx$.
(*M.D.U. Dec. 2009; K.U.K. Dec. 2010; M.T.U. 2013; P.T.U. 2011*)

11. $\int_0^\infty \int_0^x x e^{-\frac{x^2}{y}} \, dy \, dx$.
(*P.T.U. 2010; M.T.U. 2013*)

12. $\int_0^3 \int_1^{\sqrt{4 - y}} (x + y) \, dx \, dy$.

13. $\int_0^a \int_{x/a}^{\sqrt{x/a}} (x^2 + y^2) \, dy \, dx$.

Answers

1. $\frac{\pi}{16}$ 2. $\frac{1}{8}(e^{16} - 1)$ 3. π 4. $\frac{a^3\sqrt{2}}{6}$

5. $1 - \frac{1}{\sqrt{2}}$ 6. $\frac{3a^4}{8}$ 7. $\frac{1}{8}a^2b^2$ 8. $\frac{\pi a^2}{6}$

9. $\frac{2\pi}{3}$ 10. 1 11. $\frac{1}{2}$ 12. $\frac{241}{60}$

13. $\frac{a^3}{28} + \frac{a}{20}$.

12.5. TRIPLE INTEGRALS

Consider a function $f(x, y, z)$ which is continuous at every point of a finite region V of three dimensional space. Divide the region V into n sub-regions of respective volumes $\delta V_1, \delta V_2, \ldots, \delta V_n$. Let (x_r, y_r, z_r) be an arbitrary point in the rth sub-region. Consider the sum

$$\sum_{r=1}^{n} f(x_r, y_r, z_r) \, \delta V_r .$$

The limit of this sum as $n \to \infty$ and $\delta V_r \to 0$, if it exists, is called the *triple integral* of $f(x, y, z)$ over the region V and is denoted by $\iiint_V f(x, y, z) \, dV$.

For purposes of evaluation, it can be expressed as the repeated integral

$$\int_{z_1}^{z_2}\int_{y_1}^{y_2}\int_{x_1}^{x_2} f(x, y, z)\, dx\, dy\, dz \qquad ...(1)$$

the order of integration depending upon the limits.

Let x_1, x_2 be function of y, z; y_1, y_2 be function of z and z_1, z_2 be constants, *i.e.*,

Let $x_1 = f_1(y, z)$, $x_2 = f_2(y, z)$, $y_1 = \phi_1(z)$, $y_2 = \phi_2(z)$ and $z_1 = a$, $z_2 = b$.

Then the integral (*i*) is evaluated as follows:

$$\int_{z_1=a}^{z_2=b}\left[\int_{y_1=\phi_1(z)}^{y_2=\phi_2(z)}\left[\int_{x_1=f_1(y,z)}^{x_2=f_2(y,z)} f(x,y,z)\, dx\right] dy\right] dz$$

First $f(x, y, z)$ is integrated w.r.t. x (keeping y and z constant) between the limits x_1 and x_2. The resulting expression, which is a function of y and z is then integrated w.r.t. y (keeping z constant) between the limits y_1 and y_2. The resulting expression, which is a function of z only is then integrated w.r.t. z between the limits z_1 and z_2. The order of integration is from the innermost rectangle to the outermost rectangle.

Limits involving two variables are kept innermost, then the limits involving one variable and finally the constant limits.

If x_1, x_2; y_1, y_2 and z_1, z_2 are all constants, then the order of integration is immaterial, provided the limits are changed accordingly. Thus

$$\int_{z_1}^{z_2}\int_{y_1}^{y_2}\int_{x_1}^{x_2} f(x,y,z)\, dx\, dy\, dz = \int_{x_1}^{x_2}\int_{z_1}^{z_2}\int_{y_1}^{y_2} f(x,y,z)\, dy\, dz\, dx$$

$$= \int_{x_1}^{x_2}\int_{y_1}^{y_2}\int_{z_1}^{z_2} f(x,y,z)\, dz\, dy\, dx, \text{ etc.}$$

ILLUSTRATIVE EXAMPLES

Example 1. *Evaluate* $\int_0^1\int_0^{\sqrt{1-x^2}}\int_0^{\sqrt{1-x^2-y^2}} \frac{1}{\sqrt{1-x^2-y^2-z^2}}\, dz\, dy\, dx.$ (Kerala 2009)

Sol.

$$I = \int_0^1\int_0^{\sqrt{1-x^2}}\int_0^{\sqrt{1-x^2-y^2}} \frac{1}{\sqrt{(1-x^2-y^2)-z^2}}\, dz\, dy\, dx$$

$$= \int_0^1\int_0^{\sqrt{1-x^2}}\left[\sin^{-1}\frac{z}{\sqrt{1-x^2-y^2}}\right]_0^{\sqrt{1-x^2-y^2}} dy\, dx$$

$$= \int_0^1\int_0^{\sqrt{1-x^2}} (\sin^{-1}1 - \sin^{-1}0)\, dy\, dx$$

$$= \int_0^1\int_0^{\sqrt{1-x^2}} \frac{\pi}{2}\, dy\, dx = \int_0^1 \frac{\pi}{2}\Big[y\Big]_0^{\sqrt{1-x^2}} dx$$

$$= \frac{\pi}{2}\int_0^1 \sqrt{1-x^2}\, dx = \frac{\pi}{2}\left[\frac{x\sqrt{1-x^2}}{2} + \frac{1}{2}\sin^{-1}x\right]_0^1 = \frac{\pi}{4}[\sin^{-1}1] = \frac{\pi}{4}\cdot\frac{\pi}{2} = \frac{\pi^2}{8}.$$

Example 2. *Evaluate* $\int_1^e \int_1^{\log y} \int_1^{e^x} \log z \, dz \, dx \, dy$.

Sol. $$I = \int_1^e \int_1^{\log y} \left[\int_1^{e^x} \log z \, dz \right] dx \, dy$$

Since $\int_1^{e^x} \log z \, dz = \int_1^{e^x} \log z \, . \, 1 \, dz$

Integrating by parts $= \left[\log z \cdot z \right]_1^{e^x} - \int_1^{e^x} \frac{1}{z} \cdot z \, dz$

$$= e^x \log e^x - 0 - \left[z \right]_1^{e^x} = xe^x - e^x + 1 = (x - 1) e^x + 1$$

$\therefore$ $$I = \int_1^e \int_1^{\log y} [(x - 1) e^x + 1] \, dx \, dy$$

Now $\int_1^{\log y} [(x - 1) e^x + 1] \, dx = \int_1^{\log y} (x - 1) e^x \, dx + \left[x \right]_1^{\log y}$

$$= \left[(x - 1) e^x \right]_1^{\log y} - \int_1^{\log y} 1 \cdot e^x \, dx + \log y - 1$$

$$= (\log y - 1) e^{\log y} - \left[e^x \right]_1^{\log y} + \log y - 1$$

$$= y(\log y - 1) - (e^{\log y} - e) + \log y - 1 \qquad [\because \quad e^{\log y} = y]$$

$$= y(\log y - 1) - y + e + \log y - 1 = (y + 1) \log y - 2y + e - 1$$

$\therefore$ $$I = \int_1^e [\log y \cdot (y + 1) - 2y + e - 1] \, dy$$

$$= \left[\log y \cdot \left(\frac{y^2}{2} + y \right) \right]_1^e - \int_1^e \frac{1}{y} \left(\frac{y^2}{2} + y \right) dy - \left[y^2 \right]_1^e + (e - 1) \left[y \right]_1^e$$

$$= \frac{e^2}{2} + e - \int_1^e \left(\frac{y}{2} + 1 \right) dy - (e^2 - 1) + (e - 1)^2$$

$$= \frac{e^2}{2} + e - \left[\frac{y^2}{4} + y \right]_1^e - 2e + 2 = \frac{e^2}{2} + e - \left[\left(\frac{e^2}{4} + e \right) - \left(\frac{1}{4} + 1 \right) \right] - 2e + 2$$

$$= \frac{e^2}{4} - 2e + \frac{13}{4} = \frac{1}{4} (e^2 - 8e + 13).$$

EXERCISE 12.4

Evaluate the following integrals (1–11):

1. $\int_0^1 \int_0^1 \int_0^1 e^{x + y + z} \, dx \, dy \, dz$. *(Kerala 2012)*

2. $\int_0^a \int_0^a \int_0^a (yz + zx + xy) \, dx \, dy \, dz$.

3. $\int_{-1}^1 \int_0^z \int_{x - z}^{x + z} (x + y + z) \, dy \, dx \, dz$.

4. $\int_1^3 \int_{1/x}^1 \int_0^{\sqrt{xy}} xyz \, dz \, dy \, dx$.

5. $\int_0^1 \int_0^{1-x} \int_0^{1-x-y} xyz \, dz \, dy \, dx.$

6. $\int_0^1 \int_{y^2}^1 \int_0^{1-x} x \, dz \, dx \, dy.$

7. $\int_0^1 \int_0^{\sqrt{1-x^2}} \int_0^{\sqrt{1-x^2-y^2}} xyz \, dz \, dy \, dx.$ *(M.D.U. Dec. 2010)*

8. $\int_0^{\log 2} \int_0^x \int_0^{x+\log y} e^{x+y+z} \, dz \, dy \, dx.$

9. $\int_0^{\pi/2} d\theta \int_0^{a \sin \theta} dr \int_0^{(a^2-r^2)/a} r dz.$

10. $\int_0^{\pi/2} \int_0^{a \cos \theta} \int_0^{\sqrt{a^2-r^2}} r dz \, dr \, d\theta.$

11. $\int_0^a \int_0^x \int_0^{x+y} e^{x+y+z} \, dz \, dy \, dx.$ *(K.U.K. 2009)*

12. Evaluate $\iiint_V (x - 2y + z) \, dz \, dy \, dx$, where V is the region determined by $0 \le x \le 1$, $0 \le y \le x^2$, $0 \le z \le x + y$. *(U.P.T.U. 2009)*

Answers

1. $(e-1)^3$ 2. $\frac{3}{4}a^5$ 3. 0 4. $\frac{13}{9} - \frac{1}{6}\log 3$

5. $\frac{1}{720}$ 6. $\frac{4}{35}$ 7. $\frac{1}{48}$ 8. $\frac{8}{3}\log 2 - \frac{19}{9}$

9. $\frac{5a^3\pi}{64}$ 10. $\frac{a^3}{3}\left(\frac{\pi}{2} - \frac{2}{3}\right)$ 11. $\frac{1}{8}e^{4a} - \frac{3}{4}e^{2a} + e^a - \frac{3}{8}$ 12. $\frac{8}{35}$.

12.6. CHANGE OF VARIABLES

Quite often, the evaluation of a double or triple integral is greatly simplified by a suitable change of variables.

Let the variables x, y in the double integral $\iint_R f(x, y) \, dx \, dy$ be changed to u, v by means of the relations $x = \phi(u, v)$, $y = \psi(u, v)$, then the double integral is transformed into $\iint_{R'} f\{\phi(u, v), \psi(u, v)\} \mid J \mid du \, dv$, where $J = \frac{\partial(x, y)}{\partial(u, v)} = \begin{vmatrix} \frac{\partial x}{\partial u} & \frac{\partial x}{\partial v} \\ \frac{\partial y}{\partial u} & \frac{\partial y}{\partial v} \end{vmatrix}$

is the Jacobian of transformation from (x, y) to (u, v) co-ordinates and R′ is the region in the uv-plane which corresponds to the region R in the xy-plane.

(i) To change cartesian co-ordinates (x, y) to polar co-ordinates (r, θ).

Here we have $x = r \cos \theta$, $y = r \sin \theta$ so that $x^2 + y^2 = r^2$

$$J = \begin{vmatrix} \frac{\partial x}{\partial r} & \frac{\partial x}{\partial \theta} \\ \frac{\partial y}{\partial r} & \frac{\partial y}{\partial \theta} \end{vmatrix} = \begin{vmatrix} \cos\theta & -r\sin\theta \\ \sin\theta & r\cos\theta \end{vmatrix} = r(\cos^2\theta + \sin^2\theta) = r$$

$$\therefore \quad \iint_R f(x, y) \, dx \, dy = \iint_{R'} f(r\cos\theta, r\sin\theta) \, r \, dr \, d\theta$$

i.e., replace x by $r \cos \theta$, y by $r \sin \theta$ and $dx \, dy$ by $r \, dr \, d\theta$.

(ii) To change cartesian co-ordinates (x, y, z) to spherical polar co-ordinates (r, θ, φ).

Here, we have $x = r \sin\theta \cos\phi$

$$y = r \sin\theta \sin\phi$$

$$z = r\cos\theta$$

so that $x^2 + y^2 + z^2 = r^2$

$$J = \begin{vmatrix} \frac{\partial x}{\partial r} & \frac{\partial x}{\partial \theta} & \frac{\partial x}{\partial \phi} \\ \frac{\partial y}{\partial r} & \frac{\partial y}{\partial \theta} & \frac{\partial y}{\partial \phi} \\ \frac{\partial z}{\partial r} & \frac{\partial z}{\partial \theta} & \frac{\partial z}{\partial \phi} \end{vmatrix}$$

$$= \begin{vmatrix} \sin\theta\cos\phi & r\cos\theta\cos\phi & -r\sin\theta\sin\phi \\ \sin\theta\sin\phi & r\cos\theta\sin\phi & r\sin\theta\cos\phi \\ \cos\theta & -r\sin\theta & 0 \end{vmatrix} = r^2 \sin\theta$$

Z, P (x, y, z), (r, θ, φ), θ, r, z = r cos θ, θ, O, Y, x, φ, r sin θ, 90°, N, 90°, y, M, X

$$\therefore \quad \iiint_V f(x, y, z)\, dx\, dy\, dz = \iiint_{V'} f(r\sin\theta\cos\phi, r\sin\theta\sin\phi, r\cos\theta)\, r^2 \sin\theta\, dr\, d\theta\, d\phi.$$

Note. Equation of sphere $x^2 + y^2 + z^2 = a^2$ in spherical polar coordinates is $r = a$.

(*i*) If the region of integration is the whole sphere, then

$$0 \le r \le a, \quad 0 \le \theta \le \pi, \quad 0 \le \phi \le 2\pi$$

(*ii*) If the region of integration is the positive octant, then

$$0 \le r \le a, \quad 0 \le \theta \le \frac{\pi}{2}, \quad 0 \le \phi \le \frac{\pi}{2}.$$

(iii) To change cartesian co-ordinates (x, y , z) to cylindrical co-ordinates (r, φ, z).

Here we have $x = r\cos\phi$

$$y = r\sin\phi$$

$$z = z$$

$$J = \begin{vmatrix} \frac{\partial x}{\partial r} & \frac{\partial x}{\partial \phi} & \frac{\partial x}{\partial z} \\ \frac{\partial y}{\partial r} & \frac{\partial y}{\partial \phi} & \frac{\partial y}{\partial z} \\ \frac{\partial z}{\partial r} & \frac{\partial z}{\partial \phi} & \frac{\partial z}{\partial z} \end{vmatrix}$$

$$= \begin{vmatrix} \cos\phi & -r\sin\phi & 0 \\ \sin\phi & r\cos\phi & 0 \\ 0 & 0 & 1 \end{vmatrix}$$

$$= r(\cos^2\phi + \sin^2\phi) = r$$

Z, P(x, y, z), (r, φ, z), z, O, y, N, Y, 90° − φ, 90°, φ, x, r, M, X

$$\therefore \quad \iiint_V f(x, y, z)\, dx\, dy\, dz$$

$$= \iiint_{V'} f(r\cos\phi, r\sin\phi, z)\, r\, dr\, d\phi\, dz.$$

Note. For the cylinder $x^2 + y^2 = a^2$, $z = 0$, $z = h$, the limits of integration are

$$0 \le r \le a, \quad 0 \le \phi \le 2\pi, \; 0 \le z \le h$$

If the region of integration is a cylinder (or cone), change the problem to cylindrical polar coordinates.

ILLUSTRATIVE EXAMPLES

Example 1. *Evaluate* $\iint \sqrt{\dfrac{1-x^2-y^2}{1+x^2+y^2}}\, dx\, dy$ *over the positive quadrant of the circle* $x^2 + y^2 = 1$.

Sol. Changing to polar coordinates by putting $x = r\cos\theta$, $y = r\sin\theta$; $x^2 + y^2 = 1$ transforms into $r = 1$. For the region of integration R, r varies from 0 to 1 and θ varies from 0 to $\dfrac{\pi}{2}$.

$$\text{I} = \iint_{\text{R}} \sqrt{\frac{1-x^2-y^2}{1+x^2+y^2}}\, dx\, dy$$

$$= \int_0^{\pi/2}\int_0^1 \sqrt{\frac{1-r^2}{1+r^2}}\; r\, dr\, d\theta$$

| $dx\, dy$ is replaced by $r\, dr\, d\theta$

$$= \int_0^{\pi/2}\int_0^1 \frac{r(1-r^2)}{\sqrt{1-r^4}}\, dr\, d\theta$$

Now
$$\int_0^1 \frac{r(1-r^2)}{\sqrt{1-r^4}}\, dr = \int_0^1 \left(\frac{r}{\sqrt{1-r^4}} - \frac{r^3}{\sqrt{1-r^4}}\right) dr$$

$$= \frac{1}{2}\int_0^1 \frac{2r}{\sqrt{1-r^4}}\, dr + \frac{1}{4}\int_0^1 -4r^3\,(1-r^4)^{-1/2}\, dr$$

$$= \frac{1}{2}\int_0^1 \frac{dt}{\sqrt{1-t^2}} + \frac{1}{4}\cdot\left[\frac{(1-r^4)^{1/2}}{1/2}\right]_0^1, \text{ where } t = r^2$$

$$= \frac{1}{2}\Big[\sin^{-1} t\Big]_0^1 + \frac{1}{2}\,(0-1) = \frac{1}{2}\left(\frac{\pi}{2}\right) - \frac{1}{2} = \frac{\pi}{4} - \frac{1}{2}$$

$\therefore$
$$\text{I} = \int_0^{\pi/2}\left(\frac{\pi}{4} - \frac{1}{2}\right) d\theta = \left(\frac{\pi}{4} - \frac{1}{2}\right)\Big[\theta\Big]_0^{\pi/2} = \left(\frac{\pi}{4} - \frac{1}{2}\right)\frac{\pi}{2} = \frac{\pi^2}{8} - \frac{\pi}{4}.$$

Example 2. *Evaluate* $\iint \sqrt{a^2-x^2-y^2}\, dx\, dy$ *over the semi-circle* $x^2 + y^2 = ax$ *in the positive quadrant.*

Sol. Changing to polar co-ordinates, $x^2 + y^2 = ax$ transforms into $r = a\cos\theta$. For the region of integration R, r varies from 0 to $a\cos\theta$ and θ varies from 0 to $\dfrac{\pi}{2}$.

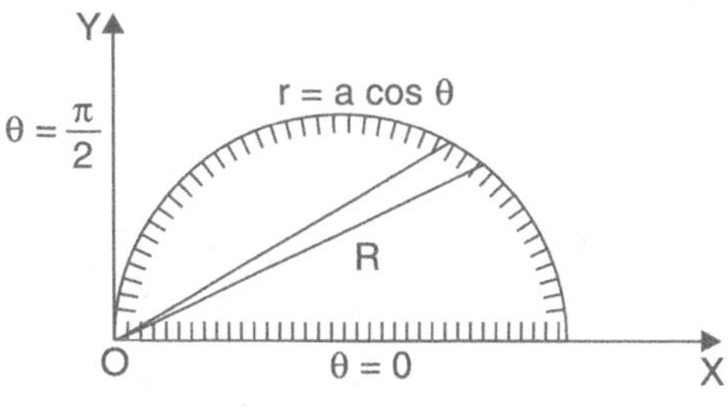

$$\therefore \iint_{\text{R}} \sqrt{a^2-x^2-y^2}\, dx\, dy$$

$$= \int_0^{\pi/2}\int_0^{a\cos\theta} \sqrt{a^2-r^2}\cdot r\, dr\, d\theta$$

$$= \int_0^{\pi/2}\int_0^{a\cos\theta} -\frac{1}{2}(a^2-r^2)^{1/2}\,(-2r)\, dr\, d\theta$$

$$= \int_0^{\pi/2} -\frac{1}{2}\cdot\left[\frac{(a^2-r^2)^{3/2}}{3/2}\right]_0^{a\cos\theta} d\theta$$

$$= -\frac{1}{3}\int_0^{\pi/2}(a^3\sin^3\theta - a^3)\,d\theta = -\frac{a^3}{3}\int_0^{\pi/2}(\sin^3\theta-1)\,d\theta$$

$$= -\frac{a^3}{3}\left[\frac{2}{3}-\frac{\pi}{2}\right] = \frac{a^3}{3}\left(\frac{\pi}{2}-\frac{2}{3}\right).$$

Example 3. *Change into polar co-ordinates and evaluate* $\int_0^\infty\int_0^\infty e^{-(x^2+y^2)}\,dy\,dx$.

Hence show that $\int_0^\infty e^{-x^2}\,dx = \frac{\sqrt{\pi}}{2}$. (K.U.K. Dec., 2013; P.T.U. May 2011, Dec. 2011)

Sol. For the region of integration in cartesian co-ordinates, y varies from 0 to ∞ and x also varies from 0 to ∞. Thus the region of integration is the plane XOY. Changing to polar co-ordinates by putting $x = r\cos\theta$, $y = r\sin\theta$ so that $x^2 + y^2 = r^2$; for the region of integration r varies from 0 to ∞ and θ varies from 0 to $\frac{\pi}{2}$.

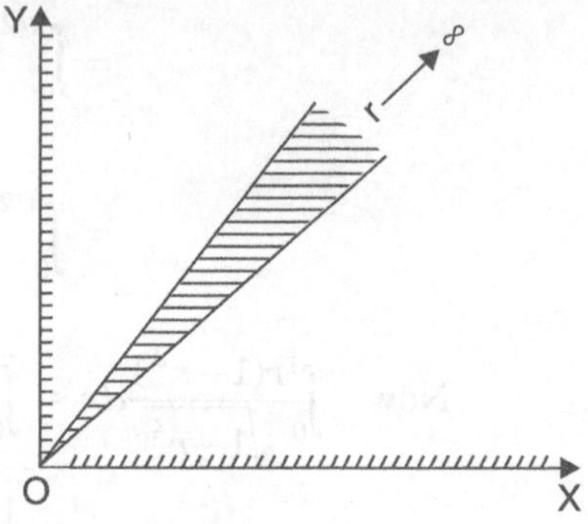

$$\therefore \quad \int_0^\infty\int_0^\infty e^{-(x^2+y^2)}\,dy\,dx = \int_0^{\pi/2}\int_0^\infty e^{-r^2}\cdot r\,dr\,d\theta$$

$$= \int_0^{\pi/2}\int_0^\infty \frac{1}{2}\cdot e^{-r^2}\cdot 2r\,dr\,d\theta$$

$$= \int_0^{\pi/2}\int_0^\infty \frac{1}{2}e^{-t}dt\,d\theta, \text{where } t = r^2$$

$$= \int_0^{\pi/2}\left[-\frac{1}{2}e^{-t}\right]_0^\infty d\theta = -\frac{1}{2}\int_0^{\pi/2}(0-1)\,d\theta = \frac{1}{2}\left[\theta\right]_0^{\pi/2} = \frac{\pi}{4}.$$

Now the above result can be written as

$$\int_0^\infty e^{-x^2}\,dx \times \int_0^\infty e^{-y^2}\,dy = \frac{\pi}{4}$$

or $$\int_0^\infty e^{-x^2}\,dx \times \int_0^\infty e^{-x^2}\,dx = \frac{\pi}{4} \qquad \text{or} \qquad \left[\int_0^\infty e^{-x^2}\,dx\right]^2 = \frac{\pi}{4}$$

$$\therefore \quad \int_0^\infty e^{-x^2}\,dx = \frac{\sqrt{\pi}}{2}.$$

Example 4. *Evaluate* $\int_0^2\int_0^{\sqrt{2x-x^2}} \frac{x\,dy\,dx}{\sqrt{x^2+y^2}}$ *by changing to polar co-ordinates.*

Sol. In the given integral, y varies from 0 to $\sqrt{2x-x^2}$ and x varies from 0 to 2.

$$y = \sqrt{2x-x^2} \;\Rightarrow\; y^2 = 2x - x^2 \;\Rightarrow\; x^2 + y^2 = 2x.$$

In polar co-ordinates, we have $r^2 = 2r \cos \theta$ or $r = 2 \cos \theta$.

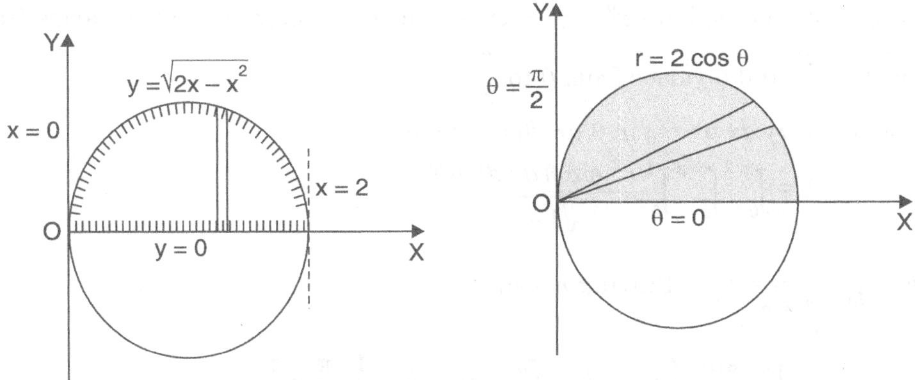

$\therefore$ For the region of integration, r varies from 0 to $2 \cos \theta$ and θ varies from 0 to $\frac{\pi}{2}$.

In the given integral, replacing x by $r \cos \theta$, y by $r \sin \theta$, $dy\, dx$ by $r\, dr\, d\theta$, we have

$$I = \int_0^{\pi/2} \int_0^{2 \cos \theta} \frac{r \cos \theta \cdot r\, dr\, d\theta}{r} = \int_0^{\pi/2} \int_0^{2 \cos \theta} r \cos \theta\, dr\, d\theta$$

$$= \int_0^{\pi/2} \cos \theta \left[\frac{r^2}{2} \right]_0^{2 \cos \theta} d\theta = \int_0^{\pi/2} 2 \cos^3 \theta\, d\theta = 2 \cdot \frac{2}{3} = \frac{4}{3}.$$

Example 5. *Evaluate* $\iiint z(x^2 + y^2 + z^2)\, dx\, dy\, dz$ *through the volume of the cylinder* $x^2 + y^2 = a^2$ *intercepted by the planes* $z = 0$ *and* $z = h$.

Sol. Changing to cylindrical co-ordinates by changing x to $r \cos \phi$, y to $r \sin \theta$ and replacing $dx\, dy\, dz$ by $r\, dr\, d\phi\, dz$

$$I = \int_0^h \int_0^{2\pi} \int_0^a z\,(r^2 + z^2)\, r\, dr\, d\phi\, dz = \int_0^h \int_0^{2\pi} \int_0^a (zr^3 + z^3 r)\, dr\, d\phi\, dz$$

$$= \int_0^h \int_0^{2\pi} \left[z \cdot \frac{r^4}{4} + z^3 \cdot \frac{r^2}{2} \right]_0^a d\phi\, dz = \int_0^h \int_0^{2\pi} \left(\frac{a^4}{4} z + \frac{a^2}{2} z^3 \right) d\phi\, dz$$

$$= \int_0^h \left(\frac{a^4}{4} z + \frac{a^2}{2} z^3 \right) \Big[\phi \Big]_0^{2\pi} dz = \int_0^h 2\pi \left(\frac{a^4}{4} z + \frac{a^2}{2} z^3 \right) dz$$

$$= 2\pi \left[\frac{a^4 z^2}{8} + \frac{a^2 z^4}{8} \right]_0^h = \frac{\pi}{4} (a^4 h^2 + a^2 h^4) = \frac{\pi}{4} a^2 h^2 (a^2 + h^2).$$

Example 6. *Evaluate* $\int_0^1 \int_0^{\sqrt{1-x^2}} \int_0^{\sqrt{1-x^2-y^2}} \frac{dz\, dy\, dx}{\sqrt{1 - x^2 - y^2 - z^2}}$, *by changing to spherical polar co-ordinates.*

(Kerala 2008; M.D.U. Dec. 2008, Dec. 2010, Dec. 2012; P.T.U. May 2008)

Sol. Here the region of integration is bounded by

$z = 0,\quad z = \sqrt{1 - x^2 - y^2}\quad$ (*i.e.*, $x^2 + y^2 + z^2 = 1$)

$y = 0,\quad y = \sqrt{1 - x^2}\quad$ (*i.e.*, $x^2 + y^2 = 1$)

$x = 0,\quad x = 1$

which is the volume of the sphere $x^2 + y^2 + z^2 = 1$ in the positive octant.

Changing to spherical polar co-ordinates by putting $x = r \sin\theta \cos\phi$, $y = r \sin\theta \sin\phi$, $z = r\cos\theta$ so that
$$x^2 + y^2 + z^2 = r^2.$$

For the volume of sphere $x^2 + y^2 + z^2 = 1$ in the positive octant, r varies from 0 to 1, θ varies from 0 to $\frac{\pi}{2}$ and ϕ varies from 0 to $\frac{\pi}{2}$.

Replacing $dz\, dy\, dx$ by $r^2 \sin\theta\, dr\, d\theta\, d\phi$, we have

$$I = \int_0^{\pi/2}\int_0^{\pi/2}\int_0^1 \frac{r^2 \sin\theta\, dr\, d\theta\, d\phi}{\sqrt{1-r^2}}$$

Now $\displaystyle\int_0^1 \frac{r^2}{\sqrt{1-r^2}}\, dr$ Putting $r = \sin t$

$$= \int_0^{\frac{\pi}{2}} \frac{\sin^2 t}{\cos t}\cos t\, dt = \int_0^{\frac{\pi}{2}} \sin^2 t\, dt = \frac{1}{2}\cdot\frac{\pi}{2} = \frac{\pi}{4}$$

$$\therefore \quad I = \int_0^{\frac{\pi}{2}}\int_0^{\frac{\pi}{2}} \frac{\pi}{4}\sin\theta\, d\theta\, d\phi = \int_0^{\frac{\pi}{2}} \frac{\pi}{4}\Big[-\cos\theta\Big]_0^{\frac{\pi}{2}} d\phi = \frac{\pi}{4}\int_0^{\frac{\pi}{2}} d\phi = \frac{\pi}{4}\cdot\frac{\pi}{2} = \frac{\pi^2}{8}.$$

Note. For the whole volume of the sphere $x^2 + y^2 + z^2 = a^2$.
$$0 \le r \le a,\ 0 \le \theta \le \pi,\ 0 \le \phi \le 2\pi.$$

Example 7. *Evaluate* $\iint_R (x + y)^2\, dx\, dy$, *where R is the parallelogram in the xy-plane with vertices (1, 0), (3, 1), (2, 2), (0, 1), using the transformation* $u = x + y$ *and* $v = x - 2y$.

Sol. The vertices A(1, 0), B(3, 1), C(2, 2), D(0, 1) of the parallelogram ABCD in the xy-plane become A′(1, 1), B′(4, 1), C′(4, –2), D′(1, –2) in the uv-plane under the given transformation.

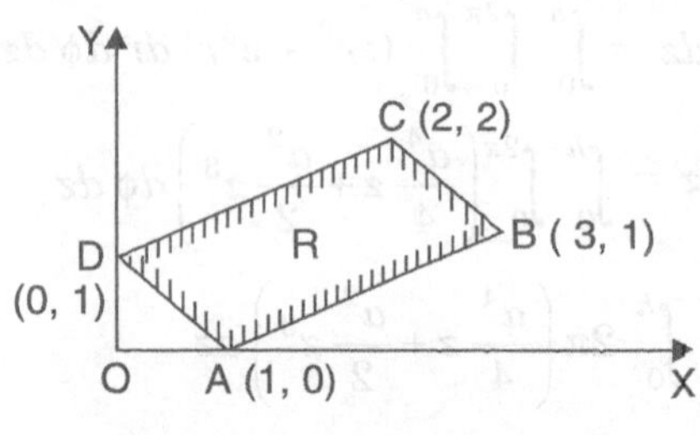

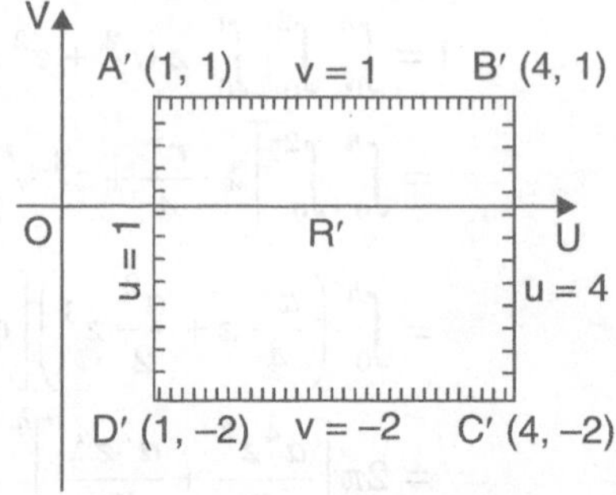

The region R in the xy-plane becomes the region R′ in the uv-plane which is a square bounded by the line $u = 1$, $u = 4$ and $v = -2$, $v = 1$.

Now
$$\frac{\partial(u, v)}{\partial(x, y)} = \begin{vmatrix} \frac{\partial u}{\partial x} & \frac{\partial u}{\partial y} \\ \frac{\partial v}{\partial x} & \frac{\partial v}{\partial y} \end{vmatrix} = \begin{vmatrix} 1 & 1 \\ 1 & -2 \end{vmatrix} = -3 \Rightarrow J = \frac{\partial(x, y)}{\partial(u, v)} = -\frac{1}{3}$$

$$\therefore \quad \iint_R (x+y)^2\, dx\, dy = \iint_{R'} u^2\, |J|\, du\, dv = \int_{-2}^{1}\int_1^4 u^2\cdot\frac{1}{3}\, du\, dv$$

$$= \int_{-2}^{1} \frac{1}{3}\left[\frac{u^3}{3}\right]_1^4 dv = \int_{-2}^{1} 7\, dv = 7\Big[v\Big]_{-2}^{1} = 7 \times 3 = 21.$$

Example 8. *Evaluate* $\iiint_R (x^2+y^2+z^2)\,dx\,dy\,dz$, *where R denotes the region bounded by* $x = 0, y = 0, z = 0$ *and* $x + y + z = a, (a > 0)$. (M.D.U. Dec. 2011)

Sol. The plane $x + y + z = a$, $(a > 0)$ meets the coordinate axes in A $(a, 0, 0)$, B$(0, a, 0)$ and C$(0, 0, a)$. On the face ABC, $z = a - x - y$. The projection of plane ABC on the xy-plane is the triangle OAB bounded by the lines OB $(x = 0)$, OA $(y = 0)$ and AB $(x + y = a)$.

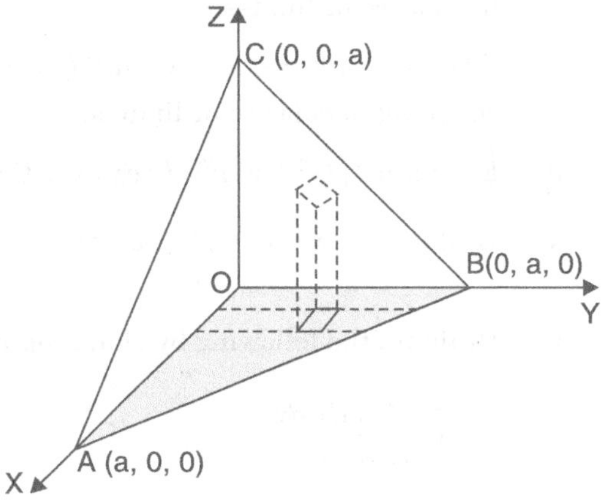

Z
C (0, 0, a)
O
B(0, a, 0)
Y
A (a, 0, 0)
X

$\therefore$ R = $\{(x, y, z): 0 \le x \le a, 0 \le y \le a - x, 0 \le z \le a - x - y\}$

$$I = \iiint_R (x^2 + y^2 + z^2)\,dx\,dy\,dz$$

$$= \int_0^a \left[\int_0^{a-x}\left[\int_0^{a-x-y}(x^2+y^2+z^2)\,dz\right]dy\right]dx$$

$$= \int_0^a \left[\int_0^{a-x}\left[(x^2+y^2)\,z+\frac{z^3}{3}\right]_0^{a-x-y} dy\right]dx$$

$$= \int_0^a \left[\int_0^{a-x}\left[(x^2+y^2)(a-x-y)+\frac{1}{3}(a-x-y)^3\right]dy\right]dx$$

$$= \int_0^a \left[\int_0^{a-x}\left[(a-x)x^2+(a-x)y^2-x^2y-y^3+\frac{1}{3}(a-x-y)^3\right]dy\right]dx$$

$$= \int_0^a \left[(a-x)x^2y+(a-x)\cdot\frac{y^3}{3}-x^2\cdot\frac{y^2}{2}-\frac{y^4}{4}+\frac{1}{3}\cdot\frac{(a-x-y)^4}{-4}\right]_0^{a-x} dx$$

$$= \int_0^a \left[x^2(a-x)^2+\frac{1}{3}(a-x)^4-\frac{1}{2}x^2(a-x)^2-\frac{1}{4}(a-x)^4-\frac{1}{12}(0)+\frac{1}{12}(a-x)^4\right]dx$$

$$= \int_0^a \left[\frac{1}{2}x^2(a-x)^2+\frac{1}{6}(a-x)^4\right]dx = \int_0^a \left[\frac{1}{2}a^2x^2-ax^3+\frac{1}{2}x^4+\frac{1}{6}(a-x)^4\right]dx$$

$$= \left[\frac{1}{2}a^2\cdot\frac{x^3}{3}-a\cdot\frac{x^4}{4}+\frac{1}{2}\cdot\frac{x^5}{5}+\frac{1}{6}\cdot\frac{(a-x)^5}{-5}\right]_0^a$$

$$= \frac{a^5}{6}-\frac{a^5}{4}+\frac{a^5}{10}-\frac{1}{30}(0)+\frac{1}{30}.a^5$$

$$= \left(\frac{10-15+6+2}{60}\right)a^5 = \frac{3}{60}a^5 = \frac{a^5}{20}$$

EXERCISE 12.5

1. Evaluate $\iint \sin \pi (x^2 + y^2)\, dx\, dy$ over the region bounded by the circle $x^2 + y^2 = 1$ by changing to polar co-ordinates.

2. Evaluate $\iint (a^2 - x^2 - y^2)\, dx\, dy$ over the semi-circle $x^2 + y^2 = ax$ in the positive quadrant by changing to polar co-ordinates.

3. Evaluate $\iint (x^2 + y^2)^{7/2}\, dx\, dy$ over the circle $x^2 + y^2 = 1$.

4. Evaluate $\iint xy\, (x^2 + y^2)^{3/2}\, dx\, dy$ over the positive quadrant of the circle $x^2 + y^2 = 1$.

(M.D.U. Dec. 2012)

5. Evaluate the following by changing into polar co-ordinates:

(*i*) $\displaystyle\int_0^a \int_y^a \frac{x\, dx\, dy}{x^2 + y^2}$ (*ii*) $\displaystyle\int_{-a}^{a} \int_{-\sqrt{a^2 - x^2}}^{\sqrt{a^2 - x^2}} dy\, dx$

(*iii*) $\displaystyle\int_0^{4a} \int_{y^2/4a}^{y} \frac{x^2 - y^2}{x^2 + y^2}\, dx\, dy$ (*iv*) $\displaystyle\int_0^a \int_0^{\sqrt{a^2 - y^2}} (x^2 + y^2)\, dx\, dy$ *(M.D.U. May 2008)*

(*v*) $\displaystyle\int_0^a \int_0^{\sqrt{a^2 - y^2}} y^2 \sqrt{x^2 + y^2}\, dx\, dy$ (*vi*) $\displaystyle\int_0^a \int_y^a \frac{x^2\, dx\, dy}{\sqrt{x^2 + y^2}}$.

(*vii*) $\displaystyle\int_0^{\frac{a}{\sqrt{2}}} \int_y^{\sqrt{a^2 - y^2}} \log (x^2 + y^2)\, dx\, dy$, $(a > 0)$

6. Evaluate $\displaystyle\iint_D e^{-(x^2 + y^2)}\, dy\, dx$, where D is the region bounded by $x^2 + y^2 = a^2$.

7. Evaluate $\displaystyle\iint xy\, (x^2 + y^2)^{n/2}\, dx\, dy$ over the positive quadrant of $x^2 + y^2 = 4$, supposing $n + 3 > 0$.

8. Evaluate $\displaystyle\iint \sqrt{\frac{1 - \frac{x^2}{a^2} - \frac{y^2}{b^2}}{1 + \frac{x^2}{a^2} + \frac{y^2}{b^2}}}\, dx\, dy$ over the positive quadrant of the ellipse $\dfrac{x^2}{a^2} + \dfrac{y^2}{b^2} = 1$.

9. Transform the following to cartesian form and hence evaluate $\displaystyle\int_0^{\pi} \int_0^a r^3 \sin\theta \cos\theta\, dr\, d\theta$.

10. Evaluate $\displaystyle\iiint (x + y + z)\, dx\, dy\, dz$ over the tetrahedron bounded by the planes $x = 0$, $y = 0$, $z = 0$ and $x + y + z = 1$.

(M.D.U. May 2009)

11. Evaluate $\displaystyle\iiint \frac{dx\, dy\, dz}{\sqrt{a^2 - x^2 - y^2 - z^2}}$,

(*i*) Over the positive octant of the sphere $x^2 + y^2 + z^2 = a^2$.

(*ii*) Taken throughout the volume of the sphere $x^2 + y^2 + z^2 = a^2$. (*M.D.U. Dec. 2009*)

12. Evaluate $\iiint \dfrac{dx\,dy\,dz}{(x+y+z+1)^3}$ over the tetrahedron bounded by the coordinate planes and the plane $x + y + z = 1$.

13. Evaluate $\iiint z(x^2 + y^2)\,dx\,dy\,dz$ over the volume of the cylinder $x^2 + y^2 = 1$ intercepted by the planes $z = 2$ and $z = 3$.

14. Evaluate the following integrals through the volume of the sphere $x^2 + y^2 + z^2 = 1$, by changing into spherical polar co-ordinates:

(*i*) $\iiint z^2\,dx\,dy\,dz$ (*ii*) $\iiint (x^2 + y^2 + z^2)^m\,dx\,dy\,dz$. $(m > 0)$

15. By using the transformation $x + y = u$, $y = uv$, show that $\int_0^1 \int_0^{1-x} e^{\frac{y}{x+y}}\,dy\,dx = \frac{1}{2}(e-1)$.

[**Hint.** Here x varies from 0 to 1 and y varies from 0 to $1 - x$. The region D of integration is the triangle OAB in xy-plane. Under the given transformation

$$x = u - uv = u(1 - v),\ y = uv$$

Now,

$$x = 0 \Rightarrow u = 0 \text{ or } v = 1$$
$$y = 0 \Rightarrow u = 0 \text{ or } v = 0$$
$$x + y = 1 \Rightarrow u = 1$$

$$J = \frac{\partial(x, y)}{\partial(u, v)} = u \text{ and } dxdy = |\,J\,|\,dudv = u\,dudv$$

The region D transforms into the region

$$D' = \{(u, v) : 0 \le u \le 1,\ 0 \le v \le 1\}$$

which is square OPQR is uv-plane.

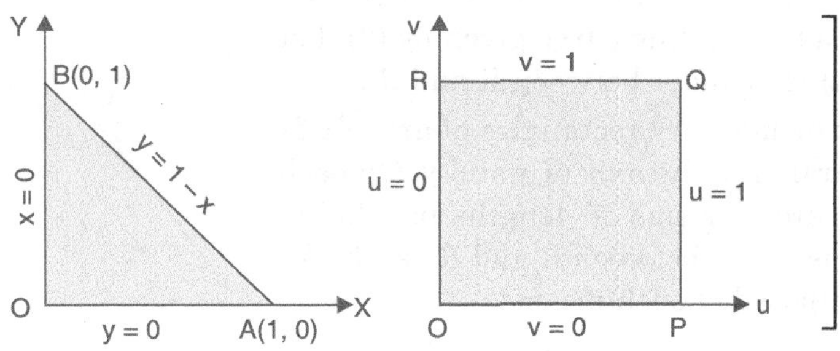

]

Answers

1. 2	**2.** $\frac{5\pi a^4}{64}$	**3.** $\frac{2\pi}{9}$	**4.** $\frac{1}{14}$
5. (*i*) $\frac{\pi a}{4}$	(*ii*) πa^2	(*iii*) $8\left(\frac{\pi}{2} - \frac{5}{3}\right)a^2$	(*iv*) $\frac{\pi a^4}{8}$

(*v*) $\dfrac{\pi a^5}{20}$ (*vi*) $\dfrac{a^3}{3}\log(\sqrt{2}+1)$ (*vii*) $\dfrac{\pi a^2}{4}\left(\log a - \dfrac{1}{2}\right)$ **6.** $\pi(1-e^{-a^2})$

7. $\dfrac{2^{n+3}}{n+4}$ **8.** $\dfrac{\pi ab}{8}(\pi-2)$ **9.** 0 **10.** $\dfrac{1}{8}$

11. (*i*) $\dfrac{\pi^2 a^2}{8}$ (*ii*) $\pi^2 a^2$ **12.** $\dfrac{1}{2}\log 2 - \dfrac{5}{16}$ **13.** $\dfrac{5\pi}{4}$

14. (*i*) $\dfrac{4\pi}{15}$ (*ii*) $\dfrac{4\pi}{2m+3}$.

12.7. AREA BY DOUBLE INTEGRATION

(*a*) *Cartesian co-ordinates.* The area A of the region bounded by the curves $y = f_1(x)$, $y = f_2(x)$ and the lines $x = a$, $x = b$ is given by $A = \int_a^b \int_{f_1(x)}^{f_2(x)} dy\, dx$

The area A of the region bounded by the curves $x = f_1(y)$, $x = f_2(y)$ and the lines $y = c$, $y = d$ is given by $A = \int_c^d \int_{f_1(y)}^{f_2(y)} dx\, dy$.

(*b*) *Polar co-ordinates.* The area A of the region bounded by the curves $f_1(\theta)$, $r = f_2(\theta)$ and the lines $\theta = \alpha$, $\theta = \beta$ is given by $A = \int_\alpha^\beta \int_{f_1(\theta)}^{f_2(\theta)} r\, dr\, d\theta$.

12.8. VOLUME AS A DOUBLE INTEGRAL

(*a*) *Cartesian co-ordinates.* Consider a surface

$z = f(x, y)$ ...(1)

Let the orthogonal projection of its region R′ on the *xy*-plane be the region R given by $\phi(x, y) = 0$. ...(2)

Now (2) represents a cylinder with generators parallel to *z*-axis and the guiding curve given by (2). Let V be the volume of this cylinder between R and R′.

Divide R into elementary rectangles of area $\delta x\, \delta y$ by drawing lines parallel to the axis of *x* and *y*. On each of these rectangles, erect prisms of lengths parallel to *z*-axis. Volume of this prism between R and R′ is $z\, \delta x\, \delta y$. The volume V is composed of such prisms.

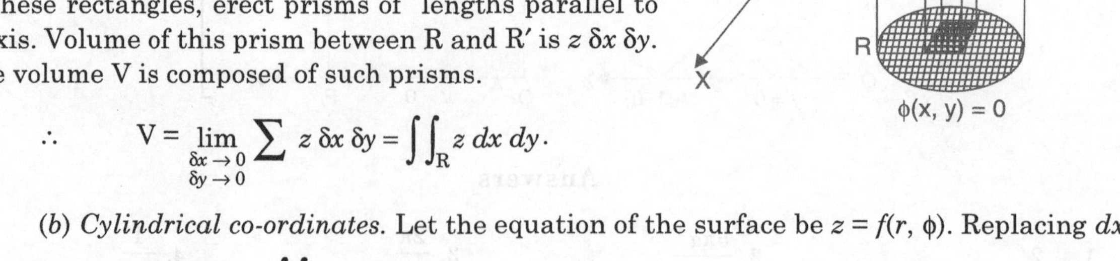

$$\therefore \quad V = \lim_{\substack{\delta x \to 0 \\ \delta y \to 0}} \sum z\, \delta x\, \delta y = \iint_R z\, dx\, dy.$$

(*b*) *Cylindrical co-ordinates.* Let the equation of the surface be $z = f(r, \phi)$. Replacing $dx\, dy$ by $r\, dr\, d\phi$, we get $V = \iint_R zr\, dr\, d\phi$.

ILLUSTRATIVE EXAMPLES

Example 1. *Find the area lying between the parabola $y = 4x - x^2$ and the line $y = x$.* (U.P.T.U. 2008)

Sol. Let two curves intersect at points whose abscissae are given by $4x - x^2 = x$

or $\quad x^2 - 3x = 0 \quad$ *i.e.,* $\quad x = 0, 3$.

Using vertical strips, the required area lies between $x = 0$, $x = 3$ and $y = x$, $y = 4x - x^2$.

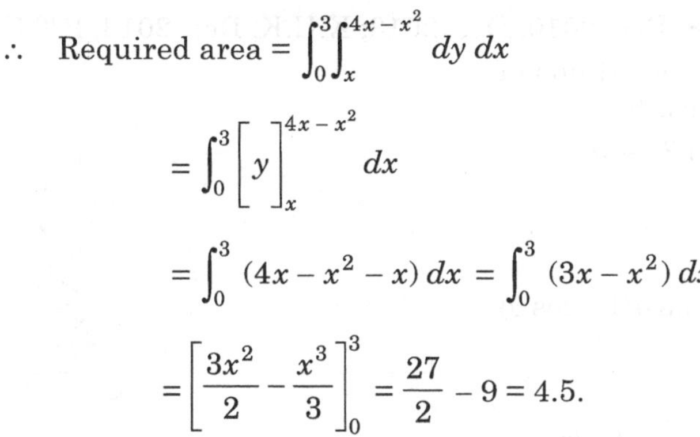

$$\therefore \quad \text{Required area} = \int_0^3 \int_x^{4x - x^2} dy\, dx$$

$$= \int_0^3 \Big[y \Big]_x^{4x - x^2} dx$$

$$= \int_0^3 (4x - x^2 - x)\, dx = \int_0^3 (3x - x^2)\, dx$$

$$= \left[\frac{3x^2}{2} - \frac{x^3}{3} \right]_0^3 = \frac{27}{2} - 9 = 4.5.$$

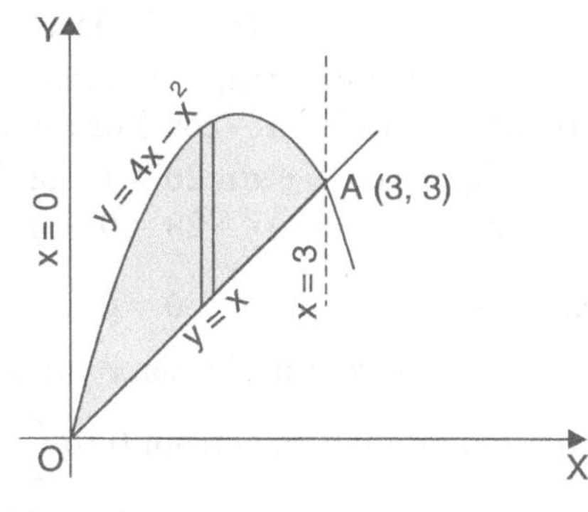

Example 2. *Find the smaller of the areas bounded by the ellipse $4x^2 + 9y^2 = 36$ and the straight line $2x + 3y = 6$.*

Sol. The equation of the ellipse is

$$\frac{x^2}{9} + \frac{y^2}{4} = 1 \quad \text{...(1)}$$

and the line is

$$\frac{x}{3} + \frac{y}{2} = 1 \quad \text{...(2)}$$

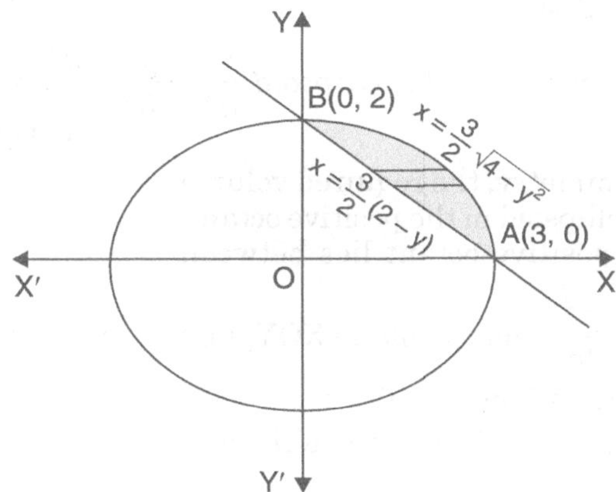

Both meet x-axis at A(3, 0) and y-axis at B(0, 2).

Using horizontal strips, the required area lies between

$$x = \frac{3}{2}(2 - y),\; x = \frac{3}{2}\sqrt{4 - y^2} \text{ and } y = 0,\, y = 2.$$

$$\therefore \quad \text{Required area} = \int_0^2 \int_{3/2(2-y)}^{3/2\sqrt{4-y^2}} dx\, dy = \int_0^2 \Big[x \Big]_{3/2(2-y)}^{3/2\sqrt{4-y^2}} dy$$

$$= \int_0^2 \frac{3}{2}\left[\sqrt{4-y^2} - (2-y)\right] dy = \frac{3}{2}\left[\frac{y\sqrt{4-y^2}}{2} + \frac{4}{2}\sin^{-1}\frac{y}{2} - 2y + \frac{y^2}{2}\right]_0^2$$

$$= \frac{3}{2}[2\sin^{-1} 1 - 4 + 2] = \frac{3}{2}\left(2 \cdot \frac{\pi}{2} - 2\right) = \frac{3}{2}(\pi - 2).$$

Example 3. *Find, by double integration, the area lying inside the circle $r = a \sin \theta$ and outside the cardioid $r = a(1 - \cos \theta)$.*

(M.D.U. Dec. 2008, Dec. 2010, Dec. 2013; K.U.K. Dec. 2013; P.T.U. 2011)

Sol. Eliminating r between the equations of two curves, $\sin \theta = 1 - \cos \theta$ or $\sin \theta + \cos \theta = 1$

Squaring $1 + \sin 2\theta = 1$ or $\sin 2\theta = 0$

$\therefore \quad 2\theta = 0$ or π

i.e., $\quad \theta = 0$ or $\frac{\pi}{2}$

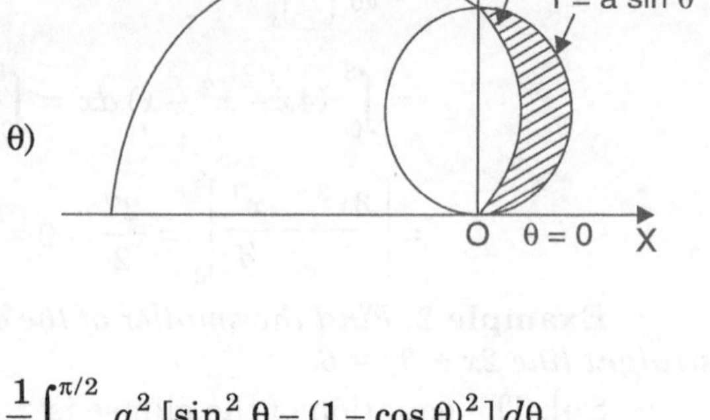

For the required area, r varies from $a(1 - \cos \theta)$ to $a \sin \theta$ and θ varies from 0 to $\frac{\pi}{2}$.

$$\therefore \quad \text{Required area} = \int_0^{\pi/2} \int_{a(1-\cos\theta)}^{a\sin\theta} r\, dr\, d\theta$$

$$= \int_0^{\pi/2} \left[\frac{r^2}{2}\right]_{a(1-\cos\theta)}^{a\sin\theta} d\theta = \frac{1}{2}\int_0^{\pi/2} a^2 [\sin^2\theta - (1-\cos\theta)^2]\, d\theta$$

$$= \frac{a^2}{2}\int_0^{\pi/2} (\sin^2\theta - 1 - \cos^2\theta + 2\cos\theta)\, d\theta$$

$$= \frac{a^2}{2}\int_0^{\pi/2} (-2\cos^2\theta + 2\cos\theta)\, d\theta = a^2\left[-\frac{1}{2}\cdot\frac{\pi}{2} + 1\right] = a^2\left(1 - \frac{\pi}{4}\right).$$

Example 4. *Find the volume of the ellipsoid $\frac{x^2}{a^2} + \frac{y^2}{b^2} + \frac{z^2}{c^2} = 1$.*

(M.D.U. Dec. 2013; Kerala 2008)

Sol. On account of symmetry, the required volume is 8 times the volume of the ellipsoid in the positive octant. The volume OABC in the positive octant lies between the ellipsoid $z = c\sqrt{1 - \frac{x^2}{a^2} - \frac{y^2}{b^2}}$ and the plane XOY, and is bounded on the sides by the planes $x = 0$, $y = 0$.

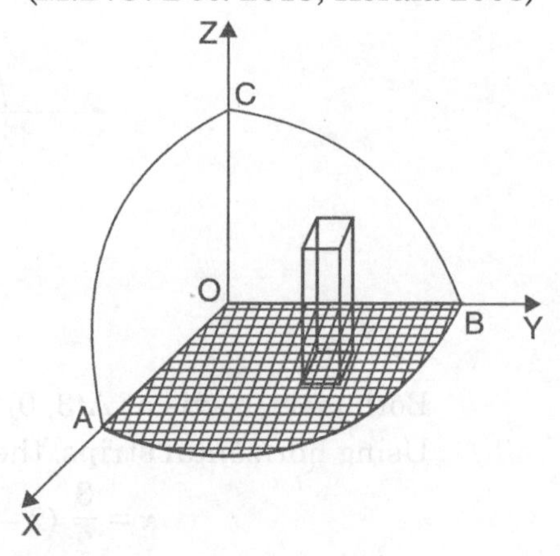

The ellipsoid cuts the plane XOY in the ellipse

$$\frac{x^2}{a^2} + \frac{y^2}{b^2} = 1, z = 0$$

$\therefore$ The region OAB above which the volume OABC lies is bounded by

$$x = 0, x = a \text{ and } y = 0, y = b\sqrt{1 - \frac{x^2}{a^2}}$$

Hence the required volume of the ellipsoid

$$= 8\int_0^a \int_0^{b\sqrt{1-x^2/a^2}} z\,dy\,dx = 8\int_0^a \int_0^{b\sqrt{1-x^2/a^2}} c\sqrt{1-\frac{x^2}{a^2}-\frac{y^2}{b^2}}\,dy\,dx$$

$$= 8\int_0^a \int_0^t c\sqrt{\frac{t^2}{b^2}-\frac{y^2}{b^2}}\,dy\,dx, \quad \text{where} \quad b\sqrt{1-\frac{x^2}{a^2}} = t$$

$$= 8\int_0^a \int_0^t \frac{c}{b}\sqrt{t^2-y^2}\,dy\,dx = 8\int_0^a \frac{c}{b}\left[\frac{y\sqrt{t^2-y^2}}{2}+\frac{t^2}{2}\sin^{-1}\frac{y}{t}\right]_0^t dx$$

$$= \frac{4c}{b}\int_0^a t^2 \sin^{-1} 1\,dx = \frac{2\pi c}{b}\int_0^c t^2\,dx$$

$$= \frac{2\pi c}{b}\int_0^a b^2\left(1-\frac{x^2}{a^2}\right)dx = 2\pi bc\left[x-\frac{x^3}{3a^2}\right]_0^a = 2\pi bc\left(a-\frac{a}{3}\right) = \frac{4}{3}\pi abc.$$

Example 5. *Find the volume of the tetrahedron bounded by the co-ordinate planes and the plane* $\frac{x}{a}+\frac{y}{b}+\frac{z}{c}=1.$ (K.U.K. Dec. 2013; Kerala 2009, 2012)

Sol. The required volume OABC lies between the plane $z=c\left(1-\frac{x}{a}-\frac{y}{b}\right)$ and the plane XOY, and is bounded on the sides by the planes $x=0$, $y=0$.

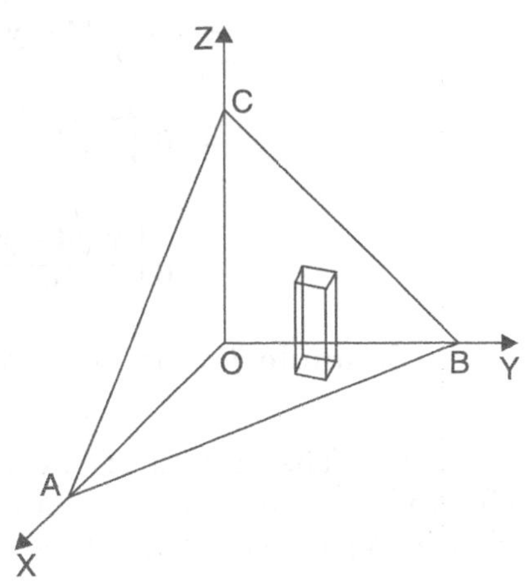

The plane $\frac{x}{a}+\frac{y}{b}+\frac{z}{c}=1$ cuts the plane XOY in the line $\frac{x}{a}+\frac{y}{b}=1, z=0$

$\therefore$ The region OAB above which the volume OABC lies is bounded by $x=0$, $x=a$ and $y=0$,

$$y = b\left(1-\frac{x}{a}\right)$$

Hence the required volume of the tetrahedron

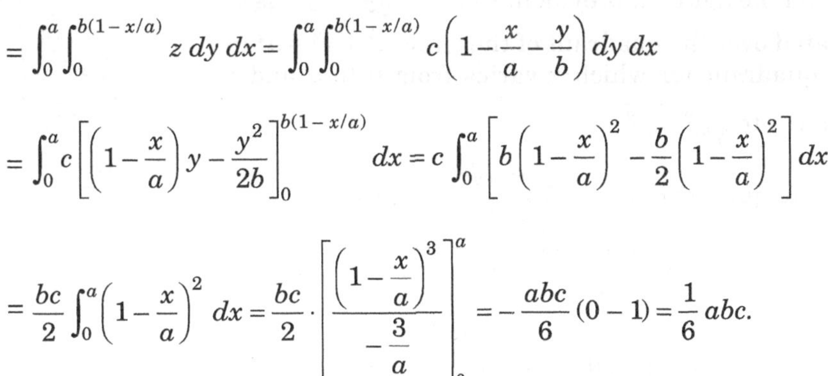

$$= \int_0^a \int_0^{b(1-x/a)} z\,dy\,dx = \int_0^a \int_0^{b(1-x/a)} c\left(1-\frac{x}{a}-\frac{y}{b}\right)dy\,dx$$

$$= \int_0^a c\left[\left(1-\frac{x}{a}\right)y-\frac{y^2}{2b}\right]_0^{b(1-x/a)} dx = c\int_0^a \left[b\left(1-\frac{x}{a}\right)^2-\frac{b}{2}\left(1-\frac{x}{a}\right)^2\right]dx$$

$$= \frac{bc}{2}\int_0^a \left(1-\frac{x}{a}\right)^2 dx = \frac{bc}{2}\cdot\left[\frac{\left(1-\frac{x}{a}\right)^3}{-\frac{3}{a}}\right]_0^a = -\frac{abc}{6}(0-1) = \frac{1}{6}abc.$$

Example 6. *Find the volume bounded by the cylinder* $x^2+y^2=4$ *and the planes* $y+z=4$ *and* $z=0.$ (M.T.U. 2011; P.T.U. 2011; K.U.K. 2012)

Sol. From the figure, it is clear that $z = 4 - y$ is to be integrated over the circle $x^2 + y^2 = 4$ in the xy-plane. To cover the shaded half of this circle, x varies from 0 to $\sqrt{4 - y^2}$ while y varies from -2 to 2.

$\therefore$ Required volume

$$= 2\int_{-2}^{2}\int_{0}^{\sqrt{4-y^2}} z\,dx\,dy$$

$$= 2\int_{-2}^{2}\int_{0}^{\sqrt{4-y^2}} (4-y)\,dx\,dy$$

$$= 2\int_{-2}^{2} (4-y)\Big[x\Big]_{0}^{\sqrt{4-y^2}} dy$$

$$= 2\int_{-2}^{2} (4-y)\sqrt{4-y^2}\,dy = 2\int_{-2}^{2} 4\sqrt{4-y^2}\,dy - 2\int_{-2}^{2} y\sqrt{4-y^2}\,dy$$

$$= 8\int_{-2}^{2} \sqrt{4-y^2}\,dy \qquad \left[\begin{array}{l}\text{The second integral is zero since}\\ y\sqrt{4-y^2} \text{ is an odd function of } y\end{array}\right]$$

$$= 16\int_{0}^{2} \sqrt{4-y^2}\,dy \qquad [\because \sqrt{4-y^2} \text{ is an even function of } y]$$

$$= 16\left[\frac{y\sqrt{4-y^2}}{2} + \frac{4}{2}\sin^{-1}\frac{y}{2}\right]_{0}^{2} = 16\,[2\sin^{-1} 1] = 32 \times \frac{\pi}{2} = 16\pi.$$

Example 7. *Find the volume common to the cylinders $x^2 + y^2 = a^2$ and $x^2 + z^2 = a^2$.* (U.P.T.U. 2009)

Sol. The section of the cylinder $x^2 + y^2 = a^2$ is the circle $x^2 + y^2 = a^2$ in the xy-plane. In the figure, only one-eighth (in positive octant) of the required volume is shown.

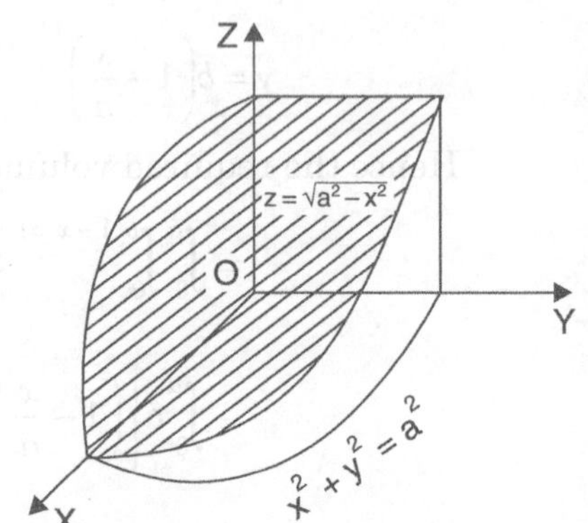

From the figure, it is evident that $z = \sqrt{a^2 - x^2}$ is to be evaluated over the quadrant of the circle $x^2 + y^2 = a^2$ in the first quadrant for which x varies from 0 to a and y varies from 0 to $\sqrt{a^2 - x^2}$.

$$\therefore \quad \text{Required volume} = 8\int_{0}^{a}\int_{0}^{\sqrt{a^2-x^2}} z\,dy\,dx$$

$$= 8\int_{0}^{a}\int_{0}^{\sqrt{a^2-x^2}} \sqrt{a^2-x^2}\,dy\,dx$$

$$= 8\int_{0}^{a} \sqrt{a^2-x^2}\Big[y\Big]_{0}^{\sqrt{a^2-x^2}} dx$$

$$= 8\int_{0}^{a} (a^2-x^2)\,dx = 8\left[a^2x - \frac{x^3}{3}\right]_{0}^{a} = 8\left(a^3 - \frac{a^3}{3}\right) = \frac{16a^3}{3}.$$

Example 8. *Find the volume common to the sphere $x^2 + y^2 + z^2 = a^2$ and the cylinder $x^2 + y^2 = ax$.*

Sol. The required volume is the part of the sphere $x^2 + y^2 + z^2 = a^2$ lying within the cylinder. On account of the symmetry of the sphere, half of it lies above the plane XOY and half below it.

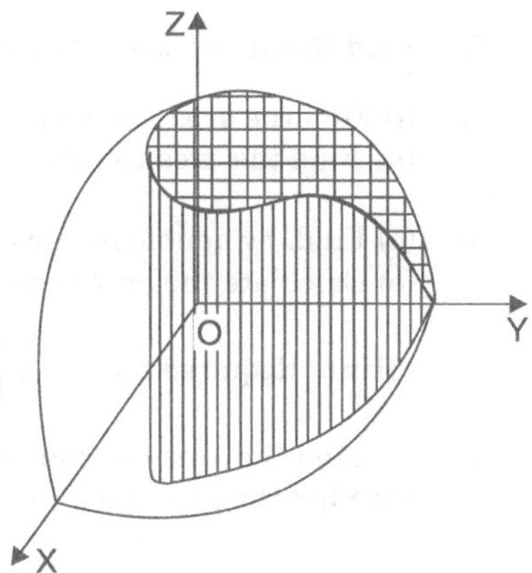

$\therefore$ Required volume $= 2 \iint z \, dy \, dx$

where $z = \sqrt{a^2 - x^2 - y^2}$, and the region of integration is the area inside the circle $x^2 + y^2 = ax$ in the xy-plane. ...(1)

On account of symmetry, the volumes above the two parts of circle (1) in the first and the fourth quadrants are equal. (The figure shows only the part in the first quadrant).

$\therefore$ Required volume $= 2 \times 2 \iint_R \sqrt{a^2 - x^2 - y^2} \, dy \, dx$

where R is the half of the circle (1) lying in the first quadrant.

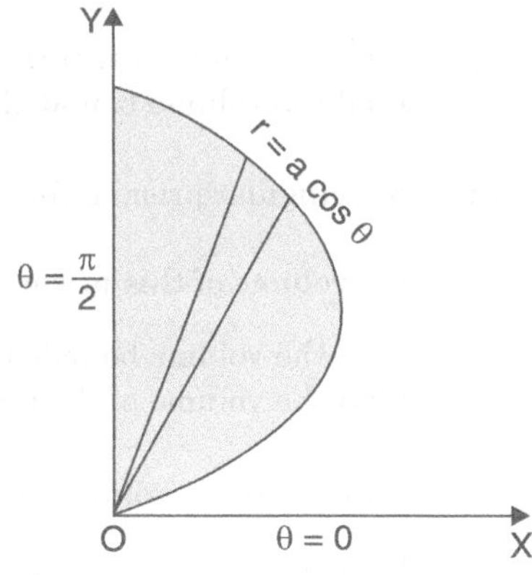

Changing to polar co-ordinates by putting $x = r \cos \theta$, $y = r \sin \theta$ so that $x^2 + y^2 = r^2$, equation (1) becomes

$$r^2 = ar \cos \theta \quad \text{or} \quad r = a \cos \theta.$$

The region of integration is bounded by

$$r = 0, r = a \cos \theta \text{ and } \theta = 0, \theta = \frac{\pi}{2}.$$

$\therefore$ Required volume

$$= 4 \int_0^{\pi/2} \int_0^{a \cos \theta} \sqrt{a^2 - r^2} \cdot r \, dr \, d\theta$$

$$= 4 \int_0^{\pi/2} \int_0^{a \cos \theta} -\frac{1}{2} (a^2 - r^2)^{1/2} (-2r) \, dr \, d\theta$$

$$= 4 \int_0^{\pi/2} -\frac{1}{2} \cdot \left[\frac{(a^2 - r^2)^{3/2}}{3/2} \right]_0^{a \cos \theta} d\theta$$

$$= -\frac{4}{3} \int_0^{\pi/2} (a^2 \sin^3 \theta - a^3) \, d\theta = -\frac{4a^3}{3} \left[\frac{2}{3} - \frac{\pi}{2} \right] = \frac{2}{9} a^3 (3\pi - 4).$$

EXERCISE 12.6

1. (*a*) Find, by double integration, the area enclosed by the ellipse $\frac{x^2}{a^2} + \frac{y^2}{b^2} = 1$. (*M.D.U. Dec. 2009*)

   (*b*) Find the area of a plate in the form of a quadrant of the ellipse $\frac{x^2}{a^2} + \frac{y^2}{b^2} = 1$.

2. Find, by double integration, the smaller of the areas bounded by the circle $x^2 + y^2 = 9$ and the line $x + y = 3$. (*Kerala 2009*)

3. Show, by double integration, that the area between the parabolas $y^2 = 4ax$ and $x^2 = 4ay$ is $\frac{16}{3} a^2$. (*G.B.T.U. 2010*)

4. Find the area bounded by the parabola $y = x^2$ and the line $y = 2x + 3$. *(Kerala 2010)*
5. Find the area bounded by the parabolas $y^2 = 4 - x$ and $y^2 = x$. *(G.B.T.U. 2013; M.D.U. Dec. 2011)*
6. Find the area enclosed by the curves $y = \dfrac{3x}{x^2 + 2}$ and $4y = x^2$. *(K.U.K. 2012)*
7. *(i)* Find the area bounded by the circles $r = 2 \sin \theta$ and $r = 4 \sin \theta$.
(ii) Find the area common to the circles $r = a \cos \theta$, $r = a \sin \theta$ by double integration. *(K.U.K. 2013)*
8. *(a)* Find, by double integration, the area of the lemniscate $r^2 = a^2 \cos 2\theta$.
(b) Calculate the area included between the curve $r = a(\sec \theta + \cos \theta)$ and its asymptote.

$$\left[\textbf{Hint.} \text{ Required area } = 2\int_0^{\pi/2}\int_{a \sec \theta}^{a (\sec \theta + \cos \theta)} r\, dr\, d\theta\right]$$

9. *(a)* Find the area outside the circle $r = 2a \cos \theta$ and inside the cardioid $r = a(1 + \cos \theta)$.
(b) Find the area lying inside the cardioid $r = 1 + \cos \theta$ and outside the parabola $r(1 + \cos \theta) = 1$. *(K.U.K. Dec. 2010)*

$$\left[\textbf{Hint.} \text{ Required area (symmetrical about } \theta = 0) \text{ is } 2\int_0^{\frac{\pi}{2}}\int_{\frac{1}{1+\cos\theta}}^{1+\cos\theta} r dr\, d\theta\right]$$

10. Find, by double integration, the volume of the sphere $x^2 + y^2 + z^2 = 9$.
11. Find the volume bounded by the *xy*-plane, the cylinder $x^2 + y^2 = 1$ and the plane $x + y + z = 3$. *(K.U.K. Dec. 2010)*
12. A triangular prism is formed by the planes whose equations are $ay = bx$, $y = 0$ and $x = a$. Prove that the volume of this prism between the plane $z = 0$ and the surface $z = c + xy$ is $\dfrac{ab}{8}(4c + ab)$.
13. Find the volume bounded by the *xy*-plane, the paraboloid $2z = x^2 + y^2$ and the cylinder $x^2 + y^2 = 4$.
14. Find the volume of the region bounded by $z = x^2 + y^2$, $z = 0$, $x = -a$, $x = a$ and $y = -a$, $y = a$.
15. Prove that the volume enclosed between the cylinders $x^2 + y^2 = 2ax$ and $z^2 = 2ax$ is $\dfrac{128a^3}{15}$.
16. Find the volume bounded by the plane $z = 0$, surface $z = x^2 + y^2 + 2$ and the cylinder $x^2 + y^2 = 4$.
17. Find the volume under the plane $z = x + y$ and above the area cut from the first quadrant by the ellipse $4x^2 + 9y^2 = 36$.
18. Find the volume of the cylinder $x^2 + y^2 - 2ax = 0$ intercepted between the paraboloid $x^2 + y^2 = 2az$ and the *xy*-plane.
19. Find the volume bounded by the cylinder $x^2 + y^2 = 4$ and the hyperboloid $x^2 + y^2 - z^2 = 1$.
20. Find the volume bounded by the paraboloid $x^2 + y^2 = az$, the cylinder $x^2 + y^2 = 2ay$ and the plane $z = 0$.

Answers

1. *(a)* πab	*(b)* $\dfrac{\pi ab}{4}$	2. $\dfrac{9}{4}(\pi - 2)$	4. $10\dfrac{2}{3}$
5. $\dfrac{16\sqrt{2}}{3}$	6. $\dfrac{3}{2}\log 3 - \dfrac{2}{3}$	7. *(i)* 3π	*(ii)* $\dfrac{a^2}{4}$
8. *(a)* a^2	*(b)* $\dfrac{5}{4}\pi a^2$	9. *(a)* $\dfrac{\pi a^2}{2}$	*(b)* $\dfrac{3\pi}{4} + \dfrac{4}{3}$
10. 36π	11. $3\pi - \dfrac{4}{3}$	13. 4π	14. $\dfrac{8a^4}{3}$
15. 16π	17. 10	18. $3\pi a^3$	19. $4\sqrt{3}\pi$
20. $\dfrac{3\pi a^3}{2}$.			

12.9. VOLUME AS A TRIPLE INTEGRAL

The volume V of a three dimensional region is given by $V = \iiint_V dx\, dy\, dz$.

If the region is bounded by $x = f_1(y, z)$, $x = f_2(y, z)$; $y = \phi_1(z)$, $y = \phi_2(z)$ and $z = a$, $z = b$, then

$$V = \int_a^b \int_{\phi_1(z)}^{\phi_2(z)} \int_{f_1(y, z)}^{f_2(y, z)} dx\, dy\, dz$$

The order of integration may be changed with a suitable change in the limits of integration.

In cylindrical co-ordinates, we have $V = \iiint_V r\, dr\, d\phi\, dz$

In spherical polar co-ordinates, we have $V = \iiint_V r^2 \sin\theta\, dr\, d\theta\, d\phi$.

ILLUSTRATIVE EXAMPLES

Example 1. *Find the volume of the solid bounded by the surface*

$$\left(\frac{x}{a}\right)^{2/3} + \left(\frac{y}{b}\right)^{2/3} + \left(\frac{z}{c}\right)^{2/3} = 1$$

Sol. Required volume $= \iiint dx\, dy\, dz$

Changing the variables x, y, z to X, Y, Z where

$$\left(\frac{x}{a}\right)^{1/3} = X, \quad \left(\frac{y}{b}\right)^{1/3} = Y, \quad \left(\frac{z}{c}\right)^{1/3} = Z$$

i.e., $x = aX^3, \quad y = bY^3, \quad z = cZ^3$ so that

$$J = \frac{\partial(x, y, z)}{\partial(X, Y, Z)} = \begin{vmatrix} \frac{\partial x}{\partial X} & \frac{\partial x}{\partial Y} & \frac{\partial x}{\partial Z} \\ \frac{\partial y}{\partial X} & \frac{\partial y}{\partial Y} & \frac{\partial y}{\partial Z} \\ \frac{\partial z}{\partial X} & \frac{\partial z}{\partial Y} & \frac{\partial z}{\partial Z} \end{vmatrix} = \begin{vmatrix} 3aX^2 & 0 & 0 \\ 0 & 3bY^2 & 0 \\ 0 & 0 & 3cZ^2 \end{vmatrix} = 27abcX^2Y^2Z^2$$

$\therefore$ Required volume $= 27abc \iiint X^2Y^2Z^2\, dX\, dY\, dZ$

taken throughout the sphere $X^2 + Y^2 + Z^2 = 1$

Changing the variables X, Y, Z to spherical polar coordinates r, θ, ϕ so that

$X = r\sin\theta\cos\phi, \quad Y = r\sin\theta\sin\phi, \quad Z = r\cos\theta$ and $J = \dfrac{\partial(X, Y, Z)}{\partial(r, \theta, \phi)} = r^2\sin\theta$.

$\therefore$ Required volume $= 27abc \int_0^{2\pi}\int_0^{\pi}\int_0^1 r^2\sin^2\theta\cos^2\phi \,.\, r^2\sin^2\theta\sin^2\phi \,.\, r^2\cos^2\theta \,.\, r^2\sin\theta\, dr\, d\theta\, d\phi$

$$= 27abc \int_0^{2\pi}\int_0^{\pi}\int_0^1 r^8\sin^5\theta\cos^2\theta\sin^2\phi\cos^2\phi\, dr\, d\theta\, d\phi$$

For the positve octant of the sphere, r varies from 0 to 1, θ varies from 0 to $\frac{\pi}{2}$ and ϕ varies from 0 to $\frac{\pi}{2}$.

$\therefore$ Required volume $= 27abc \times 8 \int_0^{\pi/2}\int_0^{\pi/2}\int_0^1 r^8\sin^5\theta\cos^2\theta\sin^2\phi\cos^2\phi\, dr\, d\theta\, d\phi$

$$= 216\, abc \int_0^{\pi/2}\int_0^{\pi/2} \left[\frac{r^9}{9}\right]_0^1 \sin^5\theta\cos^2\theta\sin^2\phi\cos^2\phi\, d\theta\, d\phi$$

$$= \frac{216abc}{9} \int_0^{\pi/2} \left[\frac{4 \cdot 2 \cdot 1}{7 \cdot 5 \cdot 3 \cdot 1} \right] \sin^2 \phi \cos^2 \phi \, d\phi$$

$$= \frac{8}{105} \times 24abc \left[\frac{1 \cdot 1}{4 \cdot 2} \cdot \frac{\pi}{2} \right] = \frac{4}{35} \pi abc$$

Note. Try example 4 before exercise 12.6 by this method.

Example 2. *Find by triple integration, the volume of the paraboloid of revolution $x^2 + y^2 = 4z$ cut off by the plane $z = 4$.*

Sol. By symmetry, the required volume is 4 times the volume in the positive octant. The volume in the positive octant is bounded on the sides by the zx and yz-planes; from above by the plane $z = 4$ and below by the curved surface $x^2 + y^2 = 4z$.

The section of the paraboloid by the plane $z = 4$ is the circle $x^2 + y^2 = 16$, $z = 4$ and its projection on the xy-plane is the circle $x^2 + y^2 = 16$, $z = 0$.

The volume in the positive octant is bounded by $z = \dfrac{x^2 + y^2}{4}$, $z = 4$, $y = 0$, $y = \sqrt{16 - x^2}$ and $x = 0$, $x = 4$.

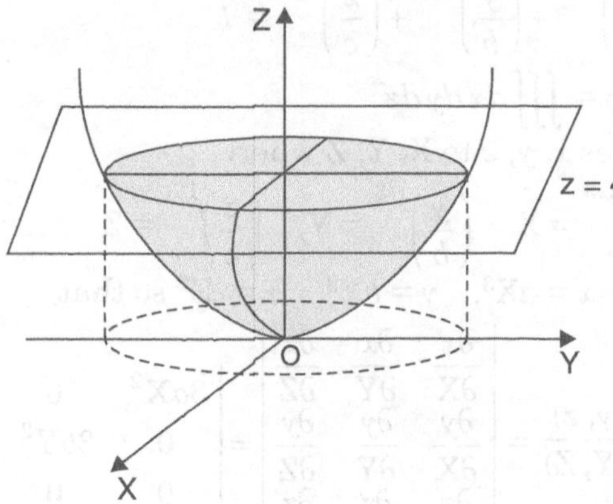

∴ Required volume

$$= 4 \int_0^4 \int_0^{\sqrt{16 - x^2}} \int_{(x^2 + y^2)/4}^4 dz \, dy \, dx = 4 \int_0^4 \int_0^{\sqrt{16 - x^2}} \left[z \right]_{x^2 + y^2/4}^4 dy \, dx$$

$$= 4 \int_0^4 \int_0^{\sqrt{16 - x^2}} \left(4 - \frac{x^2 + y^2}{4} \right) dy \, dx = 4 \int_0^4 \left[\left(4 - \frac{x^2}{4} \right) y - \frac{1}{4} \cdot \frac{y^3}{3} \right]_0^{\sqrt{16 - x^2}} dx$$

$$= 4 \int_0^4 \left[\left(4 - \frac{x^2}{4} \right) \sqrt{16 - x^2} - \frac{1}{12} (16 - x^2)^{3/2} \right] dx$$

$$= 4 \int_0^4 \left[\frac{1}{4} (16 - x^2) \sqrt{16 - x^2} - \frac{1}{12} (16 - x^2)^{3/2} \right] dx$$

$$= 4 \int_0^4 \frac{1}{6} (16 - x^2)^{3/2} \, dx = \frac{2}{3} \int_0^4 (16 - x^2)^{3/2} \, dx$$

$$= \frac{2}{3} \int_0^{\pi/2} (16)^{3/2} \cdot \cos^3 \theta \cdot 4 \cos \theta \, d\theta, \text{ where } x = 4 \sin \theta$$

$$= \frac{512}{3} \int_0^{\pi/2} \cos^4 \theta \, d\theta = \frac{512}{3} \cdot \frac{3 \cdot 1}{4 \cdot 2} \cdot \frac{\pi}{2} = 32\pi.$$

Example 3. *Find, by triple integration, the volume of the region bounded by the paraboloid $az = x^2 + y^2$ and the cylinder $x^2 + y^2 = R^2$.*

Sol. Changing to cylindrical co-ordinates, by putting $x = r \cos \theta$, $y = r \sin \theta$ the equation of the paraboloid becomes $az = r^2$ or $z = \dfrac{r^2}{a}$ and the equation of the cylinder becomes $r^2 = R^2$ or $r = R$. On account of symmetry, the required volume is four times the volume in the positive octant. Thus, in the common region, z varies from 0 to $\dfrac{r^2}{a}$, r varies from 0 to R and θ varies from 0 to $\dfrac{\pi}{2}$.

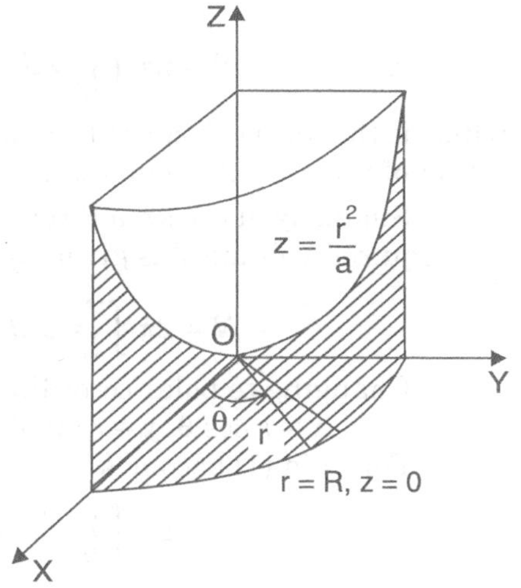

$$\therefore \quad \text{Required volume} = 4 \int_0^{\pi/2} \int_0^{R} \int_0^{r^2/a} r \, dz \, dr \, d\theta$$

$$= 4 \int_0^{\pi/2} \int_0^{R} r \Big[z \Big]_0^{r^2/a} dr \, d\theta$$

$$= 4 \int_0^{\pi/2} \int_0^{R} r \cdot \frac{r^2}{a} \, dr \, d\theta = 4 \int_0^{\pi/2} \left[\frac{r^4}{4a} \right]_0^{R} d\theta$$

$$= \frac{1}{a} \int_0^{\pi/2} R^4 \, d\theta = \frac{R^4}{a} \cdot \frac{\pi}{2} = \frac{\pi R^4}{2a}.$$

Example 4. *Find, by triple integration, the volume of a sphere of radius a.*

(M.D.U. Dec. 2010)

Sol. Changing to spherical polar co-ordinates by putting $x = r \sin \theta \cos \phi$, $y = r \sin \theta \sin \phi$, $z = r \cos \theta$ so that $x^2 + y^2 + z^2 = r^2$.

The equation of a sphere of radius a in cartesian co-ordinates is $x^2 + y^2 + z^2 = a^2$

The same equation in spherical polar coordinates is $r^2 = a^2$ or $r = a$.

On account of symmetry, the required volume is 8 times the volume of the sphere in the positive octant for which r varies from 0 to a, θ varies from 0 to $\dfrac{\pi}{2}$ and ϕ varies from 0 to $\dfrac{\pi}{2}$.

$$\therefore \quad \text{Required volume} = 8 \int_0^{\pi/2} \int_0^{\pi/2} \int_0^{a} r^2 \sin \theta \, dr \, d\theta \, d\phi$$

$$= 8 \int_0^{\pi/2} \int_0^{\pi/2} \sin \theta \cdot \left[\frac{r^3}{3} \right]_0^{a} d\theta \, d\phi = 8 \int_0^{\pi/2} \int_0^{\pi/2} \frac{a^3}{3} \sin \theta \, d\theta \, d\phi$$

$$= 8 \int_0^{\pi/2} -\frac{a^3}{3} \Big[\cos \theta \Big]_0^{\pi/2} d\phi = \frac{8}{3} a^3 \int_0^{\pi/2} d\phi = \frac{8}{3} a^3 \cdot \frac{\pi}{2} = \frac{4}{3} \pi a^3.$$

12.10. VOLUMES OF SOLIDS OF REVOLUTION

Consider an elementary area $\delta x \delta y$ at the point $P(x, y)$ of a plane area R. If this elementary area is revolved about the x-axis, it will generate a ring of radius y. The elementary volume δV of this ring is given by $\delta V = 2\pi y \cdot \delta y \, \delta x$

$$\therefore \qquad V = 2\pi \iint_R y \, dy \, dx \qquad ...(1)$$

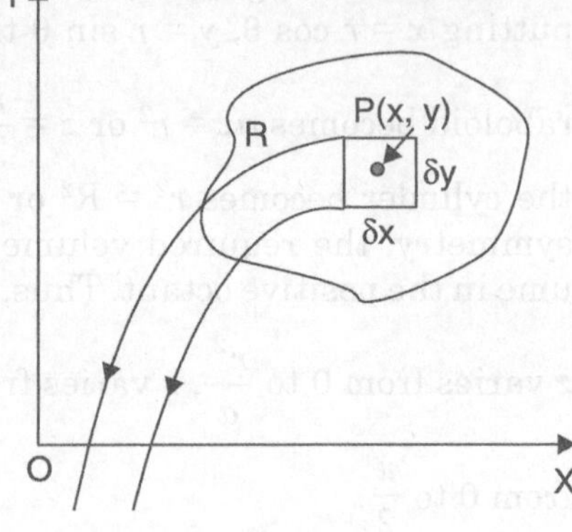

which is the total volume of the solid formed by the revolution of the area R about x-axis.

Similarly, if the area R revolves about y-axis, then the required volume V is given by

$$V = 2\pi \iint_R x \, dx \, dy \qquad ...(2)$$

Changing to polar co-ordinates by putting

$x = r \cos \theta$, $y = r \sin \theta$ and $dy \, dx = r \, dr \, d\theta$

Formula (1) becomes

$$V = 2\pi \iint_R r \sin \theta \cdot r \, dr \, d\theta = 2\pi \iint_R r^2 \sin \theta \, dr \, d\theta$$

Formula (2) becomes $V = 2\pi \iint_R r^2 \cos \theta \, dr \, d\theta$.

ILLUSTRATIVE EXAMPLES

Example 1. *Find, by double integration, the volume of the solid generated by revolving the ellipse* $\frac{x^2}{a^2} + \frac{y^2}{b^2} = 1$ *about the y-axis.*

Sol. Since the ellipse $\frac{x^2}{a^2} + \frac{y^2}{b^2} = 1$ is symmetrical about the y-axis, the volumes generated by the left and the right halves overlap. As such we consider the revolution of the right half ABB′ only for which x varies from 0 to

$$a\sqrt{1 - \frac{y^2}{b^2}} = \frac{a}{b}\sqrt{b^2 - y^2}$$

and y varies from $-b$ to b.

$\therefore$ Required volume of revolution

$$= 2\pi \int_{-b}^{b} \int_{0}^{a/b\sqrt{b^2 - y^2}} x \, dx \, dy$$

$$= 2\pi \int_{-b}^{b} \left[\frac{x^2}{2} \right]_0^{a/b\sqrt{b^2 - y^2}} dy = \pi \int_{-b}^{b} \frac{a^2}{b^2}(b^2 - y^2) \, dy = \frac{2\pi a^2}{b^2} \int_0^b (b^2 - y^2) \, dy$$

$$= \frac{2\pi a^2}{b^2} \left[b^2 y - \frac{y^3}{3} \right] = \frac{2\pi a^2}{b^2} \left[b^3 - \frac{b^3}{3} \right] = \frac{4}{3} \pi a^2 b.$$

Example 2. *Calculate, by double integration, the volume generated by the revolution of the cardioid* $r = a(1 - \cos \theta)$ *about its axis.*

Sol. Since the upper and the lower halves of the cardioid generate the same volume, we consider the revolution of the upper half only for which r varies from 0 to $a(1 - \cos \theta)$ and θ varies from 0 to π.

∴ Required volume of revolution

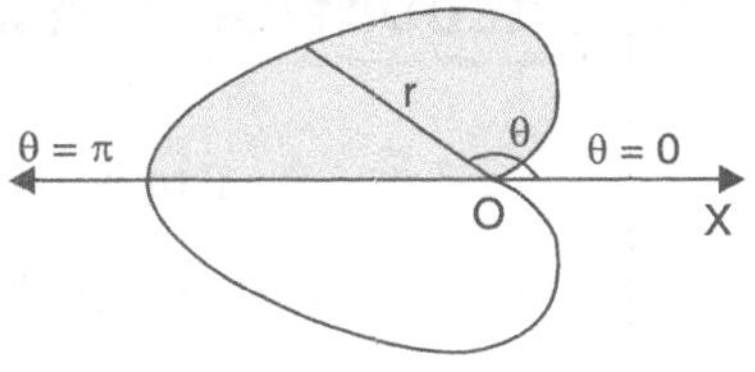

$$= 2\pi \int_0^{\pi} \int_0^{a(1-\cos\theta)} r^2 \sin\theta \, dr \, d\theta$$

$$= 2\pi \int_0^{\pi} \sin\theta \left[\frac{r^3}{3}\right]_0^{a(1-\cos\theta)} d\theta$$

$$= \frac{2\pi}{3} \int_0^{\pi} \sin\theta \cdot a^3 (1-\cos\theta)^3 \, d\theta$$

$$= \frac{2\pi a^3}{3} \int_0^{\pi} (1-\cos\theta)^3 \sin\theta \, d\theta = \frac{2\pi a^3}{3} \left[\frac{(1-\cos\theta)^4}{4}\right]_0^{\pi} = \frac{\pi a^3}{6} [(1-\cos\pi)^4]$$

$$= \frac{\pi a^3}{6} \cdot 2^4 = \frac{8}{3} \pi a^3.$$

EXERCISE 12.7

1. Find, by triple integration, the volume in the positive octant bounded by the co-ordinate planes and the plane $x + 2y + 3z = 4$. *(M.D.U. May 2008)*
2. Find the volume of the portion of the sphere $x^2 + y^2 + z^2 = a^2$ lying inside the cylinder $x^2 + y^2 = ax$.
3. Find the volume bounded above by the sphere $x^2 + y^2 + z^2 = 2a^2$ and below by the paraboloid $az = x^2 + y^2$.
4. Find the volume cut from the sphere $x^2 + y^2 + z^2 = a^2$ by the cone $x^2 + y^2 = z^2$.
5. Find, by double integration, the volume of the solid generated by revolving the ellipse $\frac{x^2}{a^2} + \frac{y^2}{b^2} = 1$ about the x-axis.
6. Find, by double integration, the volume generated by revolving the cardioid $r = a(1 + \cos\theta)$ about the initial line. *(K.U.K. 2009)*

Answers

1. $\frac{16}{9}$ 2. $\frac{2a^3}{9}(3\pi - 4)$ 3. $\left(\frac{4}{3}\sqrt{2} - \frac{7}{6}\right)\pi a^3$

4. $\frac{\pi a^3}{3}(2 - \sqrt{2})$ 5. $\frac{4}{3}\pi ab^2$ 6. $\frac{8}{3}\pi a^3$.

12.11. GAMMA FUNCTION

(G.B.T.U. 2010; U.P.T.U. 2008)

If n is positive, then the definite integral $\int_0^{\infty} e^{-x} x^{n-1} dx$,

which is a function of n, is called the Gamma function (or Eulerian integral of second kind) and is denoted by $\Gamma(n)$. Thus

$$\Gamma(n) = \int_0^{\infty} e^{-x} x^{n-1} dx, \; n > 0.$$

In particular, $\Gamma(1) = \int_0^{\infty} e^{-x} dx = \left| -e^{-x} \right|_0^{\infty} = 1.$

12.12. REDUCTION FORMULA FOR Γ (n)

$$\Gamma(n+1) = \int_0^\infty e^{-x} x^n \, dx = \int_0^\infty x^n e^{-x} \, dx$$

Integrating by parts, we have

$$\Gamma(n+1) = \left| -x^n e^{-x} \right|_0^\infty + n \int_0^\infty x^{n-1} e^{-x} \, dx$$

$$= n \int_0^\infty e^{-x} x^{n-1} \, dx \qquad \left[\because \lim_{x \to \infty} \frac{x^n}{e^x} = 0 \right]$$

$$= n\Gamma(n)$$

∴ **Γ(n + 1) = nΓ(n)**, which is the reduction formula for $\Gamma(n)$.

Note 1. If n is a positive integer, then by repeated application of above formula, we get

$$\begin{aligned} \Gamma(n+1) &= n\Gamma(n) \\ &= n(n-1)\,\Gamma(n-1) \\ &= n(n-1)(n-2)\,\Gamma(n-2) \\ &\cdots\cdots\cdots\cdots\cdots\cdots \\ &= n(n-1)(n-2)\ldots 3 \cdot 2 \cdot 1\,\Gamma(1) \\ &= n\,!, \quad \text{since } \Gamma(1) = 1 \end{aligned}$$

Hence $\Gamma(n + 1) = n\,!$ when n is a positive integer.

Note 2. If n is a positive fraction, then by repeated application of above formula, we get

$$\Gamma(n) = (n-1)(n-2) \times \text{go on decreasing by 1} \ldots$$

the series of factors being continued so long as the factors remain positive, multiplied by Γ (last factor).

Thus
$$\Gamma\left(\frac{11}{4}\right) = \frac{7}{4}\Gamma\left(\frac{7}{4}\right) = \frac{7}{4} \cdot \frac{3}{4}\Gamma\left(\frac{3}{4}\right)$$

The value of $\Gamma\left(\frac{3}{4}\right)$ can be obtained from the table of gamma functions.

Note 3. $\Gamma(n + 1) = n\,\Gamma(n)$

$$\begin{aligned} \Rightarrow \quad \Gamma(n) &= \frac{\Gamma(n+1)}{n}, \; n \neq 0 \\ &= \frac{(n+1)\Gamma(n+1)}{n(n+1)} = \frac{\Gamma(n+2)}{n(n+1)}, \; n \neq 0, -1 \\ &= \frac{(n+2)\Gamma(n+2)}{n(n+1)(n+2)} = \frac{\Gamma(n+3)}{n(n+1)(n+2)}, \; n \neq 0, -1, -2 \\ &\cdots\cdots\cdots\cdots\cdots\cdots \\ &= \frac{\Gamma(n+k+1)}{n(n+1)(n+2)\ldots(n+k)}, \; n \neq 0, -1, -2\ldots, -k \end{aligned}$$

This result defines $\Gamma(n)$ for $n < 0$, k being the least positive integer such that $n + k + 1 > 0$.

For example, to evaluate $\Gamma(-3.4)$

$$n + k + 1 > 0 \;\Rightarrow\; -3.4 + k + 1 > 0 \;\Rightarrow\; k > 2.4$$

We choose $k = 3$

$$\therefore \quad \Gamma(-3.4) = \frac{\Gamma(-3.4+3+1)}{(-3.4)(-2.4)(-1.4)(-.4)} = \frac{\Gamma(.6)}{(3.4)(2.4)(1.4)(.4)}$$

The value of $\Gamma(.6)$ can be obtained from the table of gamma functions.

Also we observe that $\Gamma(n)$ is infinite when $n = 0$ or a negative integer.

12.13. VALUE OF $\Gamma\left(\frac{1}{2}\right)$

(Kerala 2009; G.B.T.U. 2010)

$$\Gamma\left(\frac{1}{2}\right) = \int_0^\infty e^{-t}\, t^{-1/2}\, dt$$

Putting $t = x^2$ so that $dt = 2xdx$

$$\Gamma\left(\frac{1}{2}\right) = \int_0^\infty e^{-x^2} \cdot \frac{1}{x} \cdot 2xdx = 2\int_0^\infty e^{-x^2}\, dx \qquad ...(1)$$

Writing y for x, we have $\Gamma\left(\frac{1}{2}\right) = 2\int_0^\infty e^{-y^2}\, dy \qquad ...(2)$

Now we use the following result from double integrals:

If $f(x)$ and $g(y)$ are functions of x and y only, and the limits of integration are constants, then the double integral can be represented as a product of two integrals. Thus

$$\int_c^d \int_a^b f(x)\, g(y)\, dx\, dy = \int_a^b f(x)\, dx. \int_c^d g(y)\, dy$$

$\therefore$ From (1) and (2), we have

$$\left[\Gamma\left(\frac{1}{2}\right)\right]^2 = 4\int_0^\infty e^{-x^2}\, dx. \int_0^\infty e^{-y^2}\, dy = 4\int_0^\infty \int_0^\infty e^{-(x^2+y^2)}\, dx\, dy$$

Changing to polar co-ordinates with $x = r\cos\theta$, $y = r\sin\theta$, $dx\, dy = r\, dr\, d\theta$; the region of integration in this integral is the complete positive quadrant, to cover which, r must vary from 0 to ∞ and θ from 0 to $\frac{\pi}{2}$.

$$\therefore \qquad \left[\Gamma\left(\frac{1}{2}\right)\right]^2 = 4\int_0^{\pi/2}\int_0^\infty e^{-r^2}\, r\, dr\, d\theta = 4\int_0^{\pi/2}\left[-\frac{1}{2}e^{-r^2}\right]_0^\infty d\theta = 2\int_0^{\pi/2} d\theta = \pi$$

Hence $\qquad \Gamma\left(\frac{1}{2}\right) = \sqrt{\pi}$.

12.14. BETA FUNCTION

(G.B.T.U. 2010)

If m, n are positive, then the definite integral $\int_0^1 x^{m-1}(1-x)^{n-1}\, dx$, which is a function of m and n, is called the Beta Function (or Eulerian integral of first kind) and is denoted by $B(m, n)$. Thus,

$$B(m, n) = \int_0^1 x^{m-1}(1-x)^{n-1} dx,\ m > 0,\ n > 0.$$

12.15. SYMMETRY OF BETA FUNCTION *i.e.,* B(m, n) = B(n, m)

$$B(m, n) = \int_0^1 x^{m-1}(1-x)^{n-1} dx,\ m > 0,\ n > 0$$

Since $\qquad \int_0^a f(x)\, dx = \int_0^a f(a-x)\, dx$

$$\therefore \qquad B(m, n) = \int_0^1 (1-x)^{m-1}\,[1-(1-x)^{n-1}]\, dx = \int_0^1 x^{n-1}(1-x)^{m-1}\, dx = B(n, m)$$

Hence $\qquad B(m, n) = B(n, m)$.

12.16. (a) RELATION BETWEEN BETA AND GAMMA FUNCTIONS

$$\mathbf{B(m, n) = \frac{\Gamma(m)\Gamma(n)}{\Gamma(m+n)}}$$ (*M.D.U. Dec. 2013; K.U.K. 2009, 2012*)

We know that $\Gamma(m) = \int_0^\infty e^{-t}\, t^{m-1}\, dt$

Putting $t = x^2$ so that $dt = 2x\, dx$

$$\Gamma(m) = 2\int_0^\infty e^{-x^2} x^{2m-1}\, dx \quad ...(1)$$

Similarly $\Gamma(n) = 2\int_0^\infty e^{-y^2} x^{2n-1}\, dy$

With the explanation given in Art. 12.13, we have

$$\Gamma(m)\,\Gamma(n) = 4\int_0^\infty e^{-x^2} x^{2m-1}\, dx.\int_0^\infty e^{-y^2}\, y^{2n-1}\, dy$$

$$= 4\int_0^\infty\int_0^\infty e^{-(x^2+y^2)}\, x^{2m-1}\, y^{2n-1}\, dx\, dy$$

Changing to polar co-ordinates, we have

$$\Gamma(m)\,\Gamma(n) = 4\int_0^{\frac{\pi}{2}}\int_0^\infty e^{-r^2}\cdot(r\cos\theta)^{2m-1}\cdot(r\sin\theta)^{2n-1}\cdot r dr d\theta$$

$$= 4\int_0^{\pi/2}\int_0^\infty e^{-r^2}\, r^{2(m+n)-1}\, \cos^{2m-1}\theta\,\sin^{2n-1}\theta\, dr\, d\theta$$

$$= 4\int_0^\infty e^{-r^2}\, r^{2(m+n)-1}\, dr\cdot\int_0^{\pi/2}\cos^{2m-1}\theta\,\sin^{2n-1}\theta\, d\theta \quad ...(2)$$

Now $2\int_0^\infty e^{-r^2}\, r^{2(m+n)-1}\, dr = \Gamma\,(m+n)$, from (1)

and putting $\sin^2\theta = z$ so that $2\sin\theta\cos\theta\, d\theta = dz$

$$2\int_0^{\pi/2}\cos^{2m-1}\theta\,\sin^{2n-1}\theta\, d\theta = \int_0^1 (1-z)^{m-1}\, z^{n-1}\, dz = \int_0^1 z^{n-1}\,(1-z)^{m-1}\, dz$$

$= \text{B}\,(n, m) = \text{B}\,(m, n)$ by symmetry of Beta Function.

$\therefore$ From (2), we have $\Gamma(m)\,\Gamma(n) = \Gamma(m+n)\,\text{B}\,(m, n)$

Hence $$\text{B}\,(m, n) = \frac{\Gamma(m)\Gamma(n)}{\Gamma(m+n)}.$$

12.16. (b) VALUE OF $\Gamma\left(\frac{1}{2}\right)$ [Second Method]

(*M.D.U. Dec. 2013; Kerala 2009*)

Using $\text{B}(m, n) = \dfrac{\Gamma(m)\,\Gamma(n)}{\Gamma(m+n)}$, we have

$$\text{B}\left(\frac{1}{2},\frac{1}{2}\right) = \frac{\Gamma\left(\frac{1}{2}\right)\Gamma\left(\frac{1}{2}\right)}{\Gamma\left(\frac{1}{2}+\frac{1}{2}\right)} \Rightarrow \text{B}\left(\frac{1}{2},\frac{1}{2}\right) = \frac{\left[\Gamma\left(\frac{1}{2}\right)\right]^2}{\Gamma(1)}$$

$$\Rightarrow \quad \left[\Gamma\left(\frac{1}{2}\right)\right]^2 = B\left(\frac{1}{2}, \frac{1}{2}\right) \qquad [\because \quad \Gamma(1) = 1]$$

$$= \int_0^1 x^{-\frac{1}{2}} (1-x)^{-\frac{1}{2}} dx = \int_0^1 \frac{dx}{\sqrt{x(1-x)}}$$

[Put $x = \sin^2 \theta$ so that $dx = 2 \sin \theta \cos \theta \, d\theta$]

$$= \int_0^{\frac{\pi}{2}} \frac{2 \sin \theta \cos \theta \, d\theta}{\sin \theta \cos \theta} = 2\int_0^{\frac{\pi}{2}} d\theta = 2[\theta]_0^{\frac{\pi}{2}} = 2\left(\frac{\pi}{2}\right) = \pi$$

$$\therefore \quad \Gamma\left(\frac{1}{2}\right) = \sqrt{\pi}$$

For Example:

1. $\Gamma(3.5) = \Gamma\left(\frac{7}{2}\right) = \frac{5}{2} \cdot \frac{3}{2} \cdot \frac{1}{2} \Gamma\left(\frac{1}{2}\right) = \frac{15}{8}\sqrt{\pi}$

2. $B(2.5, 1.5) = B\left(\frac{5}{2}, \frac{3}{2}\right) = \dfrac{\Gamma\left(\frac{5}{2}\right)\Gamma\left(\frac{3}{2}\right)}{\Gamma\left(\frac{5}{2}+\frac{3}{2}\right)} = \dfrac{\left[\frac{3}{2} \cdot \frac{1}{2} \Gamma\left(\frac{1}{2}\right)\right]\left[\frac{1}{2} \Gamma\left(\frac{1}{2}\right)\right]}{\Gamma(4)}$

$$= \frac{\frac{3}{8}\left[\Gamma\left(\frac{1}{2}\right)\right]^2}{3!} = \frac{\frac{3}{8}(\sqrt{\pi})^2}{3 \cdot 2 \cdot 1} = \frac{\pi}{16}$$

3. $B\left(\frac{9}{2}, \frac{7}{2}\right) = \dfrac{\Gamma\left(\frac{9}{2}\right)\Gamma\left(\frac{7}{2}\right)}{\Gamma\left(\frac{9}{2}+\frac{7}{2}\right)} = \dfrac{\left[\frac{7}{2} \cdot \frac{5}{2} \cdot \frac{3}{2} \cdot \frac{1}{2} \Gamma\left(\frac{1}{2}\right)\right]\left[\frac{5}{2} \cdot \frac{3}{2} \cdot \frac{1}{2} \Gamma\left(\frac{1}{2}\right)\right]}{\Gamma(8)}$

$$= \frac{7 \cdot 5 \cdot 3 \cdot 5 \cdot 3\left[\Gamma\left(\frac{1}{2}\right)\right]^2}{16 \times 8 \times 7!} = \frac{7 \cdot 5 \cdot 3 \cdot 5 \cdot 3(\sqrt{\pi})^2}{128 \times 7 \cdot 6 \cdot 5 \cdot 4 \cdot 3 \cdot 2 \cdot 1} = \frac{5\pi}{2048}.$$

12.17. TO EVALUATE $\int_0^{\pi/2} \sin^p x \cos^q x \, dx$ *(U.K.T.U. 2011)*

Putting $\sin^2 x = t$ so that $2 \sin x \cos x \, dx = dt$

$$\int_0^{\pi/2} \sin^p x \cos^q x \, dx = \int_0^{\pi/2} (\sin^{p-1} x \cos^{q-1} x) \sin x \cos x \, dx$$

$$= \int_0^{\pi/2} (\sin^2 x)^{(p-1)/2} \cdot (1 - \sin^2 x)^{(q-1)/2} \sin x \cos x \, dx$$

$$= \frac{1}{2}\int_0^1 t^{(p-1)/2}(1-t)^{(q-1)/2} \, dt$$

$$= \frac{1}{2} B\left(\frac{p+1}{2}, \frac{q+1}{2}\right) = \frac{\Gamma\left(\frac{p+1}{2}\right)\Gamma\left(\frac{q+1}{2}\right)}{2\Gamma\left(\frac{p+q+2}{2}\right)}$$

For example: $\int_0^{\frac{\pi}{2}} \sin^2\theta\cos^4\theta\,d\theta$ (*M.D.U. Dec. 2011*)

(Here $p = 2$, $q = 4$)

$$= \frac{\Gamma\left(\frac{2+1}{2}\right)\Gamma\left(\frac{4+1}{2}\right)}{2\Gamma\left(\frac{2+4+2}{2}\right)} = \frac{\Gamma\left(\frac{3}{2}\right)\Gamma\left(\frac{5}{2}\right)}{2\Gamma(4)}$$

$$= \frac{\frac{1}{2}\Gamma\left(\frac{1}{2}\right)\cdot\frac{3}{2}\cdot\frac{1}{2}\Gamma\left(\frac{1}{2}\right)}{2[(4-1)!]} = \frac{\frac{3}{8}\left[\Gamma\left(\frac{1}{2}\right)\right]^2}{2\times 6} = \frac{(\sqrt{\pi})^2}{32} = \frac{\pi}{32}$$

Cor. 1. Putting $p = q = 0$, we have

$$\frac{[\Gamma(\frac{1}{2})]^2}{2\Gamma(1)} = \int_0^{\pi/2} dx = \frac{\pi}{2} \quad\text{or}\quad [\Gamma(\tfrac{1}{2})]^2 = \pi,\ \text{since } \Gamma(1) = 1$$

$$\therefore \qquad \Gamma(\tfrac{1}{2}) = \sqrt{\pi}$$

Cor. 2. Putting $p = n$ and $q = 0$, we have

$$\int_0^{\pi/2} \sin^n x\,dx = \frac{\Gamma\left(\frac{n+1}{2}\right)\Gamma\left(\frac{1}{2}\right)}{2\Gamma\left(\frac{n+2}{2}\right)} = \frac{\Gamma\left(\frac{n+1}{2}\right)}{\Gamma\left(\frac{n+2}{2}\right)}\cdot\frac{\sqrt{\pi}}{2}$$

Similarly, $$\int_0^{\pi/2} \cos^n x\,dx = \frac{\Gamma\left(\frac{n+1}{2}\right)}{\Gamma\left(\frac{n+2}{2}\right)}\cdot\frac{\sqrt{\pi}}{2}.$$

ILLUSTRATIVE EXAMPLES

Example 1. *Prove that* $\int_0^\infty e^{-ax}\,x^{n-1}\,dx = \frac{\Gamma(n)}{a^n}$, *where a, n are positive. Deduce that*

(*i*) $\int_0^\infty e^{-ax}\,x^{n-1}\cos bx\,dx = \frac{\Gamma(n)}{r^n}\cos n\theta$ (*ii*) $\int_0^\infty e^{-ax}\,x^{n-1}\sin bx\,dx = \frac{\Gamma(n)}{r^n}\sin n\theta$

where $r^2 = a^2 + b^2$ *and* $\theta = \tan^{-1}\frac{b}{a}$. (K.U.K. 2013)

Sol. Put $ax = z$ so that $dx = \frac{dz}{a}$

$$\therefore \qquad \int_0^\infty e^{-ax}\,x^{n-1}\,dx = \int_0^\infty e^{-z}\left(\frac{z}{a}\right)^{n-1}\cdot\frac{dz}{a} = \frac{1}{a^n}\int_0^\infty e^{-z}\,z^{n-1}\,dz = \frac{\Gamma(n)}{a^n}.$$

Deduction. Replacing a by $(a + ib)$, we have

$$\int_0^\infty e^{-(a+ib)x}\, x^{n-1}\, dx = \frac{\Gamma(n)}{(a+ib)^n} \qquad ...(1)$$

Now $e^{-(a+ib)x} = e^{-ax} \,.\, e^{-ibx} = e^{-ax}(\cos bx - i \sin bx)$

Also, putting $a = r\cos\theta$ and $b = r \sin\theta$

so that $r^2 = a^2 + b^2$ and $\theta = \tan^{-1}\dfrac{b}{a}$

$$(a+ib)^n = (r\cos\theta + ir\sin\theta)^n = r^n(\cos\theta + i\sin\theta)^n$$

$$= r^n(\cos n\theta + i \sin n\theta) \qquad \text{[De Moivre's Theorem]}$$

$\therefore$ From (1), we have

$$\int_0^\infty e^{-ax}(\cos bx - i\sin bx)\, x^{n-1}\, dx$$

$$= \frac{\Gamma(n)}{r^n(\cos n\theta + i\sin n\theta)} = \frac{\Gamma(n)}{r^n}(\cos n\theta + i \sin n\theta)^{-1} = \frac{\Gamma(n)}{r^n}(\cos n\theta - i\sin n\theta)$$

Now equate real and imaginary parts on the two sides.

Example 2. *Show that* $\displaystyle\int_0^1 y^{q-1}\left(\log\frac{1}{y}\right)^{p-1} dy = \frac{\Gamma(p)}{q^p}$, *where* $p > 0$, $q > 0$.

Sol. Put $\log\dfrac{1}{y} = x$ so that $\dfrac{1}{y} = e^x$ or $y = e^{-x}$ and $dy = -e^{-x}\,dx$

$$\therefore \int_0^1 y^{q-1}\left(\log\frac{1}{y}\right)^{p-1} dy = \int_\infty^0 e^{-(q-1)x}\cdot x^{p-1}(-e^{-x})\,dx = \int_0^\infty e^{-qx}\, x^{p-1}\, dx$$

$$= \int_0^\infty e^{-t}\cdot\left(\frac{t}{q}\right)^{p-1}\cdot\frac{dt}{q}, \qquad \text{where } qx = t$$

$$= \frac{1}{q^p}\int_0^\infty e^{-t}\, t^{p-1}\, dt = \frac{\Gamma(p)}{q^p}.$$

Example 3. *Show that* $\displaystyle B(p, q) = \int_0^\infty \frac{y^{q-1}}{(1+y)^{p+q}}\, dy = \int_0^1 \frac{x^{p-1} + x^{q-1}}{(1+x)^{p+q}}\, dx.$

(Kerala 2010; M.T.U. 2012)

Sol. $$B(p, q) = \int_0^1 x^{p-1}(1-x)^{q-1}\, dx$$

Putting $x = \dfrac{1}{1+y}$ so that $dx = -\dfrac{1}{(1+y)^2}\, dy$

$$B(p, q) = \int_\infty^0 \left(\frac{1}{1+y}\right)^{p-1}\left(\frac{y}{1+y}\right)^{q-1}\cdot\frac{-1}{(1+y)^2}\, dy = \int_0^\infty \frac{y^{q-1}}{(1+y)^{p+q}}\, dy$$

$$= \int_0^1 \frac{y^{q-1}}{(1+y)^{p+q}}\, dy + \int_1^\infty \frac{y^{q-1}}{(1+y)^{p+q}}\, dy \qquad ...(1)$$

Now, putting $y = \frac{1}{z}$ in the second integral, we have

$$\int_1^{\infty} \frac{y^{q-1}}{(1+y)^{p+q}} dy = \int_1^0 \frac{\left(\frac{1}{z}\right)^{q-1}}{\left(1+\frac{1}{z}\right)^{p+q}} \left(-\frac{1}{z^2}\right) dz = \int_0^1 \frac{z^{p-1}}{(1+z)^{p+q}} dz$$

$\therefore$ From (1), we have $B(p, q) = \int_0^1 \frac{y^{q-1}}{(1+y)^{p+q}} dy + \int_0^1 \frac{z^{p-1}}{(1+z)^{p+q}} dz = \int_0^1 \frac{x^{p-1} + x^{q-1}}{(1+x)^{p+q}} dx$.

Example 4. *Prove that* $\int_0^1 \frac{dx}{\sqrt{1-x^4}} = \frac{\sqrt{\pi}}{4} \cdot \frac{\Gamma(\frac{1}{4})}{\Gamma(\frac{3}{4})}$. (M.D.U. Dec. 2011)

Sol. Put $x^4 = t$ *i.e.,* $x = t^{1/4}$ so that $dx = \frac{1}{4} t^{-3/4} dt$

$$\therefore \quad \int_0^1 \frac{dx}{\sqrt{1-x^4}} = \frac{1}{4} \int_0^1 t^{-3/4} (1-t)^{-1/2} dt = \frac{1}{4} \int_0^1 t^{1/4-1} (1-t)^{1/2-1} dt$$

$$= \frac{1}{4} B\left(\frac{1}{4}, \frac{1}{2}\right) = \frac{1}{4} \cdot \frac{\Gamma(\frac{1}{4})\Gamma(\frac{1}{2})}{\Gamma(\frac{1}{4}+\frac{1}{2})} = \frac{\sqrt{\pi}}{4} \cdot \frac{\Gamma(\frac{1}{4})}{\Gamma(\frac{3}{4})}. \qquad | \because \Gamma(\tfrac{1}{2}) = \sqrt{\pi}$$

Example 5. *Prove that* $\int_0^{\infty} x^{p-1} e^{-kx} dx = \frac{\Gamma(p)}{k^p}$, $(k > 0)$

Sol. Put $kx = t$ so that $kdx = dt$

$$\therefore \quad \int_0^{\infty} x^{p-1} e^{-kx} dx = \int_0^{\infty} \left(\frac{t}{k}\right)^{p-1} \cdot e^{-t} \cdot \frac{dt}{k}$$

$$= \frac{1}{k^p} \int_0^{\infty} t^{p-1} e^{-t} dt = \frac{\Gamma(p)}{k^p}$$

Example 6. *Prove that* $\int_0^{\infty} \frac{x^{m-1}}{(a+bx)^{m+n}} dx = \frac{B(m, n)}{a^n b^m}$, *where m, n, a, b are positive.*

Sol. Put $bx = at$ *i.e.,* $x = \frac{at}{b}$ so that $dx = \frac{a}{b} dt$

$$\therefore \quad \int_0^{\infty} \frac{x^{m-1}}{(a+bx)^{m+n}} dx = \int_0^{\infty} \frac{\left(\frac{at}{b}\right)^{m-1}}{(a+at)^{m+n}} \cdot \frac{a}{b} dt = \frac{1}{a^n b^m} \int_0^{\infty} \frac{t^{m-1}}{(1+t)^{m+n}} dt = \frac{B(m,n)}{a^n b^m}.$$

Example 7. *Prove that* $B(m, n) = B(m+1, n) + B(m, n+1)$. (U.P.T.U. 2009)

Sol. $\quad B(m+1, n) + B(m, n+1)$

$$= \int_0^1 x^m (1-x)^{n-1}\, dx + \int_0^1 x^{m-1}(1-x)^n\, dx$$

$$= \int_0^1 x^{m-1}(1-x)^{n-1}[x+1-x]dx = \int_0^1 x^{m-1}(1-x)^{n-1}\, dx = \mathrm{B}(m,n).$$

Example 8. *Prove that* $\Gamma\,(n+\frac{1}{2}) = \dfrac{\sqrt{\pi}\cdot\Gamma(2n+1)}{2^{2n}\Gamma(n+1)}$.

Sol. Since $$\mathrm{B}(m,\, n) = \frac{\Gamma(m)\Gamma(n)}{\Gamma(m+n)}$$

$\therefore$ $$\mathrm{B}\,(n+\tfrac{1}{2},\, n+\tfrac{1}{2}) = \frac{[\Gamma(n+\frac{1}{2})]^2}{\Gamma(2n+1)} \qquad ...(1)$$

Now $$\mathrm{B}(n+\tfrac{1}{2}, n+\tfrac{1}{2}) = \int_0^1 x^{n-1/2}\,(1-x)^{n-1/2}\, dx$$

Putting $x = \sin^2\theta$ so that $dx = 2\sin\theta\cos\theta\, d\theta$

$$\mathrm{B}(n+\tfrac{1}{2}, n+\tfrac{1}{2}) = \int_0^{\pi/2} \sin^{2n-1}\theta\cos^{2n-1}\theta . 2\,\sin\theta\cos\theta\, d\theta$$

$$= 2\int_0^{\pi/2} \sin^{2n}\theta\cos^{2n}\theta\, d\theta = 2\int_0^{\pi/2}\left(\frac{\sin 2\theta}{2}\right)^{2n} d\theta = \frac{1}{2^{2n-1}}\int_0^{\pi/2} \sin^{2n} 2\theta\, d\theta$$

$$= \frac{1}{2^{2n}}\int_0^{\pi} \sin^{2n}\phi\, d\phi, \quad \text{where } \phi = 2\theta$$

$$= \frac{1}{2^{2n}}.2\int_0^{\pi/2} \sin^{2n}\phi\, d\phi \qquad \left[\because \int_0^{2a} f(x)\,dx = 2\int_0^a f(x)\,dx, \text{ if } f(2a-x) = f(x)\right]$$

$$= \frac{1}{2^{2n-1}}\cdot\frac{\sqrt{\pi}}{2}\cdot\frac{\Gamma\left(\frac{2n+1}{2}\right)}{\Gamma\left(\frac{2n+2}{2}\right)} = \frac{\sqrt{\pi}\,\Gamma\left(n+\frac{1}{2}\right)}{2^{2n}\,\Gamma(n+1)}$$

$\therefore$ From (1), we have $$\frac{\sqrt{\pi}\,\Gamma\left(n+\frac{1}{2}\right)}{2^{2n}\,\Gamma(n+1)} = \frac{\left[\Gamma\left(n+\frac{1}{2}\right)\right]^2}{\Gamma\,(2n+1)}$$

$\Rightarrow$ $$\Gamma(n+\frac{1}{2}) = \frac{\sqrt{\pi}\,\Gamma(2n+1)}{2^{2n}\,\Gamma\,(n+1)}.$$

Cor. Putting $n = \dfrac{1}{4}$, we have

$$\Gamma\,(\tfrac{3}{4}) = \frac{\sqrt{\pi}\,\Gamma\,(3/2)}{\sqrt{2}\,\Gamma\,(5/4)} = \frac{\sqrt{\pi}\cdot\frac{1}{2}\Gamma\left(\frac{1}{2}\right)}{\sqrt{2}\cdot\frac{1}{4}\Gamma\left(\frac{1}{4}\right)}$$

$\Rightarrow$ $$\Gamma\,(\tfrac{1}{4})\,\Gamma\,(\tfrac{3}{4}) = \sqrt{2}\pi.$$

Example 9. *Prove that*

$$(a)\ \iint_D x^{l-1}\, y^{m-1}\, dx\, dy = \frac{\Gamma(l)\Gamma(m)}{\Gamma(l+m+1)}\, a^{l+m}$$

where D is the domain $x \geq 0, y \geq 0$ *and* $x + y \leq a$.

(b) Establish Dirichlet's integral:

$$\iiint_V x^{l-1}\, y^{m-1}\, z^{n-1}\, dx\, dy\, dz = \frac{\Gamma(l)\Gamma(m)\Gamma(n)}{\Gamma(l+m+n+1)}$$

where V is the region $x \geq 0, y \geq 0, z \geq 0$ *and* $x + y + z \leq 1$.

Sol. (*a*) Putting $x = a\mathrm{X}$ and $y = a\mathrm{Y}$, the given integral reduces to

$$\mathrm{I} = \iint_{\mathrm{D}'} (a\mathrm{X})^{l-1}\, (a\mathrm{Y})^{m-1}\, a^2\, d\mathrm{X}\, d\mathrm{Y}$$

where D′ is the domain $\mathrm{X} \geq 0$, $\mathrm{Y} \geq 0$ and $\mathrm{X} + \mathrm{Y} \leq 1$

$$= a^{l+m} \int_0^1 \int_0^{1-\mathrm{X}} \mathrm{X}^{l-1}\, \mathrm{Y}^{m-1}\, d\mathrm{Y}\, d\mathrm{X}$$

$$= a^{l+m} \int_0^1 \mathrm{X}^{l-1} \left| \frac{\mathrm{Y}^m}{m} \right|_0^{1-\mathrm{X}} d\mathrm{X} = \frac{a^{l+m}}{m} \int_0^1 \mathrm{X}^{l-1}\, (1-\mathrm{X})^m\, d\mathrm{X}$$

$$= \frac{a^{l+m}}{m}\, \mathrm{B}(l, m+1) = \frac{d^{l+m}}{m} \cdot \frac{\Gamma(l)\,\Gamma(m+1)}{\Gamma(l+m+1)}$$

$$= a^{l+m}\, \frac{\Gamma(l)\,\Gamma(m)}{\Gamma(l+m+1)}, \text{ since } \frac{\Gamma(m+1)}{m} = \Gamma(m).$$

(*b*) Taking $y + z \leq 1 - x = a$ (say), the given integral

$$\mathrm{I} = \int_0^1 \int_0^{1-x} \int_0^{1-x-y} x^{l-1}\, y^{m-1}\, z^{n-1}\, dz\, dy\, dx$$

$$= \int_0^1 x^{l-1} \left[\int_0^a \int_0^{a-y} y^{m-1}\, z^{n-1}\, dz\, dy \right] dx = \int_0^1 x^{l-1}\, \frac{\Gamma(m)\,\Gamma(n)}{\Gamma(m+n+1)}\, a^{m+n}\, dx$$

[by part (*a*)]

$$= \frac{\Gamma(m)\,\Gamma(n)}{\Gamma(m+n+1)} \int_0^1 x^{l-1}\, (1-x)^{m+n}\, dx, \quad \text{since } a = 1 - x$$

$$= \frac{\Gamma(m)\,\Gamma(n)}{\Gamma(m+n+1)}\, \mathrm{B}(l, m+n+1) = \frac{\Gamma(m)\,\Gamma(n)}{\Gamma(l+m+n+1)} \cdot \frac{\Gamma(l)\,\Gamma(m+n+1)}{\Gamma(m+n+1)}$$

$$= \frac{\Gamma(l)\Gamma(m)\,\Gamma(n)}{\Gamma(l+m+n+1)}.$$

Example 10. *The plane* $\frac{x}{a} + \frac{y}{b} + \frac{z}{c} = 1$ *meets the axes in A, B and C. Apply Dirichlet's integral to find the volume of the tetrahedron OABC.*

Sol. Put $\frac{x}{a} = u, \frac{y}{b} = v, \frac{z}{c} = w$, then $u \geq 0$, $v \geq 0$, $w \geq 0$ and $u + v + w \leq 1$.

Also, $dx = a\, du$, $dy = b\, dv$, $dz = c\, dw$.

$$\text{Volume OABC} = \iiint_D dx\, dy\, dz$$

$$= \iiint_{D'} abc\, du\, dv\, dw, \qquad \text{where } u + v + w \leq 1$$

$$= abc \iiint_{D'} u^{1-1} v^{1-1} w^{1-1}\, du\, dv\, dw$$

$$= abc \frac{\Gamma(1)\,\Gamma(1)\,\Gamma(1)}{\Gamma(1+1+1+1)} = \frac{abc}{3!} = \frac{abc}{6}.$$

EXERCISE 12.8

Prove that:

1. $\int_{-\infty}^{\infty} e^{-k^2x^2}\, dx = \frac{\sqrt{\pi}}{k}$

2. $\int_0^{\infty} x^3 e^{-x^3}\, dx = \frac{1}{9}\Gamma\left(\frac{1}{3}\right)$

3. $\int_0^{\infty} \sqrt{x}\, e^{-x^3}\, dx = \frac{\sqrt{\pi}}{3}$

4. $\int_0^{\infty} e^{-x^2}\, x^{2n-1}\, dx = \frac{1}{2}\Gamma(n)$

5. $\int_0^{\infty} \frac{x^c}{c^x}\, dx = \frac{\Gamma(c+1)}{(\log c)^{c+1}}, c > 1$

6. (*i*) $\int_0^1 \left(\log \frac{1}{y}\right)^{n-1} dy = \Gamma(n),\ n > 0$ (*ii*) $\int_0^1 x^4 \left(\log \frac{1}{x}\right)^3 dx = \frac{6}{625}$

7. $\int_0^{\infty} x^{2n-1}\, e^{-ax^2}\, dx = \frac{\Gamma(n)}{2a^n}$

8. $\int_0^1 x^m (\log x)^n\, dx = \frac{(-1)^n\, n!}{(m+1)^{n+1}}$, where n is a positive integer and $m > 1$.

9. $\int_0^{\infty} \sqrt{x} e^{-x^2}\, dx \times \int_0^{\infty} \frac{e^{-x^2}}{\sqrt{x}}\, dx = \frac{\pi}{2\sqrt{2}}$

10. $\int_0^{\infty} \frac{e^{-x^2}}{\sqrt{x}}\, dx \times \int_0^{\infty} x^2 e^{-x^4}\, dx = \frac{\pi}{4\sqrt{2}}$

11. $\int_0^{\pi/2} \sqrt{\tan\theta}\, d\theta = \int_0^{\pi/2} \sqrt{\cot\theta} = \frac{1}{2}\Gamma\left(\frac{1}{4}\right)\Gamma\left(\frac{3}{4}\right) = \frac{\pi}{\sqrt{2}}$ (*K.U.K. Dec. 2010; U.K.T.U. 2011*)

12. $\int_0^{\pi/2} \sqrt{\sin\theta}\, d\theta \times \int_0^{\pi/2} \frac{d\theta}{\sqrt{\sin\theta}} = \pi$ (*M.D.U. May 2009*)

13. $\int_0^\infty \int_0^\infty e^{-(ax^2+by^2)}\, x^{2m-1}\, y^{2n-1}\, dx\, dy = \dfrac{\Gamma(m)\Gamma(n)}{4a^m b^n}$, where a, b, m, n are positive.

14. $\Gamma(n)\Gamma(1-n) = \dfrac{\pi}{\sin n\pi}$, given $\int_0^\infty \dfrac{x^{n-1}}{1+x}\, dx = \dfrac{\pi}{\sin n\pi}$. *(G.B.T.U. 2010)*

15. $y\text{B}(x+1, y) = x\text{B}(x, y+1)$.

16. $\int_0^2 (8-x^3)^{-1/3}\, dx = \dfrac{1}{3}\Gamma\left(\dfrac{1}{3}\right)\Gamma\left(\dfrac{2}{3}\right)$.

17. $\int_0^a x^{n-1}(a-x)^{m-1}\, dx = a^{m+n-1}\, \text{B}(m, n)$.

18. $\int_a^b (x-a)^m (b-x)^n\, dx = (b-a)^{m+n+1}\, \text{B}(m+1, n+1)$. **[Hint.** Put $x = a + (b-a)z$]

19. $\int_0^1 \dfrac{x^{m-1}(1-x)^{n-1}}{(a+bx)^{m+n}}\, dx = \dfrac{\text{B}(m,n)}{a^n(a+b)^m}$ $\left[\textbf{Hint. } \text{Put } \dfrac{x}{a+bx} = \dfrac{z}{a+b.1}\right]$

20. Express $\int_0^1 x^m(1-x^n)^k\, dx$ in terms of gamma function and show that $\int_0^1 x^5(1-x^3)^{10}\, dx = \dfrac{1}{396}$.

(M.D.U. May 2008, Dec. 2010, Dec. 2012)

21. $\dfrac{\text{B}(p, q+1)}{q} = \dfrac{\text{B}(p+1, q)}{p} = \dfrac{\text{B}(p, q)}{p+q}$. *(U.K.T.U. 2012)*

22. $\dfrac{\text{B}(m+1, n)}{\text{B}(m, n)} = \dfrac{m}{m+n}$.

23. $\dfrac{\text{B}(m+2, n-2)}{\text{B}(m, n)} = \dfrac{m(m+1)}{(n-1)(n-2)}$.

24. $\int_0^\infty x^n e^{-a^2x^2}\, dx = \dfrac{1}{2a^{n+1}}\Gamma\left(\dfrac{n+1}{2}\right)$, $n > -1$. Deduce that $\int_{-\infty}^\infty e^{-a^2x^2}\, dx = \dfrac{\sqrt{\pi}}{a}$.

25. $\int_0^{\pi/2} \sin^3 x \cos^{5/2} x\, dx = \dfrac{8}{77}$.

26. $\int_0^1 x^3(1-x)^{4/3}\, dx = \dfrac{243}{7280}$.

27. (*i*) $\text{B}(n, n) = \dfrac{\sqrt{\pi}\,\Gamma(n)}{2^{2n-1}\,\Gamma(n+\frac{1}{2})}$. *(M.D.U. Dec. 2010)*

(*ii*) $\int_0^{\pi/2} (\sqrt{\tan\theta} + \sqrt{\sec\theta})\, d\theta = \dfrac{1}{2}\Gamma(\tfrac{1}{4})\,[\Gamma(\tfrac{3}{4}) + \sqrt{\pi}/\Gamma(\tfrac{3}{4})]$

28. $2^n\,\Gamma(n+\frac{1}{2}) = 1\cdot3\cdot5 \ldots (2n-1)\sqrt{\pi}$.

29. $\int_0^1 \dfrac{dx}{\sqrt{1-x^n}} = \dfrac{\sqrt{\pi}}{n}\dfrac{\Gamma\left(\dfrac{1}{n}\right)}{\Gamma\left(\dfrac{1}{n}+\dfrac{1}{2}\right)}$. *(M.T.U. 2013)*

30. $\int_0^1 \dfrac{x^2\, dx}{\sqrt{1-x^4}} \times \int_0^1 \dfrac{dx}{\sqrt{1+x^4}} = \dfrac{\pi}{4\sqrt{2}}$ *(M.D.U. Dec. 2008; U.P.T.U. 2008)*

[Hint. Put $x^2 = \sin\theta$ and $x^2 = \tan\theta$ in the two integrals.]

31. Prove that

(*i*) $\int_0^\infty \frac{x^8(1-x^6)}{(1+x)^{24}}\,dx = 0$ (*M.T.U. 2013*) (*ii*) $\int_0^\infty \frac{x^4(1+x^5)}{(1+x)^{15}}\,dx = \frac{1}{5005}$

(*iii*) $B(m, m) \,.\, B\left(m+\frac{1}{2}, m+\frac{1}{2}\right) = \frac{\pi m^{-1}}{2^{4m-1}}$ (*iv*) $\int_0^{\pi/2} \sin^n \theta\, d\theta = \frac{\sqrt{\pi}}{2} \cdot \frac{\Gamma\left(\frac{n+1}{2}\right)}{\Gamma\left(\frac{n+2}{2}\right)}, n > -1$

(*M.D.U. Dec. 2009*)

(*v*) $\int_0^1 \frac{x\,dx}{\sqrt{1-x^5}} = \frac{1}{5} B\left(\frac{2}{5}, \frac{1}{2}\right)$ (*vi*) $\int_0^1 \frac{dx}{\sqrt{1+x^4}} = \frac{1}{4\sqrt{2}} B\left(\frac{1}{4}, \frac{1}{2}\right)$ (*G.B.T.U. 2011*)

(*vii*) $B\left(m, \frac{1}{2}\right) = 2^{2m-1} B(m, m)$ (*viii*) $\Gamma(m)\,\Gamma\left(m+\frac{1}{2}\right) = \frac{\sqrt{\pi}}{2^{2m-1}}\,\Gamma(2m)$

(*K.U.K. Dec. 2013*) (*M.T.U. 2013*)

32. Show that $\iint x^{m-1} y^{n-1}\, dx\, dy$ over the positive octant of the ellipse

$$\left(\frac{x}{a}\right)^2 + \left(\frac{y}{b}\right)^2 = 1 \text{ is } \frac{a^m b^n}{2n} \beta\left(\frac{m}{2}, \frac{n}{2}+1\right).$$

33. Show that the volume of the solid bounded by the coordinate planes and the surface

$$\sqrt{\frac{x}{a}} + \sqrt{\frac{y}{b}} + \sqrt{\frac{z}{c}} = 1 \text{ is } \frac{abc}{90}.$$

34. Show that the area in the first quadrant enclosed by the curve

$$\left(\frac{x}{a}\right)^\alpha + \left(\frac{y}{b}\right)^\beta = 1, \quad \alpha > 0, \beta > 0 \text{ is given by}$$

$$\frac{ab}{\alpha+\beta} \cdot \frac{\Gamma\left(\frac{1}{\alpha}\right)\Gamma\left(\frac{1}{\beta}\right)}{\Gamma\left(\frac{1}{\alpha}+\frac{1}{\beta}\right)}.$$

SECTION E

Vector Calculus

13*

Vector Calculus

13.1. VECTOR FUNCTIONS

If to each value of a scalar variable t, there corresponds a value of a vector $\vec{r}$, then $\vec{r}$ is called a vector function of the scalar variable t and we write $\vec{r} = \vec{r}(t)$ or $\vec{r} = \vec{f}(t)$.

For example, the position vector $\vec{r}$ of a particle moving along a curved path is a vector function of time (a scalar).

Since every vector can be uniquely expressed as a linear combination of three fixed non-coplanar vectors, therefore, we may write $\vec{f}(t) = f_1(t)\hat{i} + f_2(t)\hat{j} + f_3(t)\hat{k}$ where $\hat{i}, \hat{j}, \hat{k}$ denote unit vectors along the axes of x, y, z respectively. $f_1(t), f_2(t)$ and $f_3(t)$ are called the components of the vector $\vec{f}(t)$ along the coordinate axes.

13.2. DERIVATIVE OF A VECTOR FUNCTION WITH RESPECT TO A SCALAR

Let $\vec{r} = \vec{f}(t)$ be a vector function of the scalar variable t. Let δt be a [illegible] of t and $\delta\vec{r}$, the corresponding increment in $\vec{r}$.

Then $\vec{r} + \delta\vec{r} = \vec{f}(t + \delta t)$ so that $\delta\vec{r} = \vec{f}(t + \delta t) - \vec{f}(t)$

and
$$\frac{\delta\vec{r}}{\delta t} = \frac{\vec{f}(t + \delta t) - \vec{f}(t)}{\delta t}$$

If $\lim\limits_{\delta t \to 0} \frac{\vec{f}(t + \delta t) - \vec{f}(t)}{\delta t}$ exists, then the value of this limit is denoted by $\frac{d\vec{r}}{dt}$ [illegible]

[illegible]

[illegible] $\frac{d\vec{r}}{dt}$ [illegible] of t, its derivative [illegible] denoted by [illegible] and [illegible]

the second derivative of $\vec{r}$ with respect to t. Similarly, we can define higher order derivatives of $\vec{r}$.

* This chapter is not included in the syllabus of MDU, Rohtak.

13*

Vector Calculus

13.1. VECTOR FUNCTIONS

If to each value of a scalar variable t, there corresponds a value of a vector $\vec{r}$, then $\vec{r}$ is called a vector function of the scalar variable t and we write $\vec{r} = \vec{r}\,(t)$ or $\vec{r} = \vec{f}(t)$.

For example, the position vector $\vec{r}$ of a particle moving along a curved path is a vector function of time t, a scalar.

Since every vector can be uniquely expressed as a linear combination of three fixed non-coplanar vectors, therefore, we may write $\vec{f}(t) = f_1(t)\,\hat{i} + f_2(t)\,\hat{j} + f_3\,(t)\,\hat{k}$ where $\hat{i}, \hat{j}, \hat{k}$ denote unit vectors along the axis of x, y, z respectively, $f_1(t)$, $f_2(t)$ and $f_3(t)$ are called the components of the vector $\vec{f}\,(t)$ along the coordinate axes.

13.2. DERIVATIVE OF A VECTOR FUNCTION WITH RESPECT TO A SCALAR

Let $\vec{r} = \vec{f}(t)$ be a vector function of the scalar variable t. Let δt be a small increment in t and $\delta \vec{r}$, the corresponding increment in $\vec{r}$.

Then $\quad \vec{r} + \delta \vec{r} = \vec{f}\,(t + \delta t)$ so that $\delta \vec{r} = \vec{f}(t + \delta t) - \vec{f}(t)$

and
$$\frac{\delta \vec{r}}{\delta t} = \frac{\vec{f}(t + \delta t) - \vec{f}(t)}{\delta t}$$

If $\lim\limits_{\delta t \to 0} \dfrac{\delta \vec{r}}{\delta t} = \lim\limits_{\delta t \to 0} \dfrac{\vec{f}(t + \delta t) - \vec{f}(t)}{\delta t}$ exists, then the value of this limit is denoted by $\dfrac{d\vec{r}}{dt}$ and is called the derivative of $\vec{r}$ with respect to t.

Since $\dfrac{d\vec{r}}{dt}$ is itself a vector function of t, its derivative is denoted by $\dfrac{d^2\vec{r}}{dt^2}$ and is called the second derivative of $\vec{r}$ with respect to t. Similarly, we can define higher order derivatives of $\vec{r}$.

* This chapter is not included in the syllabus of MDU, Rohtak

13.3. GENERAL RULES FOR DIFFERENTIATION

If $\vec{a}, \vec{b}$ and $\vec{c}$ are vector functions of a scalar t and ϕ is a scalar function of t, then

(i) $\frac{d}{dt}(\vec{a} \pm \vec{b}) = \frac{d\vec{a}}{dt} \pm \frac{d\vec{b}}{dt}$ (ii) $\frac{d}{dt}(\vec{a} \cdot \vec{b}) = \vec{a} \cdot \frac{d\vec{b}}{dt} + \frac{d\vec{a}}{dt} \cdot \vec{b}$

(iii) $\frac{d}{dt}(\vec{a} \times \vec{b}) = \vec{a} \times \frac{d\vec{b}}{dt} + \frac{d\vec{a}}{dt} \times \vec{b}$ (iv) $\frac{d}{dt}(\phi \vec{a}) = \phi \frac{d\vec{a}}{dt} + \frac{d\phi}{dt} \vec{a}$

(v) $\frac{d}{dt}[\vec{a}\ \vec{b}\ \vec{c}] = \left[\frac{d\vec{a}}{dt}\ \vec{b}\ \vec{c}\right] + \left[\vec{a}\ \frac{d\vec{b}}{dt}\ \vec{c}\right] + \left[\vec{a}\ \vec{b}\ \frac{d\vec{c}}{dt}\right]$

(vi) $\frac{d}{dt}\{\vec{a} \times (\vec{b} \times \vec{c})\} = \frac{d\vec{a}}{dt} \times (\vec{b} \times \vec{c}) + \vec{a} \times \left(\frac{d\vec{b}}{dt} \times \vec{c}\right) + \vec{a} \times \left(\vec{b} \times \frac{d\vec{c}}{dt}\right)$.

Proof. (i) $\frac{d}{dt}(\vec{a} + \vec{b}) = \lim_{\delta t \to 0} \frac{\{(\vec{a} + \delta\vec{a}) + (\vec{b} + \delta\vec{b})\} - (\vec{a} + \vec{b})}{\delta t}$

$$= \lim_{\delta t \to 0} \frac{\delta\vec{a} + \delta\vec{b}}{\delta t} = \lim_{\delta t \to 0} \left(\frac{\delta\vec{a}}{\delta t} + \frac{\delta\vec{b}}{\delta t}\right)$$

$$= \lim_{\delta t \to 0} \frac{\delta\vec{a}}{\delta t} + \lim_{\delta t \to 0} \frac{\delta\vec{b}}{\delta t} = \frac{d\vec{a}}{dt} + \frac{d\vec{b}}{dt}$$

Similarly, $\frac{d}{dt}(\vec{a} - \vec{b}) = \frac{d\vec{a}}{dt} - \frac{d\vec{b}}{dt}$

(ii) $\frac{d}{dt}(\vec{a} \cdot \vec{b}) = \lim_{\delta t \to 0} \frac{(\vec{a} + \delta\vec{a}) \cdot (\vec{b} + \delta\vec{b}) - \vec{a} \cdot \vec{b}}{\delta t}$

$$= \lim_{\delta t \to 0} \frac{\vec{a} \cdot \vec{b} + \vec{a} \cdot \delta\vec{b} + \delta\vec{a} \cdot \vec{b} + \delta\vec{a} \cdot \delta\vec{b} - \vec{a} \cdot \vec{b}}{\delta t}$$

$$= \lim_{\delta t \to 0} \frac{\vec{a} \cdot \delta\vec{b} + \delta\vec{a} \cdot \vec{b} + \delta\vec{a} \cdot \delta\vec{b}}{\delta t} = \lim_{\delta t \to 0} \left\{\vec{a} \cdot \frac{\delta\vec{b}}{\delta t} + \frac{\delta\vec{a}}{\delta t} \cdot \vec{b} + \frac{\delta\vec{a}}{\delta t} \cdot \delta\vec{b}\right\}$$

$$= \vec{a} \cdot \frac{d\vec{b}}{dt} + \frac{d\vec{a}}{dt} \cdot \vec{b} + \frac{d\vec{a}}{dt} \cdot \vec{0}, \quad \text{since } \delta\vec{b} \to \vec{0} \text{ as } \delta t \to 0$$

$$= \vec{a} \cdot \frac{d\vec{b}}{dt} + \frac{d\vec{a}}{dt} \cdot \vec{b} \qquad \left[\because \frac{d\vec{a}}{dt} \cdot \vec{0} = 0\right]$$

Note. Since $\vec{a} \cdot \vec{b} = \vec{b} \cdot \vec{a}$, while evaluating $\frac{d}{dt}(\vec{a} \cdot \vec{b})$, the order of factors is immaterial.

(iii) $\frac{d}{dt}(\vec{a} \times \vec{b}) = \lim_{\delta t \to 0} \frac{(\vec{a} + \delta\vec{a}) \times (\vec{b} + \delta\vec{b}) - \vec{a} \times \vec{b}}{\delta t}$

$$= \lim_{\delta t \to 0} \frac{\vec{a} \times \vec{b} + \vec{a} \times \delta\vec{b} + \delta\vec{a} \times \vec{b} + \delta\vec{a} \times \delta\vec{b} - \vec{a} \times \vec{b}}{\delta t}$$

$$= \lim_{\delta t \to 0} \frac{\vec{a} \times \delta\vec{b} + \delta\vec{a} \times \vec{b} + \delta\vec{a} \times \delta\vec{b}}{\delta t} = \lim_{\delta t \to 0} \left\{ \vec{a} \times \frac{\delta\vec{b}}{\delta t} + \frac{\delta\vec{a}}{\delta t} \times \vec{b} + \frac{\delta\vec{a}}{\delta t} \times \delta\vec{b} \right\}$$

$$= \vec{a} \times \frac{d\vec{b}}{dt} + \frac{d\vec{a}}{dt} \times \vec{b} + \frac{d\vec{a}}{dt} \times \vec{0} \quad \text{since } \delta\vec{b} \to \vec{0} \text{ as } \delta t \to 0$$

$$= \vec{a} \times \frac{d\vec{b}}{dt} + \frac{d\vec{a}}{dt} \times \vec{b} \qquad \left[\because \frac{d\vec{a}}{dt} \times \vec{0} = \vec{0} \right]$$

Note. Since $\vec{a} \times \vec{b} \neq \vec{b} \times \vec{a}$, while evaluating $\frac{d}{dt}(\vec{a} \times \vec{b})$, the order of factors $\vec{a}$ and $\vec{b}$ must be maintained.

$$(iv) \quad \frac{d}{dt}(\phi\vec{a}) = \lim_{\delta t \to 0} \frac{(\phi + \delta\phi)(\vec{a} + \delta\vec{a}) - \phi\vec{a}}{\delta t} = \lim_{\delta t \to 0} \frac{\phi\vec{a} + \phi\delta\vec{a} + \delta\phi\vec{a} + \delta\phi\delta\vec{a} - \phi\vec{a}}{\delta t}$$

$$= \lim_{\delta t \to 0} \frac{\phi\delta\vec{a} + \delta\phi\vec{a} + \delta\phi\delta\vec{a}}{\delta t} = \lim_{\delta t \to 0} \left\{ \phi\frac{\delta\vec{a}}{\delta t} + \frac{\delta\phi}{\delta t}\vec{a} + \frac{\delta\phi}{\delta t}\delta\vec{a} \right\}$$

$$= \phi\frac{d\vec{a}}{dt} + \frac{d\phi}{dt}\vec{a} + \frac{d\phi}{dt}\vec{0}, \quad \text{since } \delta\vec{a} \to \vec{0} \text{ as } \delta t \to 0$$

$$= \phi\frac{d\vec{a}}{dt} + \frac{d\phi}{dt}\vec{a} \qquad \left[\because \frac{d\phi}{dt}\vec{0} = \vec{0} \right]$$

Note. $\phi\vec{a}$ is the product of a vector by a scalar. We usually write the scalar in the first position and the vector in the second position.

$$(v) \quad \frac{d}{dt}[\vec{a}\ \vec{b}\ \vec{c}] = \frac{d}{dt}[\vec{a} \cdot (\vec{b} \times \vec{c})] = \vec{a} \cdot \frac{d}{dt}(\vec{b} \times \vec{c}) + \frac{d\vec{a}}{dt} \cdot (\vec{b} \times \vec{c}) \qquad \text{[By rule (ii)]}$$

$$= \vec{a} \cdot \left(\vec{b} \times \frac{d\vec{c}}{dt} + \frac{d\vec{b}}{dt} \times \vec{c} \right) + \frac{d\vec{a}}{dt} \cdot (\vec{b} \times \vec{c}) \qquad \text{[By rule (iii)]}$$

$$= \vec{a} \cdot \left(\vec{b} \times \frac{d\vec{c}}{dt} \right) + \vec{a} \cdot \left(\frac{d\vec{b}}{dt} \times \vec{c} \right) + \frac{d\vec{a}}{dt} \cdot (\vec{b} \times \vec{c})$$

$$= \left[\vec{a}\ \vec{b}\ \frac{d\vec{c}}{dt} \right] + \left[\vec{a}\ \frac{d\vec{b}}{dt}\ \vec{c} \right] + \left[\frac{d\vec{a}}{dt}\ \vec{b}\ \vec{c} \right] = \left[\frac{d\vec{a}}{dt}\ \vec{b}\ \vec{c} \right] + \left[\vec{a}\ \frac{d\vec{b}}{dt}\ \vec{c} \right] + \left[\vec{a}\ \vec{b}\ \frac{d\vec{c}}{dt} \right]$$

Note. $[\vec{a}\ \vec{b}\ \vec{c}]$ is the scalar product of three vectors $\vec{a}$, $\vec{b}$ and $\vec{c}$. While evaluating $\frac{d}{dt}[\vec{a}\ \vec{b}\ \vec{c}]$, the cyclic order of factors must be maintained.

(vi) $\dfrac{d}{dt}\{\vec{a}\times(\vec{b}\times\vec{c})\} = \vec{a}\times\dfrac{d}{dt}(\vec{b}\times\vec{c}) + \dfrac{d\vec{a}}{dt}\times(\vec{b}\times\vec{c})$ [By rule (iii)]

$$= \vec{a}\times\left(\vec{b}\times\frac{d\vec{c}}{dt} + \frac{d\vec{b}}{dt}\times\vec{c}\right) + \frac{d\vec{a}}{dt}\times(\vec{b}\times\vec{c})$$

$$= \frac{d\vec{a}}{dt}\times(\vec{b}\times\vec{c}) + \vec{a}\times\left(\frac{d\vec{b}}{dt}\times\vec{c}\right) + \vec{a}\times\left(\vec{b}\times\frac{d\vec{c}}{dt}\right).$$

13.4. DERIVATIVE OF A CONSTANT VECTOR

A vector is said to be constant if both its magnitude and direction are fixed. If either of these changes, the vector is not constant.

Let $\vec{r}$ be a constant vector function of the scalar variable t.

Let $\vec{r} = \vec{f}(t)$, then $\vec{r} = \vec{f}(t+\delta t)$ so that $\vec{f}(t+\delta t) - \vec{f}(t) = \vec{0}$

$$\therefore \quad \frac{d\vec{r}}{dt} = \lim_{\delta t\to 0}\frac{\vec{f}(t+\delta t)-\vec{f}(t)}{\delta t} = \lim_{\delta t\to 0}\vec{0} = \vec{0}$$

Thus, the derivative of a constant vector is equal to the null vector.

Note. $\hat{i}, \hat{j}, \hat{k}$ being fixed unit vectors are constant vectors.

$$\therefore \quad \frac{d\hat{i}}{dt} = \frac{d\hat{j}}{dt} = \frac{d\hat{k}}{dt} = \vec{0}.$$

13.5. DERIVATIVE OF A VECTOR FUNCTION IN TERMS OF ITS COMPONENTS

Let $\vec{r}$ be a vector function of the scalar variable t.

Let $\vec{r} = x\hat{i} + y\hat{j} + z\hat{k}$, where the components x, y, z are scalar function of t.

$$\frac{d\vec{r}}{dt} = \frac{d}{dt}(x\hat{i}) + \frac{d}{dt}(y\hat{j}) + \frac{d}{dt}(z\hat{k}) = x\frac{d\hat{i}}{dt} + \frac{dx}{dt}\hat{i} + y\frac{d\hat{j}}{dt} + \frac{dy}{dt}\hat{j} + z\frac{d\hat{k}}{dt} + \frac{dz}{dt}\hat{k}$$

$$= \frac{dx}{dt}\hat{i} + \frac{dy}{dt}\hat{j} + \frac{dz}{dt}\hat{k}, \quad \text{since} \quad \frac{d\hat{i}}{dt} = \frac{d\hat{j}}{dt} = \frac{d\hat{k}}{dt} = \vec{0}.$$

If $x = f_1(t), y = f_2(t), z = f_3(t)$; then $\vec{r} = f_1(t)\,\hat{i} + f_2(t)\,\hat{j} + f_3(t)\,\hat{k}$

$$\Rightarrow \quad \frac{d\vec{r}}{dt} = f_1'(t)\,\hat{i} + f_2'(t)\,\hat{j} + f_3'(t)\,\hat{k}$$

Therefore to differentiate a vector, differentiate its components.

13.6. IF $\vec{F}(t)$ HAS A CONSTANT MAGNITUDE, THEN $\vec{F}\cdot\dfrac{d\vec{F}}{dt} = 0$ (*P.T.U. Dec. 2011*)

$\vec{F}(t)$ has a constant magnitude $\Rightarrow$ $|\vec{F}(t)|$ = constant

$$\therefore \quad \vec{F}(t)\cdot\vec{F}(t) = |\vec{F}(t)|^2 = \text{constant} \Rightarrow \frac{d}{dt}(\vec{F}\cdot\vec{F}) = 0$$

$$\Rightarrow \quad \vec{F} \cdot \frac{d\vec{F}}{dt} + \frac{d\vec{F}}{dt} \cdot \vec{F} = 0 \quad \Rightarrow \quad 2\vec{F} \cdot \frac{d\vec{F}}{dt} = 0 \quad \Rightarrow \quad \vec{F} \cdot \frac{d\vec{F}}{dt} = 0$$

Note. $\vec{F} \cdot \dfrac{d\vec{F}}{dt} = 0 \quad \Rightarrow \quad \dfrac{d\vec{F}}{dt} \perp \vec{F}$.

13.7. IF $\vec{F}$ (t) HAS A CONSTANT DIRECTION, THEN $\vec{F} \times \dfrac{d\vec{F}}{dt} = \vec{0}$

Let $|\vec{F}(t)| = f(t)$. Let $\hat{G}(t)$ be a unit vector in the direction of $\vec{F}(t)$ so that $\vec{F}(t) = f(t)\,\hat{G}(t)$

$$\therefore \qquad \frac{d\vec{F}}{dt} = f\frac{d\hat{G}}{dt} + \frac{df}{dt}\hat{G} \qquad \text{...(1)}$$

If $\vec{F}(t)$ has constant direction, so has $\hat{G}(t)$. Thus, $\hat{G}(t)$ is a constant vector and $\dfrac{d\hat{G}}{dt} = \vec{0}$

From (1),
$$\frac{d\vec{F}}{dt} = \frac{df}{dt}\hat{G}$$

$$\therefore \qquad \vec{F} \times \frac{d\vec{F}}{dt} = f\hat{G} \times \left(\frac{df}{dt}\hat{G}\right) = f\frac{df}{dt}\hat{G} \times \hat{G} = \vec{0}.$$

13.8. GEOMETRICAL INTERPRETATION OF $\dfrac{d\vec{r}}{dt}$

Let O be the origin of reference. Let the position vector of a point P be given by $\vec{r} = \vec{f}(t)$. As t varies continuously, P traces out a curve C as shown in the figure. Thus, a vector function $\vec{f}(t)$ represents a curve in space.

For example, (*i*) the vector equation $\vec{r} = at^2\hat{i} + 2at\hat{j}$ represents the parabola $y^2 = 4ax$ in the xy-plane because its parametric equations are

$$x = at^2, \; y = 2at.$$

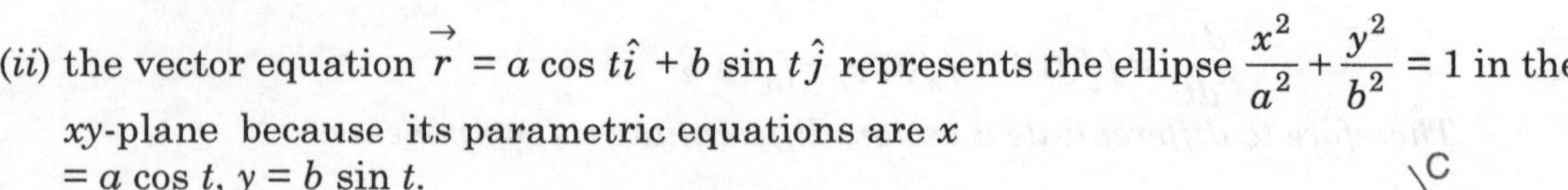

(*ii*) the vector equation $\vec{r} = a\cos t\hat{i} + b\sin t\hat{j}$ represents the ellipse $\dfrac{x^2}{a^2} + \dfrac{y^2}{b^2} = 1$ in the xy-plane because its parametric equations are $x = a\cos t$, $y = b\sin t$.

Now, let $\vec{r} = \vec{f}(t)$ be the vector equation of a curve C in space. Let $\vec{r}$ and $\vec{r} + \delta\vec{r}$ be the position vectors of two neighbouring points P and Q on this curve.

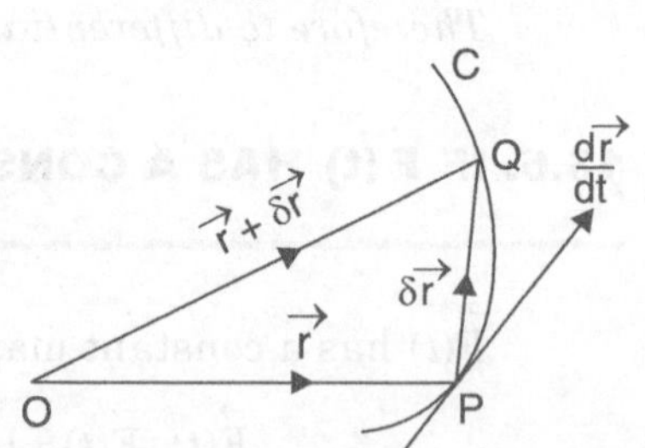

$$\overrightarrow{PQ} = \overrightarrow{OQ} - \overrightarrow{OP} = (\vec{r} + \delta\vec{r}) - \vec{r} = \delta\vec{r}$$

$\therefore$ $\dfrac{\delta\vec{r}}{\delta t}$ is directed along the chord PQ.

As $\delta t \to 0$, Q $\to$ P, chord PQ $\to$ tangent to the curve at P.

$\therefore$ $\lim\limits_{\delta t \to 0} \dfrac{\delta\vec{r}}{\delta t} = \dfrac{d\vec{r}}{dt}$ *is a vector along the tangent to the curve at P.*

Suppose the scalar parameter t is replaced by s, where s denotes the arc length from any convenient point A on the curve upto P. Thus, arc AP = s and arc AQ = $s + \delta s$ so that δs = arc PQ. In this case $\dfrac{d\vec{r}}{ds}$ will be a vector along the tangent at P. Also

$$\left|\frac{d\vec{r}}{ds}\right| = \lim_{\delta s \to 0}\left|\frac{\delta\vec{r}}{\delta s}\right| = \lim_{Q \to P}\frac{\text{chord PQ}}{\text{arc PQ}} = 1.$$

Thus, $\dfrac{d\vec{r}}{ds}$ *is the unit vector* $\hat{T}$ *along the tangent at P.*

13.9. VELOCITY AND ACCELERATION (*G.B.T.U. 2010*)

If the scalar variable t denotes the time and $\vec{r}$ is the position vector of a moving particle P, then $\delta\vec{r}$ is the displacement of the particle in time δt. The vector $\dfrac{\delta\vec{r}}{\delta t}$ is the average velocity of the particle during the interval δt. If $\vec{v}$ represents the velocity vector of the particle at P, then $\vec{v} = \lim\limits_{\delta t \to 0} \dfrac{\delta\vec{r}}{\delta t} = \dfrac{d\vec{r}}{dt}$ and its direction is along the tangent at P.

If $\delta\vec{v}$ be the change in velocity $\vec{v}$ during the time δt, then $\dfrac{\delta\vec{v}}{\delta t}$ is the average acceleration of the particle during the interval δt. If $\vec{a}$ represents the acceleration of the particle at P, then

$$\vec{a} = \lim_{\delta t \to 0}\frac{\delta\vec{v}}{\delta t} = \frac{d\vec{v}}{dt} = \frac{d}{dt}\left(\frac{d\vec{r}}{dt}\right) = \frac{d^2\vec{r}}{dt^2}.$$

ILLUSTRATIVE EXAMPLES

Example 1. *Show that if* $\vec{r} = \vec{a}$ *sin* ωt + $\vec{b}$ *cos* ωt, *where* $\vec{a}, \vec{b}, \omega$ *are constants, then*

$$\frac{d^2\vec{r}}{dt^2} = -\omega^2\vec{r} \quad \textit{and} \quad \vec{r} \times \frac{d\vec{r}}{dt} = -\omega\vec{a} \times \vec{b}. \qquad \text{(M.D.U. May 2012)}$$

Sol. We know that if $\vec{r} = \phi\vec{f}$, where ϕ is a scalar function of t, then $\dfrac{d\vec{r}}{dt} = \phi\dfrac{d\vec{f}}{dt} + \dfrac{d\phi}{dt}\vec{f}$.

But if $\vec{f}$ is a constant vector, then $\dfrac{d\vec{f}}{dt} = \vec{0}$

$$\therefore \quad \frac{d\vec{r}}{dt} = \frac{d\phi}{dt}\vec{f}$$

Now $\quad \vec{r} = \vec{a}\sin\omega t + \vec{b}\cos\omega t \quad \Rightarrow \quad \dfrac{d\vec{r}}{dt} = \vec{a}\,\omega\cos\omega t - \vec{b}\,\omega\sin\omega t$

$$\therefore \quad \frac{d^2\vec{r}}{dt^2} = -\vec{a}\,\omega^2\sin\omega t - \vec{b}\,\omega^2\cos\omega t = -\omega^2(\vec{a}\sin\omega t + \vec{b}\cos\omega t) = -\omega^2\vec{r}$$

Also $\quad \vec{r} \times \dfrac{d\vec{r}}{dt} = (\vec{a}\sin\omega t + \vec{b}\cos\omega t) \times (\vec{a}\,\omega\cos\omega t - \vec{b}\,\omega\sin\omega t)$

$$= \omega(-\vec{a}\times\vec{b}\sin^2\omega t + \vec{b}\times\vec{a}\cos^2\omega t) \qquad [\because \ \vec{a}\times\vec{a} = \vec{0} = \vec{b}\times\vec{b}\,]$$

$$= \omega(-\vec{a}\times\vec{b}\sin^2\omega t - \vec{a}\times\vec{b}\cos^2\omega t) = -\omega\,\vec{a}\times\vec{b}.$$

Example 2. *If* $\vec{r} = a\cos t\hat{i} + a\sin t\hat{j} + at\tan\alpha\hat{k}$, *find* $\left|\dfrac{d\vec{r}}{dt} \times \dfrac{d^2\vec{r}}{dt^2}\right|$ *and* $\left[\dfrac{d\vec{r}}{dt}\ \dfrac{d^2\vec{r}}{dt^2}\ \dfrac{d^3\vec{r}}{dt^3}\right]$.

(P.T.U. Jan. 2010)

Sol. $\quad \vec{r} = a\cos t\hat{i} + a\sin t\hat{j} + at\tan\alpha\hat{k}$

$$\Rightarrow \quad \frac{d\vec{r}}{dt} = -a\sin t\hat{i} + a\cos t\hat{j} + a\tan\alpha\hat{k}$$

$$\frac{d^2\vec{r}}{dt^2} = -a\cos t\hat{i} - a\sin t\hat{j}$$

$$\frac{d^3\vec{r}}{dt^3} = a\sin t\hat{i} - a\cos t\hat{j}$$

$$\therefore \quad \frac{d\vec{r}}{dt} \times \frac{d^2\vec{r}}{dt^2} = \begin{vmatrix} \hat{i} & \hat{j} & \hat{k} \\ -a\sin t & a\cos t & a\tan\alpha \\ -\cos t & -a\sin t & 0 \end{vmatrix}$$

$$= a^2\sin t\tan\alpha\hat{i} - a^2\cos t\tan\alpha\hat{j} + a^2\hat{k}$$

$$\therefore \quad \left|\frac{d\vec{r}}{dt} \times \frac{d^2\vec{r}}{dt^2}\right| = \sqrt{a^4\sin^2 t\tan^2\alpha + a^4\cos^2 t\tan^2\alpha + a^4}$$

$$= a^2\sqrt{\tan^2\alpha + 1} = a^2\sec\alpha$$

Also $\quad \left[\dfrac{d\vec{r}}{dt}\ \dfrac{d^2\vec{r}}{dt^2}\ \dfrac{d^3\vec{r}}{dt^3}\right] = \begin{vmatrix} -a\sin t & a\cos t & a\tan\alpha \\ -a\cos t & -a\sin t & 0 \\ a\sin t & -a\cos t & 0 \end{vmatrix}$ (Expanding by third column)

$$= a\tan\alpha(a^2\cos^2 t + a^2\sin^2 t) = a^3\tan\alpha.$$

Example 3. *If* $\frac{d\vec{u}}{dt} = \vec{w} \times \vec{u}$ *and* $\frac{d\vec{v}}{dt} = \vec{w} \times \vec{v}$, *prove that* $\frac{d}{dt}(\vec{u} \times \vec{v}) = \vec{w} \times (\vec{u} \times \vec{v})$.

Sol.
$$\frac{d}{dt}(\vec{u} \times \vec{v}) = \vec{u} \times \frac{d\vec{v}}{dt} + \frac{d\vec{u}}{dt} \times \vec{v} = \vec{u} \times (\vec{w} \times \vec{v}) + (\vec{w} \times \vec{u}) \times \vec{v}$$
$$= (\vec{u} \cdot \vec{v})\vec{w} - (\vec{u} \cdot \vec{w})\vec{v} + (\vec{v} \cdot \vec{w})\vec{u} - (\vec{v} \cdot \vec{u})\vec{w}$$
$$= (\vec{v} \cdot \vec{w})\vec{u} - (\vec{u} \cdot \vec{w})\vec{v} \qquad [\because \ \vec{u} \cdot \vec{v} = \vec{v} \cdot \vec{u}]$$
$$= (\vec{w} \cdot \vec{v})\vec{u} - (\vec{w} \cdot \vec{u})\vec{v} = \vec{w} \times (\vec{u} \times \vec{v}).$$

Example 4. *If* $\hat{R}$ *is a unit vector in the direction of* $\vec{r}$, *prove that* $\hat{R} \times \frac{d\hat{R}}{dt} = \frac{\vec{r}}{r^2} \times \frac{d\vec{r}}{dt}$, *where* $r = |\vec{r}|$.

Sol. We have $\vec{r} = r\hat{\mathrm{R}}$ so that $\hat{\mathrm{R}} = \frac{1}{r}\vec{r} \Rightarrow \frac{d\hat{\mathrm{R}}}{dt} = \frac{1}{r}\frac{d\vec{r}}{dt} - \frac{1}{r^2}\frac{dr}{dt}\vec{r}$

$$\therefore \quad \hat{\mathrm{R}} \times \frac{d\hat{\mathrm{R}}}{dt} = \frac{\vec{r}}{r} \times \left(\frac{1}{r}\frac{d\vec{r}}{dt} - \frac{1}{r^2}\frac{dr}{dt}\vec{r}\right) = \frac{\vec{r}}{r^2} \times \frac{d\vec{r}}{dt} - \frac{1}{r^3}\frac{dr}{dt}\vec{r} \times \vec{r}$$
$$= \frac{\vec{r}}{r^2} \times \frac{d\vec{r}}{dt} \qquad (\because \ \vec{r} \times \vec{r} = \vec{0})$$

Example 5. *If* $\vec{r}$ *is a vector function of a scalar t and* $\vec{a}$ *is a constant vector, differentiate the following with respect to t:*

(*i*) $\frac{\vec{r} \times \vec{a}}{\vec{r} \cdot \vec{a}}$ (*ii*) $\frac{\vec{r} + \vec{a}}{\vec{r}^2 + \vec{a}^2}$.

Sol. (*i*) Let $\vec{\mathrm{R}} = \frac{\vec{r} \times \vec{a}}{\vec{r} \cdot \vec{a}}$

Here $\vec{r} \cdot \vec{a}$ is a scalar function of t and $\frac{d\vec{a}}{dt} = \vec{0}$

$$\therefore \quad \frac{d\vec{\mathrm{R}}}{dt} = \frac{1}{\vec{r} \cdot \vec{a}}\frac{d}{dt}(\vec{r} \times \vec{a}) + \left\{\frac{d}{dt}\left(\frac{1}{\vec{r} \cdot \vec{a}}\right)\right\}(\vec{r} \times \vec{a})$$
$$= \frac{1}{\vec{r} \cdot \vec{a}}\left(\vec{r} \times \frac{d\vec{a}}{dt} + \frac{d\vec{r}}{dt} \times \vec{a}\right) - \frac{\frac{d}{dt}(\vec{r} \cdot \vec{a})}{(\vec{r} \cdot \vec{a})^2}(\vec{r} \times \vec{a}) \qquad \left[\because \ \frac{d}{dt}\left(\frac{1}{f(t)}\right) = -\frac{f'(t)}{(f(t))^2}\right]$$
$$= \frac{\frac{d\vec{r}}{dt} \times \vec{a}}{\vec{r} \cdot \vec{a}} - \frac{\vec{r} \cdot \frac{d\vec{a}}{dt} + \frac{d\vec{r}}{dt} \cdot \vec{a}}{(\vec{r} \cdot \vec{a})^2}(\vec{r} \times \vec{a}) = \frac{\frac{d\vec{r}}{dt} \times \vec{a}}{\vec{r} \cdot \vec{a}} - \frac{\frac{d\vec{r}}{dt} \cdot \vec{a}}{(\vec{r} \cdot \vec{a})^2}(\vec{r} \times \vec{a}).$$

(*ii*) Let $\quad \vec{R} = \dfrac{\vec{r} + \vec{a}}{\vec{r}^2 + \vec{a}^2}$

Here $\quad \vec{r}^2 = |\vec{r}|^2$ is a scalar function of t

$\vec{a}^2 = |\vec{a}|^2$ is a constant, independent of t

$\therefore \quad \vec{r}^2 + \vec{a}^2$ is a scalar function of t

Also $\quad \dfrac{d}{dt}(\vec{r}^2) = \dfrac{d}{dt}(\vec{r} \cdot \vec{r}) = \vec{r} \cdot \dfrac{d\vec{r}}{dt} + \dfrac{d\vec{r}}{dt} \cdot \vec{r} = 2\vec{r} \cdot \dfrac{d\vec{r}}{dt}$

$$\therefore \quad \frac{d\vec{R}}{dt} = \frac{1}{\vec{r}^2 + \vec{a}^2} \frac{d}{dt}(\vec{r} + \vec{a}) + \left\{ \frac{d}{dt}\left(\frac{1}{\vec{r}^2 + \vec{a}^2} \right) \right\} (\vec{r} + \vec{a})$$

$$= \frac{1}{\vec{r}^2 + \vec{a}^2} \left(\frac{d\vec{r}}{dt} + \frac{d\vec{a}}{dt} \right) - \frac{\frac{d}{dt}(\vec{r}^2 + \vec{a}^2)}{(\vec{r}^2 + \vec{a}^2)^2} (\vec{r} + \vec{a}) = \frac{\frac{d\vec{r}}{dt}}{\vec{r}^2 + \vec{a}^2} - \frac{2\vec{r} \cdot \frac{d\vec{r}}{dt}}{(\vec{r}^2 + \vec{a}^2)^2} (\vec{r} + \vec{a}).$$

Example 6. *Find*

(*i*) $\dfrac{d^2}{dt^2}\left[\vec{r} \; \dfrac{d\vec{r}}{dt} \; \dfrac{d^2\vec{r}}{dt^2} \right]$ $\qquad$ (*ii*) $\dfrac{d}{dt}\left[\vec{r} \times \left(\dfrac{d\vec{r}}{dt} \times \dfrac{d^2\vec{r}}{dt^2} \right) \right]$.

Sol. (*i*) Let $R = \left[\vec{r} \; \dfrac{d\vec{r}}{dt} \; \dfrac{d^2\vec{r}}{dt^2} \right]$, then R is the scalar triple product of three vectors $\vec{r}, \dfrac{d\vec{r}}{dt} \dfrac{d^2\vec{r}}{dt^2}$

$$\therefore \quad \frac{dR}{dt} = \left[\frac{d\vec{r}}{dt} \; \frac{d\vec{r}}{dt} \; \frac{d^2\vec{r}}{dt^2} \right] + \left[\vec{r} \; \frac{d^2\vec{r}}{dt^2} \; \frac{d^2\vec{r}}{dt^2} \right] + \left[\vec{r} \; \frac{d\vec{r}}{dt} \; \frac{d^3\vec{r}}{dt^3} \right] = \left[\vec{r} \; \frac{d\vec{r}}{dt} \; \frac{d^3\vec{r}}{dt^3} \right],$$

since scalar triple products having two equal vectors vanish.

Differentiating again, we have $\dfrac{d^2R}{dt^2} = \left[\dfrac{d\vec{r}}{dt} \; \dfrac{d\vec{r}}{dt} \; \dfrac{d^3\vec{r}}{dt^3} \right] + \left[\vec{r} \; \dfrac{d^2\vec{r}}{dt^2} \; \dfrac{d^3\vec{r}}{dt^3} \right] + \left[\vec{r} \; \dfrac{d\vec{r}}{dt} \; \dfrac{d^4\vec{r}}{dt^4} \right]$

$$= \left[\vec{r} \; \frac{d^2\vec{r}}{dt^2} \; \frac{d^3\vec{r}}{dt^3} \right] + \left[\vec{r} \; \frac{d\vec{r}}{dt} \; \frac{d^4\vec{r}}{dt^4} \right]$$

(*ii*) Let $\qquad \vec{R} = \vec{r} \times \left(\dfrac{d\vec{r}}{dt} \times \dfrac{d^2\vec{r}}{dt^2} \right)$, then $\vec{R}$ is the vector triple product of three vectors.

$$\therefore \quad \frac{d\vec{R}}{dt} = \frac{d\vec{r}}{dt} \times \left(\frac{d\vec{r}}{dt} \times \frac{d^2\vec{r}}{dt^2}\right) + \vec{r} \times \left(\frac{d^2\vec{r}}{dt^2} \times \frac{d^2\vec{r}}{dt^2}\right) + \vec{r} \times \left(\frac{d\vec{r}}{dt} \times \frac{d^3\vec{r}}{dt^3}\right)$$

$$= \frac{d\vec{r}}{dt} \times \left(\frac{d\vec{r}}{dt} \times \frac{d^2\vec{r}}{dt^2}\right) + \vec{r} \times \left(\frac{d\vec{r}}{dt} \times \frac{d^3\vec{r}}{dt^3}\right) \quad \text{since } \frac{d^2\vec{r}}{dt^2} \times \frac{d^2\vec{r}}{dt^2} = \vec{0}.$$

Example 7. *Find the unit tangent vector at any point on the curve $x = t^2 + 2$, $y = 4t - 5$, $z = 2t^2 - 6t$, where t is any variable. Also determine the unit tangent vector at the point $t = 2$.*

Sol. If $\vec{r}$ is the position vector of any point (x, y, z) on the given curve, then $\vec{r} = x\hat{i} + y\hat{j} + z\hat{k}$

$$\Rightarrow \quad \vec{r} = (t^2 + 2)\hat{i} + (4t - 5)\hat{j} + (2t^2 - 6t)\hat{k}$$

The vector $\dfrac{d\vec{r}}{dt}$ is along the tangent at the point (x, y, z) to the given curve.

Now, $$\frac{d\vec{r}}{dt} = 2t\hat{i} + 4\hat{j} + (4t - 6)\hat{k}$$

and $$\left|\frac{d\vec{r}}{dt}\right| = \sqrt{(2t)^2 + (4)^2 + (4t-6)^2} = \sqrt{20t^2 - 48t + 52} = 2\sqrt{5t^2 - 12t + 13}$$

$$\therefore \quad \text{The unit tangent vector } \hat{T} = \frac{\dfrac{d\vec{r}}{dt}}{\left|\dfrac{d\vec{r}}{dt}\right|} = \frac{t\hat{i} + 2\hat{j} + (2t-3)\hat{k}}{\sqrt{5t^2 - 12t + 13}}$$

Also the unit tangent vector at the point $t = 2$ is $\dfrac{2\hat{i} + 2\hat{j} + (2 \times 2 - 3)\hat{k}}{\sqrt{5 \times 4 - 12 \times 2 + 13}} = \dfrac{1}{3}(2\hat{i} + 2\hat{j} + \hat{k})$.

Example 8. *Find the angle between the tangents to the curve $\vec{r} = t^2\hat{i} + 2t\hat{j} - t^3\hat{k}$ at the points $t = \pm 1$.*

Sol. $\dfrac{d\vec{r}}{dt} = 2t\hat{i} + 2\hat{j} - 3t^2\hat{k}$ is a vector along the tangent at any point 't'.

If $\vec{T_1}$ and $\vec{T_2}$ are the vectors along the tangents at $t = 1$ and $t = -1$ respectively, then

$$\vec{T_1} = 2\hat{i} + 2\hat{j} - 3\hat{k} \quad \text{and} \quad \vec{T_2} = -2\hat{i} + 2\hat{j} - 3\hat{k}$$

If θ is the angle between $\vec{T_1}$ and $\vec{T_2}$, then

$$\cos\theta = \frac{\vec{T_1} \cdot \vec{T_2}}{|\vec{T_1}||\vec{T_2}|} = \frac{2(-2) + 2(2) - 3(-3)}{\sqrt{4+4+9} \cdot \sqrt{4+4+9}} = \frac{9}{17}$$

$$\therefore \quad \theta = \cos^{-1}\left(\frac{9}{17}\right).$$

Example 9. *A particle moves on the curve $x = 2t^2$, $y = t^2 - 4t$, $z = 3t - 5$ where t is the time. Find the components of velocity and acceleration at time t = 1 in the direction $\hat{i} - 3\hat{j} + 2\hat{k}$.*

Sol. If $\vec{r}$ is the position vector of any point (x, y, z) on the given curve, then

$$\vec{r} = x\hat{i} + y\hat{j} + z\hat{k} = 2t^2\hat{i} + (t^2 - 4t)\hat{j} + (3t - 5)\hat{k}$$

Velocity $\quad \vec{v} = \dfrac{d\vec{r}}{dt} = 4t\hat{i} + (2t - 4)\hat{j} + 3\hat{k} = 4\hat{i} - 2\hat{j} + 3\hat{k} \quad$ at $t = 1$

Acceleration $\quad \vec{a} = \dfrac{d^2\vec{r}}{dt^2} = 4\hat{i} + 2\hat{j} = 4\hat{i} + 2\hat{j} \quad$ at $t = 1$

Now the unit vector in the given direction $\hat{i} - 3\hat{j} + 2\hat{k}$

$$= \frac{\hat{i} - 3\hat{j} + 2\hat{k}}{|\hat{i} - 3\hat{j} + 2\hat{k}|} = \frac{\hat{i} - 3\hat{j} + 2\hat{k}}{\sqrt{14}} = \hat{n} \qquad \text{(say)}$$

$\therefore$ The component of velocity in the given direction

$$= \vec{v} \cdot \hat{n} = (4\hat{i} - 2\hat{j} + 3\hat{k}) \cdot \frac{\hat{i} - 3\hat{j} + 2\hat{k}}{\sqrt{14}}$$

$$= \frac{4(1) - 2(-3) + 3(2)}{\sqrt{14}} = \frac{16\sqrt{14}}{14} = \frac{8\sqrt{14}}{7}$$

and the component of acceleration in the given direction

$$= \vec{a} \cdot \hat{n} = (4\hat{i} + 2\hat{j}) \cdot \frac{\hat{i} - 3\hat{j} + 2\hat{k}}{\sqrt{14}} = \frac{-2}{\sqrt{14}} = -\frac{\sqrt{14}}{7}.$$

Note. Let $\hat{i}$ and $\hat{j}$ be two mutually perpendicular unit vectors in a plane. Let $\hat{a}$ and $\hat{b}$ be two unit vectors along two mutually perpendicular directions.

For examples: $\hat{T}$ and $\hat{N}$ along tangent and normal; $\hat{R}$ and $\hat{T}$ along radial and transverse directions.

Then $\qquad \hat{a} = \vec{OP} = \vec{OL} + \vec{LP}$

$$= (\cos\theta)\hat{i} + (\sin\theta)\hat{j} \qquad (\because \ OP = |\hat{a}| = 1)$$

and $\qquad \hat{b} = \cos\left(\dfrac{\pi}{2} + \theta\right)\hat{i} + \sin\left(\dfrac{\pi}{2} + \theta\right)\hat{j}$

$$= (-\sin\theta)\hat{i} + (\cos\theta)\hat{j}$$

$\therefore \qquad \dfrac{d\hat{a}}{d\theta} = (-\sin\theta)\hat{i} + (\cos\theta)\hat{j} = \hat{b}$

and $\qquad \dfrac{d\hat{b}}{d\theta} = (-\cos\theta)\hat{i} + (-\sin\theta)\hat{j} = -\hat{a}$

Hence $\dfrac{d\hat{T}}{d\theta} = \hat{N}$ and $\dfrac{d\hat{N}}{d\theta} = -\hat{T}$

$$\frac{d\hat{R}}{d\theta} = \hat{T} \text{ and } \frac{d\hat{T}}{d\theta} = -\hat{R}$$

Example 10. *Find the tangential and normal accelerations of a point moving in a plane curve.*

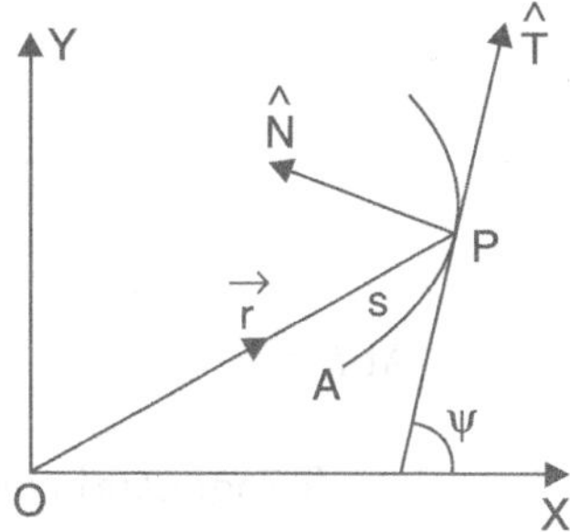

Sol. Let $\vec{r}$ be the position vector of a point P at any time t. Its velocity

$$\vec{V} = \frac{d\vec{r}}{dt} = \frac{d\vec{r}}{ds}\frac{ds}{dt}$$

Since $\dfrac{d\vec{r}}{ds} = \hat{T}$ is a unit vector along the tangent at P, we have

$$\vec{V} = \frac{ds}{dt}\hat{T}$$

$\therefore$ Acceleration is given by $\vec{a} = \dfrac{d\vec{V}}{dt} = \dfrac{d^2s}{dt^2}\hat{T} + \dfrac{ds}{dt}\dfrac{d\hat{T}}{dt}$ ...(1)

As $\hat{T}$ has constant magnitude, $\hat{T} \cdot \dfrac{d\hat{T}}{d\psi} = 0$ *i.e.,* $\dfrac{d\hat{T}}{d\psi} \perp \hat{T}$

Now $\dfrac{d\hat{T}}{dt} = \dfrac{d\hat{T}}{d\psi}\dfrac{d\psi}{dt} = \dfrac{d\psi}{dt}\hat{N}$, where $\hat{N} = \dfrac{d\hat{T}}{d\psi}$ is a unit vector along the normal at P.

$$= \frac{d\psi}{ds}\frac{ds}{dt}\hat{N} = \frac{1}{\rho}\frac{ds}{dt}\hat{N}$$

where ρ is the radius of curvature at P.

$\therefore$ From (1), $\vec{a} = \dfrac{d\vec{V}}{dt} = \dfrac{d^2s}{dt^2}\hat{T} + \dfrac{1}{\rho}\left(\dfrac{ds}{dt}\right)^2\hat{N} = \dfrac{dv}{dt}\hat{T} + \dfrac{v^2}{\rho}\hat{N}$, where $v = \dfrac{ds}{dt}$.

Hence the tangential and normal accelerations (*i.e.*, components of acceleration along $\hat{T}$ and $\hat{N}$) are

$$a_T = \frac{dv}{dt} \quad \text{or} \quad \frac{d^2s}{dt^2} \quad \text{and} \quad a_N = \frac{v^2}{\rho}.$$

Example 11. *The position vector of a moving particle at a time t is* $\vec{R}(t) = t^2\hat{i} - t^3\hat{j} + t^4\hat{k}$. *Find the tangential and normal components of its acceleration at time t = 1.*

(P.T.U. May 2010)

Sol. Let a_T and a_N be the tangential and normal components of acceleration $\vec{a}$ respectively. Then

$$\vec{a} = a_T\hat{T} + a_N\hat{N} \quad \text{...(1)}$$

Given $\qquad \vec{R}(t) = t^2\hat{i} - t^3\hat{j} + t^4\hat{k}$

$$\therefore \quad \text{Velocity } \vec{v} = \frac{d\vec{R}}{dt} = 2t\hat{i} - 3t^2\hat{j} + 4t^3\hat{k}$$

$$\Rightarrow \quad v = |\vec{v}| = \sqrt{4t^2 + 9t^4 + 16t^6}$$

$$\frac{dv}{dt} = \frac{8t + 36t^3 + 96t^5}{2\sqrt{4t^2 + 9t^4 + 16t^6}}$$

At $t = 1$, $\qquad \dfrac{dv}{dt} = \dfrac{8 + 36 + 96}{2\sqrt{4 + 9 + 16}} = \dfrac{70}{\sqrt{29}}$

$\therefore$ Tangential component of acceleration at $t = 1$ is

$$a_T = \left.\frac{dv}{dt}\right]_{t=1} = \frac{70}{\sqrt{29}}$$

Also, $\qquad \dfrac{d\vec{R}}{dt} = 2\hat{i} - 3\hat{j} + 4\hat{k} \text{ at } t = 1$

Since $\dfrac{d\vec{R}}{dt}$ is a vector along the tangent

$$\therefore \quad \hat{T} = \frac{\dfrac{d\vec{R}}{dt}}{\left|\dfrac{d\vec{R}}{dt}\right|} = \frac{2\hat{i} - 3\hat{j} + 4\hat{k}}{\sqrt{29}}$$

Acceleration $\qquad \vec{a} = \dfrac{d^2\vec{R}}{dt^2} = 2\hat{i} - 6t\hat{j} + 12t^2\hat{k}$

$$= 2\hat{i} - 6\hat{j} + 12\hat{k} \quad \text{at } t = 1$$

Now, from (1)

$$a_N\hat{N} = \vec{a} - a_T\hat{T}$$

$$= (2\hat{i} - 6\hat{j} + 12\hat{k}) - \frac{70}{\sqrt{29}}\,\frac{2\hat{i} - 3\hat{j} + 4\hat{k}}{\sqrt{29}}$$

$$= \frac{58\hat{i} - 174\hat{j} + 348\hat{k} - 140i + 210\hat{j} - 280\hat{k}}{29}$$

$$= \frac{-82}{29}\hat{i} + \frac{36}{29}\hat{j} + \frac{68}{29}\hat{k}$$

$$\Rightarrow \qquad |a_N\hat{N}| = \sqrt{\left(-\frac{82}{29}\right)^2 + \left(\frac{36}{29}\right)^2 + \left(\frac{68}{29}\right)^2}$$

$$a_N = \sqrt{\frac{12644}{29 \times 29}} = \sqrt{\frac{29 \times 436}{29 \times 29}} = \sqrt{\frac{436}{29}} = 2\sqrt{\frac{109}{29}}$$

Hence $\qquad a_T = \dfrac{70}{\sqrt{29}}$ and $a_N = 2\sqrt{\dfrac{109}{29}}$

Check. $\quad (a_T)^2 + (a_N)^2 = \dfrac{4900}{29} + \dfrac{436}{29} = \dfrac{5336}{29} = 184$

$$|\vec{a}|^2 = 4 + 36 + 144 = 184$$

$$\Rightarrow \qquad (a_T)^2 + (a_N)^2 = |\vec{a}|^2$$

Example 12. *Find the radial and transverse accelerations of a particle moving in a plane curve.* (U.P.T.U. 2008)

Sol. At any time t, let $\vec{r}$ be the position vector of the moving point P(r, θ). Let $\hat{R}$ and $\hat{T}$ be the unit vectors in radial and transverse directions respectively.

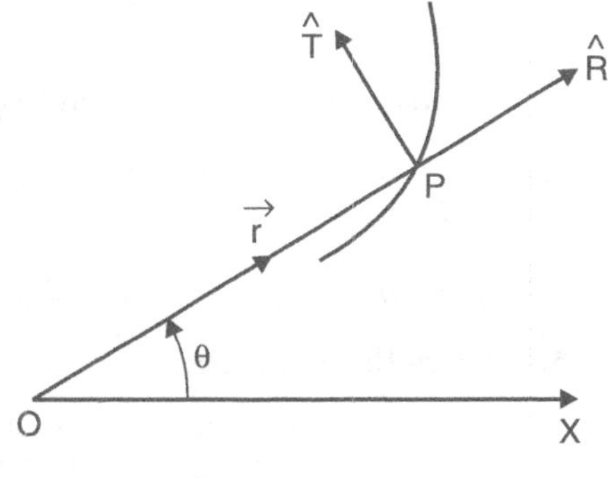

Now, $\qquad \vec{r} = r\hat{R}$

Velocity $\quad \vec{V} = \dfrac{d\vec{r}}{dt}$

$$= \frac{dr}{dt}\hat{R} + r\frac{d\hat{R}}{dt} = \frac{dr}{dt}\hat{R} + r\frac{d\hat{R}}{d\theta}\frac{d\theta}{dt}$$

$$= \frac{dr}{dt}\hat{R} + r\frac{d\theta}{dt}\hat{T}, \quad \text{since } \frac{d\hat{R}}{d\theta} = \hat{T}$$

$$\left(\text{Since } \hat{R} \text{ has fixed length, } \hat{R} \cdot \frac{d\hat{R}}{d\theta} = 0 \Rightarrow \frac{d\hat{R}}{d\theta} \perp \hat{R} \;\therefore\; \frac{d\hat{R}}{d\theta} = \hat{T}\right)$$

$\Rightarrow$ the components of velocity in the radial and transverse directions are $\dfrac{dr}{dt}$ and $r\,\dfrac{d\theta}{dt}$.

Also acceleration $\vec{A} = \dfrac{d\vec{V}}{dt} = \left(\dfrac{d^2r}{dt^2}\hat{R} + \dfrac{dr}{dt}\dfrac{d\hat{R}}{dt}\right) + \left(\dfrac{dr}{dt}\dfrac{d\theta}{dt}\hat{T} + r\dfrac{d^2\theta}{dt^2}\hat{T} + r\dfrac{d\theta}{dt}\dfrac{d\hat{T}}{dt}\right)$

$$= \frac{d^2r}{dt^2}\hat{R} + \frac{dr}{dt}\frac{d\hat{R}}{d\theta}\frac{d\theta}{dt} + \left(\frac{dr}{dt}\frac{d\theta}{dt} + r\frac{d^2\theta}{dt^2}\right)\hat{T} + r\frac{d\theta}{dt}\frac{d\hat{T}}{d\theta}\frac{d\theta}{dt}$$

$$= \frac{d^2r}{dt^2}\hat{R} + \frac{dr}{dt}\frac{d\theta}{dt}\hat{T} + \left(\frac{dr}{dt}\frac{d\theta}{dt} + r\frac{d^2\theta}{dt^2}\right)\hat{T} - r\left(\frac{d\theta}{dt}\right)^2\hat{R}$$

$$\left(\text{Since } \frac{d\hat{R}}{d\theta} = \hat{T} \text{ and } \frac{d\hat{T}}{d\theta} = -\hat{R}\right)$$

$$= \left[\frac{d^2r}{dt^2} - r\left(\frac{d\theta}{dt}\right)^2\right]\hat{R} + \left[2\frac{dr}{dt}\frac{d\theta}{dt} + r\frac{d^2\theta}{dt^2}\right]\hat{T}$$

Hence the radial and transverse components of the acceleration are

$$\frac{d^2r}{dt^2} - r\left(\frac{d\theta}{dt}\right)^2 \quad \text{and} \quad 2\frac{dr}{dt}\frac{d\theta}{dt} + r\frac{d^2\theta}{dt^2}.$$

EXERCISE 13.1

1. If $\vec{r} = \sin t\hat{i} + \cos t\hat{j} + t\hat{k}$, find $\left|\frac{d^2\vec{r}}{dt^2}\right|$.

2. If $\vec{r} = (\cos nt)\hat{i} + (\sin nt)\hat{j}$, where n is a constant and t varies, show that $\vec{r} \times \frac{d\vec{r}}{dt} = n\hat{k}$.

3. Show that $\vec{r} = \vec{a}\, e^{mt} + \vec{b}\, e^{nt}$ is the solution of the differential equation

$$\frac{d^2\vec{r}}{dt^2} - (m+n)\frac{d\vec{r}}{dt} + mn\vec{r} = \vec{0}.$$

[**Hint.** $\vec{a}$ and $\vec{b}$ are constant vectors.]

4. If $\vec{r}$ is a vector function of a scalar t and $\vec{a}$ is a constant vector, differentiate the following with respect to t:

(*i*) $\vec{r} \cdot \vec{a}$ (*ii*) $\vec{r} \times \vec{a}$ (*iii*) $\vec{r} \times \frac{d\vec{r}}{dt}$ (*iv*) $\vec{r} \cdot \frac{d\vec{r}}{dt}$.

5. Prove the following:

(*i*) $\frac{d}{dt}\left[\vec{a} \cdot \frac{d\vec{b}}{dt} - \frac{d\vec{a}}{dt} \cdot \vec{b}\right] = \vec{a} \cdot \frac{d^2\vec{b}}{dt^2} - \frac{d^2\vec{a}}{dt^2} \cdot \vec{b}$ (*ii*) $\frac{d}{dt}\left[\vec{a} \times \frac{d\vec{b}}{dt} - \frac{d\vec{a}}{dt} \times \vec{b}\right] = \vec{a} \times \frac{d^2\vec{b}}{dt^2} - \frac{d^2\vec{a}}{dt^2} \times \vec{b}$.

6. (*a*) Verify the formula, $\frac{d}{dt}(\vec{A} \cdot \vec{B}) = \vec{A} \cdot \frac{d\vec{B}}{dt} + \frac{d\vec{A}}{dt} \cdot \vec{B}$ for $\vec{A} = 5t^2\hat{i} + t\hat{j} - t^3\hat{k}$, $\vec{B} = \sin t\hat{i} - \cos t\hat{j}$.

(*b*) If $\vec{A} = 2t\hat{i} - t^2\hat{j} + t^3\hat{k}$, $\vec{B} = -t\hat{i} + t^2\hat{k}$, $\vec{C} = t^3\hat{i} - 2t\hat{k}$, find $\frac{d}{dt}(\vec{A} \cdot \vec{B} \times \vec{C})$ at $t = 1$.

(*c*) If $\vec{A} = \sin t\,\hat{i} - \cos t\,\hat{j} + t\,\hat{k}$, $\vec{B} = \cos t\,\hat{i} - \sin t\,\hat{j} - 3\,\hat{k}$ and $\vec{C} = 2\hat{i} + 3\hat{j} - \hat{k}$, find $\frac{d}{dt}[\vec{A} \times (\vec{B} \times \vec{C})]$ at $t = 0$.

7. Find the unit tangent vector at any point on the curve $x = a\cos t$, $y = a\sin t$, $z = bt$.

8. Find the angle between the tangents to the curve $x = t$, $y = t^2$, $z = t^3$, at $t = \pm 1$.

9. A particle moves along the curve $x = e^{-t}$, $y = 2\cos 3t$, $z = 2\sin 3t$, where t is the time. Determine its velocity and acceleration vectors and also the magnitudes of velocity and acceleration at $t = 0$.

10. The position vector of a particle at time t is $\vec{r} = \cos(t-1)\,\hat{i} + \sinh(t-1)\,\hat{j} + \alpha t^3\hat{k}$. Find the condition imposed on α by requiring that at time $t = 1$, the acceleration is normal to the position vector.

11. A particle moves along the curve $x = t^3 + 1$, $y = t^2$, $z = 2t + 5$ where t is the time. Find the components of its velocity and acceleration at $t = 1$ in the direction $2\hat{i} + 3\hat{j} + 6\hat{k}$. *(P.T.U. Jan. 2009)*

12. A particle moves so that its position vector is given by $\vec{r} = \cos\omega t\hat{i} + \sin\omega t\hat{j}$. Show that the velocity $\vec{v}$ of the particle is perpendicular to $\vec{r}$ and $\vec{r} \times \vec{v}$ is a constant vector.

13. The position vector of a point at time t is given by $\vec{r} = e^t(\cos t\hat{i} + \sin t\hat{j})$. Show that

    (*i*) $\vec{a} = 2(\vec{v} - \vec{r})$, where $\vec{a}, \vec{v}$ are acceleration and velocity of the particle.

    (*ii*) the angle between the radius vector and the acceleration is constant.

14. A particle moves along the curve $\vec{r} = (t^3 - 4t)\hat{i} + (t^2 + 4t)\hat{j} + (8t^2 - 3t^3)\hat{k}$ where t denotes time. Find the magnitudes of acceleration along the tangent and normal at time $t = 2$. *(P.T.U. 2011)*

15. For the curve $x = \cos t + t\sin t$, $y = \sin t - t\cos t$, find the tangential and the normal components of the acceleration at any time t. *(M.D.U. May 2011)*

    [**Hint.** $\vec{r} = x\hat{i} + y\hat{j} = (\cos t + t\sin t)\,\hat{i} + (\sin t - t\cos t)\hat{j}$]

16. A particle (position vector $\vec{r}$) is moving in a circle with constant angular velocity ω. Show by vector methods, that the acceleration is equal to $-\omega^2\vec{r}$.

Answers

1. 1   4. (*i*) $\dfrac{d\vec{r}}{dt}\cdot\vec{a}$ (*ii*) $\dfrac{d\vec{r}}{dt}\times\vec{a}$

(*iii*) $\vec{r}\times\dfrac{d^2\vec{r}}{dt^2}$ (*iv*) $\left(\dfrac{d\vec{r}}{dt}\right)^2 + \vec{r}\cdot\dfrac{d^2\vec{r}}{dt^2}$ 6. (*b*) -11 (*c*) $7\hat{i} + 6\hat{j} - 6\hat{k}$

7. $\dfrac{(-a\sin t)\,\hat{i} + (a\cos t)\,\hat{j} + b\hat{k}}{\sqrt{a^2+b^2}}$ 8. $\cos^{-1}\left(\dfrac{3}{7}\right)$ 9. $\sqrt{37}, 5\sqrt{13}$

10. $\pm\dfrac{1}{\sqrt{6}}$ 11. $\dfrac{24}{7}, \dfrac{18}{7}$ 14. $16, 2\sqrt{73}$ 15. $1, t$.

13.10. SCALAR AND VECTOR FIELDS

A variable quantity whose value at any point in a region of space depends upon the position of the point, is called a *point function*. There are two types of point functions.

(*a*) *Scalar Point Function*. Let R be a region of space at each point of which a scalar $\phi = \phi(x, y, z)$ is given, then ϕ is called a *scalar function* and R is called a *scalar field*.

The temperature distribution in a medium, the distribution of atmospheric pressure in space are examples of scalar point functions.

(*b*) *Vector Point Function*. Let R be a region of space at each point of which a vector $\vec{v} = \vec{v}(x, y, z)$ is given, then $\vec{v}$ is called a *vector point function* and R is called a *vector field*.

Each vector $\vec{v}$ of the field is regarded as a localised vector attached to the corresponding point (x, y, z).

The velocity of a moving fluid at any instant, the gravitational force are examples of vector point functions.

13.11. GRADIENT OF A SCALAR FIELD

(K.U.K. 2009, Dec. 2013; U.P.T.U. 2008; P.T.U. 2009)

Let $\phi(x, y, z)$ be a function defining a scalar field, then the vector $\hat{i}\frac{\partial \phi}{\partial x} + \hat{j}\frac{\partial \phi}{\partial y} + \hat{k}\frac{\partial \phi}{\partial z}$ is called the **gradient** of the scalar field ϕ and is denoted by grad ϕ.

Thus, $$\text{grad } \phi = \hat{i}\frac{\partial \phi}{\partial x} + \hat{j}\frac{\partial \phi}{\partial y} + \hat{k}\frac{\partial \phi}{\partial z}$$

The gradient of scalar field ϕ is obtained by operating on ϕ by the vector operator

$$\hat{i}\frac{\partial}{\partial x} + \hat{j}\frac{\partial}{\partial y} + \hat{k}\frac{\partial}{\partial z}$$

This operator is denoted by the symbol ∇, read as **del** (also called nabla).

Thus, grad $\phi = \nabla\phi$.

13.12. GEOMETRICAL INTERPRETATION OF GRADIENT

(M.D.U. 2013; K.U.K. 2009, Dec. 2013; P.T.U. May 2010)

If a surface $\phi(x, y, z) = c$ is drawn through any point P such that at each point on the surface, the function has the same value as at P, then such a surface is called a *level surface* through P. For example, if $\phi(x, y, z)$ represents potential at the point (x, y, z), the **equipotential surface** $\phi(x, y, z) = c$ is a level surface.

Through any point passes one and only one level surface. Moreover, no two level surfaces can intersect.

Consider the level surface through P at which the function has value ϕ and another level surface through a neighbouring point Q where the value is $\phi + \delta\phi$.

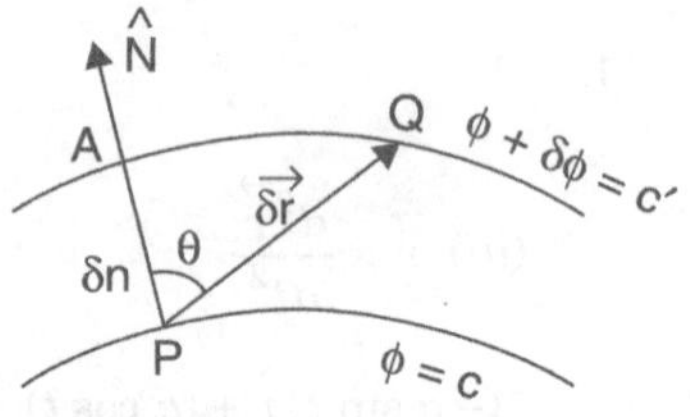

Let $\vec{r}$ and $\vec{r} + \delta\vec{r}$ be the position vectors of P and Q respectively, then $\overrightarrow{PQ} = \delta\vec{r}$.

Now $$\nabla\phi . \delta\vec{r} = \left(\hat{i}\frac{\partial \phi}{\partial x} + \hat{j}\frac{\partial \phi}{\partial y} + \hat{k}\frac{\partial \phi}{\partial z}\right) \cdot (\hat{i}\,\delta x + \hat{j}\,\delta y + \hat{k}\,\delta z)$$

$$= \frac{\partial \phi}{\partial x}\delta x + \frac{\partial \phi}{\partial y}\delta y + \frac{\partial \phi}{\partial z}\delta z = \delta\phi \qquad ...(1)$$

If Q lies on the same level surface as P, then $\delta\phi = 0$,

$\therefore$ (1) reduces to $\nabla\phi \cdot \delta\vec{r} = 0$.

Thus, $\nabla\phi$ is perpendicular to every $\delta\vec{r}$ lying in the surface.

Hence $\nabla\phi$ **is normal to the surface** $\phi(x, y, z) = c$.

Let $\nabla\phi = |\nabla\phi|\,\hat{N}$, where $\hat{N}$ is a unit vector normal to the surface. Let PA $= \delta n$ be the perpendicular distance between the two level surfaces through P and Q. Then the rate of change of ϕ in the direction of normal to the surface through P is

$$\frac{\partial \phi}{\partial n} = \lim_{\delta n \to 0} \frac{\partial \phi}{\partial n} = \lim_{\delta n \to 0} \frac{\nabla \phi \cdot \delta \vec{r}}{\delta n} \qquad \text{[by (1)]}$$

$$= \lim_{\delta n \to 0} \frac{|\nabla \phi| \hat{N} \cdot \delta \vec{r}}{\delta n} = |\nabla \phi| \qquad (\because \hat{N} \cdot \delta \vec{r} = |\hat{N}||\delta \vec{r}| \cos \theta = |\delta \vec{r}| \cos \theta = \delta n)$$

$$\therefore \quad |\nabla \phi| = \frac{\partial \phi}{\partial n}.$$

Hence the gradient of a scalar field ϕ is a vector normal to the surface $\phi = c$ and has a magnitude equal to the rate of change of ϕ along this normal.

13.13. DIRECTIONAL DERIVATIVE

(P.T.U. Jan. 2010)

Let PQ = δr, then $\lim_{\delta r \to 0} \frac{\delta \phi}{\delta r} = \frac{\partial \phi}{\partial r}$ is called the directional derivative of ϕ at P in the direction PQ.

Let $\hat{N}'$ be a unit vector in the direction PQ, then $\delta r = \frac{\delta n}{\cos \theta} = \frac{\delta n}{\hat{N} \cdot \hat{N}'}$

$$\therefore \quad \frac{\partial \phi}{\partial r} = \lim_{\delta r \to 0} \left[\hat{N} \cdot \hat{N}' \frac{\delta \phi}{\delta n} \right] = \hat{N} \cdot \hat{N}' \frac{\partial \phi}{\partial n}$$

$$= \hat{N}' \cdot \hat{N} \frac{\partial \phi}{\partial n} = \hat{N}' \cdot \hat{N} |\nabla \phi| = \hat{N}' \cdot \nabla \phi \qquad \left(\because |\nabla \phi| = \frac{\partial \phi}{\partial n} \text{ and } \hat{N} |\nabla \phi| = \nabla \phi \right)$$

Thus, the directional derivative $\frac{\partial \phi}{\partial r}$ is the resolved part of $\nabla \phi$ in the direction $\hat{N}'$.

Since $\quad \frac{\partial \phi}{\partial r} = \hat{N}' \cdot \nabla \phi = |\nabla \phi| \cos \theta \le |\nabla \phi|.$

$\therefore$ *$\nabla \phi$ gives the maximum rate of change of ϕ and the magnitude of this maximum is $|\nabla \phi|$.*

Hence the directional derivative of a scalar field ϕ at a point (x, y, z) in the direction of unit vector $\hat{a}$ is given by $(\nabla \phi) \cdot \hat{a}$.

13.14. PROPERTIES OF GRADIENT

(*a*) If ϕ is a constant scalar point function, then $\nabla \phi = \vec{0}$

(*b*) If ϕ_1 and ϕ_2 are two scalar point functions, then

(*i*) $\nabla (\phi_1 \pm \phi_2) = \nabla \phi_1 \pm \nabla \phi_2$

(*ii*) $\nabla (c_1 \phi_1 + c_2 \phi_2) = c_1 \nabla \phi_1 + c_2 \nabla \phi_2$, where c_1, c_2 are constant

(*iii*) $\nabla (\phi_1 \phi_2) = \phi_1 \nabla \phi_2 + \phi_2 \nabla \phi_1$

(*iv*) $\nabla \left(\frac{\phi_1}{\phi_2} \right) = \frac{\phi_2 \nabla \phi_1 - \phi_1 \nabla \phi_2}{\phi_2^{\,2}}, \ \phi_2 \ne 0.$

All the above results can be easily proved. For example

(*iii*) $$\nabla(\phi_1 \phi_2) = \left(\hat{i} \frac{\partial}{\partial x} + \hat{j} \frac{\partial}{\partial y} + \hat{k} \frac{\partial}{\partial z} \right) (\phi_1 \phi_2) = \hat{i} \frac{\partial}{\partial x} (\phi_1 \phi_2) + \hat{j} \frac{\partial}{\partial y} (\phi_1 \phi_2) + \hat{k} \frac{\partial}{\partial z} (\phi_1 \phi_2)$$

$$= \hat{i} \left(\phi_1 \frac{\partial \phi_2}{\partial x} + \phi_2 \frac{\partial \phi_1}{\partial x} \right) + \hat{j} \left(\phi_1 \frac{\partial \phi_2}{\partial y} + \phi_2 \frac{\partial \phi_1}{\partial y} \right) + \hat{k} \left(\phi_1 \frac{\partial \phi_2}{\partial z} + \phi_2 \frac{\partial \phi_1}{\partial z} \right)$$

$$= \phi_1 \left(\hat{i}\frac{\partial \phi_2}{\partial x} + \hat{j}\frac{\partial \phi_2}{\partial y} + \hat{k}\frac{\partial \phi_2}{\partial z} \right) + \phi_2 \left(\hat{i}\frac{\partial \phi_1}{\partial x} + \hat{j}\frac{\partial \phi_1}{\partial y} + \hat{k}\frac{\partial \phi_1}{\partial z} \right)$$

$$= \phi_1 \nabla \phi_2 + \phi_2 \nabla \phi_1.$$

(*iv*) $$\nabla \left(\frac{\phi_1}{\phi_2} \right) = \left(\hat{i}\frac{\partial}{\partial x} + \hat{j}\frac{\partial}{\partial y} + \hat{k}\frac{\partial}{\partial z} \right)\left(\frac{\phi_1}{\phi_2} \right) = \hat{i}\frac{\partial}{\partial x}\left(\frac{\phi_1}{\phi_2} \right) + \hat{j}\frac{\partial}{\partial y}\left(\frac{\phi_1}{\phi_2} \right) + \hat{k}\frac{\partial}{\partial z}\left(\frac{\phi_1}{\phi_2} \right)$$

$$= \hat{i}\,\frac{\phi_2 \frac{\partial \phi_1}{\partial x} - \phi_1 \frac{\partial \phi_2}{\partial x}}{\phi_2^2} + \hat{j}\,\frac{\phi_2 \frac{\partial \phi_1}{\partial y} - \phi_1 \frac{\partial \phi_2}{\partial y}}{\phi_2^2} + \hat{k}\,\frac{\phi_2 \frac{\partial \phi_1}{\partial z} - \phi_1 \frac{\partial \phi_2}{\partial z}}{\phi_2^2}$$

$$= \frac{1}{{\phi_2}^2}\left[\phi_2 \left(\hat{i}\frac{\partial \phi_1}{\partial x} + \hat{j}\frac{\partial \phi_1}{\partial y} + \hat{k}\frac{\partial \phi_1}{\partial z} \right) - \phi_1 \left(\hat{i}\frac{\partial \phi_2}{\partial x} + \hat{j}\frac{\partial \phi_2}{\partial y} + \hat{k}\frac{\partial \phi_2}{\partial z} \right) \right]$$

$$= \frac{\phi_2 \nabla \phi_1 - \phi_1 \nabla \phi_2}{\phi_2^2}.$$

ILLUSTRATIVE EXAMPLES

Example 1. *Find grad ϕ when ϕ is given by $\phi = 3x^2y - y^3z^2$ at the point (1, –2, –1).*

Sol. $$\text{Grad } \phi = \nabla \phi = \left(\hat{i}\frac{\partial}{\partial x} + \hat{j}\frac{\partial}{\partial y} + \hat{k}\frac{\partial}{\partial z} \right)(3x^2y - y^3z^2)$$

$$= \hat{i}\frac{\partial}{\partial x}(3x^2y - y^3z^2) + \hat{j}\frac{\partial}{\partial y}(3x^2y - y^3z^2) + \hat{k}\frac{\partial}{\partial z}(3x^2y - y^3z^2)$$

$$= \hat{i}\,(6xy) + \hat{j}(3x^2 - 3y^2z^2) + \hat{k}\,(-2y^3z)$$

$$= -12\hat{i} - 9\hat{j} - 16\hat{k} \text{ at the point } (1, -2, -1).$$

Example 2. *If $\vec{r} = x\hat{i} + y\hat{j} + z\hat{k}$, show that*

(*i*) *grad* $r = \frac{\vec{r}}{r}$ (G.B.T.U. 2011) (*ii*) *grad* $\left(\frac{1}{r}\right) = -\frac{\vec{r}}{r^3}$ (G.B.T.U. 2011)

(*iii*) $\nabla r^n = nr^{n-2}\,\vec{r}$ (G.B.T.U. 2011) (*iv*) $\nabla(\vec{a}\cdot\vec{r}) = \vec{a}$, *where $\vec{a}$ is a constant vector.* (U.P.T.U. 2008)

Sol. $r = |\vec{r}| = \sqrt{x^2 + y^2 + z^2}$, or $r^2 = x^2 + y^2 + z^2$

Differentiating partially w.r.t. x, we have $2r\frac{\partial r}{\partial x} = 2x$ or $\frac{\partial r}{\partial x} = \frac{x}{r}$

Similarly, $\frac{\partial r}{\partial y} = \frac{y}{r}$ and $\frac{\partial r}{\partial z} = \frac{z}{r}$

(*i*) $$\text{grad } r = \nabla r = \left(\hat{i}\frac{\partial}{\partial x} + \hat{j}\frac{\partial}{\partial y} + \hat{k}\frac{\partial}{\partial z} \right) r = \hat{i}\frac{\partial r}{\partial x} + \hat{j}\frac{\partial r}{\partial y} + \hat{k}\frac{\partial r}{\partial z}$$

$$= \hat{i}\left(\frac{x}{r}\right) + \hat{j}\left(\frac{y}{r}\right) + \hat{k}\left(\frac{z}{r}\right) = \frac{x\hat{i} + y\hat{j} + z\hat{k}}{r} = \frac{\vec{r}}{r}.$$

(*ii*) $$\text{grad}\left(\frac{1}{r}\right) = \nabla\left(\frac{1}{r}\right) = \left(\hat{i}\frac{\partial}{\partial x} + \hat{j}\frac{\partial}{\partial y} + \hat{k}\frac{\partial}{\partial z} \right)\left(\frac{1}{r}\right)$$

$$= \hat{i}\left(-\frac{1}{r^2}\cdot\frac{\partial r}{\partial x}\right) + \hat{j}\left(-\frac{1}{r^2}\cdot\frac{\partial r}{\partial y}\right) + \hat{k}\left(-\frac{1}{r^2}\cdot\frac{\partial r}{\partial z}\right)$$

$$= \hat{i}\left(-\frac{1}{r^2}\cdot\frac{x}{r}\right) + \hat{j}\left(-\frac{1}{r^2}\cdot\frac{y}{r}\right) + \hat{k}\left(-\frac{1}{r^2}\cdot\frac{z}{r}\right)$$

$$= -\frac{1}{r^3}(x\hat{i} + y\hat{j} + z\hat{k}) = -\frac{\vec{r}}{r^3}.$$

(*iii*) $$\nabla r^n = \left(\hat{i}\frac{\partial}{\partial x} + \hat{j}\frac{\partial}{\partial y} + \hat{k}\frac{\partial}{\partial z}\right) r^n = \hat{i}\left(nr^{n-1}\frac{\partial r}{\partial x}\right) + \hat{j}\left(nr^{n-1}\frac{\partial r}{\partial y}\right) + \hat{k}\left(nr^{n-1}\frac{\partial r}{\partial z}\right)$$

$$= \hat{i}\left(nr^{n-1}\cdot\frac{x}{r}\right) + \hat{j}\left(nr^{n-1}\cdot\frac{y}{r}\right) + \hat{k}\left(nr^{n-1}\cdot\frac{z}{r}\right) = nr^{n-2}(x\hat{i} + y\hat{j} + z\hat{k}) = nr^{n-2}\,\vec{r}.$$

(*iv*) Let $\vec{a} = a_1\hat{i} + a_2\hat{j} + a_3\hat{k}$, where a_1, a_2, a_3 are constants.

$$\vec{a}\cdot\vec{r} = a_1x + a_2y + a_3z$$

$$\therefore \quad \nabla(\vec{a}\cdot\vec{r}) = \left(\hat{i}\frac{\partial}{\partial x} + \hat{j}\frac{\partial}{\partial y} + \hat{k}\frac{\partial}{\partial z}\right)(a_1x + a_2y + a_3z)$$

$$= \hat{i}\frac{\partial}{\partial x}(a_1x + a_2y + a_3z) + \hat{j}\frac{\partial}{\partial y}(a_1x + a_2y + a_3z) + \hat{k}\frac{\partial}{\partial z}(a_1x + a_2y + a_3z)$$

$$= a_1\hat{i} + a_2\hat{j} + a_3\hat{k} = \vec{a}.$$

Example 3. *Find a unit vector normal to the surface $x^3 + y^3 + 3xyz = 3$ at the point (1, 2, –1).*

Sol. Let $\phi = x^3 + y^3 + 3xyz = 3$, then $\frac{\partial\phi}{\partial x} = 3x^2 + 3yz,\ \frac{\partial\phi}{\partial y} = 3y^2 + 3xz,\ \frac{\partial\phi}{\partial z} = 3xy$

$$\therefore \quad \nabla\phi = \hat{i}\frac{\partial\phi}{\partial x} + \hat{j}\frac{\partial\phi}{\partial y} + \hat{k}\frac{\partial\phi}{\partial z} = (3x^2 + 3yz)\hat{i} + (3y^2 + 3xz)\hat{j} + (3xy)\hat{k}$$

At (1, 2, – 1), $\nabla\phi = -3\hat{i} + 9\hat{j} + 6\hat{k}$

Which is a vector normal to the given surface at (1, 2, –1).

Hence a unit vector normal to the given surface at (1, 2, –1)

$$= \frac{-3\hat{i} + 9\hat{j} + 6\hat{k}}{\sqrt{[(-3)^2 + (9)^2 + (6)^2]}} = \frac{-3\hat{i} + 9\hat{j} + 6\hat{k}}{3\sqrt{14}} = \frac{1}{\sqrt{14}}(-\hat{i} + 3\hat{j} + 2\hat{k}).$$

Example 4. *Find the directional derivative of the function $f = x^2 - y^2 + 2z^2$ at the point P(1, 2, 3) in the direction of the line PQ, where Q is the point (5, 0, 4).* (U.K.T.U. 2011)
In what direction it will be maximum ? Find also the magnitude of this maximum.

Sol. We have $\nabla f = \hat{i}\frac{\partial f}{\partial x} + \hat{j}\frac{\partial f}{\partial y} + \hat{k}\frac{\partial f}{\partial z} = 2x\hat{i} - 2y\hat{j} + 4z\hat{k} = 2\hat{i} - 4\hat{j} + 12\hat{k}$ at P(1, 2, 3)

Also, $$\vec{PQ} = \vec{OQ} - \vec{OP} = (5\hat{i} + 4\hat{k}) - (\hat{i} + 2\hat{j} + 3\hat{k}) = 4\hat{i} - 2\hat{j} + \hat{k}$$

If $\hat{n}$ is a unit vector in the direction $\vec{PQ}$, then $\hat{n} = \frac{4\hat{i} - 2\hat{j} + \hat{k}}{\sqrt{16 + 4 + 1}} = \frac{1}{\sqrt{21}}(4\hat{i} - 2\hat{j} + \hat{k})$

$\therefore$ Directional derivative of f in the direction $\vec{PQ} = (\nabla f)\cdot\hat{n}$

$$= (2\hat{i} - 4\hat{j} + 12\hat{k}) \cdot \frac{1}{\sqrt{21}} (4\hat{i} - 2\hat{j} + \hat{k}) = \frac{1}{\sqrt{21}} [2(4) - 4(-2) + 12(1)]$$

$$= \frac{28}{\sqrt{21}} = \frac{4}{3}\sqrt{21}$$

The directional derivative of f is maximum in the direction of the normal to the given surface *i.e.*, in the direction of $\nabla f = 2\hat{i} - 4\hat{j} + 12\hat{k}$

The maximum value of this directional derivative = $|\nabla f|$

$$= \sqrt{(2)^2 + (-4)^2 + (12)^2} = \sqrt{164} = 2\sqrt{41}.$$

Example 5. *Find the directional derivative of* $\phi = 5x^2y - 5y^2z + \frac{5}{2}z^2x$ *at the point* $P(1, 1, 1)$ *in the direction of the line* $\frac{x-1}{2} = \frac{y-3}{-2} = \frac{z}{1}$. (G.B.T.U. 2010)

Sol. Here, $\phi = 5x^2y - 5y^2z + \frac{5}{2}z^2x$

$$\therefore \quad \nabla\phi = \hat{i}\frac{\partial\phi}{\partial x} + \hat{j}\frac{\partial\phi}{\partial y} + \hat{k}\frac{\partial\phi}{\partial z}$$

$$= \left(10xy + \frac{5}{2}z^2\right)\hat{i} + (5x^2 - 10yz)\hat{j} + (-5y^2 + 5zx)\hat{k}$$

$$= \frac{25}{2}\hat{i} - 5\hat{j} \qquad \text{at P(1, 1, 1)}$$

The direction of the given line is $\vec{a} = 2\hat{i} - 2\hat{j} + \hat{k}$

$$\Rightarrow \quad \hat{a} = \frac{\vec{a}}{|\vec{a}|} = \frac{2\hat{i} - 2\hat{j} + \hat{k}}{3}$$

$\therefore$ The required directional derivative

$$= (\nabla\phi)\cdot\hat{a} = \left(\frac{25}{2}\hat{i} - 5\hat{j}\right)\cdot\left(\frac{2\hat{i} - 2\hat{j} + \hat{k}}{3}\right)$$

$$= \left(\frac{25}{2}\right)\left(\frac{2}{3}\right) + (-5)\left(-\frac{2}{3}\right) + (0)\left(\frac{1}{3}\right) = \frac{35}{3}.$$

Example 6. *Find the directional derivative of* $f(x, y, z) = e^{2x}\cos(yz)$ *at (0, 0, 0) in the direction of the tangent to the curve* $x = a\sin t$, $y = a\cos t$, $z = at$ *at* $t = \pi/4$.

Sol. $f(x, y, z) = e^{2x}\cos(yz)$

$$\therefore \quad \nabla f = \hat{i}\frac{\partial f}{\partial x} + \hat{j}\frac{\partial f}{\partial y} + \hat{k}\frac{\partial f}{\partial z}$$

$$= (2e^{2x}\cos(yz))\,\hat{i} - (ze^{2x}\sin(yz))\,\hat{j} - (ye^{2x}\sin(yz))\,\hat{k}$$

$$= 2\hat{i} \quad \text{at } (0, 0, 0)$$

The given curve is $\vec{r} = x\hat{i} + y\hat{j} + z\hat{k}$

i.e., $\vec{r} = (a \sin t)\hat{i} + (a \cos t)\hat{j} + (at)\hat{k}$

Tangent to the given curve is

$$\frac{d\vec{r}}{dt} = (a \cos t)\hat{i} - (a \sin t)\hat{j} + a\hat{k}$$

$$= a\left(\frac{1}{\sqrt{2}}\hat{i} - \frac{1}{\sqrt{2}}\hat{j} + \hat{k}\right) \quad \text{at} \quad t = \frac{\pi}{4}$$

$$\therefore \quad \left|\frac{d\vec{r}}{dt}\right| = \sqrt{\frac{a^2}{2} + \frac{a^2}{2} + a^2} = a\sqrt{2}$$

$\Rightarrow$ Unit vector along the tangent at $t = \frac{\pi}{4}$ is

$$\hat{n} = \frac{\frac{d\vec{r}}{dt}}{\left|\frac{d\vec{r}}{dt}\right|} = \frac{a\left(\frac{1}{\sqrt{2}}\hat{i} - \frac{1}{\sqrt{2}}\hat{j} + \hat{k}\right)}{a\sqrt{2}} = \frac{\hat{i}}{2} - \frac{\hat{j}}{2} + \frac{\hat{k}}{\sqrt{2}}$$

$\therefore$ The required directional derivative

$$= (\nabla f) \cdot \hat{n} = 2\hat{i} \cdot \left(\frac{\hat{i}}{2} - \frac{\hat{j}}{2} + \frac{\hat{k}}{\sqrt{2}}\right) = 2\left(\frac{1}{2}\right) = 1$$

Example 7. *If $\vec{a}, \vec{b}$ are constant vectors, prove that $\nabla\, [\vec{r}\ \vec{a}\ \vec{b}\,] = \vec{a} \times \vec{b}$.*

Sol. Let $\vec{a} = a_1\hat{i} + a_2\hat{j} + a_3\hat{k}$ and $\vec{b} = b_1\hat{i} + b_2\hat{j} + b_3\hat{k}$

where $a_1, a_2, a_3, b_1, b_2, b_3$ are constants.

Also $\vec{r} = x\hat{i} + y\hat{j} + z\hat{k}$ so that

$$\left[\vec{r}\ \ \vec{a}\ \ \vec{b}\right] = \begin{vmatrix} x & y & z \\ a_1 & a_2 & a_3 \\ b_1 & b_2 & b_3 \end{vmatrix} = (a_2b_3 - a_3b_2)x + (a_3b_1 - a_1b_3)\, y + (a_1b_2 - a_2\, b_1)\, z$$

$$\therefore \quad \nabla\left[\vec{r}\ \ \vec{a}\ \ \vec{b}\right] = \left(\hat{i}\frac{\partial}{\partial x} + \hat{j}\frac{\partial}{\partial y} + \hat{k}\frac{\partial}{\partial z}\right)[(a_2b_3 - a_3b_2)x + (a_3b_1 - a_1b_3)y + (a_1b_2 - a_2b_1)\, z]$$

$$= (a_2\, b_3 - a_3\, b_2)\hat{i} + (a_3b_1 - a_1b_3)\hat{j} + (a_1b_2 - a_2\, b_1)\hat{k}$$

$$= \begin{vmatrix} \hat{i} & \hat{j} & \hat{k} \\ a_1 & a_2 & a_3 \\ b_1 & b_2 & b_3 \end{vmatrix} = \vec{a} \times \vec{b}$$

Hence $\nabla\left[\vec{r}\ \ \vec{a}\ \ \vec{b}\right] = \vec{a} \times \vec{b}$.

Example 8. *If the directional derivative of* $\phi = ax^2y + by^2z + cz^2x$ *at the point (1, 1, 1) has maximum magnitude 15 in the direction parallel to the line* $\frac{x-1}{2} = \frac{y-3}{-2} = \frac{z}{1}$, *find the values of a, b and c.* (M.D.U. Dec. 2010)

Sol. Here, $\phi = ax^2y + by^2z + cz^2x$

$$\therefore \quad \nabla\phi = \hat{i}\frac{\partial\phi}{\partial x} + \hat{j}\frac{\partial\phi}{\partial y} + \hat{k}\frac{\partial\phi}{\partial z}$$

$$= (2axy + cz^2)\hat{i} + (ax^2 + 2byz)\hat{j} + (by^2 + 2czx)\hat{k}$$

$$= (2a + c)\hat{i} + (a + 2b)\hat{j} + (b + 2c)\hat{k} \text{ at } (1, 1, 1)$$

Now, the directional derivative of ϕ is maximum in the direction of the normal to the given surface *i.e.,* in the direction of $\nabla\phi$.

But we are given that the directional derivative of ϕ is maximum in the direction parallel to the line.

$$\frac{x-1}{2} = \frac{y-3}{-2} = \frac{z}{1} \quad \textit{i.e.,} \text{ parallel to the vector } 2\hat{i} - 2\hat{j} + \hat{k}.$$

$$\therefore \quad \frac{2a+c}{2} = \frac{a+2b}{-2} = \frac{b+2c}{1}$$

[Two vectors are parallel if the corresponding scalar components are proportional].

$$\Rightarrow \quad \frac{2a+c}{2} = \frac{a+2b}{-2} \quad \text{and} \quad \frac{a+2b}{-2} = \frac{b+2c}{1}$$

$$\Rightarrow \quad 2a + c = -a - 2b \quad \text{and} \quad a + 2b = -2b - 4c$$

$$\Rightarrow \quad 3a + 2b + c = 0 \quad \text{and} \quad a + 4b + 4c = 0$$

By cross-multiplication, we have

$$\frac{a}{8-4} = \frac{b}{1-12} = \frac{c}{12-2}$$

or

$$\frac{a}{4} = \frac{b}{-11} = \frac{c}{10} = \lambda \quad \text{(say)}$$

$$\Rightarrow \quad a = 4\lambda, \ b = -11\lambda, \ c = 10\lambda$$

The maximum value of directional derivative of ϕ

$$= |\nabla\phi| = \sqrt{(2a+c)^2 + (a+2b)^2 + (b+2c)^2}$$

Since it is given to be 15, we have

$$\sqrt{(2a+c)^2 + (a+2b)^2 + (b+2c)^2} = 15$$

$$\Rightarrow \quad (8\lambda + 10\lambda)^2 + (4\lambda - 22\lambda)^2 + (-11\lambda + 20\lambda)^2 = 225$$

$$\Rightarrow \quad (324 + 324 + 81)\lambda^2 = 225 \quad \Rightarrow \quad \lambda^2 = \frac{225}{729} = \frac{25}{81}$$

$$\Rightarrow \quad \lambda = \pm 5/9$$

$$\therefore \quad a = \pm\frac{20}{9}, \ b = \mp\frac{55}{9}, \ c = \pm\frac{50}{9}$$

Example 9. *Find the values of constants a, b and c so that the maximum value of the directional derivative of* $\phi = axy^2 + byz + cz^2x^3$ *at (1, 2, –1) has a magnitude 64 in the direction parallel to z-axis.* (M.D.U. Dec. 2010)

Sol. Here, $\phi = axy^2 + byz + cz^2x^3$

$$\therefore \quad \nabla\phi = \hat{i}\frac{\partial \phi}{\partial x} + \hat{j}\frac{\partial \phi}{\partial y} + \hat{k}\frac{\partial \phi}{\partial z}$$

$$= (ay^2 + 3cz^2x^2)\,\hat{i} + (2axy + bz)\,\hat{j} + (by + 2czx^3)\,\hat{k}$$

$$= (4a + 3c)\,\hat{i} + (4a - b)\,\hat{j} + (2b - 2c)\,\hat{k} \text{ at } (1, 2, -1)$$

Now, the directional derivative of ϕ is maximum in the direction of the normal to the given surface *i.e.*, in the direction of $\nabla\phi$. But we are given that the directional derivative of ϕ is maximum in the direction parallel to z-axis *i.e.*, parallel to $\hat{k}$.

Hence co-efficients of $\hat{i}$ and $\hat{j}$ in $\nabla\phi$ should be zero and the co-efficient of $\hat{k}$ positive.

Thus, $4a + 3c = 0$ ...(1)

$4a - b = 0$ ...(2)

and $2b - 2c > 0 \quad i.e., \quad b > c$ ...(3)

Then, $\nabla\phi = 2(b - c)\,\hat{k}$

Also maximum value of directional derivative = $|\nabla\phi|$

$\therefore \quad |2\,(b - c)\,\hat{k}\,| = 64$ (Given)

$\Rightarrow \quad 2(b - c) = 64 \quad \text{or} \quad b - c = 32$ ...(4)

Solving (1), (2) and (4), we have

$$a = 6,\ b = 24,\ c = -8.$$

Example 10. *Find the angle between the surfaces* $x^2 + y^2 + z^2 = 9$ *and* $z = x^2 + y^2 - 3$ *at the point (2, – 1, 2).*

Sol. Angle between two surfaces at a point is the angle between the normals to the surfaces at that point.

Let $\phi_1 = x^2 + y^2 + z^2 = 9 \quad \text{and} \quad \phi_2 = x^2 + y^2 - z = 3$

Then $\text{grad}\,\phi_1 = 2x\hat{i} + 2y\hat{j} + 2z\hat{k} \quad \text{and} \quad \text{grad}\,\phi_2 = 2x\hat{i} + 2y\hat{j} - \hat{k}$

Let $\vec{n_1}$ = grad ϕ_1 at the point (2, –1, 2) and $\vec{n_2}$ = grad ϕ_2 at the point (2, –1, 2).

Then $\vec{n_1} = 4\hat{i} - 2\hat{j} + 4\hat{k} \quad \text{and} \quad \vec{n_2} = 4\hat{i} - 2\hat{j} - \hat{k}$

The vectors $\vec{n_1}$ and $\vec{n_2}$ are along normals to the two surfaces at the point (2, –1, 2). If θ is the angle between these vectors, then

$$\cos\theta = \frac{\vec{n_1}\cdot\vec{n_2}}{|\vec{n_1}|\,|\vec{n_2}|} = \frac{4(4) - 2(-2) + 4(-1)}{\sqrt{16+4+16}\cdot\sqrt{16+4+1}} = \frac{16}{6\sqrt{21}}$$

$$\therefore \quad \theta = \cos^{-1}\left(\frac{8}{3\sqrt{21}}\right).$$

EXERCISE 13.2

1. Find grad ϕ when ϕ is given by
   (i) $\phi = x^2 + yz$ (ii) $\phi = x^3 + y^3 + 3xyz$ (iii) $\phi = \log (x^2 + y^2 + z^2)$.
2. If $r = |\vec{r}|$, where $\vec{r} = x\hat{i} + y\hat{j} + z\hat{k}$, prove that
   (i) $\nabla f(r) = f'(r) \nabla r$ (ii) $\nabla \log r = \frac{\vec{r}}{r^2}$
   (iii) $\nabla (e^{r^2}) = 2e^{r^2} \vec{r}$ (iv) grad $|\vec{r}|^2 = 2\vec{r}$
   (v) grad$\left(\frac{1}{r^2}\right) = -\frac{2\vec{r}}{r^4}$ (*M.D.U. 2013*)
   (vi) $\nabla \phi (r) = \frac{\phi'(r)}{r} \vec{r}$ and hence show that $\nabla\left(\int r^n \, dr\right) = r^{n-1} \vec{r}$ (*M.D.U. May 2011*)
3. If $u = x + y + z$, $v = x^2 + y^2 + z^2$, $w = yz + zx + xy$, prove that:
   (i) (grad u) . [(grad v) × (grad w)] = 0
   (ii) grad u, grad v and grad w are coplanar vectors. (*M.T.U. 2013; U.K.T.U. 2010*)
   [**Hint.** Three vectors are coplanar if their scalar triple product is zero].
4. Find a unit vector normal to the surface
   (i) $xy^3z^2 = 4$ at the point (–1, –1, 2) (*U.P.T.U. 2008*)
   (ii) $x^2y + 2xz = 4$ at the point (2, –2, 3).
5. Find the directional derivative of the function
   (i) $f(x, y, z) = xy^2 + yz^3$ at the point (2, –1, 1) in the direction of the vector $\hat{i} + 2\hat{j} + 2\hat{k}$. (*P.T.U. 2011*)
   (ii) $f(x, y, z) = 2xy + z^2$ at the point (1, –1, 3) in the direction of the vector $\hat{i} + 2\hat{j} + 2\hat{k}$. (*Anna 2009*)
   (iii) $\phi = x^2yz + 4xz^2$ at the point (1, –2, –1) in the direction of the vector $2\hat{i} - \hat{j} - 2\hat{k}$. (*V.T.U. 2012*)
   (iv) $\phi = 4xz^3 - 3x^2yz^2$ at (2, –1, 2) along z-axis.
   (v) $\phi = xy^2 + yz^3$ at the point (2, –1, 1) in the direction of the normal to the surface $x \log z - y^2 + 4 = 0$ at (–1, 2, 1). ((*Anna 2009; G.B.T.U. 2011*)
   (vi) $\phi = (x^2 + y^2 + z^2)^{-\frac{1}{2}}$ at the point P(3, 1, 2) in the direction of the vector $yz\hat{i} + zx\hat{j} + xy\hat{k}$.
   (vii) $\psi(x, y, z) = 4e^{x + 5y - 13z}$ at the point (1, 2, 3) in the direction towards the point (–3, 5, 7). (*U.P.T.U. 2009*)
6. Find the directional derivative of the function $\phi = \frac{y}{x^2 + y^2}$ at the point (0, 1) making an angle 30° with the positive x-axis.
   [**Hint.** Here $\hat{a} = \cos 30° \, \hat{i} + \sin 30° \, \hat{j}$]
7. In what direction from (3, 1, –2) is the directional derivative of $\phi = x^2y^2z^4$ maximum and what is its magnitude ? (*K.U.K. 2010*)
8. What is the greatest rate of increase of $u = x^2 + yz^2$ at the point (1, – 1, 3)?
9. The temperature at a point (x, y, z) in space is given by $T(x, y, z) = x^2 + y^2 - z$. A mosquito located at (1, 1, 2) desires to fly in such a direction that it will get warm as soon as possible. In what direction should it fly?
10. Calculate the angle between the normals to the surface $xy = z^2$ at the points (4, 1, 2) and (3, 3, –3). (*U.K.T.U. 2012*)

11. If θ is the acute angle between the surfaces $xy^2z = 3x + z^2$ and $3x^2 - y^2 + 2z = 1$ at the point $(1, -2, 1)$, show that $\cos\theta = \frac{3}{7\sqrt{6}}$.

12. Find the constants a and b so that the surface $ax^2 - byz = (a+2)x$ is orthogonal to the surface $4x^2y + z^3 = 4$ at the point $(1, -1, 2)$.
[**Hint.** The point $P(1, -1, 2)$ lies on both the surfaces and $(\text{grad } \phi_1)_P \cdot (\text{grad } \phi_2)_P = 0$]

13. Find the angle between the tangent planes to the surfaces $x \log z = y^2 - 1$ and $x^2y = 2 - z$ at the point $(1, 1, 1)$.

14. Find the directional derivative of $\vec{V}^2$ where $\vec{V} = xy^2\hat{i} + zy^2\hat{j} + xz^2\hat{k}$ at the point $(2, 0, 3)$ in the direction of the outward normal to the sphere $x^2 + y^2 + z^2 = 14$ at the point $(3, 2, 1)$. *(M.T.U. 2013)*

Answers

1. (*i*) $2x\hat{i} + z\hat{j} + y\hat{k}$ (*ii*) $3(x^2 + yz)\hat{i} + 3(y^2 + xz)\hat{j} + 3xy\hat{k}$ (*iii*) $\frac{2(x\hat{i} + y\hat{j} + z\hat{k})}{x^2 + y^2 + z^2}$

4. (*i*) $-\frac{1}{\sqrt{11}}(\hat{i} + 3\hat{j} - \hat{k})$ (*ii*) $\frac{1}{3}(-\hat{i} + 2\hat{j} + 2\hat{k})$ **5.** (*i*) $-\frac{11}{3}$ (*ii*) $\frac{14}{3}$
(*iii*) $\frac{37}{3}$ (*iv*) 144 (*v*) $\frac{15}{\sqrt{17}}$ (*vi*) $\frac{-9}{49\sqrt{14}}$
(*vii*) $-4\sqrt{41}\,e^{-28}$ **6.** $-\frac{1}{2}$ **7.** $96(\hat{i} + 3\hat{j} - 3\hat{k});\ 96\sqrt{19}$

8. 11 **9.** $\frac{1}{3}(2\hat{i} + 2\hat{j} - \hat{k})$ **10.** $\cos^{-1}\left(\frac{1}{\sqrt{22}}\right)$ **12.** $a = 2.5, b = 1$

13. $\cos^{-1}\left(\frac{1}{\sqrt{30}}\right)$ **14.** $\frac{1404}{\sqrt{14}}$.

13.15. DIVERGENCE OF A VECTOR POINT FUNCTION *(U.P.T.U. 2008)*

The divergence of a differentiable vector point function $\vec{V}$ is denoted by div $\vec{V}$ and is defined as

$$\text{div } \vec{V} = \nabla \cdot \vec{V} = \left(\hat{i}\frac{\partial}{\partial x} + \hat{j}\frac{\partial}{\partial y} + \hat{k}\frac{\partial}{\partial z}\right) \cdot \vec{V} = \hat{i} \cdot \frac{\partial \vec{V}}{\partial x} + \hat{j} \cdot \frac{\partial \vec{V}}{\partial y} + \hat{k} \cdot \frac{\partial \vec{V}}{\partial z}.$$

Obviously, the divergence of a vector point function is a scalar point function.

If $$\vec{V} = V_1\hat{i} + V_2\hat{j} + V_3\hat{k}$$

then $$\text{div } \vec{V} = \nabla \cdot \vec{V} = \left(\hat{i}\frac{\partial}{\partial x} + \hat{j}\frac{\partial}{\partial y} + \hat{k}\frac{\partial}{\partial z}\right) \cdot [V_1\hat{i} + V_2\hat{j} + V_3\hat{k}] = \frac{\partial V_1}{\partial x} + \frac{\partial V_2}{\partial y} + \frac{\partial V_3}{\partial z}.$$

Since $\hat{i} \cdot \hat{i} = \hat{j} \cdot \hat{j} = \hat{k} \cdot \hat{k} = 1$ and $\hat{i} \cdot \hat{j} = \hat{j} \cdot \hat{k} = \hat{k} \cdot \hat{i} = 0$.

13.16. CURL OF A VECTOR POINT FUNCTION *(G.B.T.U. 2012; U.P.T.U. 2008)*

The curl (or rotation) of a differentiable vector point function $\vec{V}$ is denoted by curl $\vec{V}$ and is defined as

$$\text{curl } \vec{V} = \nabla \times \vec{V} = \left(\hat{i}\frac{\partial}{\partial x} + \hat{j}\frac{\partial}{\partial y} + \hat{k}\frac{\partial}{\partial z}\right) \times \vec{V} = \hat{i} \times \frac{\partial \vec{V}}{\partial x} + \hat{j} \times \frac{\partial \vec{V}}{\partial y} + \hat{k} \times \frac{\partial \vec{V}}{\partial z}.$$

Obviously, the curl of a vector point function is a vector point function.

If $\vec{V} = V_1\hat{i} + V_2\hat{j} + V_3\hat{k}$

then $$\text{curl } \vec{V} = \nabla \times \vec{V} = \left(\hat{i}\frac{\partial}{\partial x} + \hat{j}\frac{\partial}{\partial y} + \hat{k}\frac{\partial}{\partial z}\right) \times [V_1\hat{i} + V_2\hat{j} + V_3\hat{k}]$$

$$= \begin{vmatrix} \hat{i} & \hat{j} & \hat{k} \\ \frac{\partial}{\partial x} & \frac{\partial}{\partial y} & \frac{\partial}{\partial z} \\ V_1 & V_2 & V_3 \end{vmatrix} = \hat{i}\left(\frac{\partial V_3}{\partial y} - \frac{\partial V_2}{\partial z}\right) + \hat{j}\left(\frac{\partial V_1}{\partial z} - \frac{\partial V_3}{\partial x}\right) + \hat{k}\left(\frac{\partial V_2}{\partial x} - \frac{\partial V_1}{\partial y}\right).$$

ILLUSTRATIVE EXAMPLES

Example 1. *If $\vec{r} = x\hat{i} + y\hat{j} + z\hat{k}$, show that*

(i) div $\vec{r} = 3$ *(ii) curl $\vec{r} = \vec{0}$.*

Sol. (*i*) $\text{div } \vec{r} = \nabla \cdot \vec{r} = \frac{\partial}{\partial x}(x) + \frac{\partial}{\partial y}(y) + \frac{\partial}{\partial z}(z) = 1 + 1 + 1 = 3.$

(*ii*) $$\text{curl } \vec{r} = \nabla \times \vec{r} = \begin{vmatrix} \hat{i} & \hat{j} & \hat{k} \\ \frac{\partial}{\partial x} & \frac{\partial}{\partial y} & \frac{\partial}{\partial z} \\ x & y & z \end{vmatrix}$$

$$= \hat{i}\left[\frac{\partial}{\partial y}(z) - \frac{\partial}{\partial z}(y)\right] + \hat{j}\left[\frac{\partial}{\partial z}(x) - \frac{\partial}{\partial x}(z)\right] + \hat{k}\left[\frac{\partial}{\partial x}(y) - \frac{\partial}{\partial y}(x)\right]$$

$$= \hat{i}(0) + \hat{j}(0) + \hat{k}(0) = \vec{0}.$$

Example 2. *Find the divergence and curl of the vector $\vec{V} = (xyz)\hat{i} + (3x^2y)\hat{j} + (xz^2 - y^2z)\hat{k}$ at the point (2, –1, 1).*

Sol. $$\text{div } \vec{V} = \frac{\partial}{\partial x}(xyz) + \frac{\partial}{\partial y}(3x^2y) + \frac{\partial}{\partial z}(xz^2 - y^2z)$$

$$= yz + 3x^2 + 2xz - y^2 = -1 + 12 + 4 - 1 = 14 \text{ at } (2, -1, 1)$$

$$\text{curl } \vec{V} = \begin{vmatrix} \hat{i} & \hat{j} & \hat{k} \\ \frac{\partial}{\partial x} & \frac{\partial}{\partial y} & \frac{\partial}{\partial z} \\ xyz & 3x^2y & xz^2 - y^2z \end{vmatrix} = \hat{i}(-2yz - 0) + \hat{j}(xy - z^2) + \hat{k}(6xy - xz)$$

$$= 2\hat{i} - 3\hat{j} - 14\hat{k} \text{ at } (2, -1, 1).$$

Example 3. *Find div $\vec{F}$ and curl $\vec{F}$ where $\vec{F}$ = grad $(x^3 + y^3 + z^3 - 3xyz)$.* (M.D.U. 2013)

Sol. Let $\phi = x^3 + y^3 + z^3 - 3xyz$, then

$$\vec{F} = \text{grad } \phi = \left(\hat{i}\frac{\partial}{\partial x} + \hat{j}\frac{\partial}{\partial y} + \hat{k}\frac{\partial}{\partial z}\right)(x^3 + y^3 + z^3 - 3xyz)$$

$$= (3x^2 - 3yz)\hat{i} + (3y^2 - 3zx)\hat{j} + (3z^2 - 3xy)\hat{k}$$

$\therefore$ $$\text{div } \vec{F} = \frac{\partial}{\partial x}(3x^2 - 3yz) + \frac{\partial}{\partial y}(3y^2 - 3zx) + \frac{\partial}{\partial x}(3z^2 - 3xy)$$
$$= 6x + 6y + 6z = 6(x + y + z)$$

and $$\text{curl } \vec{F} = \begin{vmatrix} \hat{i} & \hat{j} & \hat{k} \\ \frac{\partial}{\partial x} & \frac{\partial}{\partial y} & \frac{\partial}{\partial z} \\ 3x^2 - 3yz & 3y^2 - 3zx & 3z^2 - 3xy \end{vmatrix}$$
$$= \hat{i}(-3x + 3x) + \hat{j}(-3y + 3y) + \hat{k}(-3z + 3z) = \vec{0}.$$

Example 4. *Find curl (curl $\vec{V}$) where $\vec{V} = (2xz^2)\hat{i} - yz\hat{j} + 3xz^3\hat{k}$, at (1, 1, 1).*

Sol. Here, $$\vec{V} = (2xz^2)\hat{i} - yz\hat{j} + 3xz^3\hat{k}$$

$\therefore$ $$\text{curl } \vec{V} = \begin{vmatrix} \hat{i} & \hat{j} & \hat{k} \\ \frac{\partial}{\partial x} & \frac{\partial}{\partial y} & \frac{\partial}{\partial z} \\ 2xz^2 & -yz & 3xz^3 \end{vmatrix}$$
$$= \hat{i}\left\{\frac{\partial}{\partial y}(3xz^3) - \frac{\partial}{\partial z}(-yz)\right\} - \hat{j}\left\{\frac{\partial}{\partial x}(3xz^3) - \frac{\partial}{\partial z}(2xz^2)\right\}$$
$$+ \hat{k}\left\{\frac{\partial}{\partial x}(-yz) - \frac{\partial}{\partial y}(2xz^2)\right\}$$
$$= \hat{i}(0 + y) - \hat{j}(3z^3 - 4xz) + \hat{k}(0 - 0) = y\hat{i} + (4xz - 3z^3)\hat{j}$$

$$\text{curl (curl } \vec{V}) = \text{curl } \{y\hat{i} + (4xz - 3z^3)\hat{j}\}$$
$$= \begin{vmatrix} \hat{i} & \hat{j} & \hat{k} \\ \frac{\partial}{\partial x} & \frac{\partial}{\partial y} & \frac{\partial}{\partial z} \\ y & 4xz - 3z^3 & 0 \end{vmatrix}$$
$$= \hat{i}\left\{\frac{\partial}{\partial y}(0) - \frac{\partial}{\partial z}(4xz - 3z^3)\right\} - \hat{j}\left\{\frac{\partial}{\partial x}(0) - \frac{\partial}{\partial z}(y)\right\}$$
$$+ \hat{k}\left\{\frac{\partial}{\partial x}(4xz - 3z^3) - \frac{\partial}{\partial y}(y)\right\}$$
$$= \hat{i}\left(0 - (4x - 9z^2)\right) - \hat{j}(0 - 0) + \hat{k}(4z - 1)$$
$$= (9z^2 - 4x)\hat{i} + (4z - 1)\hat{k} = 5\hat{i} + 3\hat{k} \text{ at } (1, 1, 1).$$

Example 5. *Let $\vec{r} = x\hat{i} + y\hat{j} + z\hat{k}$ and $\vec{a}$ be a constant vector, find the value of div $\left(\frac{\vec{a} \times \vec{r}}{r^n}\right)$.*

Sol. $$\vec{r} = x\hat{i} + y\hat{j} + z\hat{k} \quad \Rightarrow \quad r = |\vec{r}| = \sqrt{x^2 + y^2 + z^2}$$

Let $$\vec{a} = a_1\hat{i} + a_2\hat{j} + a_3\hat{k}.$$

$$\vec{a} \times \vec{r} = \begin{vmatrix} \hat{i} & \hat{j} & \hat{k} \\ a_1 & a_2 & a_3 \\ x & y & z \end{vmatrix} = (a_2 z - a_3 y)\,\hat{i} + (a_3 x - a_1 z)\,\hat{j} + (a_1 y - a_2 x)\,\hat{k}$$

$$\frac{\vec{a} \times \vec{r}}{r^n} = \frac{(a_2 z - a_3 y)\hat{i} + (a_3 x - a_1 z)\hat{j} + (a_1 y - a_2 x)\hat{k}}{(x^2 + y^2 + z^2)^{n/2}}$$

$$\therefore \quad \text{div}\left(\frac{\vec{a} \times \vec{r}}{r^n}\right) = \nabla \cdot \frac{\vec{a} \times \vec{r}}{r^n}$$

$$= \left(\hat{i}\frac{\partial}{\partial x} + \hat{j}\frac{\partial}{\partial y} + \hat{k}\frac{\partial}{\partial z}\right) \cdot \frac{(a_2 z - a_3 y)\hat{i} + (a_3 x - a_1 z)\hat{j} + (a_1 y - a_2 x)\hat{k}}{(x^2 + y^2 + z^2)^{n/2}}$$

$$= \frac{\partial}{\partial x}\left\{\frac{a_2 z - a_3 y}{(x^2 + y^2 + z^2)^{n/2}}\right\} + \frac{\partial}{\partial y}\left\{\frac{a_3 x - a_1 z}{(x^2 + y^2 + z^2)^{n/2}}\right\} + \frac{\partial}{\partial z}\left\{\frac{a_1 y - a_2 x}{(x^2 + y^2 + z^2)^{n/2}}\right\}$$

$$= (a_2 z - a_3 y) \cdot \left(-\frac{n}{2}\right)(x^2 + y^2 + z^2)^{-\frac{n}{2} - 1} \cdot 2x$$

$$+ (a_3 x - a_1 z)\left(-\frac{n}{2}\right)(x^2 + y^2 + z^2)^{-\frac{n}{2} - 1} \cdot 2y$$

$$+ (a_1 y - a_2 x)\left(-\frac{n}{2}\right)(x^2 + y^2 + z^2)^{-\frac{n}{2} - 1} \cdot 2z$$

$$= \frac{-n}{(x^2 + y^2 + z^2)^{\frac{n}{2} + 1}}\,[(a_2 z - a_3 y)\,x + (a_3 x - a_1 z)\,y + (a_1 y - a_2 x)\,z]$$

$$= \frac{-n}{(x^2 + y^2 + z^2)^{\frac{n}{2} + 1}}\,[0] = 0$$

Hence, $\text{div}\left(\frac{\vec{a} \times \vec{r}}{r^n}\right) = 0.$

Example 6. *Find the directional derivative of div $(\vec{u})$ at the point (1, 2, 2) in the direction of the outer normal to the sphere $x^2 + y^2 + z^2 = 9$ for $\vec{u} = x^4\hat{i} + y^4\hat{j} + z^4\hat{k}$.*

Sol. Here, $\vec{u} = x^4\hat{i} + y^4\hat{j} + z^4\hat{k}$

$$\therefore \quad \text{div}\,(\vec{u}) = \nabla \cdot \vec{u} = \left(\hat{i}\frac{\partial}{\partial x} + \hat{j}\frac{\partial}{\partial y} + \hat{k}\frac{\partial}{\partial z}\right) \cdot (x^4\hat{i} + y^4\hat{j} + z^4\hat{k})$$

$$= \frac{\partial}{\partial x}(x^4) + \frac{\partial}{\partial y}(y^4) + \frac{\partial}{\partial z}(z^4)$$

$$= 4\,(x^3 + y^3 + z^3)$$

Directional derivative of div $\vec{u} = \nabla (4x^3 + 4y^3 + 4z^3)$

$$= \left(\hat{i}\frac{\partial}{\partial x} + \hat{j}\frac{\partial}{\partial y} + \hat{k}\frac{\partial}{\partial z}\right)(4x^3 + 4y^3 + 4z^3)$$

$$= 12\,(x^2\hat{i} + y^2\hat{j} + z^2\hat{k})$$

$$= 12\,(\hat{i} + 4\hat{j} + 4\hat{k}) \text{ at } (1, 2, 2)$$

Outer normal to the sphere $= \nabla (x^2 + y^2 + z^2 - 9)$

$$= \left(\hat{i}\frac{\partial}{\partial x} + \hat{j}\frac{\partial}{\partial y} + \hat{k}\frac{\partial}{\partial z}\right)(x^2 + y^2 + z^2 - 9)$$

$$= \hat{i}\,(2x) + \hat{j}\,(2y) + \hat{k}\,(2z)$$

$$= 2(x\hat{i} + y\hat{j} + z\hat{k})$$

$$= 2(\hat{i} + 2\hat{j} + 2\hat{k}) \text{ at } (1, 2, 2)$$

$$= 2\hat{i} + 4\hat{j} + 4\hat{k}$$

Unit outer normal to the sphere at (1, 2, 2) is

$$\hat{n} = \frac{2\hat{i} + 4\hat{j} + 4\hat{k}}{\sqrt{4 + 16 + 16}} = \frac{2\hat{i} + 4\hat{j} + 4\hat{k}}{6}$$

$\therefore$ Directional derivative of div $\vec{u}$ at (1, 2, 2) in the direction of outer normal

$$= 12\,(\hat{i} + 4\hat{j} + 4\hat{k}) \cdot \frac{2\hat{i} + 4\hat{j} + 4\hat{k}}{6} = 2\,(2 + 16 + 16) = 68$$

13.17. PHYSICAL INTERPRETATION OF DIVERGENCE

(G.B.T.U. 2010; K.U.K. 2012; P.T.U. May 2010)

Consider a fluid having density $\rho = \rho\,(x, y, z, t)$ and velocity $\vec{v} = \vec{v}\,(x, y, z, t)$ at a point (x, y, z) at time t. Let $\vec{V} = \rho\vec{v}$, then $\vec{V}$ is a vector having the same direction as $\vec{v}$ and magnitude $\rho\,|\,\vec{v}\,|$. It is known as *flux*. Its direction gives the direction of the fluid flow, and its magnitude gives the mass of the fluid crossing per unit time a unit area placed perpendicular to the direction of flow.

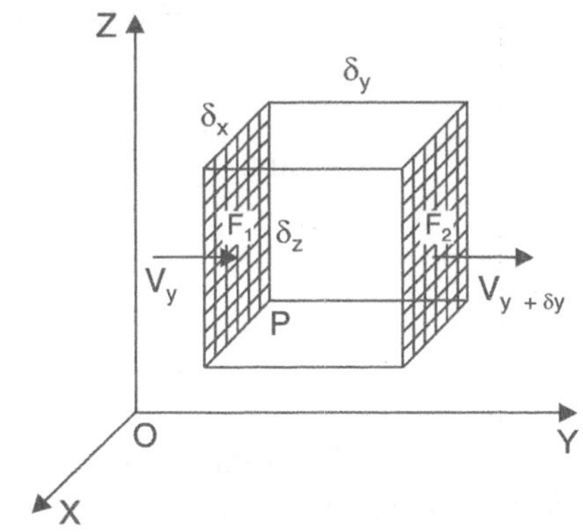

Consider the motion of the fluid having velocity $\vec{V} = V_x\hat{i} + V_y\hat{j} + V_z\hat{k}$ at a point P(x, y, z). Consider a small parallelopiped with edges δx, δy, δz parallel to the axes with one of its corners at P.

The mass of the fluid entering through the face F_1 per unit time is $V_y\ \delta x\ \delta z$ and that flowing out through the opposite face F_2 is $V_{y+\delta y}\ \delta x\ \delta z = \left(V_y + \frac{\partial V_y}{\partial y}\,\partial y\right)\delta x\ \delta z$

by using Taylor's series.

$\therefore$ The net decrease in the mass of fluid flowing across these two faces

$$= \left(V_y + \frac{\partial V_y}{\partial y}\partial y\right)\delta x\,\delta z - V_y\,\delta x\,\delta z = \frac{\partial V_y}{\partial y}\delta x\,\delta y\,\delta z$$

Similarly, considering the other two pairs of faces, we get the total decrease in the mass of fluid inside the parallelopiped per unit time $= \left(\frac{\partial V_x}{\partial x} + \frac{\partial V_y}{\partial y} + \frac{\partial V_z}{\partial z}\right)\delta x\,\delta y\,\delta z.$

Dividing this by the volume $\delta x\,\delta y\,\delta z$ of the parallelopiped, we have the rate of loss of fluid per unit time $= \frac{\partial V_x}{\partial x} + \frac{\partial V_y}{\partial y} + \frac{\partial V_z}{\partial z} = \text{div}\,\vec{V}$

Hence div $\vec{V}$ gives the rate of outflow per unit volume at a point of the fluid.

Note. If div $\vec{V} = 0$ everywhere in some region R of space, then $\vec{V}$ is called *solenoidal* vector point function.

13.18. PHYSICAL INTERPRETATION OF CURL

(*P.T.U. May 2010, Dec. 2011; G.B.T.U. 2010; M.T.U. 2011*)

Consider a rigid body rotating about a fixed axis through O with uniform angular velocity

$$\vec{\omega} = \omega_1\hat{i} + \omega_2\hat{j} + \omega_3\hat{k}$$

The velocity $\vec{V}$ of any point P(x, y, z) on the body is given by $\vec{V} = \vec{\omega} \times \vec{r}$, where $\vec{r} = x\hat{i} + y\hat{j} + z\hat{k}$ is the position vector of P.

$$\therefore \quad \vec{V} = \vec{\omega} \times \vec{r} = \begin{vmatrix} \hat{i} & \hat{j} & \hat{k} \\ \omega_1 & \omega_2 & \omega_3 \\ x & y & z \end{vmatrix}$$

$$= (\omega_2 z - \omega_3 y)\hat{i} + (\omega_3 x - \omega_1 z)\hat{j} + (\omega_1 y - \omega_2 x)\hat{k}$$

$$\text{curl}\,\vec{V} = \nabla \times \vec{V} = \begin{vmatrix} \hat{i} & \hat{j} & \hat{k} \\ \frac{\partial}{\partial x} & \frac{\partial}{\partial y} & \frac{\partial}{\partial z} \\ \omega_2 z - \omega_3 y & \omega_3 x - \omega_1 z & \omega_1 y - \omega_2 x \end{vmatrix}$$

$$= (\omega_1 + \omega_1)\hat{i} + (\omega_2 + \omega_2)\hat{j} + (\omega_3 + \omega_3)\hat{k} \quad [\because \ \omega_1, \omega_2, \omega_3 \text{ are constants}]$$

$$= 2(\omega_1\hat{i} + \omega_2\hat{j} + \omega_3\hat{k}) = 2\vec{\omega} \ \Rightarrow \ \vec{\omega} = \tfrac{1}{2}\,\text{curl}\,\vec{V}.$$

Thus, the angular velocity at any point is equal to half the curl of the linear velocity at that point of the body.

Note. If curl $\vec{V} = \vec{0}$, then $\vec{V}$ is said to be an **irrotational vector,** otherwise **rotational.**

13.19. PROPERTIES OF DIVERGENCE AND CURL

1. For a constant vector $\vec{a}$, div $\vec{a} = 0$, curl $\vec{a} = \vec{0}$

2. **div** $(\vec{\mathbf{A}} + \vec{\mathbf{B}}) = \textbf{div}\, \vec{\mathbf{A}} + \textbf{div}\, \vec{\mathbf{B}}$ **or** $\nabla \cdot (\vec{\mathbf{A}} + \vec{\mathbf{B}}) = \nabla \cdot \vec{\mathbf{A}} + \nabla \cdot \vec{\mathbf{B}}$

Proof. $\text{div}\,(\vec{A} + \vec{B}) = \nabla \cdot (\vec{A} + \vec{B}) = \left(\hat{i}\frac{\partial}{\partial x} + \hat{j}\frac{\partial}{\partial y} + \hat{k}\frac{\partial}{\partial z}\right) \cdot (\vec{A} + \vec{B})$

$$= \hat{i} \cdot \frac{\partial}{\partial x}(\vec{A} + \vec{B}) + \hat{j} \cdot \frac{\partial}{\partial y}(\vec{A} + \vec{B}) + \hat{k} \cdot \frac{\partial}{\partial z}(\vec{A} + \vec{B})$$

$$= \hat{i} \cdot \left(\frac{\partial \vec{A}}{\partial x} + \frac{\partial \vec{B}}{\partial x}\right) + \hat{j} \cdot \left(\frac{\partial \vec{A}}{\partial y} + \frac{\partial \vec{B}}{\partial y}\right) + \hat{k} \cdot \left(\frac{\partial \vec{A}}{\partial z} + \frac{\partial \vec{B}}{\partial z}\right)$$

$$= \left(\hat{i} \cdot \frac{\partial \vec{A}}{\partial x} + \hat{j} \cdot \frac{\partial \vec{A}}{\partial y} + \hat{k} \cdot \frac{\partial \vec{A}}{\partial z}\right) + \left(\hat{i} \cdot \frac{\partial \vec{B}}{\partial x} + \hat{j} \cdot \frac{\partial \vec{B}}{\partial y} + \hat{k} \cdot \frac{\partial \vec{B}}{\partial z}\right)$$

$$= \nabla \cdot \vec{A} + \nabla \cdot \vec{B} = \text{div}\, \vec{A} + \text{div}\, \vec{B}.$$

3. **curl** $(\vec{\mathbf{A}} + \vec{\mathbf{B}}) = \textbf{curl}\, \vec{\mathbf{A}} + \textbf{curl}\, \vec{\mathbf{B}}$ **or** $\nabla \times (\vec{\mathbf{A}} + \vec{\mathbf{B}}) = \nabla \times \vec{\mathbf{A}} + \nabla \times \vec{\mathbf{B}}$

Proof. $\text{curl}\,(\vec{A} + \vec{B}) = \nabla \times (\vec{A} + \vec{B}) = \left(\hat{i}\frac{\partial}{\partial x} + \hat{j}\frac{\partial}{\partial y} + \hat{k}\frac{\partial}{\partial z}\right) \times (\vec{A} + \vec{B})$

$$= \sum \hat{i} \times \frac{\partial}{\partial x}(\vec{A} + \vec{B}) = \sum \hat{i} \times \left(\frac{\partial \vec{A}}{\partial x} + \frac{\partial \vec{B}}{\partial x}\right)$$

$$= \sum \hat{i} \times \frac{\partial \vec{A}}{\partial x} + \sum \hat{i} \times \frac{\partial \vec{B}}{\partial x} = \nabla \times \vec{A} + \nabla \times \vec{B} = \text{curl}\, \vec{A} + \text{curl}\, \vec{B}.$$

4. If $\vec{A}$ is a vector function and ϕ is a scalar function, then

div $(\phi \vec{A}) = \phi\, \textbf{div}\, \vec{\mathbf{A}} + (\textbf{grad}\, \phi) \cdot \vec{A}$ **or** $\nabla \cdot (\phi \vec{A}) = \phi\,(\nabla \cdot \vec{\mathbf{A}}) + (\nabla \phi) \cdot \vec{A}$

Proof. $\text{div}\,(\phi \vec{A}) = \nabla \cdot (\phi \vec{A}) = \left(\hat{i}\frac{\partial}{\partial x} + \hat{j}\frac{\partial}{\partial y} + \hat{k}\frac{\partial}{\partial z}\right) \cdot (\phi \vec{A})$

$$= \hat{i} \cdot \frac{\partial}{\partial x}(\phi \vec{A}) + \hat{j} \cdot \frac{\partial}{\partial y}(\phi \vec{A}) + \hat{k} \cdot \frac{\partial}{\partial z}(\phi \vec{A})$$

$$= \sum \hat{i} \cdot \frac{\partial}{\partial x}(\phi \vec{A}) = \sum \hat{i} \cdot \left(\phi \frac{\partial \vec{A}}{\partial x} + \frac{\partial \phi}{\partial x}\vec{A}\right) = \phi \sum \left(\hat{i} \cdot \frac{\partial \vec{A}}{\partial x}\right) + \sum \left(\hat{i}\,\frac{\partial \phi}{\partial x}\right) \cdot \vec{A}$$

$$= \phi\,(\nabla \cdot \vec{A}) + (\nabla \phi) \cdot \vec{A} = \phi\, \text{div}\, \vec{A} + (\text{grad}\, \phi) \cdot \vec{A}.$$

5. If A is a vector function and ϕ is a scalar function, then

curl $(\phi \vec{\mathbf{A}})$ = $(\textbf{grad}\, \phi) \times \vec{A}$ + $\phi\, \textbf{curl}\, \vec{\mathbf{A}}$ **or** $\nabla \times (\phi \vec{\mathbf{A}}) = (\nabla \phi) \times \vec{\mathbf{A}} + \phi\,(\nabla \times \vec{\mathbf{A}})$

(P.T.U. Jan. 2010)

Proof. $\text{curl}\,(\phi \vec{A}) = \nabla \times (\phi \vec{A}) = \left(\hat{i}\frac{\partial}{\partial x} + \hat{j}\frac{\partial}{\partial y} + \hat{k}\frac{\partial}{\partial z}\right) \times (\phi \vec{A})$

$$= \sum \hat{i} \times \frac{\partial}{\partial x}(\phi \vec{A}) = \sum \hat{i} \times \left(\frac{\partial \phi}{\partial x} \vec{A} + \phi \frac{\partial \vec{A}}{\partial x} \right) = \sum \hat{i} \times \frac{\partial \phi}{\partial x} \vec{A} + \phi \sum \hat{i} \times \frac{\partial \vec{A}}{\partial x}$$

$$= \sum \frac{\partial \phi}{\partial x} \hat{i} \times \vec{A} + \phi \sum \hat{i} \times \frac{\partial \vec{A}}{\partial x} \qquad [\because \ \vec{a} \times (m\vec{b}) = (m\vec{a}) \times \vec{b} = m(\vec{a} \times \vec{b})]$$

$$= (\nabla \phi) \times \vec{A} + \phi\,(\nabla \times \vec{A}) = (\text{grad } \phi) \times \vec{A} + \phi \text{ curl } \vec{A}.$$

6. $\nabla (\vec{\mathbf{A}} \cdot \vec{\mathbf{B}}) = (\vec{\mathbf{A}} \cdot \nabla) \vec{\mathbf{B}} + (\vec{\mathbf{B}} \cdot \nabla) \vec{\mathbf{A}} + \vec{\mathbf{A}} \times (\nabla \times \vec{\mathbf{B}}) + \vec{\mathbf{B}} \times (\nabla \times \vec{\mathbf{A}})$

Proof. $\nabla(\vec{A} \cdot \vec{B}) = \sum \hat{i} \frac{\partial}{\partial x}(\vec{A} \cdot \vec{B}) = \sum \hat{i} \left(\vec{A} \cdot \frac{\partial \vec{B}}{\partial x} + \frac{\partial \vec{A}}{\partial x} \cdot \vec{B} \right)$

$$= \sum \left(\vec{A} \cdot \frac{\partial \vec{B}}{\partial x} \right) \hat{i} + \sum \left(\vec{B} \cdot \frac{\partial \vec{A}}{\partial x} \right) \hat{i} \qquad \text{...(1)}$$

Now, we know that $\vec{a} \times (\vec{b} \times \vec{c}) = (\vec{a} \,.\, \vec{c}) \vec{b} - (\vec{a} \,.\, \vec{b}) \vec{c}$

$\Rightarrow \qquad (\vec{a} \cdot \vec{b}) \vec{c} = (\vec{a} \cdot \vec{c}) \vec{b} - \vec{a} \times (\vec{b} \times \vec{c})$

$\therefore \qquad \left(\vec{A} \cdot \frac{\partial \vec{B}}{\partial x} \right) \hat{i} = (\vec{A} \cdot \hat{i}) \frac{\partial \vec{B}}{\partial x} - \vec{A} \times \left(\frac{\partial \vec{B}}{\partial x} \times \hat{i} \right) = (\vec{A} \cdot \hat{i}) \frac{\partial \vec{B}}{\partial x} + \vec{A} \times \left(\hat{i} \times \frac{\partial \vec{B}}{\partial x} \right)$

or $\qquad \sum \left(\vec{A} \cdot \frac{\partial \vec{B}}{\partial x} \right) \hat{i} = \left(\vec{A} \cdot \sum \hat{i} \frac{\partial}{\partial x} \right) \vec{B} + \vec{A} \times \sum \left(\hat{i} \times \frac{\partial \vec{B}}{\partial x} \right)$

$$= (\vec{A} \cdot \nabla) \vec{B} + \vec{A} \times (\nabla \times \vec{B}) \qquad \text{...(2)}$$

Interchanging $\vec{A}$ and $\vec{B}$, $\sum \left(\vec{B} \cdot \frac{\partial \vec{A}}{\partial x} \right) \hat{i} = (\vec{B} \cdot \nabla) \vec{A} + \vec{B} \times (\nabla \times \vec{A})$ ...(3)

Substituting the values from (2) and (3) in (1), we get

$$\nabla(\vec{A} \cdot \vec{B}) = (\vec{A} \cdot \nabla) \vec{B} + (\vec{B} \cdot \nabla) \vec{A} + \vec{A} \times (\nabla \times \vec{B}) + \vec{B} \times (\nabla \times \vec{A}).$$

7. $\nabla \cdot (\vec{\mathbf{A}} \times \vec{\mathbf{B}}) = \vec{\mathbf{B}} \cdot (\nabla \times \vec{\mathbf{A}}) - \vec{\mathbf{A}} \cdot (\nabla \times \vec{\mathbf{B}})$

Or

div $(\vec{\mathbf{A}} \times \vec{\mathbf{B}}) = \vec{\mathbf{B}} \cdot \text{curl } \vec{\mathbf{A}} - \vec{\mathbf{A}} \cdot \text{curl } \vec{\mathbf{B}}$ *(G.B.T.U. 2012; P.T.U. Jan. 2010)*

Proof. $\nabla \cdot (\vec{A} \times \vec{B}) = \sum \hat{i} \cdot \frac{\partial}{\partial x}(\vec{A} \times \vec{B}) = \sum \hat{i} \left(\frac{\partial \vec{A}}{\partial x} \times \vec{B} + \vec{A} \times \frac{\partial \vec{B}}{\partial x} \right)$

$$= \sum \hat{i} \cdot \left(\frac{\partial \vec{A}}{\partial x} \times \vec{B} \right) + \sum \hat{i} \cdot \left(\vec{A} \times \frac{\partial \vec{B}}{\partial x} \right) = \sum \hat{i} \cdot \left(\frac{\partial \vec{A}}{\partial x} \times \vec{B} \right) - \sum \hat{i} \,.\, \left(\frac{\partial \vec{B}}{\partial x} \times \vec{A} \right)$$

$$= \sum \left(\hat{i} \times \frac{\partial \vec{A}}{\partial x} \right) \cdot \vec{B} - \sum \left(\hat{i} \times \frac{\partial \vec{B}}{\partial x} \right) \cdot \vec{A} \qquad [\text{Since } \vec{a} \cdot (\vec{b} \times \vec{c}) = (\vec{a} \times \vec{b}) \cdot \vec{c}\,]$$

$$= (\nabla \times \vec{A}) \cdot \vec{B} - (\nabla \times \vec{B}) \cdot \vec{A} = \vec{B} \cdot (\nabla \times \vec{A}) - \vec{A} \cdot (\nabla \times \vec{B}).$$

8. $\nabla \times (\vec{\mathbf{A}} \times \vec{\mathbf{B}}) = (\nabla \cdot \vec{\mathbf{B}})\vec{\mathbf{A}} - (\nabla \cdot \vec{\mathbf{A}})\vec{\mathbf{B}} + (\vec{\mathbf{B}} \cdot \nabla)\vec{\mathbf{A}} - (\vec{\mathbf{A}} \cdot \nabla)\vec{\mathbf{B}}$

(G.B.T.U. 2013; P.T.U. May 2010; M.D.U. May 2012)

Proof. $\nabla \times (\vec{A} \times \vec{B}) = \sum \hat{i} \times \frac{\partial}{\partial x}(\vec{A} \times \vec{B}) = \sum \hat{i} \times \left(\frac{\partial \vec{A}}{\partial x} \times \vec{B} + \vec{A} \times \frac{\partial \vec{B}}{\partial x}\right)$

$$= \sum \hat{i} \times \left(\frac{\partial \vec{A}}{\partial x} \times \vec{B}\right) + \sum \hat{i} \times \left(\vec{A} \times \frac{\partial \vec{B}}{\partial x}\right)$$

$$= \sum \left[(\hat{i} \cdot \vec{B})\frac{\partial \vec{A}}{\partial x} - \left(\hat{i} \cdot \frac{\partial \vec{A}}{\partial x}\right)\vec{B}\right] + \sum \left[\left(\hat{i} \cdot \frac{\partial \vec{B}}{\partial x}\right)\vec{A} - (\hat{i} \cdot \vec{A})\frac{\partial \vec{B}}{\partial x}\right]$$

[Since $\vec{a} \times (\vec{b} \times \vec{c}) = (\vec{a} \,.\, \vec{c})\vec{b} - (\vec{a} \cdot \vec{b})\vec{c}$]

$$= \sum (\vec{B} \cdot \hat{i})\frac{\partial \vec{A}}{\partial x} - \left(\sum \hat{i} \cdot \frac{\partial \vec{A}}{\partial x}\right)\vec{B} + \left(\sum \hat{i} \cdot \frac{\partial \vec{B}}{\partial x}\right)\vec{A} - \sum (\vec{A} \cdot \hat{i})\frac{\partial \vec{B}}{\partial x}$$

$$= \left(\vec{B} \cdot \sum \hat{i}\frac{\partial}{\partial x}\right)\vec{A} - (\nabla \cdot \vec{A})\vec{B} + (\nabla \cdot \vec{B})\vec{A} - \left(A \cdot \sum \hat{i}\frac{\partial}{\partial x}\right)\vec{B}$$

$$= (\vec{B} \cdot \nabla)\vec{A} - (\nabla \cdot \vec{A})\vec{B} + (\nabla \cdot \vec{B})\vec{A} - (\vec{A} \cdot \nabla)\vec{B}$$

$$= (\nabla \cdot \vec{B})\vec{A} - (\nabla \cdot \vec{A})\vec{B} + (\vec{B} \cdot \nabla)\vec{A} - (\vec{A} \cdot \nabla)\vec{B}.$$

13.20. REPEATED OPERATIONS BY ∇

Let $\phi\ (x, y, z)$ and $\vec{V}\ (x, y, z)$ be scalar and vector point functions respectively.

Since grad ϕ and curl $\vec{V}$ are also vector point functions, we can find their divergence as well as curl, whereas div $\vec{V}$ being a scalar point function, we can find its gradient only.

1. **div (grad ϕ) = $\nabla^2\phi$ where** $\nabla^2 = \frac{\partial^2}{\partial \mathbf{x}^2} + \frac{\partial^2}{\partial \mathbf{y}^2} + \frac{\partial^2}{\partial \mathbf{z}^2}$

Proof. div (grad ϕ) $= \nabla \,.\, (\nabla\phi) = \left(\hat{i}\frac{\partial}{\partial x} + \hat{j}\frac{\partial}{\partial y} + \hat{k}\frac{\partial}{\partial z}\right) \cdot \left(\hat{i}\frac{\partial \phi}{\partial x} + \hat{j}\frac{\partial \phi}{\partial y} + \hat{k}\frac{\partial \phi}{\partial z}\right)$

$$= \frac{\partial}{\partial x}\left(\frac{\partial \phi}{\partial x}\right) + \frac{\partial}{\partial y}\left(\frac{\partial \phi}{\partial y}\right) + \frac{\partial}{\partial z}\left(\frac{\partial \phi}{\partial z}\right) = \frac{\partial^2 \phi}{\partial x^2} + \frac{\partial^2 \phi}{\partial y^2} + \frac{\partial^2 \phi}{\partial z^2} = \left(\frac{\partial^2}{\partial x^2} + \frac{\partial^2}{\partial y^2} + \frac{\partial^2}{\partial z^2}\right)\phi$$

$$= \nabla^2\phi \quad \text{where} \quad \nabla^2 = \frac{\partial^2}{\partial x^2} + \frac{\partial^2}{\partial y^2} + \frac{\partial^2}{\partial z^2}$$

∇^2 is called the *Laplacian operator* and $\nabla^2\phi = 0$ is called *Laplace's equation*.

2. **curl (grad φ) = ∇ × ∇φ = $\vec{0}$** *(G.B.T.U. 2011)*

Proof. curl (grad φ) $= \nabla \times \nabla\phi = \nabla \times \left(\hat{i}\dfrac{\partial\phi}{\partial x} + \hat{j}\dfrac{\partial\phi}{\partial y} + \hat{k}\dfrac{\partial\phi}{\partial z}\right)$

$$= \begin{vmatrix} \hat{i} & \hat{j} & \hat{k} \\ \dfrac{\partial}{\partial x} & \dfrac{\partial}{\partial y} & \dfrac{\partial}{\partial z} \\ \dfrac{\partial\phi}{\partial x} & \dfrac{\partial\phi}{\partial y} & \dfrac{\partial\phi}{\partial z} \end{vmatrix} = \sum \hat{i}\left(\frac{\partial^2\phi}{\partial y\partial z} - \frac{\partial^2\phi}{\partial z\partial y}\right) = \vec{0}$$

3. **div (curl $\vec{V}$) = ∇ · (∇ × $\vec{V}$) = 0.** *(G.B.T.U. 2011; M.D.U. May 2011)*

Proof. Let $\vec{V} = V_1\hat{i} + V_2\hat{j} + V_3\hat{k}$, then curl $\vec{V} = \nabla \times \vec{V} = \begin{vmatrix} \hat{i} & \hat{j} & \hat{k} \\ \dfrac{\partial}{\partial x} & \dfrac{\partial}{\partial y} & \dfrac{\partial}{\partial z} \\ V_1 & V_2 & V_3 \end{vmatrix}$

$$= \hat{i}\left(\frac{\partial V_3}{\partial y} - \frac{\partial V_2}{\partial z}\right) + \hat{j}\left(\frac{\partial V_1}{\partial z} - \frac{\partial V_3}{\partial x}\right) + \hat{k}\left(\frac{\partial V_2}{\partial x} - \frac{\partial V_1}{\partial y}\right)$$

∴ div (curl $\vec{V}$) $= \nabla \cdot (\nabla \times \vec{V})$

$$= \frac{\partial}{\partial x}\left(\frac{\partial V_3}{\partial y} - \frac{\partial V_2}{\partial z}\right) + \frac{\partial}{\partial y}\left(\frac{\partial V_1}{\partial z} - \frac{\partial V_3}{\partial x}\right) + \frac{\partial}{\partial z}\left(\frac{\partial V_2}{\partial x} - \frac{\partial V_1}{\partial y}\right)$$

$$= \left(\frac{\partial^2 V_3}{\partial x\partial y} - \frac{\partial^2 V_2}{\partial x\partial z}\right) + \left(\frac{\partial^2 V_1}{\partial y\partial z} - \frac{\partial^2 V_3}{\partial y\partial x}\right) + \left(\frac{\partial^2 V_2}{\partial z\partial x} - \frac{\partial^2 V_1}{\partial z\partial y}\right) = 0.$$

4. **curl (curl $\vec{V}$) = grad div $\vec{V}$ − ∇² $\vec{V}$** *(M.D.U. 2013)*

or **∇ × (∇ × $\vec{V}$) = ∇ (∇ · $\vec{V}$) − ∇²$\vec{V}$.**

Proof. Let $\vec{V} = V_1\hat{i} + V_2\hat{j} + V_3\hat{k}$

then curl $\vec{V} = \nabla \times \vec{V} = \begin{vmatrix} \hat{i} & \hat{j} & \hat{k} \\ \dfrac{\partial}{\partial x} & \dfrac{\partial}{\partial y} & \dfrac{\partial}{\partial z} \\ V_1 & V_2 & V_3 \end{vmatrix}$

$$= \hat{i}\left(\frac{\partial V_3}{\partial y} - \frac{\partial V_2}{\partial z}\right) + \hat{j}\left(\frac{\partial V_1}{\partial z} - \frac{\partial V_3}{\partial x}\right) + \hat{k}\left(\frac{\partial V_2}{\partial x} - \frac{\partial V_1}{\partial y}\right)$$

∴ curl (curl $\vec{V}$) $= \nabla \times (\nabla \times \vec{V}) = \begin{vmatrix} \hat{i} & \hat{j} & \hat{k} \\ \dfrac{\partial}{\partial x} & \dfrac{\partial}{\partial y} & \dfrac{\partial}{\partial z} \\ \dfrac{\partial V_3}{\partial y} - \dfrac{\partial V_2}{\partial z} & \dfrac{\partial V_1}{\partial z} - \dfrac{\partial V_3}{\partial x} & \dfrac{\partial V_2}{\partial x} - \dfrac{\partial V_1}{\partial y} \end{vmatrix}$

$$= \sum \hat{i}\left\{\frac{\partial}{\partial y}\left(\frac{\partial V_2}{\partial x} - \frac{\partial V_1}{\partial y}\right) - \frac{\partial}{\partial z}\left(\frac{\partial V_1}{\partial z} - \frac{\partial V_3}{\partial x}\right)\right\}$$

$$= \sum \hat{i} \left\{ \frac{\partial}{\partial x}\left(\frac{\partial V_2}{\partial y} + \frac{\partial V_3}{\partial z}\right) - \left(\frac{\partial^2 V_1}{\partial y^2} + \frac{\partial^2 V_1}{\partial z^2}\right) \right\}$$

$$= \sum \hat{i} \left\{ \frac{\partial}{\partial x}\left(\frac{\partial V_1}{\partial x} + \frac{\partial V_2}{\partial y} + \frac{\partial V_3}{\partial z}\right) - \left(\frac{\partial^2 V_1}{\partial x^2} + \frac{\partial^2 V_1}{\partial y^2} + \frac{\partial^2 V_1}{\partial z^2}\right) \right\}$$

$$\left[\because \frac{\partial}{\partial x}\left(\frac{\partial V_1}{\partial x}\right) = \frac{\partial^2 V_1}{\partial x^2}\right]$$

$$= \sum \hat{i} \left\{ \frac{\partial}{\partial x}(\nabla \cdot \vec{V}) - (\nabla^2 V_1) \right\} = \sum \hat{i} \frac{\partial}{\partial x}(\nabla \cdot \vec{V}) - \nabla^2 \Sigma \hat{i} V_1$$

$$= \nabla(\nabla \cdot \vec{V}) - \nabla^2 \vec{V} = \text{grad }(\text{div } \vec{V}) - \nabla^2 \vec{V}.$$

Note 1. The above result can also be written as grad (div $\vec{V}$) = curl (curl $\vec{V}$) + $\nabla^2 \vec{V}$

or $\nabla(\nabla \cdot \vec{V}) = \nabla \times (\nabla \times \vec{V}) + \nabla^2 \vec{V}$.

Note 2. Treating ∇ as a vector, the results of repeated application of ∇ can be easily written down. Thus

$\nabla \cdot \nabla\phi = \nabla^2\phi$ $\quad (\because \vec{a} \cdot \vec{a} = a^2)$

$\nabla \times \nabla\phi = \vec{0}$ $\quad (\because \vec{a} \times \vec{a} = \vec{0})$

$\nabla \cdot (\nabla \times \vec{V}) = 0$ $\quad (\because \vec{a} \cdot (\vec{a} \times \vec{b}) = [\vec{a}\ \vec{a}\ \vec{b}] = 0)$

$\nabla \times (\nabla \times \vec{V}) = \nabla(\nabla \cdot \vec{V}) - \nabla^2 \vec{V}$ (By expanding as a vector triple product)

ILLUSTRATIVE EXAMPLES

Example 1. *A vector field is given by* $\vec{A} = (x^2 + xy^2)\hat{i} + (y^2 + x^2y)\hat{j}$. *Show that the field is irrotational and find the scalar potential.*

Sol. Field $\vec{A}$ is irrotational if curl $\vec{A} = \vec{0}$

Now,
$$\text{curl } \vec{A} = \nabla \times \vec{A} = \begin{vmatrix} \hat{i} & \hat{j} & \hat{k} \\ \dfrac{\partial}{\partial x} & \dfrac{\partial}{\partial y} & \dfrac{\partial}{\partial z} \\ x^2 + xy^2 & y^2 + x^2y & 0 \end{vmatrix}$$

$$= \hat{i}(0 - 0) - \hat{j}(0 - 0) + \hat{k}(2xy - 2xy) = \vec{0}.$$

$\therefore$ Field $\vec{A}$ is irrotational.

If ϕ is the scalar potential, then $\vec{A}$ = grad ϕ $\quad [\because \text{curl (grad } \phi) = \vec{0}]$

$$\Rightarrow \quad (x^2 + xy^2)\hat{i} + (y^2 + x^2y)\hat{j} = \hat{i}\frac{\partial\phi}{\partial x} + \hat{j}\frac{\partial\phi}{\partial y} + \hat{k}\frac{\partial\phi}{\partial z}$$

$$\Rightarrow \quad \frac{\partial\phi}{\partial x} = x^2 + xy^2, \quad \frac{\partial\phi}{\partial y} = y^2 + x^2y, \quad \frac{\partial\phi}{\partial z} = 0$$

$$\therefore \qquad d\phi = \frac{\partial \phi}{\partial x}\, dx + \frac{\partial \phi}{\partial y} dy + \frac{\partial \phi}{\partial z}\, dz$$

$$= (x^2 + xy^2)\, dx + (y^2 + x^2y)dy$$

$$= x^2\, dx + y^2\, dy + xy(y\, dx + x\, dy)$$

$$= d\left(\frac{x^3}{3}\right) + d\left(\frac{y^3}{3}\right) + d\left(\frac{x^2y^2}{2}\right)$$

Integrating, $\qquad \phi = \frac{x^3}{3} + \frac{y^3}{3} + \frac{x^2y^2}{2} + c.$

Example 2. *A fluid motion is given by* $\vec{V} = (y \sin z - \sin x)\,\hat{i} + (x \sin z + 2yz)\hat{j} + (xy \cos z + y^2)\hat{k}$ *is the motion irrotational ? If so, find the velocity potential.*

(P.T.U. May 2010; U.K.T.U. 2011)

Sol. The motion is irrotational if curl $\vec{V} = \vec{0}$

Now, $\qquad \text{curl } \vec{V} = \begin{vmatrix} \hat{i} & \hat{j} & \hat{k} \\ \frac{\partial}{\partial x} & \frac{\partial}{\partial y} & \frac{\partial}{\partial z} \\ y \sin z - \sin x & x \sin z + 2yz & xy \cos z + y^2 \end{vmatrix}$

$$= \hat{i}\left\{\frac{\partial}{\partial y}(xy \cos z + y^2) - \frac{\partial}{\partial z}(x \sin z + 2yz)\right\}$$

$$+ \hat{j}\left\{\frac{\partial}{\partial z}(y \sin z - \sin x) - \frac{\partial}{\partial x}(xy \cos z + y^2)\right\}$$

$$+ \hat{k}\left\{\frac{\partial}{\partial x}(x \sin z + 2yz) - \frac{\partial}{\partial y}(y \sin z - \sin x)\right\}$$

$$= \hat{i}\,(x \cos z + 2y - x \cos z - 2y) + \hat{j}(y \cos z - y \cos z) + \hat{k}\,(\sin z - \sin z)$$

$$= \hat{i}\,(0) + \hat{j}\,(0) + \hat{k}\,(0) = \vec{0}$$

$\therefore$ The motion is irrotational.

Let u be the velocity potential, then $\vec{V} = \text{grad } u \qquad [\because \text{ curl (grad } u) = \vec{0}]$

$$\Rightarrow \quad (y \sin z - \sin x)\,\hat{i} + (x \sin z + 2yz)\,\hat{j} + (xy \cos z + y^2)\,\hat{k}$$

$$= \hat{i}\frac{\partial u}{\partial x} + \hat{j}\frac{\partial u}{\partial y} + \hat{k}\frac{\partial u}{\partial z}$$

$$\Rightarrow \qquad \frac{\partial u}{\partial x} = y \sin z - \sin x, \quad \frac{\partial u}{\partial y} = x \sin z + 2yz, \quad \frac{\partial u}{\partial z} = xy \cos z + y^2$$

$$\therefore \qquad du = \frac{\partial u}{\partial x}\, dx + \frac{\partial u}{\partial y}\, dy + \frac{\partial u}{\partial z}\, dz$$

$$= (y \sin z - \sin x)\, dx + (x \sin z + 2yz)\, dy + (xy \cos z + y^2)\, dz$$
$$= (y \sin z\, dx + x \sin z\, dy + xy \cos z\, dz) + (-\sin x\, dx) + (2yz\, dy + y^2\, dz)$$
$$= d(xy \sin z) + d(\cos x) + d(y^2 z)$$

Integrating, $\quad u = xy \sin z + \cos x + y^2 z + c.$

Example 3. *If the vector* $\vec{F} = (ax^2y + yz)\hat{i} + (xy^2 - xz^2)\hat{j} + (2xyz - 2x^2y^2)\hat{k}$ *is solenoidal, find the value of a. Find also the curl of this solenoidal vector.*

Sol. Here $\vec{F} = (ax^2y + yz)\hat{i} + (xy^2 - xz^2)\,\hat{j} + (2xyz - 2x^2y^2)\hat{k}$

$$\text{div } \vec{F} = \nabla \cdot \vec{F} = \left(\hat{i}\frac{\partial}{\partial x} + \hat{j}\frac{\partial}{\partial y} + \hat{k}\frac{\partial}{\partial z}\right) \cdot \left[(ax^2y + yz)\hat{i} + (xy^2 - xz^2)\hat{j} + (2xyz - 2x^2y^2)\hat{k}\right]$$

$$= \frac{\partial}{\partial x}(ax^2y + yz) + \frac{\partial}{\partial y}(xy^2 - xz^2) + \frac{\partial}{\partial z}(2xyz - 2x^2y^2)$$
$$= 2axy + 2xy + 2xy = 2(a + 2)xy$$

Since $\vec{F}$ is solenoidal, div $\vec{F} = 0$

$\Rightarrow \quad 2(a + 2)xy = 0 \qquad \therefore \quad a = -2$

Now, $\quad \vec{F} = (-2x^2y + yz)\hat{i} + (xy^2 - xz^2)\,\hat{j} + (2xyz - 2x^2y^2)\,\hat{k}$

$$\therefore \quad \text{curl } \vec{F} = \nabla \times \vec{F} = \begin{vmatrix} \hat{i} & \hat{j} & \hat{k} \\ \dfrac{\partial}{\partial x} & \dfrac{\partial}{\partial y} & \dfrac{\partial}{\partial z} \\ -2x^2y + yz & xy^2 - xz^2 & 2xyz - 2x^2y^2 \end{vmatrix}$$

$$= \hat{i}\left[\frac{\partial}{\partial y}(2xyz - 2x^2y^2) - \frac{\partial}{\partial z}(xy^2 - xz^2)\right] - \hat{j}\left[\frac{\partial}{\partial x}(2xyz - 2x^2y^2) - \frac{\partial}{\partial z}(-2x^2y + yz)\right]$$
$$+ \hat{k}\left[\frac{\partial}{\partial x}(xy^2 - xz^2) - \frac{\partial}{\partial y}(-2x^2y + yz)\right]$$

$$= \hat{i}\,(2xz - 4x^2y + 2xz) - \hat{j}\,(2yz - 4xy^2 - y) + \hat{k}\,(y^2 - z^2 + 2x^2 - z)$$
$$= 4x(z - xy)\hat{i} + (y + 4xy^2 - 2yz)\,\hat{j} + (2x^2 + y^2 - z^2 - z)\,\hat{k}.$$

Example 4. *Show that* $r^{\alpha}\vec{R}$ *is an irrotational vector for any value of* α *but it is solenoidal if* $\alpha + 3 = 0$ *where* $\vec{R} = x\hat{i} + y\hat{j} + z\hat{k}$ *and r is the magnitude of* $\vec{R}$. (M.D.U. 2010)

Sol. Let $\quad \vec{V} = r^{\alpha}\,\vec{R} = (x^2 + y^2 + z^2)^{\frac{\alpha}{2}}\,(x\hat{i} + y\hat{j} + z\hat{k})$

$$= x(x^2 + y^2 + z^2)^{\alpha/2}\,\hat{i} + y(x^2 + y^2 + z^2)^{\alpha/2}\,\hat{j} + z(x^2 + y^2 + z^2)^{\alpha/2}\,\hat{k}$$

$$\therefore \quad \text{curl } \vec{V} = \begin{vmatrix} \hat{i} & \hat{j} & \hat{k} \\ \dfrac{\partial}{\partial x} & \dfrac{\partial}{\partial y} & \dfrac{\partial}{\partial z} \\ x\,(x^2 + y^2 + z^2)^{\alpha/2} & y\,(x^2 + y^2 + z^2)^{\alpha/2} & z\,(x^2 + y^2 + z^2)^{\alpha/2} \end{vmatrix}$$

$$= \sum \hat{i}\left\{\frac{\alpha z}{2}(x^2 + y^2 + z^2)^{\frac{\alpha}{2} - 1}.\,2y - \frac{\alpha y}{2}(x^2 + y^2 + z^2)^{\frac{\alpha}{2} - 1}.\,2z\right\}$$

$$= 0\hat{i} + 0\hat{j} + 0\hat{k} = \vec{0}$$

$\Rightarrow \quad \vec{V} = r^\alpha \vec{R}$ is irrotational for any value of α.

Now, $\quad \text{div } \vec{V} = \nabla \cdot (r^\alpha \vec{R})$

$$= r^\alpha (\text{div } \vec{R}) + \text{grad } r^\alpha \cdot \vec{R} \quad ...(1)$$

$$\left[\because \ \text{div } (\phi \vec{A}) = \phi \, (\text{div } \vec{A}) + \text{grad } \phi \cdot \vec{A}\right]$$

and $\quad \text{div } (\vec{R}) = \nabla \cdot (x\hat{i} + y\hat{j} + z\hat{k})$

$$= \frac{\partial}{\partial x}(x) + \frac{\partial}{\partial y}(y) + \frac{\partial}{\partial z}(z) = 1 + 1 + 1 = 3$$

Also, $\quad r^2 = x^2 + y^2 + z^2$ so that $\dfrac{\partial r}{\partial x} = \dfrac{x}{r}, \ \dfrac{\partial r}{\partial y} = \dfrac{y}{r}, \ \dfrac{\partial r}{\partial z} = \dfrac{z}{r}$

$$\text{grad } r = \hat{i}\frac{\partial r}{\partial x} + \hat{j}\frac{\partial r}{\partial y} + \hat{k}\frac{\partial r}{\partial z} = \frac{x\hat{i} + y\hat{j} + z\hat{k}}{r} = \frac{\vec{R}}{r}$$

$$\therefore \quad \text{grad } r^\alpha = \alpha \, r^{\alpha - 1} \text{ grad } r = \alpha r^{\alpha - 1} \frac{\vec{R}}{r} = \alpha r^{\alpha - 2} \vec{R}$$

$\therefore$ From (1), we have

$$\text{div } \vec{V} = r^\alpha (3) + \alpha r^{\alpha - 2} \vec{R} \cdot \vec{R} = 3r^\alpha + \alpha \, r^{\alpha - 2} (x^2 + y^2 + z^2)$$

$$= 3r^\alpha + \alpha \, r^{\alpha - 2} (r^2) = (3 + \alpha) \, r^\alpha$$

Now, $\vec{V}$ is solenoidal if $\text{div } \vec{V} = 0$ *i.e.*, $(3 + \alpha) \, r^\alpha = 0$

$\Rightarrow \quad r^\alpha \vec{R}$ is solenoidal if $\alpha + 3 = 0$.

Note. The first part of the above example may be done using properties of curl of a vector. Thus

$$\text{curl } \vec{V} = \text{curl } (r^\alpha \vec{R})$$

$$= (\text{grad } r^\alpha) \times \vec{R} + r^\alpha (\text{curl } \vec{R}) \quad [\because \ \text{curl } (\phi \vec{A}) = (\text{grad } \phi) \times \vec{A} + \phi (\text{curl } \vec{A})]$$

$$= (\alpha r^{\alpha - 1} \text{ grad } r) \times \vec{R} + r^\alpha (\vec{0}) \quad [\because \ \text{curl } \vec{R} = \vec{0}]$$

$$= \alpha r^{\alpha - 1} \frac{\vec{R}}{r} \times \vec{R} + \vec{0}$$

$$= \alpha r^{\alpha - 2} (\vec{R} \times \vec{R}) = \alpha r^{\alpha - 2} (\vec{0})$$

$$= \vec{0}$$

Example 5. *If $\vec{a}$ is a constant vector and $\vec{r} = x\hat{i} + y\hat{j} + z\hat{k}$, prove that curl $(\vec{a} \times \vec{r}) = 2\vec{a}$.*

(Calicut 2009; K.U.K. 2009; P.T.U. May 2010; M.D.U. May 2010)

Sol. Let $\quad \vec{a} = a_1\hat{i} + a_2\hat{j} + a_3\hat{k}$, where a_1, a_2, a_3 are constants.

$$\vec{a} \times \vec{r} = \begin{vmatrix} \hat{i} & \hat{j} & \hat{k} \\ a_1 & a_2 & a_3 \\ x & y & z \end{vmatrix} = (a_2 z - a_3 y)\hat{i} + (a_3 x - a_1 z)\hat{j} + (a_1 y - a_2 x)\hat{k}$$

$$\text{curl}\,(\vec{a} \times \vec{r}) = \nabla \times (\vec{a} \times \vec{r}) = \begin{vmatrix} \hat{i} & \hat{j} & \hat{k} \\ \dfrac{\partial}{\partial x} & \dfrac{\partial}{\partial y} & \dfrac{\partial}{\partial z} \\ a_2 z - a_3 y & a_3 x - a_1 z & a_1 y - a_2 x \end{vmatrix}$$

$$= (a_1 + a_1)\hat{i} + (a_2 + a_2)\hat{j} + (a_3 + a_3)\hat{k} = 2(a_1\hat{i} + a_2\hat{j} + a_3\hat{k}) = 2\vec{a}.$$

Example 6. *Prove that*

(*i*) $\nabla\,(\vec{a} \cdot \vec{u}) = (\vec{a} \cdot \nabla)\vec{u} + \vec{a} \times (\nabla \times \vec{u})$ (*ii*) $\nabla \times (\vec{a} \times \vec{u}) = (\nabla \cdot \vec{u})\vec{a} - (\vec{a} \cdot \nabla)\vec{u}$

where $\vec{a}$ *is a constant vector.*

Sol. (*i*)
$$\nabla(\vec{a} \cdot \vec{u}) = \sum \hat{i}\frac{\partial}{\partial x}(\vec{a} \cdot \vec{u}) = \sum \hat{i}\left(\vec{a} \cdot \frac{\partial \vec{u}}{\partial x}\right) \quad ...(1)$$

Now
$$\vec{a} \times \left(\hat{i} \times \frac{\partial \vec{u}}{\partial x}\right) = \left(\vec{a} \cdot \frac{\partial \vec{u}}{\partial x}\right)\hat{i} - (\vec{a} \cdot \hat{i})\frac{\partial \vec{u}}{\partial x}$$

$$\Rightarrow \quad \left(\vec{a} \cdot \frac{\partial \vec{u}}{\partial x}\right)\hat{i} = \vec{a} \times \left(\hat{i} \times \frac{\partial \vec{u}}{\partial x}\right) + (\vec{a} \cdot \hat{i})\frac{\partial \vec{u}}{\partial x}$$

∴ From (1), we have
$$\nabla(\vec{a} \cdot \vec{u}) = \sum \vec{a} \times \left(\hat{i} \times \frac{\partial \vec{u}}{\partial x}\right) + \sum (\vec{a} \cdot \hat{i})\frac{\partial \vec{u}}{\partial x}$$

$$= \vec{a} \times (\nabla \times \vec{u}) + (\vec{a} \cdot \nabla)\vec{u} = (\vec{a} \cdot \nabla)\vec{u} + \vec{a} \times (\nabla \times \vec{u}).$$

(*ii*)
$$\nabla \times (\vec{a} \times \vec{u}) = \sum \hat{i}\frac{\partial}{\partial x} \times (\vec{a} \times \vec{u}) = \sum \hat{i} \times \left(\vec{a} \times \frac{\partial \vec{u}}{\partial x}\right)$$

$$= \sum \left(\hat{i} \cdot \frac{\partial \vec{u}}{\partial x}\right)\vec{a} - \sum (\hat{i} \cdot \vec{a})\frac{\partial \vec{u}}{\partial x} = (\nabla \cdot \vec{u})\vec{a} - (\vec{a} \cdot \nabla)\vec{u}.$$

Example 7. *Prove that* $\nabla \cdot (\phi\nabla\psi - \psi\nabla\phi) = \phi\nabla^2\psi - \psi\nabla^2\phi$.

Sol. For a scalar function f and a vector function $\vec{G}$, we know that

$$\nabla \cdot (f\vec{G}) = f(\nabla \cdot \vec{G}) + (\nabla f) \cdot \vec{G}$$

Also
$$\nabla \cdot (\vec{F} - \vec{G}) = \nabla \cdot \vec{F} - \nabla \cdot \vec{G}$$

$$\therefore \quad \nabla\cdot(\phi\nabla\psi-\psi\nabla\phi)=\nabla\cdot(\phi\nabla\psi)-\nabla\cdot(\psi\nabla\phi)$$
$$=[\phi(\nabla\cdot\nabla\psi)+\nabla\phi\cdot\nabla\psi]-[\psi(\nabla\cdot\nabla\phi)+\nabla\psi\cdot\nabla\phi]$$
$$=\phi\nabla^2\psi+\nabla\phi\cdot\nabla\psi-\psi\nabla^2\phi-\nabla\psi\cdot\nabla\phi$$
$$=\phi\nabla^2\psi-\psi\nabla^2\phi \qquad [\because \text{ dot product is commutative}]$$

Example 8. *Prove that*

(*i*) $div\left(\frac{\vec{r}}{r^3}\right)=0$ (Calicut 2009) (*ii*) $\nabla^2(r^n)=n(n+1)\,r^{n-2}$

(U.K.T.U. 2012; K.U.K. Dec. 2013; P.T.U. 2008)

(*iii*) $\nabla^2\left(\frac{x}{r^3}\right)=0$, *where r is the magnitude of* $\vec{r}=x\hat{i}+y\hat{j}+z\hat{k}$.

Sol. Here $r^2=x^2+y^2+z^2$ so that $\frac{\partial r}{\partial x}=\frac{x}{r},\frac{\partial r}{\partial y}=\frac{y}{r},\frac{\partial r}{\partial z}=\frac{z}{r}$

$$\text{grad } r=\hat{i}\frac{\partial r}{\partial x}+\hat{j}\frac{\partial r}{\partial y}+\hat{k}\frac{\partial r}{\partial z}=\frac{\vec{r}}{r}$$

(*i*) Since $\text{div}(\phi\vec{A})=\phi(\text{div }\vec{A})+\text{grad }\phi\cdot\vec{A}$

$$\therefore \quad \text{div}\left(\frac{\vec{r}}{r^3}\right)=\text{div}(r^{-3}\vec{r})=r^{-3}(\text{div }\vec{r})+(\text{grad } r^{-3})\cdot\vec{r}$$
$$=3r^{-3}+(-3r^{-4}\text{ grad } r)\cdot\vec{r} \qquad [\because \text{ div }\vec{r}=3]$$
$$=3r^{-3}+\left(-3r^{-4}\frac{\vec{r}}{r}\right)\cdot\vec{r}=3r^{-3}-3r^{-5}(\vec{r}\cdot\vec{r})=3r^{-3}-3r^{-5}(r^2)=0.$$

(*ii*)
$$\nabla^2(r^n)=\nabla\cdot(\nabla r^n)=\nabla\cdot\left(nr^{n-1}\frac{\vec{r}}{r}\right)=n\nabla\cdot(r^{n-2}\vec{r})$$
$$=n[(\nabla r^{n-2})\cdot\vec{r}+r^{n-2}(\nabla\cdot\vec{r})] \quad [\because \nabla\cdot(\phi\vec{A})=(\nabla\phi)\cdot\vec{A}+\phi(\nabla\cdot\vec{A})]$$
$$=n\left[(n-2)r^{n-3}\frac{\vec{r}}{r}\cdot\vec{r}+r^{n-2}(3)\right] \qquad [\because \nabla\cdot\vec{r}=3]$$
$$=n[(n-2)r^{n-4}(r^2)+3r^{n-2}] \qquad [\because \vec{r}\cdot\vec{r}=r^2]$$
$$=n(n+1)r^{n-2}.$$

Second Method

$$\nabla^2(r^n)=\left(\frac{\partial^2}{\partial x^2}+\frac{\partial^2}{\partial y^2}+\frac{\partial^2}{\partial z^2}\right)r^n$$
$$=\sum\frac{\partial^2}{\partial x^2}(r^n)=\sum\frac{\partial}{\partial x}\left(\frac{\partial r^n}{\partial x}\right)$$
$$=\sum\frac{\partial}{\partial x}\left(nr^{n-1}\frac{\partial r}{\partial x}\right)=\sum\frac{\partial}{\partial x}\left(nr^{n-1}\frac{x}{r}\right)=\sum n\frac{\partial}{\partial x}(r^{n-2}x)$$

$$= n \sum \left[(n-2)\, r^{n-3} \frac{\partial r}{\partial x} \cdot x + r^{n-2}\right] = n \sum \left[(n-2)\, r^{n-3} \frac{x}{r} \cdot x + r^{n-2}\right]$$

$$= n \sum [(n-2)\, r^{n-4}\, x^2 + r^{n-2}] = n[(n-2)\, r^{n-4}\, (x^2+y^2+z^2) + 3r^{n-2}]$$

$$= n\, [(n-2)\, r^{n-4}(r^2) + 3r^{n-2}] = n(n+1)r^{n-2}.$$

(Second Statement)

Prove that div (grad r^n) = $n(n+1)r^{n-2}$.

(iii) $$\nabla^2\left(\frac{x}{r^3}\right) = \left(\frac{\partial^2}{\partial x^2} + \frac{\partial^2}{\partial y^2} + \frac{\partial^2}{\partial z^2}\right)\left(\frac{x}{r^3}\right) = \frac{\partial^2}{\partial x^2}\left(\frac{x}{r^3}\right) + \frac{\partial^2}{\partial y^2}\left(\frac{x}{r^3}\right) + \frac{\partial^2}{\partial z^2}\left(\frac{x}{r^3}\right) \quad ...(1)$$

Now $$\frac{\partial}{\partial x}\left(\frac{x}{r^3}\right) = \frac{1}{r^3} - \frac{3x}{r^4}\frac{\partial r}{\partial x} = \frac{1}{r^3} - \frac{3x^2}{r^5} \qquad \left|\ \because \frac{\partial r}{\partial x} = \frac{x}{r}\right.$$

$\therefore$ $$\frac{\partial^2}{\partial x^2}\left(\frac{x}{r^3}\right) = \frac{\partial}{\partial x}\left(\frac{1}{r^3} - \frac{3x^2}{r^5}\right) = -\frac{3}{r^4}\frac{\partial r}{\partial x} - \frac{6x}{r^5} + \frac{15x^2}{r^6}\frac{\partial r}{\partial x}$$

$$= -\frac{3x}{r^5} - \frac{6x}{r^5} + \frac{15r^3}{r^7} = -\frac{9x}{r^5} + \frac{15x^3}{r^7}$$

$$\frac{\partial}{\partial y}\left(\frac{x}{r^3}\right) = -\frac{3x}{r^4}\frac{\partial r}{\partial y} = -\frac{3xy}{r^5} \qquad \left|\ \because \frac{\partial r}{\partial y} = \frac{y}{r}\right.$$

$\therefore$ $$\frac{\partial^2}{\partial y^2}\left(\frac{x}{r^3}\right) = \frac{\partial}{\partial y}\left(-\frac{3xy}{r^5}\right) = -\frac{3x}{r^5} + \frac{15xy}{r^6}\frac{\partial r}{\partial y} = -\frac{3x}{r^5} + \frac{15xy^2}{r^7}$$

Similarly, $$\frac{\partial^2}{\partial z^2}\left(\frac{x}{r^3}\right) = -\frac{3x}{r^5} + \frac{15xz^2}{r^7}$$

$\therefore$ From (1), we have

$$\nabla^2\left(\frac{x}{r^3}\right) = \left(-\frac{9x}{r^5} + \frac{15x^3}{r^7}\right) + \left(-\frac{3x}{r^5} + \frac{15xy^2}{r^7}\right) + \left(-\frac{3x}{r^5} + \frac{15xz^2}{r^7}\right)$$

$$= -\frac{15x}{r^5} + \frac{15x}{r^7}(x^2+y^2+z^2) = -\frac{15x}{r^5} + \frac{15x}{r^7}(r^2) \quad |\ \because\ x^2+y^2+z^2 = r^2$$

$$= -\frac{15x}{r^5} + \frac{15x}{r^5} = 0$$

Example 9. *Prove that the vector $f(r)\,\vec{r}$ is irrotational.* (Anna 2009)

Sol. The vector $f(r)\,\vec{r}$ will be irrotational if curl $[f(r)\,\vec{r}\,] = \vec{0}$

Since $$\text{curl}\,(\phi\vec{A}) = (\text{grad}\,\phi) \times \vec{A} + \phi\ \text{curl}\,\vec{A}$$

$\therefore$ $$\text{curl}\,[f(r)\,\vec{r}\,] = [\text{grad}\,f(r)] \times \vec{r} + f(r)\ \text{curl}\,\vec{r}$$

$$= [f'(r)\ \text{grad}\ r] \times \vec{r} + f(r)\,\vec{0} \qquad [\because\ \text{curl}\,\vec{r} = \vec{0}]$$

$$= \left[f'(r)\frac{\vec{r}}{r}\right] \times \vec{r} = \frac{f'(r)}{r}(\vec{r} \times \vec{r}) = \vec{0}, \text{ since } \vec{r} \times \vec{r} = \vec{0}.$$

$\therefore$ The vector $f(r)\,\vec{r}$ is irrotational.

Example 10. *Prove that* $\nabla^2 f(r) = f''(r) + \frac{2}{r} f'(r)$. *Hence evaluate* $\nabla^2 (\log r)$

(M.D.U. May 2011, May 2012; U.K.T.U. 2010; G.B.T.U. 2012)

Sol. $\nabla^2 f(r) = \nabla \cdot \{\nabla f(r)\} = \text{div}\,\{\text{grad}\, f(r)\} = \text{div}\,\{f'(r)\,\text{grad}\, r\} = \text{div}\left\{f'(r)\frac{\vec{r}}{r}\right\}$

$$= \text{div}\left\{\frac{1}{r} f'(r)\,\vec{r}\right\} = \frac{1}{r} f'(r)\,\text{div}\,\vec{r} + \vec{r} \cdot \text{grad}\left\{\frac{1}{r} f'(r)\right\}$$

$$= \frac{3}{r} f'(r) + \vec{r} \cdot \left[\frac{d}{dr}\left(\frac{1}{r} f'(r)\right) \text{grad}\, r\right] = \frac{3}{r} f'(r) + \vec{r} \cdot \left[\left\{-\frac{1}{r^2} f'(r) + \frac{1}{r} f''(r)\right\}\frac{\vec{r}}{r}\right]$$

$$= \frac{3}{r} f'(r) + \left[-\frac{1}{r^3} f'(r) + \frac{1}{r^2} f''(r)\right](\vec{r} \cdot \vec{r}) = \frac{3}{r} f'(r) + \left[-\frac{1}{r^3} f'(r) + \frac{1}{r^2} f''(r)\right] r^2$$

$$= \frac{3}{r} f'(r) - \frac{1}{r} f'(r) + f''(r) = f''(r) + \frac{2}{r} f'(r).$$

If $\quad f(r) = \log r$, then $f'(r) = \frac{1}{r}$ and $f''(r) = -\frac{1}{r^2}$

$$\therefore \quad \nabla^2 (\log r) = -\frac{1}{r^2} + \frac{2}{r}\left(\frac{1}{r}\right) = \frac{1}{r^2} = \frac{1}{x^2 + y^2 + z^2}.$$

Example 11. *If* $u\vec{F} = \nabla v$, *where u, v are scalar fields and* $\vec{F}$ *is a vector, field, show that* $\vec{F} \cdot \text{curl}\, \vec{F} = 0$.

Sol. $\quad \text{curl}\, \vec{F} = \nabla \times \left(\frac{1}{u} \nabla v\right) \qquad \left[\because \vec{F} = \frac{1}{u} \nabla v\right]$

$$= \nabla \frac{1}{u} \times \nabla v + \frac{1}{u} \nabla \times (\nabla v) \qquad [\because \nabla \times (\phi \vec{A}) = \nabla\phi \times \vec{A} + \phi \nabla \times \vec{A}]$$

$$= \nabla \frac{1}{u} \times \nabla v \qquad [\because \nabla \times \nabla v = \vec{0}]$$

$$\therefore \quad \vec{F} \cdot \text{curl}\, \vec{F} = \frac{1}{u} \nabla v \cdot \left(\nabla \frac{1}{u} \times \nabla v\right) \frac{1}{u}\left[\nabla v \cdot \left(\nabla \frac{1}{u} \times \nabla v\right)\right] = 0$$

being the scalar triple product in which two factors are equal.

Example 12. *Prove that* $\vec{a} \cdot \nabla\left(\vec{b} \cdot \nabla \frac{1}{r}\right) = \frac{3(\vec{a} \cdot \vec{r})(\vec{b} \cdot \vec{r})}{r^5} - \frac{\vec{a} \cdot \vec{b}}{r^3}$ *where* $\vec{a}$ *and* $\vec{b}$ *are constant vectors.*

Sol. We know that $\quad \vec{r} = x\hat{i} + y\hat{j} + z\hat{k}$ and $r = |\vec{r}| = \sqrt{x^2 + y^2 + z^2}$

$$\therefore \quad \nabla \frac{1}{r} = \left(\hat{i}\frac{\partial}{\partial x} + \hat{j}\frac{\partial}{\partial y} + \hat{k}\frac{\partial}{\partial z}\right)(x^2 + y^2 + z^2)^{-\frac{1}{2}}$$

$$= \frac{-1}{2(x^2 + y^2 + z^2)^{\frac{3}{2}}}(2x\hat{i} + 2y\hat{j} + 2z\hat{k}) = -\frac{\vec{r}}{r^3}$$

$$\Rightarrow \quad \nabla\left(\vec{b}\cdot\nabla\frac{1}{r}\right) = \nabla\left[\vec{b}\cdot\left(-\frac{\vec{r}}{r^3}\right)\right] = -\nabla\left[\frac{1}{r^3}(\vec{b}\cdot\vec{r})\right] \qquad \text{[Form } \nabla(fg)]$$

$$= -\left[\frac{1}{r^3}\nabla(\vec{b}\cdot\vec{r}) + (\vec{b}\cdot\vec{r})\nabla\frac{1}{r^3}\right]$$

$$= -\left[\frac{1}{r^3}\{(\vec{b}\cdot\nabla)\vec{r} + (\vec{r}\cdot\nabla)\vec{b} + \vec{b}\times(\nabla\times\vec{r}) + \vec{r}\times(\nabla\times\vec{b})\}\right.$$

$$\left. + (\vec{b}\cdot\vec{r})\left\{\hat{i}\frac{\partial}{\partial x} + \hat{j}\frac{\partial}{\partial y} + \hat{k}\frac{\partial}{\partial z}\right\}(x^2+y^2+z^2)^{-\frac{3}{2}}\right] \qquad \ldots(1)$$

Let $\vec{b} = b_1\hat{i} + b_2\hat{j} + b_3\hat{k}$ where b_1, b_2, b_3 are independent of x, y, z (since $\vec{b}$ is a constant vector)

$$(\vec{b}\cdot\nabla)\vec{r} = (b_1\hat{i} + b_2\hat{j} + b_3\hat{k})\cdot\left(\hat{i}\frac{\partial}{\partial x} + \hat{j}\frac{\partial}{\partial y} + \hat{k}\frac{\partial}{\partial z}\right)\vec{r}$$

$$= \left(b_1\frac{\partial}{\partial x} + b_2\frac{\partial}{\partial y} + b_3\frac{\partial}{\partial z}\right)(x\hat{i} + y\hat{j} + z\hat{k})$$

$$= b_1\hat{i} + b_2\hat{j} + b_3\hat{k} = \vec{b}$$

$$(\vec{r}\cdot\nabla)\vec{b} = (x\hat{i} + y\hat{j} + z\hat{k})\cdot\left(\hat{i}\frac{\partial}{\partial x} + \hat{j}\frac{\partial}{\partial y} + \hat{k}\frac{\partial}{\partial z}\right)\vec{b}$$

$$= \left(x\frac{\partial}{\partial x} + y\frac{\partial}{\partial y} + z\frac{\partial}{\partial z}\right)(b_1\hat{i} + b_2\hat{j} + b_3\hat{k}) = \vec{0}$$

$$\left(\hat{i}\frac{\partial}{\partial x} + \hat{j}\frac{\partial}{\partial y} + \hat{k}\frac{\partial}{\partial z}\right)(x^2+y^2+z^2)^{-\frac{3}{2}} = \frac{-3}{2(x^2+y^2+z^2)^{\frac{5}{2}}}(2x\hat{i} + 2y\hat{j} + 2z\hat{k})$$

$$= -\frac{3\vec{r}}{r^5}$$

Also $\qquad \nabla\times\vec{r} = \vec{0},\ \nabla\times\vec{b} = \vec{0}$

$\therefore$ From (1), we have

$$\nabla\left(\vec{b}\cdot\nabla\frac{1}{r}\right) = -\left[\frac{1}{r^3}\left\{\vec{b} + \vec{0} + \vec{0} + \vec{0}\right\} + (\vec{b}\cdot\vec{r})\left(\frac{-3\vec{r}}{r^5}\right)\right] = -\frac{\vec{b}}{r^3} + \frac{3(\vec{b}\cdot\vec{r})\vec{r}}{r^5}$$

$$\Rightarrow \quad \vec{a}\cdot\nabla\left(\vec{b}\cdot\nabla\frac{1}{r}\right) = \vec{a}\cdot\left[-\frac{\vec{b}}{r^3} + \frac{3(\vec{b}\cdot\vec{r})\vec{r}}{r^5}\right] = -\frac{\vec{a}\cdot\vec{b}}{r^3} + \frac{3(\vec{b}\cdot\vec{r})(\vec{a}\cdot\vec{r})}{r^5}$$

$$= \frac{3(\vec{a}\cdot\vec{r})(\vec{b}\cdot\vec{r})}{r^5} - \frac{\vec{a}\cdot\vec{b}}{r^3}$$

Example 13. *If r is the distance of a point (x, y, z) from the origin, prove that curl* $\left(\hat{k} \times grad \frac{1}{r}\right) + grad\left(\hat{k} \cdot grad \frac{1}{r}\right) = \vec{0}$, *where* $\hat{k}$ *is the unit vector in the direction of OZ.*

Sol. Here, $\quad r = \sqrt{x^2 + y^2 + z^2} \quad$ so that $r^2 = x^2 + y^2 + z^2$

and $$\frac{\partial r}{\partial x} = \frac{x}{r}, \; \frac{\partial r}{\partial y} = \frac{y}{r}, \; \frac{\partial r}{\partial z} = \frac{z}{r}$$

$$\therefore \quad \text{grad } r = \hat{i}\frac{\partial r}{\partial x} + \hat{j}\frac{\partial r}{\partial y} + \hat{k}\frac{\partial r}{\partial z} = \frac{x\hat{i} + y\hat{j} + z\hat{k}}{r}$$

$$\text{grad } \frac{1}{r} = -\frac{1}{r^2} \text{ grad } r = -\frac{x\hat{i} + y\hat{j} + z\hat{k}}{r^3}$$

$$\Rightarrow \quad \hat{k} \times \text{grad } \frac{1}{r} = -\frac{x(\hat{k} \times \hat{i}) + y(\hat{k} \times \hat{j}) + z(\hat{k} \times \hat{k})}{r^3}$$

$$= -\frac{x\hat{j} - y\hat{i}}{r^3} = \frac{y\hat{i} - x\hat{j}}{(x^2 + y^2 + z^2)^{3/2}}$$

and $$\hat{k} \cdot \text{grad } \frac{1}{r} = -\frac{\hat{k} \cdot (x\hat{i} + y\hat{j} + z\hat{k})}{r^3} = -\frac{z}{(x^2 + y^2 + z^2)^{3/2}}$$

$$\therefore \quad \text{curl}\left(\hat{k} \times \text{grad } \frac{1}{r}\right) = \begin{vmatrix} \hat{i} & \hat{j} & \hat{k} \\ \frac{\partial}{\partial x} & \frac{\partial}{\partial y} & \frac{\partial}{\partial z} \\ \frac{y}{(x^2 + y^2 + z^2)^{3/2}} & \frac{-x}{(x^2 + y^2 + z^2)^{3/2}} & 0 \end{vmatrix}$$

$$= \hat{i}\frac{\partial}{\partial z}\left\{\frac{x}{(x^2 + y^2 + z^2)^{3/2}}\right\} + \hat{j}\frac{\partial}{\partial z}\left\{\frac{y}{(x^2 + y^2 + z^2)^{3/2}}\right\}$$
$$+ \hat{k}\left[\frac{\partial}{\partial x}\left\{\frac{-x}{(x^2 + y^2 + z^2)^{3/2}}\right\} - \frac{\partial}{\partial y}\left\{\frac{y}{(x^2 + y^2 + z^2)^{3/2}}\right\}\right]$$

$$= \hat{i}\left[-\frac{3}{2}x(x^2 + y^2 + z^2)^{-\frac{5}{2}} \cdot 2z\right] + \hat{j}\left[-\frac{3}{2}y(x^2 + y^2 + z^2)^{-\frac{5}{2}} \cdot 2z\right]$$
$$+ \hat{k}\left[\left\{-(x^2 + y^2 + z^2)^{-\frac{3}{2}} + \frac{3}{2}x(x^2 + y^2 + z^2)^{-\frac{5}{2}} \cdot 2x\right\}\right.$$
$$\left. - \left\{(x^2 + y^2 + z^2)^{-\frac{3}{2}} - \frac{3}{2}y(x^2 + y^2 + z^2)^{-\frac{5}{2}} \cdot 2y\right\}\right]$$

$$= \frac{-3xz\hat{i}}{(x^2 + y^2 + z^2)^{5/2}} - \frac{3yz\hat{j}}{(x^2 + y^2 + z^2)^{5/2}}$$
$$+ \hat{k}\left[\frac{-(x^2 + y^2 + z^2) + 3x^2 - (x^2 + y^2 + z^2) + 3y^2}{(x^2 + y^2 + z^2)^{5/2}}\right]$$

$$= \frac{-3xz\hat{i} - 3yz\hat{j} + (x^2 + y^2 - 2z^2)\hat{k}}{r^5} \qquad \ldots(1)$$

Also, $\text{grad}\left(\hat{k}\cdot\text{grad}\,\frac{1}{r}\right) = \nabla\left[\frac{-z}{(x^2+y^2+z^2)^{3/2}}\right]$

$$= \left(\hat{i}\frac{\partial}{\partial x}+\hat{j}\frac{\partial}{\partial y}+\hat{k}\frac{\partial}{\partial z}\right)\left[\frac{-z}{(x^2+y^2+z^2)^{3/2}}\right]$$

$$= \hat{i}\left[\frac{3}{2}z\,(x^2+y^2+z^2)^{-\frac{5}{2}}\cdot 2x\right]+\hat{j}\left[\frac{3}{2}z\,(x^2+y^2+z^2)^{-\frac{5}{2}}\cdot 2y\right]$$

$$+\hat{k}\left[-(x^2+y^2+z^2)^{-\frac{3}{2}}+\frac{3}{2}z\,(x^2+y^2+z^2)^{-\frac{5}{2}}\cdot 2z\right]$$

$$= \frac{3xz\hat{i}}{(x^2+y^2+z^2)^{5/2}}+\frac{3yz\hat{j}}{(x^2+y^2+z^2)^{5/2}}+\frac{\{-(x^2+y^2+z^2)+3z^2\}\hat{k}}{(x^2+y^2+z^2)^{5/2}}$$

$$= \frac{3xz\hat{i}+3yz\hat{j}+(2z^2-x^2-y^2)\hat{k}}{r^5} \qquad \text{...(2)}$$

Adding (1) and (2), we have

$$\text{curl}\left(\hat{k}\times\text{grad}\,\frac{1}{r}\right)+\text{grad}\left(\hat{k}\cdot\text{grad}\,\frac{1}{r}\right)=\vec{0}.$$

Example 14. *If $\vec{V}_1$ and $\vec{V}_2$ be the vectors joining the fixed points (x_1, y_1, z_1) and (x_2, y_2, z_2) respectively to a variable point (x, y, z), prove that curl $(\vec{V}_1\times\vec{V}_2) = 2\,(\vec{V}_1-\vec{V}_2)$.*

Sol. Given

$$\vec{V}_1 = (x-x_1)\hat{i}+(y-y_1)\hat{j}+(z-z_1)\hat{k}$$

$$\vec{V}_2 = (x-x_2)\hat{i}+(y-y_2)\hat{j}+(z-z_2)\hat{k}$$

(x_1, y_1, z_1) (x_2, y_2, z_2) $\vec{V}_1$ $\vec{V}_2$ (x, y, z)

$$\therefore\ \text{Curl}\,(\vec{V}_1\times\vec{V}_2) = \nabla\times(\vec{V}_1\times\vec{V}_2)$$

$$= \sum\hat{i}\times\frac{\partial}{\partial x}(\vec{V}_1\times\vec{V}_2)$$

$$= \sum\hat{i}\times\left(\frac{\partial\vec{V}_1}{\partial x}\times\vec{V}_2+\vec{V}_1\times\frac{\partial\vec{V}_2}{\partial x}\right)$$

$$= \sum\hat{i}\times\left(\frac{\partial\vec{V}_1}{\partial x}\times\vec{V}_2\right)+\sum\hat{i}\times\left(\vec{V}_1\times\frac{\partial\vec{V}_2}{\partial x}\right)$$

$$= \sum\left[\left(\hat{i}\cdot\vec{V}_2\right)\frac{\partial\vec{V}_1}{\partial x}-\left(\hat{i}\cdot\frac{\partial\vec{V}_1}{\partial x}\right)\vec{V}_2\right]+\sum\left[\left(\hat{i}\cdot\frac{\partial\vec{V}_2}{\partial x}\right)\vec{V}_1-\left(\hat{i}\cdot\vec{V}_1\right)\frac{\partial\vec{V}_2}{\partial x}\right]$$

$[\because\ \vec{a}\times(\vec{b}\times\vec{c})=(\vec{a}\cdot\vec{c})\,\vec{b}-(\vec{a}\cdot\vec{b})\,\vec{c}]$

$$= \sum [(x - x_2)\hat{i} - (\hat{i} \cdot \hat{i})\vec{V}_2] + \sum [(\hat{i} \cdot \hat{i})\vec{V}_1 - (x - x_1)\hat{i}]$$

$$= \sum [(x - x_2)\hat{i} - \vec{V}_2] + \sum [\vec{V}_1 - (x - x_1)\hat{i}]$$

$$= [(x - x_2)\hat{i} - \vec{V}_2 + (y - y_2)\hat{j} - \vec{V}_2 + (z - z_2)\hat{k} - \vec{V}_2]$$

$$+ [\vec{V}_1 - (x - x_1)\hat{i} + \vec{V}_1 - (y - y_1)\hat{j} + \vec{V}_1 - (z - z_1)\hat{k}]$$

$$= (\vec{V}_2 - 3\vec{V}_2) + (3\vec{V}_1 - \vec{V}_1)$$

$$= -2\vec{V}_2 + 2\vec{V}_1 = 2(\vec{V}_1 - \vec{V}_2)$$

EXERCISE 13.3

1. Evaluate

   (*i*) div $(3x^2\hat{i} + 5xy^2\hat{j} + xyz^3\hat{k})$ at the point (1, 2, 3).

   (*ii*) div $[(xy \sin z)\hat{i} + (y^2 \sin x)\hat{j} + (z^2 \sin xy)\hat{k}]$ at the point $\left(0, \frac{\pi}{2}, \frac{\pi}{2}\right)$.

   (*iii*) curl $[e^{xyz}(\hat{i} + \hat{j} + \hat{k})]$

2. Find the divergence and curl of the vectors

   (*i*) $\vec{V} = (xyz)\hat{i} + (3x^2y)\hat{j} + (xz^2 - y^2z)\hat{k}$ at the point (2, –1, 1).

   (*ii*) $\vec{R} = (x^2 + yz)\hat{i} + (y^2 + zx)\hat{j} + (z^2 + xy)\hat{k}$.

   (*iii*) $\vec{F}(x, y, z) = xz^3\hat{i} - 2x^2yz\hat{j} + 2yz^4\hat{k}$

   (*iv*) $\vec{F}(x, y, z) = e^{xyz}(xy^2\hat{i} + yz^2\hat{j} + zx^2\hat{k})$ at the point (1, 2, 3).

3. If $\vec{F} = (x + y + 1)\hat{i} + \hat{j} - (x + y)\hat{k}$, show that $\vec{F} \cdot \text{Curl } \vec{F} = 0$.

4. If $\vec{V} = \dfrac{x\hat{i} + y\hat{j} + z\hat{k}}{\sqrt{x^2 + y^2 + z^2}}$, show that $\nabla \cdot \vec{V} = \dfrac{2}{\sqrt{x^2 + y^2 + z^2}}$ and $\nabla \times \vec{V} = \vec{0}$.

   (*M.T.U. 2011; Calicut 2009*)

5. (*i*) If $\vec{A} = (3xz^2)\hat{i} - (yz)\hat{j} + (x + 2z)\hat{k}$, find curl (curl $\vec{A}$).

   (*ii*) Evaluate curl curl of $\vec{V} = (2xz^2)\hat{i} - yz\hat{j} + (3xz^3)\hat{k}$ at (1, 1, 1).

6. Show that each of the following vectors are solenoidal:

   (*i*) $(x + 3y)\hat{i} + (y - 3z)\hat{j} + (x - 2z)\hat{k}$ (*P.T.U. Dec. 2010*)

   (*ii*) $(-x^2 + yz)\hat{i} + (4y - z^2x)\hat{j} + (2xz - 4z)\hat{k}$ (*P.T.U. May 2010, May 2011*)

   (*iii*) $3y^4z^2\hat{i} + 4x^3z^2\hat{j} + 3x^2y^2\hat{k}$.

7. If $u = x^2 + y^2 + z^2$ and $\vec{V} = x\hat{i} + y\hat{j} + z\hat{k}$, show that div $(u\vec{V}) = 5u$. (*Calicut 2009*)

8. (*a*) Show that the vector field $\vec{V} = (\sin y + z)\hat{i} + (x \cos y - z)\hat{j} + (x - y)\hat{k}$ is irrotational.

   (*b*) Find the value of constant '*a*' such that $\vec{A} = (ax + 4y^2z)\hat{i} + (x^3 \sin z - 3y)\hat{j} - (e^x + 4 \cos x^2y)\hat{k}$ is solenoidal. (*M.D.U. May 2012*)

(c) Show that $\vec{A} = (6xy + z^3)\hat{i} + (3x^2 - z)\hat{j} + (3xz^2 - y)\hat{k}$ is irrational. Find the velocity potential ϕ such that $\vec{A} = \nabla\phi$. *(Anna 2012; U.K.T.U. 2011)*

9. (i) Find the values of a, b, c for which the vector $\vec{V} = (x + y + az)\hat{i} + (bx + 3y - z)\hat{j} + (3x + cy + z)\hat{k}$ is irrotational. *(P.T.U. Dec. 2011; M.D.U. May 2010)*

(ii) Prove that $(y^2 - z^2 + 3yz - 2x)\hat{i} + (3xz + 2xy)\hat{j} + (3xy - 2xz + 2z)\hat{k}$ is both solenoidal and irrotational. *(U.P.T.U. 2009)*

(iii) A fluid motion is given by $\vec{V} = (y + z)\hat{i} + (z + x)\hat{j} + (x + y)\hat{k}$. Is this motion irrotational? If so, find the velocity potential. *(P.T.U. Dec. 2011)*

(iv) Show that the vector field $\vec{A}$, where $\vec{A} = (x^2 - y^2 + x)\hat{i} - (2xy + y)\hat{j}$ is irrotational and find its scalar potential. *(Anna 2010; M.T.U. 2013; P.T.U. May 2010)*

(v) Show that the vector field defined by $\vec{F} = 2xyz^3\,\hat{i} + x^2z^3\hat{j} + 3x^2yz^2\hat{k}$ is irrotational and find its scalar potential.

(vi) Show that the vector field $\vec{F} = 2x(y^2 + z^3)\,\hat{i} + 2x^2y\hat{j} + 3x^2z^2\hat{k}$ is irrotational and find its scalar potential.

(vii) Prove that $\vec{F} = (y^2 \cos x + z^3)\,\hat{i} + (2y \sin x - 4)\hat{j} + 3xz^2\hat{k}$ is irrotational and find its scalar potential.

10. (a) Show that $\vec{E} = \dfrac{\vec{r}}{r^2}$ is irrotational.

(b) Show that the vector field $\vec{F} = \dfrac{\vec{r}}{r^3}$ is irrotational as well as solenoidal.

11. If $\vec{E}$ and $\vec{H}$ are irrotational, prove that $\vec{E} \times \vec{H}$ is solenoidal. *(G.B.T.U. 2010; Calicut 2008)*

12. For a solenoidal vector $\vec{F}$, prove that curl curl curl curl $\vec{F} = \nabla^4\vec{F}$.

13. Find the directional derivative of $\nabla\cdot(\nabla\phi)$ at the point $(1, -2, 1)$ in the direction of the normal to the surface $xy^2z = 3x + z^2$, where $\phi = 2x^3y^2z^4$. *(U.P.T.U. 2009)*

14. If $\vec{V}_1$ and $\vec{V}_2$ be the vectors joining the fixed points (x_1, y_1, z_1) and (x_2, y_2, z_2) respectively to a variable point (x, y, z), prove that

(i) div $(\vec{V}_1 \times \vec{V}_2) = 0$ (ii) grad $(\vec{V}_1 \cdot \vec{V}_2) = \vec{V}_1 + \vec{V}_2$

15. Prove that $\nabla \times (\vec{r} \times \vec{u}) = \vec{r}\,(\nabla \cdot \vec{u}) - 2\vec{u} - (\vec{r} \cdot \nabla)\,\vec{u}$.

16. (a) Find curl (curl $\vec{V}$) given $\vec{V} = x^2y\hat{i} + y^2z\hat{j} + z^2y\hat{k}$

(b) If $f = (x^2 + y^2 + z^2)^{-n}$, find div grad f and determine n if div grad $f = 0$

(c) If $u = 3x^2y$, $v = xz^2 - 2y$, find

(i) $\nabla(\nabla u \cdot \nabla v)$ (ii) $\nabla\cdot(\nabla u \times \nabla v)$.

17. If $\vec{a}$ is a constant vector and $\vec{r} = x\hat{i} + y\hat{j} + z\hat{k}$, prove that

(i) div $(\vec{a} \times \vec{r}) = 0$ *(K.U.K. 2009)* (ii) curl $[(\vec{a} \cdot \vec{r})\,\vec{r}] = \vec{a} \times \vec{r}$ *(M.T.U. 2012)*

(iii) $\nabla(\vec{a}\cdot\vec{a}) = 2(\vec{a}\cdot\nabla)\vec{a} + 2\vec{a}\times(\nabla\times\vec{a})$

(iv) $\vec{a}\times(\nabla\times\vec{r}) = \nabla(\vec{a}\cdot\vec{r}) - (\vec{a}\cdot\nabla)\vec{r}$. *(U.P.T.U. 2008)*

18. **If $\vec{r} = x\hat{i} + y\hat{j} + z\hat{k}$, prove that**

(i) div $(r^n\,\vec{r}) = (n+3)\,r^n$ *(M.D.U. 2013)*

(ii) curl $(r^n\,\vec{r}) = \vec{0}$

(iii) $\dfrac{\vec{r}}{r^3}$ is solenoidal as well as irrotational

(iv) $\nabla^2\left(\dfrac{1}{r}\right) = 0$

(v) $\nabla^2(r^n\,\vec{r}) = n(n+3)r^{n-2}\,\vec{r}$

(vi) $\nabla\cdot\left\{r\nabla\left(\dfrac{1}{r^3}\right)\right\} = \dfrac{3}{r^4}$

(vii) $\nabla^2\left[\nabla\cdot\left(\dfrac{\vec{r}}{r^2}\right)\right] = 2r^{-4}$

(viii) $\nabla\left[\nabla\cdot\dfrac{\vec{r}}{r}\right] = -\dfrac{2}{r^3}\vec{r}$

(ix) $\nabla^2(r\,\vec{r}) = 4\hat{r}$

(x) $\nabla^2(xy\hat{i} + yz\hat{j} + zx\hat{k}) = 0$.

19. **Show that the vector $\nabla\phi\times\nabla\psi$ is solenoidal.**

20. **Show that div (grad r^n) $= n(n+1)r^{n-2}$, where $r^2 = x^2 + y^2 + z^2$.** *(K.U.K. 2012; U.K.T.U. 2012)*

21. **If r and $\vec{r}$ have their usual meanings and $\vec{a}$ is a constant vector, prove that**

$$\nabla\times\left(\frac{\vec{a}\times\vec{r}}{r^n}\right) = \frac{2-n}{r^n}\vec{a} + \frac{n(\vec{a}\cdot\vec{r})}{r^{n+2}}\vec{r}.$$

[Hint. $\nabla\times[r^{-n}(\vec{a}\times\vec{r})] = (\nabla r^{-n})\times(\vec{a}\times\vec{r}) + r^{-n}[\nabla\times(\vec{a}\times\vec{r})]$

{Since $\nabla\times(\phi\vec{A}) = (\nabla\phi)\times\vec{A} + \phi(\nabla\times\vec{A})$ by Art. 13.19}]

Answers

1. (i) 80 (ii) $\dfrac{\pi}{2}$ (iii) $e^{xyz}[x(z-y)\hat{i} + y(x-z)\hat{j} + z(y-x)\hat{k}]$

2. (i) 14; $2\hat{i} - 3\hat{j} - 14\hat{k}$ (ii) $2(x+y+z)$; $\vec{0}$

(iii) $z^3 - 2x^2z + 8yz^3$; $2(x^2y + z^4)\hat{i} + 3xz^2\hat{j} - 4xyz\hat{k}$

(iv) $98e^6$; $-39e^6\hat{i} - 16e^6\hat{j} + 92e^6\hat{k}$

5. (i) $-6x\hat{i} + (6z-1)\hat{k}$ (ii) $5\hat{i} + 3\hat{k}$

8. (b) 3 (c) $\phi = 3x^2y + xz^3 - zy + c$

9. (i) $a = 3$, $b = 1$, $c = -1$ (iii) $xy + yz + zx + c$

(iv) $\dfrac{1}{3}x^3 - xy^2 + \dfrac{1}{2}(x^2 - y^2) + c$ (v) $x^2yz^3 + c$

(vi) $x^2(y^2 + z^3) + c$ (vii) $y^2\sin x + xz^3 - 4y + c$

13. $\dfrac{1724}{\sqrt{21}}$

16. (a) $2(x+z)\hat{j} + 2y\hat{k}$ (b) $\dfrac{2n(2n-1)}{(x^2+y^2+z^2)^{n+1}}$; $n = \dfrac{1}{2}$

(c) (i) $(6yz^2 - 12x)\hat{i} + 6xz^2\hat{j} + 12xyz\hat{k}$ (ii) 0.

13.21. INTEGRATION OF VECTOR FUNCTIONS

Let $\vec{f}(t)$ and $\vec{F}(t)$ be two vector functions of a scalar variable t such that $\frac{d}{dt}\vec{F}(t) = \vec{f}(t)$, then $\vec{F}(t)$ is called an integral of $\vec{f}(t)$ with respect to t and we write $\int \vec{f}(t)\,dt = \vec{F}(t)$.

If $\vec{c}$ is any *arbitrary constant vector* independent of t, then $\frac{d}{dt}\{\vec{F}(t) + \vec{c}\} = \vec{f}(t)$

This is equivalent to $\int \vec{f}(t)\,dt = \vec{F}(t) + \vec{c}$

$\vec{F}(t)$ is called the *indefinite integral* of $\vec{f}(t)$. The constant vector $\vec{c}$ is called the *constant of integration* and can be determined if some initial conditions are given.

The *definite integral* of $\vec{f}(t)$ between the limits $t = a$ and $t = b$ is written as

$$\int_a^b \vec{f}(t)\,dt = \left[\vec{F}(t)\right]_a^b = \vec{F}(b) - \vec{F}(a).$$

Note 1. If $\vec{f}(t) = f_1(t)\hat{i} + f_2(t)\hat{j} + f_3(t)\hat{k}$, then

$$\int \vec{f}(t)\,dt = \hat{i}\int f_1(t)\,dt + \hat{j}\int f_2(t)\,dt + \hat{k}\int f_3(t)\,dt.$$

Thus in order to integrate a vector function, integrate its components.

Note 2. We can obtain some standard results for integration of vector functions by considering the derivatives of suitable vector functions. For example,

(*i*) $\frac{d}{dt}(\vec{r}\cdot\vec{s}) = \frac{d\vec{r}}{dt}\cdot\vec{s} + \vec{r}\cdot\frac{d\vec{s}}{dt} \quad \Rightarrow \quad \int\left(\frac{d\vec{r}}{dt}\cdot\vec{s} + \vec{r}\cdot\frac{d\vec{s}}{dt}\right)dt = \vec{r}\cdot\vec{s} + c$

Here c is a scalar quantity since the integrand is a scalar.

(*ii*) $\frac{d}{dt}(\vec{r}^{\,2}) = 2\vec{r}\cdot\frac{d\vec{r}}{dt} \quad \Rightarrow \quad \int\left(2\vec{r}\cdot\frac{d\vec{r}}{dt}\right)dt = \vec{r}^{\,2} + c$ where c is a scalar quantity.

(*iii*) $\frac{d}{dt}\left(\vec{r}\times\frac{d\vec{r}}{dt}\right) = \frac{d\vec{r}}{dt}\times\frac{d\vec{r}}{dt} + \vec{r}\times\frac{d^2\vec{r}}{dt^2} = \vec{r}\times\frac{d^2\vec{r}}{dt^2} \quad \Rightarrow \quad \int\left(\vec{r}\times\frac{d^2\vec{r}}{dt^2}\right)dt = \vec{r}\times\frac{d\vec{r}}{dt} + \vec{c}$

Here $\vec{c}$ is a vector quantity since the integrand is a vector.

(*iv*) If $\vec{a}$ is a constant vector, then $\frac{d}{dt}(\vec{a}\times\vec{r}) = \frac{d\vec{a}}{dt}\times\vec{r} + \vec{a}\times\frac{d\vec{r}}{dt} = \vec{a}\times\frac{d\vec{r}}{dt}$

$$\Rightarrow \int\left(\vec{a}\times\frac{d\vec{r}}{dt}\right)dt = \vec{a}\times\vec{r} + \vec{c},$$ where $\vec{c}$ is a vector quantity.

ILLUSTRATIVE EXAMPLES

Example 1. *The acceleration of a particle at time t is given by* $\vec{a} = 18 \cos 3t\, \hat{i} - 8 \sin 2t\, \hat{j} + 6t\, \hat{k}$.

If the velocity $\vec{v}$ *and displacement* $\vec{r}$ *be zero at* $t = 0$, *find* $\vec{v}$ *and* $\vec{r}$ *at any time t.*

Sol. Here $\quad \vec{a} = \dfrac{d^2\vec{r}}{dt^2} = 18 \cos 3t\hat{i} - 8 \sin 2t\hat{j} + 6t\hat{k}$

Integrating, we have $\vec{v} = \dfrac{d\vec{r}}{dt} = \hat{i} \int 18 \cos 3t\, dt + \hat{j} \int -8 \sin 2t\, dt + \hat{k} \int 6t\, dt$

$$= 6 \sin 3t\hat{i} + 4 \cos 2t\hat{j} + 3t^2\hat{k} + \vec{c}$$

At $\quad t = 0,\ \vec{v} = \vec{0} \quad \Rightarrow \quad \vec{0} = 4\hat{j} + \vec{c} \quad \Rightarrow \quad \vec{c} = -4\hat{j}$

$\therefore \quad \vec{v} = \dfrac{d\vec{r}}{dt} = 6 \sin 3t\hat{i} + 4 (\cos 2t - 1)\hat{j} + 3t^2\hat{k}$

Integrating again, we have $\vec{r} = \hat{i} \int 6 \sin 3t\, dt + \hat{j} \int 4 (\cos 2t - 1)\, dt + \hat{k} \int 3t^2\, dt$

$$= -2 \cos 3t\hat{i} + (2 \sin 2t - 4t)\hat{j} + t^3\hat{k} + \vec{c}$$

At $t = 0,\ \vec{r} = \vec{0} \quad \Rightarrow \quad \vec{0} = -2\hat{i} + \vec{c} \quad \Rightarrow \quad \vec{c} = 2\hat{i}$

$\therefore \quad \vec{r} = 2 (1 - \cos 3t)\hat{i} + 2 (\sin 2t - 2t)\hat{j} + t^3\hat{k}.$

Example 2. *If* $\vec{A}(t) = (3t^2 - 2t)\hat{i} + (6t - 4)\hat{j} + 4t\hat{k}$, *evaluate* $\int_2^3 \vec{A}(t)\, dt$.

Sol.
$$\int_2^3 \vec{A}(t)\, dt = \int_2^3 [(3t^2 - 2t)\hat{i} + (6t - 4)\hat{j} + 4t\hat{k}]dt$$

$$= \hat{i} \int_2^3 (3t^2 - 2t)\, dt + \hat{j} \int_2^3 (6t - 4)\, dt + \hat{k} \int_2^3 4t\, dt$$

$$= \hat{i} \left[t^3 - t^2 \right]_2^3 + \hat{j} \left[3t^2 - 4t \right]_2^3 + \hat{k} \left[2t^2 \right]_2^3 = 14\hat{i} + 11\hat{j} + 10\hat{k}.$$

Example 3. *If* $\vec{r}(t) = 5t^2\hat{i} + t\hat{j} - t^3\hat{k}$, *prove that* $\displaystyle\int_1^2 \left(\vec{r} \times \frac{d^2\vec{r}}{dt^2} \right) dt = -14\hat{i} + 75\hat{j} - 15\hat{k}$.

Sol. Since $\dfrac{d}{dt}\left(\vec{r} \times \dfrac{d\vec{r}}{dt} \right) = \dfrac{d\vec{r}}{dt} \times \dfrac{d\vec{r}}{dt} + \vec{r} \times \dfrac{d^2\vec{r}}{dt^2} = \vec{r} \times \dfrac{d^2\vec{r}}{dt^2}$

$$\therefore \quad \int_1^2 \left(\vec{r} \times \frac{d^2\vec{r}}{dt^2} \right) dt = \left[\vec{r} \times \frac{d\vec{r}}{dt} \right]_1^2 \qquad \text{...(1)}$$

Let us now find $\vec{r} \times \dfrac{d\vec{r}}{dt}$

$$\vec{r} \times \frac{d\vec{r}}{dt} = (5t^2\hat{i} + t\hat{j} - t^3\hat{k}) \times (10t\hat{i} + \hat{j} - 3t^2\hat{k})$$

$$= \begin{vmatrix} \hat{i} & \hat{j} & \hat{k} \\ 5t^2 & t & -t^3 \\ 10t & 1 & -3t^2 \end{vmatrix} = -2t^3\hat{i} + 5t^4\hat{j} - 5t^2\hat{k}$$

$\therefore$ From (1), we have $\displaystyle\int_1^2 \left(\vec{r} \times \frac{d^2\vec{r}}{dt^2}\right) dt = \left[-2t^3\hat{i} + 5t^4\hat{j} - 5t^2\hat{k}\right]_1^2$

$$= \left[-2t^3\right]_1^2 \hat{i} + \left[5t^4\right]_1^2 \hat{j} - \left[-5t^2\right]_1^2 \hat{k} = -14\hat{i} + 75\hat{j} - 15\hat{k}.$$

Example 4. *Given that* $\vec{r}(t) = \begin{cases} 2\hat{i} - \hat{j} + 2\hat{k}, & \text{when } t = 2 \\ 4\hat{i} - 2\hat{j} + 3\hat{k}, & \text{when } t = 3 \end{cases}$, *show that* $\displaystyle\int_2^3 \left(\vec{r} \cdot \frac{d\vec{r}}{dt}\right) dt = 10.$

Sol. Since $\displaystyle\frac{d}{dt}(\vec{r}^2) = 2\vec{r} \cdot \frac{d\vec{r}}{dt}$

$\therefore$ $$\int_2^3 \left(\vec{r} \cdot \frac{d\vec{r}}{dt}\right) dt = \frac{1}{2}\left[\vec{r}^2\right]_2^3 \qquad ...(1)$$

When $t = 3$, $\vec{r}^2 = (4\hat{i} - 2\hat{j} + 3\hat{k}) \cdot (4\hat{i} - 2\hat{j} + 3\hat{k}) = 16 + 4 + 9 = 29$

When $t = 2$, $\vec{r}^2 = (2\hat{i} - \hat{j} + 2\hat{k}) \cdot (2\hat{i} - \hat{j} + 2\hat{k}) = 4 + 1 + 4 = 9$

$\therefore$ From (1), we have $\displaystyle\int_2^3 \left(\vec{r} \cdot \frac{d\vec{r}}{dt}\right) dt = \frac{1}{2}[29 - 9] = 10.$

EXERCISE 13.4

1. If $\vec{f}(t) = t\hat{i} + (t^2 - 2t)\hat{j} + (3t^2 + 3t^3)\hat{k}$, find $\displaystyle\int_0^1 \vec{f}(t)\, dt$.

2. If $\vec{r} = t\hat{i} - t^2\hat{j} + (t-1)\hat{k}$ and $\vec{S} = 2t^2\hat{i} + 6t\hat{k}$, evaluate

   (*i*) $\displaystyle\int_0^2 \vec{r} \cdot \vec{S}\, dt$ (*ii*) $\displaystyle\int_0^2 \vec{r} \times \vec{S}\, dt$.

3. Find the value of $\vec{r}$ satisfying the equation $\dfrac{d^2\vec{r}}{dt^2} = 6t\hat{i} - 12t^2\hat{j} + 4\cos t\hat{k}$, given that $\vec{r} = 2\hat{i} + \hat{j}$ and $\dfrac{d\vec{r}}{dt} = -\hat{i} - 3\hat{k}$ at $t = 0$. (*M.D.U. 2013*)

4. The acceleration of a particle at any time t is given by $\vec{a} = 12\cos 2t\hat{i} - 8\sin 2t\hat{j} + 16t\hat{k}$. If the velocity $\vec{v}$ and displacement $\vec{r}$ are zero at $t = 0$, find $\vec{v}$ and $\vec{r}$ at any time t.

Answers

1. $\frac{1}{2}\hat{i} - \frac{2}{3}\hat{j} + \frac{7}{4}\hat{k}$ 2. (*i*) 12 (*ii*) $-24\hat{i} - \frac{40}{3}\hat{j} + \frac{64}{5}\hat{k}$

3. $\vec{r} = (t^3 - t + 2)\hat{i} + (1 - t^4)\hat{j} + (4 - 4\cos t - 3t)\hat{k}$

4. $\vec{v} = 6\sin 2t\hat{i} + 4(\cos 2t - 1)\hat{j} + 8t^2\hat{k}$, $\vec{r} = 3(1 - \cos 2t)\hat{i} + 2(\sin 2t - 2t)\hat{j} + \frac{8}{3}t^3\hat{k}$.

13.22. LINE INTEGRALS

Any integral which is to be evaluated along a curve is called a line integral.

Let $\vec{F}(P)$ be a continuous vector point function defined at every point of a curve C is space. Divide the curve C into n parts by the points

$$A = P_0, P_1, P_2, \ldots, P_n = B$$

B = P$_n$
P$_{n-1}$
P$_i$
C
Q$_i$
P$_{i-1}$
P$_1$
A = P$_0$

and let $\vec{R}_0, \vec{R}_1, \vec{R}_2, \ldots, \vec{R}_n$ be the position vectors of these points. Let Q_i be any point on the arc $P_{i-1}P_i$. Then the limit of the sum

$$\sum_{i=1}^{n} \vec{F}(Q_i) \cdot \delta\vec{R}_i, \quad \text{where} \quad \delta\vec{R}_i = \vec{R}_i - \vec{R}_{i-1} \qquad \ldots(1)$$

as $n \to \infty$ and every $|\delta\vec{R}_i| \to 0$, if it exists, is called a line integral of $\vec{F}$ along C and is denoted by

$$\int_C \vec{F} \cdot d\vec{R} \quad \text{or} \quad \int_C \vec{F} \cdot \frac{d\vec{R}}{dt}\, dt.$$

Clearly, it is a scalar. It is called the *tangential line integral* of $\vec{F}$ along C.

If the scalar products in (1) are replaced by vector products, then the corresponding line integral is defined as $\int_C \vec{F} \times d\vec{R}$ which is a vector.

If the vector function $\vec{F}$ is replaced by a scalar function ϕ, then the corresponding line integral is defined as $\int_C \phi\, d\vec{R}$, which is a vector.

If $\quad \vec{F}(x, y, z) = f_1\hat{i} + f_2\hat{j} + f_3\hat{k}$ and $\vec{R} = x\hat{i} + y\hat{j} + z\hat{k}$ then $d\vec{R} = \hat{i}\,dx + \hat{j}\,dy + \hat{k}\,dz$

$$\therefore \quad \int_C \vec{F} \cdot d\vec{R} = \int_C (f_1 dx + f_2 dy + f_3 dz)$$

If the parametric equations of the curve C are $x = x(t)$, $y = y(t)$, $z = z(t)$ and $t = t_1$ at A, $t = t_2$ at B, then

$$\int_C \vec{F} \cdot d\vec{R} = \int_{t_1}^{t_2} \left(f_1 \frac{dx}{dt} + f_2 \frac{dy}{dt} + f_3 \frac{dz}{dt} \right) dt$$

If C is a closed curve, then the integral sign $\int_C$ is replaced by $\oint_C$.

13.23. CIRCULATION *(G.B.T.U. 2010)*

In fluid dynamics, if $\vec{V}$ represents the velocity of a fluid particle and C is a closed curve, then the integral $\oint_C \vec{V} \cdot d\vec{R}$ is called the *circulation* of $\vec{V}$ around the curve C.

If the circulation of $\vec{V}$ around every closed curve in a region D vanishes, then $\vec{V}$ is said to be *irrotational* in D.

13.24. WORK DONE BY A FORCE

Let $\vec{F}$ represent the force acting on a particle moving along an arc AB. The work done during a small displacement $\delta\vec{R}$ is $\vec{F} \cdot \delta\vec{R}$.

$\therefore$ The total work done by $\vec{F}$ during displacement from A to B is given by $\int_A^B \vec{F} \cdot d\vec{R}$

If the force $\vec{F}$ is conservative, then there exists a scalar function ϕ such that

$$\vec{F} = \nabla\phi = \hat{i}\frac{\partial\phi}{\partial x} + \hat{j}\frac{\partial\phi}{\partial y} + \hat{k}\frac{\partial\phi}{\partial z}$$

$\therefore$ The work done by $\vec{F}$ during displacement from A to B $= \int_A^B \vec{F} \cdot d\vec{R}$

$$= \int_A^B \left(\hat{i}\frac{\partial\phi}{\partial x} + \hat{j}\frac{\partial\phi}{\partial y} + \hat{k}\frac{\partial\phi}{\partial z}\right) \cdot (\hat{i}\,dx + \hat{j}dy + \hat{k}dz)$$

$$= \int_A^B \left(\frac{\partial\phi}{\partial x}dx + \frac{\partial\phi}{\partial y}dy + \frac{\partial\phi}{\partial z}dz\right) = \int_A^B d\phi = \Big[\phi\Big]_A^B = \phi_B - \phi_A$$

Thus, in a conservative field, the work done depends on the value of ϕ at the end points A and B, and not on the path joining A and B.

ILLUSTRATIVE EXAMPLES

Example 1. *If $\vec{F} = 3xy\hat{i} - y^2\hat{j}$, evaluate $\int_C \vec{F} \cdot d\vec{r}$, where C is the arc of the parabola $y = 2x^2$ from (0, 0) to (1, 2).* (Calicut 2009)

Sol. Since the integration is performed in the *xy*-plane ($z = 0$), we take

$$\vec{r} = x\hat{i} + y\hat{j} \quad \text{so that } d\vec{r} = dx\hat{i} + dy\hat{j}$$

$$\therefore \quad \vec{F} \cdot d\vec{r} = (3xy\hat{i} - y^2\hat{j}) \cdot (dx\hat{i} + dy\hat{j}) = 3xy\,dx - y^2\,dy$$

On C : $y = 2x^2$ from (0, 0) to (1, 2)

$$\vec{F} \cdot d\vec{r} = 3x(2x^2)\,dx - (2x^2)^2\,(4x\,dx) = (6x^3 - 16x^5)\,dx$$

Also, x varies from 0 to 1.

$$\therefore \quad \int_C \vec{F} \cdot d\vec{r} = \int_0^1 (6x^3 - 16x^5)\,dx = \left[\frac{6x^4}{4} - \frac{16x^6}{6}\right]_0^1 = \frac{3}{2} - \frac{8}{3} = -\frac{7}{6}$$

Note that if the curve is traversed in the opposite sense, *i.e.*, from (1, 2) to (0, 0), the value of the integral would be $\frac{7}{6}$.

Second Method

Let $x = t$, then the parametric equations of the parabola $y = 2x^2$ are $x = t$, $y = 2t^2$.

At the point (0, 0), $x = 0$ and so $t = 0$

At the point (1, 2), $x = 1$ and so $t = 1$

If $\vec{r}$ is the position vector of any point (x, y) on C, then $\vec{r} = x\hat{i} + y\hat{j} = t\hat{i} + 2t^2\hat{j}$

Also in terms of t, $\vec{F} = (3t \times 2t^2)\hat{i} - (2t^2)^2\hat{j} = 6t^3\hat{i} - 4t^4\hat{j}$

$$\therefore \quad \int_C \vec{F}\cdot d\vec{r} = \int_C \left(\vec{F}\cdot\frac{d\vec{r}}{dt}\right) dt = \int_0^1 (6t^3\hat{i} - 4t^4\hat{j})\cdot(\hat{i} + 4t\hat{j})\, dt$$

$$= \int_0^1 (6t^3 - 16t^5)\, dt = \left[\frac{6t^4}{4} - \frac{16t^6}{6}\right]_0^1 = \frac{3}{2} - \frac{8}{3} = -\frac{7}{6}.$$

Example 2. *A vector field is given by* $\vec{F} = (\sin y)\hat{i} + x(1 + \cos y)\hat{j}$. *Evaluate the line integral over the circular path given by* $x^2 + y^2 = a^2$, $z = 0$.

Sol. The parametric equations of the circular path are $x = a\cos t$, $y = a \sin t$, $z = 0$, where t varies from 0 to 2π.

Since the particle moves in the xy-plane ($z = 0$), we can take $\vec{r} = x\hat{i} + y\hat{j}$ so that $d\vec{r} = dx\hat{i} + dy\hat{j}$

$$\therefore \quad \oint_C \vec{F}\cdot d\vec{r} = \oint_C [\sin y\hat{i} + x(1 + \cos y)\hat{j}]\cdot(dx\hat{i} + dy\hat{j})$$

$$= \oint_C [\sin y\, dx + x\,(1 + \cos y)\, dy] = \oint_C [(\sin y\, dx + x\cos y\, dy) + x\, dy]$$

$$= \oint_C d\,(x\sin y) + \oint_C x\, dy = \int_0^{2\pi} d\,[a\cos t \sin(a\sin t)] + \int_0^{2\pi} a\cos t\cdot a\cos t\, dt$$

$$= \Big[a\cos t\sin(a\sin t)\Big]_0^{2\pi} + a^2\int_0^{2\pi}\cos^2 t\, dt$$

$$= \frac{a^2}{2}\int_0^{2\pi}(1 + \cos 2t)\, dt = \frac{a^2}{2}\left[t + \frac{\sin 2t}{2}\right]_0^{2\pi} = \frac{a^2}{2}(2\pi) = \pi a^2.$$

Example 3. *If* $\vec{F} = 2y\hat{i} - z\hat{j} + x\hat{k}$, *evaluate* $\oint_C \vec{F}\times d\vec{r}$ *along the curve* $x = \cos t$, $y = \sin t$, $z = 2\cos t$ *from* $t = 0$ *to* $t = \frac{\pi}{2}$.

Sol. $\vec{F}\times d\vec{r} = \begin{vmatrix} \hat{i} & \hat{j} & \hat{k} \\ 2y & -z & x \\ dx & dy & dz \end{vmatrix} = (-z dz - x\, dy)\hat{i} + (x\, dx - 2y\, dz)\hat{j} + (2y\, dy + z\, dx)\hat{k}$

In terms of t,

$$\vec{F} \times d\vec{r} = [-2 \cos t(-2 \sin t)\, dt - \cos t(\cos t)\, dt]\hat{i}$$
$$+ [\cos t(-\sin t)\, dt - 2 \sin t(-2 \sin t)\, dt]\hat{j} + [2 \sin t(\cos t)\, dt + 2 \cos t(-\sin t)\, dt]\hat{k}$$
$$= [(4 \cos t \sin t - \cos^2 t)\hat{i} + (4 \sin^2 t - \cos t \sin t)\hat{j}]\, dt$$

$$\therefore \quad \int_C \vec{F} \times d\vec{r} = \int_0^{\pi/2} [(4 \cos t \sin t - \cos^2 t)\hat{i} + (4 \sin^2 t - \cos t \sin t)\hat{j})\, dt$$
$$= \left(4 \cdot \frac{1}{2} - \frac{1}{2} \cdot \frac{\pi}{2}\right)\hat{i} + \left(4 \cdot \frac{1}{2} \cdot \frac{\pi}{2} - \frac{1}{2}\right)\hat{j} = \left(2 - \frac{\pi}{4}\right)\hat{i} + \left(\pi - \frac{1}{2}\right)\hat{j}.$$

Example 4. *Compute* $\int_C \vec{F} \cdot d\vec{r}$, *where* $\vec{F} = \dfrac{\hat{i}y - \hat{j}x}{x^2 + y^2}$ *and C is the circle* $x^2 + y^2 = 1$ *traversed counter clockwise.*

Sol. For the circle $x^2 + y^2 = 1$ in the xy-plane, we take $\vec{r} = x\hat{i} + y\hat{j}$ so that $d\vec{r} = dx\hat{i} + dy\hat{j}$.

$$\int_C \vec{F} \cdot d\vec{r} = \int_C \left(\frac{\hat{i}y - \hat{j}x}{x^2 + y^2}\right) \cdot (dx\hat{i} + dy\hat{j})$$
$$= \int_C \frac{y\,dx - x\,dy}{x^2 + y^2} = \int_C (y\,dx - x\,dy) \qquad [\because\ x^2 + y^2 = 1]$$

Parametric equation of the circle C : $x^2 + y^2 = 1$ is $x = \cos \theta$, $y = \sin \theta$

so that $dx = -\sin \theta d\theta$, $dy = \cos \theta\, d\theta$ and θ varies from 0 to 2π.

$$\therefore \quad \int_C \vec{F} \cdot d\vec{r} = \int_0^{2\pi} \sin \theta\, (-\sin \theta\, d\theta) - \cos \theta\, (\cos \theta\, d\theta)$$
$$= -\int_0^{2\pi} (\sin^2 \theta + \cos^2 \theta)\, d\theta = -\int_0^{2\pi} d\theta$$
$$= -[\theta]_0^{2\pi} = -2\pi.$$

Example 5. *Compute the line integral* $\int_C (y^2 dx - x^2 dy)$ *about the triangle whose vertices are (1, 0), (0, 1) and (–1, 0).*

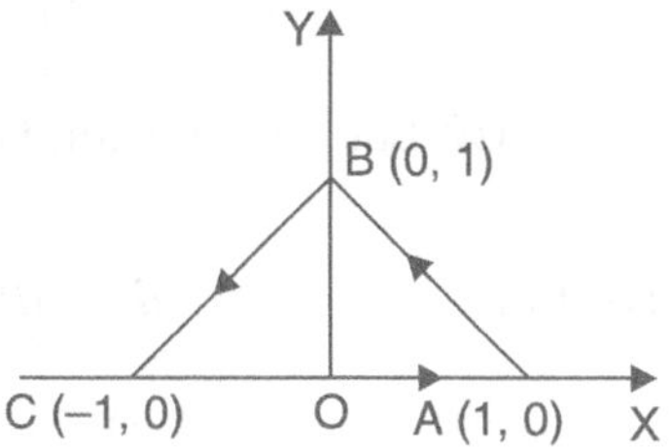

Sol. Here C is the triangle ABC.

On AB Equation of AB is

$$y - 0 = \frac{1 - 0}{0 - 1}(x - 1) \quad \text{or} \quad y = 1 - x$$

$\therefore$ $dy = -dx$ and x varies from 1 to 0.

On BC Equation of BC is

$$y - 1 = \frac{0 - 1}{-1 - 0}(x - 0) \quad \text{or} \quad y = 1 + x$$

$\therefore$ $dy = dx$ and x varies from 0 to –1.

On CA $y = 0$ $\therefore$ $dy = 0$ and x varies from –1 to 1.

$$\therefore \quad \int_C (y^2 dx - x^2 dy) = \int_{AB} (y^2 dx - x^2 dy) + \int_{BC} (y^2 dx - x^2 dy) + \int_{CA} (y^2 dx - x^2 dy)$$

$$= \int_1^0 [(1-x)^2 dx - x^2(-dx)] + \int_0^{-1} [(1+x)^2 dx - x^2 dx] + \int_{-1}^1 0\, dx$$

$$= \int_1^0 (2x^2 - 2x + 1)\, dx + \int_0^{-1} (2x+1)\, dx + 0$$

$$= \left[\frac{2x^3}{3} - \frac{2x^2}{2} + x\right]_1^0 + \left[\frac{2x^2}{2} + x\right]_0^{-1}$$

$$= \left(-\frac{2}{3} + 1 - 1\right) + (1-1) = -\frac{2}{3}.$$

Example 6. *Find the work done in moving a particle in the force field* $\vec{F} = 3x^2\hat{i} + (2xz - y)\hat{j} + z\hat{k}$ *along*

(*i*) *the straight line from (0, 0, 0) to (2, 1, 3)* (P.T.U. May 2010)

(*ii*) *the curve defined by* $x^2 = 4y$, $3x^3 = 8z$ *from* $x = 0$ *to* $x = 2$. (M.T.U. 2012)

Sol. Work done $= \int_C \vec{F}.d\vec{r} = \int_C [3x^2\hat{i} + (2xz - y)\hat{j} + z\hat{k}].(dx\hat{i} + dy\hat{j} + dz\hat{k})$

$$= \int_C [3x^2 dx + (2xz - y)dy + z dz] \qquad ...(1)$$

(*i*) Equation of straight line from (0, 0, 0) to (2, 1, 3) are

$$\frac{x-0}{2-0} = \frac{y-0}{1-0} = \frac{z-0}{3-0} \quad \text{or} \quad \frac{x}{2} = \frac{y}{1} = \frac{z}{3} = t \quad \text{(say)}$$

$\therefore$ C : $x = 2t$, $y = t$, $z = 3t$

so that $dx = 2dt$, $dy = dt$, $dz = 3dt$

The points (0, 0, 0) and (2, 1, 3) correspond to $t = 0$ and $t = 1$ respectively.

$\therefore$ From (1), we have

$$\text{Work done} = \int_0^1 [3(2t)^2\, 2dt + \{2(2t)(3t) - t\}\, dt + (3t)\, 3dt]$$

$$= \int_0^1 (24t^2 + 12t^2 - t + 9t)dt = \int_0^1 (36t^2 + 8t)dt$$

$$= \left[12t^3 + 4t^2\right]_0^1 = 16$$

(*ii*) Let $x = t$, then C: $x = t$, $y = \frac{t^2}{4}$, $z = \frac{3t^3}{8}$

so that $dx = dt$, $dy = \frac{t}{2}\, dt$, $dz = \frac{9t^2}{8}\, dt$

From $x = 0$ to $x = 2$, t varies from 0 to 2. $(\because \ t = x)$

$\therefore$ From (1), we have

$$\text{Work done} = \int_0^2 \left[3t^2 dt + \left\{2(t)\left(\frac{3t^3}{8}\right) - \frac{t^2}{4}\right\}\left(\frac{t}{2}\, dt\right) + \frac{3t^3}{8}\left(\frac{9t^2}{8}\, dt\right)\right]$$

$$= \int_0^2 \left(3t^2 + \frac{3}{8}t^5 - \frac{1}{8}t^3 + \frac{27}{64}t^5\right) dt = \int_0^2 \left(\frac{51}{64}t^5 - \frac{1}{8}t^3 + 3t^2\right) dt$$

$$= \left[\frac{51}{64} \cdot \frac{t^6}{6} - \frac{t^4}{32} + t^3 \right]_0^2 = \frac{17}{2} - \frac{1}{2} + 8 = 16.$$

Example 7. *Find the work done by the force $\vec{F} = x\hat{i} - z\hat{j} + 2y\hat{k}$ in the displacement along the closed path C consisting of the segments C_1, C_2 and C_3 where*

on C_1, $0 \le x \le 1$, $y = x$, $z = 0$
on C_2, $0 \le z \le 1$, $x = 1$, $y = 1$
on C_3, $1 \ge x \ge 0$, $y = z = x$.

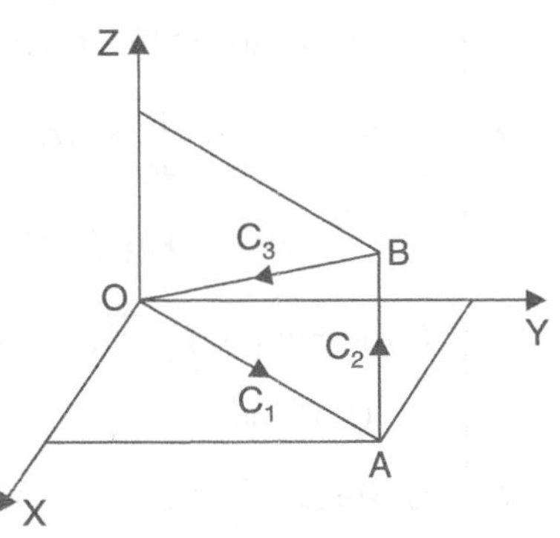

Sol. Total work done

$$= \oint_C \vec{F} \cdot \vec{dr} = \oint_C (x\hat{i} - z\hat{j} + 2y\hat{k}) \cdot (dx\hat{i} + dy\hat{j} + dz\hat{k})$$

$$= \oint_C (xdx - zdy + 2ydz)$$

$$= \oint_{C_1} (xdx - zdy + 2ydz) + \oint_{C_2} (xdx - zdy + 2ydz) + \oint_{C_3} (xdx - zdy + 2ydz)$$

$$= W_1 + W_2 + W_3$$

where W_1, W_2, W_3 denote the work done in displacement along C_1, C_2 and C_3 respectively.

On C_1, $0 \le x \le 1$, $y = x$, $z = 0$, $dy = dx$, $dz = 0$

$$\therefore \quad W_1 = \int_{C_1} x\,dx = \int_0^1 x\,dx = \frac{1}{2}$$

On C_2, $0 \le z \le 1$, $x = 1$, $y = 1$, $dx = 0$, $dy = 0$

$$\therefore \quad W_2 = \int_{C_2} 2dz = 2\int_0^1 dz = 2$$

On C_3, $1 \ge x \ge 0$, $y = z = x$, $dy = dz = dx$

$$\therefore \quad W_3 = \int_{C_3} (x\,dx - x\,dx + 2\,xdx) = 2\int_1^0 x\,dx = -1.$$

Total work done $= W_1 + W_2 + W_3 = \frac{1}{2} + 2 - 1 = \frac{3}{2}.$

EXERCISE 13.5

1. (*a*) If $\phi = 2xyz^2$, $\vec{F} = xy\hat{i} - z\hat{j} + x^2\hat{k}$, and C is the curve $x = t^2$, $y = 2t$, $z = t^3$ from $t = 0$ to $t = 1$, evaluate the line integrals

   (*i*) $\int_C \phi\, \vec{dr}$ (*ii*) $\int_C \vec{F} \times \vec{dr}$.

   (*b*) Evaluate $\int_C (x^2 + yz)\,dz$, where C is the curve defined by $x = t$, $y = t^2$, $z = 3t$ for t lying in the interval $1 \le t \le 2$. *(P.T.U. May 2008)*

2. (*a*) Evaluate the line integral $\int_C [(x^2 + xy)dx + (x^2 + y^2)dy]$, where C is the square formed by the lines $y = \pm 1$ and $x \pm 1$.

(b) If $\vec{F} = (5xy - 6x^2)\hat{i} + (2y - 4x)\hat{j}$, evaluate $\int_C \vec{F} \cdot d\vec{r}$ along the curve C in the xy-plane, $y = x^3$ from the point (1, 1) to (2, 8).

3. If $\vec{F} = (3x^2 + 6y)\hat{i} - 14yz\hat{j} + 20xz^2\hat{k}$, evaluate the line integral $\int_C \vec{F} \cdot d\vec{r}$ from (0, 0, 0) to (1, 1, 1) along the following paths C:

(i) $x = t, y = t^2, z = t^3$ (G.B.T.U. 2010) (ii) the straight line joining (0, 0, 0) to (1, 1, 1)

(iii) the straight line from (0, 0, 0) to (1, 0, 0), then to (1, 1, 0) and then to (1, 1, 1).

4. (i) Show that $\int_C \vec{F} \cdot d\vec{r} = 3\pi$, given that $\vec{F} = z\hat{i} + x\hat{j} + y\hat{k}$ and C being the arc of the curve $\vec{r} = \cos t\hat{i} + \sin t\hat{j} + t\hat{k}$ from $t = 0$ to $t = 2\pi$.

(ii) Evaluate $\int_C \vec{F} \cdot d\vec{R}$, where $\vec{F} = (2z, x, -y)$ and C is $\vec{R} = (\cos t, \sin t, 2t)$ from (1, 0, 0) to (1, 0, 4π). (K.U.K. Dec. 2012)

5. Evaluate $\int_C [(2x^2y + y + z^2)\hat{i} + 2(1 + yz^3)\hat{j} + (2z + 3y^2z^2)\hat{k}] \cdot d\vec{r}$ along the curve C: $y^2 + z^2 = a^2$, $x = 0$.

6. Evaluate $\int_C [(y + 3z)\, dx + (2z + x)\, dy + (3x + 2y)\, dz]$, where C is the arc of helix $x = a \cos \theta$, $y = a \sin \theta$, $z = \dfrac{2a\theta}{\pi}$ between the points (a, 0, 0) and (0, a, a).

7. (a) Find the circulation of $\vec{F}$ round the curve C, where $\vec{F} = y\hat{i} + z\hat{j} + x\hat{k}$ and C is the circle $x^2 + y^2 = 1, z = 0$.

(b) If $\vec{A} = (x - y)\hat{i} + (x + y)\hat{j}$, evaluate $\oint_C \vec{A} \cdot d\vec{r}$ around the curve C consisting of $y = x^2$ and $y^2 = x$. (G.B.T.U. 2012)

8. Find the circulation of $\vec{F}$ round the curve C, where $\vec{F} = e^x \sin y\hat{i} + e^x \cos y\hat{j}$ and C is the rectangle whose vertices are (0, 0), (1, 0), $\left(1, \dfrac{\pi}{2}\right)$ and $\left(0, \dfrac{\pi}{2}\right)$.

9. Find the work done when a force $\vec{F} = (x^2 - y^2 + x)\hat{i} - (2xy + y)\hat{j}$ moves a particle in the xy-plane from (0, 0) to (1, 1) along the parabola $y^2 = x$. Is the work done different when the path is the straight line $y = x$?

10. Find the work done in moving a particle once round the circle $x^2 + y^2 = 9$ in the xy-plane if the field of force is $\vec{F} = (2x - y - z)\hat{i} + (x + y - z^2)\hat{j} + (3x - 2y + 4z)\hat{k}$. If possible, find its scalar potential. (M.D.U. May 2010)

11. Compute the work done by the force $\vec{F} = (2y + 3)\hat{i} + xz\hat{j} + (yz - x)\hat{k}$ when it moves a particle from the point (0, 0, 0) to the point (2, 1, 1) along the curve $x = 2t^2, y = t, z = t^3$. (G.B.T.U. 2011)

12. (a) Suppose $\vec{F}(x, y, z) = x^3\hat{i} + y\hat{j} + z\hat{k}$ is the force field. Find the work done by $\vec{F}$ along the line from (1, 2, 3) to (3, 5, 7).

(b) Find the total work done by a force $\vec{F} = (x^2 + y^2)\hat{i} - 2xy\hat{j}$ in moving a point from (0, 0) to (a, b) along the rectangle bounded by the lines $x = 0, x = a, y = 0$ and $y = b$.

Answers

1. (a) (i) $\frac{8}{11}\hat{i}+\frac{4}{5}\hat{j}+\hat{k}$ (ii) $-\frac{9}{10}\hat{i}-\frac{2}{3}\hat{j}+\frac{7}{5}\hat{k}$ (b) $\frac{163}{4}$

2. (a) 0 (b) 35

3. (i) 5 (ii) $\frac{13}{3}$ (iii) $\frac{23}{3}$ **5.** 0

6. $2a^2$ **7.** (a) $-\pi$ (b) $\frac{2}{3}$ **8.** 0

9. $-\frac{2}{3}$; No, it is same **10.** 18π ; not possible **11.** $8\frac{8}{35}$

12. (a) $\frac{101}{2}$ (b) $\frac{a^3}{3}-ab^2$

13.25. SURFACE INTEGRALS

Any integral which is to be evaluated over a surface is called a surface integral.

Let $\vec{F}(P)$ be a continuous vector point function and S a two sided surface. Divide S into a finite number of sub-surfaces $\delta S_1, \delta S_2, ..., \delta S_k$. Let P_i be any point in δS_i and $\hat{n}_i$ be the unit vector at P in the direction of outward drawn normal to the surface at P_i. Then the limit of the sum

$$\sum_{i=1}^{k} \vec{F}(P_i)\cdot\hat{n}_i\,\delta S_i,$$ as $k \to \infty$ and each $\delta S_i \to 0$ is called the *normal surface integral* of $\vec{F}(P)$ over S and is denoted by $\iint_S \vec{F}\cdot\hat{n}\,dS$.

The surface element $\vec{\delta S}$ surrounding any point P can be regarded as a vector whose magnitude is area δS and the direction that of the outward drawn normal $\hat{n}$ *i.e.* $\vec{\delta S}=\hat{n}\,\delta S$. The surface integral may alternatively be written as $\iint_S \vec{F}\cdot d\vec{S}$.

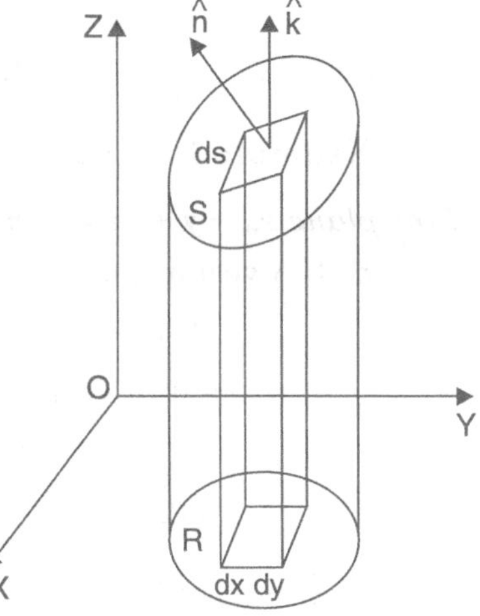

If $\vec{F}$ represents the velocity of a fluid at any point P on a closed surface S, then $\vec{F}\cdot\hat{n}$ is the normal component of $\vec{F}$ at P and $\iint_S \vec{F}\cdot\hat{n}\,dS=\iint_S \vec{F}\cdot d\vec{S}$ is a measure of volume emerging from S per unit time, *i.e.* it measures the *flux of $\vec{F}$ over S.*

Other types of surface integrals are $\iint_S \vec{F}\times d\vec{S}$, $\iint_S \phi\, d\vec{S}$.

Note. In order to evaluate surface integrals, it is convenient to express them as double integrals taken over the orthogonal projection of S on one of the coordinate planes. But this is possible only when any line perpendicular to the coordinate plane chosen meets the surface S in not more than one point.

Let R be the orthogonal projection of S on the *xy*-plane.

Let $\hat{n} = \cos\alpha\hat{i} + \cos\beta\hat{j} + \cos\gamma\hat{k}$ where $\cos\alpha$, $\cos\beta$, $\cos\gamma$ are the direction cosines of $\hat{n}$.

Now, $dxdy$ = projection of dS on the xy-plane = dS $\cos\gamma \Rightarrow d\text{S} = \dfrac{dx\,dy}{\cos\gamma}$

Also $|\hat{k}\cdot\hat{n}| = \cos\gamma \quad \therefore \quad d\text{S} = \dfrac{dxdy}{|\hat{k}\cdot\hat{n}|}$

Hence $\iint_S \vec{F}\cdot\hat{n}\,ds = \iint_R \vec{F}\cdot\hat{n}\,\dfrac{dxdy}{|\hat{k}\cdot\hat{n}|}$.

13.26. VOLUME INTEGRALS

Any integral which is to be evaluated over a volume is called a volume integral.

If V is a volume bounded by a surface S, then the triple integrals

$$\iiint_V \phi\,dV \quad \text{and} \quad \iiint_V \vec{F}\,dV$$

are called volume integrals. The first of these is a scalar and the second is a vector.

If we sub-divide the volume V into small cuboids by drawing planes parallel to the coordinate planes, then $dV = dx\,dy\,dz$.

$$\therefore \quad \iiint_V \phi\,dV = \iiint_V \phi(x, y, z)\,dx\,dy\,dz$$

If $\vec{F} = F_1\hat{i} + F_2\hat{j} + F_3\hat{k}$, then

$$\iiint_V \vec{F}\,dV = \hat{i}\iiint_V F_1(x, y, z)\,dx\,dy\,dz + \hat{j}\iiint_V F_2(x, y, z)\,dx\,dy\,dz + \hat{k}\iiint_V F_3(x, y, z)\,dx\,dy\,dz.$$

ILLUSTRATIVE EXAMPLES

Example 1. *Evaluate* $\iint_S \vec{A}\cdot\hat{n}\,dS$, *where* $\vec{A} = (x + y^2)\hat{i} - 2x\hat{j} + 2yz\hat{k}$ *and S is the surface of the plane* $2x + y + 2z = 6$ *in the first octant.*

Sol. A vector normal to the surface S is given by

$$\nabla(2x + y + 2z) = 2\hat{i} + \hat{j} + 2\hat{k}$$

$\therefore \quad \hat{n}$ = a unit vector normal to surface S

$$= \frac{2\hat{i} + \hat{j} + 2\hat{k}}{\sqrt{(2)^2 + 1^2 + (2)^2}} = \frac{2}{3}\hat{i} + \frac{1}{3}\hat{j} + \frac{2}{3}\hat{k}$$

$$\hat{k}\cdot\hat{n} = \hat{k}\cdot\left(\frac{2}{3}\hat{i} + \frac{1}{3}\hat{j} + \frac{2}{3}\hat{k}\right) = \frac{2}{3}$$

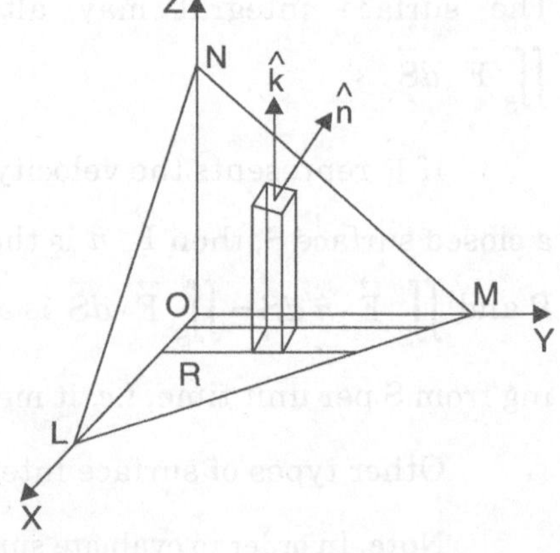

$$\therefore \quad \iint_S \vec{A}\cdot\hat{n}\,dS = \iint_R \vec{A}\cdot\hat{n}\,\frac{dxdy}{|\hat{k}\cdot\hat{n}|},$$

where R is the projection of S, *i.e.* triangle LMN on the xy-plane. The region R, *i.e.* triangle OLM is bounded by x-axis, y-axis and the line $2x + y = 6$, $z = 0$.

Now, $\vec{A} \cdot \hat{n} = [(x + y^2)\hat{i} - 2x\hat{j} + 2yz\hat{k}] \cdot \left(\frac{2}{3}\hat{i} + \frac{1}{3}\hat{j} + \frac{2}{3}\hat{k}\right)$

$$= \frac{2}{3}(x + y^2) - \frac{2}{3}x + \frac{4}{3}yz = \frac{2}{3}y^2 + \frac{4}{3}yz$$

$$= \frac{2}{3}y^2 + \frac{4}{3}y\left(\frac{6 - 2x - y}{2}\right)$$

$$\left(\because \text{ on the plane } 2x + y + 2z = 6, z = \frac{6 - 2x - y}{2}\right)$$

$$= \frac{2}{3}y(y + 6 - 2x - y) = \frac{4}{3}y(3 - x)$$

Hence, $\iint_S \vec{A} \cdot \hat{n}\, dS = \iint_R \vec{A} \cdot \hat{n} \frac{dxdy}{|\hat{k} \cdot \hat{n}|}$

$$= \iint_R \frac{4}{3}y(3 - x) \cdot \frac{3}{2}\, dxdy = \int_0^3 \int_0^{6 - 2x} 2y(3 - x)\, dydx$$

$$= \int_0^3 2(3 - x) \cdot \left[\frac{y^2}{2}\right]_2^{6 - 2x} dx = \int_0^3 (3 - x)(6 - 2x)^2\, dx$$

$$= 4\int_0^3 (3 - x)^3\, dx = 4 \cdot \left[\frac{(3 - x)^4}{4(-1)}\right]_0^3 = -(0 - 81) = 81.$$

Example 2. *Evaluate* $\iint_S \vec{A} \cdot \hat{n}\, dS$, *where* $\vec{A} = z\hat{i} + x\hat{j} - 3y^2z\hat{k}$ *and S is the curved surface of the cylinder* $x^2 + y^2 = 16$ *included in the first octant between* $z = 0$ *and* $z = 5$.

Sol. A vector normal to the surface S is given by $\nabla(x^2 + y^2) = 2x\hat{i} + 2y\hat{j}$

$\therefore$ $\hat{n}$ = a unit vector normal to surface S

$$= \frac{2x\hat{i} + 2y\hat{j}}{\sqrt{(2x)^2 + (2y)^2}} = \frac{x\hat{i} + y\hat{j}}{\sqrt{x^2 + y^2}} = \frac{x\hat{i} + y\hat{j}}{4}$$

($\because$ on the surface of cylinder, $x^2 + y^2 = 16$)

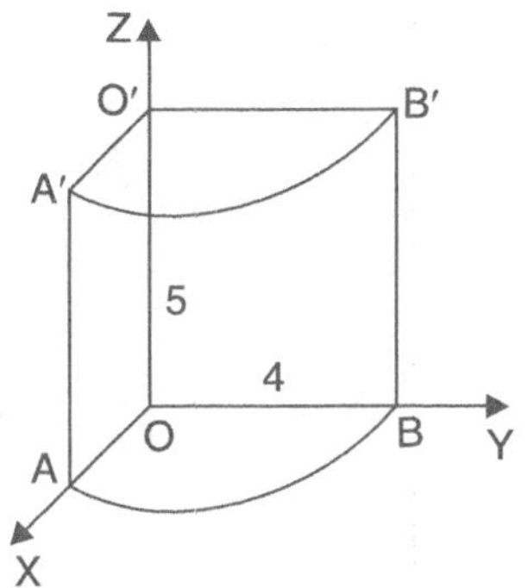

Let R be the projection of S on yz-plane, then

$$\iint_S \vec{A} \cdot \hat{n}\, dS = \iint_R \vec{A} \cdot \hat{n} \frac{dydz}{|\hat{i} \cdot \hat{n}|}$$

The region R is OBB′O′ enclosed by $y = 0$ to $y = 4$ and $z = 0$ to $z = 5$.

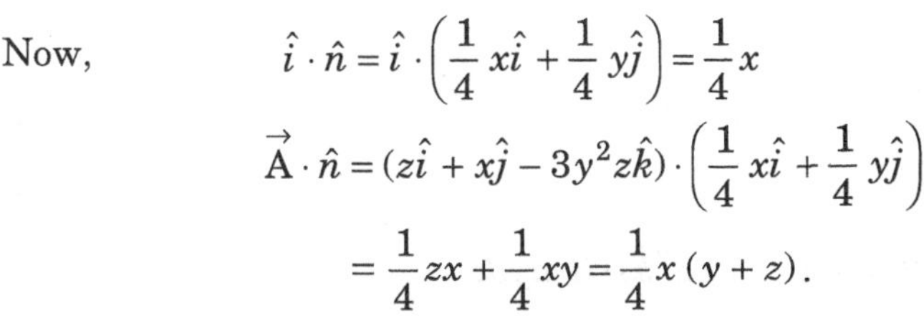

Now, $\hat{i} \cdot \hat{n} = \hat{i} \cdot \left(\frac{1}{4}x\hat{i} + \frac{1}{4}y\hat{j}\right) = \frac{1}{4}x$

$$\vec{A} \cdot \hat{n} = (z\hat{i} + x\hat{j} - 3y^2z\hat{k}) \cdot \left(\frac{1}{4}x\hat{i} + \frac{1}{4}y\hat{j}\right)$$

$$= \frac{1}{4}zx + \frac{1}{4}xy = \frac{1}{4}x(y + z).$$

$$\text{Hence } \iint_S \vec{A} \cdot \hat{n}\, dS = \iint_R \vec{A} \cdot \hat{n} \frac{dydz}{|\hat{i} \cdot \hat{n}|} = \iint_R \frac{1}{4} x(y+z) \frac{dydz}{\frac{1}{4}x} = \int_0^5 \int_0^4 (y+z)\, dy\, dz$$

$$= \int_0^5 \left[\frac{y^2}{2} + zy \right]_0^4 dz = \int_0^5 (8 + 4z)\, dz = \left[8z + 2z^2 \right]_0^5 = 40 + 50 = 90.$$

Example 3. *Evaluate* $\iint_S \vec{F} \cdot \hat{n}\, dS$, *where* $\vec{F} = 2x^2 y\hat{i} - y^2 \hat{j} + 4xz^2 \hat{k}$ *and S is the closed surface of the region in the first octant bounded by the cylinder* $y^2 + z^2 = 9$ *and the planes* $x = 0$, $x = 2$, $y = 0$ *and* $z = 0$.

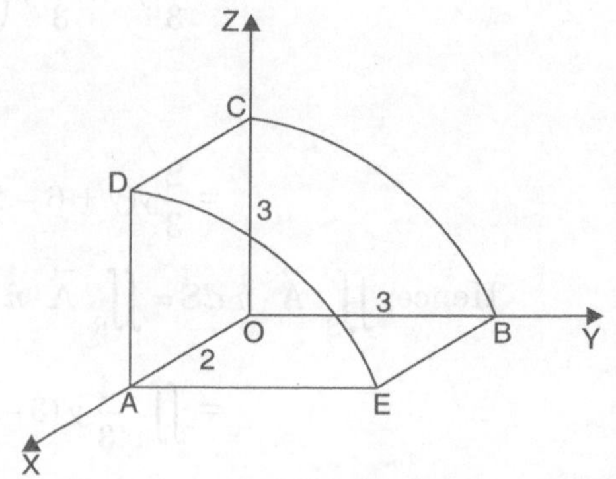

Sol. The given surface is comprised of the following five surfaces:

Surface	*Equation*	*ds*	*Outward normal*
$S_1 \to$ rectangle OAEB	$z = 0$	$dxdy$	$-\hat{k}$
$S_2 \to$ rectangle OADC	$y = 0$	$dxdz$	$-\hat{j}$
$S_3 \to$ circular quadrant OBC	$x = 0$	$dydz$	$-\hat{i}$
$S_4 \to$ circular quadrant ADE	$x = 2$	$dydz$	$\hat{i}$
$S_5 \to$ curved surface BCDE	$y^2 + z^2 = 9$, $0 \le x \le 2$	$\dfrac{dxdy}{\lvert\hat{n} \cdot \hat{k}\rvert}$	$\hat{n}$

$$\therefore \quad \iint_S \vec{F} \cdot \hat{n}\, dS = \iint_{S_1} \vec{F} \cdot \hat{n}\, dS + \iint_{S_2} \vec{F} \cdot \hat{n}\, dS + \iint_{S_3} \vec{F} \cdot \hat{n}\, dS + \iint_{S_4} \vec{F} \cdot \hat{n}\, dS + \iint_{S_5} \vec{F} \cdot \hat{n}\, dS \quad \ldots(1)$$

Now,
$$\iint_{S_1} \vec{F} \cdot \hat{n}\, dS = \iint_{S_1} (2x^2 y\hat{i} - y^2 \hat{j}) \cdot (-\hat{k})\, dxdy = 0 \qquad (\because z = 0)$$

$$\iint_{S_2} \vec{F} \cdot \hat{n}\, dS = \iint_{S_2} (4xz^2 \hat{k}) \cdot (-\hat{j})\, dxdz = 0 \qquad (\because y = 0)$$

$$\iint_{S_3} \vec{F} \cdot \hat{n}\, dS = \iint_{S_3} (-y^2 \hat{j}) \cdot (-\hat{i}) dydz = 0 \qquad (\because x = 0)$$

$$\iint_{S_4} \vec{F} \cdot \hat{n}\, dS = \iint_{S_4} (8y\hat{i} - y^2 \hat{j} + 8z^2 \hat{k}) \cdot (\hat{i}) dydz \qquad (\because x = 2)$$

$$= \iint_{S_4} 8y\,dy\,dz = \int_0^3 \int_0^{\sqrt{9-z^2}} 8y\,dy\,dz$$

$$= \int_0^3 [4y^2]_0^{\sqrt{9-z^2}} dz = \int_0^3 4(9 - z^2)dz$$

$$= 4\left[9z - \frac{z^3}{3}\right]_0^3 = 4(27 - 9) = 72$$

To find $\iint_{S_5} \vec{F} \cdot \hat{n}\, dS$, we first find $\hat{n}$.

A vector normal to $y^2 + z^2 = 9$ is $\vec{n} = \nabla(y^2 + z^2) = \left(\hat{i}\frac{\partial}{\partial x} + \hat{j}\frac{\partial}{\partial y} + \hat{k}\frac{\partial}{\partial z}\right)(y^2 + z^2)$

$$= 2(y\hat{j} + z\hat{k})$$

$\hat{n}$ = unit vector normal to S_5

$$= \frac{\vec{n}}{|\vec{n}|} = \frac{2(y\hat{j} + z\hat{k})}{2\sqrt{y^2 + z^2}} = \frac{y\hat{j} + z\hat{k}}{3} \qquad (\because\ y^2 + z^2 = 9)$$

$$|\hat{n} \cdot \hat{k}| = \left|\frac{y\hat{j} + z\hat{k}}{3} \cdot \hat{k}\right| = \frac{z}{3}$$

$$\therefore \quad ds = \frac{dxdy}{\frac{z}{3}} = \frac{3\,dxdy}{z}$$

$$\vec{F} \cdot \hat{n} = (2x^2y\hat{i} - y^2\hat{j} + 4x\,z^2\hat{k}) \cdot \frac{y\hat{j} + z\hat{k}}{3} = \frac{1}{3}(-y^3 + 4xz^3)$$

$$\therefore \quad \iint_{S_5} \vec{F} \cdot \hat{n}\, dS = \iint_{S_5} \frac{1}{3}(-y^3 + 4xz^3)\frac{3dxdy}{z} = \iint_{S_5} \left(-\frac{y^3}{z} + 4xz^2\right) dxdy$$

$$= \int_0^3 \int_0^2 \left[-\frac{y^3}{\sqrt{9 - y^2}} + 4x\,(9 - y^2)\right] dxdy$$

$$\left[\because \text{ On } S_5,\ y^2 + z^2 = 9 \Rightarrow z = \sqrt{9 - y^2}\right]$$

$$= \int_0^3 \left[-\frac{xy^3}{\sqrt{9 - y^2}} + 2x^2(9 - y^2)\right]_0^2 dy = \int_0^3 \left[-\frac{2y^3}{\sqrt{9 - y^2}} + 8(9 - y^2)\right] dy$$

$$\left[\begin{array}{l}\text{Put } y = 3\sin\theta, \text{ then } dy = 3\cos\theta\, d\theta \\ \text{when } y = 0,\ \theta = 0;\ \text{when } y = 3,\ \theta = \frac{\pi}{2}\end{array}\right]$$

$$= \int_0^{\frac{\pi}{2}} \left[\frac{-2 \times 27\sin^3\theta}{3\cos\theta} + 8 \cdot 9\cos^2\theta\right] 3\cos\theta\, d\theta$$

$$= \int_0^{\frac{\pi}{2}} (-54\sin^3\theta + 216\cos^3\theta)\, d\theta$$

$$= -54 \times \frac{2}{3} + 216 \times \frac{2}{3} = \frac{2}{3}(162) = 108$$

$\therefore$ From (1),

$$\iint_S \vec{F} \cdot \hat{n}\, dS = 0 + 0 + 0 + 72 + 108 = 180$$

Example 4. *If* $\vec{F} = (2x^2 - 3z)\hat{i} - 2xy\hat{j} - 4x\hat{k}$, *then evaluate* $\iiint_V \nabla \cdot \vec{F}\, dV$, *where V is bounded by the planes x = 0, y = 0, z = 0 and 2x + 2y + z = 4.*

Sol. $$\nabla \cdot \vec{F} = \frac{\partial}{\partial x}(2x^2 - 3z) + \frac{\partial}{\partial y}(-2xy) + \frac{\partial}{\partial z}(-4x) = 4x - 2x = 2x$$

$$\therefore \quad \iiint_V \nabla \cdot \vec{F}\, dV = \iiint_V 2x\, dx\, dy\, dz$$

$$= \int_0^2 \int_0^{2-x} \int_0^{4-2x-2y} 2x\, dz\, dy\, dx = \int_0^2 \int_0^{2-x} 2x \Big[z \Big]_0^{4-2x-2y} dy\, dx$$

$$= \int_0^2 \int_0^{2-x} 2x(4 - 2x - 2y)\, dy\, dx = \int_0^2 \int_0^{2-x} [4x(2-x) - 4xy]\, dy dx$$

$$= \int_0^2 \Big[4x(2-x)y - 2xy^2 \Big]_0^{2-x} dx = \int_0^2 [4x(2-x)^2 - 2x(2-x)^2]\, dx$$

$$= \int_0^2 2x(2-x)^2\, dx = 2\int_0^2 (4x - 4x^2 + x^3)\, dx$$

$$= 2\left[2x^2 - 4\frac{x^3}{3} + \frac{x^4}{4} \right]_0^2 = 2\left(8 - \frac{32}{3} + 4 \right) = \frac{8}{3}.$$

Example 5. *If* $\vec{A} = 2xz\hat{i} - x\hat{j} + y^2\hat{k}$, *evaluate* $\iiint_V \vec{A}\, dV$, *where V is the region bounded by the surface x = 0, y = 0, x = 2, y = 6, z = x^2, z = 4.*

Sol. $$\iiint_V \vec{A}\, dV = \iiint_V (2xz\hat{i} - x\hat{j} + y^2\hat{k})\, dx\, dy\, dz$$

$$= \int_0^2 \int_0^6 \int_{x^2}^4 (2xz\hat{i} - x\hat{j} + y^2\hat{k})\, dz\, dy\, dx$$

$$= \hat{i}\int_0^2 \int_0^6 \int_{x^2}^4 2xz dz\, dy\, dx - \hat{j}\int_0^2 \int_0^6 \int_{x^2}^4 x dz\, dy\, dx + \hat{k}\int_0^2 \int_0^6 \int_{x^2}^4 y^2\, dz\, dy\, dx$$

$$= \hat{i}\int_0^2 \int_0^6 \Big[xz^2 \Big]_{x^2}^4 dy\, dx - \hat{j}\int_0^2 \int_0^6 \Big[xz \Big]_{x^2}^4 dy\, dx + \hat{k}\int_0^2 \int_0^6 \Big[y^2 z \Big]_{x^2}^4 dy\, dx$$

$$= \hat{i}\int_0^2 \int_0^6 (16x - x^5)\, dy\, dx - \hat{j}\int_0^2 \int_0^6 (4x - x^3)\, dy\, dx + \hat{k}\int_0^2 \int_0^6 y^2(4 - x^2)\, dy\, dx$$

$$= \hat{i}\int_0^2 (16x - x^5)\Big[y \Big]_0^6 dx - \hat{j}\int_0^2 (4x - x^3)\Big[y \Big]_0^6 dx + \hat{k}\int_0^2 (4 - x^2)\left[\frac{y^3}{3} \right]_0^6 dx$$

$$= 6\hat{i}\int_0^2 (16x - x^5)\,dx - 6\hat{j}\int_0^2 (4x - x^3)\,dx + 72\hat{k}\int_0^2 (4 - x^2)\,dx$$

$$= 6\hat{i}\left[8x^2 - \frac{x^6}{6}\right]_0^2 - 6\hat{j}\left[2x^2 - \frac{x^4}{4}\right]_0^2 + 72\hat{k}\left[4x - \frac{x^3}{3}\right]_0^2$$

$$= 128\hat{i} - 24\hat{j} + 384\hat{k}.$$

EXERCISE 13.6

1. Evaluate $\iint_S \vec{F}.\hat{n}\,dS$, where $\vec{F} = 18z\hat{i} - 12\hat{j} + 3y\hat{k}$ and S is the surface of the plane $2x + 3y + 6z = 12$ in the first octant. *(M.T.U. 2012; P.T.U. May 2010)*

2. (*a*) Evaluate $\iint_S \vec{A}\cdot\hat{n}\,dS$, where $\vec{A} = 12x^2y\hat{i} - 3yz\hat{j} + 2z\hat{k}$ and S is the portion of the plane $x + y + z = 1$ included in the first octant.

   (*b*) Evaluate $\iint_S \vec{F}\cdot\hat{n}\,dS$, where $\vec{F} = 6z\hat{i} - 4\hat{j} + y\hat{k}$ and S is the portion of the plane $2x + 3y + 6z = 12$ in the first octant.

3. Show that $\iint_S \vec{F}\cdot\hat{n}\,dS = \frac{3}{2}$, where $\vec{F} = 4xz\hat{i} - y^2\hat{j} + yz\hat{k}$ and S is the surface of the cube bounded by the planes $x = 0, x = 1, y = 0, y = 1, z = 0, z = 1$.

4. Evaluate $\iint_S \vec{F}\cdot d\vec{S}$, where $\vec{F} = yz\hat{i} + zx\hat{j} + xy\hat{k}$ and S is that part of the surface $x^2 + y^2 + z^2 = 1$ which lies in the first octant.

5. Evaluate $\iint_S \vec{F}\cdot\hat{n}\,dS$, where $\vec{F} = 4x\hat{i} - 2y^2\hat{j} + z^2\hat{k}$ and S is the surface bounding the region
   $x^2 + y^2 = 4,\ z = 0,\ z = 3$. *(Calicut 2009; K.U.K. 2012)*

6. Evaluate $\iiint_V \phi\,dV$, where $\phi = 45x^2y$ and V is the closed region bounded by the planes
   $4x + 2y + z = 8,\ x = 0,\ y = 0,\ z = 0$.

7. If $\vec{F} = (2x^2 - 3z)\hat{i} - 2xy\hat{j} - 4x\hat{k}$, then evaluate $\iiint_V \nabla\times\vec{F}\,dV$ and $\iiint_V \nabla\cdot\vec{F}\,dV$, where V is the closed region bounded by the planes $x = 0, y = 0, z = 0$ and $2x + 2y + z = 4$.

8. If $\hat{F} = 2z\hat{i} - x\hat{j} + y\hat{k}$, evaluate $\iiint_V \vec{F}\,dV$, where V is the region bounded by the surface $x = 0$, $y = 0, x = 2, y = 4, z = x^2, z = 2$.

Answers

1. 24   2. (*a*) $\frac{49}{120}$ (*b*) 8   4. $\frac{3}{8}$

5. 84π 6. 128   7. $\frac{8}{3}(\hat{j} - \hat{k});\ \frac{8}{3}$ 8. $\frac{32}{15}(3\hat{i} + 5\hat{k})$.

13.27. DIVERGENCE THEOREM OF GAUSS

(Relation between surface and volume integrals) *(G.B.T.U. 2012; Calicut 2010)*

Statement. *If* $\vec{F}$ *is a vector point function having continuous first order partial derivatives in the* **region V bounded by a closed surface S,** *then* $\iiint_V \nabla \cdot \vec{F}\, dV = \iint_S \vec{F} \cdot \hat{n}\, dS$, *where* $\hat{n}$ *is the outwards drawn unit normal vector to the surface S.*

[*i.e.*, the volume integral of the divergence of a vector point function $\vec{F}$ taken over the volume V enclosed by a surface S, is equal to the surface integral of the normal component of $\vec{F}$ taken over the closed surface S].

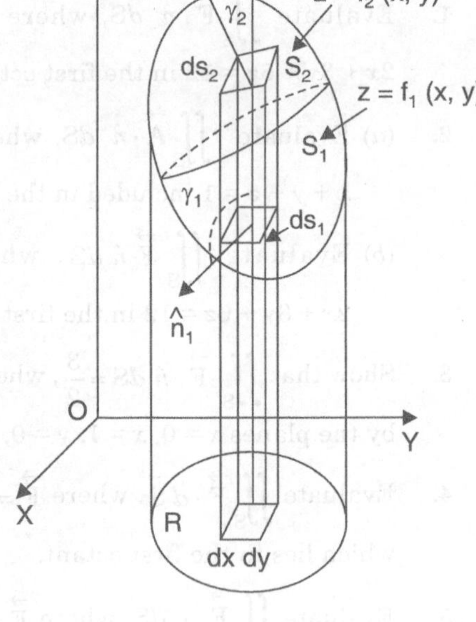

Proof. Let $\vec{F} = F_1\hat{i} + F_2\hat{j} + F_3\hat{k}$, then

$$\nabla \cdot \vec{F} = \text{div}\, \vec{F} = \frac{\partial F_1}{\partial x} + \frac{\partial F_2}{\partial y} + \frac{\partial F_3}{\partial z}.$$

Let α, β, γ be the angles which the outwards drawn unit normal vector $\hat{n}$ makes with the positive directions of *x*, *y*, *z*-axes respectively, then cos α, cos β, cos γ are the direction cosines of $\hat{n}$ and $\hat{n} = \cos\alpha\hat{i} + \cos\beta\hat{j} + \cos\gamma\hat{k}$.

$$\therefore \vec{F} \cdot \hat{n} = (F_1\hat{i} + F_2\hat{j} + F_3\hat{k}) \cdot (\cos\alpha\hat{i} + \cos\beta\hat{j} + \cos\gamma\hat{k})$$
$$= F_1\cos\alpha + F_2\cos\beta + F_3\cos\gamma$$

Therefore, the cartesian equivalent of divergence theorem is

$$\iiint_V \left(\frac{\partial F_1}{\partial x} + \frac{\partial F_2}{\partial y} + \frac{\partial F_3}{\partial z}\right) dxdydz$$
$$= \iint_S (F_1\cos\alpha + F_2\cos\beta + F_3\cos\gamma)\, dS \quad ...(1)$$
$$= \iint_S (F_1\, dydz + F_2\, dzdx + F_3\, dxdy)$$

since $\cos\alpha\, dS = dy\, dz$, etc.

Suppose that S is such a closed surface that a line parallel to the co-ordinate axes meets it in two points only. Let S_1 and S_2 denote the lower and upper portions of S with equations $z = f_1(x, y)$ and $z = f_2(x, y)$ respectively.

Let R be the projection of S on the *xy*-plane, then

$$\iiint_V \frac{\partial F_3}{\partial z}\, dx\, dy\, dz = \iint_R \left[\int_{z=f_1(x,y)}^{z=f_2(x,y)} \frac{\partial F_3}{\partial z}\, dz\right] dxdy$$

$$= \iint_R \Big[F_3(x, y, z)\Big]_{f_1}^{f_2} dxdy = \iint_R [F_3(x, y, f_2) - F_3(x, y, f_1)]\, dxdy$$

$$= \iint_R F_3(x, y, f_2)\, dxdy - \iint_R F_3(x, y, f_1)\, dxdy \quad ...(2)$$

Now for the upper portion S_2 of S, the normal $\hat{n}_2$ to S_2 makes an acute angle γ_2 with $\hat{k}$.

$$\therefore \qquad dxdy = \cos\gamma_2 \, dS_2 = \hat{k}\cdot\hat{n}_2 \, dS_2$$

For the lower portion S_1 of S, the normal $\hat{n}_1$ to S_1 makes an obtuse angle γ_1 with $\hat{k}$.

$$\therefore \qquad dxdy = -\cos\gamma_1 \, dS_1 = -\hat{k}\cdot\hat{n}_1 \, dS_1$$

$$\therefore \quad \iint_R F_3(x, y, f_2)\, dxdy = \iint_{S_2} F_3\, \hat{k}\cdot\hat{n}_2\, dS_2 \qquad ...(3)$$

and

$$\iint_R F_3(x, y, f_1)\, dxdy = -\iint_{S_1} F_3\, \hat{k}\cdot\hat{n}_1\, dS_1 \qquad ...(4)$$

Using (3) and (4), (2) becomes

$$\iiint_V \frac{\partial F_3}{\partial z}\, dx\,dy\,dz = \iint_{S_2} F_3\,\hat{k}\cdot\hat{n}_2\, dS_2 + \iint_{S_1} F_3\,\hat{k}\cdot\hat{n}_1\, dS_1$$

$$= \iint_S F_3\hat{k}\cdot\hat{n}\, dS = \iint_R F_3 \cos\gamma\, dS \qquad ...(5)$$

Similarly, by considering the projection of S on *yz* and *zx*-planes, we have

$$\iiint_V \frac{\partial F_1}{\partial x}\, dx\,dy\,dz = \iint_S F_1\hat{i}\cdot\hat{n}\, dS = \iint_S F_1 \cos\alpha\, dS \qquad ...(6)$$

and

$$\iiint_V \frac{\partial F_2}{\partial y}\, dx\,dy\,dz = \iint_S F_2\hat{j}\cdot\hat{n}\, dS = \iint_S F_2 \cos\beta\, dS \qquad ...(7)$$

Adding (5), (6) and (7), we get (1) *i.e.*, $\iiint_V \left(\frac{\partial F_1}{\partial x} + \frac{\partial F_2}{\partial y} + \frac{\partial F_3}{\partial z}\right) dx\,dy\,dz$

$$= \iint_S (F_1\cos\alpha + F_2\cos\beta + F_3\cos\gamma)\, dS \quad \text{or} \quad \iiint_V \nabla\cdot\vec{F}\, dV = \iint_R \vec{F}\cdot\hat{n}\, dS$$

In case the region be such that the lines drawn parallel to the coordinate axes meet it in more than two points, then we divide the region into various sub-regions each of which is met by a line parallel to any axis in only two points. Applying the theorem to each of these sub-regions and adding the results, we get the volume integral over the whole region.

ILLUSTRATIVE EXAMPLES

Example 1. *For any closed surface S, prove that* $\iint_S curl\,\vec{F}\cdot\hat{n}\, dS = 0$.

Sol. By the divergence theorem, we have $\iint_S \text{curl}\,\vec{F}\cdot\hat{n}\, dS = \iiint_V (\text{div curl}\,\vec{F})\, dV$,

where V is the volume enclosed by S = 0. Since div curl $\vec{F} = 0$, therefore, $\iint_S \text{curl}\,\vec{F}\cdot\hat{n}\, dS = 0$

Example 2. *Evaluate* $\iint_S \vec{r}\cdot\hat{n}\, dS$, *where S is a closed surface.* (M.D.U. May 2010)

Sol. By the divergence theorem, we have $\iint_S \vec{r}\cdot\hat{n}\, dS = \iiint_V \nabla\cdot\vec{r}\, dV$,

where V is the volume enclosed by S

$$= \iiint_V 3dV, \text{ since } \nabla \cdot \vec{r} = \text{div } \vec{r} = 3$$
$$= 3V.$$

Example 3. *Use divergence theorem to show that* $\oint_S \nabla r^2 \cdot d\vec{S} = 6V$, *where S is any closed surface enclosing a volume V.*

Sol. By the divergence theorem, we have $\oint_S \nabla r^2 \cdot d\vec{S} = \int_V \text{div}(\nabla r^2)\, dV$

$$= \int_V \nabla \cdot (\nabla r^2)\, dV = \int_V \nabla^2 r^2\, dV$$

$$= \int_V \left(\frac{\partial^2}{\partial x^2} + \frac{\partial^2}{\partial y^2} + \frac{\partial^2}{\partial z^2} \right) (x^2 + y^2 + z^2)\, dV = \int_V 6dV = 6V.$$

Example 4. *Verify divergence theorem for* $\vec{F} = (x^2 - yz)\,\hat{i} + (y^2 - zx)\,\hat{j} + (z^2 - xy)\,\hat{k}$ *taken over the rectangular parallelopiped* $0 \le x \le a$, $0 \le y \le b$, $0 \le z \le c$.

(M.T.U. 2013; M.D.U. 2013; K.U.K. Dec. 2010; P.T.U. Dec. 2010, May 2011)

Sol. For verification of divergence theorem, we shall evaluate the volume and surface integrals separately and show that they are equal.

Now, $$\text{div } \vec{F} = \frac{\partial}{\partial x}(x^2 - yz) + \frac{\partial}{\partial y}(y^2 - zx) + \frac{\partial}{\partial z}(z^2 - xy)$$
$$= 2(x + y + z)$$

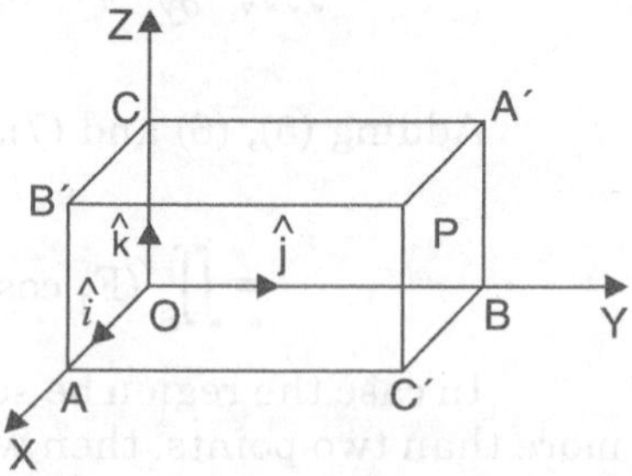

$$\therefore \quad \iiint_V \text{div } \vec{F}\, dV$$

$$= \int_0^c \int_0^b \int_0^a 2(x + y + z)\, dx dy dz$$

$$= \int_0^c \int_0^b 2\left[\left(\frac{x^2}{2} + yx + zx \right) \right]_0^a dy dz$$

$$= \int_0^c \int_0^b 2\left(\frac{a^2}{2} + ya + za \right) dy\, dz = \int_0^c 2\left[\left(\frac{a^2}{2} y + \frac{ay^2}{2} + azy \right) \right]_0^b dz$$

$$= 2\int_0^c \left(\frac{a^2 b}{2} + \frac{ab^2}{2} + abz \right) dz = 2\left[\frac{a^2 b}{2} z + \frac{ab^2}{2} z + ab\frac{z^2}{2} \right]_0^c$$

$$= a^2bc + ab^2c + abc^2 = abc\,(a + b + c) \qquad \text{...(1)}$$

To evaluate the surface integral, divide the closed surface S of the rectangular parallelopiped into 6 parts.

S_1 : the face OAC′B, S_2 : the face CB′PA′, S_3 : the face OBA′C,
S_4 : the face AC′PB′, S_5 : the face OCB′A, S_6 : the face BA′PC′.

Also $$\iint_S \vec{F} \cdot \hat{n}\, dS = \iint_{S_1} \vec{F} \cdot \hat{n}\, dS + \iint_{S_2} \vec{F} \cdot \hat{n}\, dS + \iint_{S_3} \vec{F} \cdot \hat{n}\, dS$$
$$+ \iint_{S_4} \vec{F} \cdot \hat{n}\, dS + \iint_{S_5} \vec{F} \cdot \hat{n}\, dS + \iint_{S_6} \vec{F} \cdot \hat{n}\, dS$$

On S_1 $(z = 0)$, we have $\hat{n} = -\hat{k}, \vec{F} = x^2\hat{i} + y^2\hat{j} - xy\hat{k}$

so that $\vec{F}\cdot\hat{n} = (x^2\hat{i} + y^2\hat{j} - xy\hat{k}).(-\hat{k}) = xy$

$$\therefore \quad \iint_S \vec{F}\cdot\hat{n}\,dS = \int_0^b\int_0^a xy\,dx\,dy = \int_0^b\left[y\frac{x^2}{2}\right]_0^a dy = \frac{a^2}{2}\int_0^b y\,dy = \frac{a^2b^2}{4}$$

On S_2 $(z = c)$, we have $\hat{n} = \hat{k}, \vec{F} = (x^2 - cy)\hat{i} + (y^2 - cx)\hat{j} + (c^2 - xy)\hat{k}$

so that $\vec{F}\cdot\hat{n} = [(x^2 - cy)\hat{i} + (y^2 - cx)\hat{j} + (c^2 - xy)\hat{k}]\cdot\hat{k} = c^2 - xy$

$$\iint_{S_2} \vec{F}\cdot\hat{n}\,dS = \int_0^b\int_0^a (c^2 - xy)\,dx\,dy = \int_0^b\left(c^2a - \frac{a^2}{2}y\right)dy = abc^2 - \frac{a^2b^2}{4}$$

On S_3 $(x = 0)$, we have $\hat{n} = -\hat{i}, \vec{F} = -yz\hat{i} + y^2\hat{j} + z^2\hat{k}$

so that $\vec{F}\cdot\hat{n} = (-yz\hat{i} + y^2\hat{j} + z^2\hat{k})\cdot(-\hat{i}) = yz$

$$\therefore \quad \iint_{S_3} \vec{F}\cdot\hat{n}\,dS = \int_0^c\int_0^b yz\,dy\,dz = \int_0^c \frac{b^2}{2}z\,dz = \frac{b^2c^2}{4}$$

On S_4 $(x = a)$, we have $\hat{n} = \hat{i}, \vec{F} = (a^2 - yz)\hat{i} + (y^2 - az)\hat{j} + (z^2 - ay)\hat{k}$

so that $\vec{F}\cdot\hat{n} = [(a^2 - yz)\hat{i} + (y^2 - az)\hat{j} + (z^2 - ay)\hat{k}]\cdot\hat{i} = a^2 - yz$

$$\therefore \quad \iint_{S_4} \vec{F}\cdot\hat{n}\,dS = \int_0^c\int_0^b (a^2 - yz)\,dy\,dz = \int_0^c\left(a^2b - \frac{b^2}{2}z\right)dz = a^2bc - \frac{b^2c^2}{4}$$

On S_5 $(y = 0)$, we have $\hat{n} = -\hat{j}, \vec{F} = x^2\hat{i} - zx\hat{j} + z^2\hat{k}$

so that $\vec{F}\cdot\hat{n} = (x^2\hat{i} - zx\hat{j} + z^2\hat{k})\cdot(-\hat{j}) = zx$

$$\therefore \quad \iint_{S_5} \vec{F}\cdot\hat{n}\,dS = \int_0^a\int_0^c zx\,dz\,dx = \int_0^a \frac{c^2}{2}x\,dx = \frac{a^2c^2}{4}$$

On S_6 $(y = b)$, we have $\hat{n} = \hat{j}, \vec{F} = (x^2 - bz)\hat{i} + (b^2 - zx)\hat{j} + (z^2 - bx)\hat{k}$

so that $\vec{F}\cdot\hat{n} = [(x^2 - bz)\hat{i} + (b^2 - zx)\hat{j} + (z^2 - bx)\hat{k}]\cdot\hat{j} = b^2 - zx$

$$\therefore \quad \iint_{S_6} \vec{F}\cdot\hat{n}\,dS = \int_0^a\int_0^c (b^2 - zx)\,dz\,dx = \int_0^a\left(b^2c - \frac{c^2}{2}x\right)dx = ab^2c - \frac{a^2c^2}{4}$$

$$\therefore \quad \iint_S \vec{F}\cdot\hat{n}\,dS = \frac{a^2b^2}{4} + abc^2 - \frac{a^2b^2}{4} + \frac{b^2c^2}{4} + a^2bc - \frac{b^2c^2}{4} + \frac{a^2c^2}{4} + ab^2c - \frac{a^2c^2}{4}$$

$$= abc\,(a + b + c) \qquad \ldots(2)$$

The equality of (1) and (2) verifies divergence theorem.

Example 5. *Verify divergence theorem for* $\vec{F} = 4x\hat{i} - 2y^2\hat{j} + z^2\hat{k}$ *taken over the region bounded by the cylinder* $x^2 + y^2 = 4$, $z = 0$, $z = 3$.

(M.D.U. Dec. 2010, May 2010; K.U.K. Dec. 2013; Calicut 2010)

Sol. Since div $\vec{F} = \frac{\partial}{\partial x}(4x) + \frac{\partial}{\partial y}(-2y^2) + \frac{\partial}{\partial z}(z^2) = 4 - 4y + 2z$

$$\therefore \quad \iiint_V \text{div } \vec{F}\, dV = \iiint_V (4 - 4y + 2z)\, dxdydz$$

$$= \int_{-2}^{2}\int_{-\sqrt{4-x^2}}^{\sqrt{4-x^2}}\int_0^3 (4 - 4y + 2z)\, dzdydx$$

$$= \int_{-2}^{2}\int_{-\sqrt{4-x^2}}^{\sqrt{4-x^2}} \left[4z - 4yz + z^2\right]_0^3 dydx$$

$$= \int_{-2}^{2}\int_{-\sqrt{4-x^2}}^{\sqrt{4-x^2}} (12 - 12y + 9)\, dydx$$

$$= \int_{-2}^{2}\int_{-\sqrt{4-x^2}}^{\sqrt{4-x^2}} 21\, dydx,$$

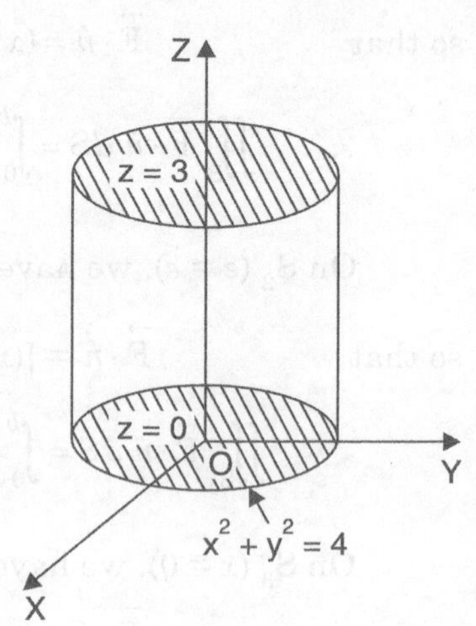

(Since $12y$ is an odd function, $\int_{-a}^{a} 12y\, dy = 0$)

$$= \int_{-2}^{2} 42\sqrt{4-x^2}\, dx = 84\int_0^2 \sqrt{4-x^2}\, dx = 84\left[\frac{x\sqrt{4-x^2}}{2} + \frac{4}{2}\sin^{-1}\frac{x}{2}\right]_0^2 = 84[2\sin^{-1} 1]$$

$$= 84\left[2 \times \frac{\pi}{2}\right] = 84\pi \qquad \ldots(1)$$

To evaluate the surface integral, divide the closed surface S of the cylinder into 3 parts.

S_1 : the circular base in the plane $z = 0$

S_2 : the circular top in the plane $z = 3$

S_3 : the curved surface of the cylinder, given by the equation $x^2 + y^2 = 4$.

Also, $\iint_S \vec{F}\cdot\hat{n}\, dS = \iint_{S_1} \vec{F}\cdot\hat{n}\, dS + \iint_{S_2} \vec{F}\cdot\hat{n}\, dS + \iint_{S_3} \vec{F}\cdot\hat{n}\, dS$

On S_1 ($z = 0$), we have $\hat{n} = -\hat{k}$, $\vec{F} = 4x\hat{i} - 2y^2\hat{j}$

so that $\vec{F}\cdot\hat{n} = (4x\hat{i} - 2y^2\hat{j})\cdot(-\hat{k}) = 0$

$$\therefore \quad \iint_{S_1} \vec{F}\cdot\hat{n}\, dS = 0$$

On S_2 ($z = 3$), we have $\hat{n} = \hat{k}$, $\vec{F} = 4x\hat{i} - 2y^2\hat{j} + 9\hat{k}$

so that $\vec{F}\cdot\hat{n} = (4x\hat{i} - 2y^2\hat{j} + 9\hat{k})\cdot\hat{k} = 9$

$$\therefore \quad \iint_{S_2} \vec{F}\cdot\hat{n}\, dS = \iint_{S_2} 9 dx\, dy = 9\iint_{S_2} dx\, dy$$

$$= 9 \times \text{area of surface } S_2 = 9\,(\pi \cdot 2^2) = 36\pi$$

On S_3, $x^2 + y^2 = 4$

A vector normal to the surface S_3 is given by $\nabla(x^2 + y^2) = 2x\hat{i} + 2y\hat{j}$

$\therefore$ $\hat{n}$ = a unit vector normal to surface S_3

$$= \frac{2x\hat{i} + 2y\hat{j}}{\sqrt{4x^2 + 4y^2}} = \frac{2x\hat{i} + 2y\hat{j}}{\sqrt{4 \times 4}}, \text{ since } x^2 + y^2 = 4$$

$$= \frac{x\hat{i} + y\hat{j}}{2}$$

$$\vec{F} \cdot \hat{n} = (4x\hat{i} - 2y^2\hat{j} + z^2\hat{k}) \cdot \left(\frac{x\hat{i} + y\hat{j}}{2}\right) = 2x^2 - y^3$$

Also, on S_3, *i.e.*, $x^2 + y^2 = 4$, $x = 2 \cos \theta$, $y = 2 \sin \theta$ and $dS = 2d\theta\, dz$.

To cover the whole surface S_3, z varies from 0 to 3 and θ varies from 0 to 2π.

$$\therefore \quad \iint_{S_3} \vec{F} \cdot \hat{n}\, dS = \int_0^{2\pi} \int_0^3 [2(2 \cos \theta)^2 - (2 \sin \theta)^3]\, 2dz\, d\theta$$

$$= \int_0^{2\pi} 16 (\cos^2 \theta - \sin^3 \theta) \times 3\, d\theta = 48 \int_0^{2\pi} (\cos^2 \theta - \sin^3 \theta)\, d\theta = 48\pi$$

$$\left(\text{since} \int_0^{2\pi} \cos^2 \theta\, d\theta = 2 \int_0^{\pi} \cos^2 \theta\, d\theta = 4 \int_0^{\pi/2} \cos^2 d\theta = 4 \times \frac{1}{2} \times \frac{\pi}{2} = \pi, \int_0^{2\pi} \sin^3 \theta\, d\theta = 0\right)$$

$$\therefore \quad \iint_S \vec{F} \cdot \hat{n}\, dS = 0 + 36\pi + 48\pi = 84\pi \qquad \text{...(2)}$$

The equality of (1) and (2) verifies divergence theorem.

Example 6. *Find* $\iint_S \vec{F} \cdot \hat{n}\, dS$, *where* $\vec{F} = (2x + 3z)\hat{i} - (xz + y)\hat{j} + (y^2 + 2z)\hat{k}$ *and S is the surface of the sphere having centre at (3, –1, 2) and radius 3.*

Sol. Let V be the volume enclosed by the surface S. Then by Gauss divergence theorem, we have

$$\iint_S \vec{F} \cdot d\vec{S} = \iiint_V \text{div}\, \vec{F}\, dV$$

$$= \iiint_V \left[\frac{\partial}{\partial x}(2x + 3z) + \frac{\partial}{\partial y}(-xz - y) + \frac{\partial}{\partial z}(y^2 + 2z)\right] dV$$

$$= \iiint_V (2 - 1 + 2)\, dV = 3 \iiint_V dV = 3V$$

But V is the volume of a sphere of radius 3.

$$\therefore \quad V = \frac{4}{3}\pi(3)^3 = 36\pi.$$

Hence $\iint_S \vec{F} \cdot d\vec{S} = 3 \times 36\pi = 108\pi$.

Example 7. *Evaluate* $\iint_S (y^2z^2\hat{i} + z^2x^2\hat{j} + z^2y^2\hat{k}) \cdot \hat{n}\, dS$, *where S is the part of the sphere* $x^2 + y^2 + z^2 = 1$ *above the xy-plane and bounded by this plane.*

Sol. Let V be the volume enclosed by the surface S. Then by divergence theorem, we have

$$\iint_S (y^2z^2\hat{i} + z^2x^2\hat{j} + z^2y^2\hat{k}) \cdot \hat{n}\, dS = \iiint_V \text{div}\, (y^2z^2\hat{i} + z^2x^2\hat{j} + z^2y^2\hat{k})\, dV$$

$$= \iiint_V \left[\frac{\partial}{\partial x}(y^2z^2) + \frac{\partial}{\partial y}(z^2x^2) + \frac{\partial}{\partial z}(z^2y^2)\right] dV$$

$$= \iiint_V 2zy^2\, dV = 2 \iiint_V zy^2\, dV$$

Changing to spherical polar coordinates by putting

$$x = r \sin \theta \cos \phi,\ y = r \sin \theta \sin \phi,\ z = r \cos \theta$$
$$dV = r^2 \sin \theta \, dr \, d\theta \, d\phi$$

To cover V, the limits of r will be 0 to 1, those of θ will be 0 to $\frac{\pi}{2}$ and those of ϕ will be 0 to 2π.

$$\therefore \quad 2 \iiint_V zy^2 \, dV = 2 \int_0^{2\pi} \int_0^{\pi/2} \int_0^1 (r \cos \theta)(r^2 \sin^2 \theta \sin^2 \phi) \, r^2 \sin \theta \, dr \, d\theta \, d\phi$$
$$= 2 \int_0^{2\pi} \int_0^{\pi/2} \int_0^1 r^5 \sin^3 \theta \cos \theta \sin^2 \phi \, dr \, d\theta \, d\phi$$
$$= 2 \int_0^{2\pi} \int_0^{\pi/2} \sin^3 \theta \cos \theta \sin^2 \phi \left[\frac{r^6}{6} \right]_0^1 d\theta \, d\phi$$
$$= \frac{2}{6} \int_0^{2\pi} \sin^2 \phi \cdot \frac{2}{4.2} \, d\phi = \frac{1}{12} \int_0^{2\pi} \sin^2 \phi \, d\phi = \frac{\pi}{12}.$$

Example 8. *Evaluate* $\iint_S \vec{F} \cdot \hat{n} \, dS$ *over the entire surface of the region above the xy-plane bounded by the cone* $z^2 = x^2 + y^2$ *and the plane* $z = 4$, *if* $\vec{F} = 4xz\hat{i} + xyz^2\hat{j} + 3z\hat{k}$.

Sol. If V is the volume enclosed by S, then V is bounded by the surfaces $z = 0$, $z = 4$, $z^2 = x^2 + y^2$.

By divergence theorem, we have $\iint_S \vec{F} \cdot \hat{n} \, dS = \iiint_V \text{div} \, \vec{F} \, dV$

$$= \iiint_V \left[\frac{\partial}{\partial x}(4xz) + \frac{\partial}{\partial y}(xyz^2) + \frac{\partial}{\partial z}(3z) \right] dV = \iiint_V (4z + xz^2 + 3) \, dV$$
$$= \int_0^4 \int_{-z}^{z} \int_{-\sqrt{z^2 - y^2}}^{\sqrt{z^2 - y^2}} (4z + xz^2 + 3) \, dx \, dy \, dz$$
$$= 2 \int_0^4 \int_{-z}^{z} \int_0^{\sqrt{z^2 - y^2}} (4z + 3) \, dx \, dy \, dz, \text{ since } \int_{-\sqrt{z^2 - y^2}}^{\sqrt{z^2 - y^2}} x \, dx = 0$$
$$= 2 \int_0^4 \int_{-z}^{z} (4z + 3) \sqrt{z^2 - y^2} \, dy \, dz = 4 \int_0^4 \int_0^z (4z + 3) \sqrt{z^2 - y^2} \, dy \, dz$$
$$= 4 \int_0^4 (4z + 3) \left[\frac{y\sqrt{z^2 - y^2}}{2} + \frac{z^2}{2} \sin^{-1} \frac{y}{z} \right]_0^z dz$$
$$= 4 \int_0^4 (4z + 3) \left[\frac{z^2}{2} \sin^{-1} 1 \right] dz = 4 \times \frac{\pi}{4} \int_0^4 (4z^3 + 3z^2) \, dz$$
$$= \pi \left[z^4 + z^3 \right]_0^4 = \pi(256 + 64) = 320\pi.$$

Example 9. *Evaluate* $\iint_S (\nabla \times \vec{A}) \cdot \hat{n} \, dS$, *where S is the surface of the cone* $z = 2 - \sqrt{x^2 + y^2}$ *above the xy-plane and* $\vec{A} = (x - z)\hat{i} + (x^3 + yz)\hat{j} - 3xy^2\hat{k}$.

Sol. Here S is not a closed surface. The surface $z = 2 - \sqrt{x^2 + y^2}$ meets the xy-plane in a circle C given by $x^2 + y^2 = 4$, $z = 0$. Let S_1 be the plane region bounded by the circle C. If S′ is the surface consisting of the surfaces S and S_1, then *S′ is a closed surface.* By divergence theorem, we have

$$\iint_{S'} \text{curl}\, \vec{A} \cdot \hat{n}\, dS = 0 \qquad \text{[See Example 1]}$$

$$\Rightarrow \iint_S \text{curl}\, \vec{A} \cdot \hat{n}\, dS + \iint_{S_1} \text{curl}\, \vec{A} \cdot \hat{n}\, dS = 0 \qquad [\because \text{ S}' \text{ consists of S and } S_1]$$

$$\Rightarrow \iint_S \text{curl}\, \vec{A} \cdot \hat{n}\, dS - \iint_{S_1} \text{curl}\, \vec{A} \cdot \hat{k}\, dS = 0 \qquad [\because \text{ on } S_1,\ \hat{n} = -\hat{k}]$$

$$\Rightarrow \int_S \text{curl}\, \vec{A} \cdot \hat{n}\, dS = \iint_{S_1} \text{curl}\, \vec{A} \cdot \hat{k}\, dS$$

Now, $$\text{curl}\, \vec{A} = \begin{vmatrix} \hat{i} & \hat{j} & \hat{k} \\ \dfrac{\partial}{\partial x} & \dfrac{\partial}{\partial y} & \dfrac{\partial}{\partial z} \\ x - z & x^3 + yz & -3xy^2 \end{vmatrix} = \hat{i}(-6xy - y) + \hat{j}(-1 + 3y^2) + \hat{k}(3x^2 - 0)$$

$$\therefore \quad \text{curl}\, \vec{A} \cdot \hat{k} = 3x^2$$

$$\therefore \iint_S \text{curl}\, \vec{A} \cdot \hat{n}\, dS = \iint_S 3x^2\, dS = \int_0^{2\pi}\int_0^2 3(r^2 \cos^2\theta)\, r\, dr\, d\theta, \text{ changing to polars}$$

$$= 3\int_0^{2\pi}\int_0^2 r^3 \cos^2\theta\, dr\, d\theta = 3\int_0^{2\pi}\left[\frac{r^4}{4}\right]_0^2 \cos^2\theta\, d\theta = 12\int_0^{2\pi} \cos^2\theta\, d\theta = 12\pi.$$

Example 10. *Use divergence theorem to evaluate* $\iint_S (xdydz + ydzdx + zdxdy)$, *where S is the portion of the plane x + 2y + 3z = 6 in the first octant.*

Sol. We know that

$$\iint_S (F_1 dydz + F_2 dzdx + F_3 dxdy) = \iiint_V \left(\frac{\partial F_1}{\partial x} + \frac{\partial F_2}{\partial y} + \frac{\partial F_3}{\partial z}\right) dxdydz$$

where V is the region bounded by closed surface S.

Comparing L.H.S. with $\iint_S (xdydz + ydzdx + zdxdy)$, we have

$F_1 = x$, $F_2 = y$, $F_3 = z$

$$\therefore \iint_S (xdydz + ydzdx + zdxdy)$$

$$= \iiint_V (1 + 1 + 1)\, dxdydz$$

$$= 3\iiint_V dxdydz$$

= 3 × volume of tetrahedron OABC

$$= 3\left\{\frac{1}{3}\,\Delta AOB \times \text{hight OC}\right\}$$

$$= \frac{1}{2}\,(OA)\,(OB)\,(OC) = \frac{1}{2}\,(6)\,(3)\,(2) = 18.$$

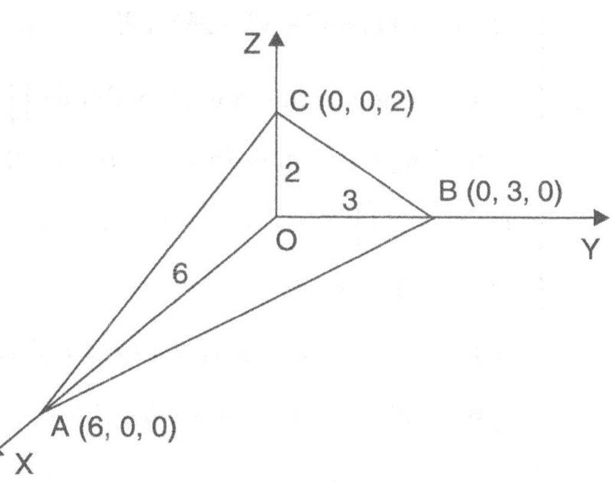

EXERCISE 13.7

1. Use divergence theorem to show that $\iint_S r^n \vec{r} \cdot d\vec{S} = (n+3) \int_V r^n \, dV \; (n \neq -3)$, where S is any closed surface enclosing a volume V.

2. Verify divergence theorem for $\vec{F} = xz\hat{i} + y^2\hat{j} + yz\hat{k}$ taken over the cube bounded by $x = 0$, $x = 1$; $y = 0$; $y = 1$; $z = 0$, $z = 1$. *(U.P.T.U. 2009; G.B.T.U. 2010)*

3. (*i*) Verify divergence theorem for $\vec{F} = 4xz\hat{i} - y^2\hat{j} + yz\hat{k}$ taken over the cube bounded by $x = 0$, $x = 1$; $y = 0$, $y = 1$; $z = 0$, $z = 1$. *(Anna 2010; Calicut 2010; U.P.T.U. 2009)*

   (*ii*) Verify Gauss divergence theorem for $\vec{F} = (x^3 - yz)\hat{i} + (y^3 - zx)\hat{j} + (z^3 - xy)\hat{k}$ taken over the cube bounded by the planes $x = 0$, $x = 1$, $y = 0$, $y = 1$, $z = 0$, $z = 1$.

4. Verify divergence theorem for $\vec{F} = (x^3 - yz)\hat{i} - 2x^2y\hat{j} + 2\hat{k}$ taken over the cube bounded by the planes $x = 0$, $x = a$; $y = 0$, $y = a$; $z = 0$, $z = a$.

5. (*a*) Verify divergence theorem for $\vec{F} = y\hat{i} + x\hat{j} + z^2\hat{k}$ over the cylindrical region bounded by $x^2 + y^2 = a^2$, $z = 0$ and $z = h$.

   (*b*) Verify Gauss divergence theorem for the function $\vec{F} = y\hat{i} + x\hat{j} + z^2\hat{k}$ over the cylindrical region bounded by $x^2 + y^2 = 9$, $z = 0$ and $z = 2$.

6. Verify divergence theorem for $\vec{F} = 2x^2y\hat{i} - y^2\hat{j} + 4xz^2\hat{k}$ taken over the region in the first octant bounded by $y^2 + z^2 = 9$ and $x = 2$.

7. If S is any closed surface enclosing a volume V and $\vec{F} = x\hat{i} + 2y\hat{j} + 3z\hat{k}$, prove that $\iint_S \vec{F} \cdot \hat{n} \, dS = 6V$.

8. If $\vec{F} = ax\hat{i} + by\hat{j} + cz\hat{k}$, a, b, c are constants, show that $\iint_S \vec{F} \cdot d\vec{S} = \frac{4}{3}\pi(a + b + c)$, where S is the surface of a unit sphere.

9. Use divergence theorem to evaluate $\iint_S \vec{F} \cdot d\vec{S}$, where $\vec{F} = x^3\hat{i} + y^3\hat{j} + z^3\hat{k}$ and S is the surface of the sphere $x^2 + y^2 + z^2 = a^2$. *(P.T.U. May 2009, Jan. 2010)*

10. Use divergence theorem to evaluate $\iint_S \vec{F} \cdot \hat{n} dS$, where $\vec{F} = x^3\hat{i} + x^2y\hat{j} + x^2z\hat{k}$ and S is the surface bounding the region $x^2 + y^2 = a^2$, $z = 0$, $z = b$. *(P.T.U. Dec. 2011)*

11. Evaluate $\iint_S (x\hat{i} + y\hat{j} + z^2\hat{k}) \cdot \hat{n} dS$, where S is the closed surface bounded by the cone $x^2 + y^2 = z^2$ and the plane $z = 1$.

12. If $\vec{F} = (x^2 + y - 4)\hat{i} + 3xy\hat{j} + (2xz + z^2)\hat{k}$, evaluate $\iint_S (\nabla \times \vec{F}) \cdot \hat{n} dS$, where S is the surface of the sphere $x^2 + y^2 + z^2 = 16$ above the xy-plane.

13. Evaluate $\iint_S [x^2\, dydz + y^2\, dzdx + 2z\,(xy - x - y)\, dxdy]$, where S is the surface of the cube $0 \le x \le 1, 0 \le y \le 1, 0 \le z \le 1$.

14. Use the divergence theorem to evaluate $\iint_S [(x^3 - yz)\, dydz - 2x^2y\, dzdx + z\, dxdy]$ over the surface of a cube bounded by the coordinate planes and the planes $x = y = z = a$.

15. Show that $\iint_S (x^2\hat{i} + y^2\hat{j} + z^2\hat{k})\cdot\hat{n}dS = 0$, where S*u* denotes the surface of the ellipsoid $\frac{x^2}{a^2} + \frac{y^2}{b^2} + \frac{z^2}{c^2} = 1$.

16. Prove that $\iiint_V \frac{1}{r^2}\, dV = \iint_S \frac{\vec{r}\cdot\hat{n}}{r^2}\, dS$.

17. Use divergence theorem to evaluate $\iint (x\, dy\, dz + y\, dz\, dx + z dx\, dy)$ over the surface of a sphere of radius a. *(K.U.K. 2009)*

Answers

9. $\frac{12}{5}\pi a^5$ 10. $\frac{5}{4}\pi a^4 b$ 11. $\frac{7}{6}\pi$

12. -16π 13. $\frac{1}{2}$ 14. $\frac{1}{3}a^5 + a^3$.

17. $4\pi a^3$

13.28. GREEN'S THEOREM IN THE PLANE

(M.T.U. 2012; P.T.U. Jan. 2009, May 2011)

Statement. *If M(x, y) and N(x, y) be continuous functions of x and y having continuous partial derivatives* $\frac{\partial M}{\partial y}$ *and* $\frac{\partial N}{\partial x}$ *in a region R of the xy-plane bounded by a closed curve C, then* $\oint_C (Mdx + Ndy) = \iint_R \left(\frac{\partial N}{\partial x} - \frac{\partial M}{\partial y}\right) dxdy$ *where C is traversed in the counterclockwise direction.*

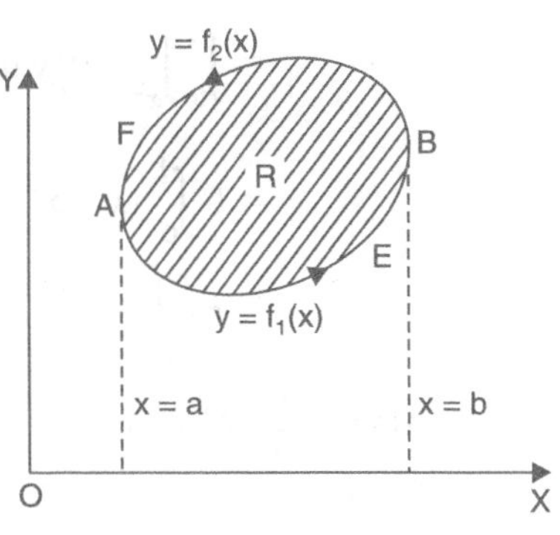

Proof. Let us assume that the region R is such that any line parallel to either axes meets the boundary curve C in at most two points.

[The proof can be easily extended to other cases.]

Suppose the region R is bounded between the lines $x = a$, $x = b$ and two arcs AEB and BFA whose equations are $y = f_1(x)$ and $y = f_2(x)$ respectively such that $f_2(x) > f_1(x)$.

Now, $$\iint_R \frac{\partial M}{\partial y}\, dxdy = \int_a^b \left[\int_{f_1(x)}^{f_2(x)} \frac{\partial M}{\partial y}\, dy\right] dx$$

$$= \int_a^b \Big[M(x, y)\Big]_{f_1(x)}^{f_2(x)} dx = \int_a^b [M\,(x, f_2) - M(x, f_1)]\, dx$$

$$= \int_a^b M(x, f_2)\, dx - \int_a^b M(x, f_1)\, dx = -\int_b^a M(x, f_2)\, dx - \int_a^b M(x, f_1)\, dx$$

$$= -\left[\int_a^b M(x, f_1)\, dx + \int_b^a M(x, f_2)\, dx\right] = -\oint_C M dx$$

or $$\oint_C M dx = -\iint_R \frac{\partial M}{\partial y} dxdy \qquad ...(1)$$

Similarly, we can show that $$\oint_C N dy = \iint_R \frac{\partial N}{\partial x} dxdy \qquad ...(2)$$

Adding (1) and (2), we have $$\oint_C (M dx + N dy) = \iint_R \left(\frac{\partial N}{\partial x} - \frac{\partial M}{\partial y}\right) dxdy$$

This theorem is useful for changing a line integral around a closed curve C into a double integral over the region R enclosed by C.

ILLUSTRATIVE EXAMPLES

Example 1. *Verify Green's theorem in the plane for* $\oint_C (3x^2 - 8y^2)\, dx + (4y - 6xy)\, dy$ *where C is the boundary of the region defined by*

(a) $y = \sqrt{x}$, $y = x^2$ (U.K.T.U. 2011) *(b)* $x = 0$, $y = 0$, $x + y = 1$.

(K.U.K. Dec. 2013; Anna 2011)

Sol. *(a)* $y = \sqrt{x}$ *i.e.,* $y^2 = x$ and $y = x^2$ are two parabolas intersecting at O(0, 0) and A(1, 1).

Here $M = 3x^2 - 8y^2$, $N = 4y - 6xy$

$$\frac{\partial M}{\partial y} = -16y, \qquad \frac{\partial N}{\partial x} = -6y$$

$$\therefore \quad \frac{\partial N}{\partial x} - \frac{\partial M}{\partial y} = 10y$$

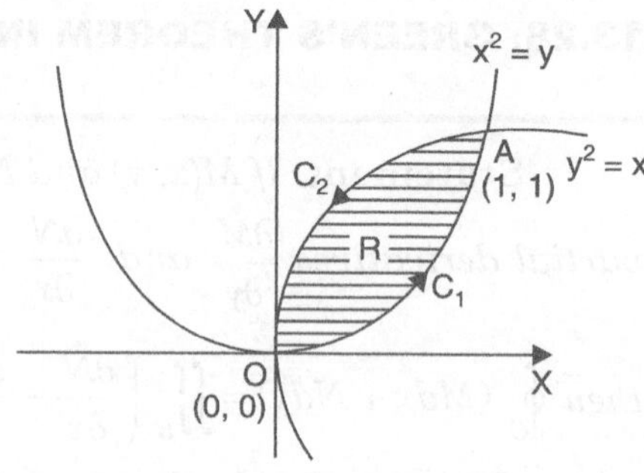

If R is the region bounded by C, then

$$\iint_R \left(\frac{\partial N}{\partial x} - \frac{\partial M}{\partial y}\right) dxdy$$

$$= \int_0^1 \int_{x^2}^{\sqrt{x}} 10y\, dydx = \int_0^1 5\left[y^2\right]_{x^2}^{\sqrt{x}} dx$$

$$= 5\int_0^1 (x - x^4)\, dx = 5\left[\frac{x^2}{2} - \frac{x^5}{5}\right]_0^1$$

$$= 5\left(\frac{1}{2} - \frac{1}{5}\right) = 5\left(\frac{3}{10}\right) = \frac{3}{2} \qquad ...(1)$$

Also, $$\oint_C (M dx + N dy) = \int_{C_1} (M dx + N dy) + \int_{C_2} (M dx + N dy)$$

Along C_1, $x^2 = y$, $\therefore$ $2x dx = dy$ and the limits of x are from 0 to 1.

$\therefore$ Line integral along $C_1 = \int_{C_1} (M dx + N dy)$

$$= \int_0^1 (3x^2 - 8x^4)\, dx + (4x^2 - 6x \cdot x^2)\, 2x\, dx = \int_0^1 (3x^2 + 8x^3 - 20x^4)\, dx$$

$$= \left[x^3 + 2x^4 - 4x^5\right]_0^1 = -1$$

Along C_2, $y^2 = x$, $\therefore$ $2ydy = dx$ and the limits of y are from 1 to 0.

$\therefore$ Line integral along $C_2 = \int_{C_2} (Mdx + Ndy)$

$$= \int_1^0 (3y^4 - 8y^2)\, 2ydy + (4y - 6y^2 \,.\, y)\, dy$$

$$= \int_1^0 (4y - 22y^3 + 6y^5)\, dy = \left[2y^2 - \frac{11}{2}y^4 + y^6\right]_1^0 = \frac{5}{2}$$

$\therefore$ Line integral along $C = -1 + \frac{5}{2} = \frac{3}{2}$ *i.e.*, $\oint_C (Mdx + Ndy) = \frac{3}{2}$ ...(2)

The equality of (1) and (2) verifies Green's theorem in the plane.

(*b*) Here $\iint_R \left(\frac{\partial N}{\partial x} - \frac{\partial M}{\partial y}\right) dxdy = \int_0^1 \int_0^{1-x} 10y\, dydx$

$$= \int_0^1 5\left[y^2\right]_0^{1-x} dx$$

$$= 5\int_0^1 (1-x)^2\, dx = 5\left[\frac{(1-x)^3}{-3}\right]_0^1$$

$$= -\frac{5}{3}(0 - 1) = \frac{5}{3} \qquad \text{...(1)}$$

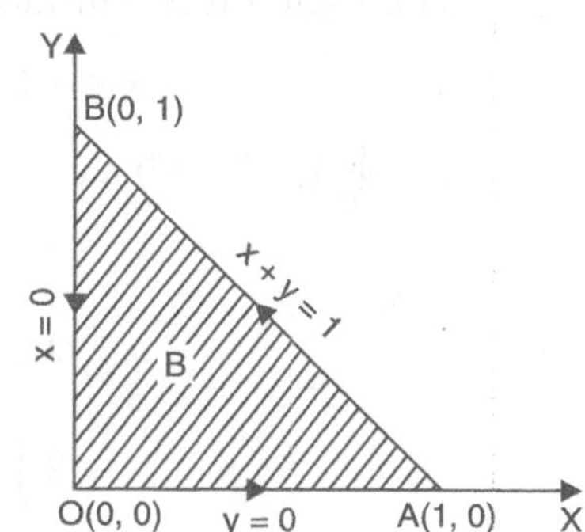

Along OA, $y = 0$ $\therefore$ $dy = 0$ and the limits of x are from 0 to 1.

$\therefore$ Line integral along OA $= \int_0^1 3x^2\, dx = \left[x^3\right]_0^1$

Along AB, $y = 1 - x$ $\therefore$ $dy = -\,dx$ and the limits of x are from 1 to 0.

$\therefore$ Line integral along AB $= \int_1^0 [3x^2 - 8\,(1 - x)^2]\, dx + [4(1 - x) - 6x\,(1 - x)]\,(-dx)$

$$= \int_1^0 (3x^2 - 8 + 16x - 8x^2 - 4 + 4x + 6x - 6x^2)\, dx = \int_1^0 (-12 + 26x - 11x^2)\, dx$$

$$= \left[-12x + 13x^2 - \frac{11}{3}x^3\right]_1^0 = -\left[-12 + 13 - \frac{11}{3}\right] = \frac{8}{3}$$

Along BO, $x = 0$ $\therefore$ $dx = 0$ and the limits of y are from 1 to 0.

$\therefore$ Line integral along BO $= \int_1^0 4y\, dy = \left[2y^2\right]_1^0 = -2$

$\therefore$ Line integral along C (*i.e.*, along OABO) $= 1 + \frac{8}{3} - 2 = \frac{5}{3}$

i.e., $$\oint_C (Mdx + Ndy) = \frac{5}{3} \qquad \text{...(2)}$$

The equality of (1) and (2) verifies Green's theorem in the plane.

Example 2. *Use Green's theorem in a plane to evaluate the integral* $\oint_C [(2x^2 - y^2)\, dx + (x^2 + y^2)\, dy]$, *where C is the boundary in the xy-plane of the area enclosed by the x-axis and the semi-circle* $x^2 + y^2 = 1$ *in the upper half xy-plane.* (K.U.K. Dec. 2010; P.T.U. Jan. 2009)

Sol. If R is the region bounded by the closed curve C, then by Green's theorem in the plane, we have

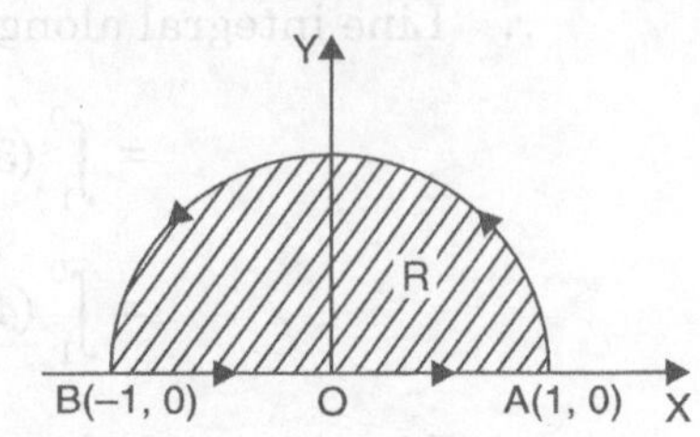

$$\oint_C (\mathrm{M}dx + \mathrm{N}dy) = \iint_R \left(\frac{\partial \mathrm{N}}{\partial x} - \frac{\partial \mathrm{M}}{\partial y}\right) dx\, dy$$

Here $\mathrm{M} = 2x^2 - y^2,\ \mathrm{N} = x^2 + y^2$

$$\frac{\partial \mathrm{M}}{\partial y} = -2y, \frac{\partial \mathrm{N}}{\partial x} = 2x$$

$$\therefore \quad \frac{\partial \mathrm{N}}{\partial x} - \frac{\partial \mathrm{M}}{\partial y} = 2(x + y)$$

The region R is bounded by

$$x = -1,\ x = 1,\ y = 0,\ y = \sqrt{1 - x^2}$$

$$\therefore \quad \oint_C [(2x^2 - y^2)\, dx + (x^2 + y^2)\, dy] = \iint_R 2(x + y)\, dxdy$$

$$= \int_{-1}^{1} \int_{0}^{\sqrt{1-x^2}} 2\,(x + y)\, dydx = 2\int_{-1}^{1} \left[xy + \frac{y^2}{2}\right]_0^{\sqrt{1-x^2}} dx$$

$$= 2\int_{-1}^{1} \left[x\sqrt{1 - x^2} + \frac{1}{2}(1 - x^2)\right] dx$$

$$= \int_{-1}^{1} (1 - x^2)\, dx, \text{ since } \int_{-1}^{1} x\sqrt{1 - x^2}\, dx = 0$$

$$= 2\int_0^1 (1 - x^2)\, dx = 2\left[x - \frac{x^3}{3}\right]_0^1 = 2\left(1 - \frac{1}{3}\right) = \frac{4}{3}.$$

EXERCISE 13.8

1. Verify Green's theorem in the plane for $\oint_C (xy + y^2)\, dx + x^2\, dy$, where C is the closed curve of the region bounded by $y = x$ and $y = x^2$. *(U.P.T.U. 2008)*

2. Verify Green's theorem in the plane for $\oint_C (2xy - x^2)\, dx + (x^2 + y^2)\, dy$, where C is the boundary of the region enclosed by $y = x^2$ and $y^2 = x$.

3. Verify Green's theorem in the xy-plane for $\oint_C (xy^2 - 2xy)\, dx + (x^2y + 3)\, dy$, around the boundary C of the region enclosed by $y^2 = 8x$ and $x = 2$.

4. Verify Green's theorem in the plane for $\int_C (x^3 - xy^3)\, dx + (y^2 - 2xy)\, dy$, where C is the square having vertices at the points (0, 0), (2, 0), (2, 2) and (0, 2). *(U.P.T.U. 2008)*

5. Evaluate by Green's theorem $\int_C (\cos x \sin y - xy)\, dx + \sin x \cos y\, dy$, where C is the circle $x^2 + y^2 = 1$.

6. Apply Green's theorem to evaluate $\oint_C [(y - \sin x)\, dx + \cos x\, dy]$, where C is the plane triangle enclosed by the lines $y = 0$, $x = \frac{\pi}{2}$ and $y = \frac{2}{\pi} x$. *(P.T.U. Dec. 2011; M.D.U. 2012)*

7. Evaluate by Green's theorem $\oint_C e^{-x} (\sin y\, dx + \cos y\, dy)$, where C is the rectangle with vertices $(0, 0)$, $(\pi, 0)$, $\left(\pi, \frac{\pi}{2}\right)$ and $\left(0, \frac{\pi}{2}\right)$. *(M.D.U. 2013)*

8. Apply Green's theorem to prove that the area enclosed by a plane curve is $\frac{1}{2}\int_C (xdy - ydx)$. Hence find the area of an ellipse whose semi major and minor axes are of lengths a and b.

9. Find the area of a circle of radius 'a' using Green's theorem.

10. A vector field $\vec{F}$ is given by $\vec{F} = \sin y\hat{i} + x(1 + \cos y)\hat{j}$. Using Green's theorem, evaluate the line integral $\int_C \vec{F} \cdot d\vec{r}$, where C is the circular path given by $x^2 + y^2 = a^2$.

11. Use Green's theorem to evaluate $\int_C (x^2 + xy)\, dx + (x^2 + y^2)\, dy$, where C is the square formed by the lines $x = \pm 1$, $y = \pm 1$. *(G.B.T.U. 2010; M.T.U. 2011; M.D.U. Dec. 2010)*

12. Using Green's theorem evaluate $\int_C (x^2 y\, dx + x^2\, dy)$, where C is the boundary described counter clockwise of the triangle with vertices (0, 0), (1, 0) and (1, 1). *(U.K.T.U. 2010)*

Answers

5. 0	6. $-\left(\frac{\pi}{4} + \frac{2}{\pi}\right)$	7. $2(e^{-\pi} - 1)$	8. πab
9. πa^2	10. πa^2	11. 0	12. $\frac{5}{12}$.

13.29. STOKE'S THEOREM *(Relation between line and surface integrals)*

(U.P.T.U. 2009; P.T.U. Jan. 2009, Dec. 2010; U.K.T.U. 2012)

Statement. *If S be an open surface bounded by a closed curve C and* $\vec{F} = F_1\hat{i} + F_2\hat{j} + F_3\hat{k}$ *be any vector point function having continuous first order partial derivatives, then* $\oint_C \vec{F} \cdot d\vec{r} = \iint_S \text{curl}\, \vec{F} \cdot \hat{n}\, dS$ *where* $\hat{n}$ *is a unit normal vector at any point of S drawn in the sense in which a right handed screw would advance when rotated in the sense of description of C.*

Proof. Let $\hat{n}$ make angles α, β, γ with positive directions of x, y, z axes, then

$$\hat{n} = \cos\alpha\hat{i} + \cos\beta\hat{j} + \cos\gamma\hat{k}.$$

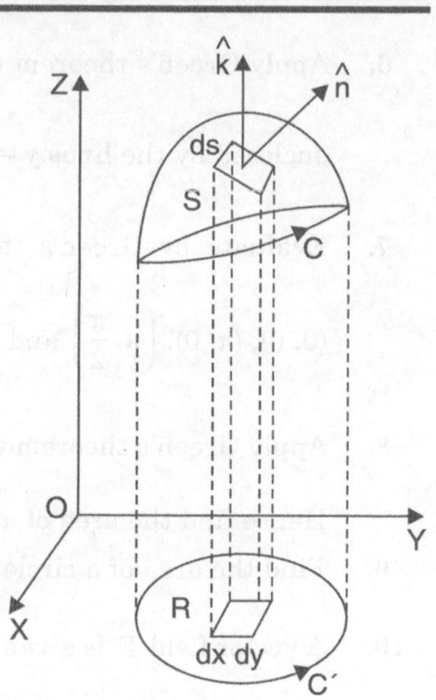

Also, $\vec{r} = x\hat{i} + y\hat{j} + z\hat{k}$ so that $d\vec{r} = \hat{i}\,dx + \hat{j}dy + \hat{k}dz$

$$\text{curl } \vec{F} = \nabla \times \vec{F} = \begin{vmatrix} \hat{i} & \hat{j} & \hat{k} \\ \frac{\partial}{\partial x} & \frac{\partial}{\partial y} & \frac{\partial}{\partial z} \\ F_1 & F_2 & F_3 \end{vmatrix}$$

$$= \left(\frac{\partial F_3}{\partial y} - \frac{\partial F_2}{\partial z}\right)\hat{i} + \left(\frac{\partial F_1}{\partial z} - \frac{\partial F_3}{\partial x}\right)\hat{j} + \left(\frac{\partial F_2}{\partial x} - \frac{\partial F_1}{\partial y}\right)\hat{k}$$

$$\therefore \quad \text{curl } \vec{F} \cdot \hat{n} = \left(\frac{\partial F_3}{\partial y} - \frac{\partial F_2}{\partial z}\right)\cos\alpha + \left(\frac{\partial F_1}{\partial z} - \frac{\partial F_3}{\partial x}\right)\cos\beta + \left(\frac{\partial F_2}{\partial x} - \frac{\partial F_1}{\partial y}\right)\cos\gamma$$

Also, $\vec{F} \cdot d\vec{r} = (F_1\hat{i} + F_2\hat{j} + F_3\hat{k}) \,.\, (\hat{i}\,dx + \hat{j}dy + \hat{k}dz)$

$$= F_1\,dx + F_2\,dy + F_3\,dz$$

$\therefore$ Stoke's theorem can be written as $\oint_C (F_1 dx + F_2 dy + F_3 dz)$

$$= \iint_S \left[\left(\frac{\partial F_3}{\partial y} - \frac{\partial F_2}{\partial z}\right)\cos\alpha + \left(\frac{\partial F_1}{\partial z} - \frac{\partial F_3}{\partial x}\right)\cos\beta + \left(\frac{\partial F_2}{\partial x} - \frac{\partial F_1}{\partial y}\right)\cos\gamma\right] dS \quad \text{...(1)}$$

Let $z = \phi(x, y)$ be the equation of the surface S whose projection on the xy-plane is R. Then the projection of C on the xy-plane is the curve C′ which bounds the region R.

$$\therefore \quad \oint_C F_1(x, y, z)\,dx = \oint_{C'} F_1(x, y, \phi)\,dx = \oint_{C'} [F_1(x, y, \phi)\,dx + 0\,dy]$$

$$= -\iint_R \frac{\partial}{\partial y} F_1(x, y, \phi)\,dxdy \qquad \text{By Green's theorem in plane for the region R}$$

$$= -\iint_R \left[\frac{\partial F_1}{\partial y} + \frac{\partial F_1}{\partial z}\frac{\partial \phi}{\partial y}\right] dxdy \quad \text{...(2)}$$

Now the direction ratios of the normal $\hat{n}$ to the surface $z = \phi(x, y)$ are $\frac{\partial \phi}{\partial x}, \frac{\partial \phi}{\partial y}, -1$

$$\Rightarrow \quad \frac{\cos\alpha}{\frac{\partial \phi}{\partial x}} = \frac{\cos\beta}{\frac{\partial \phi}{\partial y}} = \frac{\cos\gamma}{-1} \quad \Rightarrow \quad \frac{\partial \phi}{\partial y} = -\frac{\cos\beta}{\cos\gamma} \qquad [\because \; z = \phi(x, y) \Rightarrow \phi(x, y) - z = 0]$$

Since $dx\,dy$ is the projection of dS on the xy-plane

$\therefore$ $dxdy = \cos\gamma\,dS$

From (2), we have $\oint_C F_1(x, y, z)\,dx = -\iint_S \left[\frac{\partial F_1}{\partial y} - \frac{\partial F_1}{\partial z} \cdot \frac{\cos\beta}{\cos\gamma}\right] \cos\gamma\,dS$

$$= -\iint_S \left[\frac{\partial F_1}{\partial y}\cos\gamma - \frac{\partial F_1}{\partial z}\cos\beta\right] dS$$

$$= \iint_S \left[\frac{\partial F_1}{\partial z}\cos\beta - \frac{\partial F_1}{\partial y}\cos\gamma\right] dS \quad \text{...(3)}$$

Similarly, we can prove that $\oint_C F_2(x, y, z)\, dy$

$$= \iint_S \left[\frac{\partial F_2}{\partial x} \cos\gamma - \frac{\partial F_2}{\partial z} \cos\alpha \right] dS \quad ...(4)$$

and
$$\oint_C F_3(x, y, z)\, dz = \iint_S \left[\frac{\partial F_3}{\partial y} \cos\alpha - \frac{\partial F_3}{\partial x} \cos\beta \right] dS \quad ...(5)$$

Adding (3), (4) and (5), we get (1). Hence the theorem is proved. In words, Stoke's theorem states that "The line integral of the tangential component of a vector point function $\vec{F}$ taken around a closed curve C is equal to the surface integral of the normal component of $\vec{F}$ taken over any surface S having C as its boundary".

ILLUSTRATIVE EXAMPLES

Example 1. *Verify Stoke's theorem for* $\vec{F} = (x^2 + y^2)\hat{i} - 2xy\,\hat{j}$ *taken round the rectangle bounded by the lines* $x = \pm a$, $y = 0$, $y = b$.

Sol. Let C denote the boundary of the rectangle ABED, then

$$\oint_C \vec{F} \cdot d\vec{r} = \oint_C [(x^2 + y^2)\hat{i} - 2xy\hat{j}] \cdot (\hat{i}dx + \hat{j}dy)$$

$$= \oint_C [(x^2 + y^2)dx - 2xy\, dy]$$

E (– a, b) Y y = b B (a, b) x = – a x = a D (– a, 0) O y = 0 A (a, 0) X

The curve C consists of four lines AB, BE, ED and DA.

Along AB, $x = a$, $dx = 0$ and y varies from 0 to b.

$$\therefore \int_{AB} [(x^2 + y^2)\, dx - 2xydy] = \int_0^b - 2ay\, dy = -a\left[y^2\right]_0^b = -ab^2 \quad ...(1)$$

Along BE, $y = b$, $dy = 0$ and x varies from a to $-a$.

$$\therefore \int_{BE} [(x^2 + y^2)\, dx - 2xydy] = \int_a^{-a} (x^2 + b^2)\, dx = \left[\frac{x^3}{3} + b^2x\right]_a^{-a} = -\frac{2a^2}{3} - 2ab^2 \quad ...(2)$$

Along ED, $x = -a$, $dx = 0$ and y varies from b to 0.

$$\therefore \int_{ED} [(x^2 + y^2)\, dx - 2xydy] = \int_a^0 2ay\, dy = a\left[y^2\right]_b^0 = -ab^2 \quad ...(3)$$

Along DA, $y = 0$, $dy = 0$ and x varies from $-a$ to a.

$$\therefore \int_{DA} [(x^2 + y^2)\, dx - 2dxdy] = \int_{-a}^{a} x^2\, dx = \frac{2a^3}{3} \quad ...(4)$$

Adding (1), (2), (3) and (4), we get

$$\oint_C \vec{F} \cdot d\vec{r} = -ab^2 - \frac{2a^3}{3} - 2ab^2 - ab^2 + \frac{2a^3}{3} = -4ab^2 \quad ...(5)$$

Now,
$$\text{curl } \vec{F} = \begin{vmatrix} \hat{i} & \hat{j} & \hat{k} \\ \frac{\partial}{\partial x} & \frac{\partial}{\partial y} & \frac{\partial}{\partial z} \\ x^2 + y^2 & -2xy & 0 \end{vmatrix} = (-2y - 2y)\hat{k} = -4y\hat{k}$$

For the surface S, $\hat{n} = \hat{k}$

$$\text{curl } \vec{F} \cdot \hat{n} = -4y\hat{k} \cdot \hat{k} = -4y$$

$$\therefore \quad \iint_S \text{curl } \vec{F} \cdot \hat{n}\, dS = \int_0^b \int_{-a}^{a} -4y\, dx dy = \int_0^b -4y \Big[x \Big]_{-a}^{a} dy$$

$$= -8a \int_0^b y\, dy = -8a \left[\frac{y^2}{2}\right]_0^b = -4ab^2 \qquad \text{...(6)}$$

The equality of (5) and (6) verifies Stoke's Theorem.

Example 2. *Verify Stoke's Theorem for the vector field $\vec{F} = (2x - y)\hat{i} - yz^2\hat{j} - y^2z\hat{k}$ over the upper half surface of $x^2 + y^2 + z^2 = 1$, bounded by its projection on the xy-plane.* (M.D.U. 2012)

Sol. Let S be the upper half surface of the sphere $x^2 + y^2 + z^2 = 1$. The boundary C of S is a circle in the xy-plane of radius unity and centre O. The equations of C are $x^2 + y^2 = 1$, $z = 0$ whose parametric form is $x = \cos t$, $y = \sin t$, $z = 0$, $0 \le t < 2\pi$.

Z O Y C X

$$\oint_C \vec{F} \cdot d\vec{r} = \oint_C [(2x - y)\hat{i} - yz^2\hat{j} - y^2z\hat{k}] \cdot (\hat{i}\, dx + \hat{j}\, dy + \hat{k}\, dz)$$

$$= \oint_C [(2x - y)\, dx - yz^2\, dy - y^2z\, dz]$$

$$= \oint_C (2x - y)\, dx, \text{ since on C, } z = 0,\ dz = 0$$

$$= \int_0^{2\pi} (2\cos t - \sin t) \frac{dx}{dt}\, dt = \int_0^{2\pi} (2\cos t - \sin t)(-\sin t)\, dt$$

$$= \int_0^{2\pi} (-\sin 2t + \sin^2 t)\, dt = \int_0^{2\pi} \left(-\sin 2t + \frac{1 - \cos 2t}{2}\right) dt$$

$$= \left[\frac{\cos 2t}{2} + \frac{t}{2} - \frac{\sin 2t}{4}\right]_0^{2\pi} = \frac{1}{2} + \pi - \frac{1}{2} = \pi$$

Also,
$$\text{curl } \vec{F} = \begin{vmatrix} \hat{i} & \hat{j} & \hat{k} \\ \dfrac{\partial}{\partial x} & \dfrac{\partial}{\partial y} & \dfrac{\partial}{\partial z} \\ 2x - y & -yz^2 & -y^2z \end{vmatrix} = (-2yz + 2yz)\hat{i} + (0 - 0)\hat{j} + (0 + 1)\hat{k} = \hat{k}$$

$$\text{Curl } \vec{f} \cdot \hat{n} = \hat{k} \cdot \hat{n} = \hat{n} \cdot \hat{k}$$

$$\iint_S \text{curl } \vec{F} \cdot \hat{n}\, dS = \iint_S \hat{k} \cdot \hat{n}\, dS = \iint_R \hat{k} \cdot \hat{n} \frac{dxdy}{|\hat{k} \cdot \hat{n}|} = \iint_R dxdy$$

where R is the projection of S on xy-plane.

= area bounded by R

= area of a circle with radius 1

= π

Since $\oint_C \vec{F} \cdot d\vec{r} = \iint_S \text{curl } \vec{F} \cdot \hat{n}\, dS$, Stoke's theorem is verified.

Example 3. *Evaluate* $\oint_C \vec{F} \cdot d\vec{r}$ *by Stoke's Theorem, where* $\vec{F} = y^2\hat{i} + x^2\hat{j} - (x+z)\hat{k}$ *and C is the boundary of the triangle with vertices at (0, 0, 0), (1, 0, 0) and (1, 1, 0).*

(P.T.U. May 2010; U.K.T.U. 2012)

Sol. Since z-coordinates of each vertex of the triangle is zero, therefore, the triangle lies in the xy-plane and $\hat{n} = \hat{k}$

$$\text{curl } \vec{F} = \begin{vmatrix} \hat{i} & \hat{j} & \hat{k} \\ \dfrac{\partial}{\partial x} & \dfrac{\partial}{\partial y} & \dfrac{\partial}{\partial z} \\ y^2 & x^2 & -(x+z) \end{vmatrix} = \hat{j} + 2(x-y)\hat{k}$$

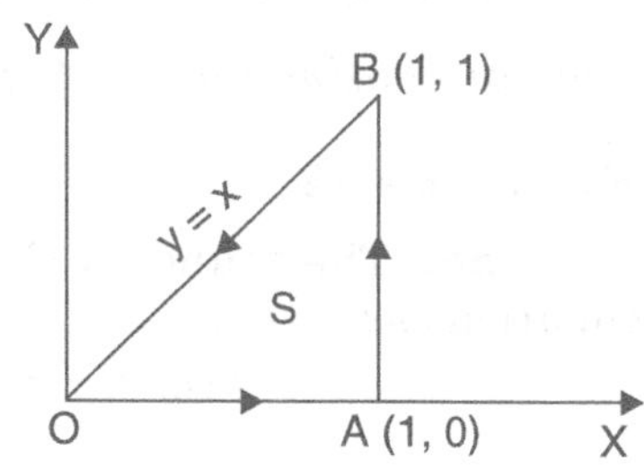

$\therefore \quad \text{curl } \vec{F} \cdot \hat{n} = [\hat{j} + 2(x-y)\hat{k}] \cdot \hat{k} = 2(x-y)$

The equation of line OB is $y = x$.

By Stoke's theorem, $\oint_C \vec{F} \cdot d\vec{r} = \iint_S \text{curl } \vec{F} \cdot \hat{n} \, dS$

$$= \int_0^1 \int_0^x 2(x-y) \, dy dx$$

$$= \int_0^1 2\left[xy - \frac{y^2}{2}\right]_0^x dx = 2\int_0^1 \left(x^2 - \frac{x^2}{2}\right) dx = \int_0^1 x^2 \, dx = \frac{1}{3}.$$

Example 4. *Using Stoke's theorem for the vector function* $\vec{F} = (2x+y-2z)\hat{i} + (2x-4y+z^2)\hat{j} + (x-2y+z^2)\hat{k}$ *evaluate the integral* $\oint_C \vec{F} \cdot d\vec{R}$, *where C is the circle with centre at (0, 0, 3) and radius 5 in the plane z = 3.*

Sol. Here $\vec{F} = (2x + y - 2z)\hat{i} + (2x - 4y + z^2)\hat{j} + (x - 2y + z^2)dz$ and C is the curve given by $(x - 0)^2 + (y - 0)^2 + (z - 3)^2 = 5^2$, $z = 3$ *i.e.*, $x^2 + y^2 + (z - 3)^2 = 25$, $z = 3$

[It is a great circle with centre at (0, 0, 3) and radius 5]

By Stoke's theorem

$$\oint_C \vec{F} \cdot d\vec{R} = \iint_S \text{curl } \vec{F} \cdot \hat{n} \, dS \qquad \ldots(1)$$

Now,
$$\text{curl } \vec{F} = \begin{vmatrix} \hat{i} & \hat{j} & \hat{k} \\ \dfrac{\partial}{\partial x} & \dfrac{\partial}{\partial y} & \dfrac{\partial}{\partial z} \\ 2x+y-2z & 2x-4y+z^2 & x-2y+z^2 \end{vmatrix}$$

$$= \hat{i}\,\{-2-2z\} - \hat{j}\,\{1-(-2)\} + \hat{k}\,\{2-1\}$$

$$= (-2-2z)\,\hat{i} - 3\hat{j} + \hat{k}$$

$\hat{n}$ = a unit vector normal to the plane $z = 3$

i.e., $\hat{n} = \hat{k}$

$\therefore$ From (1), $\oint_C \vec{F} \cdot d\vec{R} = \iint_S \left[(-2-2z)\,\hat{i} - 3\hat{j} + \hat{k}\right] \cdot \hat{k} dS = \iint_S dS = S$

$= \text{area of circle} = \pi \times 5^2 = 25\pi.$

Example 5. *Evaluate the surface integral* $\iint_S \text{curl}\ \vec{F} \cdot \hat{n}\ dS$ *by transforming it into a line integral, S being that part of the surface of the paraboloid* $z = 1 - x^2 - y^2$, *for which* $z \geq 0$ *and* $\vec{F} = y\hat{i} + z\hat{j} + x\hat{k}$.

Sol. The boundary C of the surface S is the circle $x^2 + y^2 = 1$, $z = 0$ whose parametric equations are

$$x = \cos t,\ y = \sin t,\ z = 0,\ 0 \leq t < 2\pi.$$

By Stoke's theorem, we have $\iint_S \text{curl}\ \vec{F} \cdot \hat{n}\ dS = \oint_C \vec{F} \cdot d\vec{r}$

$$= \oint_C (y\hat{i} + z\hat{j} + x\hat{k}) \cdot (\hat{i}\,dx + \hat{j}\,dy + \hat{k}\,dz)$$

$$= \oint_C (y\,dx + z\,dy + x\,dz) = \oint_C y\,dx, \text{ since on C, } z = 0,\ dz = 0$$

$$= \oint_C y\frac{dx}{dt}\,dt = \int_0^{2\pi} \sin t(-\sin t)\,dt = -\int_0^{2\pi} \sin^2 t\,dt$$

$$= -2\int_0^{\pi} \sin^2 t\,dt = -4\int_0^{\pi/2} \sin^2 t\,dt = -4 \times \frac{1}{2} \times \frac{\pi}{2} = -\pi.$$

EXERCISE 13.9

1. Verify Stoke's theorem for the vector field $\vec{F} = (x^2 - y^2)\hat{i} + 2xy\hat{j}$ integrated round the rectangle in the plane $z = 0$ and bounded by the lines $x = 0$, $y = 0$, $x = a$, $y = b$.
2. Verify Stoke's theorem for the function $\vec{F} = x^2\hat{i} + xy\hat{j}$ integrated round the square whose sides are $x = 0$, $y = 0$, $x = a$ and $y = a$ in the plane $z = 0$.
3. Verify Stoke's theorem for $\vec{F} = (x^2 - y^2)\hat{i} + 2xy\hat{j}$ over the box bounded by the planes $x = 0$, $x = a$; $y = 0$; $y = b$, $z = 0$, $z = c$; if the face $z = 0$ is cut.
4. Verify Stoke's theorem for $\vec{F} = y\hat{i} + z\hat{j} + x\hat{k}$, where S is the upper half surface of the sphere $x^2 + y^2 + z^2 = 1$ and C is its boundary.
5. Evaluate by Stoke's theorem $\oint_C (yz\,dx + zx\,dy + xy\,dz)$, where C is the curve $x^2 + y^2 = 1$, $z = y^2$.
6. Use Stoke's theorem to evaluate $\iint_S (\nabla \times \vec{F}) \cdot \hat{n}\ dS$, where $\vec{F} = y\hat{i} + (x - 2xz)\hat{j} - xy\hat{k}$ and S is the surface of the sphere $x^2 + y^2 + z^2 = a^2$ above the xy-plane.
7. Evaluate by Stoke's theorem $\oint_C (\sin z\,dx - \cos x\,dy + \sin y\,dz)$, where C is the boundary of the rectangle $0 \leq x \leq \pi$, $0 \leq y \leq 1$, $z = 3$. (*Calicut 2010*)

8. Evaluate $\int_C \vec{F} \cdot \vec{dr}$, where $\vec{F} = y\hat{i} + xz^3\hat{j} - zy^3\hat{k}$, C is the circle $x^2 + y^2 = 4$, $z = 1.5$.

9. Apply Stoke's theorem to evaluate $\int_C [(x + y)dx + (2x - z)dy + (y + z)dz]$, where C is the boundary of the triangle with vertices (2, 0, 0), (0, 3, 0) and (0, 0, 6).

(K.U.K. 2009; P.T.U. May 2010; Calicut 2010; M.T.U. 2013)

10. Apply Stoke's theorem to evaluate $\int_C (ydx + zdy + xdz)$, where C is the curve of intersection of $x^2 + y^2 + z^2 = a^2$ and $x + z = a$. *(M.D.U. May 2010)*

Answers

5. 0	6. 0	7. 2	8. $\frac{19\pi}{2}$
9. 21	10. $-\frac{\pi a^2}{\sqrt{2}}$.		

APPENDIX—I

SHORT ANSWER TYPE QUESTIONS

According to the latest syllabus of MDU, Q. 1 will be compulsory. It will cover all the four sections and will be of short answer type.

Answer the following short answer type questions by choosing the correct option or by filling up the blanks.

Infinite Series:

1. The geometric series $1 + x + x^2 + \cdots \infty$ converges in the interval .... .

2. The hyper-harmonic series $\frac{1}{1^p} + \frac{1}{2^p} + \frac{1}{3^p} + \cdots \infty$ converges if

   (*a*) $p < 1$ (*b*) $p = 1$ (*c*) $p > 1$ (*d*) $p \leq 1$.

   (M.D.U. Dec. 2013)

3. If $\sum u_n$ is convergent, then $\lim_{n\to\infty} u_n = \cdots$

4. If $\sum u_n$ is a series of positive terms and $\lim_{n\to\infty} u_n \neq 0$, then $\sum u_n$ must be $\cdots$

5. A series of positive terms either $\cdots$ or $\cdots$ but never $\cdots$.

6. The series $\sum_{n=1}^{\infty} (-1)^{n-1}$

   (*a*) converges (*b*) diverges
   (*c*) oscillates finitely (*d*) oscillates infinitely.

7. The series $1 + \frac{1}{4^{2/3}} + \frac{1}{9^{2/3}} + \frac{1}{16^{2/3}} + \cdots \infty$ is

   (*a*) convergent (*b*) divergent

8. The series $\sum_{n=0}^{\infty} (2x)^n$ converges if

   (*a*) $-2 < x < 2$ (*b*) $-1 < x < 1$ (*c*) $-\frac{1}{2} \leq x \leq \frac{1}{2}$ (*d*) $-\frac{1}{2} < x < \frac{1}{2}$

9. The series $\frac{1}{\sqrt{1.2}} + \frac{1}{\sqrt{2.3}} + \frac{1}{\sqrt{3.4}} + \cdots \infty$ is

   (*a*) convergent (*b*) divergent

10. The series $\sum \frac{\sqrt[3]{3n^2+1}}{\sqrt[4]{4n^3+2n+7}}$ is

    (*a*) convergent (*b*) divergent

11. The series $\sum \frac{1}{n^p (n+1)^p}$ converges if

(*a*) $p < \frac{1}{2}$ (*b*) $p = \frac{1}{2}$ (*c*) $p > \frac{1}{2}$ (*d*) $p \geq \frac{1}{2}$.

12. The series $\frac{2^p}{1^q} + \frac{3^p}{2^q} + \frac{4^p}{3^q} + \cdots \infty$ converges for ... (p and q are positive numbers.)

13. The series $1 - \frac{1}{\sqrt{2}} + \frac{1}{\sqrt{3}} - \frac{1}{\sqrt{4}} + \cdots \infty$ is

(*a*) convergent (*b*) divergent (*c*) oscillatory

14. The series $\sqrt{\frac{1}{4}} + \sqrt{\frac{2}{6}} + \sqrt{\frac{3}{8}} + \cdots \infty$ is

(*a*) convergent (*b*) divergent

15. The series $\sum_{n=2}^{\infty} \frac{1}{(\log n)^n}$ is

(*a*) convergent (*b*) divergent

16. The series $\sum_{n=1}^{\infty} \frac{n!}{n^n}$ is

(*a*) convergent (*b*) divergent

17. The series $\sum_{n=1}^{\infty} n(-1)^n$

(*a*) converges (*b*) diverges

(*c*) oscillates finitely (*d*) oscillates infinitely.

18. The series $5 - 4 - 1 + 5 - 4 - 1 + 5 - 4 - 1 + \cdots \infty$

(*a*) converges (*b*) diverges

(*c*) oscillates finitely (*d*) oscillates infinitely.

19. The series $\frac{1}{1.2} + \frac{2}{3.4} + \frac{3}{5.6} + \cdots \infty$ is

(*a*) convergent (*b*) divergent

20. The series $\sum_{n=1}^{\infty} \frac{1}{n} \sin \frac{1}{n}$ (*M.D.U. Dec. 2012*)

(*a*) converges (*b*) diverges (*c*) oscillates

21. Is the series $\sum n \tan \frac{1}{n}$ divergent ?

22. The series $1 - \frac{1}{3} + \frac{1}{5} - \frac{1}{7} + \cdots \infty$

(*a*) converges absolutely (*b*) converges conditionally

23. The series $1+\frac{1}{2^2}-\frac{1}{3^2}-\frac{1}{4^2}+\frac{1}{5^2}+\frac{1}{6^2}-\frac{1}{7^2}-\frac{1}{8^2}+\cdots\infty$ is

(*a*) convergent (*b*) conditionally convergent
(*c*) oscillatory (*d*) divergent

24. The series $\frac{3}{5}+\frac{4}{5^2}+\frac{3}{5^3}+\frac{4}{5^4}+\frac{3}{5^5}+\frac{4}{5^6}+\cdots\infty$ is

(*a*) convergent (*b*) divergent

25. The series $\sum\left(\sqrt{n^4+1}-\sqrt{n^4-1}\right)$ is (*M.D.U. Dec. 2012*)

(*a*) convergent (*b*) divergent (*c*) oscillatory.

26. The series $\sum \cot^{-1} n^2$ is

(*a*) convergent (*b*) divergent (*M.D.U. Dec. 2009*)

27. The series $\frac{2}{1^2}-\frac{3}{2^2}+\frac{4}{3^2}-\frac{5}{4^2}+\cdots\infty$ is

(*a*) conditionally convergent (*b*) absolutely convergent
(*c*) divergent (*d*) oscillates

28. The series $\sum \frac{(n!)\,.\,2^n}{n^n}$ is

(*a*) convergent (*b*) divergent

29. The series $\sum_{n=2}^{\infty}\frac{1}{n(\log n)^p}$, $p>0$ is convergent if

(*a*) $p<1$ (*b*) $p=1$ (*c*) $p>1$ (*d*) $p\le 1$.

30. The series $\sum_{n=1}^{\infty}\left(1+\frac{1}{n}\right)^{-n^2}$ is

(*a*) convergent (*b*) divergent (*c*) oscillatory

31. The series $\frac{\sin x}{1^3}-\frac{\sin 2x}{2^3}+\frac{\sin 3x}{3^3}-\cdots$ converges

(*a*) conditionally (*b*) absolutely

32. The series $\sum_{n=1}^{\infty}\frac{\cos n\pi}{n^2+1}$ is

(*a*) conditionally convergent (*b*) absolutely convergent

33. Which one of the following series is not convergent?

(*a*) $\frac{1}{2\sqrt{2}}+\frac{1}{3\sqrt{3}}+\frac{1}{4\sqrt{4}}+\cdots$ (*b*) $1-\frac{1}{2}+\frac{1}{3}-\frac{1}{4}+\cdots$

(*c*) $1\frac{1}{2}-1\frac{1}{3}+1\frac{1}{4}-1\frac{1}{5}+\cdots$ (*d*) $x+x^2+x^3+\cdots$, $|x|<1$

34. The series $x - \frac{x^2}{2} + \frac{x^3}{3} - \frac{x^4}{4} + \cdots \infty$ converges in the interval ⋯ .

35. The series $1 - \frac{1}{3}(x-1) + \frac{1}{9}(x-1)^2 - \frac{1}{27}(x-1)^3 + \cdots \infty$ converges in the interval ⋯ .

Matrices and Its Applications:

36. If every minor of order r of a matrix A is zero, then

(a) $\rho(A) > r$ (b) $\rho(A) = r$

(c) $\rho(A) < r$ (d) $\rho(A) \le r$.

37. The maximum value of the rank of a 3×4 matrix is ⋯ .

38. The rank of the matrix (i) $\begin{bmatrix} 2 & 3 & 4 & -1 \\ 5 & 2 & 0 & -1 \\ -4 & 5 & 12 & -1 \end{bmatrix}$ (ii) $\begin{bmatrix} 4 & 0 & 0 \\ 0 & 5 & 0 \\ 0 & 0 & 6 \end{bmatrix}$ is ⋯ .

39. The inverse of the matrix $\begin{bmatrix} 2 & 1 & -1 \\ 0 & 2 & 1 \\ 5 & 2 & -3 \end{bmatrix}$ is ⋯ .

40. For a system of non-homogeneous linear equations AX = B, if ρ [A : B] = ρ(A) < the number of unknowns, then the system has ⋯ solutions.

41. A system of non-homogeneous linear equations is consistent if and only if the rank of co-efficient matrix is equal to the rank of ⋯ .

42. By applying elementary transformations to a matrix, its rank

(a) increases (b) decreases (c) does not change.

43. The rank of the matrix $\begin{bmatrix} 6 & 1 & 3 & 8 \\ 4 & 2 & 6 & -1 \\ 10 & 3 & 9 & 7 \\ 16 & 4 & 12 & 15 \end{bmatrix}$ is ⋯ .

44. (i) If the rank of matrix A is 2, then what will be the rank of A′? (*K.U.K. 2013*)

(ii) Find the value of P for which the matrix $A = \begin{bmatrix} 3 & P & P \\ P & 3 & P \\ P & P & 3 \end{bmatrix}$ is of rank 1. (*M.T.U. 2012*)

45. A square matrix A is called orthogonal if A′ = ⋯ .

46. If A is an orthogonal matrix, then $|A| = \cdots$.

47. Is the matrix $\begin{bmatrix} \cos\theta & 0 & \sin\theta \\ 0 & 1 & 0 \\ -\sin\theta & 0 & \cos\theta \end{bmatrix}$ orthogonal?

48. The eigen values of the matrix $\begin{bmatrix} 1 & -2 \\ -5 & 4 \end{bmatrix}$ are ⋯ .

49. The sum of the eigen values of $\begin{bmatrix} 1 & 1 & 3 \\ 1 & 5 & 1 \\ 3 & -1 & 1 \end{bmatrix}$ is ⋯ . (*K.U.K. 2013*)

50. (*i*) The product of the eigen values of $\begin{bmatrix} 2 & 3 & -2 \\ -2 & 1 & 1 \\ 1 & 0 & 2 \end{bmatrix}$ is ⋯ .

(*ii*) What will be the product of eigen values of a singular matrix? (*M.D.U. Dec. 2013*)

51. The eigen values of the matrix $\begin{bmatrix} 2 & 7 & 1 \\ 0 & 4 & 3 \\ 0 & 0 & 5 \end{bmatrix}$ are ⋯ .

52. If $A = \begin{bmatrix} 1 & 0 & 0 \\ 2 & 3 & 0 \\ 3 & 4 & 5 \end{bmatrix}$, then latent roots of A^{-1} are ⋯ .

53. If the eigen values of a square matrix A are $\lambda_1, \lambda_2, \lambda_3$ then the eigen values of A^T are ⋯ .

54. If the product of two eigen values of $\begin{bmatrix} -2 & 2 & -3 \\ 2 & 1 & -6 \\ -1 & -2 & 0 \end{bmatrix}$ is – 15, then the third eigen value is ⋯ .

55. Every eigen value of A is an eigen value of B only if B = ⋯

56. The sum of the eigen values of a matrix is the ⋯ of the elements on its principal diagonal.

57. The eigen values of the matrix $\begin{bmatrix} 8 & -6 & 2 \\ -6 & 7 & -4 \\ 2 & -4 & 3 \end{bmatrix}$ are ⋯ .

58. The eigen values of a triangular matrix are ⋯ .

59. The characteristic equation of $\begin{bmatrix} 2 & -1 & 1 \\ -1 & 2 & -1 \\ 1 & -1 & 2 \end{bmatrix}$ is ⋯ .

60. If $A = \begin{bmatrix} 1 & 4 \\ 3 & 2 \end{bmatrix}$, then eigen values of A^{-1} are ⋯ .

61. If $A = \begin{bmatrix} -2 & 5 & 1 \\ 0 & 2 & 3 \\ 0 & 0 & 3 \end{bmatrix}$, then the eigen values of A^5 are ⋯ . (*M.D.U. Dec. 2009*)

62. If two eigen values of $\begin{bmatrix} -9 & 2 & 6 \\ 5 & 0 & -3 \\ -16 & 4 & 11 \end{bmatrix}$ are 1 and 2, then the third eigen value is ⋯ .

63. An example of a 3 × 3 matrix of rank one is ⋯ .

64. The product of the eigen values of $\begin{bmatrix} 2 & 0 & 0 \\ 7 & 4 & 3 \\ 15 & 6 & 5 \end{bmatrix}$ is ⋯ .

65. The eigen values of $\begin{bmatrix} 1 & -2 \\ -5 & 4 \end{bmatrix}$ are the roots of ⋯ .

66. (*i*) The sum of the squares of the eigen values of $\begin{bmatrix} 1 & 2 & -3 \\ 0 & 4 & 5 \\ 0 & 0 & 3 \end{bmatrix}$ is ⋯ .

(*ii*) Using Cayley-Hamilton theorem, find A^4 if $A = \begin{bmatrix} 2 & 1 \\ 5 & -2 \end{bmatrix}$ (*M.D.U. Dec. 2013*)

67. The matrix corresponding to the quadratic form $ax^2 + by^2 + cz^2 + 2fyz + 2gzx + 2hxy$ is ⋯ .

68. The index and signature of the quadratic form $x^2 + 2y^2 - 3z^2$ are respectively ⋯ and ⋯ .

69. A quadratic form is negative semi-definite when ⋯ .

70. The matrix of the quadratic form $2x^2 - 5y^2 + 3z^2 + 4xy - 6yz + 2zx$ is ⋯ .

Successive Differentiation:

71. If $y = \log(\cos x)$, then $y_3 = \cdots$.

72. If $y = e^{-x} \cos x$, then $y_4 + 4y = \cdots$.

73. If $y = A \sin mx + B \cos mx$, then $\dfrac{d^2y}{dx^2} = \cdots$.

74. If $\dfrac{x^2}{a^2} + \dfrac{y^2}{b^2} = 1$, then $a^2 y^3 \dfrac{d^2y}{dx^2} = \cdots$.

75. If $x = at^2$, $y = 2at$, then $\dfrac{d^2y}{dx^2} = \cdots$.

76. If $x = a(\theta + \sin\theta)$, $y = a(1 + \cos\theta)$, then $\left.\dfrac{d^2y}{dx^2}\right]_{\theta=\frac{\pi}{2}} = \cdots$.

77. If $y = (\sin^{-1} x)^2$, then $(1-x^2)\dfrac{d^2y}{dx^2} - x\dfrac{dy}{dx} = \cdots$.

78. If $\sqrt{x} + \sqrt{y} = \sqrt{a}$, then $\left.\dfrac{d^2y}{dx^2}\right]_{x=a} = \cdots$.

79. The nth derivative of $\dfrac{1}{(x+2)(x-1)}$ is ⋯ .

80. The nth derivative of $\log\sqrt{\dfrac{2x+1}{x-2}}$ is ⋯ .

81. The nth derivative of $\log(x^2 - a^2)$ is ⋯ .

82. If $y^2 = P(x)$, then $2\dfrac{d}{dx}\left(y^3 \dfrac{d^2y}{dx^2}\right)$ equals

(*a*) $P''(x)\, P'''(x)$ (*b*) $P(x)\, P'''(x)$ (*c*) $P''(x) + P'''(x)$

83. If $y = \sin^3 x$, then $y_n = \cdots$.

84. The nth derivative of $\cos^4 x$ is ⋯ .

85. The nth derivative of $\cos 2x \sin 3x$ is ⋯ .

86. The nth derivative of $\sin x \sin 2x \sin 3x$ is ⋯ .

87. If $y = e^x \sin x \cos x$, then $y_n = \cdots$.

88. If $y = x^2 e^x$, then $\frac{d^n y}{dx^n} = \cdots$.

89. If $y = x^{n-1} \log x$, then $xy_n = \cdots$.

Taylor's and Maclaurin's Series:

90. Apply Taylor's series to expand $\cos\left(\frac{\pi}{4} + h\right)$.

91. Apply Taylor's series to express the polynomial $2x^3 + 7x^2 + x - 6$ in powers of $(x - 2)$.

92. Apply Taylor's series to find the approximate value of $f\left(\frac{9}{10}\right)$, if $f(x) = x^3 + 2x^2 - 5x + 11$.

93. $\log\sqrt{\frac{1+x}{1-x}} = \cdots$.

94. $\cosh x = 1 + \cdots$.

95. $\tan x = \cdots$.

96. $\tan^{-1} x = \cdots$.

97. $\sin^{-1}\left(\frac{2x}{1+x^2}\right) = \cdots$.

Curvature:

98. The radius of curvature at the point (s, ψ) on the curve $s = a \log \tan\left(\frac{\pi}{4} + \frac{\psi}{2}\right)$ is $\cdots$

99. The radius of curvature of the curve $y = e^x$ at the point where it crosses the y-axis is $\cdots$

100. The radius of curvature of the catenary $y = c \cosh \frac{x}{c}$ at the point where it crosses the y-axis is $\cdots$

101. The curvature of the circle $x^2 + y^2 = 13$ at the point $(-2, 3)$ is $\cdots$

102. The radius of curvature of the curve $y = 4 \sin x - \sin 2x$ at $x = \frac{\pi}{2}$ is $\cdots$

103. The radius of curvature at any point of the cardioid $r = a(1 + \cos\theta)$ is $\cdots$

104. The radius of curvature at the origin for the curve $x^3 - 2x^2y + 3xy^2 - 4y^3 + 5x^2 - 6xy + 7y^2 - 8y = 0$ is $\cdots$

105. The chord of curvature parallel to y-axis for the curve $y = a \log \sec\left(\frac{x}{a}\right)$ is of $\cdots$ length.

106. If (a, b) are the coordinates of the centre of curvature whose curvature is k, then the equation of the circle of curvature is $\cdots$

Asymptote:

107. The asymptotes, parallel to the axes, of the curve $y^3 - xy^2 = x^2 + 1$ are $\cdots$

108. The asymptotes, parallel to the axes, of the curve $(x^2 - a^2)(y^2 - b^2) = a^2b^2$ are $\cdots$

109. The asymptotes of the curve $x^2y^2 = a^2(x^2 + y^2)$ form a square of side $\cdots$

110. The equation of the asymptote of the curve $x^3 + y^3 = 3axy$ is

(*a*) $x + y - a = 0$ (*b*) $x - y + a = 0$ (*c*) $x + y + a = 0$ (*d*) $x - y - a = 0$

111. The parabola $y^2 = 4ax$ has $\cdots$ asymptote.

112. The equation of the asymptote of the curve $y^3 = x^3 + ax^2$ is

(*a*) $3x = 3y + a$ (*b*) $3y = 3x + a$ (*c*) $3y = 3x - a$ (*d*) $3x = 3y - a$

113. The equation of the hyperbola having $x + y - 1 = 0$ and $x - y + 2 = 0$ as its asymptotes and passing through the origin is ···

114. The four asymptotes of the curve $xy(x^2 - y^2) + 9x^2 + 25y^2 = 114$ cut the curve in ··· points which lie on an ellipse of eccentricity ···

115. The eight points of intersection of the curve $xy(x^2 - y^2) + x^2 + y^2 = a^2$ and its asymptotes lie on a circle whose centre is ···

116. The asymptote of the polar curve $r\theta = a$ is ···

Curve Tracing:

117. If the equation of a curve remains unaltered when both x and y are changed to $-x$ and $-y$ respectively, then the curve is symmetrical ···

118. If the equation of a curve remains unchanged when x and y are interchanged, then the curve is symmetrical about ···

119. If the polar equation of a curve remains unchanged when θ is changed to $\frac{\pi}{2} - \theta$, then the curve is symmetrical about the line ···

120. If the polar equation of a curve remains unchanged when θ and r are changed to $-\theta$ and $-r$ respectively, then the curve is symmetrical about the line ···

121. The equations of tangents to a curve at the origin are obtained by equating to zero the ··· degree terms is the equation of the curve.

122. In the parametric equations $x = f(t)$, $y = g(t)$ of a curve, if x and y both are odd functions of t, then the curve is symmetrical ···

Functions of Two or More Variables:

123. If $u = x^2 + y^2 + z^2$, then $xu_x + yu_y + zu_z = \cdots$

124. If $z = \log(x^2 + xy + y^2)$, then $x\frac{\partial z}{\partial x} + y\frac{\partial z}{\partial y} = \cdots$

125. If $u = x^2y + y^2z + z^2x$, then $u_x + u_y + u_z = \cdots$

126. If $u = \frac{y}{z} + \frac{z}{x} + \frac{x}{y}$, then $x\frac{\partial u}{\partial x} + y\frac{\partial u}{\partial y} + z\frac{\partial u}{\partial z} = \cdots$

127. If $f(x, y, z) = \begin{vmatrix} x^2 & y^2 & z^2 \\ x & y & z \\ 1 & 1 & 1 \end{vmatrix}$, then $f_x + f_y + f_z = \cdots$

128. If $V = (x^2 + y^2 + z^2)^{-\frac{1}{2}}$, then $\frac{\partial^2 V}{\partial x^2} + \frac{\partial^2 V}{\partial y^2} + \frac{\partial^2 V}{\partial z^2} = \cdots$

129. If $x = r\cos\theta$, $y = r\sin\theta$, then $\left(\frac{\partial r}{\partial x}\right)^2 + \left(\frac{\partial r}{\partial y}\right)^2 = \cdots$

130. If $u = \sqrt{x^2 + y^2 + z^2}$, then $\left(\frac{\partial u}{\partial x}\right)^2 + \left(\frac{\partial u}{\partial y}\right)^2 + \left(\frac{\partial u}{\partial z}\right)^2 = \cdots$

131. If $x^2 + y^2 + z^2 = \frac{1}{u^2}$, then $\frac{\partial^2 u}{\partial x^2} + \frac{\partial^2 u}{\partial y^2} + \frac{\partial^2 u}{\partial z^2} = \cdots$

132. If u is a homogeneous function of degree n in x and y, then

$$x^2 \frac{\partial^2 u}{\partial x^2} + 2xy \frac{\partial^2 u}{\partial x \partial y} + y^2 \frac{\partial^2 u}{\partial y^2} \text{ equals}$$

(*a*) nu (*b*) $n(n+1)u$ (*c*) $n(n-1)u$ (*d*) $n^2 u$

133. $\dfrac{x+2y+3z}{\sqrt{x^8+y^8+z^8}}$ is a homogeneous function of degree ···

134. If $u = f\left(\dfrac{y}{x}\right)$, then

(*a*) $x\dfrac{\partial u}{\partial x} - y\dfrac{\partial u}{\partial y} = 0$ (*b*) $x\dfrac{\partial u}{\partial x} + y\dfrac{\partial u}{\partial y} = 0$

(*c*) $x\dfrac{\partial u}{\partial x} + y\dfrac{\partial u}{\partial y} = u$ (*d*) $x\dfrac{\partial u}{\partial x} + y\dfrac{\partial u}{\partial y} = 1$

135. If $u = \sin^{-1}\dfrac{x}{y} + \tan^{-1}\dfrac{y}{x}$, then $x\dfrac{\partial u}{\partial x} + y\dfrac{\partial u}{\partial y} = \cdots$

136. If $u = \tan^{-1}\dfrac{x^3+y^3}{x-y}$, then $x\dfrac{\partial u}{\partial x} + y\dfrac{\partial u}{\partial y} = \cdots$

137. If $u = (x^2+y^2)^{1/3}$, then $x^2\dfrac{\partial^2 u}{\partial x^2} + 2xy\dfrac{\partial^2 u}{\partial x \partial y} + y^2\dfrac{\partial^2 u}{\partial y^2} = \cdots$

138. If $x = r\cos\theta$, $y = r\sin\theta$, then $J\left(\dfrac{x, y}{r, \theta}\right) = \cdots$

139. If $J_1 = \dfrac{\partial(u, v)}{\partial(x, y)}$ and $J_2 = \dfrac{\partial(x, y)}{\partial(u, v)}$, then $J_1 J_2 = \cdots$

140. For spherical polar coordinates, $\dfrac{\partial(x, y, z)}{\partial(r, \theta, \phi)} = \cdots$

141. If $u = f(y-z, z-x, x-y)$, then $\dfrac{\partial u}{\partial x} + \dfrac{\partial u}{\partial y} + \dfrac{\partial u}{\partial z}$ equals.

(*a*) 1 (*b*) 0 (*c*) -1 (*d*) u

142. If $A = f_{xx}(a, b)$, $B = f_{xy}(a, b)$, $C = f_{yy}(a, b)$, then $f(x, y)$ will have a minimum value at (a, b) if

(*a*) $f_x = 0, f_y = 0, \quad AC < B^2, A > 0$ (*b*) $f_x = 0, f_y = 0, \quad AC > B^2, A < 0$

(*c*) $f_x = 0, f_y = 0, \quad AC > B^2, A > 0$ (*d*) $f_x = 0, f_y = 0, \quad AC = B^2, A > 0$

Partial Differentiation:

143. The volume of the solid formed by the revolution of one arch of the cycloid $x = a(\theta - \sin\theta)$, $y = a(1 - \cos\theta)$ about its base is ···

144. The volume of the solid generated by revolving the astroid $x^{2/3} + y^{2/3} = a^{2/3}$ about the x-axis is ···

Integral Calculus:

145. $\int_0^1 \int_0^1 \frac{dxdy}{\sqrt{(1-x^2)(1-y^2)}} = \cdots$.

146. $\int_1^2 \int_0^x \frac{dydx}{x^2+y^2} = \cdots$

147. $\int_0^1 \int_0^{1-x} dydx = \cdots$ *(K.U.K. 2013)*

148. $\int_0^1 \int_x^{\sqrt{x}} (x^2+y^2)\, dydx = \cdots$

149. $\int_0^{\frac{\pi}{2}} \int_0^{a\cos\theta} r \sin\theta \, dr \, d\theta = \cdots$

150. $\int_{-1}^1 \int_0^z \int_{x-z}^{x+z} (x+y+z)\, dy\, dx\, dz = \cdots$

(M.D.U. Dec. 2009)

151. If R is the region bounded by the parabolas $y^2 = 4x$ and $x^2 = 4y$, then $\iint_R y\, dx\, dy = \cdots$

152. $\iint xy\, dx\, dy$ over the positive quadrant of the circle $x^2 + y^2 = a^2$ equals $\cdots$.

153. To change cartesian coordinates (x, y) to polar coordinates (r, θ), $dx\, dy$ is replaced by $\cdots$.

154. To change cartesian coordinates (x, y, z) to spherical polar coordinates (r, θ, ϕ), $dx\, dy\, dz$ is replaced by $\cdots$.

155. $\beta(m, n) = \frac{\Gamma(m)\Gamma(n)}{\cdots}$

156. $\beta(2, 1) + \beta(1, 2) = \cdots$. *(K.U.K. 2013)*

157. $\int_0^\infty \sqrt{x}\, e^{-x^3}\, dx = \cdots$

158. $\int_0^{\frac{\pi}{2}} \sqrt{\tan\theta}\, d\theta = \cdots$

159. $\int_0^{\frac{\pi}{2}} \sin^3 x \cos^{5/2} x\, dx = \cdots$

160. $\int_0^1 x^3 (1-x)^{4/3}\, dx = \cdots$.

161. Write the formula of finding the volume of the solid generated by revolving an area about both the coordinate axis. *(M.D.U. Dec. 2013)*

Answers

1. (– 1, 1) **2.** (*c*) **3.** 0

4. divergent **5.** converges, diverges, oscillates

6. (*c*) **7.** (*a*)

8. (*d*). [The geometric series with common ratio $2x$ converges if $-1 < 2x < 1$].

9. (*b*) **10.** (*b*) **11.** (*c*)

12. $q > p + 1$ **13.** (*a*)

14. (*b*) $\left[u_n = \sqrt{\frac{n}{2(n+1)}} = \frac{1}{\sqrt{2\left(1+\frac{1}{n}\right)}} \to \frac{1}{\sqrt{2}} \neq 0 \right]$

15. (*a*) by Cauchy's root test. **16.** (*a*) by ratio test.

17. (*d*) $S_n = \begin{cases} -\left(\frac{n+1}{2}\right), & \text{if } n \text{ is odd} \\ \frac{n}{2}, & \text{if } n \text{ is even} \end{cases}$

18. (*c*) $S_n = \begin{cases} 0, & \text{if } n = 3m \\ 5, & \text{if } n = 3m+1 \\ 1, & \text{if } n = 3m+2 \end{cases}$

19. (*b*) 20. (*a*) 21. yes
22. (*b*) 23. (*a*) absolutely convergent and hence convergent
24. (*a*) 25. (*a*) 26. (*a*)
27. (*a*) 28. (*a*) 29. (*c*)
30. (*a*) 31. (*b*) 32. (*b*)
33. (*c*) 34. $-1 < x \le 1$ 35. $-2 < x < 4$

---

36. (*c*) 37. 3 38. (*i*) 2 (*ii*) 3

39. $\begin{bmatrix} 8 & -1 & -3 \\ -5 & 1 & 2 \\ 10 & -1 & -4 \end{bmatrix}$

40. an infinite number of 41. Augmented matrix 42. (*c*)
43. 2 44. (*i*) 2 (*ii*) P = 3 45. A^{-1}
46. ±1 47. yes 48. 6, –1
49. 7 50. (*i*) 21 (*ii*) 0 51. 2, 4, 5
52. $1, \frac{1}{3}, \frac{1}{5}$ 53. $\lambda_1, \lambda_2, \lambda_3$ 54. – 3
55. A or A^T 56. sum 57. 0, 3, 15
58. the elements on its principal diagonal. 59. $\lambda^3 - 6\lambda^2 + 9\lambda - 4 = 0$
60. $\frac{1}{5}, -\frac{1}{2}$ 61. –32, 32, 243 62. –1

63. $\begin{bmatrix} 1 & 2 & 3 \\ 2 & 4 & 6 \\ 3 & 6 & 9 \end{bmatrix}$ 64. 4 65. $\lambda^2 - 5\lambda - 6 = 0$

66. (*i*) 26 (*ii*) 81I 67. $\begin{bmatrix} a & h & g \\ h & b & f \\ g & f & c \end{bmatrix}$ 68. 2, 1

69. all the eigen values of A are ≤ 0 and at least one eigen value is zero.

70. $\begin{bmatrix} 2 & 2 & 1 \\ 2 & -5 & -3 \\ 1 & -3 & 3 \end{bmatrix}$

---

71. $-2 \sec^2 x \tan x$ 72. 0
73. $-m^2 y$ 74. $-b^4$ 75. $-\frac{1}{2at^3}$
76. $-\frac{1}{a}$ 77. 2 78. $\frac{1}{2a}$

79. $\frac{(-1)^n (n)!}{3}\left[\frac{1}{(x-1)^{n+1}} - \frac{1}{(x+2)^{n+1}}\right].$

80. $\frac{1}{2}(-1)^{n-1} (n-1)!\left[\frac{2^n}{(2x+1)^n} - \frac{1}{(x-2)^n}\right].$

81. $(-1)^{n-1} (n-1)!\left[\frac{1}{(x+a)^n} + \frac{1}{(x-a)^n}\right].$

82. (*b*) [$y^2 = P(x) \Rightarrow 2yy' = P'(x) \Rightarrow 2y'^2 + 2yy'' = P''(x)$
$\Rightarrow 4y'y'' + 2y'y'' + 2yy''' = P'''(x) \Rightarrow 6y'y'' + 2yy''' = P'''(x)$

$\therefore \; 2\frac{d}{dx}\left(y^3 \frac{d^2y}{dx^2}\right) = 2\frac{d}{dx}(y^3 y'') = 2(3y^2 y'y'' + y^3 y''') = y^2(6y'y'' + 2yy''') = P(x)\, P'''(x)$].

83. $\frac{3}{4}\sin\left(x + \frac{n\pi}{2}\right) - \frac{1}{4} . 3^n \sin\left(3x + \frac{n\pi}{2}\right).$

84. $2^{n-1}\cos\left(2x + \frac{n\pi}{2}\right) + 2^{2n-3}\cos\left(4x + \frac{n\pi}{2}\right).$

85. $\frac{1}{2}\left[5^n \sin\left(5x + \frac{n\pi}{2}\right) + \sin\left(x + \frac{n\pi}{2}\right)\right].$

86. $\frac{1}{4}\left[2^n \sin\left(2x + \frac{n\pi}{2}\right) + 4^n \sin\left(4x + \frac{n\pi}{2}\right) - 6^n \sin\left(6x + \frac{n\pi}{2}\right)\right].$

87. $\frac{1}{2} . 5^{\frac{n}{2}} e^x \sin(2x + n\tan^{-1} 2).$

88. $[x^2 + 2nx + n(n-1)]e^x.$

89. $(n-1)\,!$

---

90. $\frac{1}{\sqrt{2}}\left(1 - h - \frac{h^2}{2!} + \frac{h^3}{3!} + \cdots\right)$

91. $40 + 53(x-2) + 19(x-2)^2 + 2(x-2)^3$

92. 8.849

93. $x + \frac{x^3}{3} + \frac{x^5}{5} + \cdots$

94. $1 + \frac{x^2}{2!} + \frac{x^4}{4!} + \frac{x^6}{6!} + \cdots$

95. $x + \frac{x^3}{3} + \frac{2x^5}{15} + \cdots$

96. $x - \frac{x^3}{3} + \frac{x^5}{5} - \cdots$

97. $2\left(x - \frac{x^3}{3} + \frac{x^5}{5} - \frac{x^7}{7} + \cdots\right)$

---

98. $a \sec \psi$

99. $2\sqrt{2}$

100. c

101. $\frac{1}{\sqrt{13}}$, [The curvature of a circle is constant and equal to the reciprocal of its radius].

102. $\frac{5\sqrt{5}}{4}$ **103.** $\frac{4a}{3}\cos\frac{\theta}{2}$ **104.** $\frac{4}{5}$

105. Constant **106.** $(x-a)^2+(y-b)^2=\left(\frac{1}{k}\right)^2$

---

107. None **108.** $x=\pm a,\ y=\pm b$ **109.** $2a$

110. (*c*) **111.** No **112.** (*b*)

113. $x^2-y^2+x+3y=0$ **114.** $8, \frac{4}{5}$ **115.** the origin

116. $r\sin\theta=a$

---

117. in opposite quadrants **118.** $y=x$ **119.** $\theta=\frac{\pi}{4}$

120. $\theta=\frac{\pi}{2}$ **121.** lowest **122.** in opposite quadrants

---

123. $2u$ **124.** 2 **125.** $(x+y+z)^2$

126. 0 **127.** 0 **128.** 0

129. 1 **130.** 1 **131.** 0

132. (*c*) **133.** –3 **134.** (*b*)

135. 0 **136.** $\sin 2u$ **137.** $-\frac{2}{9}u$

138. r **139.** 1

140. $r^2\sin\theta$ **141.** 0 **142.** (*c*)

---

143. $5\pi^2a^3$ **144.** $\frac{32}{105}\pi a^3$

---

145. $\frac{\pi^2}{4}$ **146.** $\frac{\pi}{4}\log 2$ **147.** $\frac{1}{2}$

148. $\frac{3}{35}$ **149.** $\frac{a^2}{6}$ **150.** 0

151. $\frac{48}{5}$ **152.** $\frac{a^4}{8}$ **153.** $r\,dr\,d\theta$

154. $r^2\sin\theta\,dr\,d\theta\,d\phi$ **155.** $\Gamma(m+n)$ **156.** 1

157. $\frac{\sqrt{\pi}}{3}$ **158.** $\frac{\pi}{\sqrt{2}}$ **159.** $\frac{8}{77}$

160. $\frac{243}{7280}$ **161.** About x-axis $=\int_a^b \pi y^2dx$; about y-axis $=\int_a^b \pi x^2dy$.

APPENDIX–II

SOME IMPORTANT CURVES

Cissoid (Imp.)

$y^2(2a - x) = x^3$

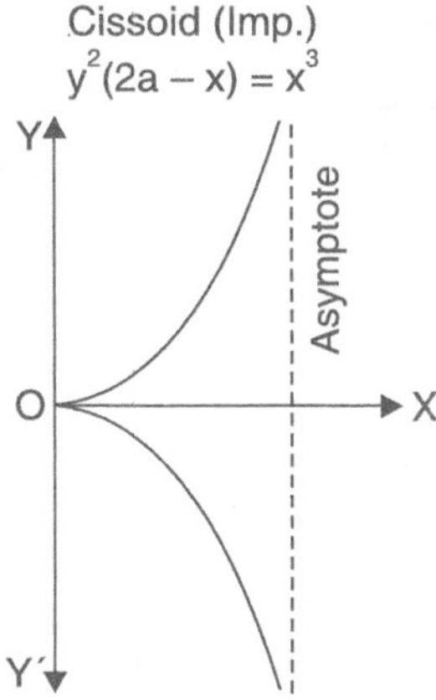

Catenary

$y = a \cosh \frac{x}{a}$

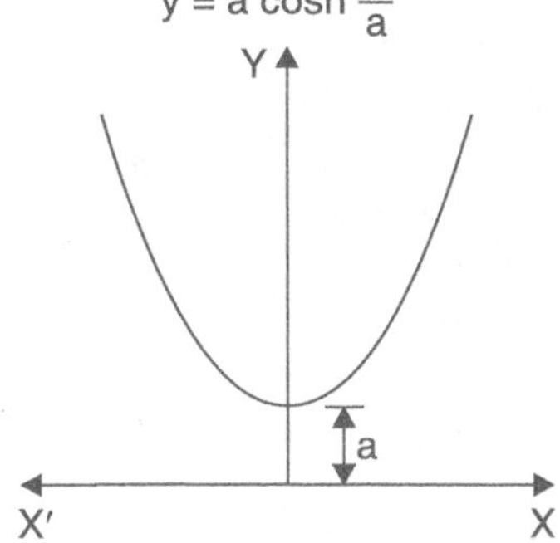

Cubical Parabola

$a^2 y = x^3$

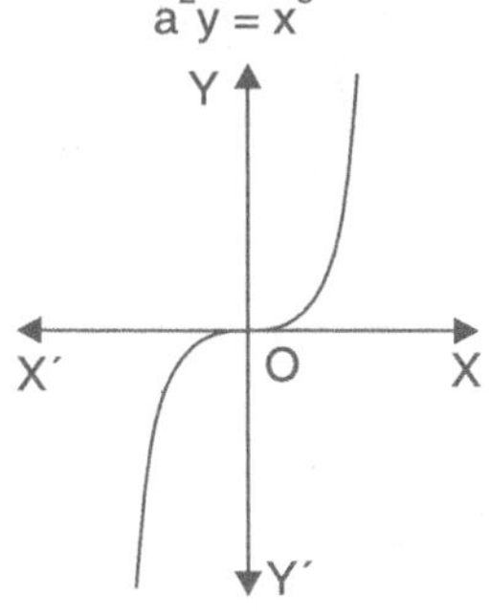

Semicubical Parabola

$ay^2 = x^3$

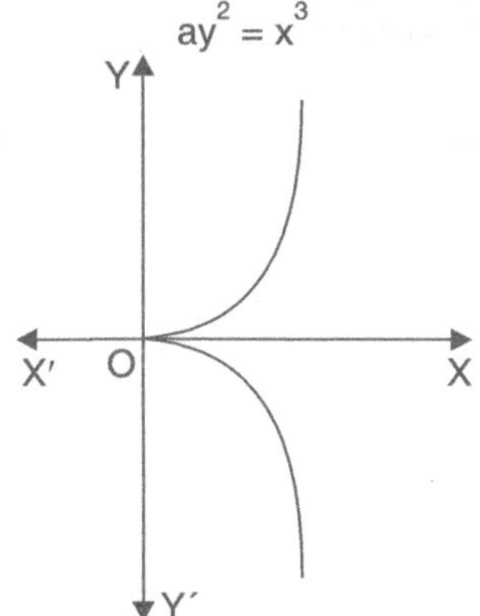

Strophoid

$y^2(a - x) = x^2(a + x)$

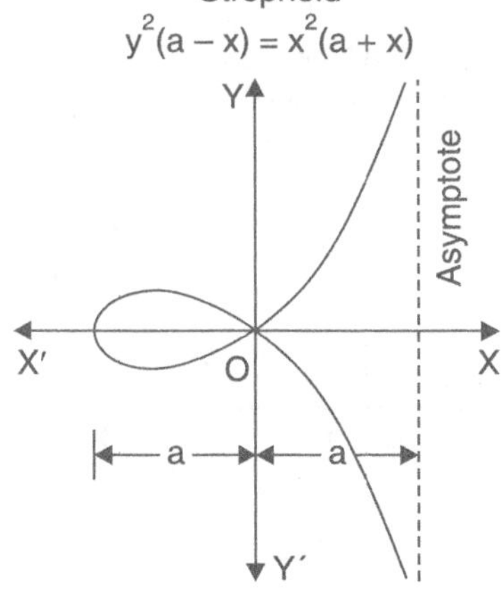

Folium of Descartes

$x^3 + y^3 = 3axy$

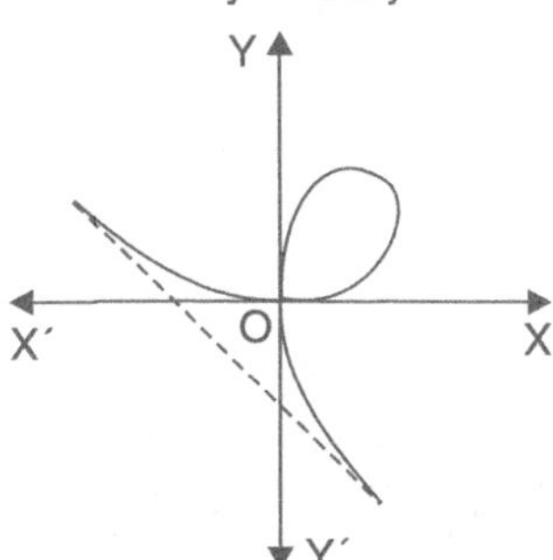

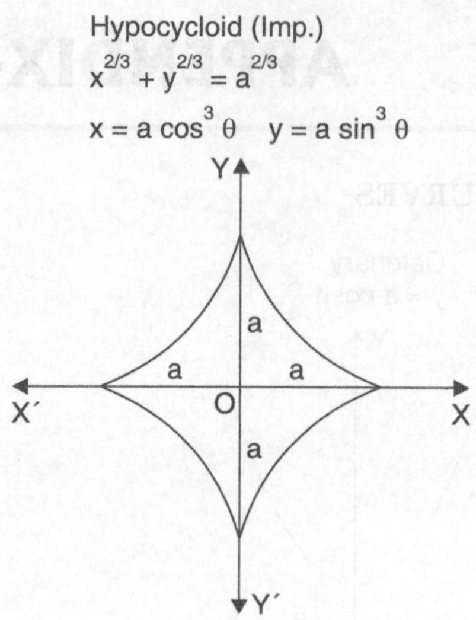
Hypocycloid (Imp.)
$x^{2/3} + y^{2/3} = a^{2/3}$
$x = a\cos^3\theta \quad y = a\sin^3\theta$
Y
a
a
a
O
a
X′
X
Y′

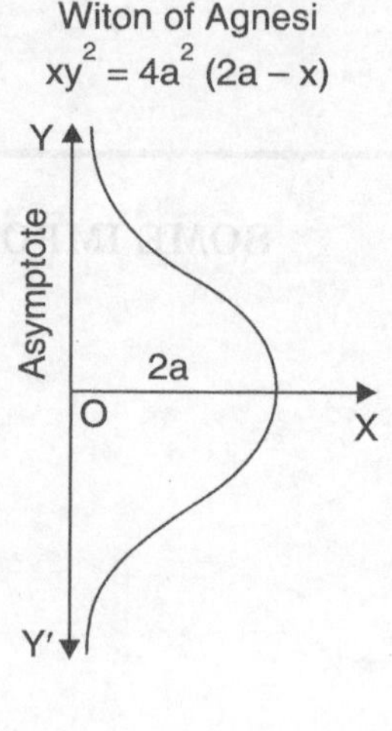
Witon of Agnesi
$xy^2 = 4a^2(2a - x)$
Y
Asymptote
2a
O
X
Y′

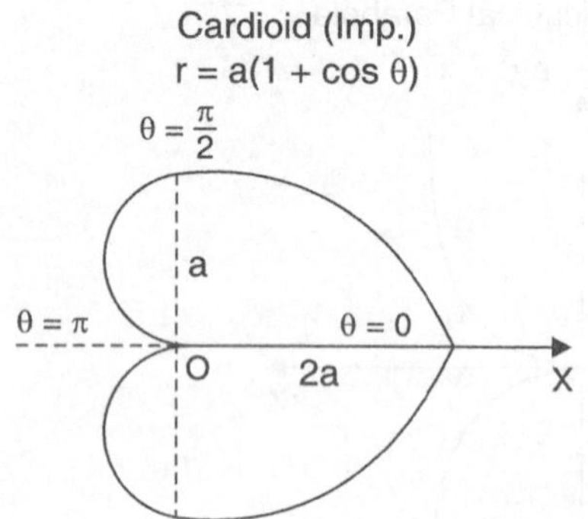
Cardioid (Imp.)
$r = a(1 + \cos\theta)$
$\theta = \frac{\pi}{2}$
a
$\theta = \pi$
$\theta = 0$
O
2a
X

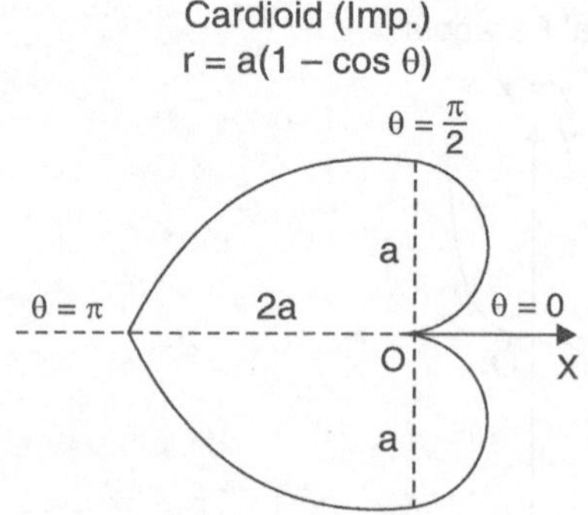
Cardioid (Imp.)
$r = a(1 - \cos\theta)$
$\theta = \frac{\pi}{2}$
a
$\theta = \pi$
2a
$\theta = 0$
O
X
a

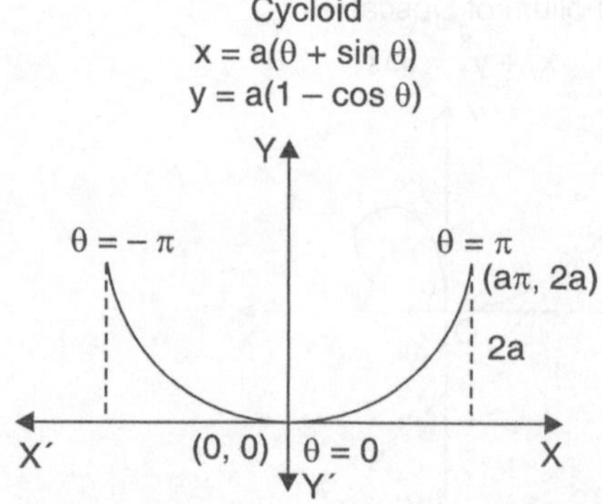
Cycloid
$x = a(\theta + \sin\theta)$
$y = a(1 - \cos\theta)$
Y
$\theta = -\pi$
$\theta = \pi$
$(a\pi, 2a)$
2a
X′
(0, 0)
$\theta = 0$
X
Y′

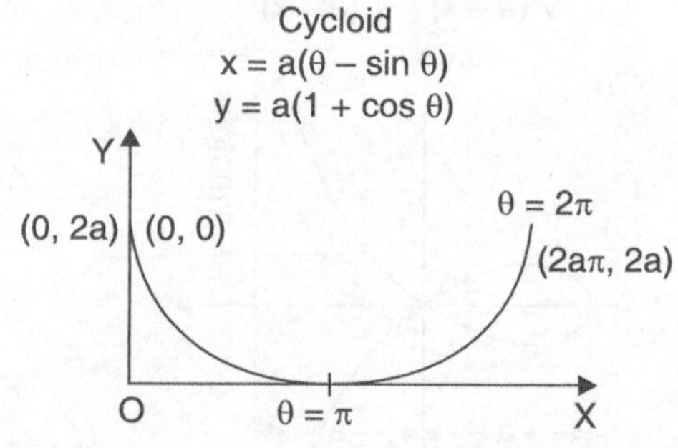
Cycloid
$x = a(\theta - \sin\theta)$
$y = a(1 + \cos\theta)$
Y
(0, 2a)
(0, 0)
$\theta = 2\pi$
$(2a\pi, 2a)$
O
$\theta = \pi$
X

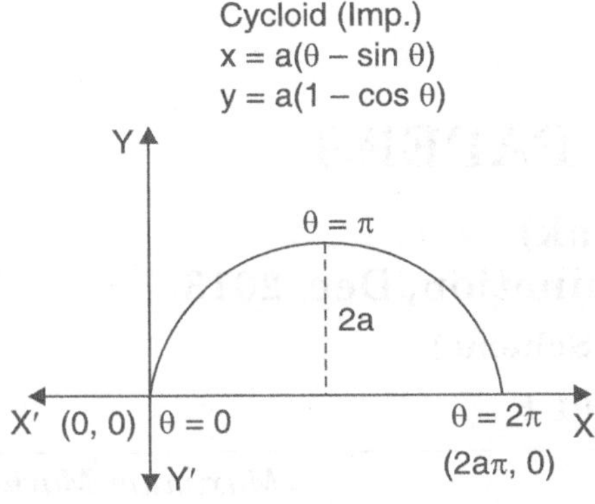
Cycloid (Imp.)
x = a(θ – sin θ)
y = a(1 – cos θ)
Y
θ = π
2a
X′ (0, 0) θ = 0
θ = 2π X
(2aπ, 0)
Y′

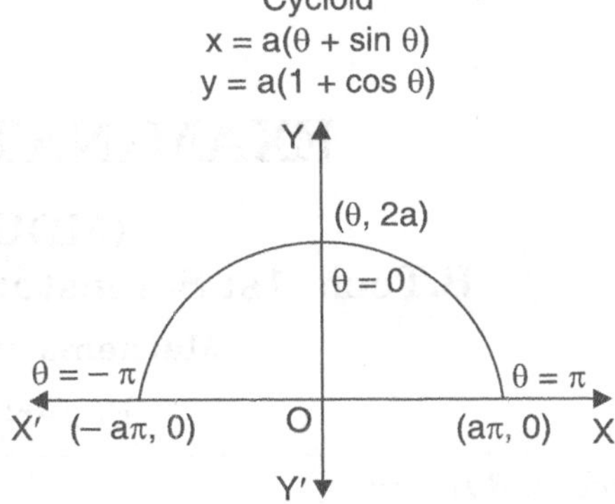
Cycloid
x = a(θ + sin θ)
y = a(1 + cos θ)
Y
(θ, 2a)
θ = 0
θ = – π
θ = π
X′ (– aπ, 0)
O
(aπ, 0) X
Y′

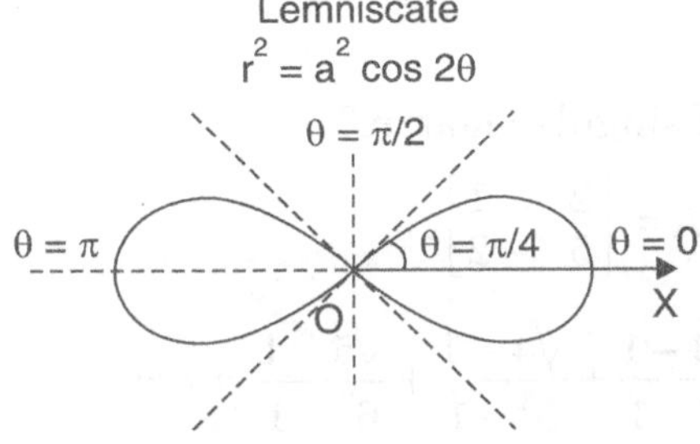
Lemniscate
r² = a² cos 2θ
θ = π/2
θ = π
θ = π/4
θ = 0
X
O

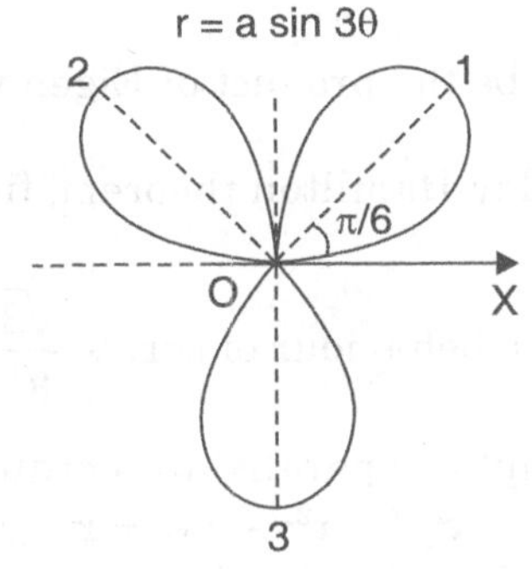
The three leaved rose
r = a sin 3θ
2
1
π/6
O
X
3

EXAMINATION PAPERS

(MDU, Rohtak)

B.Tech. 1st Semester Examination, Dec. 2013

Mathematics-I ('F' Scheme)

Paper: Math-101-F

Time allowed: 3 Hours *Maximum Marks: 100*

Note: *Question 1st is* **comulsory.** *Attempt total five questions with selecting one question from each unit. All questions carry equal marks.*

1. (*a*) What will be the product of Eigen values of singular matrix?

(*b*) Using cayley-Hamilton theorem, find A^4 if $A = \begin{bmatrix} 2 & 1 \\ 5 & -2 \end{bmatrix}$.

(*c*) Discuss the behaviour of series $\frac{\sqrt{2}-1}{3^3-1} + \frac{\sqrt{3}-1}{4^3-1} + \frac{\sqrt{4}-1}{5^3-1} + \frac{\sqrt{5}-1}{6^3-1} + \dots\dots$

(*d*) Find asymptotes parallel to coordinate axis of the curve

$$x^2y^2 - x^2y - xy^2 + x + y + 1 = 0$$

(*e*) Expand α^x by using Maclaurin's series.

(*f*) If $u = \tan^{-1}\left(\frac{x^3+y^3}{x-y}\right)$ then prove $x\frac{\partial u}{\partial x} + y\frac{\partial u}{\partial y} = \sin 2u$.

(*g*) Write the formula of finding the volume of the solid generated by revolving about both the coordinate axis.

(*h*) State *p*-Test for convergence of series.

UNIT-I

2. (*a*) Discuss the convergence of the series:

$$1 + \frac{\alpha+1}{\beta+1} + \frac{(\alpha+1)(2\alpha+1)}{(\beta+1)(2\beta+1)} + \frac{(\alpha+1)(2\alpha+1)(3\alpha+1)}{(\beta+1)(2\beta+1)(3\beta+1)} + \dots\dots \infty$$

(*b*) $x + \frac{(2x)^2}{2!} + \frac{(3x)^3}{3!} + \frac{(4x)^4}{4!} + \dots\dots$

3. (*a*) Test for absolute/conditionally convergence of the series.

$$\frac{1}{6}x^2 - \frac{2}{11}x^3 + \frac{3}{16}x^4 - \frac{4}{21}x^5 + \frac{5}{26}x^6 \;\dots\dots$$

(*b*) Discuss the convergence of the series:

$$\frac{1}{2} + \frac{2}{3}x + \left(\frac{3}{4}\right)^2 x^2 + \left(\frac{4}{5}\right)^3 x^3 + \dots\dots \infty,$$

UNIT-II

4. (*a*) Find the characteristic roots and characteristic vector of the matrix.

$$A = \begin{bmatrix} 8 & -6 & 2 \\ -6 & 7 & -4 \\ 2 & -4 & 3 \end{bmatrix}$$

(*b*) Find non-singular matrices P and Q such that PAQ is in the normal form for the matrices $A = \begin{bmatrix} 1 & 2 & 3 \\ 3 & 2 & 1 \\ 1 & 3 & 2 \\ 2 & 1 & 3 \end{bmatrix}$. Also find the rank of matrix A.

5. (*a*) Find the characteristic equation of the matrix $A = \begin{pmatrix} 1 & 3 & 7 \\ 4 & 2 & 3 \\ 1 & 2 & 1 \end{pmatrix}$. Hence obtain the inverse of the given matrix.

(*b*) Are the following vectors are linearly dependent? If so, find the relation between them.

$$x_1 = (1, 2, 1),\ x_2 = (2, 1, 4),\ x_3 = (4, 5, 6),\ x_4 = (1, 8, -3)$$

UNIT-III

6. (*a*) If $u = \sin^{-1}\left(\dfrac{x+y}{\sqrt{x}-\sqrt{y}}\right)$ then prove that $x^2\dfrac{\partial^2 u}{\partial x^2} + y^2\dfrac{\partial^2 u}{\partial y^2} + 2xy\dfrac{\partial^2 u}{\partial x \partial y} = \dfrac{-\cos 2u \sin u}{4\cos^3 u}$

(*b*) Discuss the maxima and minima of $\sin x + \sin y + \sin (x + y)$.

(*c*) By using Taylor's series prove that $\sin (x + y) = \sin x \cos y + \cos x \sin y$. Hence find the value of sin 46° correct to four places of decimals. 8, 6, 6

7. (*a*) If ρ_1 and ρ_2 be the radii of curvature at the extremities of a focal chord of a parabola whose latus rectum is '4*a*' then prove that $\rho_1^{-2/3} + \rho_2^{-2/3} = 2a^{-2/3}$.

(*b*) Prove that $\displaystyle\int_0^{\infty} \frac{\tan^{-1} ax}{(x)(1+x^2)}\,dx = \frac{\pi}{2}\log(1+a)$, where $a > 0$.

UNIT-IV

8. (*a*) Evaluate by changing the order of integration $\displaystyle\int_0^1 \int_x^{\sqrt{2-x^2}} \frac{x\,dy\,dx}{\sqrt{x^2+y^2}}$.

(*b*) Find the double integration, the area lying inside the circle $r = a \sin\theta$ and outside the cardioid $r = a(1 - \cos\theta)$. (6)

9. (*a*) Find the volume of the ellipsoid $\dfrac{x^2}{a^2} + \dfrac{y^2}{b^2} + \dfrac{z^2}{c^2} = 1$.

(*b*) Prove that: $\beta(m, n) = \dfrac{\Gamma(m)\Gamma(n)}{\Gamma(m+n)}$, hence evaluate $\Gamma(1/2)$.

B.Tech. 1st Semester Examination
(KU, Kurukshetra)
Mathematics (Dec. 2013)
Paper: Math-101-E

8101

Time allowed: 3 Hours *Maximum Marks: 100*

Note: *Attempt Five questions in all, selecting at least one question from each Unit. All questions carry equal marks.*

UNIT-I

1. (*a*) Compute to four decimal places, the value of cos 32° by use of Taylor's series.

(*b*) Define asymptotes of a polar curve. Find asymptotes of the curve $r \sin \theta = 2 \cos 2\theta$.

2. (*a*) Find the radius of curvature for the parabola $\dfrac{2a}{r} = 1 + \cos \theta$.

(*b*) Trace the curve: $y^2(a - x) = x^2 (a + x)$

UNIT-II

3. (*a*) Define homogeneous function of degree *n*. State and prove Euler's Theorem for a homogeneous function of degree *n* in *x*, *y* and *z*.

(*b*) If $x = e^u \cos v$, $y = e^u \sin v$, find $J = \dfrac{\partial(u, v)}{\partial(x, y)}$, $J' = \dfrac{\partial(x, y)}{\partial(u, v)}$ and hence show that:

$$JJ' = 1.$$

4. (*a*) Show that the rectangular solid of maximum volume that can be inscribed in a sphere is a cube.

(*b*) Evaluate $\int_0^{\pi} \log (1 + a \cos x)\, dx$ using the method of differentiation under the sign of integration.

UNIT-III

5. (*a*) Find by double integration, the area lying inside the circle $r = a \sin \theta$ and outside the cardioids $r = a (1 - \cos \theta)$.

(*b*) Evaluate $\int_0^{\infty} \int_0^{\infty} e^{-(x^2 + y^2)}\, dxdy$ by changing into polar coordinates. Hence show that:

$$\int_0^{\infty} e^{-x^2} dx = \sqrt{\frac{\pi}{2}}.$$

6. (*a*) Find the volume of the tetrahedron bounded by the coordinate planes and the plane $\dfrac{x}{a} + \dfrac{y}{b} + \dfrac{z}{c} = 1$.

(*b*) Define Beta and Gamma functions. Prove that: $\beta\left(m, \dfrac{1}{2}\right) = 2^{2m-1} \beta(m, m)$.

UNIT-IV

7. (*a*) Define Gradient of a scalar point function and hence give its geometrical interpretation.
   (*b*) With usual notations, prove that: $\nabla^2(r^n) = n(n+1)\, r^{n-2}$

8. (*a*) Verify Green Theorem for $\int_C \left[(3x^2 - 8y^2)\, dx + (4y - 6xy)dy\right]$ over C, the boundary of the region bounded by $x = 0$, $y = 0$ and $x + y = 1$.
   (*b*) Evaluate: $\int \bar{F} . d\bar{S}$
   over the surface S boundary by the region $x^2 + y^2 = 4$, $z = 0$ and $z = 3$ by taking
   $$F = 4xi - 2y^2j + z^2k.$$